The Greenland Caledonides:
Evolution of the Northeast Margin of Laurentia

edited by

A.K. Higgins
Geological Survey of Denmark and Greenland
Øster Voldgade 10
DK-1350 Copenhagen K
Denmark

Jane A. Gilotti
Department of Geoscience
University of Iowa
Iowa City, Iowa 52242
USA

M. Paul Smith
Lapworth Museum of Geology
University of Birmingham, Edgbaston
Birmingham B15 2TT
UK

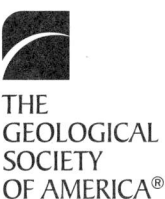

THE
GEOLOGICAL
SOCIETY
OF AMERICA®

Memoir 202

3300 Penrose Place, P.O. Box 9140 ▪ Boulder, Colorado 80301-9140, USA

2008

Copyright © 2008, The Geological Society of America (GSA). All rights reserved. GSA grants permission to individual scientists to make unlimited photocopies of one or more items from this volume for noncommercial purposes advancing science or education, including classroom use. For permission to make photocopies of any item in this volume for other noncommercial, nonprofit purposes, contact The Geological Society of America. Written permission is required from GSA for all other forms of capture or reproduction of any item in the volume, including, but not limited to, all types of electronic or digital scanning or other digital or manual transformation of articles or any portion thereof, such as abstracts, into computer-readable and/or transmittable form for personal or corporate use, either noncommercial or commercial, for-profit or otherwise. Send permission requests to GSA Copyright Permissions, 3300 Penrose Place, P.O. Box 9140, Boulder, Colorado 80301-9140, USA.

Copyright is not claimed on any material prepared wholly by government employees within the scope of their employment.

Published by The Geological Society of America, Inc.
3300 Penrose Place, P.O. Box 9140, Boulder, Colorado 80301-9140, USA
www.geosociety.org

Printed in U.S.A.

GSA Books Science Editors: Marion E. Bickford and Donald I. Siegel

The editors gratefully acknowledge the generous grant contributions of the Carlsberg Foundation, Copenhagen, to drawing of figures at the Geological Survey of Denmark and Greenland (GEUS) and printing the color illustrations in this Memoir.

Library of Congress Cataloging-in-Publication Data

The Greenland caledonides : evolution of the northeast margin of Laurentia / edited by A.K. Higgins, Jane A. Gilotti, M. Paul Smith.
    p. cm. — (Memoir (Geological Society of America) ; 202)
  Includes bibliographical references.
  ISBN 978-0-8137-1202-4 (cloth)
    1. Geology—Greenland. 2. Orogenic belts—Greenland. 3. Geology, Stratigraphic—Devonian. 4. Geology, Stratigraphic—Silurian. 5. Laurentia (Continent). I. Higgins, A. K. II. Gilotti, Jane A. III. Smith, M. Paul, 1959–.

QE70.G74 2008
551.7′3209982—dc22

2008017654

Cover: East-dipping Eleonore Bay Supergroup sediments west of Mestersvig on the south side of Kong Oscar Fjord. View looking north, with highest summits about 2000 m. Field of view is about 20 km and relief in foreground is about 1200 m. Photograph courtesy of Jakob Lautrup, Geological Survey of Denmark and Greenland (GEUS).

Contents

Introduction—The Caledonides of Greenland v
  Niels Henriksen, A.K. Higgins, Jane A. Gilotti, and M. Paul Smith

1. *Geological research and mapping in the Caledonian orogen of East Greenland, 70°N–82°N* ... 1
   Niels Henriksen and A.K. Higgins

2. *Architecture and evolution of the East Greenland Caledonides—An introduction* 29
   A.K. Higgins and A. Graham Leslie

3. *Polyorogenic history of the East Greenland Caledonides* 55
   Feiko Kalsbeek, Kristine Thrane, A.K. Higgins, Hans F. Jepsen, A. Graham Leslie,
   Allen P. Nutman, and Robert Frei

4. *Paleoproterozoic and Mesoproterozoic sedimentary and volcanic successions in the
   northern parts of the East Greenland Caledonian orogen and its foreland* 73
   John D. Collinson, Feiko Kalsbeek, Hans F. Jepsen, Stig A.S. Pedersen, and Brian G.J. Upton

5. *Neoproterozoic sedimentary basins with glacigenic deposits of the East Greenland
   Caledonides* 99
   Martin Sønderholm, Kasper S. Frederiksen, M. Paul Smith, and Henrik Tirsgaard

6. *Cambrian–Silurian development of the Laurentian margin of the Iapetus Ocean in
   Greenland and related areas* 137
   M. Paul Smith and Jan Audun Rasmussen

7. *Foreland-propagating Caledonian thrust systems in East Greenland* 169
   A. Graham Leslie and A.K. Higgins

8. *Caledonian metamorphic patterns in Greenland* 201
   Jane A. Gilotti, Kevin A. Jones, and Synnøve Elvevold

9. *Granites and granites in the East Greenland Caledonides* 227
   Feiko Kalsbeek, A.K. Higgins, Hans F. Jepsen, Robert Frei, and Allen P. Nutman

10. *Geometry, kinematics, and timing of extensional faulting in the Greenland Caledonides—
    A synthesis* 251
    Jane A. Gilotti and William C. McClelland

11. *The Devonian basin in East Greenland—Review of basin evolution and vertebrate
    assemblages* 273
    Poul-Henrik Larsen, Henrik Olsen, and Jennifer A. Clack

***12. Mineral occurrences in central East Greenland (70°N–75°N) and their relation to the
Caledonian orogeny—A Sr-Nd-Pb isotopic study of scheelite*** 293
Henrik Stendal and Robert Frei

***13. Laurentian margin evolution and the Caledonian orogeny—A template for Scotland
and East Greenland*** 307
A. Graham Leslie, Martin Smith, and N.J. Soper

***14. Caledonian orogen of East Greenland 70°N–82°N: Geological map at 1:1,000,000—
Concepts and principles of compilation*** 345
Niels Henriksen and A.K. Higgins

Introduction—The Caledonides of Greenland

Niels Henriksen
A.K. Higgins*
Geological Survey of Denmark and Greenland, Øster Voldgade 10, DK-1350 Copenhagen K, Denmark

Jane A. Gilotti
Department of Geoscience, University of Iowa, Iowa City, Iowa 52242, USA

M. Paul Smith
Lapworth Museum of Geology, University of Birmingham, Edgbaston, Birmingham B15 2TT, UK

The Caledonian–Appalachian orogenic belt extends for more than 6000 km along the eastern margin of Laurentia (Fig. 1). The northeasternmost sector of this orogenic belt in North-East Greenland has been the focus of regional geological mapping and detailed investigations by the Geological Survey of Denmark and Greenland (GEUS[1]) and its predecessor over a period of 30 yr, between 1968 and 1998. The results of these extensive investigations have been documented in a series of GEUS map sheets and in numerous articles in international journals and GEUS *Bulletins*. This volume presents an overview of the East Greenland Caledonides within a modern plate-tectonic framework, which can be summarized as "the birth, life, and death of Iapetus as recorded on the Greenland margin," and it is accompanied by a regional geological map of the entire East Greenland Caledonian orogen at a scale of 1:1,000,000. The development of North-East Greenland shows similarities with the classic British Caledonides, which were also formerly part of the Laurentian margin (Leslie et al., this volume).

The data presented in this volume are of direct significance to the tectonic development of adjacent parts of the orogen, now preserved in Scotland, Svalbard, and Scandinavia, as well as in the Appalachians farther south along the Laurentian margin. The volume is an up-to-date account of a substantial segment of the Caledonide orogen that has resulted from comprehensive research in a region that is remote and difficult of access. This volume therefore addresses all geologists with Caledonian interests, and this introduction provides the background for the geological investigations in North-East Greenland, a brief overview of the Caledonian geology in Greenland, and correlations with the British Caledonides, based on the 14 chapters that the volume contains.

The N-S–trending Caledonian orogen in East Greenland is exposed over a several-hundred-kilometer-wide stretch of ice-free land that extends for 1300 km between latitudes 70°N and 82°N. It occupies a total area of ~127,000 km² between the Inland Ice and the North Atlantic Ocean. Apart from simple regional geological maps accompanying exploration accounts, the earliest basic geological investigations of East Greenland date from the 1920s, and reconnaissance mapping of large regions was conducted in the period from 1947 to 1958 (Haller, 1971). East Greenland is generally very well exposed due to its arctic setting, sparse vegetation cover, and ice-eroded outcrops. This has made it possible to map large areas within a

---
*akh@geus.dk
[1]The Geological Survey of Greenland (GGU) was amalgamated with the Geological Survey of Denmark (DGU) in 1995 to form the present Geological Survey of Denmark and Greenland (GEUS).

Henriksen, N., Higgins, A.K., Gilotti, J.A., and Smith, M.P., 2008, Introduction—The Caledonides of Greenland, *in* Higgins, A.K., Gilotti, J.A., and Smith, M.P., eds., The Greenland Caledonides: Evolution of the Northeast Margin of Laurentia: Geological Society of America Memoir 202, p. v–xv, doi: 10.1130/2008.1202(00). For permission to copy, contact editing@geosociety.org. ©2008 The Geological Society of America. All rights reserved.

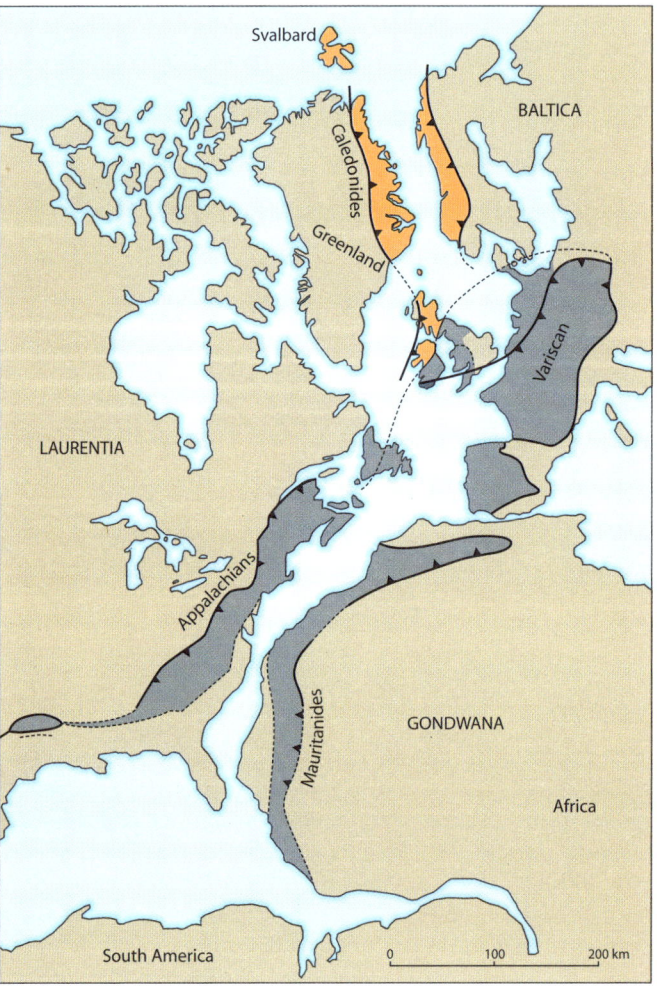

Figure 1. Map of the Paleozoic fold belts around the North Atlantic Ocean and the continents shown in their original relative position ca. 300 Ma, before seafloor spreading created the present-day Atlantic Ocean. The Caledonian fold belt in the north, shown in orange, was formed by continental collision between Laurentia and Baltica in the Silurian. This produced a doubly vergent collision geometry, resulting in eastward thrust displacement in the Scandinavian Caledonides and westward thrust displacement in the Greenland Caledonides. The other Paleozoic fold belts depicted reflect the continental collisions of Laurentia and related areas with various microcontinents and Africa, a sequence of tectonic events that lasted into the Carboniferous and created the Appalachians, the Variscan, and the Mauritanian fold belts.

limited time span, supported by extensive geological photointerpretation during drawing of new 1:100,000 topographic maps based on wide-angle aerial photographs. The geological results of the series of major expeditions have been compiled onto five 1:500,000 geological maps of the region 70°N–82°N; these are part of a set of 14 geological maps at a scale of 1:500,000 that cover all the ice-free land region of Greenland. In addition, the Scoresby Sund region (70°N–72°N) is covered by 16 detailed 1:100,000 geological maps. The 1:1,000,000 geological map published here of North-East Greenland from 70°N–82°N was compiled on the basis of the five published 1:500,000 map sheets, with modifications and updated interpretations where necessary. This map provides a tectonic overview of the Caledonian orogen in Greenland, and it includes cross sections, synoptic maps, and profiles; the exposed foreland geology is depicted, as well as post-Caledonian sedimentary rocks of Devonian–Neogene age, and Paleogene plateau basalts. This 1:1,000,000 colored geological map accompanies this volume, both as a folded printed map and as a CD-ROM.[2]

REGIONAL GEOLOGY

Four principal geological divisions are represented in the ice-free land areas of northern East Greenland: (1) The Caledonian foreland partly exposed at the margin of the Inland Ice and in tectonic windows; (2) the broad zone of the Caledonian orogen; (3) post-Caledonian, mainly sedimentary, successions found in

---

[2]The map also exists as GSA Data Repository Item 2008168 (Plate 1), which is available at www.geosociety.org/pubs/ft2008.htm, or on request from editing@geosociety.org, Documents Secretary, GSA, P.O. Box 9140, Boulder, CO 80301-9140, USA.

the coastal regions; and (4) Paleogene plateau basalts that conceal the continuation of the Caledonian orogen south of 70°N. The offshore 25–350-km-wide continental shelf region is not described in this volume; it is made up of extensive Upper Paleozoic–Neogene sedimentary basins and Paleogene plateau basalts, which are partly covered by younger sediments (Hamann et al., 2005). The continental crust passes eastward via a transition zone into oceanic crust (Henriksen et al., 2000).

The dominant geological feature of northern East Greenland is the Caledonide orogen (Fig. 2), which was created by the collision of the continents of Laurentia in the west and Baltica in the east that brought about closure of the Iapetus Ocean. West-vergent Caledonian thrust sheets characterize the East Greenland sector of the Laurentian margin, whereas east-vergent thrust sheets characterize the Caledonian orogen of Scandinavia. The basement gneiss units of Laurentia had already experienced a complex series of Archean to Neoproterozoic orogenic events prior to reworking during the Caledonian orogeny. These reworked Precambrian gneisses now form major segments of Caledonian thrust sheets, together with widespread metasedimentary rock successions of Mesoproterozoic to Neoproterozoic age. The passive-margin deposits of the west border of the Iapetus Ocean consist of thick Neoproterozoic to Ordovician or Silurian sedimentary successions, well exposed in East Greenland between latitudes 71°N and 76°N and north of 79°N. Caledonian deformation, metamorphism, thrust sheet emplacement, and granitic activity associated with continental collision took place between the Middle Ordovician and Early Carboniferous.

Results of the Geological Investigations

The geology of the Caledonian orogen in East Greenland is summarized in 13 chapters of this volume, while an additional chapter provides a description of the British Caledonides, which, prior to opening of the North Atlantic Ocean, was located immediately to the south of the East Greenland Caledonides (Leslie et al., this volume, Chapter 13). Most of the authors of individual chapters have participated in GEUS-sponsored field work in East Greenland for a number of summer seasons and have numerous publications in international journals to their credit. The chapters that make up this volume include: (1) a general introduction and a historical background of mapping and research; (2) an introduction to the structural architecture of the orogen and the principal structural domains; (3) the nature of the reworked Archean and Proterozoic crystalline complexes that made up the Greenland segment of Laurentia; (4) the Proterozoic sedimentary and volcanic successions represented in the foreland and the Caledonian thrust sheets; (5 and 6) the Neoproterozoic to Lower Paleozoic evolution of the Iapetus passive margin in East Greenland; (7) the Caledonian foreland-propagating thrust architecture; (8) Caledonian regional metamorphism; (9) Caledonian granites and migmatites, and similar early Neoproterozoic granites; (10) the geometry and timing of extensional fault systems; (11) the development of extensional continental basins in the Devonian; (12) a summary of Caledonian mineralization; (13) an account of the British Caledonides, with the focus on correlations and comparisons with East Greenland; and (14) a final chapter that describes the principles behind compilation of the 1:1,000,000 map that accompanies this volume. Figure 3 indicates the regional coverage of the different chapters, and the brief summaries that follow give an indication of the scope and focus of each chapter.

General Introduction to Geological Research and Mapping

Geological research in the East Greenland Caledonides began with the earliest exploration voyages in the early 1800s. The first small-scale geological maps accompanied expedition accounts from the early 1900s, and more detailed geological investigations resulted from Lauge Koch's long series of geological expeditions between 1926 and 1958 (Haller, 1971). In 1968, the Geological Survey of Greenland commenced a program of regional geological mapping in the Scoresby Sund region (70°N–72°N, the southernmost part of the Caledonian orogen, and during the next 30 yr, the entire 1300-km-long stretch of ice-free land making up the East Greenland Caledonides was systematically investigated in detail (Henriksen and Higgins, this volume, Chapter 1). The geological studies were carried out as a close cooperation between geologists from the GEUS and a large group of colleagues from geological institutes and universities in Europe and North America. During 15 field seasons, more than 50 geologists have been involved in the field work for one or more summer seasons. The scientific results are documented in numerous publications both in international journals and in GEUS *Bulletins* and *Reports*. The present volume presents a general summary and conclusions of this work.

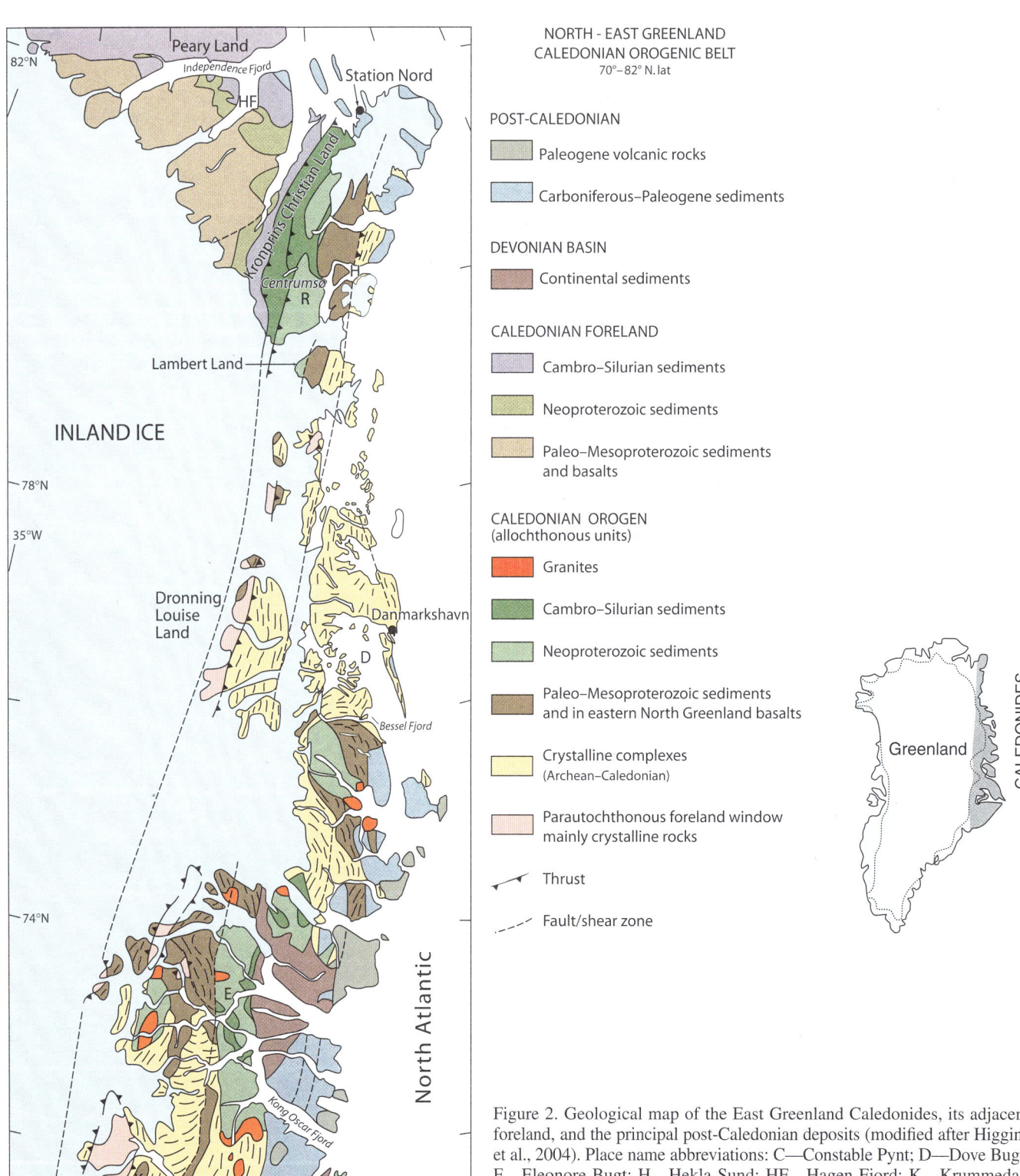

Figure 2. Geological map of the East Greenland Caledonides, its adjacent foreland, and the principal post-Caledonian deposits (modified after Higgins et al., 2004). Place name abbreviations: C—Constable Pynt; D—Dove Bugt; E—Eleonore Bugt; H—Hekla Sund; HF—Hagen Fjord; K—Krummedal; R—Rivieradal; S—Scoresbysund/Illoqqortormiut.

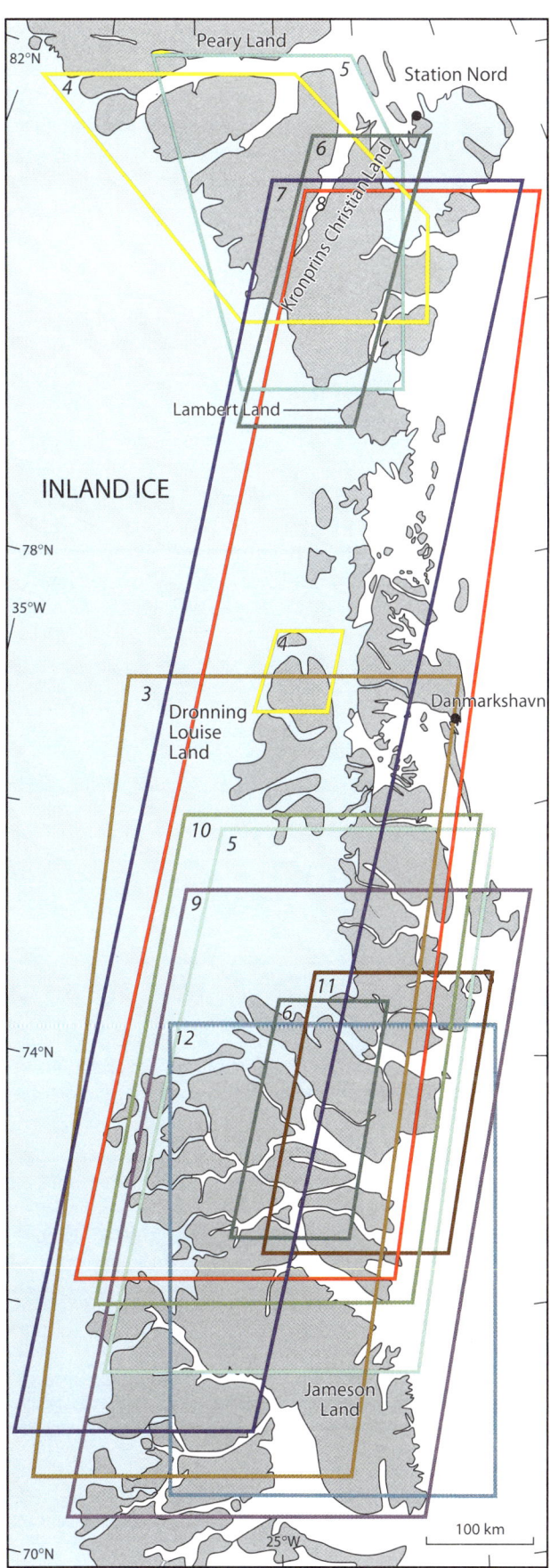

Figure 3. Map showing the location of the regions in Greenland covered by the chapters of this volume. The numbers correspond to the chapter numbers and summaries in this article. Chapters 1–2 and 14 cover the entire orogen and are not shown by a frame, while Chapter 13 is an account of the British Caledonides, and is not shown on this map.

Structural Domains

The East Greenland Caledonides can be divided into distinct structurally bounded geological domains made up of Archean to Lower Paleozoic components derived from the eastern margin of Laurentia (Higgins and Leslie, this volume, Chapter 2). These domains originally evolved as major westward-displaced thrust units during collision with Baltica. All levels of the orogen are represented, from deep-crustal segments that have undergone eclogite-facies metamorphism, to the highest nonmetamorphic units. The orogen in East Greenland has a western thrust border against the rocks of the Laurentian craton that is largely concealed beneath the Inland Ice. A foreland-propagating thrust pile is well-preserved both in the extreme north (79°N–82°N) and in the south (70°N–76°N) (Higgins and Leslie, this volume, Chapter 7), with less-well-preserved remnants in the intervening region. Between 76°N and 81°N, the deepest preserved parts of the orogen are dominated by high-grade Paleoproterozoic orthogneisses with eclogitic enclaves. Caledonian granites are confined to the upper thrust sheet in the southern orogen (70°N–76°N) (Fig. 2).

Polyorogenic History

The far-traveled thrust sheets that make up the Caledonian orogen are mainly composed of Archean to early Neoproterozoic rock units derived from the eastern margin of Laurentia that bear witness to a complex polyorogenic history of the region prior to Caledonian orogenesis (Kalsbeek et al., this volume, Chapter 3). There is evidence for Archean (ca. 2800–2600 Ma), Paleoproterozoic (2000–1750 Ma), and late "Grenvillian" (ca. 950 Ma) deformation and metamorphism, overprinted by Caledonian orogenic events, and broad descriptions of the various rock units are supported by structural, geochemical, and geochronologic data. The Archean and Paleoproterozoic gneiss complexes consist mainly of granitoid orthogneisses. Paleoproterozoic tholeiitic metabasalts are present in some of the foreland windows. A major unit of late Mesoproterozoic metasedimentary rocks (Krummedal supracrustal sequence, Smallefjord Sequence) contains early Neoproterozoic (ca. 950 Ma) as well as Caledonian granites.

Paleoproterozoic and Mesoproterozoic Sedimentary and Volcanic Successions

Paleoproterozoic sandstones are conspicuous in the Independence Fjord Basin of eastern North Greenland and northernmost East Greenland (Fig. 2). In the Caledonian thrust sheets of northernmost East Greenland, the rift sandstones are interbedded with 1740 Ma volcanic rocks (Collinson et al., this volume, Chapter 4). Similar sandstones of the Independence Fjord Group in the foreland were laid down in a continental sag basin without associated volcanics, and they are interpreted as fluvial and lacustrine sediments deposited in a semiarid environment. The sandstones are everywhere cut by abundant doleritic dikes and sills, known as the Midsommersø Dolerites (1380 Ma), and they are conformably overlain by the related Zig-Zag Dal Basalt Formation. A long hiatus (~500 m.y.) separates the basalts from the overlying Neoproterozoic successions of the Hekla Sund Basin.

In the southern half of the fold belt, a several-kilometer-thick succession of late Mesoproterozoic amphibolite-facies metasediments, known as the Krummedal supracrustal sequence or Smallefjord Sequence, makes up significant parts of major thrust sheets. The source area appears to include components with "Grenvillian" affinities.

Neoproterozoic Sedimentary Basins

Two major Neoproterozoic sedimentary basins are represented in North-East Greenland, and they may relate to an early phase of Iapetus rifting along the northeastern Laurentian margin. In the south the Eleonore Bay Basin accumulated a very thick succession, while sediments of the Hekla Sund Basin accumulated in eastern North Greenland (Sønderholm et al., this volume, Chapter 5).

The Eleonore Bay Basin accumulated a 14-km-thick Neoproterozoic succession (Eleonore Bay Supergroup) that is overlain by late Neoproterozoic sandstone-shale deposits and glacigenic diamictites (Tillite Group). The thick siliciclastic lower part of the succession evolved into a thinner succession of limestones and dolomites, overlain with slight disconformity by the Marinoan glacigenic deposits. The Eleonore Bay Supergroup is characterized by a very uniform development along the ~500-km-long south-to-north outcrop belt.

The Hekla Sund Basin is represented by sediments of the Rivieradal Group and the Hagen Fjord Group, which have a cumulative thickness of 8–11 km. The clastic Rivieradal Group sediments were deposited in a half-graben rift basin that was overlapped by the postrift, dominantly carbonate sediments of the Hagen Fjord Group.

Early Paleozoic Basin Development

Iapetus passive-margin deposits are variably preserved along the 1300-km-long northeast border of Laurentia exposed in East Greenland from 70°N to 82°N (Smith and Rasmussen, this volume, Chapter 6). North of 79°N in northernmost East Greenland, the carbonate shelf deposits are the easternmost part of the E-W–trending Franklinian Basin, which extends across North Greenland and into Arctic Canada. The Lower Cambrian sediments in this basin can be correlated with the Sauk I sequence of cratonic North America. In eastern North Greenland, a significant hiatus (the Wandel Valley unconformity) separates the Cambrian sequence from an overlying succession that extends without major breaks from the Lower Ordovician (base of Sauk IV) through to the Early Silurian. From late Llandoverian time onward, the carbonate platform was buried beneath thick, rapidly deposited sandstone turbidites. This Franklinian Basin succession differs from the Lower Paleozoic developments preserved in allochthonous thrust sheets in the southern part of the Greenland Caledonides. The upper part of the Hagar Bjerg thrust sheet, distinguished as the Franz Joseph allochthon, contains one of the thickest Cambrian-Ordovician successions known in Laurentia, a complete 4.5-km-thick succession that extends from the Lower Cambrian to Middle Ordovician (Sauk I to Tippecanoe II).

Thrust Architecture

The Greenland Caledonides display a west-verging thrust geometry that is well preserved in the southern half of the orogen and in the extreme north in Kronprins Christian Land (Leslie and Higgins, this volume, Chapter 7). The thrust architecture is illustrated by a series WNW-ESE cross sections spanning the width of the orogen. In Kronprins Christian Land, the cross sections extend from undeformed foreland through a wide fold-and-thrust belt into a series of allochthonous thrust sheets. In the extreme east, gneiss complexes of the high-grade Nørreland thrust sheet contain eclogitic enclaves that testify to exhumation from depths in excess of 50 km. In the southern half of the orogen, a pile of far-traveled major thrust sheets includes crystalline basement overlain by Mesoproterozoic to Ordovician sediments. Tectonic windows in the footwalls of the thick-skinned thrust sheets contain thin (<400 m) Neoproterozoic–Cambrian sedimentary developments, in marked contrast to very thick developments (~18 km) in the highest thrust sheet (Franz Joseph allochthon). The two major thrust sheets are difficult to restore in the absence of cutoffs of key horizons, but they have estimated displacements of several hundred kilometers, indicative of significant contraction of the Laurentian margin during collision.

Caledonian Metamorphic Patterns

Widespread regional Caledonian metamorphism accompanied the deformation of the Laurentian margin in East Greenland. Metamorphic patterns, described by pressure-temperature paths, vary as a function of structural level in the thrust sheets and are synthesized for the entire orogen by Gilotti et al. (this volume, Chapter 8). The foreland areas show unmetamorphic to low metamorphic grades. In the extreme north, conodonts in the Lower Paleozoic carbonates of the thin-skinned fold-and-thrust belt of Kronprins Christian Land show a systematic increase in color alteration indices (CAI) eastward, indicative of an increasing thickness of overburden (6–12 km). The conodont CAI in the Lower Paleozoic carbonates of the Franz Joseph allochthon in the southern orogen are very low, indicating that this structural unit is the highest level of the Caledonian thrust pile. Amphibolite-facies metamorphism characterizes the major thrust sheets in the southern orogen, as well as the thrust sheet assemblages between Lambert Land and eastern Dronning Louise Land. Extensive eclogite-facies metamorphism, locally reaching the coesite stability field, typifies the mainly gneissic terrain of the coastal region from 76°N to 79°N, an anomaly for the overriding plate of a collision zone. Two major periods of metamorphism are recognized: the first is associated with melting of the middle crust at ca. 435–415 Ma, and the second is related to the development of the high-pressure granulites and eclogites at 410–390 Ma.

Granitic Magmatism

Whereas granites are absent in the continental crust of the Baltica margin, granitic magmatism is an important element of the southern half of the Greenland Caledonides (Kalsbeek et al., this volume, Chapter 9). Both I-type and S-type granites are recognized. The less abundant, mainly Ordovician, I-type granitoids probably represent the remnants of a subduction-related magmatic arc, which accords with Laurentia being the overriding plate of the collision. In contrast, the S-type leucogranites are widespread evidence of partial melting of fertile metasedimentary rocks in the middle crust. The Caledonian S-type leucogranites are 435–415 Ma in age and are associated with regional metamorphism seen in migmatite complexes. Discrete leucogranite plutons of the same age intrude into Neoproterozoic sedimentary successions in the upper thrust sheet (Franz Joseph allochthon), which forms the roof of the orogen. The story is complicated by the presence of Mesoproterozoic (ca. 950 Ma) leucogranites that originated from the same metasedimentary units found in the migmatite complexes, and that are indistinguishable in the field from the Caledonian peraluminous granites.

Extensional Fault Systems

Regional extensional structures are common throughout the orogen. The extent and importance of extension have been debated among the different working groups, and the interpretation presented by Gilotti and McClelland (this volume, Chapter 10) offers a structural scenario that maximizes the impact of extension in the southern orogen. Two periods of synorogenic to late orogenic extensional detachment systems shuffle the thrust geometry and postdate the major periods of metamorphism. Juxtaposition of units of contrasting metamorphic grade is indicative of extensional exhumation on the order of several tens of kilometers. The superposition of one detachment system on another is responsible for the overall complexity in geometry and kinematics, along with many branching relationships and variations in displacement directions. Brittle normal faults facilitated Devonian basin formation, and extension accommodated by high-angle faulting in the Carboniferous marks the final transition to plate divergence and continental rifting.

The Devonian Basin

From the Middle Devonian onward, the Old Red Sandstone basin of North-East Greenland accumulated over 8 km of clastic, continental sediments (Larsen et al., this volume, Chapter 11). The lowermost levels in the succession consist of conglomeratic sandstones lying unconformably on folded Ordovician rocks, which is indicative of a hiatus of up to 70 m.y. The present-day Devonian deposits are exposed onshore over a region 300 km from north to south, but seismic surveys from further south demonstrate that Devonian basins are present over a distance of at least 500 km. The lower part of the onshore sedimentary succession contains thin lava flows, ash layers, flat-lying sills, and crosscutting steeply inclined dikes.

Basin initiation can be linked to the extensional collapse of the overthickened Caledonian orogen, and it was accommodated by SE-NW dip-slip faulting and subordinate N-S strike-slip faults. In addition to their importance as well-exposed, classic examples of the Old Red Sandstone molasse, the East Greenland Devonian basins preserve a rich vertebrate fauna, most notably examples of the early tetrapods *Acanthostega* and *Ichthyostega*.

Mineral Occurrences

The Caledonian orogen of East Greenland hosts numerous mineral occurrences that are related to: (1) Neoproterozoic basins (strata-bound copper); (2) hydrothermal activity along major Caledonian lineaments (vein-type gold, silver, tungsten, arsenic, and antimony); and (3) hydrothermal activity along late Caledonian lineaments, extensional and wrench faults, and dilation zones (Stendal and Frei, this volume, Chapter 12). Mineral occurrences associated with fault zones and late Caledonian veins all have a genetic relationship with Caledonian granite emplacement. The characterization of possible provenance regions for the metals presented in this volume is based on Pb, Sr, and Sm-Nd isotopic studies on scheelite, mainly from the southern half of the orogen. The Sr and Pb isotope data are compatible with Sr and Pb sources in Caledonian granites, as well as from the late Mesoproterozoic metasedimentary rocks (Krummedal supracrustal sequence). Sm-Nd isotopic data from scheelite define an errorchron with an approximate age of 380 Ma, which suggests that the formation age for scheelite perhaps postdates emplacement of most Caledonian granites.

The British Caledonides

The British Caledonides, summarized by Leslie et al. (this volume, Chapter 13), constitute only a small fragment of the Neoproterozoic to Paleozoic margin of Laurentia. This fragment was originally a southern extension of the East Greenland Caledonide orogen, but it is significant because it was located at a prominent bend in the Laurentian margin. Sequences exposed in Scottish outcrops include Mesoproterozoic, Neoproterozoic, and Cambrian-Ordovician strata that record sedimentation, volcanism, and deformation related to the amalgamation of Rodinia, the subsequent breakout of Laurentia, and growth of the Iapetus Ocean. Metamorphic and tectonic overprints then record the destruction of that ocean through Ordovician arc accretion and mid-to-late Silurian collision of Laurentia, Baltica, and Avalonia, and the final closure of Iapetus by end-Silurian time. New isotopic data and recent advances in understanding of the late Mesoproterozoic (Stenian) to Cambrian-Ordovician stratigraphic framework now better constrain the sequence and timing of events across the "Scottish Corner" and allow detailed comparison with the East Greenland Caledonides.

The 1:1,000,000 Geological Map of the Caledonian Orogen in East Greenland

The 1:1,000,000 geological map of the East Greenland Caledonides (Henriksen, 2003), reprinted as a special sheet for this volume, is the first regional overview of the entire East Greenland Caledonian orogen (Henriksen and Higgins, this volume, Chapter 14). It is based on interpretations from five published 1:500,000 map sheets produced by the Geological Survey of Denmark and Greenland that were mapped between 1968 and 1998. The 1:1,000,000 map differs from the standard regional 1:500,000 geological maps in its specific focus on the East Greenland Caledonides. The map links lithostructural units across the boundaries of the original 1:500,000 map sheets and thus provides a consistent regional overview. Geological divisions in the post-Caledonian areas have been converted from lithostratigraphic to time-stratigraphic units.

Five true-scale profiles on the map sheet illustrate the thrust sheet architecture in the northern and southern parts of the orogen. Two small-scale synoptic tectonic maps, with profiles, present a general overview of the structure.

GENERAL BACKGROUND FOR THE GEOLOGICAL INVESTIGATIONS

The region depicted on the 1:1,000,000 map is uninhabited, apart from the small town of Illoqqortormiut/Scoresbysund (79°29′N, 21°58′W) and outlying settlements (total population was 550 in 2004), the airport at Constable Pynt (70°45′N, 22°36′W), the weather station at Danmarkshavn (76°46′N, 18°39′W), and a few military outposts (Fig. 2). The almost complete lack of supporting infrastructure and logistics means that any large-scale activity such as geological field work must be carried out as self-supporting expeditions. All personnel, provisions, and camp equipment have to be transported in and out of the region each field season, and an independent local transport and communication system needs to be established. Particular consideration must be taken, and regulations followed, as the greater part of the region lies within the North-East Greenland National Park, the largest national park in the world.

North-East Greenland lies in the high arctic zone, with average temperatures of 3–5 °C in July, the warmest summer month. Winter temperatures are lowest in February, with mean temperatures between –16 °C in the south and –30 °C in the north. Conditions for field work in the summer season are generally excellent, due to 24 h of daylight, combined with low precipitation and many hours of sunshine; these factors have given rise to the characterization of the fjord region from 72°N–75°N as the "Arctic Riviera" (Hofer, 1957).

The main GEUS field work was carried out during 15 field seasons of 7–8 wk each. The expedition groups numbered 30–50 participants each year, of which 75% were scientific personnel; the remaining 25% consisted of base camp personnel, together with helicopter and aircraft pilots and mechanics. The geoscientists generally worked as teams of two from small tent camps, supported by helicopter camp moves and geological reconnaissance flights. Approximately half of the scientific group was made up of specialists from geological institutes and universities in Europe and North America, who participated as guests of GEUS and often worked in partnership with GEUS staff. The contribution of non-GEUS geoscientific personnel has been considerable and has included experts with experience in Caledonian orogenic belts outside Greenland, as well as specialists in structural, metamorphic, sedimentologic, and paleontologic disciplines. Substantial parts of the region were studied photogeologically prior to field work, and due to the exceptional high degree of exposure, the photogrammetric geological interpretations have contributed significantly to the results of the geological mapping.

The major projects were carried out in five phases: 1968–1972: Scoresby Sund region (69°N–72°N); 1979–1980: Northern Kronprins Christian Land in eastern North Greenland (81°N–82°N); 1988–1990: Dove Bugt region (75°N–78°N); 1993–1995: Lambert Land region (78°N–81°N); and 1997–1998: Kong Oscar Fjord region (72°N–75°N).

In addition to the regional geological mapping projects sponsored by GEUS, various investigations have been carried out by independent non-GEUS groups. These groups have usually concentrated on specific geological problems or particular areas, have arranged their own aircraft charter flights to reach East Greenland, and have used chartered aircraft and their own inflatable rubber boats for local transport. The published results of their investigations are referred to in the relevant chapters of this volume.

The entire Atlantic-facing coastal region of Greenland from 59°N to 82°N is officially designated "East Greenland," but except when referring to extensive regions such as the East Greenland Caledonides, this is too broad a division for geological descriptive purposes. GEUS therefore introduced unofficial but widely used subdivisions: central East Greenland (69°N–72°N), North-East Greenland (72°N–79°N), and eastern North Greenland (79°N–82°N, and east of 27°W). However, these divisions are not ideal, and they are sometimes used in a broader sense than originally intended. In this volume, the contributors often specify latitude limits where there could be confusion. Another commonly used, but unofficial designation, is the term "central fjord zone," which is used for the fjord-dissected part of East Greenland between latitudes 72°N and 75°N. Kronprins Christian Land, by its geographical location, may legitimately be regarded as both the northernmost part of East Greenland, as well as the easternmost part of North Greenland.

ACKNOWLEDGMENTS

The work described in this volume was mainly organized and funded by the Geological Survey of Greenland (GGU), part of the present-day Geological Survey of Denmark and Greenland (GEUS). The continuous and committed support provided by the Geological Survey of Denmark and Greenland (GGU/GEUS) is gratefully acknowledged, not only during the extended period of the regional mapping program in the Caledonian orogen and subsequent laboratory research, but also in respect to funding the printing of the map sheet and production of the CD-ROM that accompany this volume.

The field work and subsequent scientific studies throughout the 30 yr project period were carried out in close cooperation between GEUS staff and a large group of scientists mainly employed by geological institutes and universities in Denmark, the United Kingdom, and North America. The research of the non-GEUS participants was largely supported by their home institutions, sometimes supplemented by grants from national and private foundations. More detailed acknowledgments are given in the relevant chapters of this volume. We particularly wish to acknowledge the substantial support over many years of the following (in alphabetical order): British Geological Survey, Edinburgh, Scotland; Carlsberg Foundation, Denmark; Danish Natural Research Council (SNF); Geological Institute, University of Copenhagen, Denmark; Lapworth Museum of Geology, University of Birmingham; and New York State Geological Survey, New York.

We are indebted to the following persons for the careful and thoughtful reviews of the papers submitted for publication in this volume; they have provided valuable comments and suggestions that have in many cases greatly improved the quality of the articles: Howard A. Armstrong, Christopher J. Banks, Calvin G. Barnes, Peter A. Cawood, Winfried K. Dallman, Allen Dennis, Robert Fakundinny, Clark R.L. Friend, Peter F. Friend, Bradley R. Hacker, Michael J. Hambrey, Tekla Harms, David A.T. Harper, James Hibbard, Michael Houmark-Nielsen, Joseph M. Hull, Åke Johansson, Marian Lupulescu, Geoff M. Manby, Julian Menuge, Per Terje Osmundsen, David Peate, Karsten Piepjohn, Tony Prave, David Roberts, and Paul Tomascak.

REFERENCES CITED

Collinson, J.D., Kalsbeek, F., Jepsen, H.F., Pedersen, S.A.S., and Upton, B.G.J., 2008, this volume, Paleoproterozoic and Mesoproterozoic sedimentary and volcanic successions in the northern parts of the East Greenland Caledonian orogen and its foreland, in Higgins, A.K., Gilotti, J.A., and Smith, M.P., eds., The Greenland Caledonides: Evolution of the Northeast Margin of Laurentia: Geological Society of America Memoir 202, doi: 10.1130/2008.1202(04).

Gilotti, J.A., and McClelland, W.C., 2008, this volume, Geometry, kinematics, and timing of extensional faulting in the Greenland Caledonides—A synthesis, in Higgins, A.K., Gilotti, J.A., and Smith, M.P., eds., The Greenland Caledonides: Evolution of the Northeast Margin of Laurentia: Geological Society of America Memoir 202, doi: 10.1130/2008.1202(10).

Gilotti, J.A., Jones, K.A., and Elvevold, S., 2008, this volume, Caledonian metamorphic patterns in Greenland, in Higgins, A.K., Gilotti, J.A., and Smith, M.P., eds., The Greenland Caledonides: Evolution of the Northeast Margin of Laurentia: Geological Society of America Memoir 202, doi: 10.1130/2008.1202(08).

Haller, J., 1971, Geology of the East Greenland Caledonides: London, Interscience Publishers, 413 p.

Hamann, N.E., Whittaker, R.C., and Stemmerik, L., 2005, Structural and geological development of the North-East Greenland Shelf, in Doré, A.G., and Vining, B.A., eds., Petroleum Geology: North-West Europe and Global Perspectives: Proceedings of the 6th Conference on Petroleum: London, Geological Society, p. 887–902.

Henriksen, N., 2003, Caledonian Orogen, East Greenland 70°–82°N: Geological Map: Copenhagen, Geological Survey of Denmark and Greenland, scale 1:1,000,000.

Henriksen, N., and Higgins, A.K., 2008, this volume (Chapter 14), Caledonian orogen of East Greenland 70°N–82°N: Geological map at 1:1,000,000—Concepts and principles of compilation, in Higgins, A.K., Gilotti, J.A., and Smith, M.P., eds., The Greenland Caledonides: Evolution of the Northeast Margin of Laurentia: Geological Society of America Memoir 202, doi: 10.1130/2008.1202(14).

Henriksen, N., and Higgins, A.K., 2008, this volume (Chapter 1), Geological research and mapping in the Caledonian orogen of East Greenland 70°N–82°N, *in* Higgins, A.K., Gilotti, J.A., and Smith, M.P., eds., The Greenland Caledonides: Evolution of the Northeast Margin of Laurentia: Geological Society of America Memoir 202, doi: 10.1130/2008.1202(01).

Henriksen, N., Higgins, A.K., Kalsbeek, F., and Pulvertaft, T.C.R., 2000, Greenland from Archaean to Quaternary: Descriptive Text to the Geological Map of Greenland 1:2,500,000: Geology of Greenland Survey Bulletin, v. 185, 93 p.

Higgins, A.K., and Leslie, A.G., 2008, this volume, Architecture and evolution of the East Greenland Caledonides—An introduction, *in* Higgins, A.K., Gilotti, J.A., and Smith, M.P., eds., The Greenland Caledonides: Evolution of the Northeast Margin of Laurentia: Geological Society of America Memoir 202, doi: 10.1130/2008.1202(02).

Higgins, A.K., Elvevold, S., Escher, J.C., Frederiksen, K.S., Gilotti, J.A., Henriksen, N., Jepsen, H.F., Jones, K.A., Kalsbeek, F., Kinny, P.D., Leslie, A.G., Smith, M.P., Thrane, K., and Watt, G.R., 2004, The foreland-propagating thrust architecture of the East Greenland Caledonides 72°–75°N: Journal of the Geological Society of London, v. 161, p. 1009–1026, doi: 10.1144/0016-764903-141.

Hofer, E., 1957, Arctic Riviera, North East Greenland: Berne, Kümmerly & Frey, Geographical Publishers, 128 p.

Kalsbeek, F., Higgins, A.K., Jepsen, H.F., Frei, R., and Nutman, A.P., 2008, this volume (Chapter 9), Granites and granites in the East Greenland Caledonides, *in* Higgins, A.K., Gilotti, J.A., and Smith, M.P., eds., The Greenland Caledonides: Evolution of the Northeast Margin of Laurentia: Geological Society of America Memoir 202, doi: 10.1130/2008.1202(09).

Kalsbeek, F., Thrane, K., Higgins, A.K., Jepsen, H.F., Leslie, A.G., Nutman, A.P., and Frei, R., 2008, this volume (Chapter 3), Polyorogenic history of the East Greenland Caledonides, *in* Higgins, A.K., Gilotti, J.A., and Smith, M.P., eds., The Greenland Caledonides: Evolution of the Northeast Margin of Laurentia: Geological Society of America Memoir 202, doi: 10.1130/2008.1202(03).

Larsen, P.-H., Olsen, H., and Clack, J.A., 2008, this volume, The Devonian basin in East Greenland—Review of basin evolution and vertebrate assemblages, *in* Higgins, A.K., Gilotti, J.A., and Smith, M.P., eds., The Greenland Caledonides: Evolution of the Northeast Margin of Laurentia: Geological Society of America Memoir 202, doi: 10.1130/2008.1202(11).

Leslie, A.G., and Higgins, A.K., 2008, this volume, Foreland-propagating Caledonian thrust systems in East Greenland, *in* Higgins, A.K., Gilotti, J.A., and Smith, M.P., eds., The Greenland Caledonides: Evolution of the Northeast Margin of Laurentia: Geological Society of America Memoir 202, doi: 10.1130/2008.1202(07).

Leslie, A.G., Smith, M., and Soper, N.J., 2008, this volume, Laurentian margin evolution and the Caledonian orogeny—A template for Scotland and East Greenland, *in* Higgins, A.K., Gilotti, J.A., and Smith, M.P., eds., The Greenland Caledonides: Evolution of the Northeast Margin of Laurentia: Geological Society of America Memoir 202, doi: 10.1130/2008.1202(13).

Smith, M.P., and Rasmussen, J.A., 2008, this volume, Cambrian–Silurian development of the Laurentian margin of the Iapetus Ocean in Greenland and related areas, *in* Higgins, A.K., Gilotti, J.A., and Smith, M.P., eds., The Greenland Caledonides: Evolution of the Northeast Margin of Laurentia: Geological Society of America Memoir 202, doi: 10.1130/2008.1202(06).

Sønderholm, M., Frederiksen, K.S., Smith, M.P., and Tirsgaard, H., 2008, this volume, Neoproterozoic sedimentary basins with glacigenic deposits of the East Greenland Caledonides, *in* Higgins, A.K., Gilotti, J.A., and Smith, M.P., eds., The Greenland Caledonides: Evolution of the Northeast Margin of Laurentia: Geological Society of America Memoir 202, doi: 10.1130/2008.1202(05).

Stendal, H., and Frei, R., 2008, this volume, Mineral occurrences in central East Greenland (70°N–75°N) and their relation to the Caledonian orogeny—A Sr-Nd-Pb isotopic study of scheelite, *in* Higgins, A.K., Gilotti, J.A., and Smith, M.P., eds., The Greenland Caledonides: Evolution of the Northeast Margin of Laurentia: Geological Society of America Memoir 202, doi: 10.1130/2008.1202(12).

Manuscript Accepted by the Society 14 January 2008

Geological research and mapping in the Caledonian orogen of East Greenland, 70°N–82°N

Niels Henriksen
A.K. Higgins*
Geological Survey of Denmark and Greenland, Øster Voldgade 10, DK-1350 Copenhagen K, Denmark

ABSTRACT

The East Greenland Caledonides, which make up an ~1300-km-long stretch of North-East Greenland, were formed by the collision of Laurentia and Baltica in mid-Silurian time. Geological mapping and research in this remote and poorly accessible segment of the circum-Atlantic Caledonide orogen began in connection with geographical exploration voyages in the early part of the nineteenth century. The first regional geological mapping took place during the long series of "The Danish Expeditions to North-East Greenland" between 1926 and 1958. Modern geological research and regional mapping by the Geological Survey of Denmark and Greenland between 1968 and 1998 have resulted in the publication of a series of 1:500,000 geological maps of the orogen, and an overview geological map at 1:1,000,000 scale, which accompanies this volume.

This article reviews the history of geological research and the evolution of interpretations of the orogen. The recent systematic studies by the Geological Survey of Denmark and Greenland supplement and build on the considerable existing published literature and demonstrate that the North-East Greenland segment of the Caledonide orogen consists of a westward-propagating thrust sheet pile, with displacements estimated at 300–500 km. The thrust sheets incorporate major segments of reworked Laurentian gneiss basement, and a thick succession of Neoproterozoic to Ordovician sediments that accumulated in a major basin originally located outboard of the present coastline.

Keywords: Caledonides, Greenland, history of research, geological mapping.

INTRODUCTION

The Atlantic coast of East Greenland is almost 3000 km long, and it extends from latitude 60°N to 82°N. The ice-free stretch of land along the coast varies from a few kilometers in width in parts of South-East and North-East Greenland, to ~300 km in the central region between latitudes 70°N and 74°N. This chapter concerns the northern half of East Greenland, between latitudes 70°N and 82°N, the region of the East Greenland Caledonian orogen.

The coast of northern East Greenland is inaccessible by ship for most of the year due to the sea ice brought down from the Arctic Ocean by the East Greenland current. The floating ice forms a barrier up to several hundred kilometers wide, but

*akh@geus.dk

in the summer months, the ice thins and breaks up, and ice-strengthened ships can usually reach the coast and enter the fjords south of ~77°N. The region lies within the Arctic and High Arctic climate zones, and the ice-free land areas are snow covered and fjords are frozen for ~8–10 mo of the year. Most exploration activities and scientific investigations are therefore restricted to 2–3 mo in the middle of the summer.

Exploration of the coast of East Greenland did not begin until the early part of the nineteenth century, and the northern parts of East Greenland were not reached until the early part of the twentieth century. Even today, some parts of East Greenland remain difficult of access, and the present-day inhabitants live in the Ammassalik region around the town of Tasiilaq at ~65°30′N in South-East Greenland, and around the town of Illoqqortormiut–Scoresbysund at ~70°30′N in the Scoresby Sund region. There are no Inuit settlements farther north, and the only residents are personnel at a few military stations and at the weather station at Danmarkshavn. In the absence of any established infrastructure, all expedition activities need to be carried out by self-supporting groups. The use of STOL (short take-off and landing) aircraft and helicopters, together with modern communication systems, has considerably increased the efficiency of systematic scientific investigations in East Greenland in recent years. However, some aspects of field work methods have changed little since the first visits by explorers 100–200 yr ago. A comprehensive account of the history of geological exploration in North-East Greenland has been presented by John Haller (1971) in his book *Geology of the East Greenland Caledonides*.

EARLIEST EXPLORATION AND SCIENTIFIC INVESTIGATIONS

The first landing by Europeans on the coast in North-East Greenland (Fig. 1) was made by William Scoresby in 1822. The published journal of Scoresby's voyage (Scoresby, 1823) is notable not only for the chart of the coast between 69°N and 73°N, which shows the mouth of Scoresby Sund, but also for brief geological descriptions (by Robert Jameson) of rock samples collected by Scoresby.

The region north of Scoresby Sund was next visited by Edward Sabine and Douglas Clavering in 1823, followed in 1869–1970 by major exploration of the region from 74°N to 77°N by "The Second German North Pole Expedition," led by Karl Koldewey. In the summer of 1870, the latter expedition partially explored Kejser Franz Joseph Fjord (~73°30′N). The expedition made important scientific observations, and a reconnaissance geological map of the region 73°N–76°N was compiled by F. von Hochstetter et al. (1874). A basement complex of old crystalline rocks was shown, overlain by

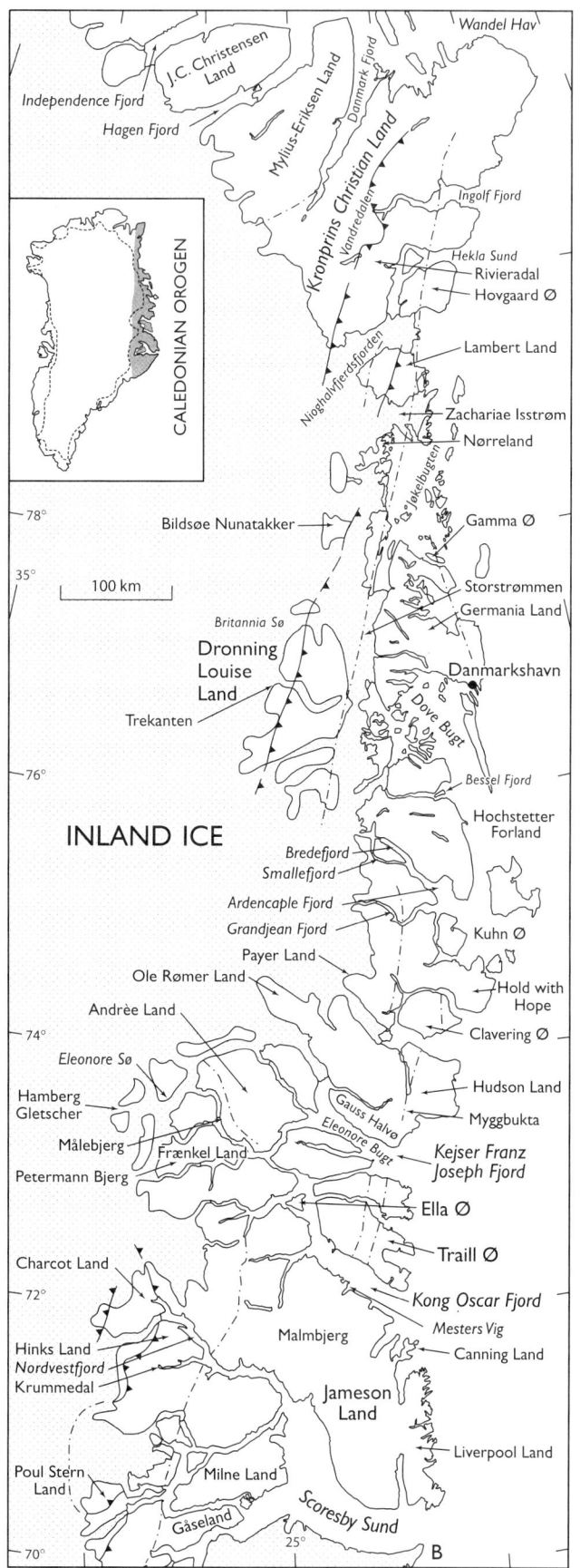

Figure 1. Map of North-East Greenland with most of the place names used in the text. B—Blosseville Kyst. Additional place names are found on the 1:1,000,000 geological map of the Caledonian orogen (Plate 1, Henriksen and Higgins, this volume, Chapter 14). Names with latitudes and longitudes are given in an appendix to Henriksen and Higgins (this volume, Chapter 14).

sedimentary rocks then believed to be Paleozoic and compared to the "Hecla Hoek Formation" of Svalbard. In the outer coastal areas, Mesozoic and Cenozoic sediments were recognized.

In 1878, the Danish State established a Scientific Commission for Greenland (Commisionen for Ledelse af de Geologiske og Geografiske Undersøgelser i Grønland), which initiated a new era of exploration and scientific work in Greenland. An expedition was sent out to explore the large Scoresby Sund fjord system in central East Greenland (Carl Ryder's Østgrønlandske Expedition 1891–1892). The two geologists (E. Bay and N. Hartz) recorded large areas of crystalline rocks, various unfolded sedimentary rocks (post-Caledonian successions), and plateau basalts. A geological sketch map of the region accompanied Bay's (1896) geological report.

In 1897, the Swedish explorer S.A. Andrée and two companions attempted to reach the North Pole from Svalbard using a balloon. The party disappeared without a trace (their bodies were not found until 1930), and in a search for them, the Swedish geologist and paleobotanist A.G. Nathorst led two Arctic expeditions, the first to Svalbard in 1898, and the second to East Greenland in 1899 (Nathorst, 1901; Liljequist, 1993). Nathorst discovered and mapped the network of interconnecting fjords between Kejser Franz Joseph Fjord and Kong Oscar Fjord (Nathorst, 1901). He was able to improve and elaborate on the scant geological information on the region given by Koldewey's 1869–1870 expedition, and, amongst other things, he recognized the Paleozoic age of the fold belt in East Greenland and recorded the unconformity between the deformed "Hecla Hoek Formation" (present-day Eleonore Bay Supergroup to Ordovician succession) and the overlying Devonian sandstones and conglomerates. On Nathorst's (1901) 1:2,000,000 colored geological map of the region from 72°N to 75°N, a basement of gneisses overlain by Silurian (and Cambrian) sediments was depicted, which were in turn overlain by Devonian and Mesozoic sediments (Keuper, Rhaetic, Jurassic), with a cover of younger volcanic rocks.

In the years 1898–1900, a major Danish expedition led by G.C. Amdrup worked in South-East Greenland (Carlsbergfondets Expedition til Øst-Grønland), mainly in the region around Ammassalik (65°36′N). In the summer of 1900, the work was extended northward to the Blosseville Kyst, Liverpool Land, and the Kong Oscar Fjord region, where the Swedish geologist and geographer Nils Otto Nordenskjöld was a participant. A colored 1:2,000,000 geological map of the region from 66°N to 75°N was compiled (Nordenskjöld, 1907), which incorporated the earlier observations of Koldewey's and Nathorst's expeditions. This map of the southern part of the Caledonian orogen distinguishes Archean gneisses, the Cambrian–Silurian succession affected by deformation, and Devonian and Mesozoic sedimentary rocks and Paleogene basalts.

Until the beginning of the twentieth century, there was a large blank area on maps of Greenland between Germania Land (~77°N) in East Greenland and Kap Clarence Wyckoff (~83°N) in easternmost North Greenland (Ventegodt, 2000). The exploration of this 700-km-long stretch of unknown land was the principal aim of the 1906–1908 "Danmark Ekspeditionen." This large and ambitious early Danish expedition sailed to Greenland on the ship *Danmark* and stayed over the winter in an excellent harbor in southern Germania Land subsequently known as Danmarkshavn (76°46′N). The expedition made numerous boat and sledge journeys in the Dove Bugt region, a long northward sledge journey in the spring and early summer of 1907 that reached northernmost East Greenland, and a journey to the interior of Independence Fjord and Danmark Fjord in eastern North Greenland (Amdrup, 1913). These journeys filled some of the major gaps on maps of Greenland. The geological observations of the Danmark Expedition reported by Nathorst (1911) and Ravn (1911) concerned the post-Caledonian sedimentary rocks and their fossils, and there are only brief remarks in respect to the older rock units.

Amongst the members of the Danmark Expedition, there was the German meteorologist Alfred Wegener, who with J.P. Koch organized an east-west crossing of the Inland Ice at its widest part in 1912–1913. Alfred Wegener subsequently became well known for his revolutionary hypothesis of continental drift. The 1912–1913 expedition began their Inland Ice crossing from the large nunatak complex of Dronning Louise Land, from which they were able to provide the first geological observations (Koch and Wegener, 1930).

MAPPING AND SCIENTIFIC INVESTIGATIONS (1920–1939)

The pre–Second World War period in East Greenland saw a series of major achievements in diverse scientific fields, including surveying, botany, zoology, archaeology, hydrography, and a wide range of geological subjects. Most of the activities were carried out during a long series of geological expeditions under the leadership of the Danish geologist Lauge Koch. After a period of mapping and exploration in North Greenland from 1916 to 1923 (Koch, 1919, 1926; Dawes, 1991, 1992), Lauge Koch was employed by the Danish State to undertake geological investigations in East Greenland north of 70°N. The pre-war "Danish Expeditions to North-East Greenland" from 1926 to 1938, were continued after the war between 1947 and 1958. Most of the results of these activities were published in the monograph series "Meddelelser om Grønland," published by the Scientific Commission for Greenland, Copenhagen.

Other expeditionary activity included the Cambridge East Greenland Expeditions led by J.M. Wordie in 1926 and 1929 and Louise A. Boyd's arctic expeditions to East Greenland in 1931 and 1933. Both Wordie's and Louise Boyd's expeditions included geologists who made brief observations in different parts of the region.

Lauge Koch's Early Expeditions (1926–1927, 1929–1930)

Lauge Koch's first East Greenland expedition in 1926–1927 consisted of three geologists and two Greenlander dog-sledge drivers. The main objective was a general geological survey of the region north of Scoresby Sund (70°30′N), and Koch, on spring sledge journeys, reached as far north as Danmarkshavn (76°46′N). In his report of the expedition's work, *The Geology of East Greenland* (Koch, 1929a), early published information was combined with his own observations into a regional 1:1,000,000 geological map (69°N–77°N). The follow-up expeditions of 1929 and 1930 included seven Danish and Swedish geologists, who worked with both Caledonian and post-Caledonian rock successions mainly in the region from 72°N to 74°N.

Important results of these early investigations included the recognition by Poulsen (1930) of a Cambrian-Ordovician succession above diamictites interpreted as tillites. The underlying, very thick, late Precambrian sedimentary succession was named the Eleonore Bay Formation by Koch (1929a). One of the notable features of Koch's report (1929a) was his speculation that the crystalline gneisses of the coastal regions might have a Caledonian origin. He envisaged an extensive geosyncline made up of Algonkian (Neoproterozoic) and Lower Paleozoic sediments covering a large region of East Greenland (Fig. 2; Koch, 1929b) that had been deformed by westward-directed pressure in Caledonian time.

J.M. Wordie's Cambridge Expeditions (1926, 1929) and Louise Boyd's Expeditions (1931, 1933)

The main objectives of the Cambridge expeditions led by J.M. Wordie were surveying and general exploration of the outer coastal regions and the inner part of Kejser Franz Joseph Fjord. Both expeditions included geologists, who described an extensive sedimentary succession named the Petermann Series from the inner fjord region (Wordie, 1930; Wordie and Whittard, 1930). In the outer fjord region, Wordie (1927) and Wordie and Whittard (1930) used the name "Franz Joseph-Beds" for the present-day Neoproterozoic–Lower Paleozoic succession corresponding to Nathorst's "Silurian." The deformation of this succession was attributed to Caledonian overfolding and thrusting across the underlying gneissic rocks. The British geologists distinguished Eastern, Central, and Western Metamorphic Complexes (Parkinson and Whittard, 1931), and compared the Precambrian rocks to the Lewisian gneisses and Moine metasediments of Scotland. Their interpretation of the structure envisaged a Neoproterozoic to Ordovician succession that was displaced across the older crystalline rocks on a major Caledonian thrust, but with little granite generation or reworking of the rock complexes below the thrust. Their interpretation of the Caledonian fold belt of East Greenland as a "superficial" orogeny, was supported by Curt Teichert, one of Koch's geologists (Teichert, 1933).

Noel Odell, who took part in the 1933 Louise Boyd Expedition, undertook a study of the Petermann Series in Knækdal, western Frænkel Land, and reported an increasing degree of metamorphism downward through the metasedimentary succession and an apparently gradual transition into the crystalline gneisses (Odell, 1939, 1944). However, he agreed with the Cambridge group in interpreting the crystalline basement as being of Archean age.

Three-Year Expedition (1931–1934), Led by Lauge Koch

The "Treårsekspedition" (three-year expedition) of 1931–1934 was one of the largest and most comprehensive expeditions ever sent to East Greenland by Denmark. The ships *Gustav Holm* and *Godthaab* were used for transport (Fig. 3), and seaplanes were used for photographic flights and geological reconnais-

Figure 3. The S.S. *Gustav Holm* in the East Greenland pack ice in 1932. The ship was used by Lauge Koch's East Greenland expeditions for a total of seven seasons in the 1930s. The Heinkel hydroplane carried up to East Greenland by the ship was used for aerial photography and geological reconnaissance. Photograph is from Koch (1940, his Fig. 33). Reproduced with permission from Meddelelser om Grønland.

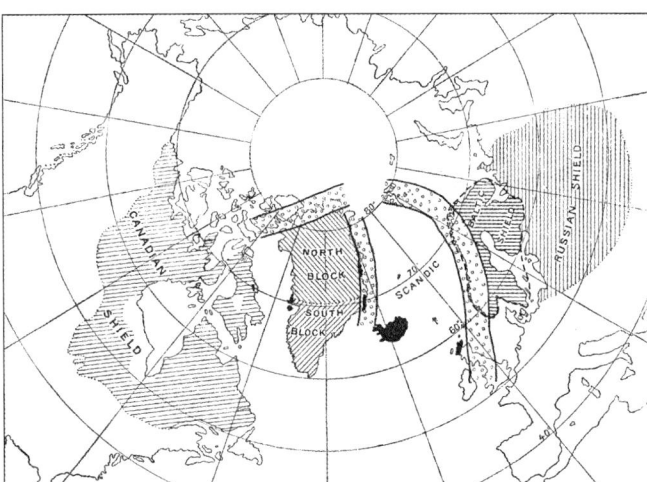

Figure 2. Lauge Koch's interpretation of the distribution of shield areas and geosynclines around the North Atlantic Ocean (from Koch, 1929b, his Fig. 60). Reproduced with permission from Meddelelser om Grønland.

sance. The expedition extended over four summers and three winters, and wintering stations were built at Ella Ø and at Eskimonæs on southern Clavering Ø. Between 60 and 100 participants took part each year, of whom 30%–50% were scientists. Much of the work undertaken was geological, but investigations included topographical surveying, biology, archaeology, and hydrography. The main region of investigation extended from Scoresby Sund (70°30′N) to Hochstetter Forland (75°30′N), with reconnaissance up as far as 77°N. The topographical surveying program produced a series of 1:250,000 maps, which became the basis for the regional geological mapping between 72°N and 76°N. General accounts of the expedition are found in Koch (1955) and Thorson (1937).

The geological studies included mapping and general investigations in both the Caledonian orogen and post-Caledonian rocks. The geologists included specialists, mainly from Danish, Swedish, German, Finnish, and Swiss universities. In the Caledonian orogen, work was carried out on the crystalline gneisses and overlying metasedimentary rocks, and a special study was made of the continental Devonian sediments. The studies of the metamorphic complexes of the inner fjord region by Backlund (1930, 1932) suggested that there was a gradual transition between the widespread metasedimentary successions and the structurally underlying crystalline gneiss complexes. Backlund proposed that Wordie's "Archaean" gneisses and granites were the result of Caledonian granitization and migmatization of a varied sedimentary succession. Wegmann's (1935) structural observations in the same region led to a division into migmatitic units, high-grade metasedimentary rocks, and marginal granites, all underlying a cover sequence of Precambrian to Lower Paleozoic sediments. Koch's early ideas, that the gneisses of the outer coastal regions were the nucleus of Caledonian orogenesis, were extended in a dramatic way by Backlund's and Wegmann's studies, who now considered the migmatitic infrastructure of the entire Caledonian fold belt to be the result of large-scale transformation of a thick geosynclinal succession.

Stratigraphical and structural studies in the continental Devonian succession were initiated by Bütler (1935). The up to 8-km-thick deposits rest with profound unconformity on folded Cambrian-Ordovician sediments. The first vertebrate fossils were found in 1929, including the earliest tetrapods—the so-called Ichthyostegalia. During the "Three-Year Expedition," detailed investigations of the tetrapods, the transitional forms between fish and land-living animals, were undertaken by vertebrate paleontologists based in Stockholm (Säve-Söderbergh, 1932, 1933, 1934).

Two Conflicts with Far-Reaching Consequences for East Greenland

In the early part of the twentieth century, Norwegian sealers began to visit the coastal areas of East Greenland to supplement their catch at sea with land-based animals such as musk ox. This practice was soon extended by small groups of Norwegian and Danish hunters, who erected small hunting stations and, in the winter, trapped foxes, and, to a lesser extent, polar bears and wolves for the European fur industry (see, e.g., Mikkelsen, 1994). The rivalry between Danish and Norwegian hunting groups culminated in 1931, when a group of Norwegian hunters raised the Norwegian flag at Myggbukta and claimed a substantial part of East Greenland between 71°30′N and 75°40′N as Norwegian territory. The occupation of this region, "Eirik Raudes Land," was supported by the Norwegian government. The Danish government immediately raised objections, and the case was referred to the International Court at The Hague. In April 1933, the International Court granted Denmark sovereignty of all of East Greenland and established that Danish political and scientific responsibilities extended to all parts of Greenland, a decision that in particular encouraged ongoing Danish scientific exploration of East Greenland. A significant factor in the Danish argument was the extensive scientific activity that had been carried out since 1926 by Lauge Koch's expeditions.

In the early 1930s, Lauge Koch was invited to contribute to an international German monograph series, *Geology of the World* (Geologie der Erde), and the result was a volume on Greenland geology, *Geologie von Grönland* (Koch, 1935). His description was considered unbalanced by some Danish geologists because it did not give appropriate credit to their own observations. A group of these Danish geologists published a detailed critical review (Böggild et al., 1935), which led to a conflict (Ries, 2003) with far-reaching consequences, not only for those directly involved, but for the course of Danish geology for many years to come. In respect of Koch's East Greenland expeditions, Lauge Koch subsequently refused to employ Danish geologists (with very few exceptions), and his later expeditions up to 1958 were largely manned by geologists from Switzerland, Sweden, and Great Britain.

Lauge Koch's Two-Year Expedition (1936–1938)

The two-year expedition of 1936–1938 had almost entirely geological objectives, and it spanned three summers and two winter seasons. The summer ships were again *Gustav Holm* and *Godthaab*, and in 1938, a seaplane was used for aerial reconnaissance. The geological studies were mainly concentrated on the post-Caledonian sedimentary rocks between Jameson Land (71°30′N) in the south and Kuhn Ø (75°N) in the north. Bütler's studies in the Middle–Upper Devonian succession around Hudson Land and Ole Rømer Land in 1936 and 1938 led to the recognition of four intra-Devonian movement phases in Hudson Land (Bütler, 1948).

RESUMPTION OF SCIENTIFIC INVESTIGATIONS AFTER WORLD WAR II

Lauge Koch's Danish Expeditions to East Greenland resumed in 1947, and a comprehensive program of mainly geological investigations was carried out every year until 1958.

During this 12 yr period, basic geological mapping of the region from 72°N to 76°N at 1:250,000 scale was completed. Reconnaissance investigations of the northern parts of the Caledonian orogen extended up to ~81°30′N, and a start was also made with field work in the Scoresby Sund region, only to be suspended when funding for Koch's expeditions was unexpectedly withdrawn after the 1958 season.

The format of the postwar expeditions was at first similar to the last prewar expeditions, based on ship transport and groups of scientists spending the winter at scientific stations in Greenland. However, Catalina flying boats soon replaced ships for transport of personnel, and after the construction of "Mestersvig airport" in 1952 (west of the bay Mesters Vig), DC-4 aircraft were used. Koch records that 691 persons took part in his postwar expeditions, but this figure includes the crews of ships in the early years, as well as the mining engineers and drilling teams involved in prospecting around Mesters Vig. In general, six to eleven geological two- to three-man teams were active each year.

From 1948, Koch's expeditions had their own single-engine Norseman seaplanes (Fig. 4), which were initially used for transport within the region. Systematic compilation of geological maps was begun in 1953 by John Haller, and to complete these maps and fill out gaps in coverage, more than 32,000 km of reconnaissance and photographic flights were carried out using Norseman aircraft in 1955, 1956, and 1958, supplementing the oblique aerial photograph routes flown for the Geodetic Institute between 1950 and 1954. Vertical aerial photograph routes were flown for the Geodetic Institute over large areas of northern East Greenland (north of 76°N) in the early 1960s and, while too late to be of use during the field work, were extensively used by John Haller for photogeological studies in connection with his regional structural and geological maps (Haller, 1970, 1983).

In 1948, a lead-zinc deposit was discovered near Mesters Vig, on the south side of Kong Oscar Fjord. Prospecting and drilling in subsequent years led to the establishment of a mining company, Nordisk Mineselskab, to exploit the deposits. The mine opened in 1954 (Mikkelsen, 1992). The presence of the airport near the mine greatly improved access for Koch's expedition personnel, and the airport subsequently became the main hub for expedition activities for the next 40 years.

In the period 1952–1954, a large British Joint Services expedition mapped and explored the large nunatak complex of Dronning Louise Land (~76°N–77°30′N), and significant new geological observations were made in this almost virgin territory (Peacock, 1956). A wintering station was established on the shore of Britannia Sø.

Systematic Geological Mapping in the Fjord Region (72°N–76°N)

The main aim of Lauge Koch's expeditions after the Second World War was to complete the geological mapping of the region from 72°N to 76°N that was covered by the 1:250,000 topographic maps prepared by the Geodetic Institute. Field map-

Figure 4. Norseman seaplane, the work-horse of the postwar Danish Expeditions to East Greenland (1947–1958). The Norseman was used for transport of geological parties to their field areas, and it was also extensively used for aerial photography and geological reconnaissance. More than 12,000 oblique photographs were taken along fjords, valleys, inland areas, and otherwise inaccessible nunataks. Photo is from the Geological Survey of Denmark and Greenland archive.

ping was mainly carried out on 1:100,000 enlargements of these maps. The field maps of the large group of geologists, mainly from Switzerland, Sweden, and Great Britain, were compiled by John Haller, who filled in gaps in the ground coverage by aerial reconnaissance from Norseman aircraft and photogeological studies. The compilation work continued after the suspension of Koch's expeditions in 1958, and the colored maps were printed as a set of 13 sheets in 1964; they were released with a short description in 1971 (Koch and Haller, 1971). The work carried out in the Caledonian orogen is briefly summarized here but is also referred to in many of the special articles in this volume.

One of the outstanding features of the central fjord zone of East Greenland (72°N–74°30′N) is the ~14.5-km-thick multicolored succession of Neoproterozoic sediments, then known as the Eleonore Bay Formation. This gently folded succession is continuously exposed over a north-south distance of 250 km, and mapping of the succession came into focus at the start of the postwar activities. A division into lower and upper parts was established, and the upper part was divided into "series," which in turn were subdivided into lithostratigraphical units (Fränkl, 1951, 1953a, 1953b; Eha, 1953; Katz, 1952a; Sommer, 1957). The Eleonore Bay Formation of Koch (1929a) was then upgraded to a "Group" (Katz, 1961; Haller 1971), and following new survey investigations in recent years, it has been upgraded to a "Supergroup," with a new formalized stratigraphy (Sønderholm and Tirsgaard, 1993).

C. Poulsen and H. Wienberg Rasmussen, working independently from Lauge Koch's expeditions, undertook mapping in 1946 of the Vendian to Ordovician succession overlying the Eleonore Bay Supergroup on Ella Ø. Studies were made of a sequence containing diamictite beds interpreted as tillites (Tillite Group), and the overlying Cambrian-Ordovician carbonate strata (Poulsen and Rasmussen, 1951). The Lower Paleozoic succession was also studied by the British geologist J.W. Cowie, working with Lauge Koch's expeditions, who established a stratigraphy for the ~4000-m-thick succession (Cowie and Adams, 1957).

Investigations in the Devonian sedimentary basins continued with H. Bütler responsible for the stratigraphy, while studies of the Devonian vertebrate fossils were undertaken by paleontologists from Stockholm (Jarvik, 1950, 1961). Bütler completed his investigations in 1957 and presented a regional correlation of the Middle and Upper Devonian deposits in central East Greenland (Bütler, 1959). Associated Devonian igneous intrusions were investigated by Dal Vesco (1954) and Graeter (1957).

The crystalline complexes of the interior fjord zone of the Caledonian orogen were mapped and investigated by a group of Swiss geologists, mainly from the University of Basel. The leader of the group was Eduard Wenk, who had first participated in Koch's expeditions in 1934, and he was accompanied by a number of younger geologists and students. One of these was John Haller, whose first summer in Greenland was in 1948. Haller stayed in Greenland in 1949–1950, and participated in a total of 10 summer field seasons up to 1958. John Haller was chief geologist of Koch's expeditions during the last part of this period and is best known for his map compilations (Haller, 1970; Koch and Haller, 1971) and his book on the East Greenland Caledonides (Haller, 1971).

The early postwar research in the crystalline complexes concentrated on structural and petrological investigations in the inner parts of Kejser Franz Joseph Fjord (Huber, 1950; Haller, 1953). Mapping was subsequently extended farther west to the nunatak region (Wenk and Haller 1953; Haller, 1956a; Katz, 1952b). The investigations of the Petermann "Series" sediments showed that the succession could be correlated with the Eleonore Bay "Group" succession. Wenk and Haller (1953) demonstrated that the metasediments of the Petermann "Series" become increasingly metamorphosed downwards with an apparent transition into the underlying gneisses and migmatites. This latter observation appeared to confirm the conclusions of Haller's (1953) work in Andrée Land, and, building on the earlier work of Backlund (1930, 1932) and Wegmann (1935), led to development of the in situ model for orogenic evolution known as the "stockwerk" concept (Fig. 5). Rising fronts of migmatization and granitization were envisaged to have transformed the lower levels of the Eleonore Bay "Group" succession into the gneisses and migmatites of the metamorphic complexes. The different levels (or stories) of the "stockwerk" and the convoluted shapes of the migmatitic domes and upwellings were illustrated in Haller's maps and cross sections (Haller, 1955, 1958).

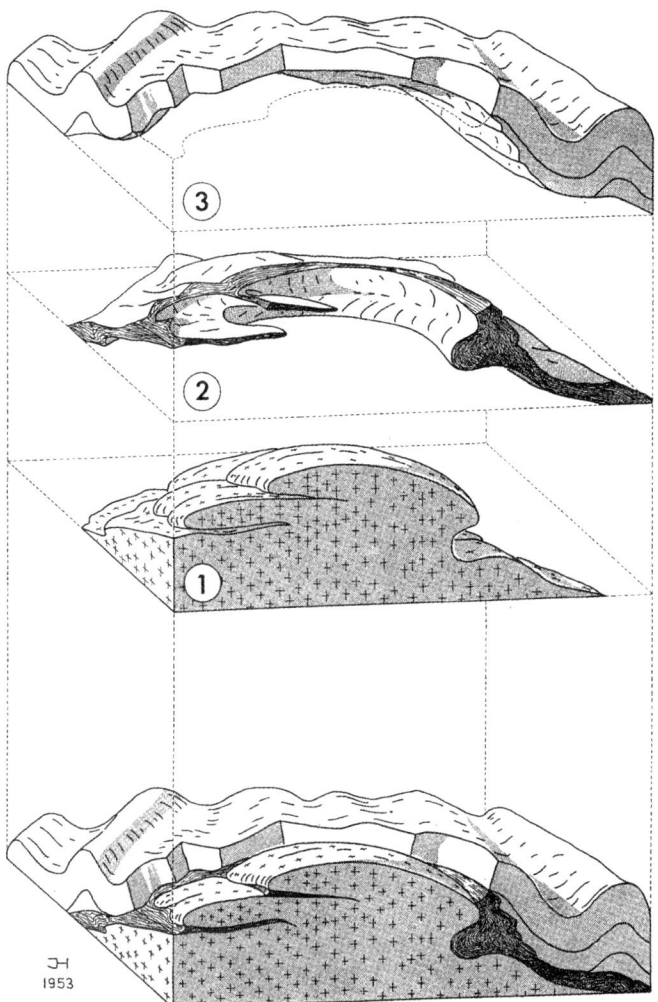

Figure 5. John Haller's interpretation of the structure in the central part of the Caledonian fold belt in East Greenland. He envisaged a "stockwerk" structure with three levels: (1) a deep-seated infrastructure with migmatitic upwellings, (2) an intermediate disharmonic detachment zone, and (3) a gently folded superstructure of low-grade metasediments. Diagram is from Haller (1971, his Fig. 52). Reproduced with permission from John Wiley & Sons Limited.

Early Investigations in the Scoresby Sund Region (70°N–72°N)

Lauge Koch had intended that his geological work would be extended to systematic mapping in the Scoresby Sund region from the 1959 field season, and he initiated reconnaissance mapping in the summers of 1957 and 1958. Previous investigations in the Scoresby Sund region had been scattered and limited. The isolated crystalline block of Liverpool Land had been visited in the 1930s (Kranck, 1935; Sahlstein 1935a, 1935b) and was one of the regions viewed by Koch (1929a) as being of Caledonian origin. The crystalline complexes of the interior part of Nordvestfjord were reached during a boat journey in 1934 (H.G. Backlund *in* Koch, 1955), and in 1957, P. Vogt and E. Wenk visited Hinks

Land, on the south side of innermost Nordvestfjord (Vogt, 1965). Vogt recognized a major thrust that displaced migmatitic paragneisses with a thick cover of metasediments westward above lower-grade metasedimentary rocks. In 1958, E. Wenk mapped part of inner Gåseland and nearby Paul Stern Land and described a significant major thrust thought to be a continuation of the thrust he had mapped in Hinks Land (Wenk, 1961). Wenk considered the underlying gneisses to be an intact region of Precambrian basement. This was the first foreland window recognized in the southern part of the East Greenland Caledonian orogen.

Reconnaissance Investigations in the Northern Part of the Orogen (76°N–81°N)

Investigations by Lauge Koch's expeditions north of 76°N were limited to observations at a few key localities, extensive aerial reconnaissance, and photogeological interpretations. The Norseman seaplanes were only able to land on ice-free lakes and the inner parts of fjords, and field work was thus limited by logistic considerations. In 1952 and 1953, a group of four geologists was able to work in southern Kronprins Christian Land. Two British geologists, J.W. Cowie and P.J. Adams, established a stratigraphy for the Proterozoic and Cambrian–Silurian sedimentary successions on the foreland west of the fold belt (Adams and Cowie, 1953; Cowie, 1961). The second party focused on the structural relationships of the deformed and displaced sedimentary sequences. Fränkl (1954, 1955) recognized the presence of nappe-scale thrust units that displaced Proterozoic sediments westward over an Ordovician–Silurian carbonate succession.

John Haller made extensive observations in the crystalline areas around Dove Bugt (76°N–78°N) in 1955, mainly from the air, and observations were extended to the region 78°N–82°N in 1958. On the basis of these observations, supplemented by occasional spot landings, and the reconnaissance work in Dronning Louise Land (Peacock, 1958) and Kronprins Christian Land (Adams and Cowie, 1953; Fränkl, 1954, 1955), a general geological map at 1:1,000,000 scale was compiled (Haller, 1983). Structural maps at 1:500,000 of the orogen north of 72°N were also published (Haller, 1970).

The aerial reconnaissance work by Haller (1961) in the northern half of the orogen seemed to indicate an angular unconformity between two Proterozoic successions, and the presence of two generations of basic intrusions. The oldest of the Proterozoic successions was assigned by Haller to the Thule Group, a sedimentary succession, ~6000 m thick, thought to have been deposited in a "Carolinidian geosyncline," and deformed during the so-called "Carolinidian" orogeny, in addition to later Caledonian deformation. The supposed "post-Carolinidian" Neoproterozoic succession (that had only suffered Caledonian deformation) was named the Hagen Fjord Group. Later, more detailed field work, which included visits to the type area, failed to find evidence for such a "Carolinidian" orogeny (Jepsen and Kalsbeek, 1985).

Despite some errors in interpretation, the reconnaissance investigations of John Haller have proved to be a valuable guide for later survey investigations, including detailed field mapping.

SURVEY INVESTIGATIONS IN THE PERIOD 1968–1998

An extended phase of compilation and publication of scientific results fortunately succeeded the suspension of Lauge Koch's Danish Expeditions to East Greenland. The chief geologist of the expeditions, John Haller, with the support of funding from Danish and Swiss sources, was enabled to produce a series of landmark publications (Haller, 1970, 1971; Koch and Haller, 1971; see also Schwarzenbach, 1993). Haller's compilation work was begun at the University of Basel, Switzerland, between 1958 and 1964, and continued from 1965 at Harvard University, United States. Lauge Koch died in 1964 and thus did not live to see the final form of the publications that provide a worthy monument to the 40 yr of East Greenland expeditions under his leadership.

In the late 1960s, the Geological Survey of Greenland (GGU) took up the unfinished work in the Scoresby Sund region of East Greenland as part of the survey's regional geological mapping program. This program, a project to cover all the land areas of Greenland with 14 geological map sheets at 1:500,000 scale, was completed in 2004. Following a small reconnaissance investigation in 1967, a series of five major expeditions was undertaken in the Scoresby Sund region (70°N–72°N) between 1968 and 1972 (Henriksen, 1986). This first survey project in the East Greenland Caledonides was followed by a series of expeditions to complete geological mapping of the Paleozoic fold belts of East and North Greenland, which, up to 1998, involved field work during a total of 18 field seasons. In addition to the major expeditionary activity, small survey groups undertook follow-up investigations throughout East Greenland. While the results of Lauge Koch's Danish Expeditions to East Greenland were largely printed in the monograph series *Meddelelser om Grønland*, the survey's investigations have been extensively published in international journals and in the GGU/GEUS *Rapport* and GGU *Bulletin* series. The regional survey activity is described in more detail in a later chapter (Henriksen and Higgins, this volume, Chapter 14).

Scoresby Sund Region (70°N–72°N)

The Scoresby Sund region encompasses the southernmost part of the Caledonian orogen, the continuation of which southward is concealed beneath a thick cover of Paleogene basalts. Mapping of the inner Scoresby Sund fjord region by geologists familiar with the old Precambrian terrains of West Greenland, led to the recognition of widespread basement crystalline gneiss complexes. Radiometric age determinations in the 1970s, based on bulk zircon analyses and Rb-Sr whole-rock determinations, led to the distinction of rock units of Archean, Paleo-, Meso-, and Neoproterozoic age, as well as of Caledonian age. This gave rise to the polyorogenic interpretation of the region (Rex and Gledhill, 1974; Steiger et al., 1979), in contrast to the exclusively Caledonian origin implied by Haller's "stockwerk" concept (Haller, 1971).

At the western margin of the Scoresby Sund region, the Gåseland foreland window, originally recognized by Wenk (1961), was remapped. The thrust system in Hinks Land reported by Vogt (1965) that borders the Charcot Land window was suggested to have a westward displacement in excess of 100 km (Higgins, 1982).

Kong Oscar Fjord Region (72°N–75°N)

In the mid-1970s, reconnaissance activities were carried out in the Kong Oscar Fjord region by small groups of geologists, with limited helicopter support. This region was well known from the activities of Lauge Koch's expeditions. Some of the main conclusions from the Scoresby Sund region that cast doubt on some of John Haller's interpretations were confirmed. Thus, some of Haller's supposed "Caledonian" granites that made up the crystalline gneiss complexes yielded Paleoproterozoic Rb-Sr whole-rock ages of ca. 1900 Ma (Higgins et al., 1978; Rex and Gledhill 1981), and a few granites intruding high-grade metasediments gave ages of ca. 1000 Ma. The high-grade metasedimentary rocks that hosted the ca. 1000 Ma granites were correlated with the Mesoproterozoic Krummedal sequence of the Scoresby Sund region (Higgins, 1988).

During systematic regional mapping in the Kong Oscar Fjord region in the summers of 1997–1998, many of the westernmost nunataks were visited for the first time, and other classic localities were reinvestigated (Henriksen, 1998, 1999; see also Henriksen and Higgins, this volume, Chapter 14). The main conclusions of the new survey mapping resulted in formulation of a new model for the tectonic architecture of the Kong Oscar Fjord region (72°N–75°N) and publication of a 1:500,000 geological map (Escher, 2001). The structural model, initially presented in a limited-circulation GGU report (Elvevold et al., 2000), and published in a revised form by Higgins et al. (2004b), is the result of recognition of two new foreland windows, and the implications for major displacements for the overlying thrust sheets (Higgins et al., 2001a, 2001b). Both windows contain thin Early Cambrian quartzite units with *Skolithos* ichnofossils (Fig. 6). Two major thrust sheets are recognized overlying the windows, the lower Niggli Spids thrust sheet, which is dominantly composed of basement gneisses and schists, and the upper Hagar Bjerg thrust sheet, which includes the very thick (up to 18 km) Neoproterozoic–Ordovician succession. These observations demonstrate that the in situ "stockwerk" concept of Haller (1970, 1971) is untenable (Higgins and Leslie, 2004).

Dove Bugt Region (75°N–78°N)

Systematic field work in the Dove Bugt region for the GGU 1:500,000 mapping project was undertaken in the summers of 1988–1990. Some early results were presented

Figure 6. Lower Cambrian quartzite with *Skolithos* ichnofossils in the Eleonore Sø window. The distortion along the bedding plane is caused by the displacement of the major thrust sheets structurally above the window. Photo is courtesy of A.K. Higgins.

as a collection of papers in the GGU *Rapport* series (Higgins, 1994), and the 1:500,000 Dove Bugt map sheet was printed in 1987 (Henriksen, 1997).

Around Ardencaple Fjord in the southern part of this region, a fault-bounded enclave of the Eleonore Bay Supergroup is preserved, surrounded by older high-grade metasedimentary rocks distinguished as the Smallefjord sequence. The Eleonore Bay Supergroup succession can be correlated at group level with the succession in the Kong Oscar Fjord region (Sønderholm and Tirsgaard, 1993). SHRIMP (sensitive high-resolution ion microprobe) U-Pb dating on zircons demonstrates that the high-grade, migmatitic sedimentary rocks of the Smallefjord sequence were deposited after ca. 1035 Ma and have been affected by a ca. 955 Ma metamorphic event (Strachan et al., 1995b). This indicates that the Smallefjord sequence can be broadly correlated with the Krummedal supracrustal sequence of the Kong Oscar Fjord and Scoresby Sund regions, where similar results have been obtained. Caledonian granites, formed by melting of the Smallefjord sequence metasediments, also invade the structurally overlying Eleonore Bay Supergroup, and they have yielded a narrow range of U-Pb zircon and monazite ages of 431–428 Ma (Strachan et al., 2001).

North of latitude 76°N, the Dove Bugt region is dominated by crystalline gneiss complexes. Apart from a small area of Archean-age gneisses near Danmarkshavn, the orthogneiss complexes have yielded Paleoproterozoic Rb-Sr and Sm-Nd model ages of ca. 2000 Ma and SHRIMP U-Pb zircon ages of 2000–1750 Ma (Kalsbeek et al., 1993; Kalsbeek, 1995). The entire region was extensively reworked during the Caledonian orogeny (Chadwick and Friend, 1994; Hull et al., 1994; Strachan et al., 1995b).

Caledonian eclogites and related high-pressure rocks are widespread within the quartzo-feldspathic orthogneisses north of Danmarkshavn (77°N), and they form part of an eclogite province that extends northward for 400 km to 81°N (Gilotti, 1993, 1994; Brueckner et al., 1998; Gilotti et al., this volume, Chapter 8). Locally, high-pressure eclogites are present in thrust sheets displaced across low-grade foreland windows, indicating that parts of the Laurentian margin were exhumed from depths in excess of 50 km prior to their westward displacement at a late stage in the Caledonian orogeny (Gilotti and Ravna, 2002).

The western thrust margin of the Caledonian orogen was recognized in Dronning Louise Land (76°N–77°N) in the early 1950s (Peacock, 1956, 1958). The later GGU work has added much new detail to the original descriptions. New observations in the foreland of Dronning Louise Land include finds of Cambrian strata below major Caledonian thrust sheets dominated by orthogneisses that were derived from the east (Strachan et al., 1992).

Lambert Land Region (78°N–82°N)

The northernmost segment of the Caledonian orogen was mapped and investigated during the 1993–1995 field seasons (Henriksen, 1996; Jepsen, 2000). The west margin of the orogen is well exposed throughout Kronprins Christian Land. The transition into the undisturbed foreland is marked by a 20–50-km-wide, thin-skinned, parautochthonous fold-and-thrust belt largely developed in Lower Paleozoic shelf limestones and dolomites. Allochthonous thrust sheets make up central Kronprins Christian Land and include the Neoproterozoic siliciclastic Rivieradal Group of the Vandredalen thrust sheet, and thrust sheets comprising imbricated sequences of Paleoproterozoic to Mesoproterozoic sandstones, basic dikes, and volcanic rocks (Pedersen et al., 2002; Collinson et al., this volume).

South of Kronprins Christian Land, thrust sheets dominated by sandstones with dolerite dikes can be traced in a more-or-less continuous zone through western Lambert Land and southward into the nunatak region. The eastern part of the orogen is dominated by Paleoproterozoic gneiss complexes that make up structurally higher thrust sheets.

On the south side of Zachariae Isstrøm, a small but significant tectonic window occurs in Nørreland (78°40′N; J.M. Hull and J.D. Friderichsen, 1995, personal commun.). Within this 20 × 8 km window, low-grade quartzitic metasedimentary rocks crop out; they are cut by Paleoproterozoic granitoids (Kalsbeek et al., 1999), and they are overlain by a thin, strongly deformed unit of Ordovician carbonates preserved immediately beneath the thrust plane (Rasmussen and Smith, 2001). The rock units structurally overlying the Nørreland window are allochthonous Paleoproterozoic gneisses that preserve Caledonian eclogitic enclaves.

INVESTIGATIONS BY NON–SURVEY RESEARCH GROUPS, 1948–2004

Greenland continues to attract the attention of geologists interested in the links between North America and northern Europe, and the Caledonian geology of East Greenland in particular has been the subject of studies by large and small research groups. The problem of transport to and within East Greenland has meant that, at times, one or two geologists have been attached to larger groups with other scientific objectives. More recently, travel possibilities have improved considerably, notably due to the logistic coordination promoted by the Danish Polar Centre.

One of the earliest expeditions was a party from Leeds University, which traveled to Greenland in 1948 with members of the Danish Peary Land Expedition aboard *Godthaab*. General geological investigations were made in A.P. Olsen Land and Th. Thomsen Land (74°30′N–75°N; Leedal, 1952). The work of the 1952–1954 British Joint Services Expedition to Dronning Louise Land has already been referred to herein (Peacock, 1956, 1958).

Following the discovery of lead and zinc mineralizations in the Mesters Vig region by geologists in Lauge Koch's expeditions in 1948, the mining and prospecting company Nordisk Mineselskab was established in 1952. The lead-zinc mine near Mesters Vig operated from 1956 to 1962, when it was worked out. Extensive diamond drilling of a molybdenum prospect at Malmbjerg was undertaken between 1958 and 1960, but due to low grades, it was not exploited. Regional mineral prospecting was carried out between 1964 and 1984 over large parts of the Nordisk Mineselskab concession region (70°N–74°30′N; Harpøth et al., 1986; Stendal and Frei, this volume).

The discovery of vertebrate fossils in the Upper Paleozoic sediments of East Greenland by Lauge Koch's expeditions was followed up in 1958–1959 and 1967, when small parties led by Svend Bendix-Almgreen of the Geological Museum, Copenhagen, made extensive collections (Larsen et al., this volume).

In the summer seasons of 1968–1970, parties from the University of Cambridge, led by Peter Friend, undertook extensive sedimentological studies of the Devonian continental sediments (Friend et al., 1983). One objective was to make comparisons with the comparable deposits of the same age in Scotland and on Svalbard. In East Greenland, 70 generalized stratigraphical columns were established, and the succession was divided into five major units with a total aggregate thickness of 10,400 m (see also Larsen et al., this volume).

In 1983 and 1984, a four-man party of American geologists led by K.G. Swett undertook detailed investigations of the uppermost Cambrian–Ordovician succession in East Greenland. Comparisons with the equivalent successions in Newfoundland and northwest Scotland with respect to stratigraphic sequences, sedimentary structures, paleocurrent patterns, and geochemical anomalies showed a very close correlation. It was concluded that during Cambrian and Early Ordovician time, these areas were closely juxtaposed on the western margin of the proto-Atlantic (Swett and Smit, 1972). The uppermost Riphean to Ordovician sedimentary developments on the Laurentian margin are at present the subject of new investigations by a research group that includes Danish geologists and geologists from the Geological Survey of Newfoundland and Labrador (Stouge et al., 2001, 2002). An independent study of Cambrian non-trilobite faunas

(Skovsted, 2003) has demonstrated faunal similarities between East Greenland and Australia.

As part of a study of North Atlantic Precambrian diamictites (tillites), a four-man party led by M.J. Hambrey visited the classic localities of the Vendian Tillite Group in the fjord zone in 1984. In 1985, studies were continued and included visits to the foreland diamictites of Gåseland and Charcot Land (Fig. 7), which were correlated with the Tillite Group (Hambrey and Spencer, 1987; Hambrey et al., 1989; Moncrieff, 1989; Fairchild and Hambrey, 1995). The very close correlation with the diamictite sequence of the same age in Svalbard suggests that they were deposited in a contiguous basin with a common source area.

In 1987, a party of British and Danish geologists put together extensive collections of Upper Devonian tetrapods in the Gauss Halvø area (Bendix-Almgreen et al., 1988, 1990). Gauss Halvø was revisited in 1998 by a party from the Universities of Cambridge and Bristol, led by Jennifer Clack, in cooperation with Svend Bendix-Almgreen, when further collections were made from the same strata (Clack, 1994, 1999; Larsen et al., this volume). The vertebrate fossils in the slightly younger Triassic sediments of the Jameson Land area have been the focus of a series of expeditions since 1988 led by Farish A. Jenkins of Harvard University.

Studies of Devonian stratigraphy and relationships to the collapse of the Caledonian orogenic belt were initiated by a group of geologists based at Oslo University in 1995. Field activities have been carried out nearly every year from 1995 onward, supported by geochronological work undertaken at the Massachusetts Institute of Technology and Brown University, Rhode Island. Their early interpretations of the Caledonian orogeny appeared to support the classic "stockwerk" concept of Haller (1971), but their most significant work has concerned the extensional history of a major fault system (Hartz and Andresen, 1995; Andresen et al., 1998; Hartz et al., 2000, 2001).

CHANGING CONCEPTS—GEOSYNCLINES TO PLATE TECTONICS

Regional geological investigations within the Caledonian orogen in North-East Greenland have taken place, with intervals, since the late eighteenth century, and more systematic surveys were conducted after 1926, during "The Danish East Greenland Expeditions" led by Lauge Koch. The revolutionary plate tectonic concept of the late 1960s was thus gaining general acceptance ten years after the end of Lauge Koch's last field season (1958), and at a time when John Haller had almost completed the map compilations and regional descriptions (Koch and Haller, 1971; Haller, 1970, 1971). The science of radiometric age determinations was still in its infancy in the 1960s, and the only dates available to John Haller were ~20 K-Ar mineral ages from widely scattered localities along the length of the orogen (Haller and Kulp, 1962). With hindsight, it is not surprising that these mineral ages reflected the Caledonian thermal overprint and, with the exception of a ca. 1800 Ma age from the gneisses of the Gåseland window, gave no hint of the protolith age of the orthogneisses making up a large proportion of the orogen. More reliable radiometric methods developed from the mid-1970s onward have provided evidence of a complex sequence of Archean to Neoproterozoic orogenic events (Rex and Gledhill, 1974; Hansen et al., 1978; Steiger et al., 1979). Increasingly sophisticated isotopic methods have since helped to further unravel this complex geological history.

Geosynclines and Early Views of Paleogeographic Positions

In East Greenland, the Paleozoic age of the fold belt was first recognized by Nathorst (1901), who described the unconformity between the folded and slightly metamorphic "Silurian" succession and overlying undeformed Devonian sediments. Wordie (1927) described the folding and thrusting as the result of "Caledonian movements"; while Koch (1929a) recognized Cambrian-Ordovician fossiliferous sediments and concluded that Caledonian folding in East Greenland was younger than the Middle Ordovician and older than the Upper Devonian (Koch, 1929b). Koch described two Lower Paleozoic geosynclines in North and East Greenland, respectively. The geosyncline in East Greenland consisted of a thick late "Algonkian" succession (present Eleonore Bay Supergroup) overlain by Cambrian-Ordovician limestones. The approximately contemporaneous "Smith Sound Geosyncline" in North Greenland included some Proterozoic sediments but was dominated by Cambrian-Ordovician carbonate rocks overlain by Silurian

Figure 7. The diamictite at Tillit Nunatak in the Charcot Land window, which has been correlated with the Vendian Tillite Group of the central fjord zone. Photo is courtesy of N. Henriksen.

sandstones and shales. The deformation of the deposits in both the N-S–trending East Greenland and the E-W–trending North Greenland geosynclines was compared by Koch (1929b) with the Caledonian folding of Western Europe.

The belief in a genetic relationship between geosynclines and the formation of mountain belts (orogenies) is reflected in many papers published by Lauge Koch's geologists. Thus, a "Carolinidian geosyncline," consisting of thick sandstone successions of the "Thule Group" (the Paleoproterozoic–Mesoproterozoic Independence Fjord Group; Collinson et al., this volume), was viewed as having been folded during a pre-Caledonian orogeny to form a fold belt called the "Carolinides" (Haller, 1961). This event was viewed as predating formation of the two "Caledonian geosynclines."

The East Greenland geosyncline encompassed the very thick succession of late Precambrian (Groenlandian; Koch, 1930) and Cambrian to Middle Ordovician sediments (up to Upper Silurian in Kronprins Christian Land) that were subsequently deformed during the Caledonian orogeny. Since volcanic rocks were absent, the succession was characterized as miogeosynclinal. Haller (1971) allotted the volcanic rocks around Eleonore Sø in the nunatak region to a supposed "Basal Series" of the Eleonore Bay "Group" and interpreted them as ophiolites. However, the volcanic complex around Eleonore Sø is cut by quartz porphyry bodies dated at pre–1950 Ma (Kalsbeek et al., this volume, Chapter 3), which cannot be part of the Eleonore Bay "Group."

Structural Interpretations Arising from Lauge Koch's Expeditions from 1926 to 1958

The evolution of structural interpretations of East Greenland has been summarized by Haller (1970, 1971), who developed the early observations of Helge Backlund and Emil Wegmann about the crystalline complexes and combined them with his own investigations (Haller, 1958). The "stockwerk" concept explained the occurrence of deep-seated and highly deformed crystalline rocks (central metamorphic complex) beneath a cover of relatively weakly deformed and little-altered sediments in terms of "vertical ductility contrasts" (Haller, 1985). Rising fronts of Caledonian migmatization and metasomatism emanating from a deep-seated orogenic core were considered to have caused in situ transformation of a widespread metasedimentary succession (the Eleonore Bay "Group"). This interpretation was primarily based on field observations, and it developed from the fact that at many localities these appeared to indicate a complete and gradational transition from little-altered sediments into highly altered metasediments, and into the gneissic infrastructure. Some of the early published descriptions of the supposed "Caledonized" gneiss complexes included statements that the "Archean basement of the Upper Algonkian–Ordovician series of deposits has hitherto not been found anywhere in central East Greenland" (Haller, 1956b, p. 160). However, after some years, these interpretations were modified significantly, such that a "basement" to the metasedimentary rocks was in fact recognizable: "Inside the Caledonian domain, rock units which were originally from the ancient basement, represent substantial ingredients of the fold belt" (Haller and Kulp, 1962, p. 18, their Fig. 3b). The sketch map included in his later regional description (Haller, 1971, their Fig. 15b) clearly indicates the existence of widespread basement gneisses in the inner parts of the Scoresby Sund region and Kong Oscar Fjord region. However, due to their extensive Caledonian reworking, they were assigned to the Caledonian crystalline complex. Farther north, the Precambrian basement gneisses were interpreted as mainly reworked "Carolinidian" rock units.

The lower levels of the "central metamorphic complex" were made up of three units (Gletscherland migmatite complex, Hagar migmatite sheet, Niggli Spids dome), which were folded together with the overlying supracrustal rocks into dome-like, mushroom, or tongue shapes (Fig. 8). While the Niggli Spids dome and Gletscherland migmatite complex were in Haller's later publications viewed as Caledonian reworked basement rocks (Haller, 1971), an entirely Caledonian origin was still envisaged for the nappe-like convolutions of the Hagar migmatite sheet. In contrast to the strongly deformed infracrustal levels, the sedimentary successions of the Eleonore Bay "Group" above the so-called "zone of detachment" were only moderately deformed. The pattern of generally N-S–trending open folds with steep axial planes has a limited regional E-W shortening, estimated at 5.3% by Eha (1953).

Haller (1971) distinguished three Caledonian orogenic phases. A Caledonian "main orogeny" was dated as Silurian and placed at 420–400 Ma on the basis of the very few K-Ar mineral ages available (Haller and Kulp, 1962). A second phase, termed "late Caledonian spasms," was considered to be Devonian and placed at 400–350 Ma. A third series of deformations, referred to as "minor succeeding episodes," was considered to be Carboniferous (350–270 Ma) in age. The main orogeny was considered to be a short and violent episode that encompassed the whole of the fold belt accompanied by widespread regional metamorphism. The late "spasms" were viewed as an extended period of more local structures undergoing minor folding associated with the developing Devonian basins. Around Grandjean Fjord (75°N), a distinctive NW-SE–trending "young" Caledonian mountain belt was viewed as part of this phase.

Major regional thrusts were not an important element in John Haller's structural interpretations of the southern parts of the Caledonian orogen in East Greenland. He recognized that the edge of the Caledonian fold belt was marked by a thrust belt traceable for 600 km between Kronprins Christian Land and Dronning Louise Land, and that this was evidence of tangential crustal movements. Apart from the Gåseland window in the south (Wenk, 1961), no further outcrops of the foreland were then known, and he stated: "However, the main Caledonian structures displayed in the well-explored fjord region are definitely not far traveled; on the contrary, they appear to be autochthonous, initiated and caused by the rise of the migmatite front resulting in a 'stockwerk' folded belt" (Haller, 1971, p. 218).

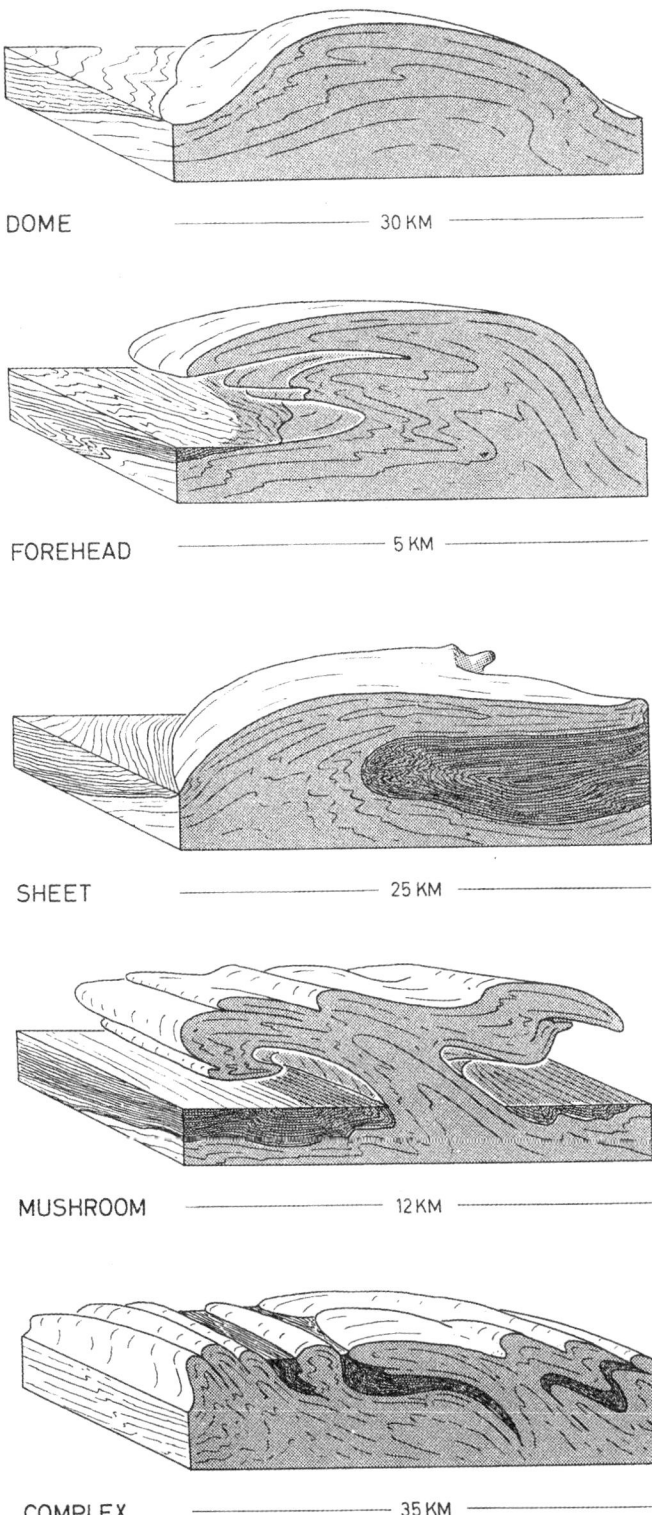

Figure 8. John Haller's interpretations of various types of infrastructural upwellings in the deep levels of the orogen. The vertical ascent of mobilized rock complexes generated in the deep levels of the orogen was envisaged to form major recumbent fold structures at shallower levels. Contractional deformation was considered to be unimportant. This diagram is from Haller (1971, his Fig. 64). Reproduced with permission from John Wiley & Sons Limited.

NEWER SCIENTIFIC RESULTS

Most of the recent studies in the East Greenland Caledonian orogen since 1968 stem from the initiation of systematic geological mapping by the GGU and the succession of major mapping projects that continued up to 1998 (Plate 1, Henriksen and Higgins, this volume, Chapter 14). These mapping expeditions typically involved a large international group of specialists working in cooperation with GGU staff members, and they covered a wide range of geological subjects, from the genesis of high-grade infracrustal gneiss regions to sedimentological, stratigraphical, and paleontological studies of the nonmetamorphic sedimentary successions. Local and regional structural analyses were supplemented by geochronological studies and interpretations. A brief description of some of the main results is given next, and more detail will be found in the relevant chapters of this volume.

Biostratigraphical Data

The stratigraphical divisions of the Lower Paleozoic in East Greenland up to the mid-1960s were largely based on work by Poulsen (1956), Cowie and Adams (1957), and Cowie (1961, 1963). Paleontological work, almost exclusively on body fossils, had led to a subdivision of the succession into formations, which in general correspond to present usage. The first studies of trace fossils were reported by Cowie and Spencer (1970).

The results of GGU work in East Greenland are summarized by Smith and Rasmussen (this volume). The northern part of the region includes parts of the Lower Paleozoic Franklinian Basin in North Greenland, for which a detailed sedimentological and stratigraphical division is available based on field work from 1978 to 1985 (Higgins et al., 1991). The Franklinian Basin extends to the easternmost parts of North Greenland, and parts of the succession crop out in the Caledonian foreland as well as in Caledonian thrust sheets in Kronprins Christian Land. Early stratigraphical results from the Franklinian Basin were reported by Peel and Cowie (1979) and Peel (1982), who compared the succession with sequences of similar age in East Greenland and revised the stratigraphical nomenclature, in particular, restricting usage of formation names formerly employed in both regions. The Ordovician system in Greenland has been described by Smith and Bjerreskov (1994), who presented a correlation chart and stratigraphic lexicon for the Ordovician strata in both the North Greenland Franklinian Basin and strata in East Greenland. Studies of Early Ordovician conodonts from East and North Greenland have been presented by Smith (1991). The Upper Ordovician and Silurian carbonate shelf stratigraphy of eastern North Greenland has been described by Hurst (1984), and the Silurian turbidite stratigraphy of North Greenland has been described by Hurst and Surlyk (1982). A review of the Lower Paleozoic stratigraphy of East Greenland by Smith et al. (2004) amended the nomenclature for several units and formally defined the Kong Oscar Fjord Group for the Cambrian–Ordovician succession in the central fjord region.

Basin Analysis

Several major sedimentary basins in East Greenland, ranging in age from the Paleoproterozoic to early Paleozoic, have been affected by Caledonian metamorphism and deformation. These basins developed differently in different parts of East Greenland, and the successions preserved in them vary from a few kilometers to ~18.5 km in thickness. A description of the sedimentary basins of North Greenland, edited by Peel and Sønderholm (1991), includes accounts of the Neoproterozoic sag and rift basins (Sønderholm and Jepsen, 1991), the Lower Paleozoic Franklinian Basin (Higgins et al., 1991), as well as the Carboniferous to Cenozoic deposits of the post-Caledonian Wandel Sea Basin (Stemmerik and Håkansson, 1991; Håkansson et al., 1991; Stemmerik 2000). These basins are all represented in Kronprins Christian Land, which forms the northernmost segment of the East Greenland Caledonian orogen. In the southern part of the Caledonian orogen in East Greenland, the Neoproterozoic–Ordovician basin succession includes the Eleonore Bay Supergroup, the Tillite Group, and the Kong Oscar Fjord Group. The development of the intramontane continental Devonian basins was closely related to the collapse and extension of the Caledonian orogen. The major basins are briefly described in chronological order next.

Independence Fjord Group and Zig-Zag Dal Basalt Formation. The earliest sedimentary basin phase in North-East and eastern North Greenland is represented by the thick sandstone-dominated succession of the Independence Fjord Group (Fig. 9). The sandstones are intracratonic, mainly ephemeral stream and eolian deposits, with thin widespread intervals of lacustrine sedimentation. In Kronprins Christian Land, the lower levels of the sandstone succession, which occur in Caledonian thrust sheets, are interbedded with basalts dated at 1740 Ma (Collinson et al., this volume). At ca. 1380 Ma, these deposits were intruded by doleritic dikes and sills (Midsommersø dolerites) and overlain by associated basalts (Zig-Zag Dal Basalt Formation; Upton et al., 2005). Representatives of the dolerites are widely exposed in Kronprins Christian Land and the Caledonian foreland to the west, and they extend southward through the nunatak region of East Greenland as far as Dronning Louise Land.

Eleonore Sø Complex, Hamberg Gletscher Complex, and Charcot Land Supracrustal Sequence. Rift-related pre–1900 Ma volcanic and sedimentary rocks are preserved in the foreland areas of Eleonore Sø (74°N), Hamberg Gletscher (73°30′N), and Charcot Land (72°N) (see Higgins et al., 2001a; Leslie and Higgins, this volume). These scattered developments may be part of a widespread succession concealed in the foreland areas beneath the Inland Ice. No equivalent developments are known in the structurally overlying Caledonian thrust sheets.

Krummedal Supracrustal Sequence and Smallefjord Sequence. Thick, latest Mesoproterozoic to earliest Neoproterozoic metasedimentary rocks (Krummedal supracrustal sequence, Smallefjord sequence; Watt and Thrane, 2001; Kalsbeek et al., this volume, Chapter 3) are widely distributed in Caledonian

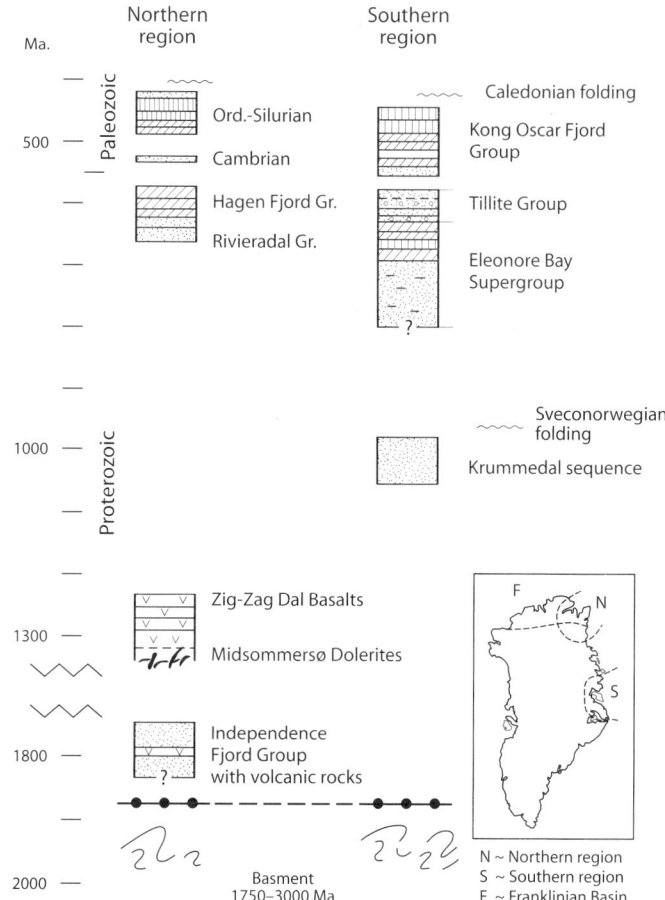

Figure 9. Proterozoic and Lower Paleozoic stratigraphy and sedimentary basins in the Caledonian orogen of East Greenland and correlative successions in the foreland to the west. Lower Neoproterozoic Sveconorwegian orogenic events have only been recorded in the southern half of the Caledonian orogen. The Lower Paleozoic Franklinian Basin that extends across North Greenland has its easternmost representatives in Kronprins Christian Land, where they were involved in Caledonian deformation.

thrust sheets in the southern half of the Caledonian orogen, but there is no evidence that they were ever deposited in the foreland areas to the west. Their wide distribution suggests that the original sedimentary basin may have been at least 600 km long, from south to north, and up to 300 km wide. The succession of dominantly siliciclastic sediments is at least 4 km thick. Successions of similar development and comparable age are also found in other Caledonian provinces bordering the North Atlantic (e.g., Svalbard; Gee and Teben'kov, 2004) and NW Scotland.

Rivieradal Group; the Hekla Sund Basin. The Rivieradal Group consists of a siliciclastic succession of Neoproterozoic age that is confined to the Caledonian Vandredalen thrust sheet in Kronprins Christian Land (Higgins et al., 2001b; Smith et al., 2004). The succession is up to 10 km thick and represents the fill of a half-graben rift basin known as the Hekla Sund Basin (Fränkl, 1954, 1955), which was at least 200 km long and 50 km wide. The Rivieradal Group postdates the Independence Fjord Group, Midsommersø Dolerite Formation, and Zig-Zag Dal Basalt Formation, all of which are represented by clasts within conglomeratic units. It is conformably overlain by formations of the Hagen Fjord Group.

Hagen Fjord Group. The Hagen Fjord Group is a transgressive shallow-marine Neoproterozoic succession, up to 1 km thick, that is widely exposed in eastern North Greenland, including Kronprins Christian Land. Basal clastic formations are overlain by reddish limestones, mudstones, and yellow-orange stromatolitic dolostones. West of Danmark Fjord, in the foreland, the succession unconformably overlies the Independence Fjord Group or Zig-Zag Dal Basalt Formation, while in the Caledonian Vandredal thrust sheet, it conformably overlies the Rivieradal Group. A significant hiatus, spanning most the Vendian, separates it from the Kap Holbæk Formation, which has now been formally excluded from the Hagen Fjord Group by Smith et al. (2004).

Franklinian Basin. The Lower Paleozoic E-W–trending Franklinian Basin, which extends ~2000 km across North Greenland and farther westward into the Canadian Arctic islands, is divided into a southern carbonate shelf and a northern clastic trough (Higgins et al., 1991; Ineson and Peel, 1997; Henriksen et al., 2000). In Kronprins Christian Land, the carbonate sediments of the shelf are well represented in the foreland, as well as in the marginal Caledonian fold-and-thrust belt. Observations during mapping of the Caledonian orogen in 1994–1995 have added further detail about the significance of the Early Ordovician Wandel Valley Formation unconformity, which affects a region of some 100,000 km^2 in eastern North and North-East Greenland. While the Laurentian passive margin to the west and south was undergoing uninterrupted thermal subsidence, this region was experiencing major uplift and westward tilting (Smith and Rasmussen, this volume). In Kronprins Christian Land, the Wandel Valley Formation unconformably overlies the Kap Holbæk Formation, whereas to the south in Lambert Land, it rests directly on the Paleoproterozoic-Mesoproterozoic Independence Fjord Group.

Eleonore Bay Supergroup. The most conspicuous succession of the southern half of the Caledonian orogen in East Greenland is formed by the brightly colored clastic and carbonate formations that make up the Neoproterozoic Eleonore Bay Supergroup, the Vendian Tillite Group, and the Cambrian–Ordovician Kong Oscar Fjord Group. These sequences make up the uppermost allochthonous unit of the Caledonian orogen, distinguished as the "Franz Joseph allochthon," which appears to have been carried passively westward as part of the Hagar Bjerg thrust sheet. The succession reaches a maximum thickness of 18.5 km, in contrast to the few hundred meters of equivalent age found on the foreland.

New studies of the Neoproterozoic Eleonore Bay Supergroup have confirmed the great thickness of the lower part reported by earlier workers, and they have demonstrated a firm correlation between the western exposures in the Petermann Bjerg region and exposures in the central fjord zone. These studies also have revealed an unexpected significant hiatus within the lower part of the Eleonore Bay Supergroup (Nathorst Land Group) in Andrée Land (Smith and Robertson, 1999). Investigations of the carbonate rocks of the Andrée Land Group, the uppermost unit of the Eleonore Bay Supergroup, suggest a correlation with the carbonates of the Hagen Fjord Group, which would imply that parts of the basin extended for at least 1000 km along the Laurentian margin (Frederiksen, 2000; Sønderholm et al., this volume).

Tillite Group. Detailed studies of the 800–1000-m-thick Tillite Group, with its two diamictite levels, suggest a very close correlation with the equivalent sequence at Svalbard (Hambrey and Spencer, 1987). The foreland diamictites, which are preserved in erosional depressions in an eroded gneiss surface in the Gåseland, Charcot Land, and Målebjerg windows, have all been correlated with the diamictites of the Tillite Group (Moncrieff, 1989; Smith et al., 2004).

Kong Oscar Fjord Group. The Kong Oscar Fjord Group (Cambrian–Ordovician) has a maximum thickness of 4 km in the fjord zone (Smith et al., 2004; Smith and Rasmussen, this volume). The great similarities in early Paleozoic platform developments along thousands of kilometers of the Laurentian margin of the Iapetus Ocean are well known from the studies of Swett and Smit (1972). New investigations of the East Greenland succession have been conducted by Stouge et al. (2001, 2002).

Devonian Continental Basins. The up to ~8-km-thick continental Devonian succession is composed mainly of conglomerates and sandstones, the accumulation of which was related to collapse of the Caledonian orogen. Survey mapping and investigations have led to introduction of a new lithostratigraphical division (Olsen and Larsen, 1993; Larsen et al., this volume). An analysis of basin development has distinguished four tectonostratigraphic basin stages defined by drainage patterns and bounding unconformities (Olsen, 1993).

Geochronological Studies

When the GGU initiated regional geological mapping in the Scoresby Sund region in 1968, the age of the widespread basement gneiss complexes was uncertain. While they showed many similarities with the gneiss complexes known from the Precambrian shield in West Greenland, Haller (1970, 1971) considered them to represent granitized Eleonore Bay Group sediments transformed during the Caledonian orogeny. In order to solve this problem, the GGU established a cooperative project with the geochronological laboratory of the Federal Institute of Technology (ETH) in Zürich, Switzerland; extensive K/Ar, Rb/Sr mineral, and whole-rock analyses were carried out, as well as U/Pb dating on multigrain zircon populations. The gneiss basement in the inner Scoresby Sund region yielded Archean ages of ca. 2500 Ma, while analyses of granitic rocks from the broad zone of migmatites and granites farther east yielded ages of ca. 1000 Ma, as well as Caledonian ages of ca. 475 Ma and 425–400 Ma (Steiger et al., 1979). Mineral ages confirmed a regional Caledonian metamorphic event, but the intensity of Caledonian orogenic reworking was unclear. Confirmation of the

Archean age of the gneiss complexes was provided by Rex and Gledhill (1974), who also showed that the widespread metasedimentary rocks (Krummedal supracrustal sequence) hosted ca. 1000 Ma augen granites, and thus could not be part of the Eleonore Bay "Group" as had been assumed by Lauge Koch's geologists (Rex and Gledhill, 1981). Furthermore, orthogneisses in the metamorphic complexes dated by Rb-Sr whole-rock methods yielded both Archean and Paleoproterozoic ages (Rex et al., 1976, 1977; Rex and Gledhill, 1981). A related K/Ar mineral age study on a variety of rock types yielded ages from 440 to 365 Ma, testifying to a widespread Caledonian regional metamorphism (Rex and Higgins, 1985).

Improved and increasingly sophisticated isotopic methods were utilized in the early 1990s, when northern parts of the orogen were mapped. These included SHRIMP analyses on single-zircon grains and Rb-Sr and Sm-Nd model age studies. The gneiss complexes that dominate the Caledonian orogen north of 76°N, and extend to 80°N, have yielded protolith ages between 1730 and 2000 Ma almost everywhere (Kalsbeek et al., 1993, 1999, this volume, Chapter 3; Nutman and Kalsbeek, 1994; Kalsbeek, 1995). This extensive region, up to 900 km from south to north, represents part of a major province of Paleoproterozoic crust accretion, with a small enclave of Archean rocks near Danmarkshavn. Migmatitic metasedimentary rocks that structurally overlie gneisses around Ardencaple Fjord, known as the Smallefjord sequence, were dated by SHRIMP studies on zircons, and the results showed that they were deposited after ca. 1035 Ma and had been affected by a ca. 955 Ma metamorphic event (Strachan et al., 1995a). Caledonian granites developed within the Smallefjord sequence that invade the structurally overlying Eleonore Bay Supergroup have yielded a narrow range of U-Pb zircon and monazite ages of 431–428 Ma (Strachan et al., 2001).

Zircons from a rhyolitic ignimbrite within a basalt sequence preserved in Caledonian thrust sheets in Kronprins Christian Land have yielded a SHRIMP age of ca. 1740 Ma (Pedersen et al., 2002). The basalt sequence known as the Hekla Sund Formation is interbedded with sandstones and conglomerates correlated with the Independence Fjord Group, while the overlying Midsommersø Dolerite Formation has been dated at ca. 1380 Ma (Collinson et al., this volume; Upton et al., 2005). There appear, therefore, to be two distinct basalt successions in northern East Greenland, one Paleoproterozoic and one Mesoproterozoic; the implication of the older age is that the quartzites referred to the Independence Fjord Group extend back into the Paleoproterozoic.

The $^{40}Ar/^{39}Ar$ mineral age analyses carried out on a wide range of rock types in the northern part of the orogen have all yielded Caledonian ages, ranging between 438 and 370 Ma (Early Silurian–Late Devonian; Dallmeyer and Strachan, 1994). They are interpreted as dating cooling following regional Caledonian metamorphism.

The 1997–1998 remapping of the Kong Oscar Fjord region (72°N–75°N) included extensive isotopic studies that have confirmed and revised conclusions based on the more primitive isotopic work 20–30 yr earlier. Ion microprobe studies on individual zircon grains from a suite of augen granite intrusions, which had previously yielded ages of ca. 1000 Ma from Rb-Sr and bulk zircon analyses (Steiger et al., 1979; Rex and Gledhill, 1981), have yielded protolith ages of between 940 and 910 Ma (Kalsbeek et al., 2000; Watt et al., 2000; Watt and Thrane, 2001; Leslie and Nutman, 2003). Furthermore, ion microprobe studies of detrital zircons from the Krummedal supracrustal sequence show that they must have been deposited after ca. 1050 Ma (the youngest detrital zircons), and before ca. 940 Ma (the oldest augen granites that intrude the metasediments). The thermal event that gave rise to the 940–910 Ma granite suite can be compared with the late Grenvillian tectono-thermal activity on Svalbard (Gee and Teben'kov, 2004) and the Sveconorwegian event of southern Scandinavia. Both the 940–910 Ma granite suite and the later (ca. 435 Ma) Caledonian granites have been shown to be S-type intrusions produced by melting of the Krummedal sequence metasediments (Kalsbeek et al., 2001a, 2001b, this volume, Chapter 9).

North-East Greenland Eclogite Province

The regional mapping of the Dove Bugt region resulted in the discovery of widespread eclogitic lenses and enclaves within the Paleoproterozoic gneisses north of ~77°N (Gilotti, 1993, 1994). The North-East Greenland eclogite province has subsequently been shown to extend over a distance of 400 km from south to north and ~100 km in width. The Paleoproterozoic basement gneisses are dominated by heterogeneous gray orthogneisses that contain bands of paragneisses and later granitoids, and these have yielded emplacement ages of between 2000 and 1750 Ma (Kalsbeek et al., 1993). The protoliths of the eclogites are considered to have been mafic and ultramafic pods and basic dikes of Paleoproterozoic age. The crystalline gneiss complexes were extensively deformed during the Caledonian orogeny, and the eclogites were formed under medium- to high-pressure and high-temperature conditions ca. 440–360 Ma (Brueckner et al., 1998; Gilotti et al., 2004, this volume). Eclogite-facies assemblages are best preserved in the cores of the pods and layers, while the rims have amphibolite-facies assemblages. The host gneisses lack the eclogite-facies paragenesis. The eclogite-facies assemblages were formed during the Caledonian orogeny at crustal depths of up to ~60 km, while some ultrahigh-pressure eclogites were perhaps formed at depths of 100 km (Gilotti and Ravna, 2002). Farther south in the orogen, in Payer Land (74°40′N), high-grade metamorphism has been dated by SHRIMP zircon analyses to ca. 404 Ma (McClelland and Gilotti, 2003; Elvevold et al., 2003), and it is considered to have been related to the same crustal thickening event that formed the eclogites.

Structural Interpretations

The East Greenland Caledonian orogen consists of a westward foreland-propagating pile of major thrust sheets (Higgins et al., 2004b). The west border of the orogen is largely concealed by the Inland Ice, but the contact with the foreland is well exposed in Kronprins Christian Land and Dronning Louise Land, and in a series of large and small tectonic windows. The southern half of the orogen, south of Bessel Fjord (76°N), is composed of two major thrust sheets. The lower (Niggli Spids) thrust sheet is made up of Archean–Paleoproterozoic gneiss complexes and a cover of medium- to high-grade metasediments (Krummedal sequence), and it has been thrust westward across the foreland windows. The higher (Hagar Bjerg) thrust sheet is made up of largely migmatitic high-grade metasediments (Krummedal sequence), which host both early Neoproterozoic (ca. 930 Ma) and Caledonian granites; the uppermost level of this thrust sheet is the thick Neoproterozoic–Ordovician succession, distinguished as the Franz Joseph allochthon, which is also the uppermost structural level in the orogen. Between Bessel Fjord (76°N) and Lambert Land (79°15′N), individual thrust sheets are generally not well defined. Neoproterozoic sedimentary or metasedimentary rocks are unknown, and Caledonian granites have not been recorded; these structural levels may have been removed by erosion, or they are perhaps preserved offshore North-East Greenland in the wide shelf areas. The onshore region of the orogen north of 76°N is dominated by Paleoproterozoic gneisses, which contain Caledonian eclogitic enclaves over a wide region that testify to rapid uplift prior to westward displacement across the foreland. In the extreme north, in Kronprins Christian Land, a complete transition, from foreland through a thin-skinned thrust belt to allochthonous thrust sheets, is well exposed.

Autochthonous Foreland. The main area of autochthonous foreland lies west of Kronprins Christian Land, while the western part of Dronning Louise Land is interpreted to be autochthonous to parautochthonous. A small region of nunataks in the upper western part of Hamberg Gletscher is interpreted as probably intact foreland (Fig. 10).

The west border of the Caledonian orogen is well exposed in western Kronprins Christian Land. The undisturbed foreland to the west is made up of a generally flat-lying undeformed and unmetamorphosed succession that is widely exposed in Mylius-Erichsen Land and J.C. Christensen Land and that extends to the northwest across southern parts of North Greenland. It includes a succession of Paleoproterozoic to Lower Paleozoic sediments and volcanic rocks (see previous).

In western Dronning Louise Land, Paleoproterozoic gneisses are unconformably overlain by Paleo- to Mesoproterozoic sandstones (Trekant "series"; Collinson et al., this volume), which in turn are unconformably overlain by Cambrian sandstones (Zebra "series"). Similar Cambrian sandstones have been identified in the Eleonore Sø and Målebjerg windows (Smith et al., 2004).

A group of small nunataks in the western part of Hamberg Gletscher exposes thick sheets of columnar-jointed, coarse-grained gabbros that have been emplaced into and thermally metamorphosed a sequence of lavas and calcareous sediments. No comparable rock types are known in the structurally overlying thrust sheets, and a broad correlation with Paleoproterozoic lavas and basic rocks in the Eleonore Sø and Charcot Land foreland windows has been suggested.

Foreland Windows. Large and small tectonic windows exposing parautochthonous foreland rocks are sporadically exposed in the western marginal zone of the orogen (Fig. 10). Windows are lacking in the eastern zone of the orogen, which is dominated by allochthonous basement gneiss complexes and thick metasedimentary successions, and which has been characterized as thick-skinned. As noted already, the Gåseland window was discovered in 1958 (Wenk, 1961), and the extent of the Charcot Land window was determined during GGU mapping in 1968 (Steck, 1971). The small window in Nørreland (78°40′N) was located in 1995 (J.M. Hull and J.D. Friderichsen, 1995, personal commun.), and the Eleonore Sø and Målebjerg windows were recognized during the 1997–1998 field seasons (Higgins and Leslie, 2000).

The successions in the various parautochthonous windows and those in the autochthonous foreland were compared and correlated by Higgins et al. (2001a). In all the foreland areas except Hamberg Gletscher and Charcot Land, the uppermost preserved units are carbonates of early Paleozoic age. Underlying these carbonates in Kronprins Christian Land, Dronning Louise Land, Eleonore Sø, and Målebjerg, there are Lower Cambrian quartzites that contain *Skolithos* ichnofossils (Smith et al., 2004). Diamictites interpreted as tillites have been found in the Gåseland, Charcot Land, and Målebjerg windows; in all cases, they can be correlated with the diamictites of the Vendian Tillite Group. The most conspicuous feature of the foreland areas in the southern part of the orogen is the very thin Neoproterozoic to Lower Paleozoic succession (<500 m), which shows great contrast to the up to 18.5-km-thick equivalent succession in the structurally overlying thrust sheets.

Foreland-Propagating Thrust Sheets (70°N–76°N). Recognition of the Eleonore Sø and Målebjerg windows in the Kong Oscar Fjord region in 1997–1998 demonstrated major displacements on the order of hundreds of kilometers for the overlying thrust sheets (Higgins and Leslie, 2004). The distinction of two major thrust sheets in the Kong Oscar Fjord region (72°N–75°N) allowed close comparisons and correlation to be made with the thrust sheets distinguished almost 30 yr earlier in the Scoresby Sund region (70°N–72°N; Henriksen, 1986; Elvevold et al., 2000; Higgins et al., 2004b).

The Niggli Spids thrust sheet is the lowest of the two major thrust sheets, and it is made up of crystalline gneiss complexes structurally overlain by high-grade metasedimentary successions; these were variably reworked during Caledonian orogenesis. Ion-microprobe analyses from the orthogneisses (Thrane, 2002; F. Kalsbeek and A.P. Nutman, personal commun., 2001) have yielded Archean and Paleoproterozoic protolith ages that confirm earlier less-precise isotopic ages (e.g., Rex and Gledhill, 1974, 1981). The high-grade metasedimentary rocks in the Kong Oscar Fjord region have been correlated with the Krummedal supracrustal sequence of the Scoresby Sund region (Higgins, 1988).

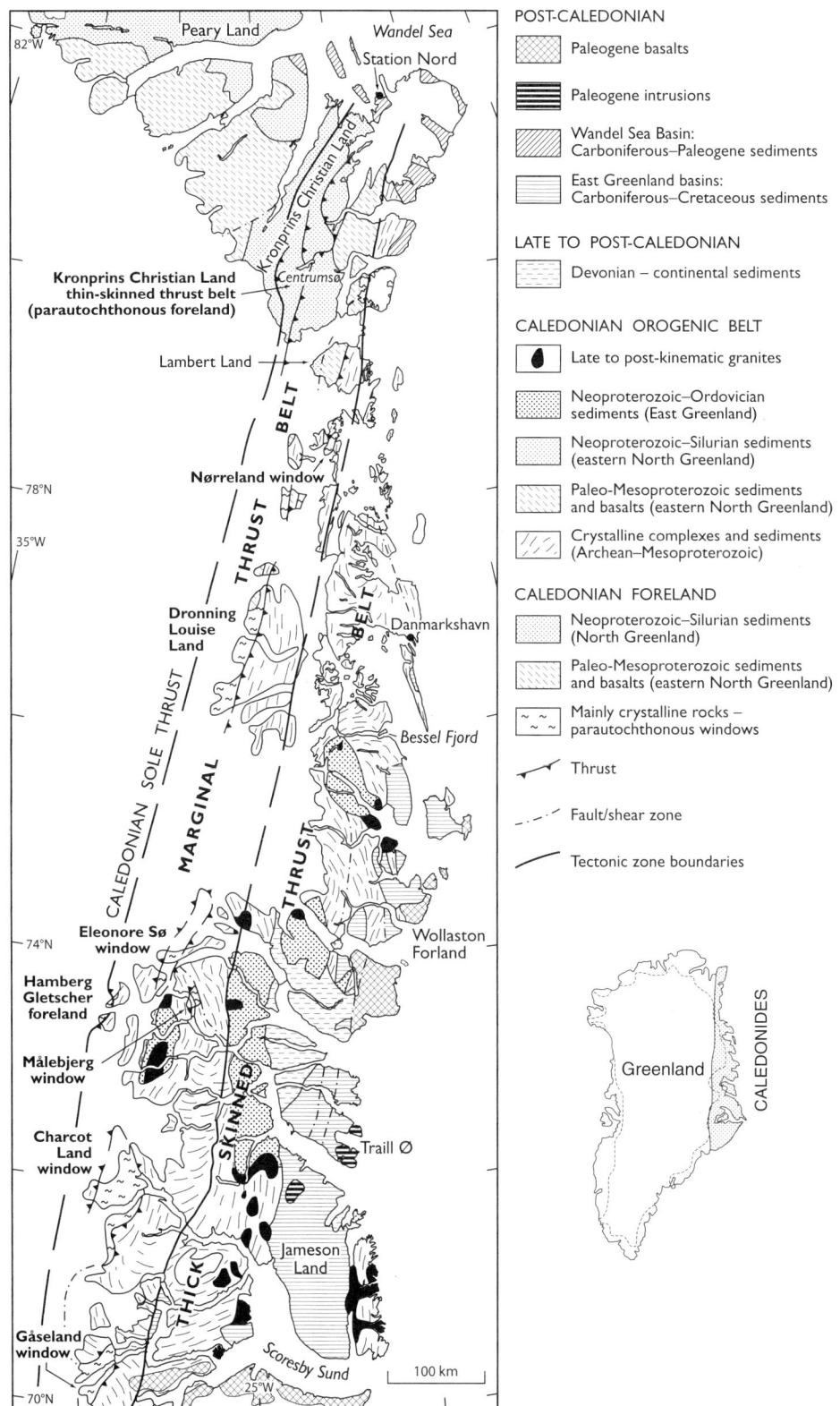

Figure 10. Simplified sketch map of the East Greenland Caledonian orogen showing the subdivision into three major tectonic zones. The western foreland is composed of Archean–Paleoproterozoic crystalline rocks of the Greenland Shield overlain by undeformed Paleoproterozoic–Silurian sedimentary and volcanic rocks. The Caledonian orogen is divided into a broad western marginal zone, and a thick-skinned central zone, both of which are characterized by westward-propagating major thrust sheets. Post-Caledonian sediments (Middle Devonian to Mesozoic) and Lower Paleogene volcanic rocks are extensively exposed in the eastern coastal regions of North-East Greenland.

As mentioned previously, the upper Hagar Bjerg thrust sheet is characterized by abundant migmatitic metasedimentary rocks correlated with the Krummedal supracrustal sequence, and it hosts two generations of leucocratic granites. The earliest granite suite, typically characterized by large feldspar phenocrysts, has yielded ion-microprobe ages on zircons of 940–910 Ma (Kalsbeek et al., 2000, this volume, Chapter 9; Watt et al., 2000; Watt and Thrane, 2001), while the younger granite suite has yielded 435–425 Ma ages (Watt et al., 2000; Hartz et al., 2001; Kalsbeek et al., 2001a, 2001b, this volume, Chapter 9). Both granite suites are S-type granites, generated by melting of the metasediments that host them.

The upper part of the Hagar Bjerg thrust sheet, distinguished as the Franz Joseph allochthon, is made up of the 18.5-km-thick Neoproterozoic–Ordovician sedimentary succession of the Eleonore Bay Supergroup, Tillite Group, and Kong Oscar Fjord Group. This succession is widely exposed in the central fjord zone of the Kong Oscar Fjord region, but it is only locally represented in the northeast part of the Scoresby Sund region in Canning Land (71°50′N). The contact between the lowest level of the Eleonore Bay Supergroup and the underlying Krummedal sequence metasedimentary rocks is a shear zone, in which both extensional and contractional strain have been recorded. In western Frænkel Land around Petermann Bjerg, the contact zone is known as the Petermann Detachment, whereas in the east, it is termed the Franz Joseph Detachment. Conodonts extracted from the highest exposed Ordovician strata of the Franz Joseph allochthon have very low color alteration indices (Smith, 1991; Stouge et al., 2002), which demonstrate that the strata were never buried beneath higher thrust sheets. The Franz Joseph allochthon would thus appear to be the highest structural unit in the Caledonian thrust pile.

Around Ardencaple Fjord and Bredefjord Fjord (75°30′N), a broad area of medium- to high-grade Eleonore Bay Supergroup sediments is preserved in a fault-bounded enclave. The generally steeply inclined bounding detachment zones preserve evidence of both contractional and extensional deformation, which were interpreted by Soper and Higgins (1993) and Higgins and Soper (1994) as indicative of an early extensional phase succeeded by Caledonian compression. Surrounding the fault-bounded enclave of Eleonore Bay Supergroup, there occurs a thick, migmatized, high-grade succession of metasediments, the Smallefjord sequence, which has been correlated with the Krummedal sequence further south (Jones and Strachan, 2000). Caledonian granites produced by melting of the Smallefjord sequence have migrated upward into the structurally overlying Eleonore Bay Supergroup, in the same way as granites produced by melting of the Krummedal supracrustal sequence in the Kong Oscar Fjord region. The close parallels of orogenic developments in the two regions suggest that the same foreland-propagating thrust sheet scenario can be extended into the Ardencaple Fjord region.

On the basis of the distribution of the outcrops of the Niggli Spids and Hagar Bjerg thrust sheets above the Eleonore Sø and Målebjerg windows in the Kong Oscar Fjord region, Higgins and Leslie (2000) estimated that total westward displacement was at least 200 km and possibly as much as 400 km. Restoration of the thrust sheets to their approximate relative locations implies that the site of collision of Laurentia and Baltica was several hundred kilometers off the coast of present-day East Greenland.

Foreland-Propagating Thrust Sheets (76°N–82°N). A prominent E-W–trending shear zone along Bessel Fjord (76°N) divides the Caledonian orogen into northern and southern parts (Fig. 10). While there are similarities between the two parts, in that both are characterized by westward-propagating thrust sheets, there are also significant differences. No representatives of the Eleonore Bay Supergroup or Tillite Group are known within the orogenic segment preserved onshore north of 76°, nor are there any significant bodies of Caledonian granitic rocks. All these elements are well represented south of 76°N. It is possible that they were formerly present to the north but have been removed by erosion, or that they are preserved in those parts of the orogen now hidden offshore beneath younger sediments.

The relationships between the geological development in the eastern half of Dronning Louise Land and the foreland areas of western Dronning Louise Land indicate a close correlation between them (Strachan et al., 1992). The implication is that the rock units making up eastern Dronning Louise Land comprise parts of a parautochthonous thrust sheet that was probably displaced only a limited distance from its original location.

The western margin of the Caledonian orogen in Kronprins Christian Land is characterized by a thin-skinned thrust belt with displacements of ~18 km, while the overlying allochthonous Vandredalen thrust sheet has been displaced ~40 km (Higgins et al., 2004a).

The higher thrust sheets of Kronprins Christian Land consist of a thick succession of sandstones correlated with the Independence Fjord Group, with interbedded volcanic units (Pedersen et al., 2002). These units are interleaved in major imbricate thrust zones and are particularly well exposed along the cliffs of central Ingolf Fjord. These conspicuous zones of white quartzites and black doleritic rocks can be traced southward through Hovgaard Ø, Lambert Land, and the western nunatak region as far as Bildsøe Nunatakker (78°N), and they no doubt represent a series of thrust sheets. Where exposed, the eastern limit of these units is a major thrust set against structurally overlying gneisses.

The structurally highest Caledonian thrust sheets in the onshore northern part of the orogen are gneiss-dominated, and, as noted already, they contain eclogitic enclaves over a wide region. It has not been possible to distinguish individual thrust sheets within the gneiss complexes. Locally, gneisses with eclogitic enclaves are displaced across significantly lower-grade foreland windows, as at Nørreland, whereas elsewhere, the gneisses structurally overlie lower thrust sheets.

Late-Orogenic Extensional Collapse and Devonian Basins. Detailed work by the GGU and other research groups has substantiated the close links between extensional collapse of the Caledonian orogen and the formation of the continental Devonian basins. Within the Caledonian orogen, a series of major lineaments preserves evidence of shear displacement related to late-orogenic extension.

The Storstrømmen shear zone is a NNE-SSW–trending belt of steep, heterogeneously deformed gneisses and mylonites that transects the Paleoproterozoic gneiss complexes of the Caledonian orogen between southeast Dronning Louise Land (76°N) and eastern Lambert Land (79°15′N). A splay of this structure appears to continue northward into Kronprins Christian Land, but, here, the latest movements are top down to the east, possible reactivating an earlier westward-directed thrust. The shear zone is up to 8 km wide, and shear criteria indicate a consistent sinistral sense of displacement. Mineral assemblages and fabrics indicate that mylonitization was initiated under low amphibolite-facies conditions and continued within the greenschist facies (Strachan and Tribe, 1994).

The Germania Land deformation zone consists of two subparallel NW-SE–striking strands of mylonites and cataclasites that can be traced for more than 150 km from eastern Germania Land across Gamma Ø to Jøkelbugten. Displacements on the Germania Land deformation zone, and parallel zones near Danmarkshavn, were predominantly dextral strike-slip (Hull and Gilotti, 1994). The Germania Land deformation zone was also the locus of Carboniferous normal faulting and basin development.

The Fjord Region fault, also known as the Fjord Zone fault or Fjord Region detachment zone, is a major N-S–trending extensional fault best known from detailed studies in the central fjord zone between 72°N and 74°N (Hartz and Andresen, 1995; Andresen et al., 1998; Hartz et al., 2000, 2001). This lineament continues southward through the inner part of the Scoresby Sund region to latitude 70°N, where it bounds the west side of a Carboniferous succession known as the Røde Ø conglomerate (Stemmerik and Piasecki, 2004). The lineament has a strike length of ~400 km. Higgins et al. (2004b) considered the Fjord Region fault (the main N-S lineament) to be a late Caledonian feature that postdated thrusting and cut through the thrust pile. However, other investigators have interpreted the Fjord Region detachment zone to be a long-lived extensional fault system that played a central role in Caledonian orogenic development (Hartz et al., 2001).

Extensional collapse of the overthickened orogen, which essentially postdated the main Caledonian compressional deformation, led to formation of intramontane continental basins that accumulated an up to ~8-km-thick succession of Devonian conglomerates and sandstones (Larsen et al., this volume). This Devonian succession hosts large quantities of vertebrate fossils, among which the most significant are the first land-living tetrapods (Clack and Neininger, 2000).

PLATE-TECTONIC SETTING

Greenland has formed part of the North American Precambrian shield at the northeastern corner of Laurentia since the late Paleoproterozoic. During the latest Cretaceous, a spreading axis from the central Atlantic propagated into the Labrador Sea between Greenland and Canada, but in the early Paleogene (ca. 55 Ma), spreading was diverted east of Greenland, and the opening of the North Atlantic and Arctic Oceans led to the separation of Greenland from Scandinavia (Fig. 11). The post-Caledonian rifting events in East Greenland that preceded final separation are well documented in the late Carboniferous to Mesozoic successions (Coward et al., 2003; Ineson and Surlyk, 2003).

East Greenland preserves evidence for at least three older cycles of continental accretion and subsequent breakup with formation of oceanic crust. The best known of these is closure of the Iapetus Ocean, which formed the Caledonian orogeny, but there is also evidence for a Mesoproterozoic to earliest Neoproterozoic event related to the Rodinia supercontinent and the Grenvillian-Sveconorwegian orogeny. In earlier times, parts of the Greenland Precambrian shield appear to have had close links with the Paleoproterozoic and Archean shield areas of Scotland and Scandinavia.

In the context of the present volume, only the plate-tectonic configurations related to Caledonian orogeny are relevant. It should be noted here that evidence for the paleogeographic position of East Greenland between early Neoproterozoic and late Paleozoic time is entirely based on data from other parts of Laurentia, and on the assumption that Greenland formed the northeastern segment of Laurentia. There are limited data from North Greenland with respect to the Franklinian Basin, but a comparison with other data from North America and Britain indicates that North America–Greenland and Great Britain were assembled in a supercontinent in Late Silurian to Early Devonian time (Stearns et al., 1989). Plate-tectonic reconstructions are based on the assumption that Greenland and the remainder of Laurentia shared the same drift history with respect to both movements and rotations.

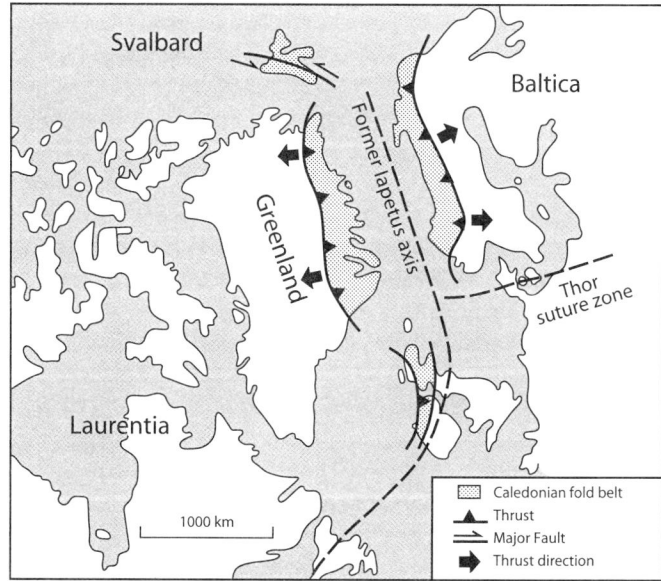

Figure 11. The Caledonian orogen is generally assumed to have formed in the collision of the northeast margin of Laurentia (including Greenland) and the west margin of Baltica (Norway). The map illustrates the approximate predrift locations of Greenland, Scandinavia, and the British Isles.

It is generally accepted that the Caledonian orogen of East Greenland was formed by the collision between NE Laurentia and Baltica and the closure of the Iapetus Ocean in Silurian to Early Devonian time. However, the two sides of the orogen are very different, both in their structural makeup and in the composition of the lithological units of which they are composed. It is therefore not possible to determine which continental segments were originally juxtaposed. Using the present-day predrift configuration, Scandinavia and Greenland are placed almost exactly opposite to each other, but on many reconstructions of the closure of the Iapetus Ocean and the collision between Scandinavia and Greenland, the two blocks are depicted as being affected by an important phase of left-lateral strike-slip faulting along the collision zone (Larsen and Bengaard, 1991; Soper et al., 1992; Strachan et al., 1995b; Coward et al., 2003; Dewey and Strachan, 2003). Comparable major faults may well be hidden in the offshore shelf areas, but the onshore orogen-parallel faults and shear zones mapped in East Greenland do not appear to have large-scale displacements.

Caledonian Plate-Tectonic Features in East Greenland

Recent investigations in the East Greenland Caledonides have yielded important results that need to be taken into consideration when considering plate-tectonic reconstructions. These are:

(1) There are major differences between the two sides of the Iapetus Ocean and between the compositions of the Scandinavian Caledonides and the East Greenland Caledonides (Fig. 12). For example, in Scandinavia, there are widespread ocean-derived terranes in some of the higher allochthons above the level of the Baltoscandian miogeosyncline (Roberts, 2003; Andréasson et al., 2003); similar oceanic and basic intrusive complexes are absent in the thrust sheets along the Laurentian margin in East Greenland.

(2) The Greenland Caledonides, in contrast to Scandinavia, are entirely ensialic, and the west side of Iapetus here developed as a passive margin.

(3) The present onshore segment of the Caledonian orogen preserved in East Greenland represents a westward-directed pile of thrust sheets. Parts of the orogen are concealed beneath younger rocks offshore, in the 100–300-km-wide shelf areas beyond the present coast. The original Laurentian continental margin, and the border of Iapetus, was probably positioned a few hundred kilometers east of the present coastline (Higgins and Leslie, 2000; Higgins et al., 2004b; Hamann et al., 2005).

(4) Sedimentation in the basins of southern parts of North-East Greenland began in the early Neoproterozoic (900–800 Ma) and continued as the supercontinent Rodinia began to break up at ca. 750–725 Ma (Torsvik et al., 1996). Accumulation of the ~14.5-km-thick Eleonore Bay Supergroup reflects a long period of steady subsidence in a shallow-marine environment

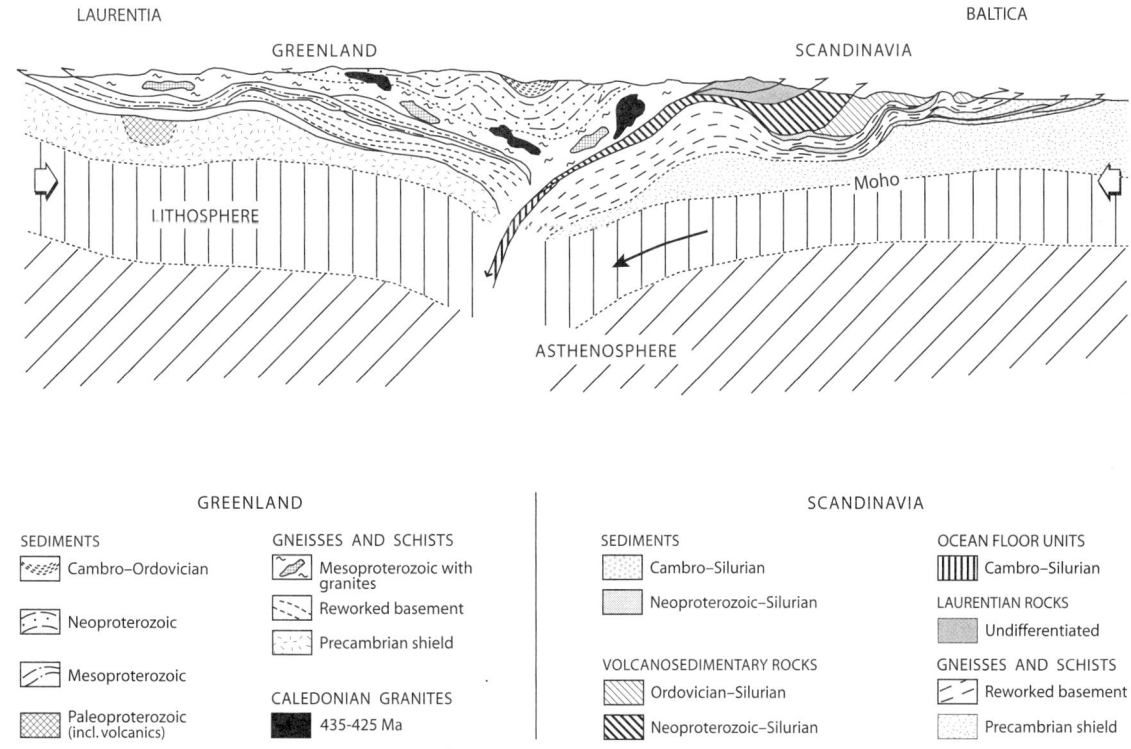

Figure 12. Sketch profile through the Caledonian orogen depicting the western thrusting in Greenland and the eastward thrusting in Scandinavia in Silurian time.

(Sønderholm and Tirsgaard, 1993), which continued through the Vendian with deposition of the sediments and diamictites of the Tillite Group (Hambrey and Spencer, 1987). This was succeeded by the accumulation of an up to 4-km-thick succession of Cambrian–Ordovician limestones and dolomites (Smith et al., 2004; Smith and Rasmussen, this volume). There are very close similarities in development with the equivalent Neoproterozoic–Ordovician succession of Svalbard (Gee and Teben'kov, 2004), which, prior to the closure of Iapetus, can be presumed to have been positioned close to this part of East Greenland.

(5) Sedimentary developments in the northernmost part of the East Greenland Caledonian orogen differ from those in the southern parts. In the north, the sedimentary rocks affected by Caledonian deformation can be related to three successive basinal developments, ranging in age from late Paleoproterozoic to Late Silurian. The Lower Paleozoic Franklinian Basin is the youngest of these basins, and it extended from the Canadian Arctic islands across North Greenland to Kronprins Christian Land. There are no counterparts along the margins of Baltica.

(6) The East Greenland Caledonian orogen preserves little evidence to constrain the timing of the opening of Iapetus. The Upper Neoproterozoic to Lower Paleozoic successions are not cut by basic intrusive rocks, apart from scattered thin lamprophyres (Haller, 1971; Hartz et al., 2001).

(7) The Iapetus margin sediments preserved in central East Greenland indicate continuous, almost undisturbed sedimentation (with minor hiatuses) until at least 462 Ma. Farther north, sedimentation continued until the Late Silurian (ca. 422 Ma).

(8) Restoration of the foreland-propagating thrust pile in the southern part of the East Greenland Caledonian orogen indicates that the site of the Eleonore Bay Supergroup–Kong Oscar Fjord Group sedimentary basin lay some distance to the east of the present coastline (Higgins et al., 2004b).

(9) No evidence for the site of the original subduction zones is preserved onshore East Greenland; these would probably have been located some hundred kilometers east of the present coastline. Basic rocks formerly interpreted as ophiolites (Haller, 1971) are now known to be of Archean or Proterozoic age. However, the occurrence of a few 466 ± 9 Ma I-type granites in the southern part of the orogen (Kalsbeek et al., this volume, Chapter 9), may be indicative of emplacement related to westward subduction. It has been suggested (Roberts et al., 2004, 2007) that these plutons may represent the northernmost tip of a Taconic magmatic arc.

Pre-Caledonian Plate-Tectonic Features from East Greenland

Evidence of Grenvillian-Sveconorwegian events within the rock units making up the Caledonian thrust sheets of East Greenland has been clarified by a succession of SHRIMP isotopic studies of zircons. These investigations have demonstrated that a widespread late Mesoproterozoic sedimentary succession (Krummedal sequence) was affected by high-grade metamorphism, migmatization, and granite formation ca. 930 Ma (e.g., Kalsbeek et al., 2000, this volume, Chapter 3). In the Renland area of the Scoresby Sund region (71°N–72°N), isoclinal folding appears to have preceded the emplacement of the granite suite; however, so far this, is the only known evidence suggestive of orogenic deformation associated with the ca. 930 Ma thermal episode (Leslie and Nutman, 2003). Studies of the detrital zircon populations in the Krummedal sequence indicate that the source area was not the Precambrian shield of Greenland, and it has been suggested that the detrital material may have been derived by long-distance transport from the Grenvillian orogen of North America, supplemented by material from the Fennoscandian Shield (Watt and Thrane, 2001).

In South-East Greenland, south of the Caledonian orogen, the Precambrian shield includes an Archean core bordered both to the north and to the south by Paleoproterozoic orogens formed at ca. 1900–1700 Ma. Comparisons have been made with Archean–Paleoproterozoic complexes in the Caledonian foreland of northwest Scotland, and it has been suggested that northwest Scotland and South-East Greenland were probably linked in the later part of the Paleoproterozoic (Kalsbeek, 1989).

REFERENCES CITED

Adams, P.J., and Cowie, J.W., 1953, A Geological Reconnaissance of the Region around the Inner Part of Danmarks Fjord, Northeast Greenland: Meddelelser om Grønland, v. 111, no. 7, 24 p.

Amdrup, G., 1913, Report on the Danmark Expedition to the North-East Coast of Greenland 1906–1908: Meddelelser om Grønland, v. 41, no. 1, 270 p.

Andréasson, P.G., Gee, D.G., Whitehouse, M.J., and Schoberg, H., 2003, Subduction-flip during Iapetus Ocean closure and Baltica-Laurentia collision, Scandinavian Caledonides: Terra Nova, v. 15, p. 362–369, doi: 10.1046/j.1365-3121.2003.00486.x.

Andresen, A., Hartz, E.[H.], and Vold, J., 1998, A late orogenic extensional origin for the infrastructural gneiss domes of the East Greenland Caledonides (72°–74°N): Tectonophysics, v. 285, p. 353–369.

Backlund, H.G., 1930, Contributions to the geology of Northeast Greenland: Meddelelser om Grønland, v. 74, no. 11, p. 207–296.

Backlund, H.G., 1932, Das Alter des 'Metamorphen Komplexes' von Franz Josef Fjord in Ost-Grönland: Meddelelser om Grønland, v. 87, no. 4, 119 p.

Bay, E., 1896, Den østgrønlandske Ekspedition, udført i Aarene 1891–1892 under Ledelse af C. Ryder, 3. Del. Geologi: Meddelelser om Grønland v. 19, no. 6, p. 147–187.

Bendix-Almgreen, S.E., Clack, J.A., and Olsen, H., 1988, Upper Devonian and Upper Permian vertebrates collected in 1987 around Kejser Franz Joseph Fjord, central East Greenland: Rapport Grønlands Geologiske Undersøgelse, v. 140, p. 95–102.

Bendix-Almgreen, S.E., Clack, J.A., and Olsen, H., 1990, Upper Devonian tetrapod palaeontology in the light of new discoveries in East Greenland: Terra Nova, v. 2, p. 131–137, doi: 10.1111/j.1365-3121.1990.tb00053.x.

Böggild, O.B., Bøgvad, R., Callisen, K., Frebold, H., Gry, H., Jessen, H., Madsen, V., Noe-Nygaard, A., Poulsen, C., Rosenkrantz, A., and Teichert, C., 1935, Bemærkninger til Lauge Koch: Geologie von Grönland: Meddelelser fra Dansk Geologisk Forening, v. 8, no. 5, p. 483–496 (English version, p. 497–512).

Brueckner, H.K., Gilotti, J.A., and Nutman, A.P., 1998, Caledonian eclogite-facies metamorphism of Early Proterozoic protoliths from the North-East Greenland eclogite province: Contributions to Mineralogy and Petrology, v. 130, p. 103–120, doi: 10.1007/s004100050353.

Bütler, H., 1935, Some New Investigations of the Devonian Stratigraphy and Tectonics of East Greenland: Meddelelser om Grønland, v. 103, no. 2, 35 p.

Bütler, H., 1948, Die Westgrenze des Devones am Kejser Franz Joseph Fjord in Ostgröland: Mitteilungen Naturforschungs Gesellschaft: Schaffhausen, v. 22, no. 3, p. 73–152.

Bütler, H., 1959, Das Old Red Gebiet am Moskusoksefjord. Attempt at a Correlation of the Series of Various Devonian Areas in Central East Greenland: Meddelelser om Grønland, v. 160, no. 5, 188 p.

Chadwick, B., and Friend, C.R.L., 1994, Reaction of Precambrian high-grade gneisses to mid-crustal ductile deformation in western Dove Bugt, North-East Greenland: Rapport Grønlands Geologiske Undersøgelse, v. 162, p. 53–70.

Clack, J.A., 1994, *Acanthostega gunnari*, a Devonian Tetrapod from Greenland; The Snouth, Palate and Ventral Parts of the Braincase, with a Discussion of Their Significance: Meddelelser om Grønland: Geoscience, v. 31, 24 p.

Clack, J.A., 1999, 1999 report on the vertebrate paleontological expedition to East Greenland, summer 1998: Danmarks og Grønlands Geologiske Undersøgelse Rapport, v. 19, p. 183–187.

Clack, J.A., and Neininger, S.L., 2000, Fossils from the Celsius Bjerg Group, Upper Devonian sequence, East Greenland: Significance and sedimentological distribution, *in* Friend, P.F., and Williams, B., eds., New Perspectives on the Old Red Sandstone: Geological Society of London Special Publication 180, p. 557–566.

Collinson, J.D., Kalsbeek, F., Jepsen, H.F., Pedersen, S.A.S., and Upton, B.G.J., 2008, this volume, Paleoproterozic and Mesoproterozoic sedimentary and volcanic successions in the northern parts of the East Greenland Caledonian orogen and its foreland, *in* Higgins, A.K., Gilotti, J.A., and Smith, M.P., eds., The Greenland Caledonides: Evolution of the Northeast Margin of Laurentia: Geological Society of America Memoir 202, doi: 10.1130/2008.1202(04).

Coward, M.P., Dewey, J., Hempton, M., and Holroyd, J., 2003, Tectonic evolution, *in* Evans, D., Graham, C., Armour, A., and Bathurst, P., eds., The Millennium Atlas: Petroleum Geology of the Central and Northern North Sea: Geological Society of London, p. 17–33.

Cowie, J.W., 1961, Contributions to the Geology of North Greenland: Meddelelser om Grønland, v. 164, no. 3, 47 p.

Cowie, J.W., 1963, The Cambro–Ordovician geology of East Greenland: Experientia, v. 19, p. 281–284, doi: 10.1007/BF02150410.

Cowie, J.W., and Adams, P.J., 1957, The Geology of the Cambro-Ordovician Rocks of Central East Greenland: Part 1. Stratigraphy and Structure: Meddelelser om Grønland, v. 153, no. 1, 193 p.

Cowie, J.W., and Spencer, A.M., 1970, Trace fossils from the late Precambrian/Lower Cambrian of East Greenland, *in* Crimes, T.P., and Harper, J.C., eds., Trace Fossils: Geological Journal, Special Issue, v. 3, p. 91–100.

Dallmeyer, R.D., and Strachan, R.A., 1994, ^{40}Ar/^{39}Ar mineral age constraints on the timing of deformation and metamorphism, North-East Greenland Caledonides: Rapport Grønlands Geologiske Undersøgelse, v. 162, p. 153–162.

Dal Vesco, E., 1954, Vulkanismus, Magmatismus und Metamorphose im Gebiet des nordostgrönländischen Devons: Meddelelser om Grønland, v. 72, no. 2, 38 p.

Dawes, P.R., 1991, Lauge Koch: Pioneer geo-explorer of Greenland's far north: Earth Sciences History, v. 10, no. 2, p. 130–153.

Dawes, P.R., 1992, Lauge Koch som polarforsker: 50 år i Grønlands tjeneste, *in* Olsen, A., ed., Tusaat—Forskning i Grønland, Volume 3: Copenhagen, Dansk Polarcenter, p. 9–15.

Dewey, J.F., and Strachan, R.A., 2003, Changing Silurian–Devonian relative plate motion in the Caledonides: Sinistral transpression to sinistral transtension: Journal of the Geological Society of London, v. 160, p. 219–229.

Eha, S., 1953, The Pre-Devonian Sediments of Ymers Ø, Suess Land, and Ella Ø (East Greenland) and Their Tectonics: Meddelelser om Grønland, v. 111, no. 2, 105 p.

Elvevold, S., and Gilotti, J.A., 2000, Pressure-temperature evolution of retrogressed kyanite eclogites, Weinschenk Island, North-East Greenland Caledonides: Lithos, v. 53, p. 127–147, doi: 10.1016/S0024-4937 (00)00014-1.

Elvevold, S., Escher, J.C., Frederiksen, K.S., Friderichsen, J.D., Gilotti, J.A., Henriksen, N., Higgins, A.K., Jepsen, H.F., Jones, K.A., Kalsbeek, F., Kinny, P.D., Leslie, A.G., Robertson, S., Smith, M.P., Thrane, K., and Watt, G.R., 2000, Tectonic Architecture of the East Greenland Caledonides 72°–74°30′N: Danmarks og Grønlands Geologiske Undersøgelse Rapport 2000/88, 34 p.

Elvevold, S., Thrane, K., and Gilotti, J.A., 2003, Metamorphic history of high-pressure granulites in Payer Land, Greenland Caledonides: Journal of Metamorphic Geology, v. 21, p. 49–63, doi: 10.1046/j.1525-1314. 2003.00419.x.

Escher, J.C., 2001, Geological Map of Greenland, Kong Oscar Fjord, Sheet 11: Copenhagen, Geological Survey of Denmark and Greenland, scale 1:500,000.

Fairchild, I.J., and Hambrey, M.J., 1995, Vendian basin evolution in East Greenland and NE Svalbard: Precambrian Research, v. 73, p. 217–233, doi: 10.1016/0301-9268(94)00079-7.

Fränkl, E., 1951, Die untere Eleonore Bay Formation im Alpefjord: Meddelelser om Grønland, v. 151, no. 6, 15 p.

Fränkl, E., 1953a, Geologische Untersuchungen in Ost-Andrées Land (Nordostgrönland): Meddelelser om Grønland, v. 113, no. 4, 160 p.

Fränkl, E., 1953b, Die Geologische Karte von Nord-Scoresby Land (Nordostgrönland): Meddelelser om Grønland, v. 113, no. 6, 56 p.

Fränkl, E., 1954, Vorläufige Mitteilung über die Geologie von Kronprins Christians Land (NE-Grönland, zwischen 80°–81° und 19°–23°W: Meddelelser om Grønland, v. 116, no. 2, 85 p.

Fränkl, E., 1955, Weitere Beiträge zur Geologie von Kronprins Christian Land (NE-Grönland): Meddelelser om Grønland, v. 103, no. 7, 35 p.

Frederiksen, K.S., 2000, A Neoproterozoic carbonate ramp and base-of-slope succession, the Andrée Land Group, Eleonore Bay Supergroup, North-East Greenland: Sedimentary facies, stratigraphy and basin evolution (Ph.D. thesis): Copenhagen, University of Copenhagen, Denmark, unpaginated.

Friend, P.F., Alexander-Marrack, P.F., Allen, K.C., Nicholson, J., and Yeats, A.K., 1983, Devonian Sediments of East Greenland: VI. Review of Results: Meddelelser om Grønland, v. 206, no. 6, 96 p.

Gee, D.G., and Teben'kov, A.M., 2004, Svalbard: A fragment of the Laurentian margin, *in* Gee, D.G., and Pease, V., eds., The Neoproterozoic Timanide Orogen of Eastern Baltica: Geological Society of London Memoir 30, p. 191–206.

Gilotti, J.A., 1993, Discovery of a medium-temperature eclogite province in the Caledonides of North-East Greenland: Geology, v. 21, p. 523–526, doi: 10.1130/0091-7613(1993)021<0523:DOAMTE>2.3.CO;2.

Gilotti, J.A., 1994, Eclogites and related high-pressure rocks from North-East Greenland: Rapport Grønlands Geologiske Undersøgelse, v. 162: Copenhagen, Denmark, Geological Survey of Denmark and Greenland, p. 77–90.

Gilotti, J.A., and McClelland, W.C., 2008, this volume, Geometry, kinematics, and timing of extensional faulting in the Greenland Caledonides—A synthesis, *in* Higgins, A.K., Gilotti, J.A., and Smith, M.P., eds., The Greenland Caledonides: Evolution of the Northeast Margin of Laurentia: Geological Society of America Memoir 202, doi: 10.1130/2008.1202(10).

Gilotti, J.A., and Ravna, E.J.K., 2002, First evidence for ultrahigh-pressure metamorphism in the North-East Greenland Caledonides: Geology, v. 30, p. 551–554, doi: 10.1130/0091-7613(2002)030<0551:FEFUPM> 2.0.CO;2.

Gilotti, J.A., Nutman, A.P., and Brueckner, H.K., 2004, Devonian to Carboniferous collision in the Greenland Caledonides: U-Pb zircon and Sm-Nd ages of polyphase high-pressure and ultrahigh-pressure metamorphisms: Contributions to Mineralogy and Petrology, v. 148, p. 216–235, doi: 10.1007/s00410-004-0600-4.

Gilotti, J.A., Jones, K.A., and Elvevold, S., 2008, this volume, Caledonian metamorphic patterns in Greenland, *in* Higgins, A.K., Gilotti, J.A., and Smith, M.P., eds., The Greenland Caledonides: Evolution of the Northeast Margin of Laurentia: Geological Society of America Memoir 202, doi: 10.1130/2008.1202(08).

Graeter, P., 1957, Die Sauren Devonischen Eruptivgesteine des Kap Franklin Gebiets am Kejser Franz Josephs Fjord in Zentral-Ostgrönland: Meddelelser om Grønland, v. 155, no. 3, 102 p.

Håkansson, E., Heinberg, C., and Stemmerik, L., 1991, Mesozoic and Cenozoic history of the Wandel Sea Basin area, North Greenland, *in* Peel, J.S., and Sønderholm, M., eds., Sedimentary Basins of North Greenland: Bulletin Grønlands Geologisle Undersøgelse, v. 160, p. 153–164.

Haller, J., 1953, Geologie und Petrographie von West-Andrées Land und Ost-Frænkels Land (NE-Grönland): Meddelelser om Grønland, v. 113, no. 5, 196 p.

Haller, J., 1955, Der "Zentrale Metamorphe Komplex" von NE-Grönland. Teil I. Der Geologische Karte von Suess Land, Gletscherland und Goodenoughs Land: Meddelelser om Grønland, v. 73, I, no. 3, 174 p.

Haller, J., 1956a, Geologie der Nunatakker Region von Zentral-Ostgrönland: Meddelelser om Grønland, v. 154, no. 1, 172 p.

Haller, J., 1956b, Probleme der Tiefentektonik: Bauformen im Migmatit Stockwerk der Ostgrönländischen Kaledoniden: Geologische Rundschau, v. 45, p. 159–167, doi: 10.1007/BF01802002.

Haller, J., 1958, Der "Zentrale Metamorphe Komplex" von Nordostgrønland. Teil 2. Die Geologische Karte der Staunings Alper und des Forsblads Fjordes: Meddelelser om Grønland, v. 154, no. 3, 153 p.

Haller, J., 1961, The Carolinides: An orogenic belt of late Precambrian age in Northeast Greenland, in Raasch, G.O., ed., Geology of the Arctic, Volume 1: Toronto, Toronto University Press, p. 155–159.

Haller, J., 1970, Tectonic Map of East Greenland (1:500,000): An Account of Tectonism, Plutonism and Volcanism in East Greenland: Meddelelser om Grønland, v. 171, no. 5, 286 p.

Haller, J., 1971, Geology of the East Greenland Caledonides: London, Interscience, 413 p.

Haller, J., 1983, Geological Map of Northeast Greenland 75°–82° N. Lat. 1:1,000,000: Meddelelser om Grønland, v. 200, no. 5, 22 p. (+ 1 map sheet).

Haller, J., 1985, The East Greenland Caledonides—Reviewed, in Gee, D.G., and Sturt, B.A., eds., The Caledonide Orogen—Scandinavia and Related Areas: Chichester, John Wiley & Sons, p. 1031–1046.

Haller, J., and Kulp, J.L., 1962, Absolute age determinations in East Greenland: Meddelelser om Grønland, v. 171, no. 1, 77 p.

Hamann, N.E., Whittaker, R.C., and Stemmerik, L., 2005, Structural and geological development of the North-East Greenland shelf, in Doré, A.G., and Vining, B.A., eds., Petroleum Geology: North-West Europe and Global Perspectives: Proceedings of the 6th Petroleum Geology Conference: London, Geological Society of London, p. 887–902.

Hambrey, M.J., and Spencer, A.M., 1987, Late Precambrian Glaciation of Central East Greenland: Meddelelser om Grønland, Geoscience, v. 19, 50 p.

Hansen, B.T., Higgins, A.K., and Bär, M.-T., 1978, Rb-Sr and U-Pb age patterns in polymetamorphic sediments from the southern part of the East Greenland Caledonides: Bulletin of the Geological Society of Denmark, v. 27, p. 55–62.

Harpøth, O., Pedersen, J.L., Schønwandt, H.K., and Thomassen, B., 1986, The Mineral Occurrences of Central East Greenland: Meddelelser om Grønland, Geoscience, v. 17, 139 p.

Hartz, E.H., and Andresen, A., 1995, Caledonian sole thrust of central East Greenland: A crustal-scale Devonian extensional detachment?: Geology, v. 23, p. 637–640, doi: 10.1130/0091-7613(1995)023<0637:CSTOCE>2.3.CO;2.

Hartz, E.H., Andresen, A., Martin, M.W., and Hodges, K.V., 2000, U-Pb and $^{40}Ar/^{39}Ar$ constraints on the Fjord Region detachment zone: A long-lived extensional fault in the central East Greenland Caledonides: Journal of the Geological Society of London, v. 157, p. 795–809.

Hartz, E.H., Andresen, A., Hodges, K.V., and Martin, M.W., 2001, Syncontractional extension and exhumation of deep crustal rocks in the East Greenland Caledonides: Tectonics, v. 20, p. 58–77, doi: 10.1029/2000TC900020.

Henriksen, N., 1986, Geological Map of Greenland, 1:500,000, Sheet 12, Scoresby Sund, Descriptive Text: Copenhagen, Geological Survey of Greenland, 27 p.

Henriksen, N., 1996, Conclusion of the 1:500,000 field mapping in eastern North Greenland: Bulletin Grønlands Geologiske Undersøgelse, v. 172, p. 42–48.

Henriksen, N., 1997, Geological Map of Greenland, Dove Bugt, Sheet 10: Copenhagen, Geological Survey of Denmark and Greenland, scale 1:500,000.

Henriksen, N., 1998, North-East Greenland 1997–1998: A new 1:500 000 mapping project in the Caledonian fold belt (72°–75°N): Geology of Greenland Survey Bulletin, v. 180, p. 119–127.

Henriksen, N., 1999, Conclusion of the 1:500,000 mapping project in the Caledonian fold belt in North-East Greenland: Geology of Greenland Survey Bulletin, v. 183, p. 10–22.

Henriksen, N., and Higgins, A.K., 2008, this volume, Chapter 14, Caledonian Orogen of East Greenland 70°–82°N: Geological Map 1:1,000,000—Concepts and principles of compilation, in Higgins, A.K., Gilotti, J.A., and Smith, M.P., eds., The Greenland Caledonides: Evolution of the Northeast Margin of Laurentia: Geological Society of America Memoir 202, doi: 10.1130/2008.1202(14).

Henriksen, N., Higgins, A.K., Kalsbeek, F., and Pulvertaft, T.C.R., 2000, Greenland from Archaean to Quaternary: Descriptive Text to the Geological Map of Greenland 1:2,500,000: Geology of Greenland Survey Bulletin, v. 185, 93 p.

Higgins, A.K., 1982, Geological Map of Greenland 1:100,000, Charcot Land 71 Ø.4 Nord, Krummedal 71 Ø 4 Syd: Descriptive Text: Copenhagen, Geological Survey of Greenland, p. 26.

Higgins, A.K., 1988, The Krummedal supracrustal sequence in East Greenland, in Winchester, J.A., ed., Later Proterozoic Stratigraphy of the Northern Atlantic Regions: Glasgow and London, Blackie and Son, p. 86–96.

Higgins, A.K., ed., 1994, Geology of North-East Greenland: Rapport Grønlands Geologiske Undersøgelse, v. 162, 209 p.

Higgins, A.K., and Leslie, A.G., 2000, Restoring thrusting in the East Greenland Caledonides: Geology, v. 28, p. 1019–1022, doi: 10.1130/0091-7613(2000)28<1019:RTITEG>2.0.CO;2.

Higgins, A.K., and Leslie, A.G., 2004, The Eleonore Sø and Målebjerg foreland windows, East Greenland Caledonides, and the demise of the 'stockwerke concept': Geological Survey of Denmark and Greenland Bulletin, v. 6, p. 77–93.

Higgins, A.K., and Soper, N.J., 1994, Structure of the Eleonore Bay Supergroup at Ardencaple Fjord, North-East Greenland, in Higgins, A.K., ed., Geology of North-East Greenland: Rapport Grønlands Geologiske Undersøgelse, v. 162, p. 91–101.

Higgins, A.K., Friderichsen, J.D., Rex, D.C., and Gledhill, A.R., 1978, Early Proterozoic isotopic ages in the East Greenland Caledonian fold belt: Contributions to Mineralogy and Petrology, v. 67, p. 87–94, doi: 10.1007/BF00371636.

Higgins, A.K., Ineson, J.R., Peel, J.S., Surlyk, F., and Sønderholm, M., 1991, Lower Palaeozoic Franklinian Basin of North Greenland, in Peel, J.S., and Sønderholm, M., eds., Sedimentary Basins of North Greenland: Bulletin Grønlands Geologiske Undersøgelse, v. 160, p. 71–139.

Higgins, A.K., Leslie, A.G., and Smith, M.P., 2001a, Neoproterozoic–Lower Palaeozoic stratigraphical relationships in the marginal thin-skinned thrust belt of the East Greenland Caledonides: Comparisons with the foreland of Scotland: Geological Magazine, v. 138, p. 143–160, doi: 10.1017/S0016756801005076.

Higgins, A.K., Smith, M.P., Soper, N.J., Leslie, A.G., Rasmussen, J.A., and Sønderholm, M., 2001b, The Neoproterozoic Hekla Sund Basin, eastern North Greenland: A pre-Iapetan extensional sequence thrust across its rift shoulders during the Caledonian orogeny: Journal of the Geological Society of London, v. 158, p. 487–499.

Higgins, A.K., Soper, N.J., Smith, M.P., and Rasmussen, J.A., 2004a, The Caledonian parautochthonous fold and thrust belt of Kronprins Christian Land, eastern North Greenland: Geological Survey of Denmark and Greenland Bulletin, v. 6, p. 41–56.

Higgins, A.K., Elvevold, S., Escher, J.C., Frederiksen, K.S., Gilotti, J.A., Henriksen, N., Jepsen, H.F., Jones, K.A., Kalsbeek, F., Kinny, P.D., Leslie, A.G., Smith, M.P., Thrane, K., and Watt, G.R., 2004b, The foreland-propagating thrust architecture of the East Greenland Caledonides 72°–75°N: Journal of the Geological Society of London, v. 161, p. 1009–1026, doi: 10.1144/0016-764903-141.

Hofer, E., 1957, Arctic Riviera, North East Greenland: Berne, Kümmerly & Frey, Geographical Publishers, 127 p.

Huber, W., 1950, Geologisch-Petrographische Untersuchungen in der inneren Fjordregion des Kejser Franz Josephs Fjordsystems in Nordostgrønland: Meddelelser om Grønland, v. 151, no. 3, 83 p.

Hull, J.M., and Gilotti, J.A., 1994, The Germania Land deformation zone and related structures, North-East Greenland: Rapport Grønlands Geologiske Undersøgelse, v. 162, p. 113–127.

Hull, J.M., Friderichsen, J.D., Gilotti, J.A., Henriksen, N., Higgins, A.K., and Kalsbeek, F., 1994, Gneiss complex of the Skærfjorden region (76°–78°N), North-East Greenland: Rapport Grønlands Geologiske Undersøgelse, v. 162, p. 35–51.

Hurst, J.M., 1984, Upper Ordovician and Silurian Carbonate Shelf Stratigraphy, Facies and Evolution, Eastern North Greenland: Bulletin Grønlands Geologiske Undersøgelse, v. 148, 73 p.

Hurst, J.M., and Surlyk, F., 1982, Stratigraphy of the Silurian Turbidite Sequence of North Greenland: Bulletin Grønlands Geologiske Undersøgelse, v. 145, 121 p.

Ineson, J.R., and Peel, J.S., 1997, Cambrian Shelf Stratigraphy of North Greenland: Geology of Greenland Survey Bulletin, v. 173, 120 p.

Ineson, J.R., and Surlyk, F., eds., 2003, The Jurassic of Denmark and Greenland: Geological Survey of Denmark and Greenland Bulletin, v. 1, 948 p.

Jarvik, E., 1950, Note on Middle Devonian Crossopterygians from the Eastern Part of Gauss Halvø, East Greenland, with an Appendix: An attempt at a

Correlation of the Upper Old Red Sandstone of East Greenland with the Marine Sequence: Meddelelser om Grønland, v. 149, no. 6, 20 p.

Jarvik, E., 1961, Devonian vertebrates, in Raasch, G.O., ed., Geology of the Arctic, Volume 1: Toronto, Toronto University Press, p. 197–204.

Jepsen, H.F., 2000, Geological Map of Greenland, Lambert Land, Sheet 9: Copenhagen, Geological Survey of Denmark and Greenland, scale 1:500,000.

Jepsen, H.F., and Kalsbeek, F., 1985, Evidence for non-existence of a Carolinidian fold belt in eastern North Greenland, in Gee, D.G., and Sturt, B.A., eds., The Caledonide Orogen: Scandinavia and Related Areas: London, Wiley & Sons, p. 1071–1076.

Jones, K.A., and Strachan, R.A., 2000, Crustal thickening and ductile extension in the NE Greenland Caledonides: A metamorphic record from anatectic pelites: Journal of Metamorphic Geology, v. 18, p. 719–735, doi: 10.1046/j.1525-1314.2000.00282.x.

Kalsbeek, F., ed., 1989, Geology of the Ammassalik Region, South-East Greenland: Rapport Grønlands Geologiske Undersøgelse, v. 146, 106 p.

Kalsbeek, F., 1995, Geochemistry, tectonic setting, and poly-orogenic history of Palaeoproterozoic basement rocks from the Caledonian fold belt of North-East Greenland: Precambrian Research, v. 72, p. 301–315, doi: 10.1016/0301-9268(94)00097-B.

Kalsbeek, F., Nutman, A.P., and Taylor, P.N., 1993, Palaeoproterozoic basement province in the Caledonian fold belt of North-East Greenland: Precambrian Research, v. 63, p. 163–178, doi: 10.1016/0301-9268(93)90010-Y.

Kalsbeek, F., Nutman, A.P., Escher, J.C., Friderichsen, J.D., Hull, J.M., Jones, K.A., and Pedersen, S.A.S., 1999, Geochronology of granitic and supracrustal rocks from the northern part of the East Greenland Caledonides: Ion microprobe U-Pb zircon ages: Geology of Greenland Survey Bulletin, v. 184, p. 31–48.

Kalsbeek, F., Thrane, K., Nutman, A.P., and Jepsen, H.F., 2000, Late Mesoproterozoic to early Neoproterozoic history of the East Greenland Caledonides: Evidence for Grenvillian orogenesis?: Journal of the Geological Society of London, v. 157, p. 1215–1225.

Kalsbeek, F., Jepsen, H.F., and Nutman, A.P., 2001a, From source migmatites to plutons: Tracking the origin of c. 435 Ma granites in the East Greenland Caledonian orogen: Lithos, v. 57, p. 1–21, doi: 10.1016/S0024-4937(00)00071-2.

Kalsbeek, F., Jepsen, H.F., and Jones, K.A., 2001b, Geochemistry and petrogenesis of S-type granites in the East Greenland Caledonides: Lithos, v. 57, p. 91–109, doi: 10.1016/S0024-4937(01)00038-X.

Kalsbeek, F., Higgins, A.K., Jepsen, H.F., Frei, R., and Nutman, A.P., 2008, this volume (Chapter 9), Granites and granites in the East Greenland Caledonides, in Higgins, A.K., Gilotti, J.A., and Smith, M.P., eds., The Greenland Caledonides: Evolution of the Northeast Margin of Laurentia: Geological Society of America Memoir 202, doi: 10.1130/2008.1202(09).

Kalsbeek, F., Thrane, K., Higgins, A.K., Jepsen, H.F., Leslie, A.G., Nutman, A.P., and Frei, R., 2008, this volume (Chapter 3), Polyorogenic history of the East Greenland Caledonides, in Higgins, A.K., Gilotti, J.A., and Smith, M.P., eds., The Greenland Caledonides: Evolution of the Northeast Margin of Laurentia: Geological Society of America Memoir 202, doi: 10.1130/2008.1202(03).

Katz, H.R., 1952a, Zur Geologie von Strindbergs Land (Nordostgrönland): Meddelelser om Grønland, v. 111, no. 1, 150 p.

Katz, H.R., 1952b, Ein Querschnitt durch die Nunatakzone Ostgrönlands (ca. 74° N.B.). Ergebnisse einer Reisen vom Inlandseis (in Zusammenarbeit mit den Expéditions Polaires Françaises von P.-E. Victor) ostwarts bis in die Fjord-region, ausgeführt im Sommer 1951: Meddelelser om Grønland, v. 144, no. 8, 65 p.

Katz, H.R., 1961, Late Precambrian to Cambrian stratigraphy in East Greenland, in Raasch, G.O., ed., Geology of the Arctic, Volume 1: Toronto, Toronto University Press, p. 299–328.

Koch, J.P., and Wegener, A., 1930, Wissenschaftlische Ergebnisse der Dänischen Expedition nach Dronning Louise Land und quer über das Inlandeis von Nordgrönland 1912–1913 unter leitung von Hauptmann J.P. Koch: Meddelelser om Grønland, v. 75, no. I and II, 676 p.

Koch, L., 1919, De geologiska resultaten af den andra Thule expeditionen till Grönland: Geologiska Föreningens Stockholm Förhandlingar, v. 41, p. 109–112.

Koch, L., 1926, Report on the Danish Bicentenary Jubilee Expedition North of Greenland 1920–1923: Meddelelser om Grønland, v. 70, no. 1, 232 p.

Koch, L., 1929a, The Geology of East Greenland: Meddelelser om Grønland, v. 73, II, no. 1, 204 p.

Koch, L., 1929b, Stratigraphy of Greenland: Meddelelser om Grønland, v. 73, II, no. 2, p. 205–320.

Koch, L., 1930, Report on the Geological Expedition to East Greenland 1926–1927: Meddelelser om Grønland, v. 76, no. 6, p. 225–282.

Koch, L., 1935, Geologie von Grönland, in Krenkel, E., ed., Geologie der Erde: Berlin, Gebrüder Bornträger, 159 p.

Koch, L., 1940, Survey of North Greenland: Meddelelser om Grønland, v. 130, no. 1, 364 p.

Koch, L., 1955, Report on the Expeditions to Central East Greenland 1926–1939, Conducted by Lauge Koch, Part II: Meddelelser om Grønland, v. 143, no. 2, 642 p.

Koch, L., and Haller, J., 1971, Geological Map of East Greenland 72°–76° N. Lat. (1:250,000): Meddelelser om Grønland, v. 183, 26 p. (+ 13 map sheets).

Kranck, E.H., 1935, On the Crystalline Complex of Liverpool Land: Meddelelser om Grønland, v. 95, no. 7, 122 p.

Larsen, P.-H., and Bengaard, H.-J., 1991, Devonian basin initiation in East Greenland: A result of sinistral wrench faulting and Caledonian extensional collapse: Journal of the Geological Society of London, v. 148, p. 355–368, doi: 10.1144/gsjgs.148.2.0355.

Larsen, P.-H., Olsen, H., and Clack, J.A., 2008, this volume, The Devonian Basin in East Greenland—Review of basin evolution and vertebrate assemblages, in Higgins, A.K., Gilotti, J.A., and Smith, M.P., eds., The Greenland Caledonides: Evolution of the Northeast Margin of Laurentia: Geological Society of America Memoir 202, doi: 10.1130/2008.1202(11).

Leedal, G.P., 1952, The Crystalline Rocks of East Greenland between Latitudes 74°30′N and 75°N: Meddelelser om Grønland, v. 142, no. 6, 80 p.

Leslie, A.G., and Higgins, A.K., 2008, this volume, Foreland-propagating Caledonian thrust systems in East Greenland, in Higgins, A.K., Gilotti, J.A., and Smith, M.P., eds., The Greenland Caledonides: Evolution of the Northeast Margin of Laurentia: Geological Society of America Memoir 202, doi: 10.1130/2008.1202(07).

Leslie, A.G., and Nutman, A.P., 2003, Evidence for Neoproterozoic orogenesis and early high temperature Scandian deformation events in the southern East Greenland Caledonides: Geological Magazine, v. 140, p. 309–333, doi: 10.1017/S0016756803007593.

Liljequist, G.H., 1993, High Latitudes: A History of Swedish Polar Travels and Research: Stockholm, Swedish Polar Research Secretariat and Streiffert Förlag AB, 607 p.

McClelland, W.C., and Gilotti, J.A., 2003, Late stage extensional exhumation of high-pressure granulites in the Greenland Caledonides: Geology, v. 31, p. 259–262, doi: 10.1130/0091-7613(2003)031<0259:LSEEOH>2.0.CO;2.

Mikkelsen, A., 1992, Blyfundet i Mesters Vig: Forskning i Grønland: Tusaat, v. 3/92, p. 60–62.

Mikkelsen, P.S., 1994, Nordøstgrønland 1908–60, Fangstmandsperioden: Copenhagen, Dansk Polarcenter, 408 p.

Moncrieff, A.M., 1989, The Tillite Group and related rocks of East Greenland: Implications for Late Proterozoic palaeogeography, in Gayer, R.A., ed., The Caledonide Geology of Scandinavia: London, Graham and Trotman, p. 285–297.

Nathorst, A.G., 1901, Bidrag till nordöstra Grönlands geologi: Geologiska Föreningens Stockholm Förhandlingär, v. 23, p. 275–306.

Nathorst, A.G., 1911, Contributions to the Carboniferous flora of North-East Greenland: Meddelelser om Grønland v. 43, no. 12, p. 339–346.

Nordenskjöld, O., 1907, On the geology and physical geography of East Greenland: Meddeleser om Grønland, v. 28, no. 1(5), p. 151–284.

Nutman, A.P., and Kalsbeek, F., 1994, Search for Archean basement in the Caledonian fold belt of North-East Greenland: Rapport Grønlands Geologiske Undersøgelse, v. 162, p. 129–133.

Odell, N.E., 1939, The Structure of the Kejser Franz Josephs Fjord Region, North-East Greenland: Meddelelser om Grønland, v. 119, no. 6, 51 p.

Odell, N.E., 1944, The petrography of the Frans Josef Fjord region, North-East Greenland, in relation to its structures: Transactions of the Royal Society of Edinburgh, v. 61, no. 1, p. 221–246.

Olsen, H., 1993, Sedimentary Basin Analysis of the Continental Devonian Basin in East Greenland: Bulletin Grønlands Geologiske Undersøgelse, v. 168, 80 p.

Olsen, H., and Larsen, P.-H., 1993, Lithostratigraphy of the Continental Devonian Sediments in North-East Greenland: Bulletin Grønlands Geologiske Undersøgelse, v. 165, 108 p.

Parkinson, M.M.L., and Whittard, W.F., 1931, The geological work of the Cambridge Expedition to East Greenland in 1929: Quarterly Journal of the Geological Society of London, v. 87, p. 650–674.

Peacock, J.D., 1956, The Geology of Dronning Louise Land, N.E. Greenland: Meddelelser om Grønland, v. 137, no. 7, 38 p.

Peacock, J.D., 1958, Some Investigations into the Geology and Petrography of Dronning Louise Land, N.E. Greenland: Meddelelser om Grønland, v. 157, no. 4, 139 p.

Pedersen, S.A.S., Craig, L.E., Upton, B.G.J., Rämö, O.T., Jepsen, H.F., and Kalsbeek, F., 2002, Palaeoproterozoic (1740 Ma) rift-related volcanism in the Hekla Sund region, eastern North Greenland: Field occurrence, geochemistry and tectonic setting: Precambrian Research, v. 114, p. 327–346, doi: 10.1016/S0301-9268(01)00234-0.

Peel, J.S., 1982, The Lower Paleozoic of Greenland, in Embry, A.F., and Balkwill, R., eds., Arctic Geology and Geophysics: Canadian Society of Petroleum Geology Memoir 8, p. 309–330.

Peel, J.S., and Cowie, J.W., 1979, New names for Ordovician formations in Greenland: Rapport Grønlands Geologiske Undersøgelse, v. 91, p. 117–124.

Peel, J.S., and Sønderholm, M., eds., 1991, Sedimentary Basins of North Greenland: Bulletin Grønlands Geologiske Undersøgelse, v. 160, 164 p.

Poulsen, C., 1930, Contributions to the stratigraphy of the Cambro–Ordovician of East Greenland (Preliminary Report): Meddelelser om Grønland v. 74, no. 12, p. 297–316.

Poulsen, C., 1956, The Cambrian of the East Greenland geosyncline, in Rodgers, J., ed., El Sistema Cámbrico, paleogeografia y el problema de su base, in 20th International Geological Congress, Mexico, 1956: International Commission for Stratigraphy, Symposium 1, p. 59–69.

Poulsen, C., and Rasmussen, H.W., 1951, Geological Map (scale 1:50,000) and Description of Ella Ø: Meddelelser om Grønland, v. 151, no. 5, p. 1–25.

Rasmussen, J.A., and Smith, M.P., 2001, Conodont geothermometry and tectonic overburden in the northernmost East Greenland Caledonides: Geological Magazine, v. 138, p. 687–698.

Ravn, J.P.J., 1911, On Jurassic and Cretaceous fossils from North-East Greenland: Meddelelser om Grønland, v. 45, no. 10, p. 477–500.

Rex, D.C., and Gledhill, A., 1974, Reconnaissance geochronology of the infracrustal rocks of Flyverfjord, Scoresby Sund, East Greenland: Bulletin of the Geological Society of Denmark, v. 23, p. 49–54.

Rex, D.C., and Gledhill, A., 1981, Isotopic studies in the East Greenland Caledonides (72°–74°N)—Precambrian and Caledonian ages: Rapport Grønlands Geologiske Undersøgelse, v. 104, p. 47–72.

Rex, D.C., and Higgins, A.K., 1985, Potassium-argon ages from the East Greenland Caledonides between 72° and 74°N, in Gee, D.G., and Sturt, B.A., eds., The Caledonide Orogen—Scandinavia and Related Areas: Chichester, John Wiley and Sons, p. 1115–1124.

Rex, D.C., Gledhill, A.R., and Higgins, A.K., 1976, Progress report on geochronological investigations in the crystalline complexes of the East Greenland Caledonian fold belt between 72° and 74°N: Rapport Grønlands Geologiske Undersøgelse, v. 80, p. 127–133.

Rex, D.C., Gledhill, A.R., and Higgins, A.K., 1977, Precambrian Rb-Sr isochron ages from the crystalline complexes of inner Forsblads Fjord, East Greenland fold belt: Rapport Grønlands Geologiske Undersøgelse, v. 85, p. 122–126.

Ries, C.J., 2003, Retten, magten og æren; Lauge Koch sagen, en strid om Grønlands geologiske udforskning: København, Lindhardt og Ringhof, 366 p.

Roberts, D., 2003, The Scandinavian Caledonides: Event chronology, palaeogeographic setting and likely modern analogues: Tectonophysics, v. 365, p. 283–299, doi: 10.1016/S0040-1951(03)00026-X.

Roberts, D., Nordgulen, Ø., and Melezhik, V.A., 2004, The uppermost allochthon in the Scandinavian Caledonides: A terrane collage of Laurentian ancestry transported onto Baltica in Siluro-Devonian time (abs.), in Hatcher, R.D., Jr., Whisner, J.B., Thigpen, J.R., Whitner, N.E., and Whisner, S.C., eds., Abstracts of the 17th International Basement Tectonics Conference 2004, Oak Ridge, Tennessee, USA: p. 28–29.

Roberts, D., Nordgulen, Ø., and Melezhik, V., 2007, The Uppermost Allochthon in the Scandinavian Caledonides: From a Laurentian ancestry through Taconian orogeny to Scandian crustal growth on Baltica, in Hatcher, R.D., Jr., Carlson, M.P., McBride, J.H. and Martínez-Catalán, J.R., eds., 4-D Framework of Continental Crust: Geological Society of America Memoir 200, p. 357–377, doi: 10.1130/2007.1200(18).

Sahlstein, T.G., 1935a, Petrographie der Eklogiteinschlüsse in den Gneisen des südwestlichen Liverpool Landes in Ostgrönland. Nebst Anhang: Granulitartiger Gneis nordöstlich von Kap Hope: Meddelelser om Grønland, v. 95, no. 5, 43 p.

Sahlstein, T.G., 1935b, Zur Regelung der Gesteine im Kristallin von Liverpool Land in Ostgrönland: Meddelelser om Grønland, v. 95, no. 6, 25 p.

Säve-Söderbergh, G., 1932, Preliminary Note on Devonian Stegocephalians from East Greenland: Meddelelser om Grønland, v. 94, no. 7, 107 p.

Säve-Söderbergh, G., 1933, Further Contributions to the Devonian Stratigraphy of East Greenland: I. Results from the Summer Expeditions 1932: Meddelelser om Grønland, v. 96, no. 1, 40 p.

Säve-Söderbergh, G., 1934, Further Contributions to the Devonian Stratigraphy of East Greenland: II. Investigations on Gauss Peninsula during the Summer of 1933, with an Appendix: Notes on the Geology of the Passage Hills (East Greenland): Meddelelser om Grønland, v. 96, no. 2, 74 p.

Schwarzenbach, F.H., 1993, Towards new horizons. John Haller 1927–1984: Verlag der Fachvereine Zürich: Zürich, Schweizerische Stiftung für Alpine Forschungen, 128 p.

Scoresby, W., 1823, Journal of a Voyage to the Northern Whale-Fishery; Including Researches and Discoveries on the Eastern Coast of West Greenland, Made in the Summer of 1822, in the Ship Baffin of Liverpool: Edinburgh, Constable & Co, 472 p.

Skovsted, C.B., 2003, The Early Cambrian Fauna of North-East Greenland [dissertation]: Uppsala, Uppsala University, Department of Earth Science, Palaeobiology, 233 p.

Smith, M.P., 1991, Early Ordovician Conodonts of East and North Greenland: Meddelelser om Grønland, Geoscience, v. 26, 81 p.

Smith, M.P., and Bjerreskov, M., 1994, The Ordovician System in Greenland: International Union of Geological Sciences Special Publication 29A, 46 p.

Smith, M.P., and Rasmussen, J.A., 2008, this volume, Cambrian–Silurian development of the Laurentian margin of the Iapetus Ocean in Greenland and related areas, in Higgins, A.K., Gilotti, J.A., and Smith, M.P., eds., The Greenland Caledonides: Evolution of the Northeast Margin of Laurentia: Geological Society of America Memoir 202, doi: 10.1130/2008.1202(06).

Smith, M.P. and Robertson, S., 1999, The Nathorst Land Group (Neoproterozoic) of East Greenland—Lithostratigraphy, basin geometry and tectonic history: Danmarks og Grønlands Geologiske Undersøgelse Rapport, v. 19, p. 127–143.

Smith, M.P., Rasmussen, J.A., Robertson, S., Higgins, A.K., and Leslie, A.G., 2004, Lower Palaeozoic stratigraphy of the East Greenland Caledonides: Geological Survey of Denmark and Greenland Bulletin, v. 6, p. 5–28.

Sommer, M., 1957, Geologie von Lyell Land (Nordostgrönland): Meddelelser om Grønland, v. 155, no. 2, 157 p.

Sønderholm, M., and Jepsen, H.F., 1991, Proterozoic basins of North Greenland: Bulletin Grønlands Geologiske Undersøgelse, v. 160, p. 49–69.

Sønderholm, M., and Tirsgaard, H., 1993, Lithostratigraphic Framework of the Upper Proterozoic Eleonore Bay Supergroup of East and North-East Greenland: Bulletin Grønlands Geologiske Undersøgelse, v. 167, 38 p.

Sønderholm, M., Frederiksen, K.S., Smith, M.P., and Tirsgaard, H., 2008, this volume, Neoproterozoic sedimentary basins with glacigenic deposits of the East Greenland Caledonides, in Higgins, A.K., Gilotti, J.A., and Smith, M.P., eds., The Greenland Caledonides: Evolution of the Northeast Margin of Laurentia: Geological Society of America Memoir 202, doi: 10.1130/2008.1202(05).

Soper, N.J., and Higgins, A.K., 1993, Basement–cover relationships in the East Greenland Caledonides: Evidence from the Eleonore Bay Supergroup at Ardencaple Fjord: Transactions of the Royal Society of Edinburgh, Earth Sciences, v. 84, p. 103–115.

Soper, N.J., Strachan, R.A., Holdsworth, R.E., Gayer, R.A., and Greiling, R.O., 1992, Sinistral transpression and the closure of Iapetus: Journal of the Geological Society of London, v. 149, p. 871–880, doi: 10.1144/gsjgs.149.6.0871.

Stearns, C., Van der Voo, R., and Abrahamsen, N., 1989, A new Siluro–Devonian paleopole from early Paleozoic rocks of the Franklinian Basin, North Greenland fold belt: Journal of Geophysical Research, v. 94, no. B8, p. 10,669–10,683.

Steck, A., 1971, Kaledonische Metamorphose der Praekambrischen Charcot Land Serie, Scoresby Sund, Ost-Grönland: Bulletin Grønlands Geologiske Undersøgelse, v. 97, 69 p.

Steiger, R.H., Hansen, B.T., Schuler, C., Bär, M.T., and Henriksen, N., 1979, Polyorogenic nature of the southern Caledonian fold belt in East Greenland: The Journal of Geology, v. 87, p. 475–495.

Stemmerik, L., ed., 2000, Palynology and Deposition in the Wandel Sea Basin, Eastern North Greenland: Geology of Greenland Survey Bulletin, v. 187, 101 p.

Stemmerik, L., and Håkansson, E., 1991, Carboniferous and Permian history of the Wandel Sea Basin, North Greenland, in Peel, J.S., and Sønderholm, M., eds., Sedimentary Basins of North Greenland: Bulletin Grønlands Geologiske Undersøgelse, v. 160, p. 141–151.

Stemmerik, L., and Piasecki, S., 2004, Isotopic evidence for the age of the Røde Ø Conglomerate, inner Scoresby Sund, East Greenland: Bulletin of the Geological Society of Denmark, v. 51, p. 137–140.

Stendal, H., and Frei, R., 2008, this volume, Mineral occurrences in central East Greenland (70°–75°N) and their relation to the Caledonian orogeny—A Sr-Nd-Pb isotopic study of scheelite, in Higgins, A.K., Gilotti, J.A., and Smith, M.P., eds., The Greenland Caledonides: Evolution of the Northeast Margin of Laurentia: Geological Society of America Memoir 202, doi: 10.1130/2008.1202(12).

Stouge, S., Boyce, D.W., Christiansen, J., Harper, D.A.T., and Knight, I., 2001, Vendian–Lower Ordovician stratigraphy of Ella Ø, North-East Greenland: New investigations: Geology of Greenland Survey Bulletin, v. 189, p. 107–114.

Stouge, S., Boyce, D.W., Christiansen, J.L., Harper, D.A.T., and Knight, I., 2002, Lower–Middle Ordovician stratigraphy of North-East Greenland: Geology of Greenland Survey Bulletin, v. 191, p. 117–125.

Strachan, R.A., and Tribe, I.R., 1994, Structure of the Storstrømmen shear zone, eastern Hertugen of Orléans Land, North-East Greenland: Rapport Grønlands Geologiske Undersøgelse, v. 162, p. 103–112.

Strachan, R.A., Holdsworth, R.E., Friderichsen, J.D., and Jepsen, H.F., 1992, Regional Caledonian structure within an oblique convergence zone, Dronning Louise Land, NE Greenland: Journal of the Geological Society of London, v. 149, p. 359–371, doi: 10.1144/gsjgs.149.3.0359.

Strachan, R.A., Nutman, A.P., and Friderichsen, J.D., 1995a, SHRIMP U-Pb geochronology and metamorphic history of the Smallefjord sequence, NE Greenland Caledonides: Journal of the Geological Society of London, v. 152, p. 779–784, doi: 10.1144/gsjgs.152.5.0779.

Strachan, R.A., Chadwick, B., Friend, C.R.L., and Holdsworth, R.E., 1995b, New perspectives on the Caledonian orogeny in Northeast Greenland, in Hibbard, J.P., van Staal, C.R., and Cawood, P.A., eds., Perspectives in the Appalachian-Caledonian Orogen: Geological Association of Canada Special Paper 41, p. 303–332.

Strachan, R.A., Martin, M.W., and Friderichsen, J.D., 2001, Evidence for contemporaneous yet contrasting styles of granite magmatism during extensional collapse of the northeast Greenland Caledonides: Tectonics, v. 20, p. 458–473, doi: 10.1029/2000TC001206.

Swett, K., and Smit, D.E., 1972, Cambro–Ordovician shelf sedimentation of western Newfoundland, northwest Scotland and central East Greenland, in Aitken, J.D., Basset, H.G., and deWit, R., eds., Proceedings of the 24th International Geological Congress, Canada, 1972, v. 6: Montreal, Canada, International Geological Congress, p. 33–41.

Teichert, C., 1933, Untersuchungen zum Bau des Kaledonischen Gebirges in Ostgrönland: Meddelelser om Grønland, v. 95, no. 1, 121 p.

Thorson, G., ed., 1937, Med Treaarsekspeditionen til Christian X's Land: Copenhagen, Gyldendalsk Boghandel Nordisk Forlag, 281 p.

Thrane, K., 2002, Relationships between Archaean and Palaeoproterozoic crystalline basement complexes in the southern part of the East Greenland Caledonides: An ion microprobe study: Precambrian Research, v. 113, p. 19–42, doi: 10.1016/S0301-9268(01)00198-X.

Torsvik, T.H., Smethurst, M.A., Meert, J.G., Van der Voo, R., McKerrow, W.S., Brasier, M.D., Sturt, B.A., and Walderhaug, H.J., 1996, Continental break-up and collision in the Neoproterozoic and Palaeozoic—A tale of Baltica and Laurentia: Earth-Science Reviews, v. 40, p. 229–258, doi: 10.1016/0012-8252(96)00008-6.

Upton, B.G.J., Rämö, O.T., Heaman, L.M., Blichert-Toft, J., Kalsbeek, F., Barry, T.L., and Jepsen, H.F., 2005, The Mesoproterozoic Zig-Zag Dal basalts and associated intrusions of eastern North Greenland: Mantle plume–lithosphere interaction: Contributions to Mineralogy and Petrology, v. 149, p. 40–56, doi: 10.1007/s00410-004-0634-7.

Ventegodt, O., 2000, Den sidste brik: Mylius-Erichsens Danmark-ekspedition til Nordøstgrønland 1906–1908: Copenhagen, Gyldendal, 428 p.

Vogt, P., 1965, Zur Geologie von Südwest-Hinks Land (Ostgrönland, 71°30′N): Meddelelser om Grønland, v. 154, no. 5, 24 p.

von Hochstetter, F., Lenz, O., Toula, F., Bauer, A., and Heer, O., 1874, Geologie, in Hartlaub, G., and Lindemann, M., eds., Geographisches Gesellschaft in Bremen: Die Zweite Deutsche Nordpolarfahrt in den Jahren 1869 und 1870, v. 2: Leipzig, Germany, F.A. Brockhaus, p. 471–517.

Watt, G.R., and Thrane, K., 2001, Early Neoproterozoic events in East Greenland: Precambrian Research, v. 110, p. 165–184, doi: 10.1016/S0301-9268(01)00186-3.

Watt, G.R., Kinny, P.D., and Friderichsen, J.D., 2000, U-Pb geochronology of Neoproterozoic and Caledonian tectonothermal events in the East Greenland Caledonides: Journal of the Geological Society of London, v. 157, p. 1031–1048.

Wegmann, C.E., 1935, Preliminary Report on the Caledonian Orogeny in Christian X's Land (North-East Greenland): Meddelelser om Grønland, v. 103, no. 3, 59 p.

Wenk, E., 1961, On the Crystalline Basement and the Basal Part of the Pre-Cambrian Eleonore Bay Group in the Southwestern Part of Scoresby Sund: Meddelelser om Grønland, v. 168, no. 1, 54 p.

Wenk, E., and Haller, J., 1953, Geological Explorations in the Petermann Region, Western Part of Frænkels Land, East Greenland: Meddelelser om Grønland, v. 111, no. 3, 48 p.

Wordie, J.M., 1927, The Cambridge expedition to East Greenland in 1926: Appendix V. Geology: The Geographical Journal, v. 70, p. 225–253, doi: 10.2307/1781943.

Wordie, J.M., 1930, Cambridge East Greenland Expedition 1929: Ascent of the Petermann Peak: The Geographical Journal, v. 75, no. 6, p. 481–495, doi: 10.2307/1784482.

Wordie, J.M., and Whittard, W.F., 1930, A contribution to the geology of the country between Petermann Peak and Kjerulf Fjord: East Greenland: Geological Magazine, v. 67, p. 145–158.

MANUSCRIPT ACCEPTED BY THE SOCIETY 14 JANUARY 2008

Architecture and evolution of the East Greenland Caledonides— An introduction

A.K. Higgins*
Geological Survey of Denmark and Greenland, Øster Voldgade 10, DK-1350 Copenhagen K, Denmark

A. Graham Leslie
British Geological Survey, Murchison House, Edinburgh EH9 3LA, UK

ABSTRACT

The East Greenland Caledonian orogen can be divided into distinct structurally bound geological domains composed of Archean to Lower Paleozoic lithostratigraphic and lithodemic components derived from the eastern margin of Laurentia. These domains originally evolved as major westward-displaced thrust units in the overriding plate during the collision with Baltica. The western border of the 1300-km-long and up to 300-km-wide segment of the orogen preserved onshore in East Greenland is thrust against the rocks of the Laurentian craton and is largely concealed beneath the Inland Ice. A foreland-propagating thrust pile is well-preserved in the extreme north of the orogen (79°N–82°N), and in the southern half (70°N–76°N), with less-well-preserved remnants in the western nunataks of the intervening region. Between 76°N and 81°N, the outer coastal region is dominated by high-grade Paleoproterozoic orthogneisses that were reworked during the Caledonian orogeny; most of this region is characterized by the presence of eclogitic mafic enclaves, which testify to exhumation from depths in excess of 50 km in late Caledonian time. Caledonian granites are confined to the southern orogen (70°N–76°N), where they intrude rock units now contained within the upper thrust sheet. Devonian continental basins are conspicuous in the southern part of the orogen and occur offshore farther north; their deposition can be linked to syn- to late-orogenic extension. Carboniferous and younger rocks are exposed onshore in the extreme north of the orogen (80°N–81°N) and are widespread in the south between 71°N and 75°N.

Keywords: Caledonian, structural domains, foreland-directed thrusting, extensional faulting.

---

*akh@geus.dk

A REVOLUTION OF UNDERSTANDING

The N-S–trending Caledonian orogenic belt in East Greenland was created by continent-continent collision and thickening in the early Paleozoic as Laurentia (North America and Greenland) converged with Baltica. With continued convergence, the subduction of Baltica beneath Laurentia led to development of the westward-directed thrust belts now preserved in Greenland, whereas eastward-directed thrust systems are preserved in Scandinavia. The East Greenland elements of the Caledonian orogen were all derived from the Laurentian margin, and there is little evidence in East Greenland for the short-lived early Paleozoic marginal arcs and basins that must have been associated with final convergence with Baltica and the subduction of Iapetan oceanic crust. These disrupted terranes appear to have been mainly incorporated as Scandian thrust sheets in the upper structural levels of the Scandinavian Caledonides (Roberts, 2003). No exotic terranes are found in the East Greenland sector of the Caledonides.

The wealth of new observations and data accumulated between 1968 and 1998 during regional mapping by the Geological Survey of Greenland and other organizations has radically changed our understanding of virtually all aspects of the East Greenland Caledonides (Fig. 1). This segment of the orogen extends for 1300 km between 70°N (Scoresby Sund) and 82°N (Kronprins Christian Land). While the existence of westward-directed thrust systems had been demonstrated locally at widely scattered locations in the 1950s and early 1960s, John Haller's (1971) book, *Geology of the East Greenland Caledonides*, promoted a "stockwerk" concept, in which widespread in situ vertical granitization was considered to be responsible for all features of the orogen. New mapping by the Geological Survey of Greenland, together with increasingly sophisticated isotopic dating carried out in collaboration with a number of international laboratories, gradually established the presence of a wide range of Archean and Proterozoic granitoid and sedimentary rock units within the orogen and evidence for westward-directed thrusting at different structural levels over a wide region. Recognition of major extensional faulting added a new dimension to regional interpretations of the orogen (Strachan et al., 1992) and, in central East Greenland (72°N–74°30′N), led to a brief revival of the "stockwerk" concept within an extensional framework that denied any contribution from thrusting (Hartz and Andresen, 1995; Andresen et al., 1998); this concept was undermined by discovery of new foreland windows and proof of hundreds of kilometers of major thrusting in the same region (Higgins and Leslie, 2000, 2004). Extension and thrusting may well have been broadly synchronous, where thrusting occurred at deep levels and extension occurred at higher structural levels (Gilotti and McClelland, this volume). Evidence for high-pressure Devonian to Mississippian metamorphism has now been recognized throughout the length of the orogen: 415–390 Ma eclogite-facies rocks occur in the coastal region from 76°N to 81°N (Gilotti et al., 2004), 405 Ma granulites have been found in Payer Land (74°30′N) (McClelland and Gilotti, 2003), and ca. 400 Ma eclogite relics occur in Liverpool Land (71°N) (Hartz et al., 2005). The Devonian continental basins in the southern orogen are now regarded as synorogenic rather than the traditional "postorogenic" interpretation, and there is increasing evidence that plate convergence may have continued into the Mississippian.

The Paleogene opening of the North Atlantic Ocean disrupted the Caledonian orogen, and the widely separated parts now occur in East Greenland, Svalbard, Scandinavia, and the NW British Isles. The contributions to this volume present evidence from East Greenland for at least 80 m.y. of Caledonian orogenic activity, stretching from 466 Ma (earliest granitoid intrusions, Middle Ordovician; Kalsbeek et al., this volume, Chapter 9) to the Mississippian (360–347 Ma ages for the ultrahigh-pressure metamorphism in parts of the eclogite province; McClelland et al., 2006).

The orogenic belt in East Greenland can be divided into a number of regionally extensive but discrete structural domains (Fig. 2); these domains and the essential characteristics, lithostratigraphy, and tectonic framework of the orogenic belt are briefly introduced in this chapter along with references to more detailed descriptions contained in the succeeding chapters of this volume. High structural levels are preserved in the extreme north in Kronprins Christian Land, as well as in the southern half of the orogen. Exhumed deep levels of the orogen occupy a wide swath of the coastal region north of 76°N.

The various chapters of this volume review the composition of the reworked segments of the Laurentian Shield preserved within the Caledonian orogen, dealing with the Paleoproterozoic siliciclastic sedimentary rocks and Neoproterozoic to Silurian heterolithic sedimentary successions deposited on the Laurentian margin of Iapetus, as well as assessments of the structural evolution, regional metamorphism, migmatization, and granite formation within the East Greenland Caledonian orogen. The continental Devonian basins associated with synorogenic extension, and some aspects of mineralization, are the subject of separate chapters, and these are followed by a review of the British Caledonides, which places particular emphasis on comparisons and relationships with East Greenland.

CALEDONIAN OROGENESIS IN EAST GREENLAND

The onshore East Greenland Caledonian orogen preserves a relict collisional geometry; it is composed of far-traveled foreland-propagating thrust sheets that were derived from the Laurentian margin and were translated westward across the orogenic foreland (Higgins and Leslie, 2000; Higgins et al., 2004a). Restoration of thrusting indicates that the site of collision was probably several hundred kilometers east of the present-day part of the orogen preserved onshore. The overthickened orogen suffered extensional reactivation of many of the original contractional shear zones, which had previously defined the major thrust sheets (Gilotti and McClelland, this volume).

The record of sedimentation and magmatic activity provides constraints on the timing and evolution of orogenic activity. In

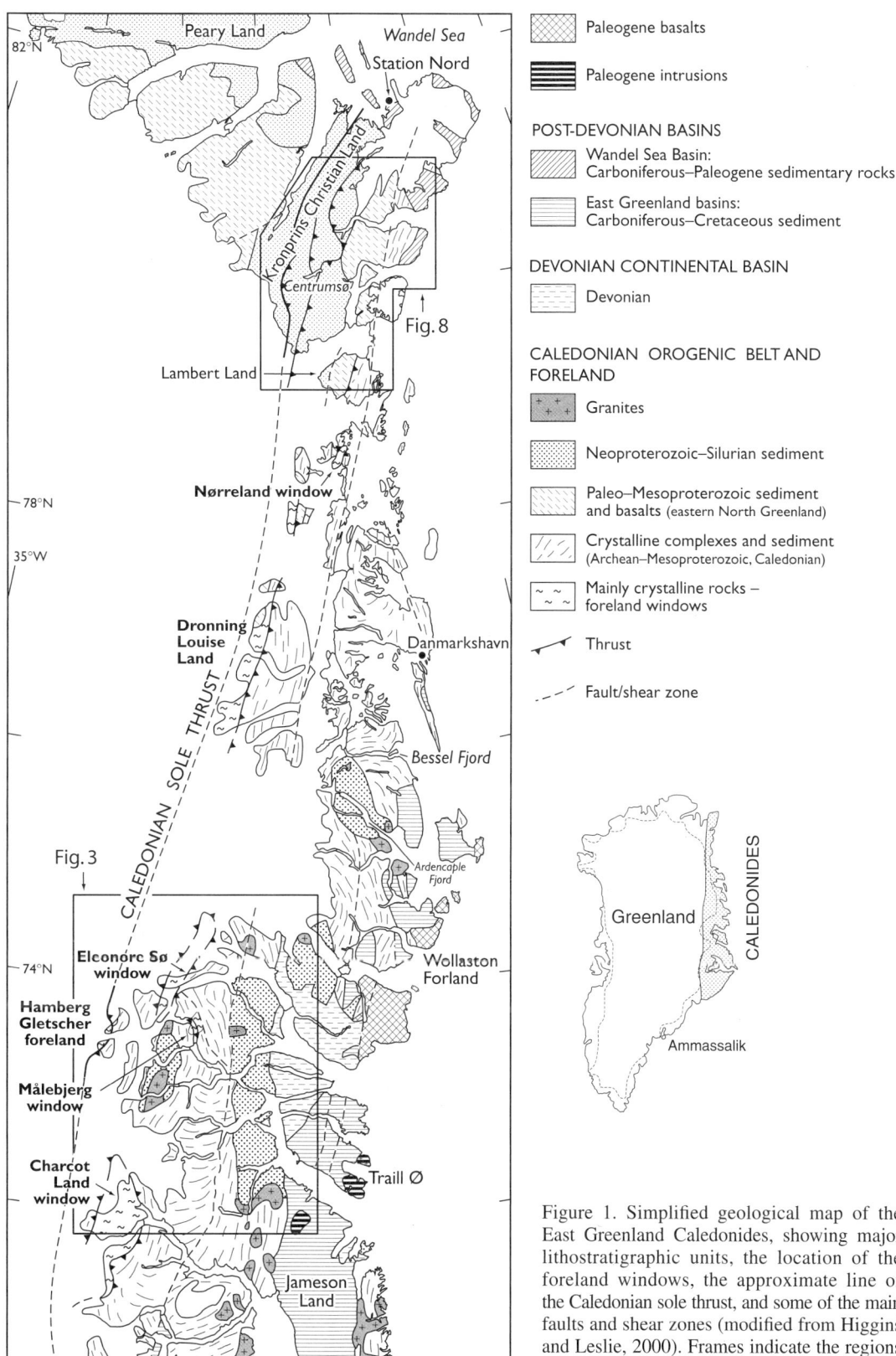

Figure 1. Simplified geological map of the East Greenland Caledonides, showing major lithostratigraphic units, the location of the foreland windows, the approximate line of the Caledonian sole thrust, and some of the main faults and shear zones (modified from Higgins and Leslie, 2000). Frames indicate the regions shown at larger scales in Figures 3 and 8.

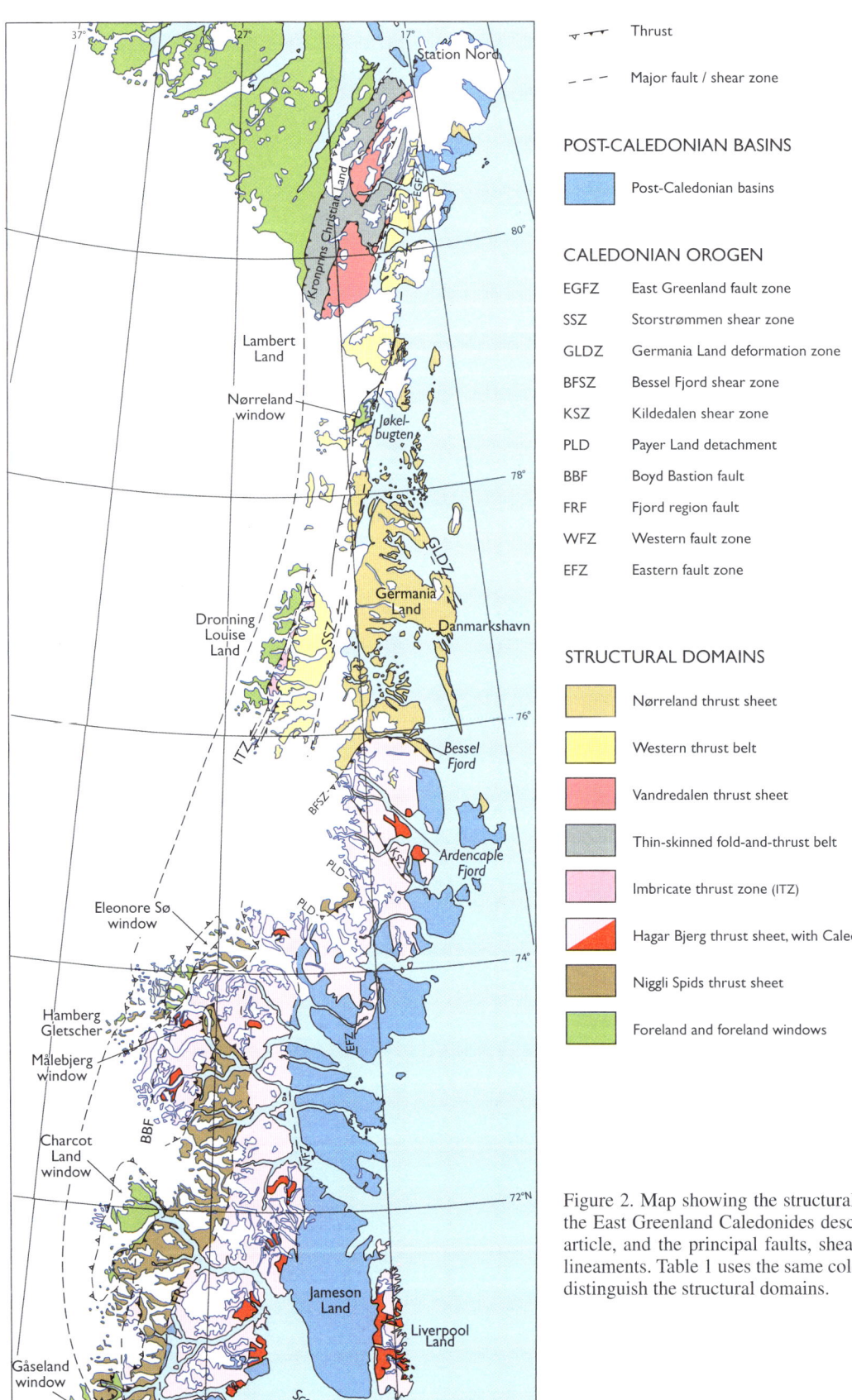

Figure 2. Map showing the structural domains of the East Greenland Caledonides described in this article, and the principal faults, shear zones, and lineaments. Table 1 uses the same color scheme to distinguish the structural domains.

central parts of East Greenland (~71°45′N–75°N), the youngest sedimentary rocks preserved in thrust sheets are found in the upper part of the Hagar Bjerg thrust sheet (Figs. 2 and 3), and they have Middle Ordovician ages (ca. 460 Ma). In this region, there are no breaks in the depositional record between the Lower Cambrian and the Middle Ordovician, and there is no evidence of tectonic activity corresponding to the Finnmarkian or Trondheim phases of Scandinavia (cf. Roberts, 2003). In the Caledonian foreland of northernmost East Greenland (north of 79°N), carbonate sedimentation on the Iapetus margin persisted into the late Llandoverian (430 Ma). Silurian turbidites that brought carbonate deposition to a close are interpreted as debris flows derived from erosion of the rising Caledonian mountain chain (Hurst et al., 1983). The youngest turbiditic sedimentary rocks over-ridden by Caledonian thrusts are shales of the middle Wenlockian (ca. 426 Ma) Profilfjeldet Member (Lauge Koch Land Formation; Table 1), which provide a maximum age for these frontal thrusts.

Restoration of the thrust sheets in the southern part of the orogen indicates that the Hagar Bjerg thrust sheet, with its abundant Caledonian migmatites and granites, was originally a substantial distance east of the present-day coast. The earliest known Caledonian granitoid rocks in the orogenic belt crop out in the southeastern part of the Hagar Bjerg thrust sheet; these are I-type calc-alkaline granodiorites and quartz diorites in the Scoresby Sund region, dated by sensitive high-resolution ion microprobe (SHRIMP) U-Pb analyses of zircons to 466 ± 9 Ma, with several ages ca. 432 Ma (A.P. Nutman and F. Kalsbeek, 2005, personal commun.; Kalsbeek et al., this volume, Chapter 9). These granitoids may represent that part of the Laurentian continent closest to the site of collision; the older I-type granitoids can be interpreted as parts of an arc that formed during subduction of Iapetus beneath Laurentia, and that correspond in time to the Taconian-Grampian phase of arc accretion on the Laurentian margin.

The thrust architecture is best preserved in the extreme north of the orogen (79°N–81°30′N), and in the southern half of the orogen (70°N–76°N). Both regions are described by Leslie and Higgins (this volume), where they are illustrated by long west to east cross sections. In the north, measured displacement on the Vandredalen thrust is 35–50 km, but higher thrust sheets may have displacements of 100 km or more (Higgins et al., 2001b). In the southern orogen, net thrust displacement is estimated at 200–400 km, corresponding to 40%–60% E-W contraction (Higgins and Leslie, 2000; Higgins et al., 2004a).

Caledonian metamorphic patterns in the orogen are described by Gilotti et al. (this volume), and they are variously superimposed on Archean, Paleoproterozoic, and early Neoproterozoic metamorphic histories in Precambrian crystalline complexes and Neoproterozoic to Lower Paleozoic sedimentary cover successions along the length of the orogen. North of Bessel Fjord (76°N), metamorphic grade increases eastward in progressively higher thrust sheets, suggesting that Caledonian metamorphic patterns are likely to have evolved from the onset of collision to the time of major Scandian (mid-Silurian) thrusting and crustal thickening. South of 76°N (Figs. 1 and 3), evidence for very low-grade to nonmetamorphic conditions is preserved in foreland windows at the lowest structural levels beneath the Niggli Spids thrust sheet, as well as in the highest stratigraphic levels of the Hagar Bjerg thrust sheet. Metasedimentary units in the Niggli Spids thrust sheet record amphibolite-facies conditions and, in some places, high-pressure granulite-facies conditions and local anatexis; however, the Krummedal metasedimentary sequence in the upper part of this thrust sheet lacks the Caledonian granites and migmatization seen in the Hagar Bjerg thrust sheet. The highest structural levels south of Bessel Fjord are dominated by the Hagar Bjerg thrust sheet; here, the Krummedal metasedimentary rocks are characterized by abundant Caledonian granites in a midcrustal-level migmatite complex, which records high-temperature amphibolite- to granulite-facies conditions. North of 76°N, and of particular note, Mississippian–age ultrahigh-pressure metamorphic conditions (eclogite facies) are preserved in parts of the Nørreland thrust sheet in Jøkelbugten (Fig. 2; McClelland et al., 2006).

Crustal-thickening processes led to the widespread formation of S-type Caledonian granites at 435–425 Ma by melting of metapelitic units of the Krummedal supracrustal sequence in what was to become the Hagar Bjerg thrust sheet. These leucogranites cut across high-grade fabrics, presumed to be of both pre-Caledonian and early Caledonian age. However, while some granites are massive, others are strongly deformed in top-to-the-west shearing, with distinct foliation and lineation. Granite emplacement apparently occurred both before and coeval with thrusting; many of the swarms of more foliated sheets show geometry consistent with foreland-propagating transport. A narrow range of ages, indicative of essentially synchronous granite emplacement at different structural levels, is evident in isotopic studies of foliated and unfoliated granites around both Ardencaple Fjord (431–428 Ma; Strachan et al., 2001) and in eastern Andrée Land (432–425 Ma; Andresen et al., 2007). Geochemical and petrological studies of the voluminous migmatites and associated leucogranites in the Hagar Bjerg thrust sheet suggest that granite formation was aided by the introduction of externally derived H_2O-rich fluids into the metapelitic host rocks (Kalsbeek et al., 2001b).

Caledonian leucogranites and migmatites are abundant in the Hagar Bjerg thrust sheet, but, as noted already, they are conspicuously absent in the Krummedal metasedimentary rocks of the structurally underlying Niggli Spids thrust sheet. Thrusting clearly continued after granite formation at 435–425 Ma, and all structures associated with the granites were passively transported during the final westward displacement of the major thrust sheets. The very latest movements of the Niggli Spids thrust sheet across the low-grade rocks preserved in the foreland windows are likely to have postdated the ca. 405 Ma high-pressure metamorphism in the high-grade gneiss basement recorded in Payer Land (McClelland and Gilotti, 2003), perhaps accompanying exhumation of these high-grade rocks.

The outcrop pattern of domains in the Caledonian orogen is clearly transected and displaced by a regional system of major extensional faults (Fig. 2; Gilotti and McClelland, this volume).

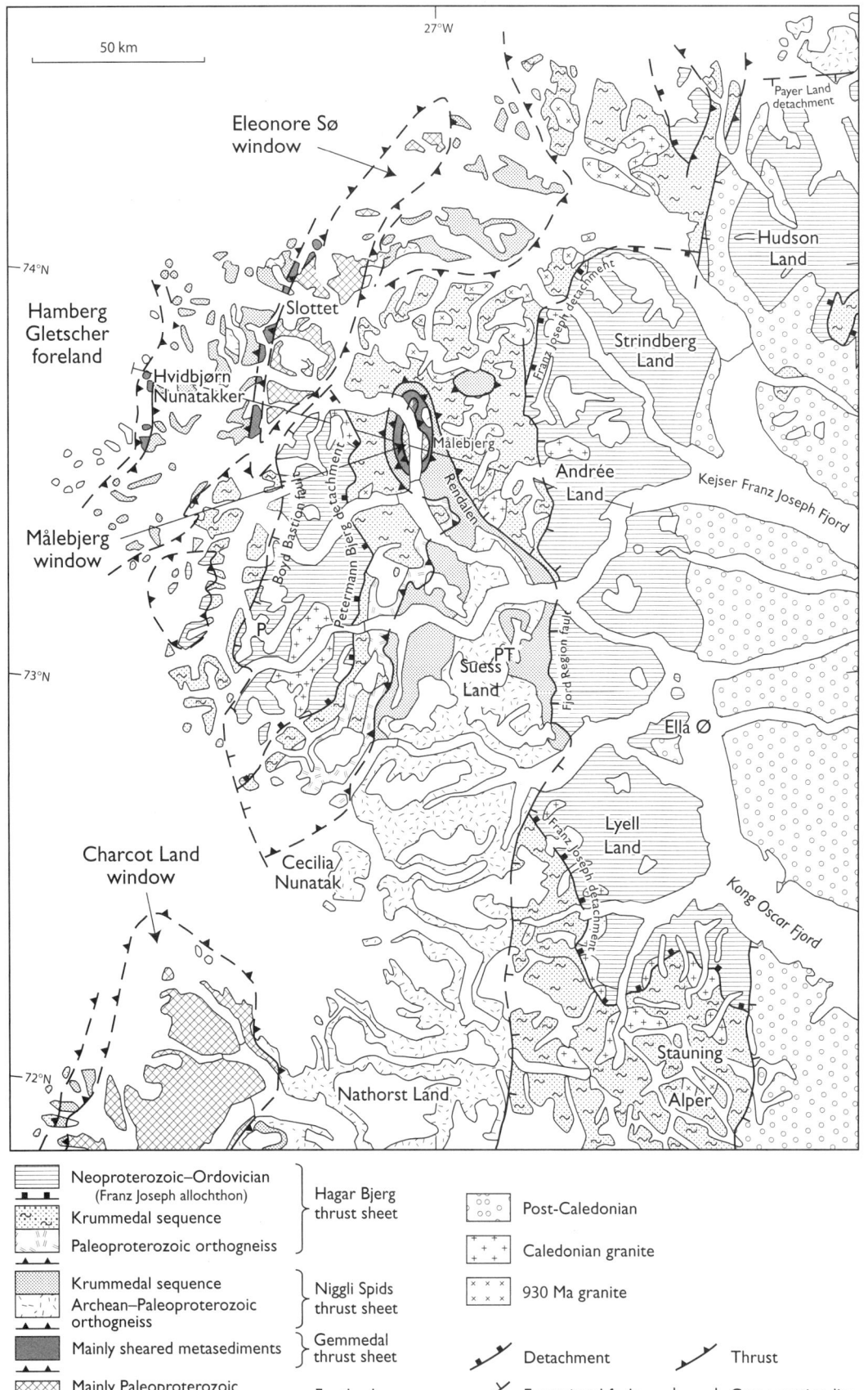

Figure 3. Geological map of North-East Greenland from 71°50′N to 74°30′N; see Figure 1 for location (modified after Higgins and Leslie, 2004). The parautochthonous Eleonore Sø, Charcot Land, and Målebjerg windows are shown, as is the location of the autochthonous Hamberg Gletscher foreland. The legend depicts the structural division into two major thrust sheets (Niggli Spids and Hagar Bjerg thrust sheets) and one minor thrust sheet (Gemmedal thrust sheet). The Franz Joseph allochthon forms the upper detached part of the Hager Bjerg thrust sheet. P—Petermann Bjerg; PT—Payer Tinde.

While many faults appear to postdate thrusting, they may also have been synorogenic, with thrusting ongoing at deep levels, while extensional structures formed at higher levels. In the southern segment of the orogen, ESE-directed extension and a N-S–directed sinistral wrench component played a major role in the initiation of the Middle to Upper Devonian continental basins (Larsen and Bengaard, 1991; Larsen et al., this volume). Top-to-the-SW displacements around Bessel Fjord (76°N; Strachan et al., 2001) and in Payer Land (74°30′N; Gilotti and Elvevold, 2002) seem to indicate orogen-parallel extension, but the geometry of the extensional detachments is complex and incompletely understood. The most important faults and lineaments are briefly described here in a separate section.

LITHOSTRATIGRAPHY AND PRE-CALEDONIAN GEOLOGICAL HISTORY

The principal lithostratigraphic and lithodemic components identified in the various domains recognized in East Greenland Caledonides are summarized in Table 1 (cf. Figs. 1 and 3) and are briefly described from oldest to youngest next.

Archean to Paleoproterozoic Basement Gneiss Complexes

Caledonian reworked basement gneiss complexes are present throughout the Caledonian orogen in East Greenland, and they contain the same broad range and ages of gneiss complexes known from the Greenland Shield of western and southern Greenland (Henriksen et al., 2000; Kalsbeek et al., this volume, Chapter 3). The Archean gneisses that make up large segments of the Niggli Spids thrust sheet south of 72°50′N (Figs. 1–3) are broadly equivalent to the autochthonous Archean gneisses of the Ammassalik region of South-East Greenland (Fig. 1, 65°30′N; Kalsbeek, 1989). Paleoproterozoic gneisses and granitoid rocks are present in the foreland windows and the Niggli Spids and Hagar Bjerg thrust sheets (Figs. 1–4), and they are widespread in the thrust complexes north of 76°N (Figs. 1 and 2). They were accreted in juvenile calc-alkaline arcs at ca. 2000–1800 Ma (Kalsbeek et al., 1993), and they are in many respects similar to the somewhat younger (and better preserved) rock units in the Ketilidian orogenic belt of South Greenland (Garde et al., 1998). Prior to the Caledonian orogeny, these Precambrian basement complexes within the Caledonian orogenic belt all formed part of the northeast margin of Laurentia.

These basement gneiss complexes have a complex history that witnessed one or more episodes of orogenic deformation and granitoid emplacement (e.g., Hull et al., 1994). Crosscutting, deformed amphibolitic dikes are common, and mafic to ultrabasic bands and lenses are present in many areas. Thin units of mica schists and marbles appear to represent infolded relics of once-more-extensive metasedimentary successions that are probably significantly older than the Paleoproterozoic–Mesoproterozoic volcano-sedimentary complexes and the latest Mesoproterozoic to earliest Neoproterozoic Krummedal supracrustal sequence.

Paleoproterozoic–Mesoproterozoic Volcanic and Sedimentary Complexes

In the foreland region west of Kronprins Christian Land (79°N–81°N), a >3-km-thick succession of Proterozoic sedimentary and volcanic rocks (Independence Fjord Group, Zig-Zag Dal Basalt Formation) overlies crystalline basement (Fig. 2; Table 1). The lower part of the Independence Fjord Group is represented within the Caledonian thrust sheets of Kronprins Christian Land (Fig. 5) as highly deformed quartzitic and feldspathic sandstone units cut by mafic dikes and interbedded with volcanic rocks; rhyolites within the volcanic rocks have been dated at 1740 Ma (Kalsbeek et al., 1999; Pedersen et al., 2002; Collinson et al., this volume). The white quartzitic sandstones of the Independence Fjord Group in the foreland areas are intruded by abundant dikes and sills (Midsommersø Dolerites) thought to be feeders to the overlying Zig-Zag Dal Basalt Formation (see Collinson et al., this volume). The Independence Fjord Group sandstones and associated dolerites are conspicuous lithologies easily recognized within the Western thrust belt, and the southernmost occurrences (known as the Trekant "series") are in the parautochthonous foreland of western Dronning Louise Land (Table 1). Rift-related, pre–1900 Ma volcanic and sedimentary rocks are preserved locally in the southern foreland areas (Charcot Land window, Hamberg Gletscher foreland, Eleonore Sø window), but they are not represented in either the Niggli Spids or Hagar Bjerg thrust sheets (Figs. 1–3; Table 1).

Latest Mesoproterozoic to Earliest Neoproterozoic Sedimentary Rocks

Thick latest Mesoproterozoic to earliest Neoproterozoic metasedimentary rocks described as the Krummedal supracrustal sequence (including the so-called Smallefjord supracrustal sequence at 75°N; Fridcrichsen et al., 1994) are widely distributed in both the Niggli Spids and Hagar Bjerg thrust sheets (Figs. 3, 4, and 6; Higgins, 1988; Leslie and Higgins, this volume); there is no evidence that they were ever deposited in the foreland areas to the west, or that contemporaneous accumulations developed in the northern parts of the orogen. The lowest preserved levels include thin carbonate units, but the bulk of the succession consists of siliciclastic units that are locally >4 km thick; in parts of the Scoresby Sund region, apparent thicknesses of up to 8 km have been recorded (Henriksen et al., 1980; Higgins 1988). Ion-microprobe studies of detrital zircons indicate that the sediments were deposited after ca. 1050 Ma (the youngest detrital grains; Kalsbeek et al., 2000; Watt et al., 2000). In East Greenland, the Krummedal metasedimentary rocks and equivalent successions are exposed over a N-S distance of at least 600 km, while thrust restoration suggests that the original depositional basin was at least 300 km wide; the western and eastern limits are undefined (Higgins and Leslie, 2000; Higgins et al., 2004a).

TABLE 1. LITHOSTRATIGRAPHY OF STRUCTURAL DOMAINS AND FORELAND AREAS

Northern Orogen 76°N to 82°N			
Stratigraphy/Chronostratigraphy	Lithostratigraphy	Age (Ma)	Lithology
POST-OROGENIC SEDIMENTATION			
Kronprins Christian Land			
Gzhelian	Foldedal Formation	ca. 300	260 m carbonates
Moscovian	Kap Jungersen Formation	ca. 310	300 m conglomerates, carbonates & sandstones
Visean	Sortebakker Formation	ca. 330–340	600 m fluviatile sandstones — *ca. 20 Ma hiatus*
Germania Land			
Late Carboniferous	Chatham Elv	ca. 305	75 m conglomerate and sandstone
ALLOCHTHONOUS LITHOTECTONIC UNITS – WNW-directed translation			
Nørreland thrust sheet (eclogite facies rocks) – Devonian to Carboniferous exhumation			
Paleoproterozoic	Independence Fjord Group	ca. 1740	alluvial white-weathering quartzitic sandstones
Paleoproterozoic	Basement	1800–2000	gneisses and metasediments
Western thrust belt (with Lambert Land and west of Jøkelbugten)			
Lower Ordovician	Amdrup & Danmark Fjord Mb.	475–480	carbonates
Paleoproterozoic	Independence Fjord Group with Hekla Sund Formation	ca. 1740	2800 m white-weathering quartzitic sandstones and 1100 m tholeiitic volcanic rocks
Paleoproterozoic	Basement	1800–2000	gneisses and metasediments
Vandredalen thrust sheet			
Cryogenian	Hagen Fjord Group	ca. 650–750	ca. 700 m, calcitic/dolomitic limestones overlie siliciclastic rocks
Cryogenian–Tonian	Rivieradal Group	750–ca. 900	7.5 to 10 km conglomerates, sandstones and mudstones — *Vandredalen thrust at base*
FORELAND AND PARAUTOCHTHONOUS THIN-SKINNED FOLD-AND-THRUST BELT			
Kronprins Christian Land			
Middle-Silurian (Wenlock)	Lauge Koch Land Formation	ca. 426	up to 400 m, black shales and limestones at base, shaly turbiditic siltstones and sandstones above
Lower Silurian (Llandovery)	Samuelsen Høj Formation		conspicuous reef limestones, up to 20 m high
	Odins Fjord Formation		up to 320 m limestone and dolostone
	Tureso Formation		up to 350 m limestone and dolostone
to	Børglum River Formation		ca. 430 m dark burrow-mottled limestone
	Sjælland Fjelde Formation		ca. 100 m burrow-mottled limestone and dolostone
Lower Ordovician	Wandel Valley Formation	ca. 480	ca. 335 m limestone and dolostone — *40 Ma hiatus*
Lower Cambrian	Kap Holbæk Formation	ca. 520	0–180 m mudstones and sandstones — *125 Ma hiatus*
Cryogenian	Hagen Fjord Group	ca. 650–750	ca. 740 m siliciclastic rocks overlain by calcitic and dolomitic limestones
Mesoproterozoic	Zig-Zag Dal Basalt Formation	ca. 1350–1380	ca. 1350 m basaltic lavas: extrusive equivalents of the Midsommersø dolerites
Paleoproterozoic	Independence Fjord Group with Aage Bertelsen Gl. Formation	ca. 1740	ca. 2 km quartzitic sandstones with ca. 400 m tholeiitic volcanic rocks
PARAUTOCHTHONOUS FORELAND			
Western Dronning Louise Land (with imbricate thrust zone)			
Lower Ordovician	Zebra "series" 2	ca. 480	— *40 Ma hiatus*
Lower Cambrian	Zebra "series" 1	ca. 520	quartz sandstones — *hiatus*
Paleoproterozoic	Trekant "series" with dolerites	ca. 1740	sandstones
Paleoproterozoic and Archean	Basement	1800–2000	orthogneisses and mafic dikes, local Archean enclaves
Nørreland window			
Lower Ordovician	Danmark Fjord Member	ca. 480	carbonates — *hiatus*
Paleoproterozoic	Independence Fjord Group	ca. 1740	quartzitic sandstones

(continued)

TABLE 1. LITHOSTRATIGRAPHY OF STRUCTURAL DOMAINS AND FORELAND AREAS (continued)

Southern Orogen 70°N to 76°N			
Stratigraphy/Chronostratigraphy	Lithostratigraphy	Age (Ma)	Lithology
SYN- to POST-OROGENIC SEDIMENTATION			
Molasse deposits			
Lower Carboniferous (Devonian)	Harder Bjerg Fm.	ca. 355	8 km intramontane continental sedimentation, mainly sandstones and siltstones, extrusive volcanism, locally faulted
Famennian to Frasnian (Devonian)	Celcius Bjerg Group, Kap Graah Group, Kap Kolthoff Group, Vilddal Group		
Givetian to Eifelian (Devonian)		ca. 395	*ca. 65 Ma hiatus*
ALLOCHTHONOUS LITHOTECTONIC UNITS – Silurian WNW-directed translation			
Franz Joseph allochthon (hosts Caledonian granites)			
Middle Ordovician to Lower Cambrian	Kong Oscar Fjord Group (comprising Heimbjerge, Narwhale Sound, Cape Weber, Antiklinalbugt, Dolomite Point, Hylolithus Creek, Ella Island, Bastion and Kløftelv Fms.)	ca. 460 to ca. 525	4.5 km carbonates, siliciclastics at base
			ca. 50 Ma hiatus
Marinoan	Tillite Group	575–660	
Cryogenian to Tonian	Eleonore Bay Supergroup (comprising Andrée Land, Ymer Ø, Lyell Land and Nathorst Land Groups)	ca. 660 to ca. 900	5.4 km carbonates, mudstones and sandstones, overlying up to 9 km sandstones and mudstones
Detachment at base			
Hagar Bjerg thrust sheet (Krummedal sequence hosts both Caledonian and ca. 930 Ma granites)			
Lower Neoproterozoic		920–950	'Older Granites'
Lower Neoproterozoic to Mesoproterozoic	Krummedal (and Smallefjord) supracrustal sequences	ca. 950–1100	metasedimentary rocks (sandstone, mudstone, local calcareous mudstone)
Paleoproterozoic	Basement	1800–2000	orthogneisses (with mafic dikes)
Hagar Bjerg thrust at base			
Niggli Spids thrust sheet (granites insignificant in Krummedal sequence, ca. 930 Ma granites unknown)			
Lower Neoproterozoic to Mesoproterozoic	Krummedal supracrustal sequence	ca. 950–1100	metasedimentary rocks (sandstone, mudstone, local calcareous mudstone)
Paleoprot.–Archean	Basement	1800–2800	orthogneisses and infolded supracrustals
Niggli Spids thrust at base			
Gemmedal thrust sheet (present locally)			
FORELAND AND PARAUTOCHTHONOUS WINDOWS			
Målebjerg window			
Lower Ordovician	Målebjerg Formation	ca. 480	32 m dolostones and burrow-mottled limestones — hiatus
Lower Cambrian	Slottet Formation	ca. 520	ca. 200 m quartz sandstones with *Skolithos* — hiatus
Marinoan	Tillite Group	575–660	31 m diamictites and sandstones — hiatus
Paleoproterozoic	Basement	ca. 2000	gneisses
Eleonore Sø window			
Lower Ordovician	Målebjerg Formation	ca. 480	45–50 m laminated dolostones and limestones — hiatus
Lower Cambrian	Slottet Formation	ca. 520	350 m quartz sandstones with *Skolithos* — hiatus
Paleoproterozoic	Eleonore Sø supracrustal rocks	>1970	metamorphosed lavas and hyaloclastites, carbonates, mudstones and sandstones
Paleoproterozoic	Basement	ca. 2000	gneisses
Charcot Land window			
Marinoan	Tillite Group	575–660	diamictites — hiatus
Paleoproterozoic	Charcot Land granite	ca. 1850	granite
Paleoproterozoic	Charcot Land supracrustal rocks	1850–1900	metasedimentary rocks and pillow lavas
Paleoproterozoic	Basement	1915–1930	gneisses
Gåseland window			
age unknown	thrust zone	?	sheared quartzite and carbonate
Marinoan		575–660	diamictites — hiatus
Paleoproterozoic	Basement	ca. 2000	gneisses
Hamberg Gletscher foreland			
unknown age			gabbroic sill complex, lavas and carbonates
Paleoproterozoic	Basement	ca. 1900	orthogneisses, granitic veins

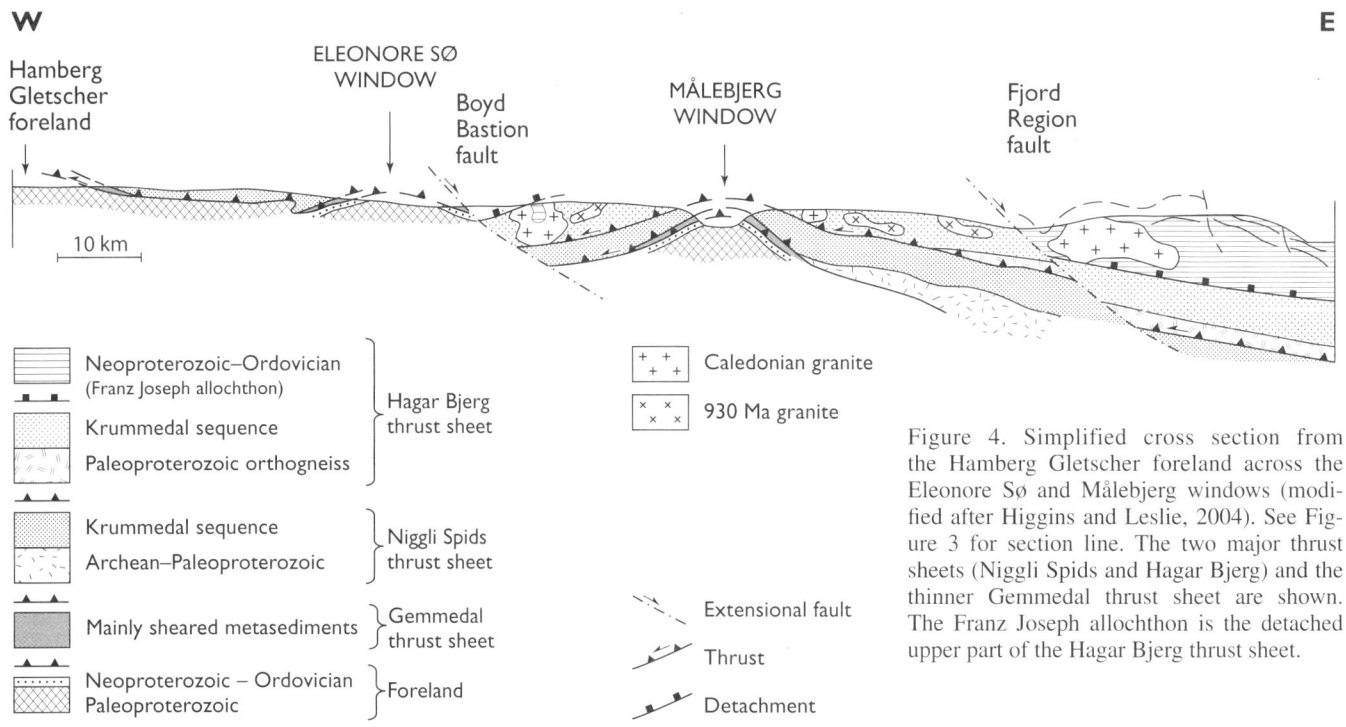

Figure 4. Simplified cross section from the Hamberg Gletscher foreland across the Eleonore Sø and Målebjerg windows (modified after Higgins and Leslie, 2004). See Figure 3 for section line. The two major thrust sheets (Niggli Spids and Hagar Bjerg) and the thinner Gemmedal thrust sheet are shown. The Franz Joseph allochthon is the detached upper part of the Hagar Bjerg thrust sheet.

Figure 5. Part of central Ingolf Fjord, looking northward, showing spectacular cliffs of folded light-colored quartzites of the Independence Fjord Group cut by dark-colored dikes and sills, part of the Western thrust belt. Cliff height is ~1350 m.

The widespread occurrence of successions of broadly comparable age in the eastern province of Spitsbergen (Brennevinsfjorden Group, Helvetsflya Formation; Gee and Teben'kov, 1996) and in the NW Highlands of Scotland in the British Caledonides (Moine Supergroup; Holdsworth et al., 1994; Leslie et al., this volume) suggests that the Krummedal succession basin may have been deposited in one of a widespread system of late Mesoproterozoic to early Neoproterozoic sedimentary basins. Detrital zircon populations of the Scottish Moine Supergroup (Friend et al., 2003) share similar spectra with the Krummedal succession and, like them, cannot have been deposited earlier than ca. 1050 Ma.

Figure 6. Looking southeast across Kejser Franz Joseph Fjord to Payer Tinde (2320 m), the snow-covered peak in the center of the photograph. The two parts of the Niggli Spids thrust sheet are clearly seen: light-colored Paleoproterozoic orthogneisses form the steep cliffs and are overlain by dark-colored Mesoproterozoic to lower Neoproterozoic Krummedal sequence metasedimentary rocks. Field of view at sea level ~12 km.

Early Neoproterozoic Granitoids

A suite of augen granites and leucogranites (Fig. 7) that has yielded protolith ages of 940–910 Ma (Jepsen and Kalsbeek, 1998; Kalsbeek et al., 2000, this volume, Chapter 9; Watt and Thrane, 2001) is widely distributed within the Krummedal supracrustal sequence in the Hagar Bjerg thrust sheet over a N-S distance of more than 400 km. These magmatic events are contemporaneous with high-grade metamorphism recorded in overgrowths on detrital zircon in the Krummedal sequence (Kalsbeek et al., 2000; Watt et al., 2000; Watt and Thrane, 2001) and with ductile deformation that resulted in nappe-scale folds (Leslie and Nutman, 2003). The absence of Neoproterozoic granites in the metasedimentary rocks of the underlying Niggli Spids thrust sheet suggests that such high-grade thermal events affected only the eastern part of the Krummedal sequence sedimentary basin preserved in East Greenland (now preserved only in the Hagar Bjerg thrust sheet) and, thus, only the easternmost border of the Laurentian margin. Lower-grade effects cannot be discounted in the Krummedal metasedimentary rocks of the Niggli Spids thrust sheet, which would have lain to the west.

In the eastern province of Spitsbergen, 970–940 Ma events are recorded, and augen granites are reported to have been emplaced synchronous with an episode of deformation (Johansson et al., 2000), a scenario resembling that recorded in the Hagar Bjerg thrust sheet in the Scoresby Sund region of East Greenland (Leslie and Nutman, 2003). In Scotland, relationships seem less clear, since zircon dating studies appear to record a number of high-pressure–high-temperature

Figure 7. Typical appearance of the ca. 930 Ma augen granites that occur as large and small sheets within the Hagar Bjerg thrust sheet. Black pen is 15 cm long.

tectonothermal events between 840 and 730 Ma (see Leslie et al., this volume). East Greenland may therefore record a phase of orogenic activity that succeeded the Grenville (Elziviran) events of North America, comparable in age at least with the Sveconorwegian magmatic events of southern Scandinavia (Möller et al., 1997; Söderlund et al., 1999). This could be interpreted to record the gradual northward movement of Baltica along the Laurentian margin during the final amalgamation of the components of Rodinia (Dalziel, 1997; Weil et al., 1998).

Upper Neoproterozoic Sedimentary Successions

Basin development in the Neoproterozoic was related to the disintegration of the Rodinia supercontinent and creation of the Iapetus Ocean. In East Greenland, thick and well-preserved successions of siliciclastic and carbonate sedimentary rocks are preserved in the Caledonian orogen in two distinct regions (Sønderholm et al., this volume; Table 1): the Rivieradal Group and Hagen Fjord Group in the north (78°N–81°N; see Figs. 8 and 9), and the Eleonore Bay Supergroup and Tillite Group in the south (71°50′N–76°N; Figs. 2 and 3).

The Rivieradal Group is a pre-Iapetus, synrift, deep-water succession that was deposited in an east-facing half-graben, which is now exclusively preserved in the Vandredalen thrust sheet. The sedimentary rocks preserved include successions of proximal coarse conglomerates, well-bedded sandstone units (Fig. 10), a thick turbiditic sandstone succession, and distal shale and carbonate units (Smith et al., 2004b). The overlying Hagen Fjord Group is extensively represented in the wide foreland region west of Kronprins Christian Land, and also within the Thin-skinned fold-and-thrust belt and in the Vandredalen thrust sheet of Kronprins Christian Land. It consists of fluvial and marine sandstones overlain by carbonate platform deposits, representing postrift thermal reequilibrium, where the youngest sediments overstepped the half-graben in which the Rivieradal Group accumulated. The Hagen Fjord Group is overlain by sandstones of the Lower Cambrian Kap Holbæk Formation.

In the southern half of the orogen, the Upper Neoproterozoic succession (Eleonore Bay Supergroup, Tillite Group) is almost confined to the upper part of the Hagar Bjerg thrust sheet; only a few tens of meters of diamictites correlated with the Tillite Group occur in foreland windows. In dramatic contrast to that reduced foreland succession, the lower part of the Eleonore Bay Supergroup is composed of ~12 km of mainly shallow-water siliciclastic sedimentary rocks, while the upper 2–3 km are dominated by carbonate platform sedimentary rocks. The overlying Tillite Group is composed of ~800–1300 m of dolomitic mudstones and sandstones, with two diamictite units interpreted as tillites. The age of the glacial episode has been much debated (see Sønderholm et al., this volume), but increasing evidence supports a Marinoan age. In the field, the contact at the base of the Eleonore Bay Supergroup is always tectonic, and it takes the form of a shear zone with evidence of both westward and eastward displacement. However, the simple Caledonian fold pattern seen in the Upper Neoproterozoic to Ordovician succession may indicate that the amount of movement along the shear zone was limited, and that this thick sedimentary package was transported more-or-less passively on top of underlying parts of the Hagar Bjerg thrust sheet during major westward thrust displacement; Higgins et al. (2004a) placed the Neoproterozoic–Ordovician successions in the Franz Joseph allochthon (Table 1).

No definite representatives of the Neoproterozoic Eleonore Bay Supergroup are preserved in the Niggli Spids thrust sheet, and no representative is seen in the foreland areas. However, clasts derived from the upper formations of the Eleonore Bay Supergroup basin are an important component of the diamictites of the Tillite Group (Hambrey and Spencer, 1987), indicating uplift and erosion in the unknown source region.

The Eleonore Bay Supergroup basin in East Greenland must have been at least 600 km long and more than 200 km wide, and if the Hagen Fjord Group carbonates of Kronprins Christian Land can be correlated with the Andrée Land Group, as suggested by Frederiksen (2000), the depositional framework can be extended for at least 1000 km along this sector of the Laurentian margin. The broad similarities between development of the Eleonore Bay Supergroup of East Greenland and the Murchisonfjorden Supergroup of Spitsbergen are well known and were recognized by the earliest explorers (e.g., Nathorst, 1901). The broadly equivalent sedimentation in the Scottish Central Highlands is represented by the Grampian Group (Harris et al., 1994), while strong parallels can be drawn between the Appin Group and the upper Eleonore Bay Supergroup (see Leslie et al., this volume), further expanding this late Neoproterozoic record of pericontinental deposition along the Laurentian margin.

Diamictites interpreted as tillites have been recognized in the Gåseland, Charcot Land, and Målebjerg foreland windows (Table 1), and, in all cases, they are preserved in erosional depressions in an eroded crystalline gneiss basement. All have been correlated with the Tillite Group of the Franz Joseph allochthon in the central fjord zone (72°N–75°N; Moncrieff, 1989; Smith et al., 2004a). Detailed work in Svalbard and East Greenland has demonstrated a correlation of the glacial horizons unit by unit, and even bed by bed (e.g., Fairchild and Hambrey, 1995). It is evident that the two regions must have been part of the same basin prior to Caledonian collision and strike-slip disruption of the Laurentian margin.

Lower Paleozoic Sedimentary Successions

Lower Paleozoic sedimentary rocks, like the underlying Neoproterozoic successions, are present in two distinct regions of East Greenland. In the north, they make up a large part of the Thin-skinned fold-and-thrust belt of Kronprins Christian Land, and they are widespread in the undeformed foreland areas to the west (Fig. 2; see also Fig. 8). In the south, Cambrian-Ordovician sedimentary rocks are well exposed in the so-called central fjord zone of East Greenland (72°N–75°N), where they make up the highest levels (Franz Joseph allochthon) of the Hagar Bjerg thrust sheet (Fig. 2). Both successions represent deposition on the Laurentian passive margin of the Iapetus Ocean (see Smith and Rasmussen, this volume).

Figure 8. Geological map of Kronprins Christian Land and Lambert Land (modified from Smith et al., 2004a). See inset map and Figure 1 for location. **BS**—Brede Spærregletscher; **M**—Marmorvigen. The post-Caledonian Wandel Sea Basin sedimentary rocks of Holm Land are Carboniferous (Visean to Gzhelian) in age (see Table 1).

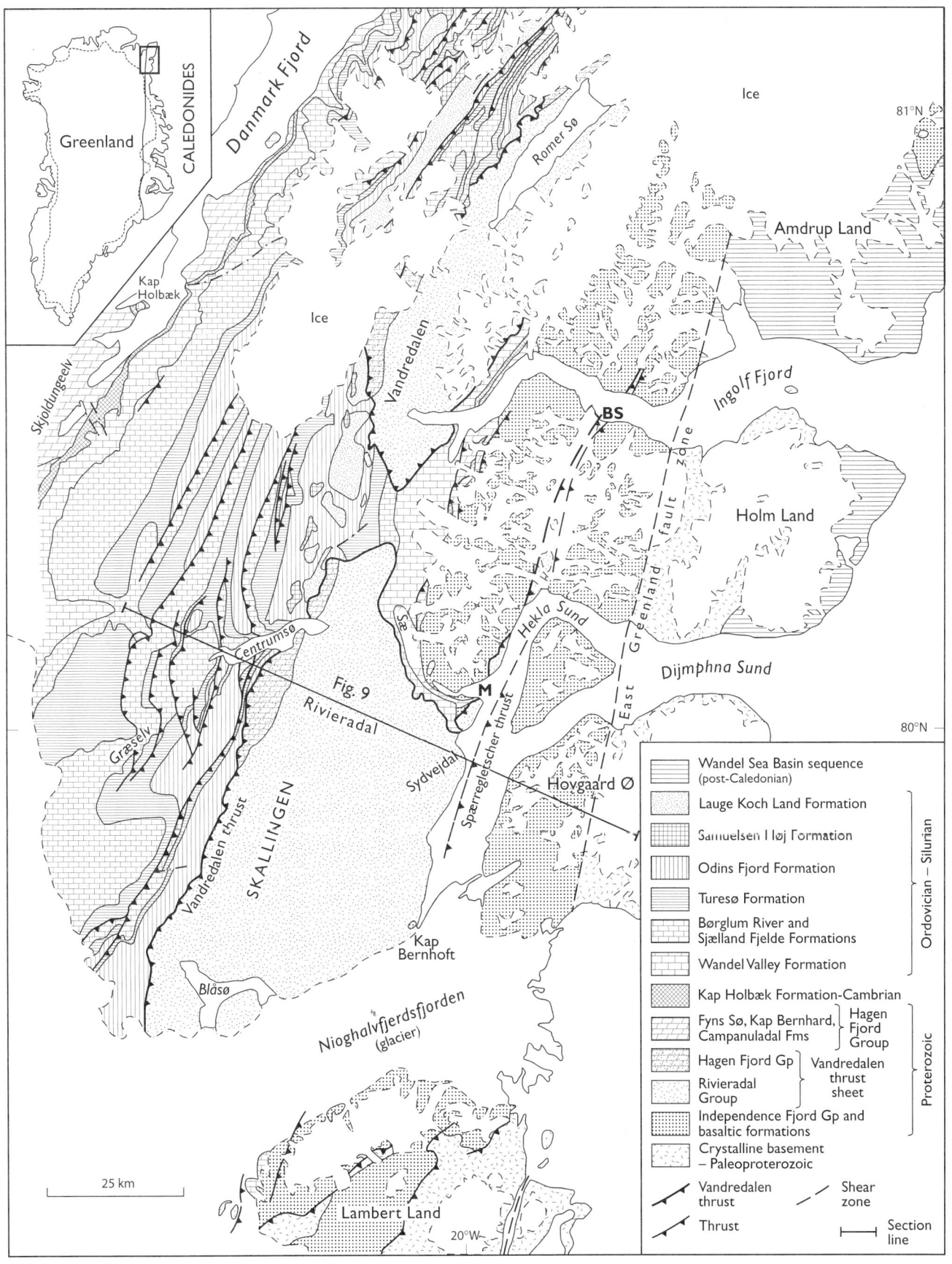

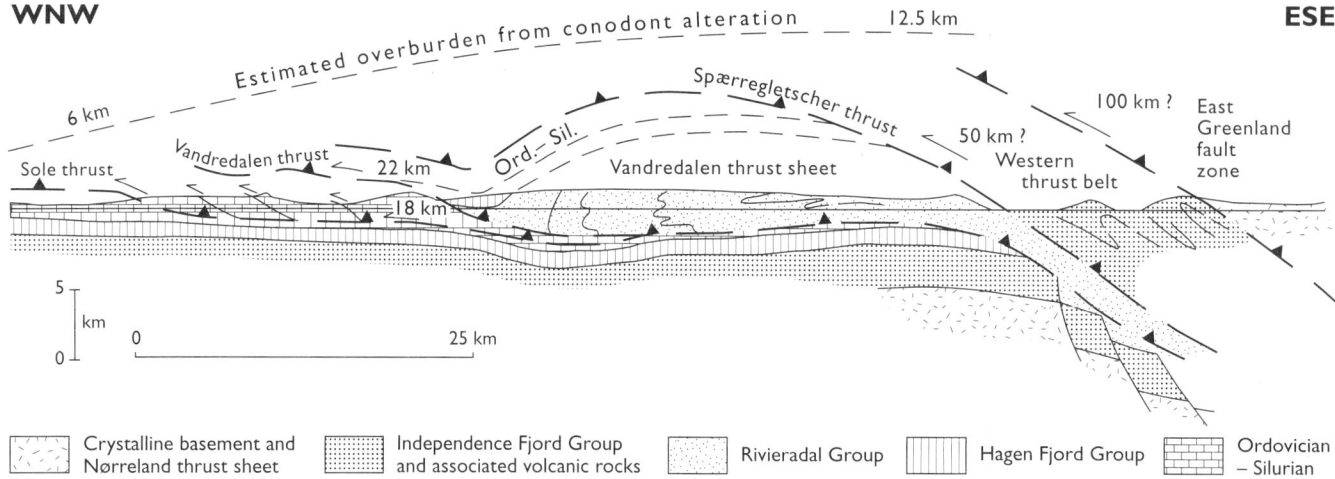

Figure 9. Cross section of Kronprins Christian Land (modified from Higgins et al., 2004a); see Figure 8 for section line. The Thin-skinned fold-and-thrust belt, between the sole thrust and the Vandredalen thrust, is mainly developed in Ordovician and Silurian shelf carbonates. The clastic Rivieradal Group sedimentary rocks were deposited in an east-facing half-graben and have been displaced westward across the faulted western margin as the Vandredalen thrust sheet (bounded by the Vandredalen thrust below and Spærregletscher thrust above). The thickness of the thrust overburden, determined from conodont alteration colors, increases from 6 km to 12.5 km across the section and demonstrates that higher thrust sheets must have been present above the Spærregletscher thrust (part of the Western thrust belt).

The Lower Paleozoic sedimentary rocks in the northern part of the orogen are the easternmost representatives of the Franklinian Basin, a major E-W–trending basin that extends for 900 km across North Greenland, and farther westward into Arctic Canada (Higgins et al., 1991; Trettin, 1991). The succession in Kronprins Christian Land differs from that in the central fjord zone to the south, but it also differs from the typical shelf stratigraphy of the Franklinian Basin in that uplift and erosion between the Early Cambrian and mid–Early Ordovician produced the sub–Wandel Valley Formation unconformity, which represents a hiatus of 40 m.y. (Table 1). The oldest sedimentary rocks are the sandstones of the Lower Cambrian Kap Holbæk Formation, which are overlain unconformably by the lower Ordovician Wandel Valley Formation, here the lowest unit of the carbonate-dominated shelf succession (Table 1; see also Figs. 8 and 9). In Kronprins Christian Land, the Silurian carbonate shelf deposits are abruptly overlain by mid-Silurian (Wenlockian) black mudstones and turbidites.

The Cambrian-Ordovician succession of the central fjord region (72°N–75°N) is up to 4 km thick and is placed in the Kong Oscar Fjord Group (Table 1); it occurs within the uppermost part of the Hagar Bjerg thrust sheet (Franz Joseph allochthon). The stratigraphy of the Kong Oscar Fjord Group (Smith and Rasmussen, this volume) can be closely correlated with the formerly contiguous terranes of NE Spitzbergen, NW Scotland, and, farther to the south, western Newfoundland (e.g., Swett and Smit, 1972). The Kong Oscar Fjord Group begins with a 70–75-m-thick quartz-rich sandstone unit (Kløftelv Formation), which is succeeded by a series of limestone and dolomite formations. The succession is truncated by the sub-Devonian unconformity, and the youngest unit (Heimbjerge Formation), which extends up to the Whiterockian (early Middle Ordovician), is only present in the northern part of the outcrop belt.

Scattered outcrops of an attenuated Lower Paleozoic succession occur in some of the foreland windows (Figs. 2 and 3). The Målebjerg and Eleonore Sø windows preserve an up to 350-m-thick quartzite unit known as the Slottet Formation (Fig. 11), which can be correlated with the lower part of the Zebra "series" of Dronning Louise Land (Table 1; Zebra "series" 1) and the Kap Holbæk Formation of Kronprins Christian Land (Table 1). *Skolithus* ichnofossils are common, and this unit is clearly the source of the abundant glacial-erratic boulders of *Skolithus*-bearing quartzites that occur in East Greenland (Haller, 1971, their Fig. 48). A thin carbonate unit (Målebjerg Formation) that overlies the Slottet Formation in the Målebjerg and Eleonore Sø windows is of Early Ordovician age. Thin carbonate units located in Dronning Louise Land (Table 1; Zebra "series" 2), in the Nørreland window, and in western Lambert Land (Smith and Rasmussen, this volume) have been correlated with the Danmark Fjord Member of the Wandel Valley Formation.

STRUCTURAL DOMAINS

The structural domains identified here (Fig. 2) comprise regions that generally extend over a considerable distance, are characterized by a coherent assemblage of elements of the lithostratigraphic and lithodemic units outlined already, and have been subjected to a similar degree of ductile deformation and metamorphism.

Two distinctive domains occur only in Kronprins Christian Land, the Thin-skinned fold-and-thrust belt that has a

Figure 10. Tight to isoclinal westward-overturned folds developed in well-bedded sandstones and shales in the upper part of the Rivieradal Group. View is from west side of Vandredalen, looking north; profile height is ~300 m. Photo is courtesy of Jakob Lautrup.

Figure 11. The Lower Cambrian Slottet Formation, made up of white- and red-weathering quartzites, resting unconformably on Paleoproterozoic metasedimentary rocks (black) at Slottet in the foreland of the Eleonore Sø window (after Higgins and Leslie, 2004). Neoproterozoic sedimentary rocks are absent in the Caledonian foreland. View is looking northward. The thickness of the quartzites of the Slottet Formation has been determined photogrammetrically to be ~350 m. The highest summit of Slottet is 600 m above the glacier in the foreground.

transitional contact with undisturbed foreland, and the structurally overlying Vandredalen thrust sheet (Figs. 2, 8, and 9). The Western thrust belt consists of a group of thrust sheets involving distinctive black and white lithologies (Fig. 5; dikes and quartzites) that can be traced southward through the alpine region of central Kronprins Christian Land, through western Hovgaard Ø and Lambert Land (Fig. 8), and farther south through the large scattered nunataks west of Jøkelbugten (Fig. 2). The Nørreland thrust sheet dominates the region north of Bessel Fjord and extends throughout the coastal region of Dove Bugt and Jøkelbugten northwards to ~80°N. Throughout the southern half of the orogen (70°N–76°N), two major thrust sheets are recognized, namely the Niggli Spids thrust sheet and the Hagar Bjerg thrust sheet (Figs. 2–4); their northern limit is defined by the Bessel Fjord shear zone at ~76°N. Caledonian anticlinal structures elevate both autochthonous and parautochthonous lower-grade foreland rocks to current exposure beneath overriding Caledonian thrust sheets (Fig. 4). A minor but discontinuous thrust unit is locally present between the foreland and the Niggli Spids thrust sheet; it is known as the Gemmedal thrust sheet. From the brief descriptions of these domains given next, it will be appreciated that crustal thickening and the development of major thrust sheets underpin the early orogenic framework. At the present-day outcrop, the boundaries of the different domains are defined by ductile thrusts or shear zones; however, in some cases, these are the expression of orogenic extension accommodating exhumation of the deeper parts of the orogen, and, in other cases, they may have controlled the Devonian to Carboniferous accumulation of sedimentary successions derived from rapid erosion of the orogenic welt. The characteristics of the principal faults and lineaments are presented in a later section.

The Foreland

The western boundary of the orogen is largely concealed by the present-day continental ice cap (Inland Ice; Figs. 1 and 2). At the northern extremity of the orogen, the Caledonian sole thrust, which corresponds to the west limit of Caledonian deformation, is the westernmost thrust in the Thin-skinned fold-and-thrust belt of Kronprins Christian Land. In some older descriptions of the East Greenland Caledonides, the Caledonian sole thrust has been linked with the imbricate thrust zone in central Dronning Louise Land (76°N–77°N; Haller, 1971; Henriksen and Higgins, 1976), but since western Dronning Louise Land exhibits considerable Caledonian deformation (Strachan et al., 1992), that region must be made of parautochthonous, not autochthonous foreland. The sole thrust is thus most likely to be a hidden splay off the imbricate thrust zone that ends as a blind thrust just west of present-day Dronning Louise Land (Figs. 1 and 2). Farther south, intact autochthonous foreland is seen in the extreme western part of Hamberg Gletscher, where the sole thrust emerges as an east-dipping shear zone. In the extreme south, the Caledonian sole thrust outlines the Gåseland window, and, here, it is a very gentle east-dipping shear zone between undisturbed Precambrian gneisses below and the Niggli Spids thrust sheet above. The Niggli Spids thrust is well exposed in the several other foreland windows between 70°N and 74°30′N, but since the rock units within these windows show a varying degree of Caledonian deformation, they are considered to be parautochthonous, and thus this thrust cannot always be strictly identified as the sole thrust.

The intermittent outcrops of foreland and foreland windows along the west margin of the Caledonian orogen provide an incomplete impression of the makeup of the Greenland craton and its sedimentary cover. The similarities and correlations between the different foreland areas in East Greenland have been discussed by Higgins et al. (2001a, 2004a), and they are referred to in the lithostratigraphical outline provided herein. The most notable contrast between the foreland areas and the overlying allochthonous thrust sheets is seen in the southern part of the orogen. The allochthonous Neoproterozoic–Ordovician succession spectacularly exposed in the central fjord zone (72°N–75°N) has an apparent thickness of up to 18.5 km, but it is represented in the foreland by a partly equivalent succession less than 400 m thick (Table 1; Fig. 4; see Sønderholm et al., this volume; Smith and Rasmussen, this volume).

Thin-Skinned Fold-and-Thrust Belt

Kronprins Christian Land is notable for the well-exposed and complete transition from undisturbed foreland in the west through a Thin-skinned fold-and-thrust belt to allochthonous thrust sheets farther to the east (Vandredalen thrust sheet, Western thrust belt, Nørreland thrust sheet). The Thin-skinned fold-and-thrust belt is mainly developed in Ordovician–Silurian platform carbonate rocks of the Iapetus passive margin (Figs. 8 and 9).

This parautochthonous, Thin-skinned fold-and-thrust belt is characterized by a series of NNE-SSW–striking and east-dipping Caledonian thrusts, which have westward displacements of generally a few kilometers each. The belt passes westward into undisturbed autochthonous foreland; eastward, the thrusts penetrate to deeper levels, and the Independence Fjord Group underlying the carbonate platform rocks makes up a fold-and-thrust duplex that is well exposed in the western part of central Ingolf Fjord (Fig. 12). Total displacement along a well-exposed WNW-ESE section through the western part of the fold-and-thrust belt, on the basis of a line and area restoration, amounts to ~17 km (Higgins et al., 2004b); this represents a shortening of ~45% in the line of section.

Biostratigraphical control in the carbonate succession is based on conodonts and macrofossils, and a by-product of the conodont studies is that their alteration colors provide an estimate of maximum burial temperatures. The thickness of the overlying thrust sheets, deduced from the estimated burial temperatures, ranges from ~6 km to 12.5 km from west to east across the width of the Thin-skinned fold-and-thrust belt (Rasmussen and Smith, 2002). Since the estimated former thickness of the Vandredalen thrust sheet above the Thin-skinned fold-and-thrust belt is insufficient to yield the observed temperature records (Fig. 9), higher thrust sheets (presumably parts of the Western thrust belt) must once have extended across the region (Rasmussen and Smith, 2002; Higgins et al., 2004b).

Vandredalen Thrust Sheet

The Vandredalen thrust sheet is a distinctive and well-defined thrust sheet in Kronprins Christian Land (Figs. 8 and 9). It is largely made up of a Neoproterozoic siliciclastic succession

Figure 12. Looking south across central Ingolf Fjord at the folded eastern part of the Thin-skinned fold-and-thrust belt. Part of Brede Spærregletscher is visible to the left. Light-colored rocks are Independence Fjord Group sandstones, and the dark units are dolerite dikes and sills. The generally darker-colored anticlinally folded rocks on the left side of the photograph lie in the hanging wall of the westernmost significant thrust in an ~10-km-wide zone of imbrication beneath the Spærregletscher thrust. The pale-colored rocks toward the right of the photograph are arranged in westward-overturned fold pairs. This large-scale thrust duplex, and the associated fold deformation, raised the Vandredalen thrust sheet above the mountaintop exposures in this view. Section height is ~1500 m.

(the Rivieradal Group) that was deposited in an east-facing half-graben (Smith et al., 2004b). During the Caledonian orogeny, the fill of the half-graben was displaced westward across its rift shoulders, and rocks of the Rivieradal Group now structurally overlie the Thin-skinned fold-and-thrust belt to the west (Higgins et al., 2001b). The Rivieradal Group grades upward into distinctive red and yellow weathering carbonate units (Kap Bernhard and Fyns Sø Formations) of the Hagen Fjord Group that overstepped the earlier half-graben structure, and these units are also well developed in the foreland to the west. A correlation of cut-offs of the Fyns Sø Formation in the hanging wall and footwall of the Vandredalen thrust demonstrates that westward displacement during the Caledonian orogeny ranged from 35 to 50 km (see Leslie and Higgins, this volume).

The Vandredalen thrust sheet has a clearly defined frontal ramp that can be traced over a N-S distance of ~200 km. The present-day extent of the Rivieradal Group within the thrust sheet indicates that the half-graben in which the Rivieradal Group was deposited must have been at least 50 km wide. The westernmost preserved outcrops of the Rivieradal Group include coarse conglomerates and sandstones evidently derived from a source region to the west, and the conglomerate clasts include white quartzites and dark dolerites that can be matched with the Independence Fjord Group and the Midsommersø Dolerites. The root zone of the Vandredalen thrust sheet is preserved as a narrow strip of metasedimentary rocks traceable northward along the west side of Hekla Sund, from Marmorvigen across the alpine region to Brede Spærregletscher and Ingolf Fjord (Fig. 8); the Spærregletscher thrust is the roof thrust to the Vandredalen thrust sheet, and it forms the contact with the structurally overlying Western thrust belt.

Imbricate Thrust Zone in Dronning Louise Land

Dronning Louise Land is divided into two parts by a N-S–trending and up to 10-km-wide east-dipping imbricate thrust zone (Holdsworth and Strachan, 1991). It separates parautochthonous foreland gneisses and quartzitic sedimentary rocks in the footwall to the west from part of the Western thrust belt in the hanging wall to the east. It is composed of a series of east-dipping thrust sheets, made up of variably deformed basement gneisses, metasedimentary cover rocks, and metadolerites. The three main rock units within the imbricate thrust zone have all been correlated with their less-deformed equivalents in the foreland to the west. Quartzitic sedimentary rocks cut by basic dike swarms have been correlated with the Trekant "series" of the foreland (which is also cut by dolerites), and other quartzites associated with shales that preserve Cruziana trace fossils (Strachan et al., 1994) have been correlated with the Lower Cambrian Zebra "series" 1 of the foreland. The rock units that make up the eastern half of Dronning Louise Land (Western thrust belt) include Paleoproterozoic gneisses, and also quartzitic rocks that have been correlated with the Trekant and Zebra "series." The implication of the close correlation is that the thrust sheet that overlies the imbricate thrust zone and makes up eastern Dronning Louise Land probably has only moderate displacement relative to the foreland (Strachan et al., 1992).

Western Thrust Belt

The Western thrust belt is a general term used here for a group of poorly defined thrust sheets that structurally overlie the Vandredalen thrust sheet in Kronprins Christian Land, and west of Jøkelbugten, lie structurally beneath the Nørreland thrust sheet. The Western thrust belt extends between latitudes 76°N and 81°N (Fig. 2), is of variable width, and is composed of amphibolite-facies basement gneisses interleaved with Paleoproterozoic to Mesoproterozoic quartzites. Gneisses dominate in the southern part of the thrust zone in eastern Dronning Louise Land, whereas northward, the proportion of quartzitic rocks increases. In Lambert Land, a complex series of thrust slices involving both quartzites and underlying gneisses has been recognized; still farther northward, the thrust belt narrows and consists of thrust imbricates of quartzites and associated volcanic rocks. In Dronning Louise Land, the rock units of the Western thrust belt are separated from the parautochthonous foreland of western Dronning Louise Land by the narrow imbricate zone referred to previously (and see Fig. 2). No equivalents of the Vandredalen thrust sheet or the Thin-skinned fold-and-thrust belt are present in Dronning Louise Land.

The northern part of the Western thrust belt has been studied in central Kronprins Christian Land in the well-exposed cliff exposures along central Ingolf Fjord, Hekla Sund, and western Hovgaard Ø (Figs. 5 and 8). Here, a zone up to 25 km wide is dominated by white quartzites (Independence Fjord Group) and black doleritic sills and dikes, along with a number of interbedded volcanic units (Hekla Sund Formation, 1740 Ma; Pedersen et al., 2002; Collinson et al., this volume). All of these units are interleaved in a thrust duplex and are tightly folded in places; the contrasting black and white weathering colors provide the dramatic sections in Ingolf Fjord described by Leslie and Higgins (this volume). Basement gneiss lithologies are not seen at exposure level in this region. The Western thrust belt here lies in the hanging wall of the Spærregletscher thrust, which is also the roof thrust to the Vandredalen thrust sheet (Fig. 9). The Spærregletscher thrust dips eastward at steep to moderate angles and can be traced across Ingolf Fjord, and southward along Brede Spærregletscher to Hekla Sund (Fig. 8). The southward continuation of the thrust is hidden beneath the waters of Hekla Sund, apart from a small outlier that caps the summit of Kap Bernhoft (Fig. 8), where quartzites are in thrust contact with the Rivieradal Group of the Vandredalen thrust sheet. It is envisaged that the southward continuation of the thrust forms the west margin of the Western thrust belt (Fig. 2), which is hidden beneath the Inland Ice, and probably merges with the imbricate thrust zone in Dronning Louise Land (~77°N).

In the well-exposed region of Lambert Land, J.C. Escher and K.A. Jones (1995, personal commun.) distinguished two distinct thrust events within the Western thrust belt, namely, early N-directed thrusting and later westward thrusting. A pile of seven, thin N-directed thrust sheets separated by ductile thrusts forms an in-sequence northward-propagating stack. A later series of six westward-directed foreland-propagating thrust sheets carried large segments of the early thrust stack in the structurally highest west-directed thrust sheet. In Lambert Land, the thrust sheets mainly involve the Independence Fjord Group quartzite succession and crosscutting basic dikes, but the earliest and highest thrust sheets also incorporate slices of basement gneisses.

Similar relationships probably prevail in the poorly exposed terrain of large nunataks west of Jøkelbugten that lie structurally beneath the Nørreland thrust sheet. These nunataks mainly expose quartzitic sedimentary rocks correlated with the Paleoproterozoic Independence Fjord Group that are cut by dike swarms.

Eastern Dronning Louise Land makes up the southern part of the Western thrust belt, and it is a region dominated by amphibolite-facies gneisses, with only few exposures of quartzitic sedimentary rocks correlated with the Trekant "series" and Zebra "series" of the foreland to the west (Table 1; Strachan et al., 1994).

Nørreland Thrust Sheet

The Bessel Fjord shear zone (Fig. 2; 76°N) marks the northern limit of the Hagar Bjerg thrust sheet (described later) and the southern border of the extensive Nørreland thrust sheet. The present-day exposure north of 76°N is dominated by high-grade Paleoproterozoic gneiss complexes, which, north of Danmarkshavn (76°40′N), contain abundant Caledonian eclogitic enclaves (Gilotti, 1993). The region where eclogites are found has also been described as the "North-East Greenland eclogite province" and extends northward to at least 80°N. There are rare enclaves of Precambrian metasedimentary rocks in this extensive region, but there is a conspicuous lack of thick metasedimentary successions comparable to the Krummedal sedimentary sequence. There are no known exposures of either Neoproterozoic–Ordovician sedimentary rocks or Caledonian granitoid intrusions.

The Nørreland thrust sheet makes up the deepest preserved level of the East Greenland Caledonides where continental crust has been subjected to medium-temperature (600–750 °C), high-pressure (1.5–2.2 GPa) metamorphism (Brueckner et al., 1998; Elvevold and Gilotti, 2000). These conditions are attributed to thickening of the East Greenland continental margin during crustal imbrication associated with west-directed Caledonian thrusting, probably with a component of pure shear (Gilotti et al., this volume). Coesite-bearing ultrahigh-pressure eclogites are exposed in the easternmost part of the region (McClelland et al., 2006); their large pressure difference requires a tectonic contact (currently unrecognized) to separate them from other parts of the Nørreland thrust sheet.

The western limit of the Nørreland thrust sheet is not well defined due to the scattered exposures in the nunatak region west of Jøkelbugten. Eclogite-bearing orthogneisses structurally overlie lower-grade gneiss complexes imbricated with epidote- to amphibolite-facies Paleoproterozoic quartzites 15 km west of the Storstrømmen shear zone in Lambert Land and Nørreland. Description of the relationships around the Nørreland window led to usage of the "Nørreland thrust sheet" for the overlying high-grade rock units (Hull et al., 1995; Gilotti et al., this volume), and this term is expanded here to include the entire high-grade domain; however, Gilotti et al. (this volume) report that despite the absence of mapped internal thrust contacts, it is unlikely that this extensive region represents a single continuous crustal slab. The Nørreland thrust sheet must have been exhumed from depths in excess of 50 km and displaced westward across higher-level rock units. Conodonts isolated from the Ordovician limestones in the footwall of the thrust that bounds the Nørreland window have alteration indices indicative of an overburden of less than 15 km (Rasmussen and Smith, 2002). This is considerably less than the thicknesses required for eclogite formation, and it demonstrates that significant displacement on the Nørreland thrust must be younger than the Devonian-age (415–390 Ma) eclogites in the hanging wall (Gilotti et al., 2004).

Gemmedal Thrust Sheet

The thrust architecture of the southern half of the Caledonian orogen in East Greenland includes two major thrust sheets (Niggli Spids thrust sheet and Hagar Bjerg thrust sheet; Higgins and Leslie, 2000; Higgins et al., 2004a), and the Neoproterozoic–Ordovician segment of the Hagar Bjerg thrust sheet is known as the Franz Joseph allochthon (Table 1). Here, a relatively thin assemblage of diverse lithologies at the base of the Niggli Spids thrust sheet is separately described as the Gemmedal thrust sheet (Figs. 3 and 4). The division into two major thrust units and one lowermost minor thrust sheet is best seen in central parts of East Greenland (72°N–74°30′N) and is described and illustrated by Leslie and Higgins (this volume).

The lithologies represented in the Gemmedal thrust sheet include low-grade black shales in the upper Hamberg Gletscher area, dark mylonitic metasedimentary units along the west side of the Eleonore Sø window, and mica schists associated with quartzites around the Målebjerg window. There appear to be no equivalents of the Gemmedal thrust sheet around the Charcot Land window. However, analogous occurrences may be present in the Gåseland window in the extreme south of the orogen, where a several-hundred-meter-thick succession of strongly sheared carbonates and quartzites occurs above the thrust contact with the foreland (Table 1; Phillips et al., 1973).

Niggli Spids Thrust Sheet

The Niggli Spids thrust sheet (Figs. 2 and 6) is the lower of the two major thrust sheets recognized in the southern half of the Caledonian orogen. Archean and Paleoproterozoic basement gneiss complexes form the lower part of the thrust sheet and are overlain by a thick succession of upper Mesoproterozoic to lower Neoproterozoic metasedimentary rocks known as the Krummedal supracrustal sequence.

In some areas (e.g., around the Charcot Land window), the Niggli Spids thrust sheet directly overlies the foreland, and the contact is sharp and marked only by a few meters of sheared gneiss or metasedimentary rocks. Elsewhere, the Niggli Spids thrust sheet is separated from the foreland by a variable thickness of low-grade mica schist and dark mylonitic shales ascribed to the Gemmedal thrust sheet.

The Archean and Paleoproterozoic basement gneiss complexes that form the lower part of the Niggli Spids thrust sheet are widely exposed in the inner fjord region between 70°N and 75°N. The Archean gneisses in the Scoresby Sund region exhibit complex deformation patterns and commonly incorporate enclaves and thin bands of amphibolites and metasedimentary rocks. In the Kong Oscar Fjord region, thick units of Paleoproterozoic orthogneisses form spectacular cliffs (Fig. 6). The basement gneiss complexes directly overlie the eastern margins of both the Gåseland and Charcot Land windows but wedge out westward, such that the Krummedal metasedimentary rocks of upper parts of the Niggli Spids thrust sheet overlie the western sides of both windows. Around the Målebjerg window, the lower basement gneiss unit is only a few hundred meters thick.

The Krummedal supracrustal sequence is dominated by siliciclastic sedimentary rocks that have a distinctive brownish red weathering color that contrasts with the underlying gneisses. This often rather uniform metasedimentary succession is generally 2–4 km in thickness and is locally up to 8 km thick (Higgins, 1988). The contact with the underlying gneisses is commonly sheared, but at a few localities, it has the appearance of a modified unconformity (Higgins et al., 1981; Escher and Jones, 1998).

Hagar Bjerg Thrust Sheet (with Franz Joseph Allochthon)

The Hagar Bjerg thrust sheet is the upper of the two major thrust sheets distinguished in the southern half of the orogen (70°N–76°N), and it overlies the Niggli Spids thrust sheet everywhere (Fig. 4). It incorporates a few major Paleoproterozoic orthogneiss units, but it is dominated by a 2–4-km-thick succession of high-grade metasedimentary rocks of the Krummedal supracrustal sequence that show spectacular migmatitic features and are cut by abundant granitic sheets and plutons. Two granite-forming episodes are recognized, an early Neoproterozoic phase that produced leucocratic augen granites dated at 950–920 Ma (Fig. 7), and a Caledonian phase dated at ca. 435–425 Ma; the two different granites are almost indistinguishable in the field (Kalsbeek et al., 2001a, 2001b, this volume, Chapter 9).

The migmatitic metasedimentary rocks of the Hagar Bjerg thrust sheet are interpreted to have formed from the Krummedal supracrustal sequence and are correlated with the equivalent units of the Niggli Spids thrust sheet. However, there is often a marked contrast across the thrust boundary between the two thrust sheets, where migmatitic metasedimentary rocks with abundant granite sheets in the hanging wall structurally overlie metasedimentary rocks lacking granite veins and sheets in the footwall. It seems likely that the migmatitic and granitic developments seen in the Hagar Bjerg thrust sheet were formed prior to Caledonian thrusting, and, furthermore, that the region from which the Hagar Bjerg thrust sheet was derived experienced two high-grade events (ca. 930 Ma and ca. 425 Ma) that the otherwise similar metasedimentary rocks of the Niggli Spids thrust sheet seem to have escaped.

The very thick Neoproterozoic–Ordovician sedimentary package (18.5 km apparent thickness) of the Eleonore Bay Supergroup, Tillite Group, and Kong Oscar Fjord Group makes up a conspicuous upper segment of the Hagar Bjerg thrust sheet and has been distinguished as the Franz Joseph allochthon (Higgins et al., 2004a). The contrast in thickness between this Neoproterozoic–Ordovician succession and the partly equivalent units (250–350 m thick) found in the structurally underlying foreland windows testifies to the substantial displacements on the Hagar Bjerg and Niggli Spids thrusts.

Although far-traveled from its original site of deposition, the Franz Joseph allochthon displays simple large-scale anticlines and synclines with only limited shortening, and it has been suggested that this allochthonous segment was transported almost passively westward, riding on the underlying part of the Hagar Bjerg thrust sheet (Higgins et al., 2004a). For this reason, it has not been distinguished as a separate thrust sheet, but rather a slightly displaced upper segment of the Hagar Bjerg thrust sheet. The contact of the base of the Eleonore Bay Supergroup against the underlying migmatitic Krummedal supracrustal sequence is a shear zone that shows evidence of both compressional and extensional movement, and it might be viewed as a modified unconformity. In the eastern exposures of the Eleonore Bay Supergroup, this contact is referred to as the Franz Joseph detachment, whereas in the western exposures around Petermann Bjerg, it has been referred to as the Petermann Bjerg detachment (Fig. 3; Higgins et al., 2004a; Leslie and Higgins, this volume).

The large-scale N-S–trending fold system that developed in the sedimentary rocks of the Franz Joseph allochthon has been illustrated by Haller (1971). The simple fold patterns led Haller and others (e.g., Fränkl, 1953) to view the structures as the result of vertical movements, with lateral forces playing a minor roll; it was this view that led Haller to develop the "stockwerk" concept for the Caledonian orogeny. Subsequently, Higgins et al. (1981) estimated the E-W shortening due to folding of the Eleonore Bay Supergroup at ~15% and observed that many of the major folds in Andrée Land and in the Petermann Bjerg region had a box-fold style; the contact zone with the underlying migmatitic metasedimentary rocks of the Hagar Bjerg thrust sheet was interpreted to be a décollement. The recent demonstration of hundreds of kilometers of westward displacement for the Niggli Spids and Hagar Bjerg thrust sheets proves Haller's "stockwerk" concept to be untenable (Higgins and Leslie, 2000; Higgins and Leslie, 2004), but the amount of displacement on the contact zone (Franz Joseph detachment/Petermann Bjerg detachment) remains controversial. The view presented by Higgins et al. (2004a), that the Franz Joseph detachment was transported almost passively as the upper part of the Hagar Bjerg thrust sheet, with perhaps 10–15 km of westward displacement, is not held by all workers in the region; other workers consider it possible for the Franz Joseph allochthon to have been a major thrust sheet in its own right (e.g., J. Gilotti, 2007, personal commun.), and that as a dominant thrust sheet, it suffered little internal deformation.

FAULTS AND LINEAMENTS

Prominent faults and lineaments (see Fig. 2), some of which are primarily related to extension and collapse of the overthickened orogenic thrust stack, are described in this section, from north to south. The major thrusts and detachments that floor the main structural domains have been described already.

East Greenland Fault Zone (79°30′N–81°N)

The East Greenland fault zone is the prominent NNE-SSW–trending lineament in eastern Kronprins Christian Land that forms the boundary between the alpine mountains of imbricated units of Independence Fjord Group quartzites cut by dikes to the west (Western thrust belt) and the low-lying terrain of high-grade gneisses to the east (Nørreland thrust sheet), which are unconformably overlain by Upper Paleozoic sedimentary rocks in Holm Land and Amdrup Land. Stemmerik and Håkansson (1991) considered post-Caledonian, normal, down-to-the-east displacement on this fault line to have controlled deposition in a Carboniferous basin that developed as a result of rifting between Greenland and Norway.

While the East Greenland fault zone is a prominent topographic feature, the best exposures of the presumed contact zone are on the south side of Hovgaard Ø, for which the only description is an unpublished cross section (K.A. Jones and J.C. Escher, 1995, personal commun.). At this locality, a several-kilometer-wide, steeply east-dipping to near-vertical shear zone exposes mylonitic Independence Fjord Group quartzites to the west in contact with mylonitic granitic gneisses containing sheared basic sheets to the east. Displacement sense is dominantly vertical, east-side-up, which is the opposite of the normal displacement on the East Greenland fault zone that controlled the Carboniferous basin, suggesting that this shear zone contact has a different origin. To the east of the shear zone, the granitic and tonalitic gneisses contain basic sheets and pods metamorphosed in the eclogite facies, which indicate that they are part of the Nørreland thrust sheet. K.A. Jones and J.C. Escher recorded mylonitic fabric in the shear zone that developed under amphibolite- to epidote-amphibolite-facies metamorphism, and they suggested that the shear zone may have played a role in the exhumation of the eclogite-facies rocks of the Nørreland thrust sheet. In this case, the structural setting would be closely similar to that at the east margin of the Western thrust belt, west of Jøkelbugten, where the Nørreland thrust sheet is displaced upward and across lower-grade units of the Western thrust belt. Higgins et al. (2001b) speculated that the gneiss areas of easternmost Kronprins Christian Land might represent the roots of deep-seated thrust sheets that formerly projected westward above the units of the Western thrust belt, and that may have had displacements of as much as 100 km (Fig. 9; see also Leslie and Higgins, this volume).

The assumed normal downthrow on the East Greenland fault zone, which is considered to have controlled the formation of the Carboniferous basin in eastern Kronprins Christian

Land (Stemmerik and Håkansson, 1991), is presumed to have been part of the system of fault-controlled basins known offshore of North-East Greenland (Hamann et al., 2005).

Storstrømmen Shear Zone (76°N–79°N)

The Storstrømmen shear zone is a major NNE-SSW–trending shear zone that is best exposed between Dronning Louise Land and Jøkelbugten, where it is characterized by a several-kilometer-wide belt of steeply dipping mylonites that have a dominantly sinistral sense of displacement (Strachan and Tribe, 1994). Mineral assemblages and fabrics indicate that mylonitization was initiated under low-amphibolite-facies conditions and continued within the greenschist facies. The continuation to the south is concealed for 150 km beneath a major lobe of the Inland Ice, and it may dissipate into several shear or fault zones as it reaches higher structural levels of the orogen. Larsen and Bengaard (1991) speculated that the so-called Western fault zone (see later section), a sinistral fault that follows the west margin of the Devonian continental basins from 72°N to 74°N, might represent a high-level continuation of the Storstrømmen shear zone. While some workers suspect that the northward continuation of the Storstrømmen shear zone trends offshore north of Lambert Land (~79°N), a link with the pronounced topographic lineament in Kronprins Christian Land known as the East Greenland fault zone (see previous) seems most likely. Holdsworth and Strachan (1991) and Strachan et al. (1992) interpreted the imbricate thrust zone in Dronning Louise Land and the Storstrømmen shear zone to be products of Caledonian sinistral transpression, perhaps indicative of a significant component of oblique collision between Laurentia and Baltica in the Silurian (cf. Soper et al., 1992). However, Gilotti et al. (2004) demonstrated that significant movement on the Storstrømmen shear zone must have been younger than the Devonian eclogites of the Nørreland thrust sheet because the amphibolite-facies fabrics are superimposed on the high-pressure fabrics associated with the eclogites.

Germania Land Deformation Zone (77°N–78°N)

The Germania Land deformation zone is a NNW-SSE–striking shear zone that can be traced from northeast Germania Land northward to Jøkelbugten (Fig. 2). It consists of two subparallel lineaments, ~1–2 km apart, associated with protomylonites, mylonites, and ultramylonites, while the eastern strand is marked by a feature known as the Chatham Elv fault (Hull and Gilotti, 1994). Outcrops of Carboniferous sedimentary rocks on the east side of the Chatham Elv fault (Piasecki et al., 1994) indicate a component of normal dip-slip movement and a geometry similar to the normal faults imaged in seismic sections just offshore (Hamann et al., 2005). The predominant movement on the ductile mylonites is dextral strike-slip, with some local sinistral strike-slip on subsidiary shear zones. Small-scale dextral mylonite zones, inferred to be comparable with the ductile structures of the Germania Land deformation zone, have been studied in the vicinity of the Danmarkshavn weather station (Sartini-Rideout et al., 2006). The mylonites were active under amphibolite- to greenschist-facies conditions from ca. 370 to 340 Ma, based on U-Pb SHRIMP zircon geochronology of pegmatites in the necks of eclogite boudins, and pegmatites that crosscut the mylonites.

Bessel Fjord Shear Zone (~76°N) and Kildedalen Shear Zone (75°15′N–75°25′N)

The Bessel Fjord shear zone is a S-dipping, approximately E-W–trending structure that separates two contrasting regions of the Caledonian orogen (Fig. 2; Friderichsen et al., 1994). The shear zone is well exposed on the south side of Bessel Fjord, where it is characterized by an up to 130-m-wide zone of heterogeneous ductile shear; belts of mylonitic gneisses and schists dip gently to moderately to the south. In the footwall, to the north of the shear zone, high-grade Proterozoic gneiss complexes are widely exposed, while in the hanging wall to the south of the Bessel Fjord shear zone, structurally higher levels of the orogen are well preserved. The latter include the Smallefjord supracrustal sequence (correlated with the Krummedal supracrustal sequence) and a fault-bounded enclave of the Eleonore Bay Supergroup centered on Ardencaple Fjord (Soper and Higgins, 1993; Higgins and Soper, 1994).

The Kildedalen shear zone (Friderichsen et al., 1994) is characterized by a 1.2–3.5-km-wide belt of reworked gneisses that follows the lower, southern boundary of the Smallefjord supracrustal sequence. Thin section study has demonstrated that reworking occurred within the amphibolite facies in the gneissic rocks, whereas the Smallefjord metasedimentary rocks carry a penetrative quartz-mica fabric overgrown by fibrolite mats, indicative of mid- to upper-amphibolite-facies metamorphic conditions. Since the Bessel Fjord and Kildedalen shear zones occupy similar structural positions (along the basal Smallefjord supracrustal sequence contact) and were formed under similar grades of metamorphism, Friderichsen et al. (1994) suggested that they developed contemporaneously during Caledonian extension.

Payer Land Detachment (74°30′N)

The Payer Land detachment is a prominent SE-dipping shear zone that developed between high-pressure granulite-facies orthogneisses and paragneisses in the footwall and low-grade quartzitic rocks of the Eleonore Bay Supergroup in the hanging wall (Fig. 2; Gilotti and Elvevold, 2002). This mylonitic shear zone is at least 1500 m thick and primarily derived from metapelitic rocks comparable to the Krummedal supracrustal sequence. Displacement, based on abundant kinematic indicators, is top-to-the-southwest. The high-pressure granulite-facies metamorphism has been dated at 403 ± 5 and 404 ± 4 Ma (McClelland and Gilotti, 2003), which carries with it the implication that the extensional displacement that juxtaposed the contrasting metamorphic terranes along the Payer Land detachment cannot be younger than ca. 405 Ma.

Boyd Bastion Fault (73°N–73°40′N)

The Boyd Bastion fault is a prominent N-S–trending extensional fault with substantial downthrow to the east (Figs. 2–4). The Eleonore Bay Supergroup is downthrown against part of the Niggli Spids thrust sheet, suggesting a vertical displacement of ~10–15 km. It was named the Nunatak zone fault by Larsen and Bengaard (1991), who put forward the concept that accumulation in the Devonian basins was caused by collapse of the overthickened Caledonian orogen, and extension was accommodated along two major east-dipping normal faults (Nunatak zone fault and Fjord zone fault; see also Larsen et al., this volume).

Fjord Region Fault (Fjord Zone Fault or Fjord Region Detachment; 70°N–74°N)

Slightly different names have been applied to this major NNE-SSW–trending fault that can be traced continuously for 400 km from 70°N to 74°N. In the inner Scoresby Sund region, it separates two contrasting structural levels, the migmatitic metasedimentary rocks and granites of the Hagar Bjerg thrust sheet to the east and the Archean gneisses of the lower Niggli Spids thrust sheet to the west. Although initially interpreted as a thrust in the Scoresby Sund region (e.g., Henriksen and Higgins, 1976; Henriksen, 1986), down-to-the-east movement that accommodated deposition of the post-Caledonian clastic sedimentary rocks of the Røde Ø conglomerate has also been recognized. In the central fjord zone of East Greenland (72°N–75°N), high levels of the Eleonore Bay Supergroup are locally juxtaposed against low levels of the Niggli Spids thrust sheet along the fault, indicating downthrow to the east of the order of 15–30 km. Larsen and Bengaard (1991) envisaged their "Fjord zone fault" to be a major extensional structure linked to initiation of the Devonian basins. Renamed the "Fjord region detachment" or "Fjord region detachment zone," this feature, and additional structures interpreted as splays, has featured prominently in extensional models of the East Greenland Caledonides (Hartz and Andresen, 1995; Andresen et al., 1998).

Leslie and Higgins (this volume) employ the term Fjord region fault for the NNE-SSW–striking steep lineament that shows substantial down-to-the-east extensional displacement and is traceable continuously from 70°N to 74°N (Figs. 2–4). In contrast, Gilotti and McClelland (this volume) employ the term Fjord region detachment system in the wider sense of Hartz and Andresen (1995), which includes not only the main NNE-SSW–trending fault but also incorporates as splays many of the detachments and thrust structures of Leslie and Higgins (this volume). In comparing the profiles presented by different authors in this volume, it should be remembered that many structures depicted as thrusts were subsequently reactivated during extension, just as some structures shown as major extensional faults may have been initiated as major thrusts.

Western Fault Zone (71°N–74°N) and Eastern Fault Zone (72°N–74°N)

These two NNE-SSW–trending sinistral wrench faults were named for their relationships to the syn- to late Caledonian Devonian basin, of which they mark the western and eastern borders, respectively. The series of deformational phases that affected the Devonian succession and continued to be active into Mississippian time were attributed by Larsen and Olesen (1991) to strike-slip movements on the two faults (see also Larsen et al., this volume).

TOWARD A BIGGER PICTURE

Eighty years have elapsed since initiation in 1926 of the first phase of regional geological mapping in the East Greenland Caledonides (see Haller, 1971) to publication of the series of modern 1:500,000-scale geological maps of the orogen (see Henriksen and Higgins, this volume, Chapter 14) and of the geological overview map at 1:1,000,000 scale that accompanies this volume (Henriksen, 2003; Henriksen and Higgins, this volume, Chapter 14). The authors of the succeeding chapters in this volume expand greatly upon the geological composition and evolution of the Caledonian orogenic belt of East Greenland provided here, and have, although the region is remote and represents a significant challenge to geological exploration, been able to draw upon a considerable volume of new research. The reward won is that we are able for the first time to place a modern understanding of the East Greenland Caledonian orogen alongside current frameworks of North America, Svalbard, Scandinavia, and northern Britain. The East Greenland strata record Tonian, Cryogenian, Marinoan, Ediacaran, and Cambrian-Ordovician sedimentation and deformation related to the latter stages of the amalgamation of Rodinia, together with the subsequent breakout of Laurentia and growth of the Iapetus Ocean. Metamorphic and tectonic overprints then provide a perspective upon the destruction of that ocean, culminating in the mid- to late Silurian collision of Laurentia, Baltica, and Avalonia. New isotopic data and recent advances in our understanding of Neoproterozoic to Cambrian-Ordovician stratigraphic framework now better constrain the sequence and timing of events and invite a dynamic, and potentially enormously rewarding, comparison of the current research into the East Greenland Caledonides with the entire North Atlantic region.

ACKNOWLEDGMENTS

We would like to thank Niels Henriksen, leader of the Geological Survey of Greenland's expeditions to North-East Greenland, and our colleagues during the mapping project for helpful discussions in the field and subsequently. Special thanks are due to the two reviewers, Jim Hibbard and Karsten Piepjohn, for their many constructive comments, and last but not least to editor Jane Gilotti, whose suggestions and insights considerably helped the shape of the final manuscript.

REFERENCES CITED

Andresen, A., Hartz, E., and Vold, J., 1998, A late orogenic extensional origin for the infrastructural gneiss domes of the East Greenland Caledonides (72°–74°N): Tectonophysics, v. 285, p. 353–369.

Andresen, A., Rehnström, E.F., and Holte, M.K., 2007, Evidence for simultaneous contraction and extension at different crustal levels during the Caledonian orogeny in NE Greenland: Journal of the Geological Society of London, v. 164, p. 869–880, doi: 10.1144/0016-76492005-056.

Brueckner, H.K., Gilotti, J.A., and Nutman, A.P., 1998, Caledonian eclogite facies metamorphism of Early Proterozoic protoliths from the North-East Greenland eclogite province: Contributions to Mineralogy and Petrology, v. 130, p. 103–120, doi: 10.1007/s004100050353.

Collinson, J.D., Kalsbeek, F., Jepsen, H.F., Pedersen, S.A.S., and Upton, B.G.J., 2008, this volume, Paleoproterozoic and Mesoproterozoic sedimentary and volcanic successions in the northern parts of the East Greenland Caledonian orogen and its foreland, in Higgins, A.K., Gilotti, J.A., and Smith, M.P., eds., The Greenland Caledonides: Evolution of the Northeast Margin of Laurentia: Geological Society of America Memoir 202, doi: 10.1130/2008.1202(04).

Dalziel, I.W.D., 1997, Neoproterozoic-Paleozoic geography and tectonics: Review, hypothesis, environmental speculation: Geological Society of America Bulletin, v. 109, p. 16–42, doi: 10.1130/0016-7606(1997)109<0016:ONPGAT>2.3.CO;2.

Elvevold, S., and Gilotti, J.A., 2000, Pressure-temperature evolution of retrogressed kyanite eclogites, Weinschenk Island, North-East Greenland Caledonides: Lithos, v. 53, p. 127–147, doi: 10.1016/S0024-4937(00)00014-1.

Escher, J.C., and Jones, K.A., 1998, Caledonian thrusting and extension in Frænkel Land, East Greenland (73°–73°30′N): Preliminary results: Danmarks og Grønlands Geologiske Undersøgelse Rapport, v. 1998/28, p. 29–42.

Fairchild, I.J., and Hambrey, M.J., 1995, Vendian basin evolution in East Greenland and NE Svalbard: Precambrian Research, v. 73, p. 217–223, doi: 10.1016/0301-9268(94)00079-7.

Fränkl, E., 1953, Geologische untersuchungen in ost-Andrées Land (NE-Grønland): Meddelelser om Grønland, v. 113, no. 4, p. 1–160.

Frederiksen, K.S., 2000, A Neoproterozoic Carbonate Ramp and Base-of-Slope Succession, the Andrée Land Group, Eleonore Bay Supergroup, North-East Greenland: Sedimentary Facies, Stratigraphy and Basin Evolution [Ph.D. thesis]: Copenhagen, University of Copenhagen, 242 p.

Friderichsen, J.D., Henriksen, N., and Strachan, R.A., 1994, Basement-cover relationships and regional structure in the Grandjean Fjord–Bessel Fjord region (75°–76°N), North-East Greenland: Rapport Grønlands Geologiske Undersøgelse, v. 162, p. 17–33.

Friend, C.R.L., Strachan, R.A., Kinny, P.D., and Watt, G.R., 2003, Provenance of the Moine Supergroup of NW Scotland: Evidence from geochronology of detrital and inherited zircons from (meta)sedimentary rocks, granites and migmatites: Journal of the Geological Society of London, v. 160, p. 247–257.

Garde, A.A., Chadwick, B., Grocott, J., Hamilton, M., McCaffrey, K., and Swager, C.P., 1998, An overview of the Paleoproterozoic Ketilidian orogen, South Greenland, in Wardle R.J., and Hall, J., compilers, Eastern Canadian Shield Onshore–Offshore Transect (ESCOOT), Report of 1998 Transect Meeting: University of British Columbia, Lithoprobe Report, v. 68, p. 50–66.

Gee, D.G., and Teben'kov, A.M., 1996, Two major unconformities beneath the Neoproterozoic Murchisonfjorden Supergroup in the Caledonides of central Nordaustlandet, Svalbard: Polar Research, v. 15, no. 1, p. 81–91, doi: 10.1111/j.1751-8369.1996.tb00460.x.

Gilotti, J.A., 1993, Discovery of a medium-temperature eclogite province in the Caledonides of North-East Greenland: Geology, v. 21, p. 523–526, doi: 10.1130/0091-7613(1993)021<0523:DOAMTE>2.3.CO;2.

Gilotti, J.A., and Elvevold, S., 2002, Extensional exhumation of a high-pressure granulite terrane in Payer Land, Greenland Caledonides: Structural, petrologic and geochronologic evidence from metapelites: Canadian Journal of Earth Sciences, v. 39, p. 1169–1187, doi: 10.1139/e02-019.

Gilotti, J.A., and McClelland, W.C., 2008, this volume, Geometry, kinematics, and timing of extensional faulting in the Greenland Caledonides—A synthesis, in Higgins, A.K., Gilotti, J.A., and Smith, M.P., eds., The Greenland Caledonides: Evolution of the Northeast Margin of Laurentia: Geological Society of America Memoir 202, doi: 10.1130/2008.1202(10).

Gilotti, J.A., Nutman, A.P., and Brueckner, H.K., 2004, Devonian to Carboniferous collision in the Greenland Caledonides: U-Pb zircon and Sm-Nd ages of high-pressure and ultrahigh-pressure metamorphism: Contributions to Mineralogy and Petrology, v. 148, p. 216–235, doi: 10.1007/s00410-004-0600-4.

Gilotti, J.A., Jones, K.A., and Elvevold, S., 2008, this volume, Caledonian metamorphic patterns in Greenland, in Higgins, A.K., Gilotti, J.A., and Smith, M.P., eds., The Greenland Caledonides: Evolution of the Northeast Margin of Laurentia: Geological Society of America Memoir 202, doi: 10.1130/2008.1202(08).

Haller, J., 1971, Geology of the East Greenland Caledonides: London, Interscience, 413 p.

Hamann, N.E., Whittaker, R.C., and Stemmerik, L., 2005, Structural and geological development of the North-East Greenland shelf, in Doré, A.G., and Vining, B., eds., Petroleum Geology: North-West Europe and Global Perspectives: Proceedings of the 6th Petroleum Conference: London, Geological Society of London, p. 887–902.

Hambrey, M.J., and Spencer, A.M., 1987, Late Precambrian glaciation of central East Greenland: Meddelelser om Grønland: Geoscience, v. 19, p. 1–50.

Harris, A.L., Haselock, P.J., Kennedy, M.J., and Mendum, J.R., 1994, The Dalradian Supergroup in Scotland, Shetland and Ireland, in Gibbins, W., and Harris, A.L., eds., A Revised Correlation of Precambrian Rocks in the British Isles: Geological Society of London Special Report 22, p. 33–53.

Hartz, E., and Andresen, A., 1995, Caledonian sole thrust of central East Greenland: A crustal-scale Devonian extensional detachment?: Geology, v. 23, p. 637–640, doi: 10.1130/0091-7613(1995)023<0637:CSTOCE>2.3.CO;2.

Hartz, E.H., Condon, D., Austrheim, H., and Erambert, M., 2005, Re-discovery of the Liverpool Land eclogites (Central East Greenland): A post- and supra-subduction UHP province: Mitteilungen der Österreichischen Mineralogischen Gesellschaft, v. 150, p. 50.

Henriksen, N., 1986, Geological Map of Greenland 1:500,000, Sheet 12, Scoresby Sund: Descriptive Text: Copenhagen, Geological Survey of Greenland, 27 p.

Henriksen, N., 2003, Geological Map of the Caledonian Orogen, East Greenland 70°–82°N: Copenhagen, Geological Survey of Denmark and Greenland, scale 1:1,000,000.

Henriksen, N., and Higgins, A.K., 1976, East Greenland Caledonides, in Escher, A., and Watt, W.S., eds., Geology of Greenland: Copenhagen, Grønlands Geologiske Undersøgelse, p. 182–246.

Henriksen, N., and Higgins, A.K., 2008, this volume, Caledonian orogen of East Greenland 70°–82°N: Geological map 1:1,000,000—Concepts and principles of compilation, in Higgins, A.K., Gilotti, J.A., and Smith, M.P., eds., The Greenland Caledonides: Evolution of the Northeast Margin of Laurentia: Geological Society of America Memoir 202, doi: 10.1130/2008.1202(14).

Henriksen, N., Perch-Nielsen, K., and Andersen, C., 1980, Geological Map of Greenland, 1:100,000 (Sydlige Stauning Alper 71 Ø.2 Nord, Frederiksdal 71 Ø.3 Nord): Descriptive Text: Copenhagen, Geological Survey of Greenland, 46 p.

Henriksen, N., Higgins, A.K., Kalsbeek, F., and Pulvertaft, T.C.R., 2000, Greenland from Archaean to Quaternary: Descriptive Text to the Geological Map of Greenland 1:2,500,000: Geology of Greenland Survey Bulletin, v. 185, 93 p.

Higgins, A.K., 1988, The Krummedal supracrustal sequence in East Greenland, in Winchester, J.A., ed., Later Proterozoic Stratigraphy of the Northern Atlantic Regions: Glasgow and London, Blackie and Son, p. 86–96.

Higgins, A.K., and Leslie, A.G., 2000, Restoring thrusting in the East Greenland Caledonides: Geology, v. 28, p. 1019–1022, doi: 10.1130/0091-7613(2000)28<1019:RTITEG>2.0.CO;2.

Higgins, A.K., and Leslie, A.G., 2004, The Eleonore Sø and Målebjerg foreland windows, East Greenland Caledonides, and the demise of the 'stockwerke' concept: Geological Survey of Denmark and Greenland Bulletin, v. 6, p. 77–93.

Higgins, A.K., and Soper, N.J., 1994, Structure of the Eleonore Bay Supergroup at Ardencaple Fjord, North-East Greenland: Rapport Grønlands Geologiske Undersøgelse, v. 162, p. 91–101.

Higgins, A.K., Friderichsen, J.D., and Thyrsted, T., 1981, Precambrian metamorphic complexes in the East Greenland Caledonides (72°–74°N): Their relationships to the Eleonore Bay Group and Caledonian orogenesis: Rapport Grønlands Geologiske Undersøgelse, v. 104, p. 1–46.

Higgins, A.K., Ineson, J.R., Peel, J.S., Surlyk, F., and Sønderholm, M., 1991, Lower Palaeozoic Franklinian Basin of North Greenland, in Peel, J.S., and Sønderholm, M., eds., Sedimentary Basins of North Greenland: Bulletin Grønlands Geologiske Undersøgelse, v. 160, p. 71–139.

Higgins, A.K., Leslie, A.G., and Smith, M.P., 2001a, Neoproterozoic–Lower Palaeozoic stratigraphical relationships in the marginal thin-skinned thrust belt of the East Greenland Caledonides: Comparisons with the foreland of Scotland: Geological Magazine, v. 138, p. 143–160, doi: 10.1017/S0016756801005076.

Higgins, A.K., Smith, M.P., Soper, N.J., Leslie, A.G., Rasmussen, J.A., and Sønderholm, M., 2001b, The Neoproterozoic Hekla Sund Basin, eastern North Greenland: A pre-Iapetan extensional sequence thrust across its rift shoulders during the Caledonian orogeny: Journal of the Geological Society of London, v. 158, p. 487–499.

Higgins, A.K., Elvevold, S., Escher, J.C., Frederiksen, K.S., Gilotti, J.A., Henriksen, N., Jepsen, H.F., Jones, K.A., Kalsbeek, F., Kinny, P.D., Leslie, A.G., Smith, M.P., Thrane, K., and Watt, G.R., 2004a, The foreland-propagating thrust architecture of the East Greenland Caledonides 72°–75°N: Journal of the Geological Society of London, v. 161, p. 1009–1026, doi: 10.1144/0016-764903-141.

Higgins, A.K., Soper, N.J., Smith, M.P., and Rasmussen, J.A., 2004b, The Caledonian parautochthonous fold and thrust belt of Kronprins Christian Land, eastern North Greenland: Geological Survey of Denmark and Greenland Bulletin, v. 6, p. 41–56.

Holdsworth, R.E., and Strachan, R.A., 1991, Interlinked system of ductile strike-slip and thrusting formed by Caledonian sinistral transpression in northeastern Greenland: Geology, v. 19, p. 510–513, doi: 10.1130/0091-7613(1991)019<0510:ISODSS>2.3.CO;2.

Holdsworth, R.E., Strachan, R.A., and Harris, A.L., 1994, Precambrian rocks in northern Scotland east of the Moine thrust: The Moine Supergroup, in Gibbons, W., and Harris, A.L., eds., A Revised Correlation of the Precambrian Rocks in the British Isles: Geological Society of London Special Report 22, p. 23–32.

Hull, J.M., and Gilotti, J.A., 1994, The Germania Land deformation zone and related structures: Rapport Grønlands Geologiske Undersøgelse, v. 162, p. 113–127.

Hull, J.M., Friderichsen, J.D., Gilotti, J.A., Henriksen, N., Higgins, A.K., and Kalsbeek, F., 1994, Gneiss complex of the Skærfjord region (76°–78°N), North-East Greenland: Rapport Grønlands Geologiske Undersøgelse, v. 162, p. 35–51.

Hull, J.M., Gilotti, J.A., and Friderichsen, J.D., 1995, A window through a basement-involved thrust sheet into sub-thrust cover, northeast sector, East Greenland Caledonides: Geological Society of America Abstracts with Programs, v. 27, no. 6, p. A-225.

Hurst, J.M., McKerrow, W.S., Soper, N.J., and Surlyk, F., 1983, The relationship between Caledonian nappe tectonics and Silurian turbidite deposition in North Greenland: Journal of the Geological Society of London, v. 140, p. 123–132, doi: 10.1144/gsjgs.140.1.0123.

Jepsen, H.F., and Kalsbeek, F., 1998, Granites in the Caledonian fold belt of East Greenland, in Higgins, A.K., and Frederiksen, K.S., eds., Caledonian Geology of East Greenland 72°–74°N: Preliminary Reports from the 1997 Expedition: Danmarks og Grønlands Geologiske Undersøgelse Rapport, v. 1998/28, p. 73–82.

Johansson, Å., Larionov, A.N., Tebenkov, A.M., Gee, D.G., Whitehouse, M.J., and Vestin, J., 2000, Grenvillian magmatism of western and central Nordaustlandet, northeastern Svalbard: Transactions of the Royal Society of Edinburgh, Earth Sciences, v. 90, p. 221–254.

Kalsbeek, F., 1989, Geology of the Ammassalik Region, South-East Greenland: Rapport Grønlands Geologiske Undersøgelse, v. 146, 106 p.

Kalsbeek, F., Nutman, A.P., and Taylor, P.N., 1993, Palaeoproterozoic basement province in the Caledonian fold belt of North-East Greenland: Precambrian Research, v. 63, p. 163–178, doi: 10.1016/0301-9268(93)90010-Y.

Kalsbeek, F., Nutman, A.P., Escher, J.C., Friderichsen, J.D., Hull, J.M., Jones, K.A., and Pedersen, S.A.S., 1999, Geochronology of granitic and supracrustal rocks from the northern part of the East Greenland Caledonides: Ion microprobe U-Pb zircon ages: Geological Survey of Denmark and Greenland Bulletin, v. 184, p. 31–48.

Kalsbeek, F., Thrane, K., Nutman, A.P., and Jepsen, H.F., 2000, Late Mesoproterozoic to early Neoproterozoic history of the East Greenland Caledonides: Evidence for Grenvillian orogenesis?: Journal of the Geological Society of London, v. 157, p. 1215–1225.

Kalsbeek, F., Jepsen, H.F., and Nutman, A.P., 2001a, From source migmatites to plutons: Tracking the origin of c. 435 Ma granites in the East Greenland Caledonian orogen: Lithos, v. 57, p. 1–21, doi: 10.1016/S0024-4937(00)00071-2.

Kalsbeek, F., Jepsen, H.F., and Jones, K.A., 2001b, Geochemistry and petrogenesis of S-type granites in the East Greenland Caledonides: Lithos, v. 57, p. 91–109, doi: 10.1016/S0024-4937(01)00038-X.

Kalsbeek, F., Higgins, A.K., Jepsen, H.F., Frei, R., and Nutman, A.P., 2008, this volume (Chapter 9), Granites and granites in the East Greenland Caledonides, in Higgins, A.K., Gilotti, J.A., and Smith, M.P., eds., The Greenland Caledonides: Evolution of the Northeast Margin of Laurentia: Geological Society of America Memoir 202, doi: 10.1130/2008.1202(09).

Kalsbeek, F., Thrane, K., Higgins, A.K., Jepsen, H.F., Leslie, A.G., Nutman, A.P., and Frei, R., 2008, this volume (Chapter 3), Polyorogenic history of the East Greenland Caledonides, in Higgins, A.K., Gilotti, J.A., and Smith, M.P., eds., The Greenland Caledonides: Evolution of the Northeast Margin of Laurentia: Geological Society of America Memoir 202, doi: 10.1130/2008.1202(03).

Larsen, P.-H., and Bengaard, H.-J., 1991, Devonian basin initiation in East Greenland: A result of sinistral wrench faulting and Caledonian extensional collapse: Journal of the Geological Society of London, v. 148, p. 355–368, doi: 10.1144/gsjgs.148.2.0355.

Larsen, P.-H., and Olesen, H., 1991, The Devonian basin project, North-East Greenland—A summary: Rapport Grønlands Geologiske Undersøgelse, v. 152, p. 108–113.

Larsen, P.-H., Olsen, H., and Clack, J.A., 2008, this volume, The Devonian basin in East Greenland—Review of basin evolution and vertebrate assemblages, in Higgins, A.K., Gilotti, J.A., and Smith, M.P., eds., The Greenland Caledonides: Evolution of the Northeast Margin of Laurentia: Geological Society of America Memoir 202, doi: 10.1130/2008.1202(11).

Leslie, A.G., and Higgins, A.K., 2008, this volume, Foreland-propagating Caledonian thrust systems in East Greenland, in Higgins, A.K., Gilotti, J.A., and Smith, M.P., eds., The Greenland Caledonides: Evolution of the Northeast Margin of Laurentia: Geological Society of America Memoir 202, doi: 10.1130/2008.1202(07).

Leslie, A.G., and Nutman, A.P., 2003, Evidence for Neoproterozoic orogenesis and early high temperature Scandian deformation events in the southern East Greenland Caledonides: Geological Magazine, v. 140, p. 309–333, doi: 10.1017/S0016756803007593.

Leslie, A.G., Smith, M., and Soper, N.J., 2008, this volume, Laurentian margin evolution and the Caledonian orogeny—A template for Scotland and East Greenland, in Higgins, A.K., Gilotti, J.A., and Smith, M.P., eds., The Greenland Caledonides: Evolution of the Northeast Margin of Laurentia: Geological Society of America Memoir 202, doi: 10.1130/2008.1202(13).

McClelland, W.C., and Gilotti, J.A., 2003, Late-stage extensional exhumation of high-pressure granulites in the Greenland Caledonides: Geology, v. 31, p. 259–262, doi: 10.1130/0091-7613(2003)031<0259:LSEEOH>2.0.CO;2.

McClelland, W.C., Power, S.-E., Gilotti, J.A., Mazdab, F., and Wopenka, B., 2006, U-Pb SHRIMP geochronology and trace element geochemistry of coesite-bearing zircons, North-East Greenland Caledonides, in Hacker, B., McClelland, W.C., and Liou, J.G., eds., Ultrahigh-Pressure Metamorphism: Deep Continental Subduction: Geological Society of America Special Paper 404, p. 23–43.

Möller, C., Andersson, J., Söderlund, U., and Johansson, L., 1997, A Sveconorwegian deformation zone within the eastern segment, Sveconorwegian orogen of SW Sweden—A first report: GFF, v. 119, p. 73–78.

Moncrieff, A.C.M., 1989, The Tillite Group and related rocks of East Greenland: Implications for late Proterozoic palaeogeography, in Gayer, R.A., ed., The Caledonide Geology of Scandinavia: London, Graham & Trotman, p. 285–297.

Nathorst, A.G., 1901, Bidrag til nordöstra Grönlands geologi: Geologisk Föreningens i Stockholm Förhandlingar, v. 23, p. 275–306.

Pedersen, S.A.S., Craig, L.E., Upton, B.G.J., Rämö, O.T., Jepsen, H.F., and Kalsbeek, F., 2002, Palaeoproterozoic (1740 Ma) rift-related volcanism in the Hekla Sund region, eastern North Greenland: Field occurrence, geochemistry and tectonic setting: Precambrian Research, v. 114, p. 327–346, doi: 10.1016/S0301-9268(01)00234-0.

Phillips, W.E.A., Stillman, C.J., Friderichsen, J.D., and Jemelin, L., 1973, Preliminary results of mapping in the western gneiss and schist zone around Vestfjord and inner Gåsefjord, south-west Scoresby Sund: Rapport Grønlands Geologiske Undersøgelse, v. 58, p. 17–32.

Piasecki, S., Stemmerik, L., Friderichsen, J.D., and Higgins, A.K., 1994, Stratigraphy of the post-Caledonian sediments in the Germania Land area,

North-East Greenland: Rapport Grønlands Geologiske Undersøgelse, v. 162, p. 177–184.

Rasmussen, J.A., and Smith, M.P., 2002, Conodont geothermometry and tectonic overburden in the northernmost East Greenland Caledonides: Geological Magazine, v. 138, p. 687–698.

Roberts, D., 2003, The Scandinavian Caledonides: Event chronology, palaeogeographic settings and likely modern analogues: Tectonophysics, v. 365, p. 283–299, doi: 10.1016/S0040-1951(03)00026-X.

Sartini-Rideout, C., Gilotti, J.A., and McClelland, W.C., 2006, Geology and timing of dextral strike-slip shear zones in Danmarkshavn, North-East Greenland Caledonides: Geological Magazine, v. 143, p. 431–446, doi: 10.1017/S0016756806001968.

Smith, M.P., and Rasmussen, J.A., 2008, this volume, Cambrian–Silurian development of the Laurentian margin of the Iapetus Ocean in Greenland and related areas, in Higgins, A.K., Gilotti, J.A., and Smith, M.P., eds., The Greenland Caledonides: Evolution of the Northeast Margin of Laurentia: Geological Society of America Memoir 202, doi: 10.1130/2008.1202(06).

Smith, M.P., Rasmussen, J.A., Robertson, S., Higgins, A.K., and Leslie, A.G., 2004a, Lower Palaeozoic stratigraphy of the East Greenland Caledonides: Geological Survey of Denmark and Greenland Bulletin, v. 6, p. 5–28.

Smith, M.P., Higgins, A.K., Soper, N.J., and Sønderholm, M., 2004b, The Neoproterozoic Rivieradal Group of Kronprins Christian Land, eastern North Greenland: Geological Survey of Denmark and Greenland Bulletin, v. 6, p. 29–39.

Söderlund, U., Jarl, L.G., Persson, P.O., Stephens, M.B., and Wahlgren, C.H., 1999, Protolith ages and timing of deformation in the eastern, marginal part of the Sveconorwegian orogen, south-western Sweden: Precambrian Research, v. 94, p. 29–48, doi: 10.1016/S0301-9268(98)00104-1.

Sønderholm, M., Frederiksen, K.S., Smith, M.P., and Tirsgaard, H., 2008, this volume, Neoproterozoic sedimentary basins with glacigenic deposits of the East Greenland Caledonides, in Higgins, A.K., Gilotti, J.A., and Smith, M.P., eds., The Greenland Caledonides: Evolution of the Northeast Margin of Laurentia: Geological Society of America Memoir 202, doi: 10.1130/2008.1202(05).

Soper, N.J., and Higgins, A.K., 1993, Basement–cover relationships in the East Greenland Caledonides: Evidence from the Eleonore Bay Supergroup at Ardencaple Fjord: Transactions of the Royal Society of Edinburgh, Earth Sciences, v. 84, p. 103–115.

Soper, N.J., Strachan, R.A., Holdsworth, R.E., Gayer, R.A., and Greiling, R.O., 1992, Sinistral transpression and the Silurian closure of Iapetus: Journal of the Geological Society of London, v. 149, p. 871–880, doi: 10.1144/gsjgs.149.6.0871.

Stemmerik, L., and Håkansson, E., 1991, Carboniferous and Permian history of the Wandel Sea Basin, North Greenland, in Peel, J.S., and Sønderholm, M., eds., Sedimentary Basins of North Greenland: Bulletin Grønlands Geologiske Undersøgelse, v. 160, p. 141–151.

Strachan, R.A., and Tribe, I.R., 1994, Structure of the Storstrømmen shear zone, eastern Hertugen af Orléans Land, North-East Greenland: Rapport Grønlands Geologiske Undersøgelse, v. 162, p. 103–112.

Strachan, R.A., Holdsworth, R.E., Friderichsen, J.D., and Jepsen, H.F., 1992, Regional Caledonian structure within an oblique convergence zone, Dronning Louise Land, NE Greenland: Journal of the Geological Society of London, v. 149, p. 359–371, doi: 10.1144/gsjgs.149.3.0359.

Strachan, R.A., Friderichsen, J.D., Holdsworth, R.E., and Jepsen, H.F., 1994, Regional geology and Caledonian structure, Dronning Louise Land, North-East Greenland, in Higgins, A.K., ed., Geology of North-East Greenland: Rapport Grønlands Geologiske Undersøgelse, v. 162, p. 71–76.

Strachan, R.A., Martin, M.W., and Friderichsen, J.D., 2001, Evidence for contemporaneous yet contrasting styles of granite magmatism during extensional collapse of the northeast Greenland Caledonides: Tectonics, v. 20, p. 458–473, doi: 10.1029/2000TC001206.

Swett, K., and Smit, D.E., 1972, Cambro-Ordovician shelf sedimentation of western Newfoundland, northwest Scotland and central East Greenland, in Aitken, J.D., Basset, H.G., and de Witt, R., eds., Proceedings of the 24th International Geological Congress, Canada, 1972, v. 6: International Geological Congress, Montreal, Canada, p. 33–41.

Trettin, H.P., ed., 1991, Geology of the Innuitian Orogen and Arctic Platform of Canada and Greenland: Geology of Canada, Volume 3: Calgary, Geological Survey of Canada, 569 p.

Watt, G.R., and Thrane, K., 2001, Early Neoproterozoic events in East Greenland: Precambrian Research, v. 110, p. 165–184, doi: 10.1016/S0301-9268(01)00186-3.

Watt, G.R., Kinny, P.D., and Friderichsen, J.D., 2000, U-Pb geochronology of Neoproterozoic and Caledonian tectonothermal events in the East Greenland Caledonides: Journal of the Geological Society of London, v. 157, p. 1031–1048.

Weil, A.B., Van der Voo, R., Mac Niocaill, R., and Meert, J.G., 1998, The Proterozoic supercontinent Rodinia: Paleomagnetically derived reconstructions for 1100–800 Ma: Earth and Planetary Science Letters, v. 154, p. 13–24, doi: 10.1016/S0012-821X(97)00127-1.

Manuscript Accepted by the Society 14 January 2008

Polyorogenic history of the East Greenland Caledonides

**Feiko Kalsbeek*
Kristine Thrane[†]
A.K. Higgins
Hans F. Jepsen**
Geological Survey of Denmark and Greenland, Øster Voldgade 10, DK-1350 Copenhagen K, Denmark

A. Graham Leslie
British Geological Survey, Murchison House, Edinburgh, EH9 3LA, UK

Allen P. Nutman[§]
Research School of Earth Sciences, Australian National University, Canberra, ACT 0200, Australia

Robert Frei
Geological Institute, University of Copenhagen, Øster Voldgade 10, DK-1350 Copenhagen, Denmark

ABSTRACT

The Caledonian orogen of East Greenland contains remnants of Archean, Paleoproterozoic, late Mesoproterozoic, and early Neoproterozoic rocks that occur within far-traveled thrust sheets, and bear witness to a complex polyorogenic history of the region prior to Caledonian orogenesis. Archean and Paleoproterozoic complexes consist mainly of granitoid orthogneisses. A succession of Paleoproterozoic tholeiitic metabasalts is present in some of the foreland windows. A major unit of late Mesoproterozoic metasedimentary rocks (Krummedal supracrustal sequence) contains early Neoproterozoic (ca. 950 Ma) as well as Caledonian granites. There is evidence for Archean (ca. 2800–2600 Ma), Paleoproterozoic (2000–1750 Ma), and late Grenvillian (ca. 950 Ma) deformation and metamorphism, but Caledonian overprinting complicates the study of these events. This paper presents a broad overview of the various rock units with structural, geochemical, and geochronologic data. The Paleoproterozoic metabasaltic rocks from the foreland windows are described in more detail.

Keywords: Caledonian orogen, East Greenland, Paleoproterozoic basement, Mesoproterozoic supracrustal rocks, Grenvillian granites.

---

*fkalsbeek@gmail.com
[†]Current address: Geological Institute, University of Copenhagen, Øster Voldgade 10, DK-1350 Copenhagen, Denmark.
[§]Current address: Beijing SHRIMP Center, Institute of Geology, Chinese Academy of Geological Sciences, 26 Baiwangzhuang Road, Beijing 100037, China.

Kalsbeek, F., Thrane, K., Higgins, A.K., Jepsen, H.F., Leslie, A.G., Nutman, A.P., and Frei, R., 2008, Polyorogenic history of the East Greenland Caledonides, in Higgins, A.K., Gilotti, J.A., and Smith, M.P., eds., The Greenland Caledonides: Evolution of the Northeast Margin of Laurentia: Geological Society of America Memoir 202, p. 55–72, doi: 10.1130/2008.1202(03). For permission to copy, contact editing@geosociety.org. ©2008 The Geological Society of America. All rights reserved.

INTRODUCTION

Study of the Precambrian basement of the East Greenland Caledonides (Fig. 1) has revealed evidence for Archean, Paleoproterozoic, and Grenvillian orogenic events prior to the formation of the present Caledonian orogen (e.g., Steiger et al., 1979). Archean rocks, mainly orthogneisses, are best preserved in the southernmost part of the orogen, south of ~72°50′N; further north, most of the crystalline basement has yielded Paleoproterozoic ages. South of ~76°N, the basement is overlain by late Mesoproterozoic high-grade metasedimentary rocks, collectively termed the "Krummedal supracrustal sequence," which retain evidence of large-scale late Grenvillian (ca. 950 Ma) folding, metamorphism, and granite formation. All these rocks were reworked during the Caledonian orogeny, giving rise to very complex structural relationships, which are as yet only partly understood.

It is important to note that most of the rocks do not occur in situ but form parts of far-traveled thrust sheets, which further complicate their interpretation. Granitoid rocks similar to those in the thrust sheets also underlie large parts of the foreland windows. However, the geology of the windows differs from that of the thrust units in several respects: (1) Archean rocks have not (yet) been recorded in the windows; (2) metasedimentary rocks equivalent to the Krummedal sequence are restricted to the thrust sheets; and (3) some of the foreland windows contain major units of Paleoproterozoic basaltic pillow lavas and associated sedimentary rocks that have not been found in the thrust sheets.

ARCHEAN ORTHOGNEISSES

Strongly deformed quartzofeldspathic orthogneisses of Archean age form a large proportion of the thrust sheets that make up the southern part of the East Greenland Caledonides. Amphibolitic dikes (metadolerites) are locally common in the gneisses (Fig. 2). They are discordant to the foliation and early fold structures in the gneisses, but they are themselves deformed and metamorphosed. The presence of these dikes suggests correlation with Archean terrains in West and South-East Greenland, where similar gneisses with mafic dikes are common. Rex and Gledhill (1974) obtained a poorly fitted Rb-Sr whole-rock isochron on gneisses from the Scoresby Sund region (70°N–71°N) that yielded an age of ca. 3000 Ma, which supports this correlation. Steiger et al. (1979) obtained multigrain zircon U-Pb ages of ca. 2300 and ca. 2500 Ma on late granite sheets within basement gneisses, and a K-Ar age of ca. 2500 Ma on hornblende from one of the discordant amphibolite dikes, consistent with an Archean age of the gneisses they cut. Later investigations have shown that, because of later disturbance, the isotope systems in the rocks are complex, and the ages obtained by these early studies must be viewed with reservation. Nevertheless, the Archean age of the basement gneisses had been established beyond any reasonable doubt.

An extensive Rb-Sr whole-rock isotope study of basement rocks in the Kong Oscar Fjord region (71°N–72°N) by Rex and Gledhill (1981) showed that Archean gneisses continue northward until just south of 73°N, and that lithologically similar gneisses further north yield Paleoproterozoic ages. Many sample suites, however, displayed so much scatter in isochron diagrams that it was not possible to obtain reliable age information from the data.

The nature of the boundary between the Archean and Paleoproterozoic basement complexes around 73°N was investigated by Thrane (2002). It proved impossible to differentiate between Archean and Paleoproterozoic rocks in the field, and age distinctions could only be made using radiometric age determinations. U-Pb analyses carried out by ion microprobe on single zircons from Archean rocks yielded imprecise ages on the order of 2650–2800 Ma, but the data scattered widely in concordia diagrams, suggesting disturbance of the isotope systems during at least two later tectonothermal events. Sm-Nd isotope data suggest involvement of older crustal components in the petrogenesis of the gneisses; this is supported by the presence of ca. 3000 Ma zircons in some of the samples.

North of 72°50′N, nearly all samples investigated have yielded Paleoproterozoic ages (see later section), but at two localities within the Paleoproterozoic terrain, Archean rocks have been found. The first of these was at Danmarkshavn (~77°N, Fig. 1), where Steiger et al. (1976) obtained multigrain U-Pb zircon and Rb-Sr whole-rock isochron ages of ca. 3000 Ma. Later reinvestigation of these Archean gneisses by SHRIMP (sensitive high-resolution ion microprobe) zircon U-Pb analyses (Nutman and Kalsbeek, 1994) revealed that, apart from Archean zircons, a large number of ca. 2000 Ma zircons was also present. The presence of two distinct age groups can be interpreted in two ways: either the rock represents an Archean protolith heavily overprinted around 2000 Ma, or it is of Paleoproterozoic origin with numerous inherited Archean zircons. The Rb-Sr data (Steiger et al., 1976) support the first of these interpretations.

Archean rocks have also been described from Payer Land (Fig. 1, 74°30′N), where there are also large numbers of Paleoproterozoic zircons present (Elvevold et al., 2003). Because of the lithological similarity between Archean and Paleoproterozoic orthogneisses, it has not been possible to map out the extent of these Archean outcrops.

PALEOPROTEROZOIC SUPRACRUSTAL ROCKS IN THE ELEONORE SØ WINDOW: FIELD OBSERVATIONS, AGE, AND CHEMICAL COMPOSITION

A several-kilometer-thick succession of nearly undeformed sedimentary and volcanic rocks crops out near the Inland Ice in the Eleonore Sø and Charcot Land windows (Fig. 1; Leslie and Higgins, this volume); these rocks have not been described in any detail previously, and new information is provided here. The volcanosedimentary succession of the Eleonore Sø window was first noted by Katz (1952) and later investigated in more detail during expeditions by the Geological Survey of Denmark and Greenland (GEUS) in 1997 and 1998 (Leslie and Higgins, 1998, 1999). The

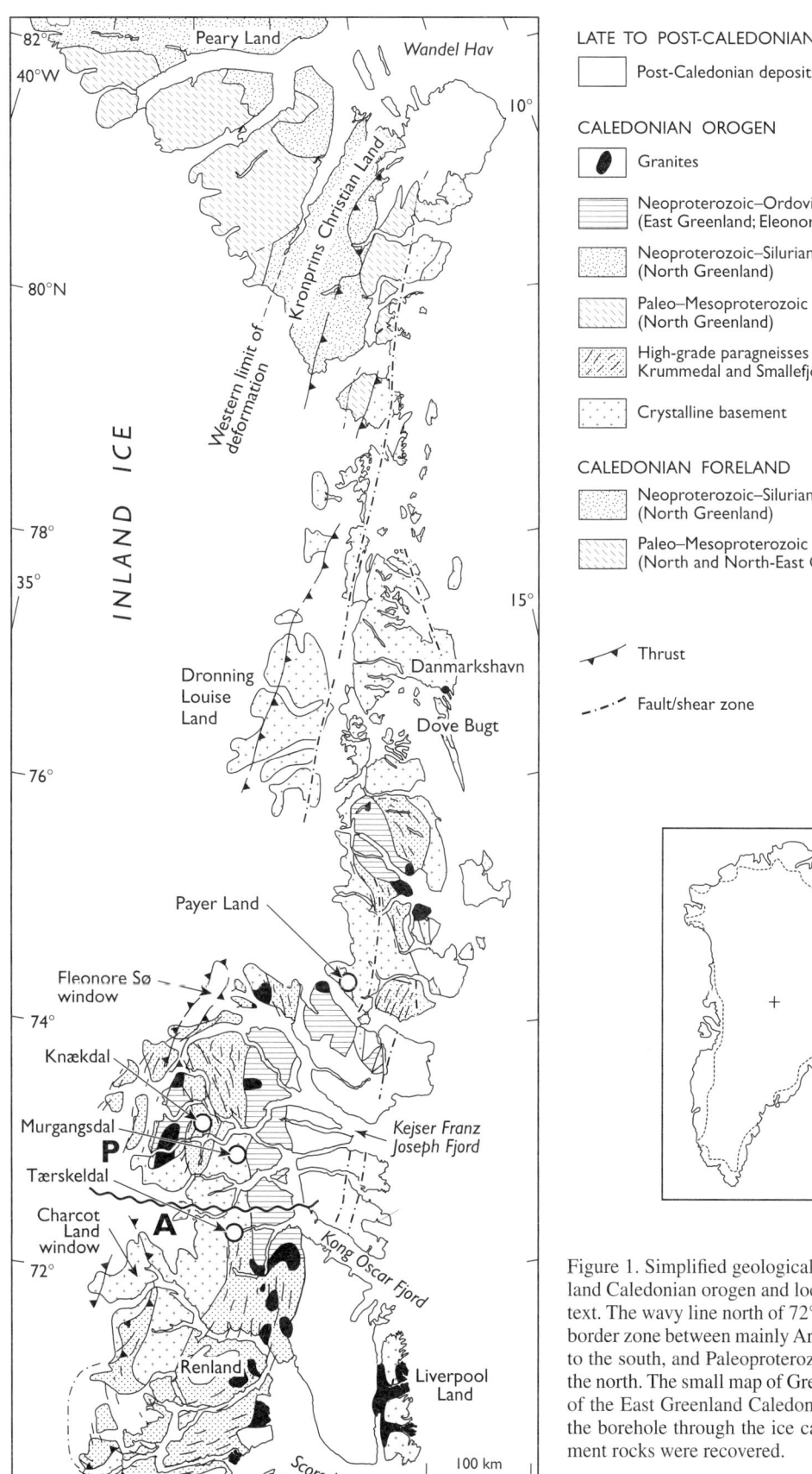

Figure 1. Simplified geological map of the East Greenland Caledonian orogen and locations mentioned in the text. The wavy line north of 72°N represents the diffuse border zone between mainly Archean orthogneisses (A) to the south, and Paleoproterozoic orthogneisses (P) to the north. The small map of Greenland shows the extent of the East Greenland Caledonides and the location of the borehole through the ice cap where Archean basement rocks were recovered.

Figure 2. Archean orthogneiss and folded discordant metadolerite dike in Tærskeldal (Fig. 1). Hammer shaft is ~55 cm long. Photograph is by J.D. Friderichsen, reproduced from Higgins et al. (1981).

rocks are unconformably overlain by the early Cambrian Slottet Formation (Leslie and Higgins, this volume), and a Precambrian age is therefore certain.

Contacts between the succession and the underlying basement were not found. The lower part of the succession consists of arkosic psammites and semipelites overlain by a major unit of carbonates, which, in turn, are overlain by basaltic pillow lavas, now largely transformed into greenschists. On the basis of field observations, it is envisaged that these rocks accumulated in an epicontinental rift setting. The occurrence of synsedimentary olistostromes in the carbonate sequence suggests tectonic instability in the region during deposition of the succession. The supracrustal units are cut by sheets of quartz porphyry.

Age

In order to constrain the age of the supracrustal rocks, SHRIMP zircon U-Pb data were acquired from a quartz porphyry (GGU 417880) that cuts sediments of the Eleonore Sø sequence. Zircons from this sample are euhedral to subhedral, up to a few hundred micrometers long, and they have aspect ratios mainly in the range from 2 to 4. Most crystals are deep brown in transmitted light. Cathodoluminescence (CL) images do not reveal distinct internal structures, but vague euhedral zoning is locally visible. No obvious cores are present. Fifteen spots were analyzed (Table 1; for analytical procedures, see, e.g., Williams, 1998), but the results are complex and do not provide an indisputable age. Four grains yielded Archean $^{207}Pb/^{206}Pb$ ages (Table 1; Fig. 3A); we interpret those as inherited zircons from Archean rocks that may be present at depth. Data for one nearly opaque grain plot near concordia just above 400 Ma. All other analyses yielded Paleoproterozoic $^{207}Pb/^{206}Pb$ ages between ca. 1900 and 2050 Ma (Fig. 3B). A best-fit discordia line for all non-Archean zircons gives an upper intersection age of 1962 ± 38 Ma and a lower intercept at 535 ± 170 Ma ($N = 11$; mean square of weighted deviates [MSWD] = 7.2; 2σ errors). This age could represent the time of emplacement of the quartz porphyry, but the poor fit of the data points on the discordia renders this interpretation dubious. A well-fitted discordia with upper and lower intercepts at 1915 ± 16 Ma and 452 ± 16 Ma ($N = 7$; MSWD = 0.73) is obtained if the four analyses giving $^{207}Pb/^{206}Pb$ ages between 1950 and

TABLE 1. SHRIMP U-Pb ZIRCON DATA FOR #417880, A QUARTZ PORPHYRY THAT CUTS THE ELEONORE SØ SUPRACRUSTAL SEQUENCE

Spot	U (ppm)	Th/U	f_{206} (%)	$^{206}Pb/^{238}U$	$^{207}Pb/^{206}Pb$	Age (Ma)	Discordance (%)
1.1	285	0.43	0.07	0.343 ± 9	0.1164 ± 6	1902 ± 6	0
2.1	920	0.14	2.48	0.082 ± 4	0.0645 ± 17	757 ± 55	−32
3.1	120	0.46	0.02	0.507 ± 15	0.1837 ± 10	2686 ± 9	−2
4.1	272	0.33	0.22	0.312 ± 11	0.1192 ± 9	1944 ± 13	−10
4.2	227	0.20	0.16	0.335 ± 9	0.1175 ± 8	1918 ± 13	−3
5.1	236	0.57	0.08	0.365 ± 11	0.1223 ± 7	1990 ± 10	1
6.1	230	0.42	0.16	0.335 ± 10	0.1203 ± 7	1960 ± 10	−1
7.1	253	0.39	4.16	0.311 ± 8	0.1179 ± 21	1925 ± 31	−9
8.1	153	1.51	0.90	0.414 ± 12	0.1809 ± 15	2661 ± 14	−16
9.1	275	0.61	0.56	0.246 ± 7	0.1109 ± 10	1814 ± 17	−22
10.1	100	0.51	0.17	0.345 ± 10	0.1267 ± 19	2052 ± 27	−7
11.1	75	0.56	0.28	0.511 ± 16	0.1712 ± 25	2569 ± 25	4
12.1	250	0.43	0.74	0.450 ± 12	0.1861 ± 19	2708 ± 17	−11
13.1	261	0.46	0.27	0.350 ± 10	0.1171 ± 8	1912 ± 13	1
14.1	247	0.65	0.09	0.336 ± 11	0.1256 ± 8	2038 ± 11	3

Note: Sample no. is from the files of the Geological Survey of Denmark and Greenland. Samples were analyzed at the Research School of Earth Sciences, Australian National University, Canberra. f_{206} is the proportion of ^{206}Pb that is not radiogenic. Age is the apparent $^{207}Pb/^{206}Pb$ age. Discordance is the difference between the $^{207}Pb/^{206}Pb$ and $^{206}Pb/^{238}U$ ages in percent of the $^{207}Pb/^{206}Pb$ age. All errors are quoted at the 1σ level.

2050 Ma are not included in the age calculation (Fig. 3B). The 1950–2050 Ma zircons may represent inheritance, similar to the Archean zircons, but there is no independent evidence that this is indeed the case. So, even though the 1915 ± 16 Ma date does not necessarily represent the age of the quartz porphyry, it is a minimum age for its emplacement, and it shows that the metavolcanic and sedimentary rocks must be Paleoproterozoic or older in age. The scatter of the data points along a discordia toward ca. 400 Ma is evidence for partial loss of radiogenic Pb from the zircons during Caledonian thermal events.

Geochemistry

Thirty-seven samples from the Eleonore Sø supracrustal sequence were analyzed for major elements (24 metavolcanics; 13 metasediments), as well as four samples from the metavolcanic rocks from the Charcot Land window. For twelve samples from the Eleonore Sø sequence, trace-element data were also acquired (9 metabasaltic rocks; 1 more acidic lava, and 2 metasediments). Representative analyses are listed in Table 2. In the Jensen (1976) diagram (Fig. 4A), most samples fall in the field of tholeiitic basalts; a few plot as komatiitic basalts or as komatiites. Tholeiitic samples fall in two groups: high-Fe and high-Mg tholeiites. The high-Fe samples mostly represent distinct pillow lavas, whereas the high-Mg samples come from more massive lithologies, massive flows or sills. Komatiitic samples contain up to 28% MgO; one of these samples represents a large (~1 m) inclusion in metabasalt. The metavolcanic rocks vary widely in TiO_2 contents (Fig. 4B). Most high-Mg samples have 0%–1% TiO_2, while high-Fe samples have 1.5%–3.4% TiO_2 (on a volatile-free basis). The four samples from the Charcot Land window belong to the high-Fe, high-Ti group.

Trace-element concentrations show relatively flat patterns in chondrite-normalized spider diagrams (Fig. 4C). Concentrations of Rb, Ba, and K are not shown because they scatter widely as a result of metasomatic alteration during low-grade metamorphism. This is illustrated by the erratic behavior of another mobile element, Sr, which yields variable positive and negative spikes in the otherwise regular patterns. There is a wide range in the concentrations of the most incompatible elements (Th, Nb, La), from ~5 to 50 times chondrite, and Ti-rich samples have higher Th, Nb, and La than Ti-poor rocks; for less incompatible elements (Y, Yb), the range in concentrations is more restricted, ~6–10 times chondrite. For most low-Ti samples, there is a slight increase in normalized concentrations from Th to Yb, whereas samples with higher TiO_2 have more-or-less horizontal spectra or spectra that slope down to the right. The spider diagrams do not show the negative anomalies for Nb, P, and Ti that are typical for arc-related magmatic rocks or rocks that contain components derived from sialic crust.

Rare earth element (REE) patterns (Fig. 4D) are similar to those of the other incompatible trace elements. Low-Ti samples are slightly to strongly light (L) REE depleted relative to heavy (H) REEs, whereas high-Ti samples are enriched in LREEs; the

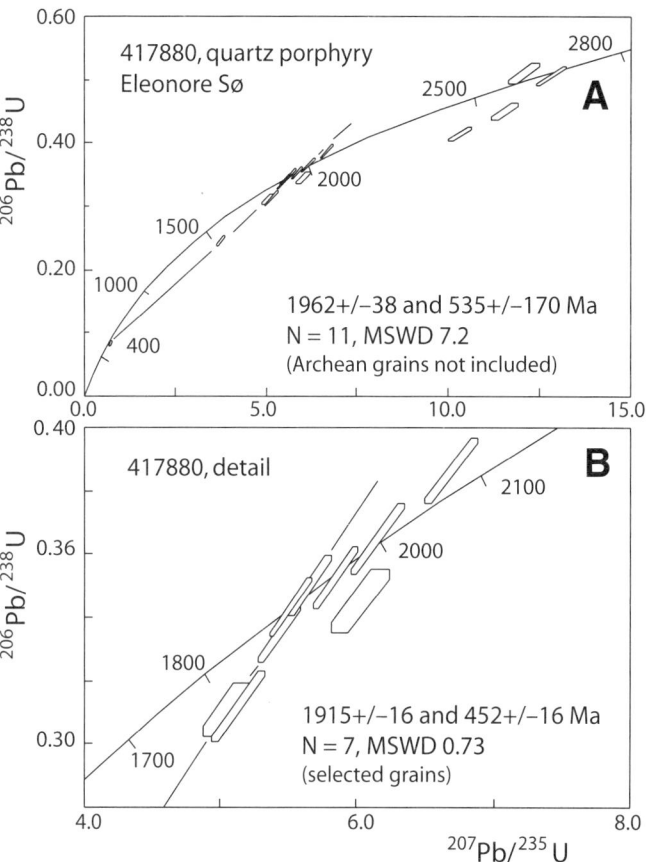

Figure 3. Concordia diagrams for zircons from GGU 417880, a quartz porphyry that cuts metasedimentary rocks in the Eleonore Sø window. Error boxes are 1σ; ages are given with 2σ errors. For discussion, see text. MSWD—mean square of weighted deviates.

most LREE-depleted sample is the Mg-rich inclusion mentioned previously. None of the samples exhibits an Eu anomaly.

Unfortunately, no detailed information is available on the tectonic setting of the Eleonore Sø lavas. Field observations suggest volcanism in a rift setting, while the primitive chemistry of the Ti-poor metabasalts indicates minimum involvement of continental crustal material in the petrogenesis of the lavas.

PALEOPROTEROZOIC ORTHOGNEISS TERRAINS

North of ~73°N and up to ~80°30′N, nearly all basement areas, both in the foreland and in the thrust sheets, consist of variously deformed juvenile Paleoproterozoic quartzofeldspathic orthogneisses with sheets of late metagranitoid rocks (Fig. 5) and metadolerite dikes (Hull et al., 1994). They represent fragments of the largest Paleoproterozoic basement province in Greenland, measuring more than 800 km along strike (Kalsbeek et al., 1993). The area north of 77°N is remarkable because of the common occurrence of Caledonian eclogites formed by high-pressure metamorphism of Paleoproterozoic protoliths (Gilotti, 1993; Brueckner et al., 1998;

TABLE 2. REPRESENTATIVE CHEMICAL ANALYSES OF METAVOLCANIC ROCKS FROM THE ELEONORE SØ WINDOW AND ASSOCIATED ROCKS

	Sample no.						
	420212	420213	420215	420218	420222	103391	420219
SiO_2 (%)	39.70	38.70	49.16	46.40	47.59	48.47	66.57
TiO_2	2.75	0.15	1.18	2.14	0.55	1.70	0.53
Al_2O_3	11.68	6.05	9.20	10.78	13.30	14.19	16.20
Fe_2O_3	1.63	2.54	1.69	1.16	1.35	1.37	1.68
FeO	11.02	7.22	8.33	9.86	7.60	10.36	2.55
MnO	0.21	0.16	0.15	0.16	0.16	0.18	0.04
MgO	8.21	28.14	12.64	10.95	10.79	7.28	1.89
CaO	10.41	3.52	11.35	8.69	11.93	9.12	0.33
Na_2O	0.90	0.03	2.27	2.89	1.45	2.21	2.14
K_2O	1.27	0.01	0.04	0.95	0.45	1.11	4.66
P_2O_5	0.25	0.02	0.08	0.19	0.06	0.16	0.13
LOI	11.64	11.79	3.43	5.23	4.39	3.52	2.73
Sum	99.66	98.33	99.53	99.40	99.61	99.68	99.46
Cs* (ppm)	10	1.2	n.d.	10	n.d.	0.5	9.9
Rb	45	1.0	<0.5	44	7.5	30	179
Ba	611	9	59	452	151	400	908
Pb	3	<1	<1	<1	<1	3	16
Sr	180	6.8	56	164	133	220	43
Y	27	7	15	22	16	22	32
Th*	1.4	n.d.	0.3	1.4	n.d.	1.9	13
Zr	177	6	60	134	31	88	213
Hf*	5.4	n.d.	1.8	3.7	1.0	2.7	6.1
Nb	15	<0.5	3.7	17	1.7	10	14
Ni	147	1430	360	438	268	108	17
Sc	43	20	30	38	42	38	13
V	480	108	284	343	211	389	55
Cr	304	3960	915	814	743	218	38
Ga	23	5	13	16	12	20	18
La* (ppm)	14.7	0.10	2.46	12.0	1.33	10.9	38.5
Ce*	38.7	0.20	7.5	29.8	3.80	24.1	79.4
Pr**	5.71	0.04	1.30	4.26	0.63	3.16	9.70
Nd*	26.1	0.33	6.88	19.1	3.49	13.5	36.7
Sm*	6.33	0.24	2.13	4.76	1.22	3.14	6.84
Eu*	2.22	0.12	0.85	1.68	0.56	1.20	1.27
Gd**	6.49	0.64	2.75	4.94	1.81	3.61	6.06
Tb**	1.02	0.16	0.49	0.82	0.39	0.65	0.96
Dy**	5.22	1.12	2.75	4.40	2.38	3.71	5.37
Ho**	0.95	0.25	0.52	0.81	0.53	0.75	1.08
Er**	2.45	0.76	1.43	2.12	1.58	2.13	3.24
Tm**	0.32	0.11	0.20	0.28	0.24	0.32	0.51
Yb*	1.99	0.72	1.25	1.78	1.62	2.05	3.29
Lu*	0.26	0.10	0.16	0.24	0.23	0.29	0.49

Note: Sample numbers are from the files of the Geological Survey of Denmark and Greenland. #103391 is from the Charcot Land window; #420219 is a metasedimentary rock from the Eleonore Sø window. Major elements were analyzed at the Geological Survey of Denmark and Greenland by X-ray fluorescence spectrometry (XRF) on glass disks; Na_2O was analyzed by atomic absorption spectrometry; and FeO was analyzed by titration; LOI (loss on ignition) represents the sum of volatiles (Kystol and Larsen, 1999). Unmarked trace elements were analyzed by XRF on powder tablets at the Geological Institute, University of Copenhagen.

*Analyzed by instrumental neutron activation analysis at Activation Laboratories Ltd., Ontario, Canada.

**Analyzed by inductively coupled plasma–mass spectrometry at Activation Laboratories Ltd., Ontario, Canada.

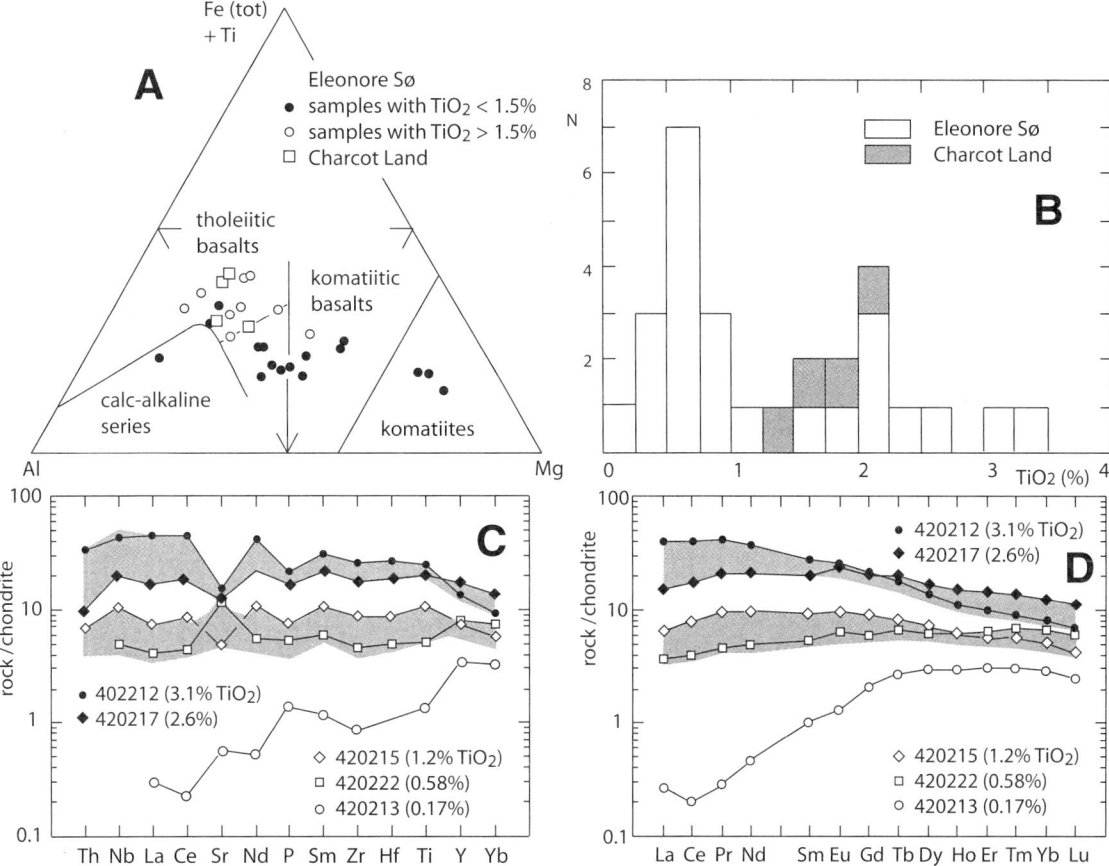

Figure 4. Diagrams illustrating the chemical compositions of Paleoproterozoic metavolcanic rocks from the Eleonore Sø window. (A) Jensen (1976) diagram of Al, Fe^{tot} + Ti, and Mg in cation percentages with the fields of various types of subalkalic volcanic rocks. (B) Histogram illustrating the distinction between low-Ti ($TiO_2 < 1.5\%$) and high-Ti ($TiO_2 > 1.5\%$) samples (TiO_2 relative to a volatile-free sum of 100%). (C) Chondrite-normalized trace-element spectra for selected samples of metabasaltic rocks of the Eleonore Sø sequence. Chondritic values are from Thompson (1982). (D) Chondrite-normalized rare earth element (REE) spectra for metavolcanic rocks of the Eleonore Sø sequence. Chondritic values are from Taylor and McLennan (1985). The shaded fields in C and D illustrate the range of trace-element and REE concentrations of all analyzed metabasaltic rocks subdivided into samples with >1.5% TiO_2 (N = 3) and <1.5 TiO_2 (N = 5). For discussion, see text.

Gilotti and Ravna, 2002; Gilotti et al., this volume). In the Caledonian foreland of Dronning Louise Land (76°N–77°20′N, Fig. 1), the gneisses and granitoid rocks are unconformably overlain by sandstones and conglomerates correlated with the Paleoproterozoic or early Mesoproterozoic Independence Fjord Group of North Greenland (Collinson et al., this volume).

Because of lithological similarity of the Archean and Paleoproterozoic orthogneisses, it has not been possible to map the boundary between the two complexes around 73°N in detail. However, Thrane (2002) found that the Archean basement is cut by Paleoproterozoic tonalitic to granodioritic intrusions that vary in size from thin dikes to kilometer-sized plutons. The western boundary of Paleoproterozoic rocks in the Caledonian orogen and the foreland, and their contact with possible Archean rocks to the west, is covered by the Inland Ice, and its nature therefore open to speculation.

Ages and Isotope Data

Detailed age determination programs on the Paleoproterozoic basement have only been carried out in the southern part of the orogen, around 73°N (Rex and Gledhill, 1981; Thrane, 2002). Further north (74°N–80°N), only spot checks have been carried out because of the large extent of the crystalline basement complexes and the reconnaissance nature of the mapping. Ages obtained here fall in three groups: (1) U-Pb zircon ages, mainly by ion microprobe, (2) Sm-Nd model ages, and (3) single-sample Rb-Sr model ages (Kalsbeek et al., 1993).

Paleoproterozoic U-Pb Zircon Ages

U-Pb zircon data were acquired for 30 samples of Paleoproterozoic rocks. One age was obtained by classic isotope-dilution–thermal ionization mass spectrometry (TIMS;

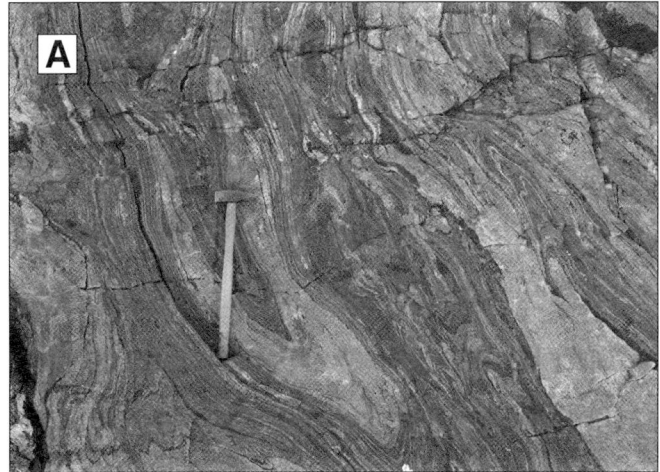

Figure 5. Paleoproterozoic granitoid basement rocks. (A) Strongly deformed orthogneiss near Dove Bugt (Fig. 1), from Henriksen (1994). Hammer shaft is ~55 cm long. (B) Little-deformed meta-augen granite with undeformed felsic granitoid dikes in Murgangsdal (Fig. 1). Hammer shaft is ~60 cm long. The augen gneiss has been dated at 1973 ± 15 Ma (F. Kalsbeek and A.P. Nutman, unpublished data).

1909 ± 6 Ma; Tucker et al., 1993), and zircons from all other samples were studied by SHRIMP. Only eight of the samples were studied in some detail; for the other 22 samples, only reconnaissance data were acquired (3–5 spot analyses per sample). Even though the resulting age determinations are much less precise, a larger number of samples could be investigated. This was regarded as useful in view of the reconnaissance nature of the study. Figure 6A gives a summary of the ages obtained. Most ages fall between 1900 and 2000 Ma, with a minority around 1750 Ma. There is good agreement between the more precise age determinations and the reconnaissance data.

In many cases, zircon U-Pb isotope data plot along well-fitted discordia plots with lower intercepts at ca. 400 Ma (e.g., Kalsbeek et al., 1993). This has been interpreted as the result of partial loss of radiogenic Pb during Caledonian thermal events. In the area south of 76°N, there is convincing evidence of earlier Grenvillian (ca. 950 Ma) metamorphism in the metasedimentary rocks of the Krummedal supracrustal sequence (Strachan et al., 1995; see following discussion). The good fit of the data on the discordias with lower intersections at ca. 400 Ma, however, indicates that this Grenvillian event did not significantly affect the zircon U-Pb systems in the granitoid rocks.

Rb-Sr and Sm-Nd Model Ages

In their Rb-Sr isotope study of the area between 72°N and 73°30′N, Rex and Gledhill (1981) acquired both Archean and Paleoproterozoic whole-rock isochron ages. However, the poor fit on most isochrons demonstrated that the Rb-Sr isotope systems had been disturbed, resulting in large errors in the age determinations. For a number of sample localities, the analytical data scattered so much in the isochron diagram that no useful ages were obtained. Rex and Gledhill's (1981) data were reinterpreted by Kalsbeek et al. (1993) by plotting mean $^{87}Rb/^{86}Sr$ and $^{87}Sr/^{86}Sr$ values for samples from individual localities in an isochron diagram. Most of these "locality means" fall along a reference isochron that has a slope corresponding to an age of ca. 2000 Ma and an initial $^{87}Sr/^{86}Sr$ ratio (Sr_i) of ~0.7025 (Fig. 6B). Two locality means plot far above the 2000 Ma isochron; these rocks are apparently of Archean origin. A locality mean for the samples from Danmarkshavn, where an Archean age had been obtained by Steiger et al. (1976), also plots far above the 2000 Ma reference isochron (Fig. 6C).

The relatively good fit of the locality means on a Paleoproterozoic isochron demonstrates that the disturbance of Rb-Sr isotope relations was on a local scale only (within each locality). An attempt was therefore made to obtain age information from large (~30 kg) single samples, in the hope that such large samples could replace locality means (Kalsbeek et al., 1993). In the isochron diagram, these samples also scatter about the 2000 Ma reference isochron, where $Sr_i = 0.7025$ (Fig. 6B). Model ages (T_{Sr}) for individual samples, calculated assuming a common initial $^{87}Sr/^{86}Sr$ ratio of 0.7025, scatter between 1800 and 2280 Ma. Clearly, none of these samples represents Archean rocks, and it is evident that, despite later disturbance, Rb-Sr data can be used with confidence to differentiate between Archean and Paleoproterozoic rocks. Rb-Sr isotope data were also obtained for 18 normal (1–2 kg) samples scattered over a large area between 75°N and 77°N (Kalsbeek, 1995). Several of these samples have very high $^{87}Rb/^{86}Sr$ ratios (Fig. 6C). In the isochron diagram (Fig. 6C), these high-$^{87}Rb/^{86}Sr$ samples scatter about a 1750 Ma reference isochron, and they apparently belong to the younger group of granitoid rocks identified by zircon U-Pb geochronology (Fig. 6A).

The Sm-Nd data from this latter collection of 1–2 kg samples yielded depleted mantle (DM) model ages (T_{DM}; DePaolo, 1981) between 2.1 and 3.4 Ga, which are much higher than the ages obtained by U-Pb zircon or single-sample Rb-Sr dating (Kalsbeek et al., 1993). The sample dated at 1909 ± 6 Ma by Tucker et al. (1993), for example, has a T_{DM} age of 2.49 Ga. The discrepancy between the Sr and Nd model ages is most marked

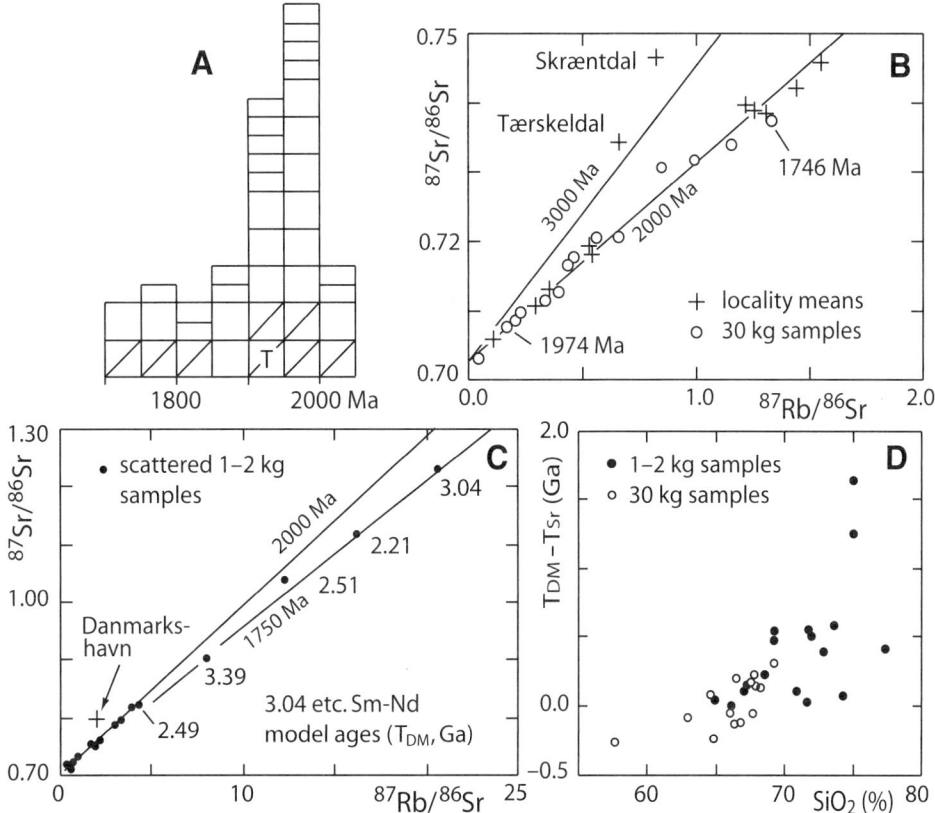

Figure 6. Diagrams illustrating age determinations and isotopic compositions of Paleoproterozoic gneisses and metagranitoid rocks from the crystalline basement. (A) Histogram of ages obtained by U-Pb zircon dating (data are from Tucker et al., 1993; Kalsbeek et al., 1993, 1999; Nutman and Kalsbeek, 1994). Each age determination is shown by a square; crossed squares represent proper age determinations, and open squares represent reconnaissance data. The square marked T represents the age of 1909 ± 6 Ma obtained by Tucker et al. (1993). Age estimates that straddle the 50 Ma borderlines are shown by two half-squares on either side of this line. (B) Rb-Sr isochron diagram for locality means and 30 kg samples illustrating the distinction between Archean and Paleoproterozoic rocks; see text. The 2000 Ma and 3000 Ma isochrons are shown for reference only. The ages of 1746 (±20) Ma and 1974 (±17) Ma were obtained from ion microprobe U-Pb zircon data (Kalsbeek et al., 1993). (C–D) Comparison between Rb-Sr data and Sm-Nd model ages. The 2000 Ma and 1750 Ma isochrons in C are shown for reference only. For further discussion, see text. Parts B and C are reproduced in modified form from similar diagrams in Kalsbeek et al. (1993) and Kalsbeek (1995) with permission from Elsevier.

for granitic (high-^{87}Rb/^{86}Sr) samples (Fig. 6C). These results have been interpreted as the result of incorporation of components derived from Archean crustal rocks into the magmas from which the protoliths of the Paleoproterozoic gneisses were formed (Kalsbeek et al., 1993; Kalsbeek, 1995).

In order to further test this interpretation, new Sm-Nd data were acquired for the collection of 30 kg samples (Table 3). In Figure 6D, the differences between the Sm-Nd (T_{DM}; DePaolo, 1981) and Rb-Sr (T_{Sr}) model ages are plotted against SiO_2, assuming that T_{Sr} is a fair approximation of the true age of the rocks in question. Even discarding the two samples where $T_{DM} - T_{Sr} > 1$ Ga, there is a distinct (and statistically highly significant) positive correlation between $T_{DM} - T_{Sr}$ and SiO_2. This is consistent with the view that many of the samples contain a component derived from older (probably Archean) crust. In view of the limited influence that the proposed Archean contributions had on the Sr-isotopic composition of the rocks, it is likely that these contaminants were derived from deep (granulite facies?) crustal sources with low Rb concentrations.

Geochemistry

Most of the Paleoproterozoic orthogneisses and metagranitoid rocks have tonalitic to granitic compositions and a surprisingly large proportion of true granitic rocks (Fig. 7A; Table 4). Apart from a number of samples with very high FeO*/MgO ratios, the rocks represent a normal calc-alkaline plutonic suite (Fig. 7B), and chondrite-normalized spider

TABLE 3. Sm-Nd ISOTOPE DATA FOR GRANITOID BASEMENT SAMPLES

Sample no.	Sm (ppm)	Nd (ppm)	$^{147}Sm/^{144}Nd$	$^{143}Nd/^{144}Nd$	T_{DM}	T_{Sr}	$T_{DM}-T_{SR}$ (Ga)	SiO_2 (wt%)
344820	9.8	69.0	0.0858	0.511372	1.98	1.83	0.15	67.84
344868	8.8	73.0	0.0728	0.511320	1.85	1.90	−0.05	67.67
344898	0.9	3.7	0.1444	0.512179	1.89	2.00	−0.11	66.74
344937	4.3	27.0	0.0965	0.511594	1.87	2.10	−0.23	64.80
344963	4.7	34.0	0.0842	0.511241	2.11	2.23	−0.12	66.30
344965	10.1	62.6	0.0981	0.511277	2.32	2.23	0.09	64.56
344966	8.8	52.9	0.1013	0.511607	1.94	2.20	−0.26	57.66
344967	1.4	6.6	0.1274	0.511889	2.03	2.03	0.00	62.90
363138	5.8	38.7	0.0899	0.511226	2.23	1.91	0.32	69.21
363184	16.6	109.2	0.0922	0.511436	2.01	1.80	0.21	66.39
364644	7.9	50.8	0.0937	0.511296	2.21	2.07	0.14	68.19
364646	4.7	32.2	0.0879	0.511197	2.23	2.28	−0.05	66.06
365368	3.5	18.7	0.1125	0.511645	2.09	1.92	0.17	67.56
365390	23.0	158.2	0.0878	0.511326	2.07	1.84	0.23	67.71

Note: Sample numbers are from the files of the Geological Survey of Denmark and Greenland. Samples were analyzed at the Geological Institute, University of Copenhagen, Denmark. Rock powders were spiked with a ^{149}Sm-^{150}Nd mixed spike, and the bulk rare earth elements (REEs) were separated over 15 mL glass-stem columns charged with AG 50W cation resin. REEs were further separated over HDEHP-coated biobeads (BioRad*) loaded in 6 mL glass-stem columns. Sm and Nd isotopic analyses were performed on a VG Sector 54 IT mass spectrometer, and both static and multidynamic routines were used for the collection of the isotopic ratios. The mean value for our internal JM Nd standard (referenced against La Jolla) during the period of measurement was 0.511115 for $^{143}Nd/^{144}Nd$, with a 2σ external reproducibility of ±0.000013 (five measurements). T_{DM}—Sm-Nd model ages according to the depleted mantle model of DePaolo (1981). T_{Sr}—Rb-Sr model ages calculated for individual samples assuming an initial $^{87}Sr/^{86}Sr$ ratio of 0.7025 (Kalsbeek et al., 1993; Kalsbeek, 1995).

diagrams for incompatible trace elements display the typical negative anomalies for Nb, P, and Ti of calc-alkaline rocks from volcanic arcs (Fig. 7C).

Among the true granitic rocks, two types can be distinguished: (1) granites with low FeO*/MgO ratios, mainly 1–5, and (2) granites with much higher FeO*/MgO ratios, up to ~50. Most granites of the first group are corundum-normative, while those of the second group are diopside-normative (Kalsbeek, 1995). The high-FeO*/MgO granites are rich in a number of trace elements, such as Ba, Sr, REE, Zr, Y, and Nb (Table 4; Figs. 7C and 7D), and most of them plot in the field of within-plate granites in the Rb versus Nb + Y diagram of Pearce et al. (1984). These granites have the composition of A-type (post- or anorogenic) granites, which occur as late intrusions in many orogenic belts. However, in North-East Greenland, there is no clear age distinction between the two types of granitic rocks. Some of the high FeO*/MgO metagranites belong to the younger (1750 Ma) age group, but the sample from which Tucker et al. (1993) obtained an age of 1909 ± 6 Ma also belongs to the high-FeO*/MgO group.

Deformation

Most Paleoproterozoic gneisses and metagranitoid rocks are strongly deformed (Fig. 5A). It is only rarely possible to assess how much of this deformation is due to Caledonian tectonism and how much is older. Chadwick and Friend (1994) investigated an unusually well-exposed area at ~76°30′N near Dove Bugt (Fig. 1), where a pink porphyritic granite, dated at 1739 ± 11 Ma by Kalsbeek et al. (1993), could be followed through complex nappe-scale folds, several kilometers in size (Fig. 8). These folds were interpreted by Chadwick and Friend (1994) to have been formed during Caledonian deformation. However, large-scale folds that predate the metagranite in question are also present (Fig. 8), and these must be Paleoproterozoic in age.

Near the Inland Ice in Dronning Louise Land (76°N–77°20′N, Fig. 1), in the foreland of the Caledonian orogen, Paleoproterozoic basement gneisses are unconformably overlain by sandstones and conglomerates of the Paleoproterozoic or early Mesoproterozoic "Trekant series" (correlated with the Independence Fjord Group of Eastern North Greenland; see Collinson et al., this volume) and by the early Cambrian "Zebra series." Gneissic fabrics and minor isoclinal folds in the gneisses predate deposition of the Trekant series and must therefore be of Paleoproterozoic age. Further east in Dronning Louise Land, east of the Caledonian boundary thrust (Fig. 1), both the Trekant series and the early Cambrian Zebra series have been affected by large-scale folding; it is clear that these folds must have been formed during Caledonian orogenesis (Strachan et al., 1992). There is no evidence for orogenic activity between deposition of the Trekant series and the Zebra series, and it appears, therefore, that the Grenvillian orogenic event recognized further south (see later discussion) was not active in Dronning Louise Land.

Paleoproterozoic Postorogenic Sedimentation and Volcanism

In Kronprins Christian Land (~80°N), sandstones and conglomerates are interbedded with basaltic and andesitic lavas and rhyolites that have been dated by SHRIMP at 1740 ± 6 Ma (Kals-

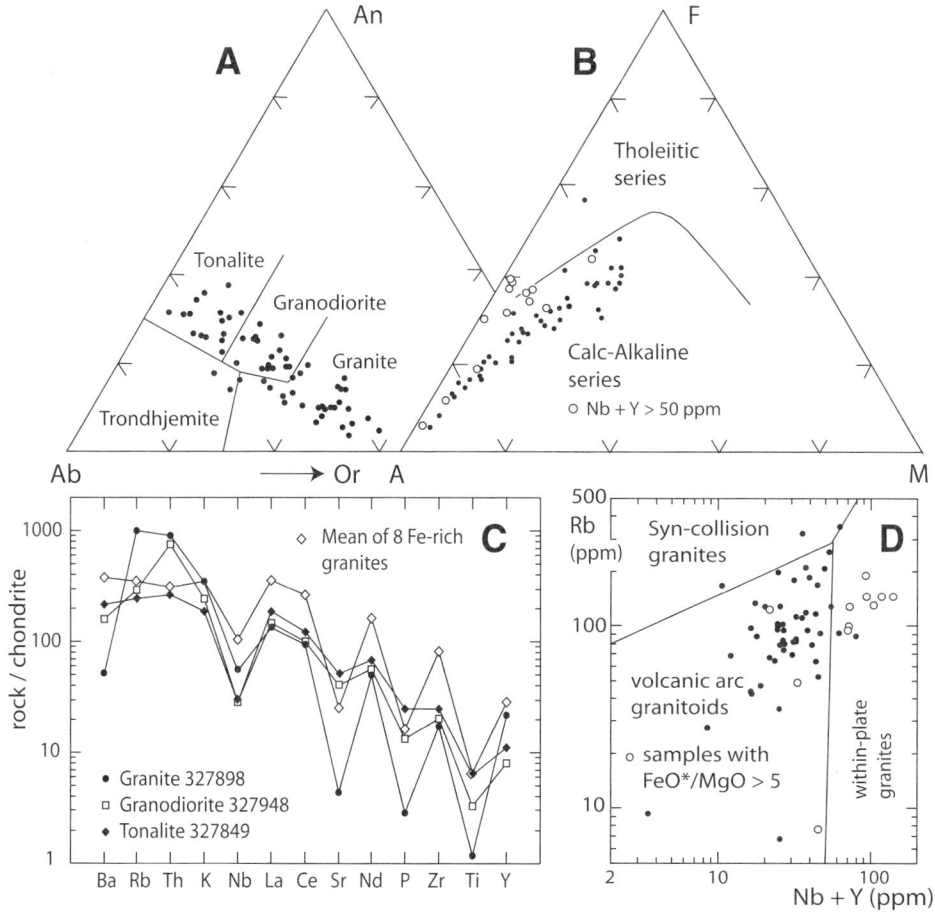

Figure 7. Diagrams illustrating the chemical compositions of gneisses and metagranitoid rocks from the crystalline basement. (A) Classification of the investigated samples in the cation-normative Ab-An-Or diagram (O'Connor, 1965; modified by Barker, 1979). (B) AFM diagram (A = $Na_2O + K_2O$; F = FeO^{tot}; M = MgO). Division between the fields of calc-alkaline and tholeiitic rock series is after Irvine and Baragar (1971). High-(Nb + Y) samples are shown separately. (C) Chondrite-normalized trace-element spectra; chondritic values are from Thompson (1982). Note high Ba, Nb, REE, Zr, and Y concentrations in high-FeO*/MgO samples. (D) Discrimination diagram for granitic rocks (Pearce et al., 1984) illustrating high Nb + Y for high-FeO*/MgO samples. For discussion, see text. Parts B and D are reproduced in modified form from similar diagrams in Kalsbeek (1995) with permission from Elsevier.

beek et al., 1999; Pedersen et al., 2002; see Collinson et al., this volume). This is more or less contemporaneous with the youngest Paleoproterozoic granitoid rocks in the region, including some of the aforementioned A-type granites. The sandstones in Kronprins Christian Land can thus be interpreted as molasse deposits that formed immediately after Paleoproterozoic orogenic activity had ceased, and they may reflect extensional orogenic collapse.

Geological Setting

The Paleoproterozoic basement complexes within the East Greenland Caledonides have not been investigated in any detail, and, as a consequence, interpretations as to their petrogenesis and geological setting must be of a provisional nature. It follows from their relatively low initial $^{87}Sr/^{86}Sr$ ratios (Figs. 6B and 6C) that the rocks cannot represent reworked Archean crust but were newly accreted at around 2000 Ma (Rex and Gledhill, 1981). Nevertheless, Sm-Nd isotope data suggest that significant proportions of Archean crustal material were involved in the petrogenesis of the rocks.

At present, most of the Paleoproterozoic crystalline rocks occur in major thrust sheets, and they may have originated several hundreds of kilometers east of their present location. Nevertheless, similarities in age and chemistry suggest that they once formed a coherent basement province together with similar rocks in the foreland. The western boundary of this province is hidden beneath the Inland Ice, and its nature is therefore uncertain.

Available evidence suggests that much of the central part of Greenland beneath the Inland Ice is underlain by Archean rocks that have been strongly reworked during Paleoproterozoic processes. Detrital zircons in sandstones from the Trekant series in Dronning Louise Land and ca. 1740 Ma sandstones in eastern North Greenland have yielded mainly Archean ages (Tucker et al., 1993; Kalsbeek et al., 1999), and a granite sample obtained from a deep drill core through the Inland Ice in central Greenland has yielded an Archean age (Weis et al., 1997; Henriksen et al., 2000). Since Archean crust was involved in the petrogenesis of the Paleoproterozoic granitoids, it appears likely that Archean complexes may not be far distant. This view is supported by the presence of inherited Archean zircons within the Paleoproterozoic granite GGU 417880 from the Eleonore Sø window (see previous section). Based on the chemical and isotopic evidence described here, we tentatively conclude that the orthogneisses were originally accreted as arcs along the eastern margin of this Archean continental terrain during Paleoproterozoic orogenic processes.

TABLE 4. REPRESENTATIVE CHEMICAL ANALYSES OF PALEOPROTEROZOIC GRANITOID ROCKS FROM THE CALEDONIAN BASEMENT, EAST GREENLAND

	Sample #					
	327849 tonalite	327948 granodiorite	327898 granite	344820 granite	Granites, mean 1	Granites, mean 2
SiO_2 (%)	60.01	66.15	75.06	67.84	72.85	68.41
TiO_2	0.65	0.34	0.12	0.65	0.20	0.68
Al_2O_3	16.46	15.31	13.72	13.43	13.82	13.14
FeO(total)	5.76	3.74	1.18	5.22	1.60	4.71
MnO	0.09	0.07	0.03	0.09	0.03	0.08
MgO	3.20	1.71	0.25	0.26	0.49	0.48
CaO	5.37	3.59	0.97	2.21	1.09	2.34
Na_2O	3.72	4.03	3.57	3.25	3.16	2.86
K_2O	2.70	3.50	4.98	5.16	5.55	5.09
P_2O_5	0.26	0.14	0.03	0.11	0.06	0.17
LOI	0.92	0.66	0.57	0.39	0.52	0.55
Sum	99.14	99.24	100.48	98.61	99.37	98.51
$FeO^{(tot)}$/MgO	1.80	2.19	4.72	19.69	3.91	19.31
Rb (ppm)	85	102	352	100	213	122
Ba	1500	1100	352	3230	733	2644
Pb	9	13	42	24	32	25
Sr	609	474	51	216	131	301
La	62	48	44	51	49	119
Ce	107	86	81	107	90	228
Nd	43	36	31	60	35	103
Y	22	16	43	34	23	57
Th	11	32	38	3	30	13
Zr	169	139	119	704	139	557
Nb	10	10	19	37	18	36
Ni	25	9	3	4	7	4
Sc	17	10	4	10	5	11
V	113	77	10	8	16	19
Cr	44	21	5	<3	13	4
Ga	18	16	20	21	18	20

Note: Sample numbers are from the files of the Geological Survey of Denmark and Greenland. #344820 is a high-FeO*/MgO granite. Granite means 1 and 2 represent the mean composition of 12 normal granitic rocks and eight high-FeO*/MgO granites, respectively. Major elements were analyzed at the Geological Survey of Denmark and Greenland by X-ray fluorescence spectrometry (XRF) on glass disks; Na_2O was analyzed by atomic absorption spectrometry; and FeO was analyzed by titration; LOI (loss on ignition) represents the sum of volatiles (Kystol and Larsen, 1999). Trace elements were analyzed by XRF on powder tablets at the Geological Institute, University of Copenhagen.

In several respects, the geology of the East Greenland Paleoproterozoic province resembles that of the somewhat younger Paleoproterozoic Ketilidian orogen of South Greenland (Garde et al., 2002). Widespread calc-alkaline magmatism (1850–1750 Ma) in South Greenland was preceded by deposition of sedimentary and basaltic rocks in a northern border zone. The supracrustal sequence of the Eleonore Sø window is broadly similar to that in the northern border zone of the Ketilidian orogen. Isotope data from Ketilidian granites intruded into Archean gneisses of the northern border zone give information on significant contamination with Archean crustal components (Kalsbeek and Taylor, 1985). Ketilidian granites in the border zone are also more granitic in composition than granitoid rocks from the central parts of the Ketilidian orogen, which are mainly granodiorites. Similarly, a large proportion of true granites in the East Greenland basement may be the result of incorporation of components from Archean crust. Finally, late high-FeO*/MgO "rapakivi" granites are present in the Ketilidian orogen; these are similar in age and chemistry to the ca. 1740 Ma A-type granites in East Greenland.

THE LATE MESOPROTEROZOIC KRUMMEDAL SUPRACRUSTAL SEQUENCE

High-grade late Mesoproterozoic metasedimentary rocks referred to as the Krummedal supracrustal sequence are widely distributed in the thrust sheets south of 76°N (Higgins, 1988). In the area between 75°N and 76°N, the term Smallefjord sequence has previously been used for comparable rocks, but because of similarities in composition and age (see following), we include them here in the Krummedal sequence. The lowest preserved levels of the sequence locally include carbonates, but the bulk of the succession consists of siliciclastic sediments locally more than 4 km thick; in parts of the Scoresby Sund region, apparent thicknesses of up to 8 km have been recorded.

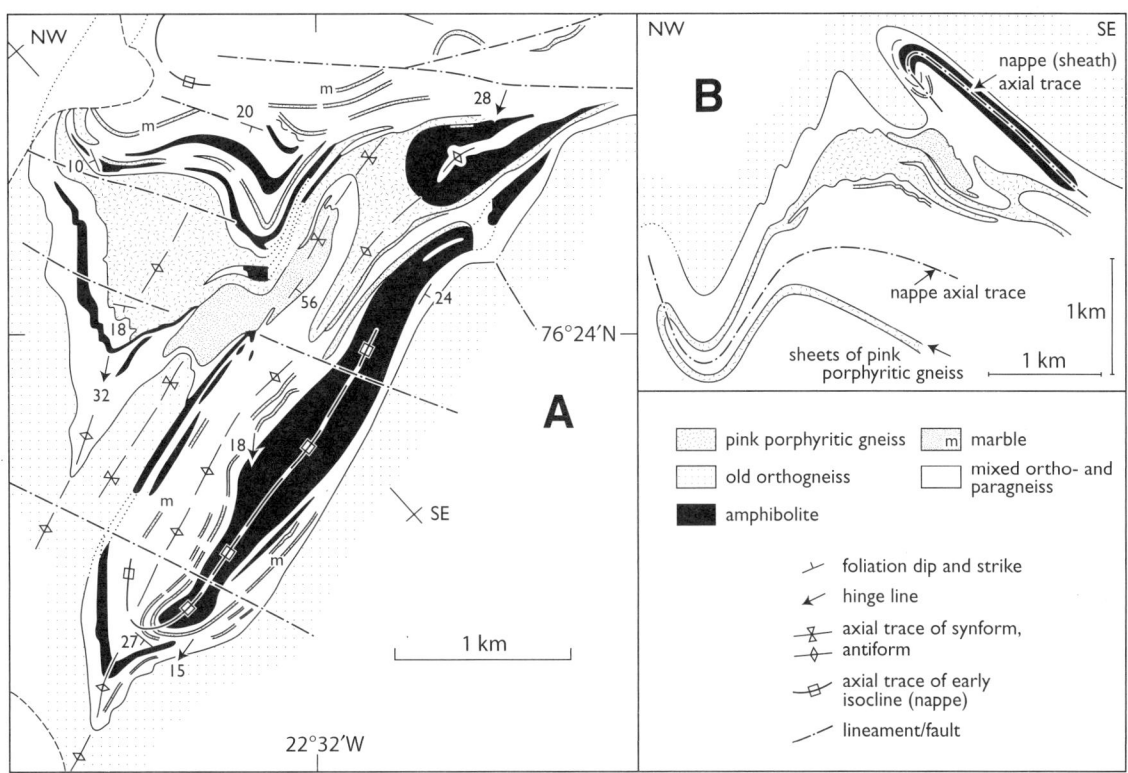

Figure 8. Geological map (A) and section (B) of an area southwest of Dove Bugt (modified from Chadwick and Friend, 1994). The pink porphyritic gneiss has been dated at 1739 ± 11 Ma by Kalsbeek et al. (1993), and the nappe-scale fold outlined by this gneiss is interpreted by Chadwick and Friend (1994) to be due to Caledonian deformation. The amphibolites were folded into large-scale folds prior to emplacement of the pink porphyritic gneiss as the result of Paleoproterozoic deformation.

Contacts with the underlying basement complexes are commonly tectonic, but at a few localities, a depositional unconformity has been recognized (Fig. 9; Higgins et al., 1981; Andresen et al., 1998; H.F. Jepsen and F. Kalsbeek, unpublished observations). Equivalents of the Krummedal sequence have not been recorded in the foreland windows.

Age and Provenance

The age of deposition of the Krummedal sequence is constrained by the age of the youngest detrital zircons in the rocks and the timing of the earliest metamorphism and granite emplacement. Age determinations on detrital zircons from the Krummedal (and Smallefjord) supracrustal sequences (SHRIMP U-Pb data) yielded a wide spectrum of ages from ca. 1800 to 1000 Ma, with a few Archean zircons also present (Fig. 10; Strachan et al., 1995; Kalsbeek et al., 2000; Watt et al., 2000; Leslie and Nutman, 2003). The youngest detrital zircons gave ages just over 1000 Ma (Kalsbeek et al., 2000), but because the rocks have subsequently experienced high-grade metamorphism, the zircons may have suffered a minor degree of Pb loss, and their true ages may be somewhat older. Metamorphism of the Krummedal sequence and emplacement of granites took place during the early Neoproterozoic around 950 Ma (Hansen et al., 1978; Steiger et al., 1979; Strachan et al., 1995; Kalsbeek et al., 2000; Watt et al., 2000), and deposition of the Krummedal sequence can therefore be bracketed between ca. 1100–1000 Ma and ca. 950 Ma, spanning the Mesoproterozoic-Neoproterozoic boundary.

There is a major difference in the age distribution of detrital zircons from the Krummedal sequence and that of zircons separated from basement gneisses (Fig. 10). It is evident, therefore, that the sedimentary material that formed the Krummedal sequence was not derived from rocks similar to those of the underlying crystalline basement. Ages of 1200–1800 Ma are common for rocks from the Grenvillian and Sveconorwegian terrains in North America and southern Scandinavia, and it has been suggested that the Krummedal sequence was formed by erosion of such a terrain, with subsequent long-distance transport (Kalsbeek et al., 2000; Watt et al., 2000; Watt and Thrane, 2001).

EARLY NEOPROTEROZOIC ("OLDER") GRANITES

Granitic rocks are common in the southern parts of the East Greenland Caledonide orogen. In the Kong Oscar Fjord region, they are prominent in the upper Hagar Bjerg thrust sheet, whereas the lower Niggli Spids thrust sheet is devoid of granites (see Kalsbeek et al., this volume).

Figure 9. Tectonically modified unconformity between metasedimentary rocks of the Krummedal sequence (right) and moderately deformed Paleoproterozoic orthogneiss (left) in Knækdalen (Fig. 1). Hammer (near arrow point) is ~50 cm long. Reproduced from Higgins et al. (1981).

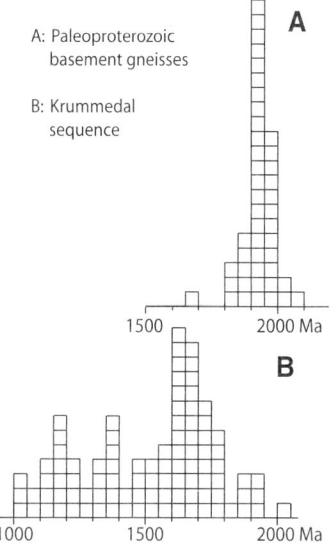

Figure 10. Age distribution of detrital zircons from the Krummedal supracrustal sequence compared to that of zircons from the Paleoproterozoic basement (each square represents one zircon date; only zircon ages <10% discordant and with error <50 Ma are shown). Data for Krummedal zircons are from Strachan et al. (1995), Kalsbeek et al. (2000), Watt et al. (2000), and Leslie and Nutman (2003). Data for the Paleoproterozoic basement are from Thrane (2002) and F. Kalsbeek and A.P. Nutman (unpublished data).

Early dating of the granites demonstrated that two age groups are present. Most granites yielded Caledonian ages, mainly 435–425 Ma, but a minority proved to be around 1000 Ma (e.g., Steiger et al., 1979; Rex and Gledhill, 1981). The latter, termed "older granites," form large sheets of granodioritic augen gneiss, hundreds of meters across, as well as wide sheets, sometimes more than 100 m, of muscovite-biotite leucogranites. "Older" and Caledonian granites are lithologically very similar, and it is impossible to differentiate them in the field.

Age

Precise age determination of the "older" granites is complicated for several reasons. The leucogranites were formed by partial melting of Krummedal sequence lithologies and often contain biotite-rich schlieren, probably incompletely digested restite material from the source (Kalsbeek et al., 2001a, 2001b, this volume). A variable proportion of the zircons in the granites is also inherited from the source; these commonly occur as cores overgrown by magmatic zircon. In earlier U-Pb age determinations, bulk zircon fractions were used, and it was not possible entirely to avoid inherited zircon components. Age determination by the Rb-Sr whole-rock isochron method suffered from incomplete Sr-isotopic homogenization of the granites, as well as from later disturbance during Caledonian metamorphic events. As a consequence, the early age determinations, although they served their purpose of demonstrating the presence of ca. 1000 Ma granites, lacked precision.

More recent age determinations have been carried out by ion microprobe (SHRIMP) methods, in which zircon cores and rims, recognized by cathodoluminescence imaging, can be analyzed separately (Kalsbeek et al., 2000; Watt et al., 2000; Leslie and Nutman, 2003). Most zircon rims, as well as a number of whole grains, yielded $^{207}Pb/^{206}Pb$ ages between 900 and 1000 Ma. A number of analyses on zircon cores gave higher $^{207}Pb/^{206}Pb$ ages, mostly in excess of the depositional age of the Krummedal sequence (ca. 1100–1000 Ma), and must represent inherited grains. A few zircons yielded ages of 400–450 Ma and were apparently formed during Caledonian metamorphism. Finally, some zircons had intermediate ages, interpreted as resulting from Pb loss from older zircon during Caledonian metamorphic events. Excluding such older and younger ages from the age calculations, the older granites could be dated at ca. 920–950 Ma in the Kong Oscar Fjord region (Kalsbeek et al., 2000; Watt et al., 2000) and at ca. 915 Ma in the Scoresby Sund region (Leslie and Nutman, 2003).

Caledonian and older granites can also be differentiated on the basis of Rb-Sr isotope data (Kalsbeek et al., 2000). Samples for which an "older" age has been established by

SHRIMP dating scatter about a 1000 Ma reference isochron in an $^{86}Sr/^{87}Sr$ versus $^{86}Rb/^{87}Sr$ diagram, while Caledonian granites (ca. 435–425 Ma) have significantly lower $^{86}Sr/^{87}Sr$ ratios at comparable $^{86}Rb/^{87}Sr$ (Fig. 11).

"GRENVILLIAN" DEFORMATION

The proposed late Precambrian supercontinent of Rodinia is considered to have been formed by the amalgamation of older cratons during the Grenvillian orogeny, ca. 1200–1000 Ma (e.g., Hoffman, 1991; Dalziel, 1992). Most reconstructions of Rodinia (e.g., Condie, 2003) show an arm of a global Grenvillian orogenic belt extending along the eastern margin of Greenland, an area now occupied by the Caledonian orogen. The presence of a Grenvillian orogen here was proposed by Park (1992) based on the older chronological data referred to already. Formation of the "older" granites in East Greenland at 950–920 Ma, however, postdates Grenvillian orogenic activity in North America and most Sveconorwegian activity in southern Scandinavia (1200–1000 Ma; Rivers, 1997; Åhäll and Connelly, 1998; Bingen and van Breemen, 1998), although postkinematic Sveconorwegian intrusions range down in age to ca. 915 Ma (see Table 1 in Hellström et al., 2004). Despite these chronological differences, the term "Grenvillian" is used here in a wide sense for lack of a better designation.

In eastern North Greenland, the presence of pre-Caledonian "Carolinidian" orogenic activity was proposed by Haller (1961), mainly based on air-photo interpretation and reconnaissance from the air. Detailed investigations of the area by Jepsen and Kalsbeek (1985), however, failed to find evidence for this event. On western Dronning Louise Land (see previous section), there is no evidence for any orogenic activity between deposition of the Paleoproterozoic or Mesoproterozoic Trekant series and the early Cambrian Zebra series (Strachan et al., 1992); Grenvillian activity clearly did not extend to that area. In the Kong Oscar Fjord area, Kalsbeek et al. (2000) documented local early Neoproterozoic deformation but failed to find evidence of large-scale "Grenvillian" structures or other proof of true orogenic activity.

In order to further study the presence or absence of late "Grenvillian" orogenic activity in the southernmost part of the East Greenland Caledonides, Leslie and Nutman (1999, 2003) reinvestigated an area of older granites in Renland (Fig. 1, ~71°N). Here, a 915 Ma augen granite within the Krummedal sequence occupies nappe-scale isoclinal folds (Fig. 12). Field observations were interpreted to demonstrate in situ generation of the augen granite by partial melting of fertile units within previously folded metasediments of the Krummedal sequence (Leslie and Nutman, 2003). This interpretation requires the isoclinal folds to be older than the 915 Ma augen granite but younger than deposition of the Krummedal sequence around or slightly before 1000 Ma. This is the strongest evidence yet available for Neoproterozoic orogenic activity in the East Greenland Caledonide orogen. Evidence for late Grenvillian orogenic activity has also been documented from Svalbard, where sedimentary rocks similar in age to the Krummedal sequence were isoclinally folded before deposition of Neoproterozoic sediments (Gee et al., 1995; Gee

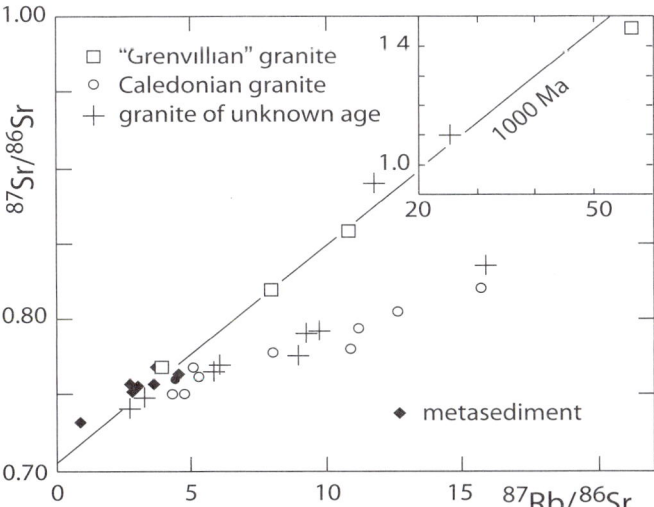

Figure 11. Rb-Sr isochron diagram showing the distinction between "Grenvillian" and Caledonian granites. The 1000 Ma isochron line is for reference only. Modified from Kalsbeek et al. (2001a) and used with permission from Elsevier.

Figure 12. Large-scale recumbent folds in metasedimentary rocks of the Krummedal sequence in Renland. The white sheets are ca. 915 Ma augen gneisses, interpreted to have been formed by in situ partial melting after folding. Height of mountain is ~1900 m. Photograph is by N. Henriksen.

and Tebenkov, 1996). A continuation of this "Grenvillian" belt from the southernmost part of the East Greenland Caledonides to Svalbard is indicated.

POLYOROGENIC NATURE OF THE EAST GREENLAND CALEDONIDES

As described in this paper, the area at present occupied by the North-East Greenland Caledonides appears to have been the site of several previous orogenic cycles. Earlier structures and mineral parageneses have been overprinted to such a degree during Caledonian tectonic and metamorphic events, however, that much of the evidence for earlier orogenic activity has been destroyed, and several aspects of the following discussion are open to alternative interpretations.

Correlation of the discordant amphibolite dikes in the Archean gneisses with Paleoproterozoic dike swarms in the Archean craton of Greenland (see Henriksen et al., 2000, p. 28–29) would seem to imply that the structures they cut are of Archean age. Although this dike correlation is plausible, it cannot be proven. Nevertheless, since nearly all gneisses in the Archean craton were strongly deformed during Archean orogenic activity, the same is likely to be true in East Greenland, even if correlation of the various dikes would prove to be incorrect.

Chemical and isotopic data suggest that the Paleoproterozoic gneisses and metagranitic rocks north of 72°50′N represent arcs that were accreted ca. 2000–1850 Ma along the eastern margin of an Archean crustal block now hidden by the Inland Ice. This interpretation would indicate subduction of Paleoproterozoic oceanic lithosphere from the east (in present-day coordinates). The presence of an ocean to the east of Greenland is implied. The volcanic rocks in the Eleonore Sø window might be related to the opening of this ocean.

Paleoproterozoic orthogneiss complexes elsewhere in Greenland underwent multiple phases of deformation and metamorphism before final stabilization. There is no reason to believe that this was not also the case in the Paleoproterozoic basement in North-East Greenland. However, with the few exceptions described previously here, Caledonian overprinting makes the recognition of older structures and mineral parageneses nearly impossible.

The Krummedal sequence of North-East Greenland has traditionally been correlated with the Moine Supergroup of NW Scotland (e.g., Winchester, 1988). The similarity in the age spectra for detrital zircons strengthens this correlation (for Scotland, see Friend et al., 1997; Kinny et al., 1999; Cawood et al., 2004). Similar metasedimentary sequences are also present in the Caledonian thrust units of Scandinavia (Williams and Claesson, 1987) and in Svalbard (Gee et al., 1995; for an overview, see Watt and Thrane, 2001). Apparently, the Krummedal sequence was deposited in part of a widespread system of late Mesoproterozoic to early Neoproterozoic sedimentary basins. Detritus for these sediments appears to have been derived from the Grenvillian-Sveconorwegian orogen (e.g., Watt and Thrane, 2001), where granitoid rocks with ages between 1700 and 1000 Ma are prominent (e.g., Rivers, 1997; Gower et al., 2002). The Krummedal and similar sedimentary sequences are thus, strictly speaking, post-Grenvillian, and their Neoproterozoic deformation cannot be related to the Grenvillian orogeny sensu stricto. Nevertheless, since diachroneity of events in global orogenic belts is common, a broad correlation with Grenvillian orogenesis appears plausible. Most reconstructions of Rodinia place Scandinavia and Svalbard opposite central East Greenland around 1000 Ma, though at somewhat different positions (e.g., Condie, 2003; Pisarevsky et al., 2003). Deformation of the Krummedal sequence around 950 Ma in the southernmost part of the East Greenland Caledonides may have been the result of a late stage of collision, related to the closure of a pre–Iapetan ocean between Baltica and Laurentia at that time.

After this "Grenvillian" interlude, new sedimentary basins were formed, and the very thick Eleonore Bay Supergroup was deposited (Sønderholm et al., this volume). The Iapetus Ocean opened, and a thick sequence of early Paleozoic sedimentary rocks was deposited on the passive margin of Iapetus. Closure of Iapetus in the Silurian led to the formation of the Caledonian orogen, followed by new sedimentary basins in the Devonian and Carboniferous. After continued deposition of sedimentary rocks during the Mesozoic and basaltic volcanism in the early Tertiary, the Atlantic Ocean was formed.

Therefore, in North-East Greenland, there is evidence—admittedly fragmentary and, in many respects, open to discussion—of at least three stages of orogenic activity related to the opening and closure of oceans, while a new stage may have started with the opening of the North Atlantic Ocean.

ACKNOWLEDGMENTS

We thank the staffs of the laboratories involved in this study for their help, Helle Zetterwall and Benny Schark for assistance with the figures, Clark Friend and Åke Johansson for helpful reviews, Elsevier BV (The Netherlands) for permission to reuse diagrams published earlier in *Precambrian Research* and *Lithos*, and the Geological Survey of Denmark and Greenland for permission to publish this paper.

REFERENCES CITED

Åhäll, K.-I., and Connelly, J., 1998, Intermittent 1.53–1.13 Ga magmatism in western Baltica: Age constraints and correlations within a postulated supercontinent: Precambrian Research, v. 92, p. 1–20, doi: 10.1016/S0301-9268(98)00064-3.

Andresen, A., Hartz, E.H., and Vold, J., 1998, A late orogenic extensional origin for the infracrustal gneiss domes of the East Greenland Caledonides (72–74°N): Tectonophysics, v. 285, p. 353–369, doi: 10.1016/S0040-1951(97)00278-3.

Barker, F., 1979, Trondhjemite: Definition, environment and hypotheses of origin, *in* Barker, F., ed., Trondhjemites, Dacites and Related Rocks: Amsterdam, Elsevier, p. 1–12.

Bingen, B., and van Breemen, O., 1998, Tectonic regimes and terrane boundaries in the high-grade Sveconorwegian belt of SW Norway, inferred from U-Pb zircon geochronology and geochemical signature of augen gneiss suites: Journal of the Geological Society of London, v. 155, p. 143–154, doi: 10.1144/gsjgs.155.1.0143.

Brueckner, H.K., Gilotti, J.A., and Nutman, A.P., 1998, Caledonian eclogite-facies metamorphism of early Proterozoic protoliths from the North-East Greenland eclogite province: Contributions to Mineralogy and Petrology, v. 130, p. 103–120, doi: 10.1007/s004100050353.

Cawood, P.A., Nemchin, A.A., Strachan, R.A., Kinny, P.D., and Loewy, S., 2004, Laurentian provenance and an intracratonic tectonic setting for the Moine Supergroup, Scotland, constrained by detrital zircons from the Loch Eil and Glen Urquhart successions: Journal of the Geological Society of London, v. 161, p. 861–874, doi: 10.1144/16-764903-117.

Chadwick, B., and Friend, C.R.L., 1994, Reaction of Precambrian high-grade gneisses to mid-crustal ductile deformation in western Dove Bugt, North-East Greenland: Rapport Grønlands Geologiske Undersøgelse, v. 162, p. 53–70.

Collinson, J.D., Kalsbeek, F., Jepsen, H.F., Pedersen, S.A.S., and Upton, B.G.J., 2008, this volume, Paleoproterozic and Mesoproterozoic sedimentary and volcanic successions in the northern parts of the East Greenland Caledonian orogen and its foreland, in Higgins, A.K., Gilotti, J.A., and Smith, M.P., eds., The Greenland Caledonides: Evolution of the Northeast Margin of Laurentia: Geological Society of America Memoir 202, doi: 10.1130/2008.1202(04).

Condie, K.C., 2003, Supercontinents, superplumes and continental growth: The Neoproterozoic record, in Yoshida, M., Windley, B.F., and Dasgupta, S., eds., Proterozoic East Gondwana: Supercontinent Assembly and Breakup: Geological Society of London Special Publication 206, p. 1–21.

Dalziel, I.W.D., 1992, On the organization of American plates in the Neoproterozoic and the breakout of Laurentia: GSA Today, v. 2, no. 11, p. 237, 240–241.

DePaolo, D.J., 1981, Neodymium isotopes in the Colorado Front Range and crust-mantle evolution in the Proterozoic: Nature, v. 291, p. 193–196, doi: 10.1038/291193a0.

Elvevold, S., Thrane, K., and Gilotti, J.A., 2003, Metamorphic history of high-pressure granulites in Payer Land, Greenland Caledonides: Journal of Metamorphic Geology, v. 21, p. 49–63, doi: 10.1046/j.1525-1314.2003.00419.x.

Friend, C.R.L., Kinny, P.D., Rogers, G., Strachan, R.A., and Paterson, B.A., 1997, U-Pb zircon geochronological evidence for Neoproterozoic events in the Glenfinnan Group (Moine Supergroup): The formation of the Ardgour granite gneiss, north-west Scotland: Contributions to Mineralogy and Petrology, v. 128, p. 101–113, doi: 10.1007/s004100050297.

Garde, A.A., Hamilton, M.A., Chadwick, B., Grocott, J., and McCaffrey, K.J.W., 2002, The Ketilidian orogen of South Greenland: Geochronology, tectonics, magmatism, and fore-arc accretion during Palaeoproterozoic oblique convergence: Canadian Journal of Earth Sciences, v. 39, p. 765–793, doi: 10.1139/e02-026.

Gee, D.G., and Tebenkov, A.M., 1996, Two major unconformities beneath the Neoproterozoic Murchisonfjorden Supergroup in the Caledonides of central Nordaustlandet, Svalbard: Polar Research, v. 15, p. 81–91, doi: 10.1111/j.1751-8369.1996.tb00460.x.

Gee, D.G., Johansson, Å., Ohta, Y., Tebenkov, A.M., Krasil'ščikov, A.A., Balashov, Y.A., Larionov, A.N., Gannibal, L.F., and Ryungenen, G.I., 1995, Grenvillian basement and a major unconformity within the Caledonides of Nordaustlandet, Svalbard: Precambrian Research, v. 70, p. 215–234, doi: 10.1016/0301-9268(94)00041-O.

Gilotti, J.A., 1993, Discovery of a medium temperature eclogite province in the Caledonides of North-East Greenland: Geology, v. 21, p. 523–526, doi: 10.1130/0091-7613(1993)021<0523:DOAMTE>2.3.CO;2.

Gilotti, J.A., and Ravna, E.J.K., 2002, First evidence of ultrahigh-pressure metamorphism in the North-East Greenland Caledonides: Geology, v. 30, p. 551–554, doi: 10.1130/0091-7613(2002)030<0551:FEFUPM>2.0.CO;2.

Gilotti, J.A., Jones, K.A., and Elvevold, S., 2008, this volume, Caledonian metamorphic patterns in Greenland, in Higgins, A.K., Gilotti, J.A., and Smith, M.P., eds., The Greenland Caledonides: Evolution of the Northeast Margin of Laurentia: Geological Society of America Memoir 202, doi: 10.1130/2008.1202(08).

Gower, C.F., Krogh, T.E., and James, D.T., 2002, Correlation chart of the eastern Grenville Province and its northern foreland: Canadian Journal of Earth Sciences, v. 39, p. 897 + 2 charts.

Haller, J., 1961, The Carolinides: An orogenic belt of late Precambrian age in Northeast Greenland, in Raasch, G.O., ed., Geology of the Arctic, Volume 1: Toronto, Toronto University Press, p. 155–159.

Hansen, B.T., Higgins, A.K., and Bär, M.-T., 1978, Rb-Sr and U-Pb age patterns in polymetamorphic sediments from the southern part of the East Greenland Caledonides: Bulletin of the Geological Society of Denmark, v. 27, p. 55–62.

Hellström, F.A., Johansson, Å., and Larson, S.Å., 2004, Age and emplacement of late Sveconorwegian monzogabbroic dykes, SW Sweden: Precambrian Research, v. 128, p. 39–55, doi: 10.1016/S0301-9268(03)00194-3.

Henriksen, N., 1994, Geology of North-East Greenland (75°–78°N): The 1988–90 mapping project: Rapport Grønlands Geologiske Undersøgelse, v. 162, p. 5–16.

Henriksen, N., Higgins, A.K., Kalsbeek, F., and Pulvertaft, T.C.R., 2000, Greenland from Archaean to Quaternary: Descriptive Text to the Geological Map of Greenland, 1:2,500,000: Geology of Greenland Survey Bulletin, v. 185, 93 p.

Higgins, A.K., 1988, The Krummedal supracrustal sequence in East Greenland, in Winchester, J.A., ed., Later Proterozoic Stratigraphy of the Northern Atlantic Regions: Glasgow, Blackie and Son Ltd., p. 86–96.

Higgins, A.K., Friderichsen, J.D., and Thyrsted, T., 1981, Precambrian metamorphic complexes in the East Greenland Caledonides (72°–74°N): Their relationships to the Eleonore Bay Group, and Caledonian orogenesis: Rapport Grønlands Geologiske Undersøgelse, v. 104, p. 5–46.

Hoffman, P.F., 1991, Did the breakout of Laurentia turn Gondwanaland inside-out?: Science, v. 252, p. 1409–1412, doi: 10.1126/science.252.5011.1409.

Hull, J.M., Friderichsen, J.D., Gilotti, J.A., Henriksen, N., Higgins, A.K., and Kalsbeek, F., 1994, Gneiss complex of the Skærfjorden region (76°–78°N), North-East Greenland: Rapport Grønlands Geologiske Undersøgelse, v. 162, p. 35–51.

Irvine, T.N., and Baragar, W.R.A., 1971, A guide to the chemical classification of the common volcanic rocks: Canadian Journal of Earth Sciences, v. 8, p. 523–548.

Jensen, L.S., 1976, A New Cation Plot for Classifying Subalkalic Volcanic Rocks: Ontario Division of Mines Miscellaneous Papers 66, 22 p.

Jepsen, H.F., and Kalsbeek, F., 1985, Evidence for non-existence of a Carolinidian fold belt in eastern North Greenland, in Gee, D.G., and Sturt, B.A., eds., The Caledonide Orogen—Scandinavia and Related Areas, Part 2: New York, John Wiley and Sons Ltd., p. 1071–1076.

Kalsbeek, F., 1995, Geochemistry, tectonic setting, and poly-orogenic history of Palaeoproterozoic basement rocks from the Caledonian fold belt of North-East Greenland: Precambrian Research, v. 72, p. 301–315, doi: 10.1016/0301-9268(94)00097-B.

Kalsbeek, F., and Taylor, P.N., 1985, Isotopic and chemical variation in granites across a Proterozoic continental margin—The Ketilidian mobile belt of South Greenland: Earth and Planetary Science Letters, v. 73, p. 65–80, doi: 10.1016/0012-821X(85)90035-4.

Kalsbeek, F., Nutman, A.P., and Taylor, P.N., 1993, Palaeoproterozoic basement province in the Caledonian fold belt of North-East Greenland: Precambrian Research, v. 63, p. 163–178, doi: 10.1016/0301-9268(93)90010-Y.

Kalsbeek, F., Nutman, A.P., Escher, J.C., Friderichsen, J.D., Hull, J.M., Jones, K.A., and Pedersen, S.A.S., 1999, Geochronology of granitic and supracrustal rocks from the northern part of the East Greenland Caledonides: Ion microprobe U-Pb zircon ages: Geology of Greenland Survey Bulletin, v. 184, p. 31–48.

Kalsbeek, F., Thrane, K., Nutman, A.P., and Jepsen, H.F., 2000, Late Mesoproterozoic to early Neoproterozoic history of the East Greenland Caledonides: Evidence for Grenvillian orogenesis?: Journal of the Geological Society of London, v. 157, p. 1215–1225.

Kalsbeek, F., Jepsen, H.F., and Nutman, A.P., 2001a, From source migmatites to plutons: Tracking the origin of ca. 435 Ma S-type granites in the East Greenland Caledonian orogen: Lithos, v. 57, p. 1–21, doi: 10.1016/S0024-4937(00)00071-2.

Kalsbeek, F., Jepsen, H.F., and Jones, K.A., 2001b, Geochemistry and petrogenesis of S-type granites in the East Greenland Caledonides: Lithos, v. 57, p. 91–109, doi: 10.1016/S0024-4937(01)00038-X.

Kalsbeek, F., Higgins, A.K., Jepsen, H.F., Frei, R., and Nutman, A.P., 2008, this volume, Granites and granites in the East Greenland Caledonides, in Higgins, A.K., Gilotti, J.A., and Smith, M.P., eds., The Greenland Caledonides: Evolution of the Northeast Margin of Laurentia: Geological Society of America Memoir 202, doi: 10.1130/2008.1202(09).

Katz, H.R., 1952, Ein Querschnitt durch die Nunatakzone Östgrönlands: Meddelelser om Grønland, v. 144/8, 65 p.

Kinny, P.D., Friend, C.R.L., Strachan, R.A., Watt, G.R., and Burns, I.M., 1999, U-Pb geochronology of regional migmatites in East Sutherland, Scotland: Evidence for crustal melting during the Caledonian orogeny: Journal of the Geological Society of London, v. 156, p. 1143–1152, doi: 10.1144/gsjgs.156.6.1143.

Kystol, J., and Larsen, L.M., 1999, Analytical procedures in the rock geochemical laboratory of the Geological Survey of Denmark and Greenland: Geology of Greenland Survey Bulletin, v. 184, p. 59–62.

Leslie, A.G., and Higgins, A.K., 1998, On the Caledonian geology of Andrée Land, Eleonore Sø and adjacent nunataks (73°30′–74°N), East Greenland, in Higgins, A.K., and Frederiksen, K.S., eds., Caledonian Geology of East Greenland 72°–74°N: Preliminary Reports from the 1997 Expedition: Danmarks og Grønlands Geologiske Undersøgelse Rapport, v. 1998/28, p. 11–27.

Leslie, A.G., and Higgins, A.K., 1999, On the Caledonian (and Grenvillian) geology of Bartholin Land, Ole Rømer Land and adjacent nunataks, East Greenland, in Higgins, A.K., and Frederiksen, K.S., eds., Geology of East Greenland 72°–75°N, Mainly Caledonian: Preliminary Reports from the 1998 Expedition: Danmarks og Grønlands Geologiske Undersøgelse Rapport, v. 1999/19, p. 11–26.

Leslie, A.G., and Higgins, A.K., 2008, this volume, Foreland-propagating Caledonian thrust systems in East Greenland, in Higgins, A.K., Gilotti, J.A., and Smith, M.P., eds., The Greenland Caledonides: Evolution of the Northeast Margin of Laurentia: Geological Society of America Memoir 202, doi: 10.1130/2008.1202(07).

Leslie, A.G., and Nutman, A.P., 1999, The evidence for episodic tectonothermal activity in southern Renland, East Greenland Caledonides: Danmarks og Grønlands Geologiske Undersøgelse Rapport 1999/80, 12 p. + 12 plates.

Leslie, A.G., and Nutman, A.P., 2003, Evidence for Neoproterozoic orogenesis and early high temperature Scandian deformation events in the southern East Greenland Caledonides: Geological Magazine, v. 140, p. 309–333, doi: 10.1017/S0016756803007593.

Nutman, A.P., and Kalsbeek, F., 1994, Search for Archaean basement in the Caledonian fold belt of North-East Greenland: Rapport Grønlands Geologiske Undersøgelse, v. 162, p. 129–133.

O'Connor, J.T., 1965, A classification of quartz-rich igneous rocks based on feldspar ratios: U.S. Geological Survey Professional Paper 525B, p. B79–B84.

Park, R.G., 1992, Plate kinematic history of Baltica during the middle to late Proterozoic: A model: Geology, v. 20, p. 725–728, doi: 10.1130/0091-7613(1992)020<0725:PKHOBD>2.3.CO;2.

Pearce, J.A., Harris, N.B.W., and Tindle, A.G., 1984, Trace element discrimination diagrams for the tectonic interpretation of granitic rocks: Journal of Petrology, v. 25, p. 956–983.

Pedersen, S.A.S., Craig, L.E., Upton, B.G.J., Rämö, O.T., Jepsen, H.F., and Kalsbeek, F., 2002, Palaeoproterozoic (1740 Ma) rift-related volcanism in the Hekla Sund region, eastern North Greenland: Field occurrence, geochemistry and tectonic setting: Precambrian Research, v. 114, p. 327–346, doi: 10.1016/S0301-9268(01)00234-0.

Pisarevsky, S.A., Wingate, M.T.D., Powell, C.McA., Johnson, S., and Evans, D.A.D., 2003, Models of Rodinia assembly and fragmentation, in Yoshida, M., Windley, B.F., and Dasgupta, S., eds., Proterozoic East Gondwana: Supercontinent Assembly and Breakup: Geological Society of London Special Publication 206, p. 35–55.

Rex, D.C., and Gledhill, A.R., 1974, Reconnaissance geochronology of the infracrustal rocks of Flyverfjord, Scoresby Sund, East Greenland: Bulletin of the Geological Society of Denmark, v. 23, p. 49–54.

Rex, D.C., and Gledhill, A.R., 1981, Isotopic studies in the East Greenland Caledonides (72°–74°N)—Precambrian and Caledonian ages: Rapport Grønlands Geologiske Undersøgelse, v. 104, p. 47–72.

Rivers, T., 1997, Lithotectonic elements of the Grenville province: Review and tectonic implications: Precambrian Research, v. 86, p. 117–154, doi: 10.1016/S0301-9268(97)00038-7.

Sønderholm, M., Frederiksen, K.S., Smith, M.P., and Tirsgaard, H., 2008, this volume, Neoproterozoic sedimentary basins with glacigenic deposits of the East Greenland Caledonides, in Higgins, A.K., Gilotti, J.A., and Smith, M.P., eds., The Greenland Caledonides: Evolution of the Northeast Margin of Laurentia: Geological Society of America Memoir 202, doi: 10.1130/2008.1202(05).

Steiger, R.H., Harnik-Šoptrajanova, G., Zimmermann, E., and Henriksen, N., 1976, Isotopic age and metamorphic history of the banded gneiss at Danmarkshavn, East Greenland: Contributions to Mineralogy and Petrology, v. 57, p. 1–24, doi: 10.1007/BF00392849.

Steiger, R.H., Hansen, B.T., Schuler, Ch., Bär, M.T., and Henriksen, N., 1979, Polyorogenic nature of the southern Caledonian fold belt in East Greenland: An isotopic age study: The Journal of Geology, v. 87, p. 475–495.

Strachan, R.A., Holdsworth, R.E., Friderichsen, J.D., and Jepsen, H.F., 1992, Regional Caledonian structure within an oblique convergence zone, Dronning Louise Land, NE Greenland: Journal of the Geological Society of London, v. 149, p. 359–371, doi: 10.1144/gsjgs.149.3.0359.

Strachan, R.A., Nutman, A.P., and Friderichsen, J.D., 1995, SHRIMP U-Pb geochronology and metamorphic history of the Smallefjord sequence, NE Greenland Caledonides: Journal of the Geological Society of London, v. 152, p. 779–784, doi: 10.1144/gsjgs.152.5.0779.

Taylor, S.R., and McLennan, S.M., 1985, The Continental Crust: Its Composition and Evolution: Oxford, Blackwell, 312 p.

Thompson, R.N., 1982, British Tertiary volcanic province: Scottish Journal of Geology, v. 18, p. 49–107.

Thrane, K., 2002, Relationships between Archaean and Palaeoproterozoic crystalline basement complexes in the southern part of the East Greenland Caledonides: An ion microprobe study: Precambrian Research, v. 113, p. 19–42, doi: 10.1016/S0301-9268(01)00198-X.

Tucker, R.D., Dallmeyer, R.D., and Strachan, R.A., 1993, Age and tectonothermal record of Laurentian basement, Caledonides of NE Greenland: Journal of the Geological Society of London, v. 150, p. 371–379, doi: 10.1144/gsjgs.150.2.0371.

Watt, G.R., and Thrane, K., 2001, Early Neoproterozoic events in East Greenland: Precambrian Research, v. 110, p. 165–184, doi: 10.1016/S0301-9268(01)00186-3.

Watt, G.R., Kinny, P.D., and Friderichsen, J.D., 2000, U-Pb geochronology of Neoproterozoic and Caledonian tectonothermal events in the East Greenland Caledonides: Journal of the Geological Society of London, v. 157, p. 1031–1048.

Weis, D., Demaiffe, D., Souchez, A.J., Gow, A.J., and Meese, D.A., 1997, Ice sheet development in central Greenland: Implications from the Nd, Sr and Pb isotopic compositions of basal material: Earth and Planetary Science Letters, v. 150, p. 161–169, doi: 10.1016/S0012-821X(97)00073-3.

Williams, I.S., 1998, U-Th-Pb geochronology by ion microprobe, in McKibben, M.A., Shanks, W.C., III, and Ridley, W.I., eds., Applications of Microanalytical Techniques to Understanding Mineralizing Processes: Society of Economic Geologists Reviews in Economic Geology, v. 7, p. 1–35.

Williams, I.S., and Claesson, S., 1987, Isotopic evidence for the Precambrian provenance and Caledonian metamorphism of high grade paragneisses from the Seve nappes, Scandinavian Caledonides: II. Ion microprobe U-Th-Pb: Contributions to Mineralogy and Petrology, v. 97, p. 205–217, doi: 10.1007/BF00371240.

Winchester, J.A., 1988, Later Proterozoic environments and tectonic evolution in the northern Atlantic lands, in Winchester, J.A., ed., Later Proterozoic Stratigraphy of the Northern Atlantic Regions: Glasgow, Blackie and Son Ltd., p. 253–270.

MANUSCRIPT ACCEPTED BY THE SOCIETY 14 JANUARY 2008

Paleoproterozoic and Mesoproterozoic sedimentary and volcanic successions in the northern parts of the East Greenland Caledonian orogen and its foreland

John D. Collinson
Delos, Knowl Wall, Beech, Staffordshire ST4 8SE, UK

Feiko Kalsbeek*
Hans F. Jepsen
Stig A.S. Pedersen
Geological Survey of Denmark and Greenland, Øster Voldgade 10, DK-1350 Copenhagen K, Denmark

Brian G.J. Upton
School of Geosciences, University of Edinburgh, Edinburgh EH9 3JW, UK

ABSTRACT

The crystalline basement within the northern parts of the Caledonian orogen, and in the adjacent foreland, is overlain by a several-kilometer-thick succession of sedimentary and volcanic rocks, the Paleoproterozoic–Mesoproterozoic Independence Fjord Group and the Mesoproterozoic Zig-Zag Dal Basalt Formation. The lowermost strata of the Independence Fjord Group, composed of quartzitic and feldspathic sandstones and conglomerates with interbedded volcanic rocks, occur within the Caledonian orogen and are strongly deformed. These strata were deposited around 1740 Ma ago, and they were associated with a period of rifting that succeeded a long sequence of Paleoproterozoic orogenic events. Similar sandstones, interbedded with siltstone units but without volcanic rocks, are widespread in the Caledonian foreland, where they are virtually undeformed. These foreland deposits were laid down in a continental sag basin under semiarid conditions. Sedimentary structures indicate a largely fluvial origin, with intermittent eolian transport. The siltstones were deposited in extensive shallow lakes. Desiccated bedding surfaces show that these periodically dried out.

The sandstones of the Independence Fjord Group are cut by a multitude of doleritic sheets and dikes, the ca. 1380 Ma Midsommersø Dolerites, and more silicic intrusions, most of which show evidence of hydrothermal alteration and variable contamination with components derived from the crystalline basement and the sandstones. Some intrusions consist almost entirely of crustally derived material. The Zig-Zag Dal Basalt Formation conformably overlies the Independence Fjord Group.

---

*Corresponding author: fkalsbeek@gmail.com.

Compositional similarities suggest a genetic relationship with the Midsommersø Dolerites, but the basalts appear to be less crustally contaminated. The basalts were deposited within a basin that underwent subsidence during and after volcanic activity. The Zig-Zag Dal Basalt Formation is unconformably overlain by Neoproterozoic sedimentary successions. The unconformity represents a stratigraphic hiatus of some 500 m.y., for which no information is available from North Greenland.

Keywords: North-East Greenland, Caledonian foreland, Paleoproterozoic, Mesoproterozoic, sandstones, sedimentology, basalts, dolerites, geochemistry, ages.

INTRODUCTION

The stratigraphic units described in this review occur in three separate areas (Fig. 1): (A) in the Caledonian foreland west of Kronprins Christian Land in eastern North Greenland (80°N–83°N); (B) in the western part of Dronning Louise Land, 400 km further south (77°N, 25°W); and (C) in the thrust sheets that form the northernmost parts of the Caledonian orogen, between Dronning Louise Land and Kronprins Christian Land. The following discussion shows that correlations of very similar rocks in the foreland and orogen are uncertain.

The lowest stratigraphic unit in the first area, west of Kronprins Christian Land, is the Independence Fjord Group, which is a succession of sandstones and extensive but thin siltstones, the base of which is hidden beneath the Inland Ice (Collinson, 1980, 1983; Sønderholm and Jepsen, 1991). The sandstones are undeformed and unmetamorphosed, and they postdate the Paleoproterozoic orogenic events recorded in North-East Greenland (Kalsbeek et al., 1993, this volume). They are conformably overlain by a succession of continental flood basalts, the Zig-Zag Dal Basalt Formation (Kalsbeek and Jepsen, 1984; Upton et al., 2005), and are cut by numerous sheets and dikes of dolerites, the Midsommersø Dolerites, and associated more siliceous intrusive rocks (Fig. 2; Kalsbeek and Jepsen, 1983; Kalsbeek and Frei, 2006).

Less extensive outcrops of the Caledonian foreland are present near the ice margin on Dronning Louise Land (area B). Here, Paleoproterozoic orthogneisses are unconformably overlain by sandstones of the Trekant "series" (Fig. 2; Peacock, 1956, 1958; Friderichsen et al., 1990; Strachan et al., 1994), which are lithologically similar to those of the Independence Fjord Group west of Kronprins Christian Land. Although the two areas are ~400 km apart, there is no reason to doubt correlation with the Independence Fjord Group farther north. Like the Independence Fjord Group, the Trekant series (as well as the underlying crystalline basement) is cut by numerous dolerite intrusions, but basalts similar to those of the Zig-Zag Dal Basalt Formation are absent.

Sandstones with dolerite intrusions, very similar to those just mentioned, but strongly deformed and tectonically interleaved with crystalline basement gneisses, are also present within the thrust sheets that form the northern parts of the Caledonian orogen (area C). These sandstones are locally interbedded with conglomerates, and the presence of syndepositional faults suggests that at least some of the sediments were deposited in rift basins. Within this succession of sandstones and conglomerates, two units of basaltic and andesitic lavas and tuffs are present, referred to as the Hekla Sund Formation and the Aage Berthelsen Gletscher Formation (Pedersen et al., 2002; see below). On the 1:2,500,000 *Geological Map of Greenland* (Escher and Pulvertaft, 1995), the sandstones and conglomerates in area C were correlated with the Independence Fjord Group of the foreland (area A), and the volcanic rocks (Hekla Sund and Aage Berthelsen Gletscher Formations) were correlated with the Zig-Zag Dal Basalt Formation. Modern age determinations, however, have shown that at least the latter correlation is not correct.

The Independence Fjord Group exposed in area A was deposited between the end of a long period of Paleoproterozoic orogenic events ca. 1750 Ma ago and the emplacement of the Midsommersø Dolerites, which has been dated at 1380 Ma (Upton et al., 2005). Because of a close geochemical similarity between the dolerites and the Zig-Zag Dal Basalt Formation, it has been assumed that the two were more or less contemporaneous (Upton et al., 2005; Kalsbeek and Frei, 2006). The basalts may be marginally younger because they are not cut by dolerites. Rhyolitic rocks within the Hekla Sund Formation of area C, on the other hand, have yielded a significantly older age, 1740 Ma (Kalsbeek et al., 1999), and it is apparent that the volcanic rocks within the Caledonian orogen (Hekla Sund Formation) cannot correlate with those of the foreland (Zig-Zag Dal Basalt Formation). Chemical differences between the two volcanic formations, as well as the observation that the lavas of the Hekla Sund and Aage Berthelsen Gletscher Formations are cut by numerous dolerite intrusions confirm this conclusion (see below).

Correlation of the sandstones and conglomerates in eastern Kronprins Christian Land (area C) with the Independence Fjord Group of the foreland (area A) is also questionable, for two reasons. First, whereas the Independence Fjord Group was deposited in an intracratonic basin during a period of slow subsidence (Collinson, 1980, 1983; see below), the sedimentary successions of Kronprins Christian Land were formed during a period of active rifting (Pedersen et al., 2002). Second, while deposition of the sediments in eastern Kronprins Christian Land was interrupted by volcanic activity, in part explosive, no ash beds have been encountered in the foreland.

Since the sandstones and conglomerates in Kronprins Christian Land are interbedded with the lavas, they must have been

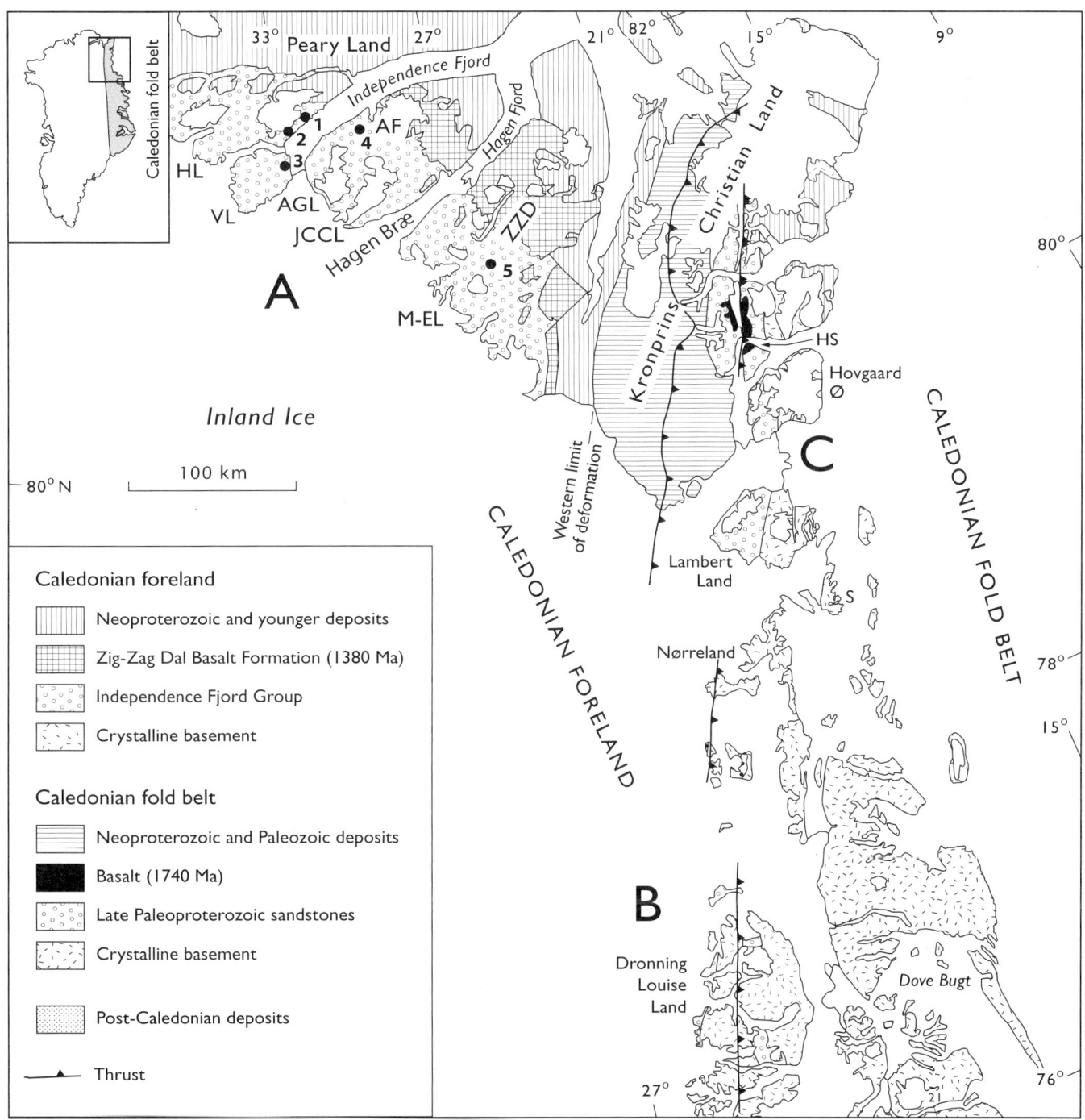

Figure 1. Simplified geological map of eastern North Greenland and localities mentioned in the text. From NW to SE along the ice margin, HL—Heilprin Land; VL—Vildtland; AGL—Academy Gletscher; JCCL—J.C. Christensen Land; M-EL—Mylius-Erichsen Land; AF (on J.C. Christensen Land)—Astrup Fjord, a little side-fjord to Independence Fjord; ZZD (on Mylius-Erichsen Land)—Zig-Zag Dal; HS (near the east coast)—Hekla Sund. Intrusive rocks (Midsommersø Dolerites), which occupy a significant part of the outcrop area of the Independence Fjord Group, are not shown on this map; localities where these intrusions were studied in detail are marked with numbers 1–5 in boldface; the locality marked **1** is Kap Einar Mikkelsen. For discussion of areas A, B, and C, see text.

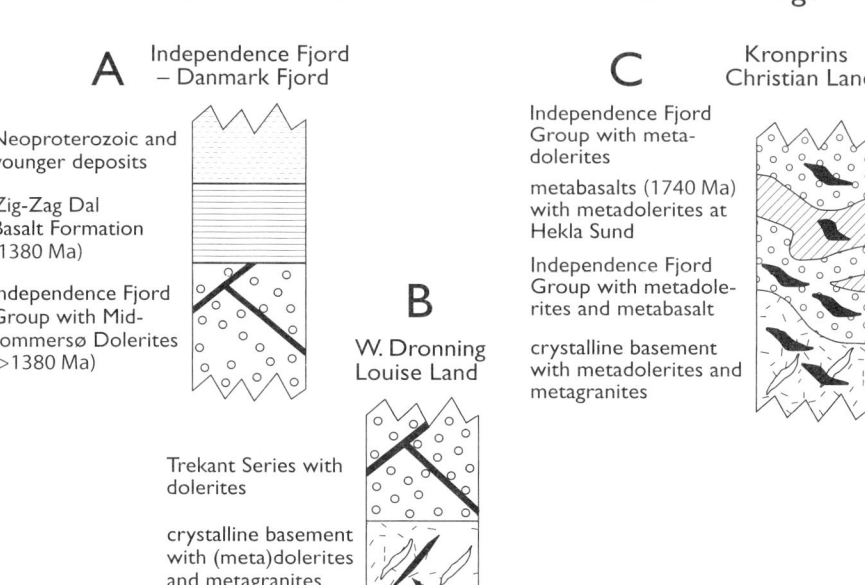

Figure 2. Simplified stratigraphies for areas A, B, and C of Figure 1 (modified from Kalsbeek et al., 1999).

deposited around 1740 Ma ago. This is immediately after the end of Paleoproterozoic orogenic and igneous events in East Greenland, where the youngest granitic rocks have been dated at ca. 1740 Ma (Kalsbeek et al., 1993, this volume). In this paper, we adopt the view that they represent the oldest units of a sedimentary sequence that grades upward into the sandstones of the foreland. In accordance with the 1:2,500,000 *Geological Map of Greenland* (Escher and Pulvertaft, 1995), all these sandstones and associated conglomerates and volcanic rocks, in the Caledonian orogen as well as in the foreland, are included in the Independence Fjord Group. Arguably, deposition of the succession may be interpreted in terms of the McKenzie (1978) model, starting with rapid extension and sedimentation in rift basins, followed by a period of thermal relaxation during which deposition in a more extensive sag basin took place.

In the following discussion, the different stratigraphic successions are described in their (assumed) chronological order, starting with the sedimentary and volcanic formations within the northernmost parts of the Caledonian orogen (area C), followed by the Independence Fjord Group of areas A and B, the Midsommersø Dolerites, and the Zig-Zag Basalt Formation (area A).

INDEPENDENCE FJORD GROUP WITH VOLCANIC ROCKS WITHIN THE NORTHERN PARTS OF THE CALEDONIAN OROGEN

The northernmost parts of the East Greenland Caledonide orogen were mapped by the Geological Survey of Greenland in the early 1990s and documented on the 1:500,000 *Geological Map of Greenland*, sheet 9, Lambert Land (Jepsen, 2000). Most of the sandstones and volcanic rocks relevant to the present review occur in eastern Kronprins Christian Land and Lambert Land. The rocks are particularly well exposed in the area around Hekla Sund in eastern Kronprins Christian Land (Fig. 3), but the severity of this alpine terrain, together with extensive thrusting and folding, makes detailed mapping difficult. The stratigraphy of the Hekla Sund area was described in an internal survey report by Pedersen et al. (1995), and their subdivisions were used in modified form on the geological map referred to above.

The rocks in question occur in two different thrust complexes that are separated by major thrusts from the intervening "Hekla Sund basin" (Fig. 3). Because correlation of formations between the two thrust complexes is not clear, different stratigraphic designations were used for the areas west and east of the Hekla Sund basin (Fig. 4). All these rocks are cut by doleritic intrusions, to such an extent that 30%–50% of the area is underlain by dolerites.

The lowest sandstone strata recognized in the western area are referred to as the Ingolf Fjord Formation (IG on the map), and they consist of ~200 m of buff-colored, thick-bedded quartzitic to arkosic sandstones with thin intercalations of conglomerate. Sedimentary structures are dominated by large-scale trough cross-bedding and local intercalations of heavy mineral placer deposits typical for a fluvial environment. Toward the top, thin beds of volcanic ash as well as volcanic bombs are increasingly common. The Ingolf Fjord Formation is interpreted as a shallow-water deposit; disturbed sandstone beds display slump features, suggesting an unstable environment close to a zone of tectonic activity.

The Ingolf Fjord Formation is overlain by an ~400-m-thick succession dominated by pillow lavas and hyaloclastites, defined as the Aage Berthelsen Gletscher Formation (BG; Pedersen et al.,

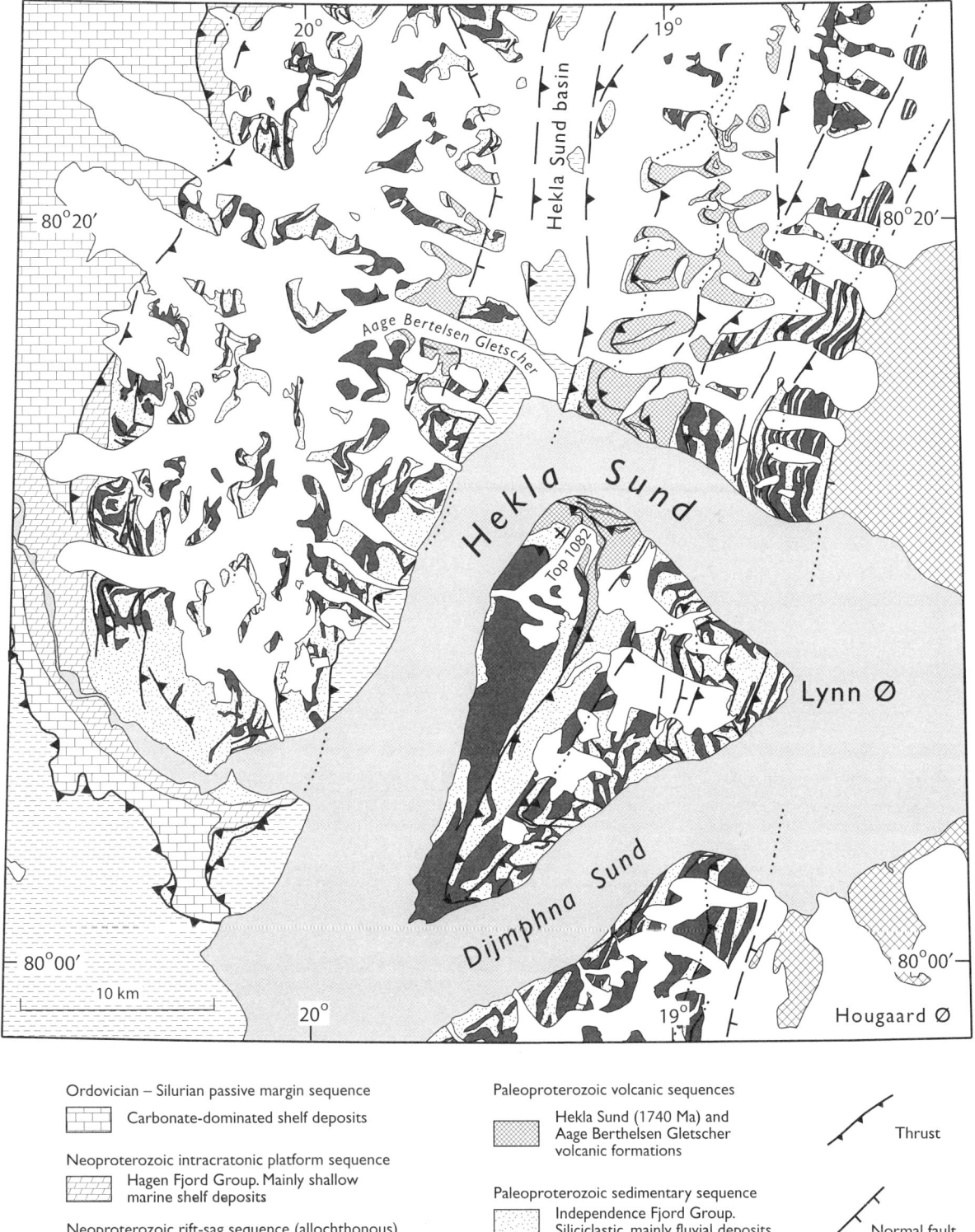

Figure 3. Geological map of the Hekla Sund area, Kronprins Christian Land, within the Caledonian orogen, after the 1:500,000 geological map of Greenland (Jepsen, 2000, sheet 9, Lambert Land).

Independence Fjord Group in the Caledonian orogen

West of Hekla Sund Basin		East of Hekla Sund Basin
Caroline Mathilde Alper Formation (CM) c. 800 m, mainly sandstones	Hekla Sund basin	Lynn Ø Formation (LY) c. 250 m, sandstones and conglomerates
Aage Berthelsen Gletscher Formation (BG) c. 400 m, basalts and andesites		Hekla Sund Formation (HS) c. 1100 m, basalts and andesites
Ingolf Fjord Formation (IG) c. 200 m, sandstones with local conglomerates		Hovgaard Ø Formation (HG) c. 2500 m, sandstones with local conglomerates (includes dolerites)
base not known		base not known

Figure 4. Stratigraphic scheme for the Independence Fjord Group in the area around Hekla Sund, Kronprins Christian Land. Note that correlation of formations east and west of the Hekla Sund basin is not possible. The codes CM, etc., are those used on the 1:500,000 geological map of Greenland (Jepsen, 2000, sheet 9, Lambert Land).

2002; see below). Above this volcanic unit, there is a succession of ~800 m of cross-bedded, pale yellow– or red-weathering sandstones, referred to as the Caroline Mathilde Alper Formation (CM). Near the base, conglomerate horizons are present. Beds enriched in heavy minerals locally display swash-wash-back separation structures. The sandstones are interpreted as fluvial or nearshore shallow-water deposits.

The lowermost stratigraphic unit in the eastern thrust complex consists of a succession of cross-bedded quartzitic sandstones with numerous dolerite sheets, termed the Hovgaard Ø Formation (HG). It is ~2500 m thick, about half of which is accounted for by dolerites. Toward the top, trough cross-bedded conglomerates are common.

The Hovgaard Ø Formation is overlain by the volcanic Hekla Sund Formation (HS), described in some detail below, which is, in turn, overlain by the ~250-m-thick Lynn Ø Formation (LY, Fig. 4). The latter consists of arkoses and trough cross-bedded, matrix- and clast-supported, polymict conglomerates with pebbles (up to 20 cm across) of gneiss, sandstone, granite, vein quartz, and basalt. Impact structures around volcanic bombs are locally preserved and indicate a terrestrial setting. The formation is interpreted to be mainly fluvial in origin.

Throughout the sedimentary successions in eastern Kronprins Christian Land, there is evidence of syndepositional deformation: sandstone beds truncated by minor listric faults are draped by sandstone unaffected by this faulting. This led Pedersen et al. (1995) to the conclusion that sedimentation, at least intermittently, took place in rift basins subject to active faulting. The occurrence of conglomerates, locally with large boulders of granite and gneiss, is consistent with deposition in a rift basin where fault scarps locally exposed crystalline basement rocks. All the rocks briefly described here are cut by numerous dolerite dikes broadly correlated with the Midsommersø Dolerites of the foreland, although some may have been associated with the volcanic event that produced the Hekla Sund and Aage Berthelsen Gletscher Formations.

The Independence Fjord Group in the Caledonian orogen is unconformably overlain by the Neoproterozoic Hagen Fjord Group. Equivalents of the Zig-Zag Dal Basalt Formation have not been recognized within the orogen.

Hekla Sund and Aage Berthelsen Gletscher Formations

The stratigraphy of the Hekla Sund and Aage Berthelsen Gletscher Formations is depicted in Figure 5. Because of problems due to accessibility and tectonic breaks, logging of the Hekla Sund Formation was carried out in three sections, and relationships between these are not entirely certain. The three sections, termed, from the base upward, HS-1, HS-2, and HS-3 (Fig. 5A; for details, see Pedersen et al., 2002), were sampled in detail, and chemical data are reported below. Within the HS-2 section, two marker horizons of conglomerate and arkosic sandstone occur within the lava succession (Fig. 5A). The second conglomerate in particular is characterized by the presence of large boulders (up to 25 cm) of granitic rocks, sandstones, vein quartz, and basalt. Volcanic sediments are prominent in an ~20-m-wide interval ~50 m above the second conglomerate.

Petrography and Geochemistry

The volcanic rocks of the Hekla Sund and Aage Berthelsen Gletscher Formations have experienced greenschist-facies metamorphism, and all original minerals have been replaced by chlorite, sericite, and epidote. Porphyritic textures are commonly preserved, and pseudomorphs of plagioclase are the most prominent phenocryst mineral.

A detailed treatment of the geochemistry of the Hekla Sund lavas, including Nd isotope data, has been presented by Pedersen et al. (2002), and only some major issues are addressed in this review. Major- and trace-element data were acquired for 52 samples from the Hekla Sund Formation and 8 from the Aage Berthelsen Gletscher Formation. Representative analyses are presented in Table 1, and the chemical variation across the Hekla Sund succession is shown in Figure 6.

Compared to the basalts of the Zig-Zag Dal Basalt Formation, the Hekla Sund lavas are significantly less magnesian: mean Mg# values for the two formations are 39.5 and 27.0, respectively, where Mg# = $100 \times MgO/(MgO + Fe_2O_3^{tot})$. Strong correlations between the concentrations of element pairs such as Ti and Zr indicate relative immobility of these and a number of other elements (Nb, Y, P, rare earth elements [REEs]) during metamorphism. Variations in the compositions of samples from the continuous section HS-2 (and to a lesser extent, the other sec-

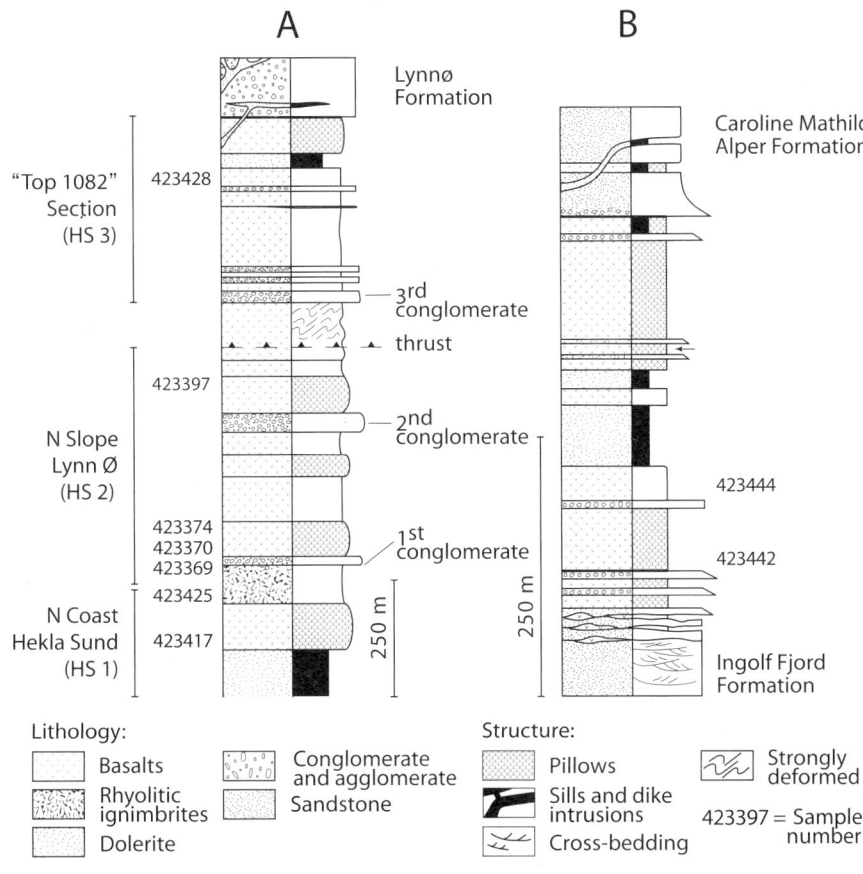

Figure 5. Schematic stratigraphic sections for the Hekla Sund Formation (A) and the Aage Berthelsen Gletscher Formation (B), Independence Fjord Group, modified from Pedersen et al. (2002). The location of analyzed samples given in Table 1 is shown.

tions) can be related to fractional crystallization involving olivine, clinopyroxene, and, probably, plagioclase. A chemical discontinuity occurs within HS-2 across the first conglomerate (Fig. 6). Discontinuities in element ratios such as Zr/Ti and Zr/Nb occur across the second conglomerate and the interval dominated by volcanic sediments 50 m above the second conglomerate (Pedersen et al., 2002). It is envisaged that these sedimentary intervals represent periods of volcanic quiescence, during which fractional crystallization took place at depth. Simultaneously, assimilation of components derived from crustal sources took place, as illustrated by the sudden increase of SiO_2 above the first conglomerate and the interval with volcanic sediments (for detailed discussion, see Pedersen et al., 2002).

The Aage Berthelsen Gletscher Formation has not been geochemically investigated in detail. However, reconnaissance data (Table 1) show that it is somewhat more magnesian than the Hekla Sund Formation (mean Mg# = 29.5), and trace-element data indicate that the two formations were not derived from similar mantle sources (Pedersen et al., 2002).

Accumulation of the Hekla Sund Formation took place ca. 1740 Ma ago (Kalsbeek et al., 1999), shortly after the termination of a major period of Paleoproterozoic crustal accretion and orogenesis in East Greenland (Kalsbeek et al., 1993, this volume). Ca. 1740 Ma A-type granites are among the youngest plutonic rocks emplaced during that period. Such granites are commonly related to late- to postorogenic extension. Because of this similarity in age, and the setting of the volcanic rocks in rifted basins, it appears plausible that the volcanism represented by the Hekla Sund and Aage Berthelsen Gletscher Formations was related to the same event of postorogenic extension. Both the late A-type granites and the volcanic rocks may have been the result of crustal underplating during extension and concomitant pressure release on underlying mantle sources.

THE INDEPENDENCE FJORD GROUP IN THE FORELAND

The foreland of the East Greenland Caledonides in North Greenland was mapped by the Geological Survey of Greenland in the late 1970s. The geology is documented on the 1:500,000 *Geological Map of Greenland*, sheet 8, Peary Land (Bengaard and Henriksen, 1986). The lowest exposed stratigraphic unit in this region is the Independence Fjord Group, which is composed of a succession of clastic sediments on the order of 2 km thick that outcrops across a wide tract of North Greenland between Heilprin Land in the west and Mylius Erichsen Land in the east (Fig. 1). Its area of outcrop and its observed lower stratigraphic boundary are determined by the margin of the Inland Ice, and the outcrop belt is fragmented by major glaciers, particularly Academy Gletscher and Hagen Bræ. The succession is only slightly disturbed by

TABLE 1. REPRESENTATIVE ANALYSES OF SAMPLES FROM THE HEKLA SUND AND AAGE BERTHELSEN GLETSCHER FORMATIONS

	\multicolumn{9}{c}{Sample numbers}								
	423417 HS-1	423425 HS-1	423369 HS-2	423370 HS-2	423374 HS-2	423397 HS-2	423428 HS-3	423442 ABG	423444 ABG
SiO_2 (wt%)	49.11	72.26	49.49	54.28	49.37	57.16	52.79	50.27	46.98
TiO_2	1.21	0.38	1.06	2.68	1.36	1.19	1.88	1.57	1.54
Al_2O_3	17.06	12.29	18.29	12.74	16.30	13.86	14.94	14.47	15.48
Fe_2O_3	12.42	4.29	11.27	13.87	12.06	9.56	10.46	13.19	14.10
MnO	0.24	0.06	0.15	0.18	0.16	0.12	0.13	0.14	0.18
MgO	6.71	0.32	5.21	2.02	4.73	3.10	5.37	5.62	6.34
CaO	4.97	1.09	6.37	6.64	9.08	9.58	6.54	8.47	9.09
Na_2O	4.52	4.35	4.52	2.10	3.05	1.59	4.54	1.89	1.91
K_2O	0.62	4.12	0.37	1.37	1.04	0.62	0.25	0.03	0.54
P_2O_5	0.28	0.04	0.19	0.90	0.31	0.32	0.56	0.25	0.23
LOI*	2.57	0.42	3.33	2.82	2.49	2.43	2.01	3.71	3.31
Sum	99.71	99.62	100.24	99.61	99.95	99.52	99.46	99.62	99.70
Mg#	35.1	6.9	31.6	12.7	28.2	24.5	33.9	29.9	31.0
Nb (ppm)	7.6	34	6.6	32	9.3	9.6	18	3.3	3.4
Zr	132	721	123.2	380	156	151	297	73	78
Y	30	79	25	53	31	32	43	24	23
Sr	195	62	291	681	614	568	788	524	262
Rb	16	86	9.0	42	19	13	5.1	0.7	11
Th	2.5	16	2.4	6.9	2.6	2.6	3.7	1.1	0.9
Pb	20	4.5	4.5	20	8.3	13	4.7	3.9	1.6
La	27	122	20	88	28	34	44	11	11
Ce	58	274	50	202	68	79	111	27	26
Nd	32	112	26	90	34	37	53	18	17
Ni	44	n.d.	30	4.4	26	20	62	104	154
Cr	94	n.d.	42	n.d.	30	29	117	147	98
V	124	n.d.	146	165	175	224	185	337	307
Ba	451	1324	275	1378	697	539	229	38	229
Sc	35	1.9	24	21	27	22	31	26	31

Note: Samples were analyzed by X-ray fluorescence spectrometry at the University of Edinburgh, major elements on fused glass discs, trace elements on pressed powder tablets (Fitton et al., 1998). HS-1, HS-2, and HS-3 represent three sampled sections of the Hekla Sund Formation in ascending order. #423425 is the dated rhyolite from HS-1. Samples marked ABG are from the Aage Berthelsen Gletscher Formation. For stratigraphic positions of the analyzed samples, see Figure 5.
*LOI—loss on ignition.

tectonic deformation, so the stratigraphy can be traced with a high level of confidence over long distances. The great continuity of stratigraphic units, the very gradual changes in their thickness, and the nonmarine nature of the sediments all suggest that the succession is the fill of a widespread intracratonic sag basin, probably similar to those of the present-day Chad Basin of Africa and the Eyre Basin of Australia. Probable stratigraphic equivalents can be locally recognized in the zone of Ellesmerian deformation to the north and, more extensively, in the belt of more intense Caledonian deformation to the east and south.

Structural Setting

The Independence Fjord Group occurs principally in a virtually undeformed cratonic setting, inboard from zones of intense Caledonian and Ellesmerian deformation to the east and north, respectively. In spite of the intensity of this nearby deformation, disturbance within the main outcrop of the Independence Fjord Group is limited to widespread, uniform dips, usually on the order of 1°–2°. Steeper dips only occur close to large dolerite intrusions. Across Mylius Erichsen Land and J.C. Christensen Land, the gentle regional dip is toward the northeast, and the sediments are overlain conformably by the Zig-Zag Dal Basalt Formation. In the west, in Heilprin Land and southern Peary Land, the Independence Fjord Group is overlain unconformably by the widespread Lower Cambrian Portfjeld Formation. Locally, however, the late Precambrian Moræneso Formation is preserved beneath the flat-lying base-Cambrian unconformity in paleovalleys incised into the Independence Fjord Group (Jepsen, 1971; Collinson et al., 1989).

The sedimentary succession is inferred to lie nonconformably on Archean crystalline basement (erratic blocks of which are found throughout the region), where the contact lies beneath the Inland Ice. At the head of Victoria Fjord, 100 km west of the Independence Fjord Group outcrop, Cambrian sediments lie nonconformably on such basement, with older sediments cut out and overlapped in the intervening area, so that the lowest parts of the succession are unknown.

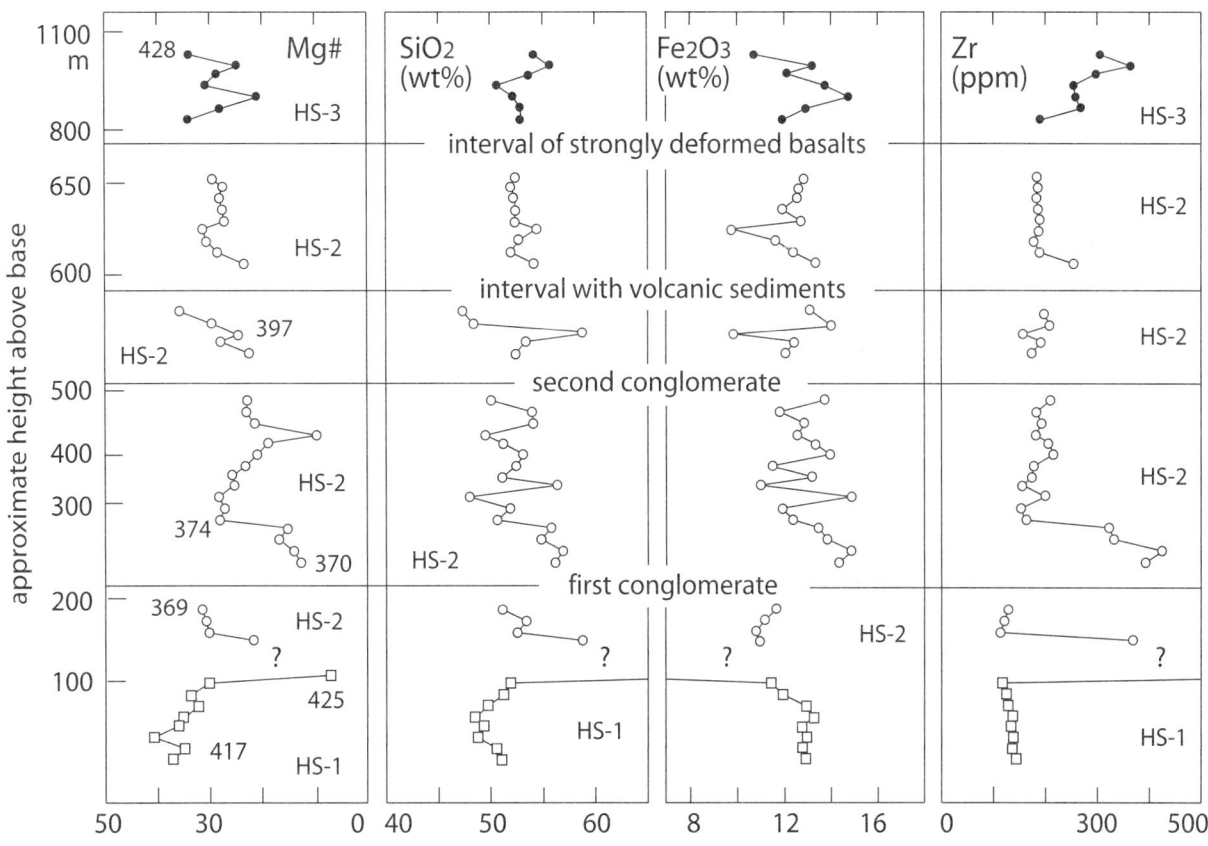

Figure 6. Chemical variation with height within the Hekla Sund Formation. Mg# = 100 × MgO/(MgO + $Fe_2O_3^{tot}$). Concentrations of SiO_2, Fe_2O_3, and Zr were recalculated on a volatile-free basis of 100%. To promote readability, the samples are not shown at their precise height, but with equal distances; see uneven scale at left-hand margin. Sample numbers (only last three digits shown) are for samples listed in Table 1.

Stratigraphic Framework

Collinson (1980) reviewed the internal stratigraphy of the Independence Fjord Group, and a formal system of formations and members was proposed (Fig. 7). The area of outcrop of the group was divided into two subareas, each of which is nominally occupied by a different formation. The precise stratigraphic relationship between these two formations is not known, but general mapping considerations suggest that there is some overlap between them. The intensity with which the sediments were disturbed by dolerite intrusions in the northwest of the area means that is impossible to propose detailed correlations across the Academy Gletscher, and it is also very difficult to map and correlate units west of the glacier. Accordingly, all the sediments in the west are assigned to the Inuiteq Sø Formation (Jepsen, 1971), within which no clear succession has been established. Two units

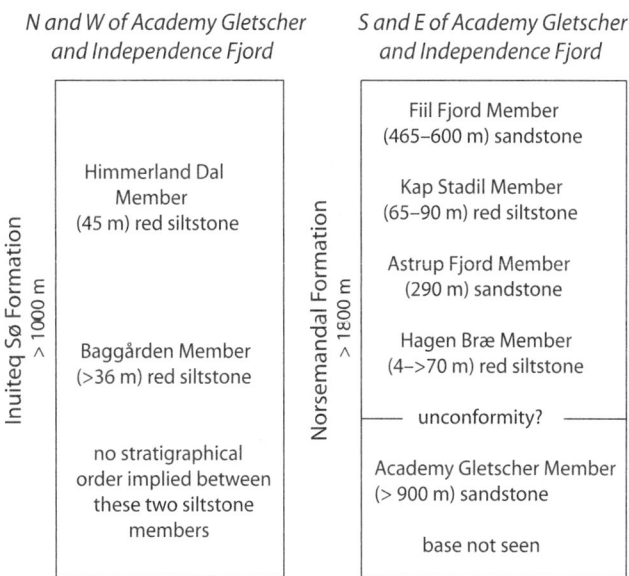

Figure 7. Stratigraphic subdivision of the Independence Fjord Group in the Caledonian foreland of eastern North Greenland (area A, Fig. 1). The relationships between the Inuiteq Sø Formation and the Norsemandal Formation are not known, and there may be overlap between them.

of siltstone are designated as members within the Inuiteq Sø Formation, but their stratigraphic relationships are unknown.

By contrast, in the east, the succession is confidently subdivided into five members, which constitute the Norsemandal Formation (cf. Norsemandal Sandstones of Adams and Cowie, 1953) (Fig. 8). These members have been mapped with a high level of confidence from Independence Fjord to the southern part of Mylius Erichsen Land, a distance of some 170 km. The five members comprise three major sandstone units, each several hundreds of meters thick, separated by two relatively thin, fine-grained members that are dominated by red siltstones, each a few tens of meters thick. In spite of their relatively small thickness, these siltstone units maintain their continuity across the area and are the main means by which the Norsemandal Formation is mapped.

The relationship between the Inuiteq Sø Formation and the Norsemandal Formation is unclear. On general structural and geographic grounds, the Inuiteq Sø Formation should lie somewhat lower in the overall succession than the majority of the Norsemandal Formation. It is likely that the lowest unit of the latter, the Academy Gletscher Member, is a partial equivalent of the Inuiteq Sø Formation because there is no evidence of any siltstones in the intensely intruded area around the head of Independence Fjord. On that basis, it is unlikely that either of the siltstone members within the Inuiteq Sø Formation is equivalent to siltstone members of the Norsemandal Formation. If that is the case, there are four siltstone units within the exposed Independence Fjord Group succession.

The top surfaces of the sandstone members directly beneath the red siltstone members in both the Inuiteq Sø Formation and in the Norsemandal Formation show evidence of prolonged nondeposition and, in the case of the Hagen Bræ Member, of an interval of significant erosion that included the development of several tens of meters of relief (Collinson, 1983; Fig. 9). While

Figure 8. Parts of the Norsemandal Formation on the north side of Hagen Bræ. The lowest unit is the Astrup Fjord Member (AF, sandstones), while the top of the cliff is the Fiil Fjord Member (FF, sandstones). KS represents the siltstones of the Kap Stadil Member. The thick, dark unit (d) is a dolerite sill intruded between the Astrup Fjord Member and the siltstones of the Kap Stadil Member. The cliff is ~1000 m high.

the base of the Hagen Bræ Member along the northwest side of Hagen Bræ is the only example that shows direct evidence of erosional downcutting associated, other surfaces that lack relief show evidence of long-term emergence.

Beyond the main areas of outcrop, a probable equivalent of the Independence Fjord Group occurs within an area of more intense deformation in northeastern Peary Land, where quartzites with dolerite intrusions and basalts occur (Christie and Ineson, 1979). These include a 40-m-thick interval of red siltstone with halite pseudomorphs, similar to red siltstone occurrences in the main outcrop. The occurrence of basalts directly overlying the sandstones suggests correlation with the uppermost part of the Norsemandal Formation, the Fiil Fjord Member, while the siltstone unit may be equivalent to the upper siltstone unit, the Kap Stadil Member. If so, the lateral extent of that unit is at least 220 km.

The age of the Independence Fjord Group is partially constrained by a U-Pb baddeleyite age of 1382 ± 2 Ma on one of the intruding dolerites (Upton et al., 2005). Rb-Sr isotope data on clay minerals from one of the siltstones suggest diagenetic ages of around 1380 Ma, around the time of dolerite intrusion (Larsen and Graff-Petersen, 1980). Together with the 1740 Ma age of the lavas from the Hekla Sund Formation, these dates constrain the depositional age between ca. 1750 and ca. 1400 Ma.

Sedimentology

The sedimentology of the Independence Fjord Group can be divided into two fairly distinct facies associations, the sandstones that make up the major part of the succession and the red siltstone intervals that provide a basis for the limited stratigraphic subdivisions and mapping within the group. The bases of the red siltstone intervals are generally much more sharply defined than their tops, which typically show a more gradational passage as the succession coarsens upward into the overlying sandstone member.

Figure 9. Inclined erosion surface with small steps on top of the horizontally bedded Academy Gletscher Member (AG) at Hagen Bræ. The wedge of dipping sandstone beds may record eolian deposition prior to flooding of the surface. The sandstones are overlain by red siltstones of the Hagen Bræ Member (HB). Maximum topography on this surface is more than 70 m. Person (in box) for scale.

Siltstones

The sedimentology of the red siltstones and their basal surfaces has been described in detail elsewhere (Collinson, 1983), and here we present a brief précis of that description. The great lateral extent (>150 km) of the red siltstone units, at least in the Norsemandal Formation, is important in establishing their sedimentologic context. Their thicknesses are typically tens of meters. Lateral variations in thickness can only be demonstrated within the Norsemandal Formation, since the examples within the Inuiteq Sø Formation cannot be mapped and correlated with any confidence. The lower red siltstone of the Norsemandal Formation, the Hagen Bræ Member, shows the most dramatic variation in thickness. On the northwest side of Hagen Bræ, it is over 70 m thick, where it fills the deep topographic lows eroded into the top of the underlying Academy Gletscher Member. The "hills" and "valleys" that characterize this paleotopography are on the order of hundreds of meters in width. The bases of the deepest incisions are not visible. By contrast, on the southeast side of Hagen Bræ, some 10 km distant, the Hagen Bræ Member is only 4 m thick.

The upper siltstone unit, the Kap Stadil Member, shows a more gradual thickness variation, thinning from 90 m at Independence Fjord to 65 m at Hagen Bræ, a distance of 65 km. This thinning may continue farther to the southeast, but the 45 m thickness recorded in central Mylius Erichsen Land is only a minimum. In spite of their widespread occurrence, accurate thicknesses of the siltstone units can only be established in areas of steep topography. Over most of the outcrop, they have been mapped by tracing reddish scree across frost-shattered regolith on plateaus. The stratigraphically unconstrained siltstone members in the Inuiteq Sø Formation cannot be traced sufficiently far to establish systematic thickness changes.

The siltstone intervals are, in reality, interbedded successions of red siltstones and thin beds of typically fine-grained sandstone (Fig. 10). The siltstones themselves are quite coarse grained and tend to be rather homogeneous. They are strongly stained by hematite and locally have irregular concretions of dolomite. The siltstones also include intraformational breccias of slightly darker red mudstone clasts, suggesting local rip-up of surface mud veneers. Thin (usually less than 50 cm) units of dolomite, with fine stromatolitic lamination and small-scale dome form, are less common.

The sandstone beds have sharp bases and parallel margins and are typically between a few millimeters and a few tens of centimeters thick (Fig. 11A). Rather more rarely, sandstone beds up to 2 m thick occur (Fig. 11B), most commonly in the upper parts of the siltstone intervals, as part of an upward coarsening into a sandstone member. Sandstone beds may be quite widely spaced within the siltstones, or they may be closely stacked, with only thin siltstone partings. While the sandstone beds do not show conspicuous graded bedding, the lower parts of some have concentrations of mud clasts. Upper bedding surfaces of the sandstones commonly show ripples, both current ripples and, more commonly, wave ripples, which frequently have interference patterns (Fig. 11C). Lower bedding surfaces commonly show desiccation cracks and halite pseudomorphs (Fig. 11D). Paleocurrent data are quite scarce in the red siltstone units, but, in central Mylius Erichsen Land, a small sample ($n = 7$) of cross-bedding and current ripple directions from the Kap Stadil Member gives a rather consistent direction toward ENE (~070°). A smaller sample ($n = 4$) from the same member at Independence Fjord gives rather consistent directions around 250°.

The whole sedimentary assemblage of the siltstone units suggests widespread, shallow, and frequently desiccated saline lakes with internal drainage and very low relief, probably similar to present-day Lake Chad or Lake Eyre (Fig. 12). An arid or semiarid setting is implicit. The sandstones are products of episodic flood events that inundated the lake. Following a flood, a body of shallow water persisted for a time and allowed locally generated waves to rework the upper surfaces of sand layers into wave ripples. Interference patterns attest to shallow water and a variable wind regime. The paleocurrents suggest that sheet floods extended into the lakes from more than one direction.

The siltstones themselves probably result mainly from suspension during periods of flooding when sandy sheet floods did not reach the area. In saline lakes, river flows would have led to extensive freshwater plumes that distributed suspended load over large areas. The thin silt partings that separate closely stacked sandstones may be products of the late stages of the preceding sand-bearing flood. It is also likely that some of the silt was delivered or reworked by wind. Stromatolitic dolomites are probably products of prolonged lake highstands when terrigenous supply was excluded, probably because the shoreline was far away at the time.

Sandstones

The sandstone members of both the Norsemandal and Inuiteq Sø Formations can largely be described together. They are typically hundreds of meters thick and commonly have rather gradational contacts with the underlying siltstone members. Their upper boundaries are generally sharp, and there is evidence of nondeposition and prolonged emergence. They are medium to coarse grained and typically show moderate to good sorting. Compositionally, sandstones range from quartzites to arkoses. The Astrup Fjord Member is the most conspicuously feldspathic, though it has an upward trend of diminishing feldspar content. Separate from these compositional differences, some of the sandstone members show diagenetically determined, stratiform color differences, usually more or less red coloration. These color differences maintain continuity over many kilometers but do not lead to any more refined sedimentologic understanding.

Internally, sandstone members consist of interbedded units of cross-bedded sandstone and horizontally bedded sandstone (Fig. 13) with no obvious patterns or trends. The cross-bedded sandstones include both trough sets, typically 20–50 cm thick, as well as thicker tabular sets, typically 1 m but up to 2.5 m thick (Fig. 14A). The smaller trough sets are inferred to be of fluvial origin, where co-sets up to a few meters thick are

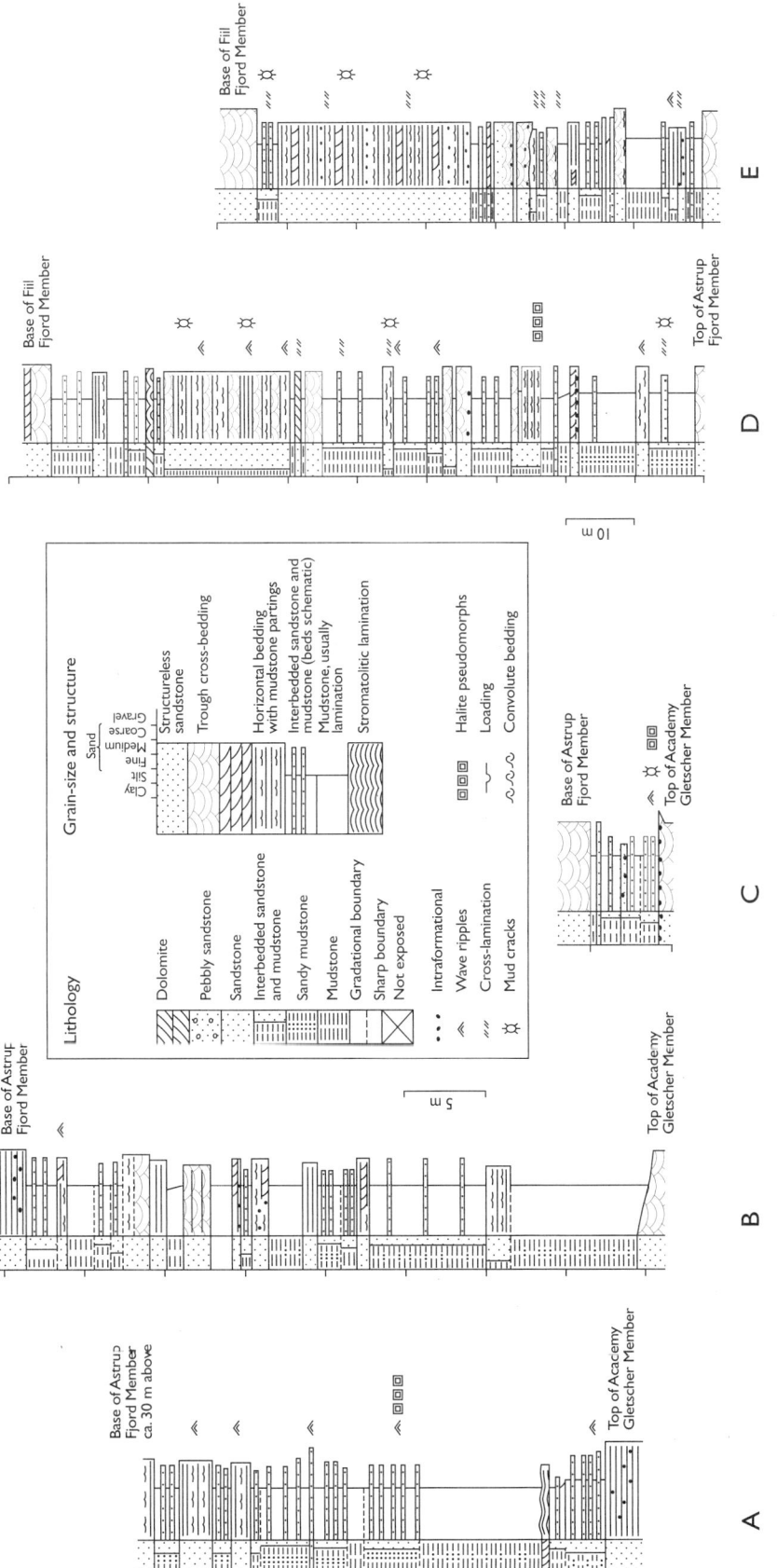

Figure 10. Measured sections in the siltstone members of the Independence Fjord Group (modified from Collinson, 1983). Note the sharp bases to the intervals and the more gradational passage to the overlying sandstones. (A–C) Hagen Bræ Member: (A) south side of Hagen Bræ; (B) north side of Hagen Bræ where it is associated with topographic relief at the lower contact; and (C) central J.C. Christensen Land. (D–E) Kap Stadil Member: (D) south coast of Independence Fjord; and (E) north side of Hagen Bræ. Legend in box.

Figure 11. Sedimentary features of the siltstone members. (A) Thin sandstone and siltstone interbedded with compaction-folded desiccation crack fills in the siltstones, Inuiteq Sø Formation. The object shown for scale is 5 cm wide. (B) Meter-scale interbedding of siltstones and sandstones in the Hagen Bræ Member. Height of section is ~20 m. (C) Wave ripples on upper bedding surfaces of thin sandstone beds in the Hagen Bræ Member. Hammer for scale. (D) Halite pseudomorphs on the base of thin sandstone bed, Kap Stadil Member. Width of photograph is ~10 cm.

products of channel migration. Rare pebble horizons may be lag deposits associated with shifting channel bases. Larger tabular sets are less easily interpreted. Some could be the products of larger fluvial bed forms, but some could equally be eolian in origin. Significant proportions of sand grains are well rounded, suggesting abrasion during eolian transport. The discrimination of eolian strata based on detailed observation of lamination types was not possible during the wide-ranging mapping study carried out in 1978 and 1979.

Horizontally bedded sandstones are subordinate to the cross-bedded units but occur scattered amongst them in units ranging in thickness from a few centimeters to several meters. The sandstones occur as essentially parallel-sided beds separated by thin silty partings, typically millimeters in thickness. The upper sur-

Figure 12. The margin of Lake Eyre, central Australia, showing the extension of a low-relief fluvial distributive system into the evaporitic lake.

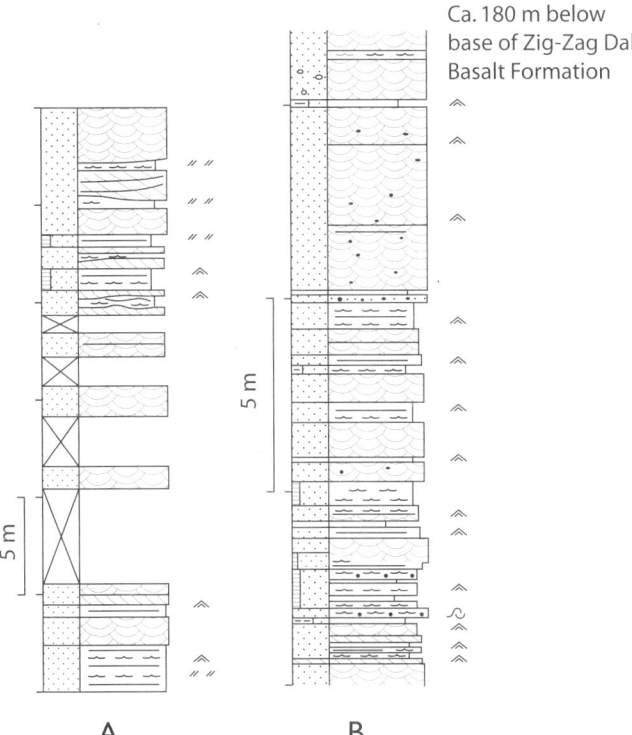

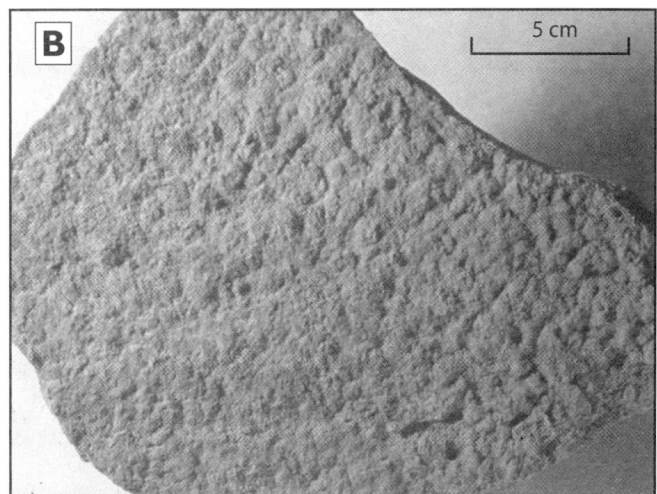

Figure 13. Measured sections in the sandstone members of the Norsemandal Formation. (A) Fragmentary section in the upper part of the Academy Gletscher Member, valley south of Astrup Fjord. The interbedding of cross-bedded and more thinly bedded sandstone is fairly typical of the Member. (B) Representative section in the middle of the Fiil Fjord Member, north of Hagen Bræ. Legend as in Figure 10.

Figure 14. Sedimentary features of the sandstone members. (A) Large-scale cross-bedding with multiple reactivation surfaces, possibly of eolian origin, Inuiteq Sø Formation. The sandstone block in the foreground is ~60 cm high. (B) Inferred wind adhesion warts on the upper surface of a sandstone, Astrup Fjord Member.

faces of the sandstones commonly show ripples, in some cases current ripples, but more usually wave ripples. More rarely, upper surfaces show structures interpreted as wind adhesion ripples (Fig. 14B). The bases of sandstones commonly show patterns of polygonal desiccation cracks and also shorter, irregular cracks that may record synaeresis or evaporite pseudomorphs.

Paleocurrent data from the sandstones show rather variable directions, both within and between the major sandstone members (Fig. 15). Some of this variation possibly stems from a lack of systematic discrimination between fluvial and eolian sets, but some also reflects changing paleogeography. In the Inuiteq Sø Formation, the dominant cross-bedding direction is toward the northeast, though there are minor modes to other directions except the northwest. This trend also dominates in the Academy Gletscher Member, which may be a partial equivalent to the Inuiteq Sø Formation. In this member, there is a rather strong unimodal trend amongst data that were collected mainly from J.C. Christensen Land. In the Astrup Fjord Member, there appear to be regional differences: there is a strong mode toward the northeast in central Mylius Erichsen Land and a less well-defined trend toward the south amongst a small data set from J.C. Christensen Land. In the uppermost sandstone unit, the Fiil Fjord Member, a strong trend toward the north and northeast is apparent in Mylius Erichsen Land. Data from elsewhere in this member are insufficient to identify clear trends. Overall, directions suggest that most sand was deposited by flows from the south or southwest but that other directions were active from time to time. From this, it is inferred that the center of the basin lay to the northeast of the present outcrop. Wave ripple crests show very variable orientations and no clearly preferred directions in most stratigraphic units. This suggests a variable wind regime.

On the basis of this evidence, the sandstone units are thought to be products of extensive and weakly channelized fluvial distributive systems that delivered sand during floods across low-

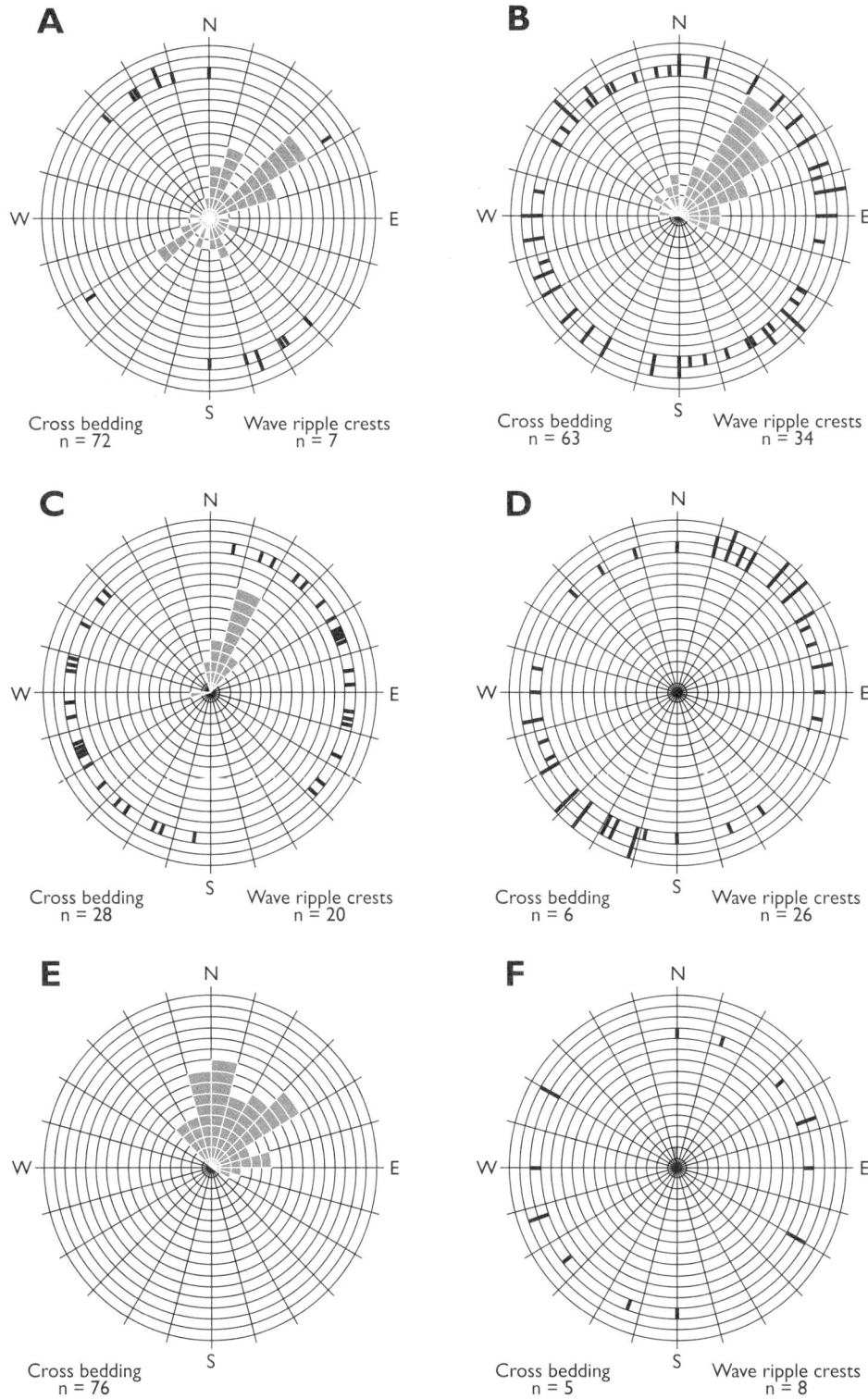

Figure 15. Paleocurrent roses for various sandstone units of the Independence Fjord Group. These show paleoflow directions derived from cross-bedding (dark gray) and the orientations of the crests of wave ripples (black). The cross-beds are dominantly fluvial, and no attempt has been made to separate the eolian cross-bedding. (A) Inuiteq Sø Formation. (B) Academy Gletscher Member, mainly J.C. Christensen Land. (C) Astrup Fjord Member, central Mylius Erichsen Land. (D) Astrup Fjord Member, J.C. Christensen Land. (E) Fiil Fjord Member, Mylius Erichsen Land. (F) Fiil Fjord Member, outside Mylius Erichsen Land.

gradient surfaces. Following flood events, large areas were inundated as shallow lakes across which winds blew to raise waves capable of reworking sand into ripples. Quiet periods allowed thin veneers of mud to be laid down, which, on emergence, became desiccated and cracked. Evaporites were locally precipitated. Between flood events, emergent sandy areas were subjected to reworking by wind. Although possible eolian dune units are scattered and quite thin, the occurrence of abundant rounded sand grains suggests that eolian transport and reworking was common. Sand probably went through many cycles of fluvial and eolian transport before coming to its final resting place, in a similar way to the situation in inland basins in Australia (Fig. 16). The occurrence of wind adhesion ripples suggests that the water table was at times sufficiently high to give a damp sediment surface.

Sandstone-Siltstone Boundaries

The boundaries between the siltstone and sandstone members record the major facies shifts that occurred during the life of the basin, and, as such, they may hold clues to the understanding of larger-scale controls, such as basin tectonics or paleoclimate. There is a clear asymmetry in the character of these basinwide facies transitions. Upward changes from sandstones to siltstone units are very sharp, and upward changes from siltstone to sandstone are gradational. At first sight, this suggests that the bases of the siltstones are flooding surfaces, recording widespread expansion of the lake and retreat of the systems that delivered abundant sand, presumably back toward the basin margins. However, the situation appears to be more complex than simply a succession of lake expansions. The upper boundaries of sandstone members show evidence of having been subaerially exposed for significant periods of time prior to deposition of the overlying siltstones (Collinson, 1983). There is clear evidence of at least partial lithification of the sands prior to flooding. Layers and pockets of conglomerate, made up of lithified sandstone clasts and silica-cemented concretions of inferred pedogenic origin, occur on top of a sandstone unit within the Inuiteq Sø Formation (Fig. 17A). A steep-sided, conglomerate-filled fissure occurs on top of another sandstone interval in the same formation (Fig. 17B). Most spectacularly, the top surface of the Academy Gletscher Member on the north side of Hagen Bræ shows relief of at least 70 m (Fig. 18). The paleotopography is locally steep, implying lithification, and overlying wedges of sediment close to inclined surfaces include blocks of sandstone like those below the erosion surface. Some of these inclined flanking beds are bedded sandstones, suggesting that lithification was not intense and that grain-by-grain erosion also took place. It seems possible that some inclined sandstone beds close to the steep unconformity could have resulted from eolian deposition in the lee of the topography. Prior to the onset

Figure 16. Part of the Strzelecki Desert, central Australia, showing the interaction of eolian and fluvial processes. Older eolian linear dunes in the darker area at the top of the photograph are eroded and reworked by a fluvial channel, and the fluvial sands are in turn reworked by the wind to create the dunes in the pale sands in the foreground. A similar interaction is envisaged for some of the sandstone members of the Independence Fjord Group.

Figure 17. Small-scale features of the top surfaces of sandstone members in the Inuiteq Sø Formation. (A) Breccia of lithified sandstone clasts derived from the underlying unit. (B) Steep-sided fissure in the top of a sandstone unit, filled with poorly sorted conglomerate. The object shown for scale is 5 cm wide.

of silt deposition, a thin unit of laminated dolomite and chert, with evidence of evaporites, was deposited, suggesting starvation of clastic supply following submergence.

These patterns of deposition and diagenesis suggest that a decrease in sand supply coincided with or was closely followed by a phase of early cementation. This was followed by a period of erosion, which, on at least one occasion, generated significant relief. When deposition resumed, it consisted of fine-grained sediments, in one case preceded by an episode of chemical deposition.

Paleogeography and Basin Tectonics

The wide extent of the various stratigraphic units within the Independence Fjord Group and the nature of the constituent facies suggest that deposition took place in a very extensive, low-relief, fluvio-lacustrine basin in a cratonic setting, probably with no connection to the ocean. Cratonic sag basins such as the Lake Chad or Lake Eyre basins provide at least partial analogues. Sand was probably supplied from several directions from flanking highlands, although flow from the southwest appears to have dominated. The center of the lake basin probably lay to the northeast of the present outcrop area. The paleolatitude is uncertain, but the climatic regime appears to have been arid with episodic floods, similar to that prevailing in central Australia at the present day. Sand supplied initially by rivers during flood events was subjected to eolian reworking by winds with rather variable directions. While it is possible that the large-scale, apparently basinwide changes in facies could reflect a long-term climatic cyclicity, we think it more likely that the variation resulted from tectonic controls stemming from thermal processes beneath the sag basin. In particular, it is difficult to envisage the generation of over 70 m of erosional relief entirely as a result of climatically induced changes in lake level. A succession of thermal events in the lower crust and upper mantle seems most likely to have driven periods of alternating uplift and subsidence within the basin and may also have been linked to uplift of surrounding source areas. The extrusion of the overlying Zig-Zag Dal Basalt Formation and the associated emplacement of the Midsommersø Dolerites may have been further manifestations of these processes.

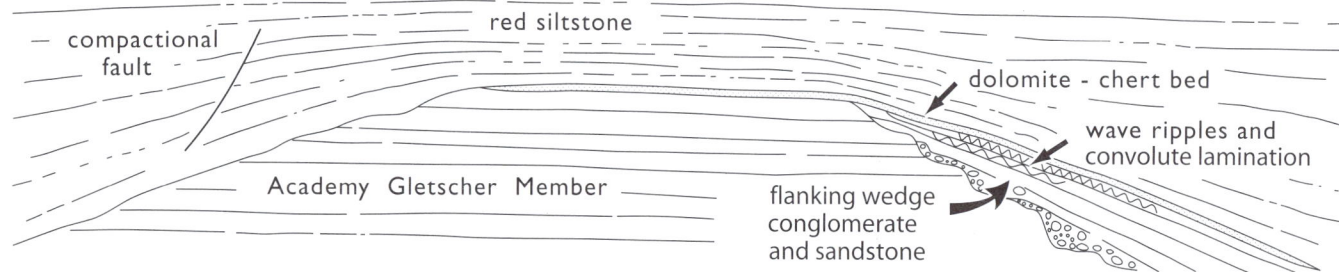

Figure 18. Schematic section showing the main features of the topography at the boundary between the Academy Gletscher Member and the Hagen Bræ member at Hagen Bræ. Typical relief is 8–10 m, and horizontal extent of the "paleohigh" is in the range of 50–200 m.

THE TREKANT SERIES IN DRONNING LOUISE LAND

The Trekant series (Peacock, 1956, 1958) is widely exposed in the westernmost parts of Dronning Louise Land. It is composed of an up to ~500-m-thick succession of sandstones, siltstones, and conglomerates, which unconformably overlie the crystalline basement (Fig. 19). The lower parts of the succession consist of irregularly bedded conglomerates with angular clasts of orthogneiss. The conglomerates pass upward into cross-bedded quartzitic and feldspathic sandstones, which are locally interlayered with beds of siltstone. Desiccation cracks and ripple marks are commonly preserved at the top of siltstone beds. The Trekant series is interpreted to have been deposited in a fluvial environment (Friderichsen et al., 1990).

MIDSOMMERSØ DOLERITES

The sandstones of the Independence Fjord Group are cut by a multitude of sheets of dolerite and associated more siliceous rocks (Fig. 20). Some sills are more than 100 m thick and can be followed along fjord walls for tens of kilometers. In many areas, intrusions form the largest component in the stratigraphy (Fig. 20A), and their aggregate width is locally >1000 m. The intrusions were termed the "Midsommersø Dolerites," after Midsommersøerne (= Midsummer lakes, in southern Peary Land) by Jepsen (1971). U-Pb analyses on baddeleyite from one dolerite have yielded an age of 1382 ± 2 Ma (weighted mean of three ^{206}Pb/^{207}Pb ages; Upton et al., 2005). Earlier Rb-Sr whole-rock isotope data (Kalsbeek and Jepsen, 1983) yielded an erroneous

Figure 19. Sandstones and conglomerates of the Trekant series (correlated with the Independence Fjord Group of North Greenland) unconformably overlying crystalline basement rocks, Dronning Louise Land. Cliff section is ~400 m high.

Figure 20. Intrusions of the Midsommersø Dolerite suite. (A) Sheets and dikes of dolerites, which make up a large proportion of an ~800-m-high cliff, inner Independence Fjord. The country rocks are sandstones of the Independence Fjord Group. (B) Doleritic and rheopsammitic intrusions in sandstones of the Independence Fjord Group. Cliff section is ~400 m high, Vildtland (locality marked 3 in Fig. 1). The lowermost subhorizontal intrusion is a dolerite, while most other intrusions are rheopsammites with dark doleritic borders. (C) Felsic granophyric rock with quartzitic inclusions at Kap Einar Mikkelsen (locality marked 1 in Fig. 1).

isochron age of 1230 ± 20 Ma, probably due to later disturbance of the Rb-Sr isotope systems (Kalsbeek and Frei, 2006).

The presence of mafic intrusions in eastern North Greenland was first reported by members of the "First Thule Expedition" in 1912, led by Knud Rasmussen (Freuchen, 1915). Ellitsgaard-Rasmussen (1950) presented the first detailed description of their field relationships, and Jepsen (unpublished thesis, 1969) presented petrographic and geochemical data. Comprehensive investigations that included large numbers of major- and trace-element analyses, as well as Sr, Nd, and Pb isotope data, have been presented by Kalsbeek and Jepsen (1983) and Kalsbeek and Frei (2006).

The intrusive rocks can be divided into three general groups: (1) normal, dark gray to black dolerites, (2) totally altered, red-brown to brick-red or red and greenish mottled rocks, grading from dolerites to fine-grained granophyric rocks, and (3) very siliceous (up to ~90 wt% SiO_2), fine-grained, yellow, pink, or red rocks, termed "rheopsammites" by Kalsbeek and Jepsen (1983). Intrusions of categories 2 and 3 are always bordered by dark dolerite (Fig. 20B). The different rock types are more or less contemporaneous; intersections are common but do not reveal any systematic differences in age.

Petrography, Geochemistry, and Petrogenesis

Eight intrusions, from five localities (Fig. 1), were sampled and studied in detail. They show a very wide variation in compositions: SiO_2 = 50%–90% and $Na_2O + K_2O$ = 2%–9% (Fig. 21). This variation has been interpreted by Kalsbeek and Jepsen (1983) as the result of three main processes: (1) mixing of magmas derived from mantle and crustal sources; (2) fractional crystallization, both before and after emplacement into individual intrusions; and (3) metasomatic changes related to intense hydrothermal alteration. It should be noted that, despite intense alteration, all these rocks appear totally fresh in hand specimen.

A well-preserved dolerite sill at Kap Einar Mikkelsen (KEM-1, Fig. 21; locality 1 on Fig. 1) consists mainly of plagioclase, augite, and pigeonite; pseudomorphs of olivine are present in the lower part of the sheet, and magnetite becomes abundant higher up in the intrusion. Samples from KEM-1 have the lowest concentrations of SiO_2 (~50%) and $Na_2O + K_2O$ (2%–3%) of the investigated rocks (Table 2; Fig. 21), and they most closely represent a mantle-derived magma. The 1382 ± 2 Ma U-Pb baddeleyite age (Upton et al., 2005) is from a sample from this intrusion.

Another intrusion at Kap Einar Mikkelsen (KEM-2, Fig. 21), consists of very fine-grained and strongly altered granophyric rock (Fig. 20C). It has a granitic composition, close to the composition of the granitoid basement underlying the Independence Fjord Group (Table 2). This intrusion has been interpreted to represent a crustally derived magma. An attempt to recover zircon from the rock for age determination failed.

The upper parts of a variably altered composite dolerite intrusion in upper Zig-Zag Dal (ZZD-1, locality 5 on Fig. 1) are enriched both in SiO_2 and $Na_2O + K_2O$ relative to most other dolerites (Fig. 21; Table 2); these are interpreted to be the result of mixing of basic magma and crustally derived melts.

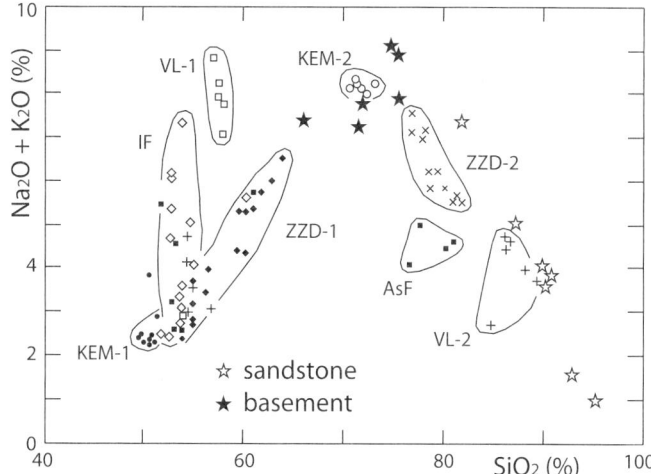

Figure 21. $Na_2O + K_2O$ versus SiO_2 diagram for the Midsommersø Dolerites and associated intrusions, showing their wide chemical variation (adapted from Kalsbeek and Frei, 2006). For discussion, see text. The codes VL-1, KEM-2, etc., refer to the different intrusions studied. Note that the acidic intrusions, such as those of VL-2 and ZZD-2, have dark doleritic borders that plot far off the fields of the siliceous samples. Sandstone samples were collected from outcrops of the Independence Fjord Group; basement samples were taken from ice-transported blocks, which are of widespread occurrence in the valleys.

An intrusion on Vildtland (VL-1; locality 3 on Fig. 1) consists of a very fine-grained brick-red rock. Intense hydrothermal alteration has strongly affected the magmatic mineralogy; some fresh augite is preserved, but the plagioclase is totally sericitized, and the rock contains large proportions of fine-grained secondary minerals. During alteration, a large proportion of Ca was removed and replaced by Na and K, resulting in high $Na_2O + K_2O$ (Table 2; Fig. 21). The border of this intrusion is a doleritic rock with SiO_2 ~ 55% and $Na_2O + K_2O$ ~ 3%. Parts of a dolerite sheet near Independence Fjord (IF; locality 2 on Fig. 1) consist of a similar red dolerite that has undergone hydrothermal alteration, resulting in increased $Na_2O + K_2O$ at more or less constant SiO_2 (Fig. 21).

Large sheets of very fine-grained rheopsammite (85–89 wt% SiO_2; Table 2; Fig. 21) are prominent on Vildtland. They are invariably mantled by normal dolerite (Fig. 20B). One of these intrusions (VL-2) was studied in detail. In thin section, quartz occurs both as pseudomorphs after tridymite and as corroded rounded grains. These rocks were interpreted to represent mobilized feldspathic sandstones by Jepsen (1969). It is envisaged that during hydrothermal activity at high temperatures, enough melt was formed from quartz and K-feldspar to mobilize the rock to such an extent that it could intrude together with the basic magma. The tridymite was interpreted to have been crystallized from the melt, whereas the corroded rounded quartz grains were believed to represent remnants of quartz grains derived from the original sandstone. The close correspondence in composition between the rheopsammites of VL-2 and feldspathic sandstones of the Independence Fjord Group (Table 2) supports this interpretation. The

TABLE 2. REPRESENTATIVE ANALYSES OF MIDSOMMERSØ DOLERITES AND ASSOCIATED INTRUSIONS

	273240	273250	273210	273482	273493	273394	273390	273226	273231	273369	273532	197408	Sandstones (mean)
	Dolerites, KEM-1		Granophyre, KEM-2 (mean)	Dolerites, ZZD-1 (273493 near top)		VL-1, red dolerite and its dark border (-390)		Dolerites, IF		"Rheopsammites," VL-2, ZZD-2 and AsF, compared to mean sandstone			
SiO$_2$ (wt%)	49.62	48.97	67.82	53.75	60.73	55.64	52.44	52.60	50.59	88.50	75.30	78.67	87.37
TiO$_2$	1.29	2.18	0.41	0.80	0.58	0.81	0.94	0.83	0.97	0.04	0.37	0.24	0.05
Al$_2$O$_3$	13.62	11.36	13.43	14.27	12.90	12.92	13.58	15.22	15.24	4.71	10.41	6.91	6.16
Fe$_2$O$_3$	5.04	6.69	1.31	2.71	2.82	2.33	2.59	2.55	2.79	0.27	1.35	1.61	0.14
FeO	8.38	11.63	1.32	6.36	3.77	5.94	7.36	6.40	6.94	0.66	1.09	1.36	0.40
MnO	0.22	0.28	0.09	0.13	0.08	0.19	0.21	0.19	0.20	0.01	0.04	0.04	0.00
MgO	7.62	5.22	3.03	6.98	4.63	5.83	7.06	6.38	7.32	0.91	1.74	3.10	0.13
CaO	10.07	9.70	0.59	9.57	5.23	4.70	9.72	10.06	5.76	0.08	0.15	0.61	0.09
Na$_2$O	1.88	2.03	4.44	1.60	3.18	3.37	1.83	1.96	3.66	0.05	0.10	0.00	0.04
K$_2$O	0.36	0.40	5.79	1.49	2.64	4.09	0.97	1.55	2.24	3.61	7.30	4.51	4.72
P$_2$O$_5$	0.11	0.14	0.10	0.09	0.09	0.11	0.12	0.12	0.13	0.04	0.08	0.05	0.04
volat.	1.88	1.35	2.05	1.80	2.46	2.56	1.95	1.95	3.52	0.72	1.46	2.13	0.48
Sum	100.09	99.95	97.99	99.55	99.11	98.49	98.77	99.81	99.36	99.60	99.39	99.23	99.61
Mg#	34.8	21.1		39.7	39.8	41.8	39.8	39.9	41.3				

Note: Samples were analyzed at the Geological Survey of Greenland (GGU); now incorporated into the Geological Survey of Denmark and Greenland, GEUS), mainly by X-ray fluorescence spectrometry on glass discs; Na$_2$O was analyzed by Atomic Absorption Spectrometry, and FeO was analyzed by titration (Kystol and Larsen, 1999). Sample numbers are according to GGU files. KEM-1, etc., are the investigated intrusions; for locations, see Figure 1.

occurrence of rheopsammites as sheets with doleritic borders suggests that the intrusions originated from zoned magma chambers (Kalsbeek and Frei, 2006).

Similar intrusions in upper Zig-Zag Dal (ZZD-2) have somewhat lower SiO$_2$ (75%–82%) and higher Na$_2$O + K$_2$O values (Table 2; Fig. 21); these rocks may represent crustal melts composed of components derived from both basement gneisses and sandstones. Rheopsammitic rocks from outcrops south of Astrup Fjord (AsF; locality 4 on Fig. 1) also have lower SiO$_2$ values than VL-2, but they do not show the same enrichment in Na$_2$O + K$_2$O. They might represent a crustal melt containing a minor proportion of mantle-derived components.

Midsommersø Dolerites in Kronprins Christian Land

As mentioned already, the dolerites that cut the sandstones and lavas of the Independence Fjord Group in Kronprins Christian Land (Fig. 22) can be broadly correlated with the Midsommersø Dolerites. This correlation is based on lithological similarities and supported by Sm-Nd data on one sample, which are similar to those obtained for Midsommersø Dolerites from the foreland, but which are clearly distinct from Sm-Nd data for samples from the Hekla Sund and Aage Berthelsen Gletscher Formations in Kronprins Christian Land (Pedersen et al., 2002; Upton et al., 2005). However, the possibility that older dolerites, similar in age to the 1740 Ma lavas, also occur, cannot be excluded.

THE ZIG-ZAG DAL BASALT FORMATION

The Zig-Zag Dal Basalt Formation (Figs. 1 and 23; Kalsbeek and Jepsen, 1984; Upton et al., 2005) consists of a succession of tholeiitic flood basalts and covers an area of at least 10,000 km^2; it has a maximum thickness of ~1350 m. It directly overlies the Independence Fjord Group with its mafic intrusions, and it is therefore natural to assume a genetic relationship between the intrusions and the lavas. This relationship is supported by a close geochemical similarity between the lavas and the intrusions (Upton et al., 2005; Kalsbeek and Frei, 2006). However, no feeders to the lavas have been seen in the sandstones, and, in fact, the upper few hundred meters of the sandstones are nearly devoid of intrusions. Attempts to date the basalts have not been successful, and contemporaneity has therefore not been proven. However, the similarity in paleomagnetic pole positions for the dolerites and the basalts (Marcussen and Abrahamsen, 1983) further supports the view that they are expressions of the same magmatic event.

Thirteen sections through the basalt sequence were measured between Independence Fjord and Danmark Fjord (Fig. 24; Kalsbeek and Jepsen, 1984), and the two sections in Zig-Zag Dal, where the Zig-Zag Dal Basalt Formation is thickest, were sampled in detail. The basalt sequence occupies a trough-shaped basin that has been peneplaned at the top. The central part of the basin apparently underwent subsidence during and after termination of volcanic activity. After cessation of volcanism, the area was subjected to a long period of erosion, during which all basalts east and west of the present area of

Figure 22. Strongly deformed Midsommersø Dolerites in sandstones of the Independence Fjord Group, Caledonian orogen, Kronprins Christian Land. Cliff section is ~1000 m high.

Figure 23. The Zig-Zag Dal Basalt Formation in Zig-Zag Dal. Height of section is ~800 m.

distribution were removed before deposition of the overlying Hagen Fjord Group (Sønderholm et al., this volume).

The Zig-Zag Dal Basalt Formation has been subdivided into three main units, a basal unit, an aphyric unit, and a porphyritic unit (Fig. 24). The basal unit (100–120 m) consists of small, macroscopically aphyric, 1–10-m-thick basalt flows. Pillow lavas are locally present in the lower parts of the unit and indicate subaqueous effusion. A thin horizon of sedimentary rocks, sandstones, and dolomites locally overlies the basal unit and marks a break in volcanic activity. The aphyric and porphyritic units (390–440 and up to 750 m thick, respectively) are together composed of some thirty flows; individual flows are up to 120 m thick. Most flows have amygdaloidal or flow-brecciated tops, suggesting subaerial effusion, and corded pahoehoe surfaces are locally preserved. Preservation of such delicate structures suggests very short periods of repose, i.e., a high rate of volcanic productivity. Thin sedimentary units are locally present, implying longer periods of dormancy. Copper mineralization (malachite staining and rare presence of native Cu) has locally been observed in connection with amygdaloidal flow tops.

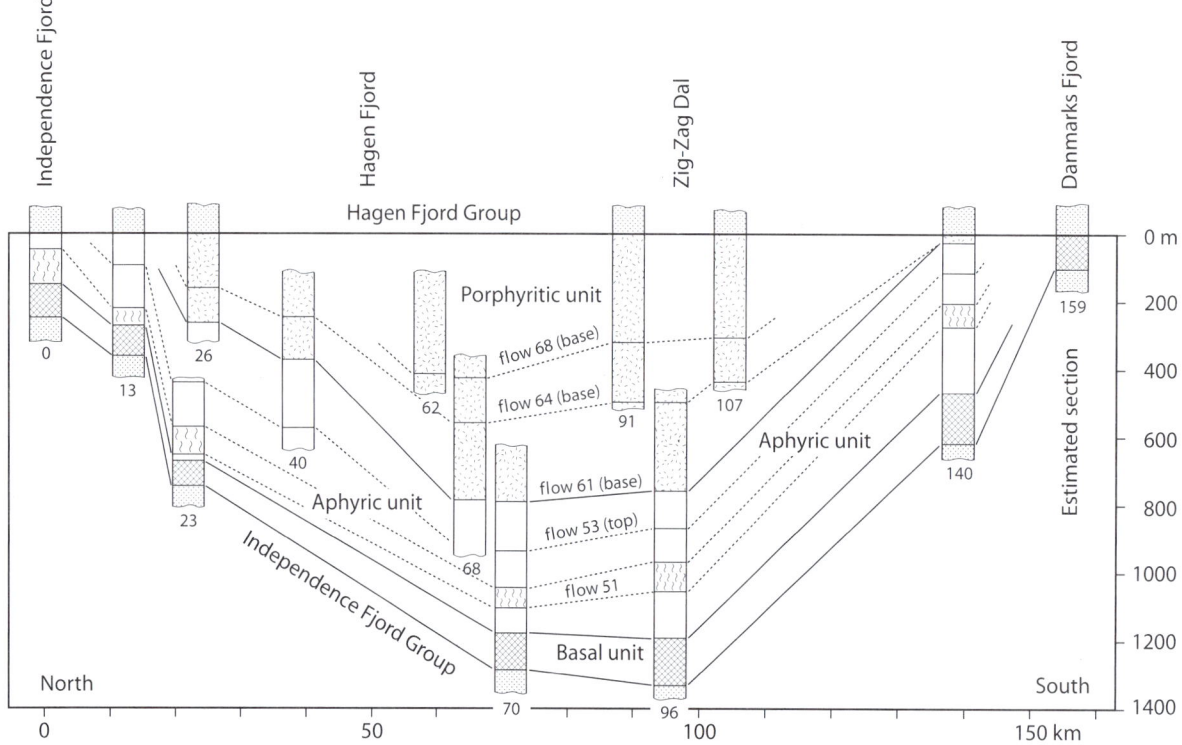

Figure 24. Synoptic correlation diagram of sections through the Zig-Zag Dal Basalt Formation (from Kalsbeek and Jepsen, 1984). The sections are arranged in order of distance from Independence Fjord in southeastward direction, and the numbers underneath each section are the distance (in km) from Independence Fjord. The numbering of the flows is semi-arbitrary, counting upward and starting with #50 at the base of the aphyric unit and #61 at the base of the porphyritic unit.

Flow characteristics such as thickness, weathering color, type of columnar jointing, and presence or absence of phenocrysts or vesicular cognate inclusions permit correlation of individual flows from section to section (Fig. 24). Some of the flows can be mapped over long distances, e.g., #51 in the aphyric unit, which has an estimated volume of at least 600 km³. The overlying flow (#52; the "brown marker") is characterized by a distinctive brown color.

At the base of section 26 in J.C. Christensen Land, more acidic lavas (SiO_2 = 64.3–65.6 wt%) occur near the top of the aphyric unit; they are exposed over an area of ~70 km² and attain a thickness of ~100 m. These rocks are very strongly altered and have not been studied in any detail. An attempt to recover zircons for age determination failed.

Petrography and Geochemistry

The Zig-Zag Dal Basalt Formation has been affected by low-temperature hydrothermal metamorphism at zeolite-facies to lowermost prehnite-pumpellyite-facies conditions, with temperatures up to 200 °C, shortly after or coeval with volcanism (Bevins et al., 1991). The basalts of the basal unit in particular have been strongly altered. Fine-grained carbonate, sericite, chlorite, and opaque minerals form a large proportion of some of the samples. In less-altered samples, remnants of plagioclase, augite, and pseudomorphs after olivine are present. The lavas of the aphyric and porphyritic units are often better preserved. In the aphyric unit, there are microphenocrysts of augite and plagioclase, whereas the porphyritic unit is characterized by larger euhedral phenocrysts of plagioclase and porphyritic aggregates of plagioclase and augite in a majority of the samples.

Representative chemical analyses of the basalts are given in Table 3, and the chemical variation across the basalt succession is illustrated in Figure 25. Most samples have normal basaltic compositions, SiO_2 around 50 wt% and MgO > 6%, with the exception of the Brown Marker, which, with values of ~55% SiO_2 and 4.7% MgO (#273426, Table 3), has the composition of a tholeiitic andesite.

As is the case for most continental flood basalt provinces, the most magnesian lavas occur at the base of the succession (Fig. 25); Mg numbers (Mg# = $100 \times MgO/[Fe_2O_3^{tot} + MgO]$) decrease from ~50 in the basal unit to ~35–40 in the porphyritic

TABLE 3. REPRESENTATIVE ANALYSES OF SAMPLES FROM THE ZIG-ZAG DAL BASALT FORMATION

	273404	273407	273414	273422	273426	273432	273436	273446	273453	273457	273302
	Basal unit		Aphyric unit					Porphyritic unit			
SiO_2 (wt%)	47.69	51.49	49.51	50.35	54.57	49.74	50.87	48.85	48.66	49.34	65.06
TiO_2	0.81	0.83	0.70	0.91	1.27	0.79	1.04	1.00	1.03	1.16	0.82
Al_2O_3	14.83	14.96	14.39	13.73	13.33	15.02	13.87	14.09	13.59	13.69	11.64
$Fe_2O_3^{tot}$	9.42	9.55	10.10	10.92	12.24	10.43	11.68	12.66	12.80	14.21	6.21
MnO	0.18	0.17	0.16	0.19	0.20	0.17	0.22	0.23	0.19	0.22	0.08
MgO	9.59	9.44	7.90	8.54	4.69	7.91	6.84	7.48	6.94	6.66	3.31
CaO	5.38	2.76	9.01	8.00	5.50	12.79	7.83	9.13	9.21	10.47	0.52
Na_2O	2.37	4.62	3.15	3.22	3.48	2.10	4.04	3.32	3.26	2.75	0.97
K_2O	1.42	0.87	1.54	1.40	2.43	0.12	0.93	0.45	0.74	0.23	7.69
P_2O_5	0.13	0.13	0.10	0.13	0.19	0.10	0.13	0.09	0.11	0.12	0.16
LOI*	7.78	4.76	3.42	2.50	1.94	0.87	2.43	2.51	3.03	1.49	2.11
Sum	99.58	99.59	99.99	99.89	99.83	100.04	99.87	99.81	99.55	100.34	98.57
Mg#	50.4	49.7	43.9	43.9	27.7	43.1	36.9	37.1	35.2	31.9	
Nb (ppm)	6.4	7.0	6.0	9.1	19	3.9	7.0	2.7	3.5	3.6	
Zr	75	85	78	122	227	66	106	59	71	81	
Y	17	16	16	25	40	18	24	20	22	27	
Sr	136	106	143	258	161	206	326	380	278	200	40
Rb	23	21	28	48	63	5.4	18	7.4	8.3	2.7	125
Th	1.3	0.6	0.8	2.4	6.5	1.0	2.1	0.8	0.5	0.7	
Pb	4.4	2.2	12	4.6	5.6	1.1	9.6	2.9	9.3	2.9	
La	11	9.8	12	15	34	7.2	14	3.9	5.7	6.9	46
Ce	23	25	23	33	68	14	29	13	9.3	16	94
Nd	14	13	12	16	33	9.3	16	7.1	7.3	11	42
Ni	99	90	89	148	30	137	75	139	100	83	21
Cr	421	384	406	361	4	378	148	306	260	106	
V	249	247	226	255	326	276	312	366	358	405	
Ba	314	77	365	869	583	69	309	438	1383	115	1410
Sc	43	40	34	40	39	43	41	44	49	47	

Note: Samples were analyzed by X-ray fluorescence spectrometry at the University of Edinburgh, major elements on fused glass discs, trace elements on pressed powder tablets (Fitton et al., 1998). Sample #273426 in the aphyric unit represents the "Brown Marker," a tholeiitic andesite. With the exception of #273302, an icelanditic lava from J.C. Christensen Land, all samples are from the type sections in Zig-Zag Dal. #273302 was analyzed by similar methods at the Geological Survey of Greenland and the University of Copenhagen; see caption to Table 2. For stratigraphic positions of the analyzed samples, see Figure 25.
*LOI—loss on ignition.

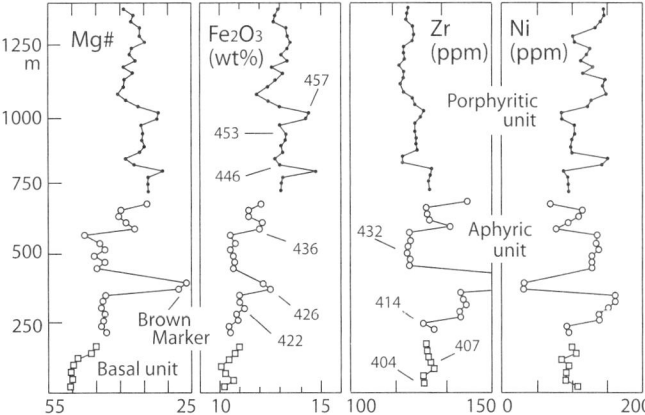

Figure 25. Chemical variation with height through the Zig-Zag Dal Basalt Formation in Zig-Zag Dal. Note decreasing Mg# ($100 \times MgO/[MgO + Fe_2O_3^{tot}]$) with height and Fe enrichment in the porphyritic unit. Sample numbers (only last three digits shown) are for the samples listed in Table 3.

unit. Samples from the porphyritic unit are rich in Fe_2O_3 (total), with 13–14 wt% Fe_2O_3 compared to 10%–12% for samples from the lower units.

Hydrothermal alteration of the basalts has had a profound influence on the chemical compositions of a large number of samples (Fig. 26). Loss on ignition (LOI) increases from <1% to ~8%; simultaneously, CaO decreases from ~12% to 2%, and Na_2O increases from ~2% to 5% (Figs. 26A and 26B). Concentrations of all other elements, including immobile elements, are passively affected by these changes. For example, the variation in SiO_2 in samples from the basal unit (47.7–51.9 wt%) is mainly caused by wide variations in LOI (3.4%–7.8%) and CaO (1.8%–6.2%).

Further variations in chemical composition are the result of fractional crystallization, differences in mantle sources, extent of melting in the source, interaction with wall rocks in the lithospheric mantle, and, perhaps, a minor degree of crustal contamination (Upton et al., 2005). Based on a combination of major- and trace-element data, Kalsbeek and Jepsen (1984) concluded that the magmas underwent low-pressure fractionation, which involved precipitation of plagioclase, clinopyroxene, and olivine, before eruption. Increasing values of Zr, and decreasing Ni, with decreasing Mg# within the upper part of the aphyric unit (above the Brown Marker, flow #52) and within the porphyritic unit (Figs. 26C and 26D) are the result of fractional crystallization; however, differences in Zr concentrations between the upper part of the aphyric unit and the porphyritic unit at similar Mg# (Fig. 26C) cannot be explained by fractional crystallization because of the incompatibility of Zr with the minerals involved. Also, differences in Ni values at similar Mg# between the aphyric and porphyritic units (Fig. 26D) are unlikely to have been caused by variations in the degree of fractional crystallization. Such disparities are interpreted as mainly due to differences in mantle source compositions.

A more comprehensive description of the Zig-Zag Dal Basalts has been presented in a recent paper by Upton et al. (2005). Based on new trace-element and isotope data, these authors concluded that magma generation took place in an upwelling mantle plume, at increasingly shallow depths, underneath an attenuating continental lithosphere. Lithospheric contamination is believed to have diminished with time, and the lavas of the porphyritic unit are considered to represent essentially uncontaminated (but variably fractionated) plume-source melts.

SUMMARY AND CONCLUSIONS

Quartzitic sandstones of the Independence Fjord Group are widely exposed in the northern parts of the Caledonian orogen and in its western foreland. The succession unconformably overlies crystalline basement complexes; the unconformity, however, is only well exposed in Dronning Louise Land.

The oldest strata in the Independence Fjord Group occur within the Caledonian orogen in Kronprins Christian Land. They consist of sandstones and conglomerates, together with interleaved volcanic rocks that have been dated at 1740 Ma. This age is equivalent to that of the youngest granites within the crystalline basement, 1740–1750 Ma (Kalsbeek et al., 1993, this volume), and the sedimentary rocks can be regarded as molasse-type deposits related to the breakdown of the Paleoproterozoic orogen in North-East Greenland. These rocks have been strongly affected by Caledonian thrusting and folding, as well as low-grade Caledonian metamorphism.

The main outcrop area of the Independence Fjord Group is in the Caledonian foreland west of Kronprins Christian Land, where the sandstones are virtually undeformed and attain a thickness of ~2000 m. Correlation between the orogen and the foreland is not straightforward. Whereas deposition of sandstones and conglomerates in Kronprins Christian Land appears to have taken place during active rifting, Independence Fjord sandstones in the foreland were deposited into a continental sag basin. Moreover, no vestiges of volcanic activity, evident within the orogen, are apparent in the foreland. It is therefore envisaged that the sandstones in the foreland represent a somewhat younger succession than those exposed within the Caledonian orogen. Strong deformation within the orogen and major tectonic dislocations between the orogen and the foreland preclude following individual strata from the orogen to the foreland.

In the foreland, the Independence Fjord Group is composed of a near-horizontal succession of quartzitic and feldspathic sandstones, with intercalated siltstone horizons a few tens of meters thick. The latter can be followed over large areas and have been used to describe the stratigraphy of the group in the foreland. Sedimentary structures in the sandstones indicate a largely fluvial origin, and there is local evidence of eolian deposition; the sedimentary environment can be compared with recent inland basins in Australia, where sands undergo repeated cycles of fluvial and eolian transport before final deposition. Fluvial transport was in general toward the northeast; wind

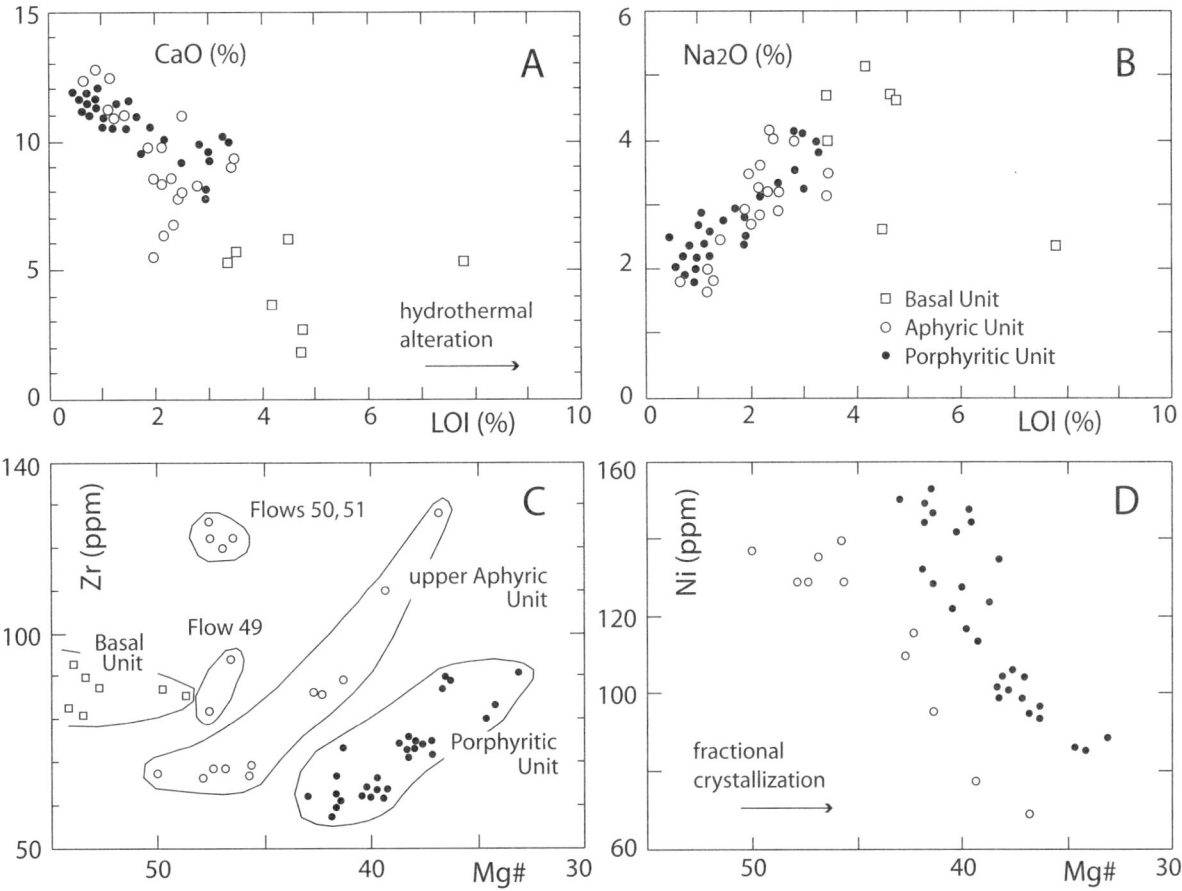

Figure 26. Chemical variations of basaltic rocks within the Zig-Zag Dal Basalt Formation (in part from Kalsbeek and Jepsen, 1984). (A–B) Diagrams illustrating the influence of hydrothermal alteration (partial spilitization) on the composition of the basalts. (C–D) Variation in chemistry related to fractional crystallization and differences in mantle sources. For discussion, see text. LOI—loss on ignition.

directions varied widely. The siltstones appear to have been deposited in extensive shallow lakes that were frequently desiccated under arid or semiarid conditions. Pseudomorphed halite crystals indicate intermittent formation of evaporites. Deposition of the Independence Fjord Group in the foreland must have taken place between 1740 and 1380 Ma.

The sandstones and siltstones of the Independence Fjord Group are dissected by numerous dolerites and associated intrusive rocks, collectively called the Midsommersø Dolerites. Acidic members of this suite are believed to represent remelted crystalline basement and partially melted feldspathic sandstones. Most of the rocks have undergone strong hydrothermal alteration. U-Pb dating of baddeleyite from a well-preserved dolerite yielded an age of 1382 Ma. The Independence Fjord Group is conformably overlain by a succession of tholeiitic flood basalts, up to 1350 m thick, the Zig-Zag Dal Basalt Formation, that is believed to be contemporaneous with and genetically related to the Midsommersø Dolerites. The basalts were deposited into a basin that underwent subsidence during and after termination of volcanic activity.

The oldest deposits overlying the Zig-Zag Dal Basalt Formation belong to the Neoproterozoic (ca. 800 Ma?) Jyske Ås Formation of the Hagen Fjord Group (Clemmensen and Jepsen, 1992; Sønderholm et al., this volume). There is thus a stratigraphic hiatus of some 500 m.y. for which no information is available from North Greenland. There is evidence for Grenvillian orogenic activity in northern Ellesmere Island within this period (e.g., Trettin, 1987), and it has been suggested that this orogeny may have caused uplift and erosion prior to deposition of the Hagen Fjord Group (Clemmensen and Jepsen, 1992). Direct evidence for Grenvillian activity, however, has not been documented in North Greenland.

ACKNOWLEDGMENTS

The investigations reported in this paper were carried out in relationship with the former Geological Survey of Greenland (GGU) regional mapping programs in northern Greenland under the inspiring direction of Niels Henriksen. We thank Julian Menuge and Tony Prave for helpful reviews, Helle Zetterwall and Benny Schark for help with the figures, Elsevier BV (the Netherlands) for permission to reuse figures that were previously published in *Precambrian Research*, and GEUS (the Geological Survey of Denmark and Greenland) for permission to publish this paper.

REFERENCES CITED

Adams, P.J., and Cowie, J.W., 1953, A Geological Reconnaissance of the Region around the Inner Part of Danmarks Fjord, Northeast Greenland: Meddelelser om Grønland, v. 111, no. 7, 24 p.

Bengaard, H.-J., and Henriksen, N., 1986, Geological Map of Greenland, Peary Land, Sheet 8: Copenhagen, Geological Survey of Greenland, scale 1:500,000.

Bevins, R.E., Rowbotham, G., and Robinson, D., 1991, Zeolite to prehnite-pumpellyite facies metamorphism of the late Proterozoic Zig-Zag Dal Basalt Formation, eastern North Greenland: Lithos, v. 27, p. 155–165, doi: 10.1016/0024-4937(91)90010-I.

Christie, R.L., and Ineson, J.R., 1979, Precambrian-Silurian geology of the G.B. Schley Fjord region, eastern Peary Land, North Greenland: Rapport Grønlands Geologiske Undersøgelse, v. 88, p. 63–71.

Clemmensen, L.B., and Jepsen, H.F., 1992, Lithostratigraphy and Geological Setting of Upper Proterozoic Shoreline-Shelf Deposits, Hagen Fjord Group, Eastern North Greenland: Rapport Grønlands Geologiske Undersøgelse, v. 157, 27 p.

Collinson, J.D., 1980, Stratigraphy of the Independence Fjord Group (Proterozoic) of eastern North Greenland: Rapport Grønlands Geologiske Undersøgelse, v. 99, p. 7–23.

Collinson, J.D., 1983, Sedimentology of unconformities within a fluvio-lacustrine sequence; Middle Proterozoic of eastern North Greenland: Sedimentary Geology, v. 34, p. 145–166, doi: 10.1016/0037-0738(83)90084-2.

Collinson, J.D., Bevins, R.E., and Clemmensen, L.B., 1989, Post-glacial mass flow and associated deposits preserved in palaeovalleys: The late Precambrian Morænesø Formation, North Greenland: Meddelelser om Grønland: Geoscience, v. 21, 27 p.

Ellitsgaard-Rasmussen, K., 1950, Preliminary report on the geological field work carried out by the Danish Peary Land Expedition in the year 1949–1950: Meddelelser fra Dansk Geologisk Forening, v. 11, p. 589–595.

Escher, J.C., and Pulvertaft, T.C.R., 1995, Geological Map of Greenland: Copenhagen, Geological Survey of Greenland, scale 1:2,500,000.

Fitton, J.G., Saunders, A.D., Larsen, L.M., Hardason, B.S., and Norry, M.J., 1998, Volcanic rocks from the South-East Greenland margin at 63°N: Composition, petrogenesis and mantle sources, *in* Saunders, A.D., Larsen, H.C., and Wise, S.H., eds., Proceedings of the Ocean Drilling Program, Scientific Results, Volume 152: College Station, Texas, Ocean Drilling Program, p. 331–350.

Freuchen, P., 1915, General observations as to natural conditions in the country traversed by the expedition: Meddelelser om Grønland, v. 51, p. 341–370.

Friderichsen, J.D., Holdsworth, R.E., Jepsen, H.F., and Strachan, R.A., 1990, Caledonian and pre-Caledonian geology of Dronning Louise Land, North-East Greenland: Rapport Grønlands Geologiske Undersøgelse, v. 148, p. 133–141.

Jepsen, H.F., 1969, Basiske intrusiver i det sydlige Pearyland og deres geologiske miljø [master's thesis]: Århus, University of Århus, Denmark, 66 p.

Jepsen, H.F., 1971, The Precambrian, Eocambrian and Early Palaeozoic Stratigraphy of the Jørgen Brønlund Fjord Area, Peary Land, North Greenland: Meddelelser om Grønland, v. 192, no. 2, 42 p.

Jepsen, H.F., 2000, Geological Map of Greenland, Lambert Land, Sheet 9: Copenhagen, Geological Survey of Denmark and Greenland, scale 1:500,000.

Kalsbeek, F., and Frei, R., 2006, The Mesoproterozoic Midsommersø Dolerites and associated high-silica intrusions, North Greenland: Crustal melting, contamination and hydrothermal alteration: Contributions to Mineralogy and Petrology, v. 152, p. 89–110, doi: 10.1007/s00410-006-0096-1.

Kalsbeek, F., and Jepsen, H.F., 1983, The Midsommersø Dolerites and associated intrusions in the Proterozoic platform of eastern North Greenland—A study of the interaction between intrusive basic magma and sialic crust: Journal of Petrology, v. 24, p. 605–634.

Kalsbeek, F., and Jepsen, H.F., 1984, The late Proterozoic Zig-Zag Dal Basalt Formation of eastern North Greenland: Journal of Petrology, v. 25, p. 644–664.

Kalsbeek, F., Nutman, A.P., and Taylor, P.N., 1993, Palaeoproterozoic basement province in the Caledonian fold belt of North-East Greenland: Precambrian Research, v. 63, p. 163–178, doi: 10.1016/0301-9268(93)90010-Y.

Kalsbeek, F., Nutman, A.P., Escher, J.C., Friderichsen, J.D., Hull, J.M., Jones, K.A., and Pedersen, S.A.S., 1999, Geochronology of granitic and supracrustal rocks from the northern part of the East Greenland Caledonides: Ion microprobe U-Pb zircon ages: Geology of Greenland Survey Bulletin, v. 184, p. 31–48.

Kalsbeek, F., Thrane, K., Higgins, A.K., Jepsen, H.F., Leslie, A.G., Nutman, A.P., and Frei, R., 2008, this volume, Polyorogenic history of the East Greenland Caledonides, *in* Higgins, A.K., Gilotti, J.A., and Smith, M.P., eds., The Greenland Caledonides: Evolution of the Northeast Margin of Laurentia: Geological Society of America Memoir 202, doi: 10.1130/2008.1202(03).

Kystol, J., and Larsen, L.M., 1999, Analytical procedures in the rock geochemical laboratory of the Geological Survey of Denmark and Greenland: Geology of Greenland Survey Bulletin, v. 184, p. 59–62.

Larsen, O., and Graff-Petersen, P., 1980, Sr-isotopic studies and mineral composition of the Hagen Bræ Member in the Proterozoic clastic sediments at Hagen Bræ, eastern North Greenland: Rapport Grønlands Geologiske Undersøgelse, v. 99, p. 111–118.

Marcussen, C., and Abrahamsen, N., 1983, Palaeomagnetism of the Proterozoic Zig-Zag Dal Basalt and the Midsommersø Dolerites, eastern North Greenland: Geophysical Journal of the Royal Astronomical Society, v. 73, p. 367–387.

McKenzie, D., 1978, Some remarks on the development of sedimentary basins: Earth and Planetary Science Letters, v. 40, p. 25–32, doi: 10.1016/0012-821X(78)90071-7.

Peacock, J.D., 1956, The Geology of N.E. Greenland: Meddelelser om Grønland, v. 137, no. 7, 38 p.

Peacock, J.D., 1958, Some Investigations into the Geology and Petrology of Dronning Louise Land, N.E. Greenland: Meddelelser om Grønland, v. 157, no. 4, 139 p.

Pedersen, S.A.S., Leslie, A.G., and Craig, L.E., 1995, Proterozoic and Caledonian geology of the Prinsesse Caroline Mathilde Alper, eastern North Greenland, *in* Higgins, A.K., ed., Express Report: Eastern North Greenland and North-East Greenland 1995: Copenhagen, Geological Survey of Greenland, p. 71–85.

Pedersen, S.A.S., Craig, L.E., Upton, B.G.J., Rämö, O.T., Jepsen, H.F., and Kalsbeek, F., 2002, Palaeoproterozoic (1740 Ma) rift-related volcanism in the Hekla Sund region, eastern North Greenland: Field occurrence, geochemistry and tectonic setting: Precambrian Research, v. 114, p. 327–346, doi: 10.1016/S0301-9268(01)00234-0.

Sønderholm, M., and Jepsen, H.F., 1991, Proterozoic basins of North Greenland: Bulletin Grønlands Geologiske Undersøgelse, v. 160, p. 49–69.

Sønderholm, M., Frederiksen, K.S., Smith, M.P., and Tirsgaard, H., 2008, this volume, Neoproterozoic sedimentary basins with glacigenic deposits of the East Greenland Caledonides, *in* Higgins, A.K., Gilotti, J.A., and Smith, M.P., eds., The Greenland Caledonides: Evolution of the Northeast Margin of Laurentia: Geological Society of America Memoir 202, doi: 10.1130/2008.1202(05).

Strachan, R.A., Friderichsen, J.D., Holdsworth, R.E., and Jepsen, H.F., 1994, Regional geology and Caledonian structure, Dronning Louise Land, North-East Greenland: Rapport Grønlands Geologiske Undersøgelse, v. 162, p. 71–76.

Trettin, H.P., 1987, Pearya: A composite terrane with Caledonian affinities in northern Ellesmere Island: Canadian Journal of Earth Sciences, v. 24, p. 224–245.

Upton, B.G.J., Rämö, O.T., Heaman, L.M., Blichert-Toft, J., Kalsbeek, F., Barry, T.L., and Jepsen, H.F., 2005, The Mesoproterozoic Zig-Zag Dal Basalts and associated intrusions of eastern North Greenland: Mantle plume–lithosphere interaction: Contributions to Mineralogy and Petrology, v. 149, p. 40–56, doi: 10.1007/s00410-004-0634-7.

MANUSCRIPT ACCEPTED BY THE SOCIETY 14 JANUARY 2008

Neoproterozoic sedimentary basins with glacigenic deposits of the East Greenland Caledonides

Martin Sønderholm*
Geological Survey of Denmark and Greenland, Øster Voldgade 10, DK-1350 Copenhagen K, Denmark

Kasper S. Frederiksen
DONG Energy, Agern Alle 24-26, DK-2970 Hørsholm, Denmark

M. Paul Smith
Lapworth Museum of Geology, University of Birmingham, Edgbaston, Birmingham B15 2TT, UK

Henrik Tirsgaard
Mærsk Oil & Gas, Esplanaden 50, DK-1098 Copenhagen K, Denmark

ABSTRACT

Two major Neoproterozoic sedimentary basins that probably formed in response to an early pulse of Iapetan rifting along the Laurentian margin are well exposed in the East Greenland Caledonides. The Hekla Sund Basin is exposed at the northern termination of the East Greenland Caledonides, and it is represented by the Rivieradal and Hagen Fjord Groups, which attain a cumulative thickness of 8–11 km. The evolution of this basin reflects deposition during active rifting and a postrift thermal equilibration stage. The Eleonore Bay Basin of East Greenland includes the deposits of the Eleonore Bay Supergroup of early Neoproterozoic age overlain by Cryogenian (mid-Neoproterozoic) glacial deposits of the Tillite Group, which have a combined thickness in excess of 14 km. Four stages of basin evolution may be distinguished based on paleogeographic reorganizations of the shelf and a change from siliciclastic to carbonate deposition, and the final stage was dominated by glacigenic deposition. Major regional stratigraphic breaks seem to be absent, as is other evidence of rift-related sedimentation, suggesting deposition in one or a series of connected ensialic basins. A comparison with other Neoproterozoic basins along the Laurentian margin of the Iapetus Ocean shows similarities between the Eleonore Bay Basin and coeval deposits on Svalbard and the Central Highlands of Scotland. The development of an extensive carbonate platform during the later stages of both the Eleonore Bay and Hekla Sund Basins testifies to a period of tectonic stability prior to onset of Iapetus rifting. The extent of this carbonate platform may have been even larger, since similar successions are present in the Caledonides of Scotland and Ireland.

Keywords: Neoproterozoic sedimentary basin evolution, sequence stratigraphy, Marinoan tillites, Caledonides, East Greenland, Laurentia.

---

*Current address: DONG Energy, Agern Alle 24-26, DK-2970 Hørsholm, Denmark; e-mail: mason@dongenergy.dk.

Sønderholm, M., Frederiksen, K.S., Smith, M.P., and Tirsgaard, H., 2008, Neoproterozoic sedimentary basins with glacigenic deposits of the East Greenland Caledonides, in Higgins, A.K., Gilotti, J.A., and Smith, M.P., eds., The Greenland Caledonides: Evolution of the Northeast Margin of Laurentia: Geological Society of America Memoir 202, p. 99–136, doi: 10.1130/2008.1202(05). For permission to copy, contact editing@geosociety.org. ©2008 The Geological Society of America. All rights reserved.

INTRODUCTION

In the North Atlantic region, extensive basin development was initiated during the Neoproterozoic, and it continued until Ordovician times. This basin formation was associated with the disintegration of the Neoproterozoic supercontinent Rodinia (Piper, 1982) and the subsequent creation of the Iapetus Ocean (e.g., Harland and Gayer, 1972; Winchester, 1988). The actual opening of the Iapetus Ocean occurred in the Ediacaran (late Neoproterozoic) around 570–535 Ma (Cawood et al., 2001). Evidence of Neoproterozoic basin formation comes from thick successions of siliciclastic and carbonate sediments, locally overlain by Marinoan tillites, preserved in the Caledonian terrane within Svalbard, western Scandinavia, the British Isles, and from North and East Greenland, where the well-exposed Hekla Sund and Eleonore Bay Basins form some of the major architectural elements of the East Greenland Caledonides (Fig. 1). In general, the sediments of these basins are only weakly metamorphosed, and only their lower parts show high grades of metamorphism and tectonic overprint. This paper reviews the present knowledge relating to these two sedimentary successions exposed in East and North-East Greenland, and we finally present a comparison with other Neoproterozoic successions along the Laurentian margin of the Iapetus Ocean.

HEKLA SUND BASIN

Sediments of the Neoproterozoic Hekla Sund Basin are exposed at the northern termination of the East Greenland Caledonides. The region constitutes a key area for studies of the western border zone of the orogen (Fig. 2), since it exposes continuous sections from the undisturbed foreland in the west, across parautochthonous foreland affected by folding and thin-skinned thrusting, to allochthonous thrust sheets in the east (Higgins et al., 2001a, 2001b).

The Hekla Sund Basin as used in this paper incorporates both the deposits of the Rivieradal Group (Smith et al., 2004a) and the Hagen Fjord Group (Clemmensen and Jepsen, 1992). The Rivieradal Group has a cumulative thickness of up to 7.5–10 km and is exposed in a major Caledonian thrust sheet (the Vandredalen thrust sheet) in eastern Kronprins Christian Land. The Hagen Fjord Group, which has a maximum thickness of 1000–1100 m, is mainly exposed in the Caledonian foreland to the northwest in western Kronprins Christian Land, Mylius-Erichsen Land, J.C. Christensen Land, and Heilprin Land, but it also occurs within the Vandredalen thrust sheet (Figs. 2 and 3). The Hagen Fjord Group unconformably overlies the Mesoproterozoic Independence Fjord Group (including the Midsommersø Dolerite Formation) and the Zig-Zag Dal Basalt Formation (Clemmensen and Jepsen, 1992; Collinson et al., this volume).

The Rivieradal Group represents a pre-Iapetus, synrift, deep-water succession deposited in an eastward-facing extensional half-graben that was originally situated at least 40 km east of the present outcrop area. The Hagen Fjord Group is composed of fluvial and shallow-marine sandstone overlain by a succession of carbonate platform deposits. It represents a period of postrift thermal reequilibration during the Neoproterozoic, where the youngest units in the group overstep the original rift shoulder (Higgins et al., 2001a, 2001b). The present-day distribution of remnants of the Rivieradal and Hagen Fjord Groups indicates that the original depositional basin was more than 200 km long (parallel to bounding rift faults). During the early stage of basin development (Rivieradal Group), the basin was at least 50 km wide (perpendicular to the rift faults), and it was >300 km wide during the later stages.

Deposits of the Fyns Sø Formation, representing the youngest part of the Hagen Fjord Basin, are at most localities unconformably overlain by sandstone of the Lower Cambrian Kap Holbæk Formation. These are in turn unconformably overlain by the Lower Ordovician Wandel Valley Formation (see Peel and Sønderholm, 1991; Smith et al., 2004b).

Previous Work

The Neoproterozoic sediments in eastern North Greenland were examined by several geological expeditions in the period from 1947 to 1958 (Troelsen, 1949; Adams and Cowie, 1953; Fränkl, 1954, 1955), combined with systematic aerial reconnaissance (e.g., Haller, 1971). During 1978–1980, sedimentologic and stratigraphic studies were carried out on the Hagen Fjord Group sediments as part of Geological Survey of Greenland mapping (Clemmensen and Jepsen, 1992). The Rivieradal Group succession was studied as part of reconnaissance field work in 1980 by Hurst and McKerrow (1981), who recognized three discrete thrust sheets in the Kronprins Christian Land region, and it was suggested that the allochthonous units had been transported ~150 km to the west during the Caledonian orogeny (Hurst and McKerrow, 1985).

Detailed studies on the sedimentologic, stratigraphic, and structural setting of the Rivieradal and Hagen Fjord Groups were carried out as part of regional mapping of the southern Kronprins Christian Land area in 1994 and 1995 (Sønderholm and Tirsgaard, 1998; M.P. Smith et al., 1999, 2004a, 2004b; Higgins et al., 2001a, 2001b). During this work, the main concepts developed by Fränkl (1954, 1955), Hurst and McKerrow (1980, 1981, 1985), and Hurst et al. (1985) were shown to be accurate and largely valid, although the Rivieradal Group succession was confirmed to be confined to a single thrust sheet, the Vandredalen thrust sheet (Higgins et al., 2001b).

Lithostratigraphy

The concept of the Hagen Fjord Group was introduced by Haller (1961), mainly on the basis of aerial reconnaissance work. As a result of field work carried out in 1978–1980, the group was redefined by Clemmensen and Jepsen (1992) to include only Neoproterozoic sediments of mainly shallow-marine origin in the region between Heilprin Land and Kronprins Christian Land (Fig. 2).

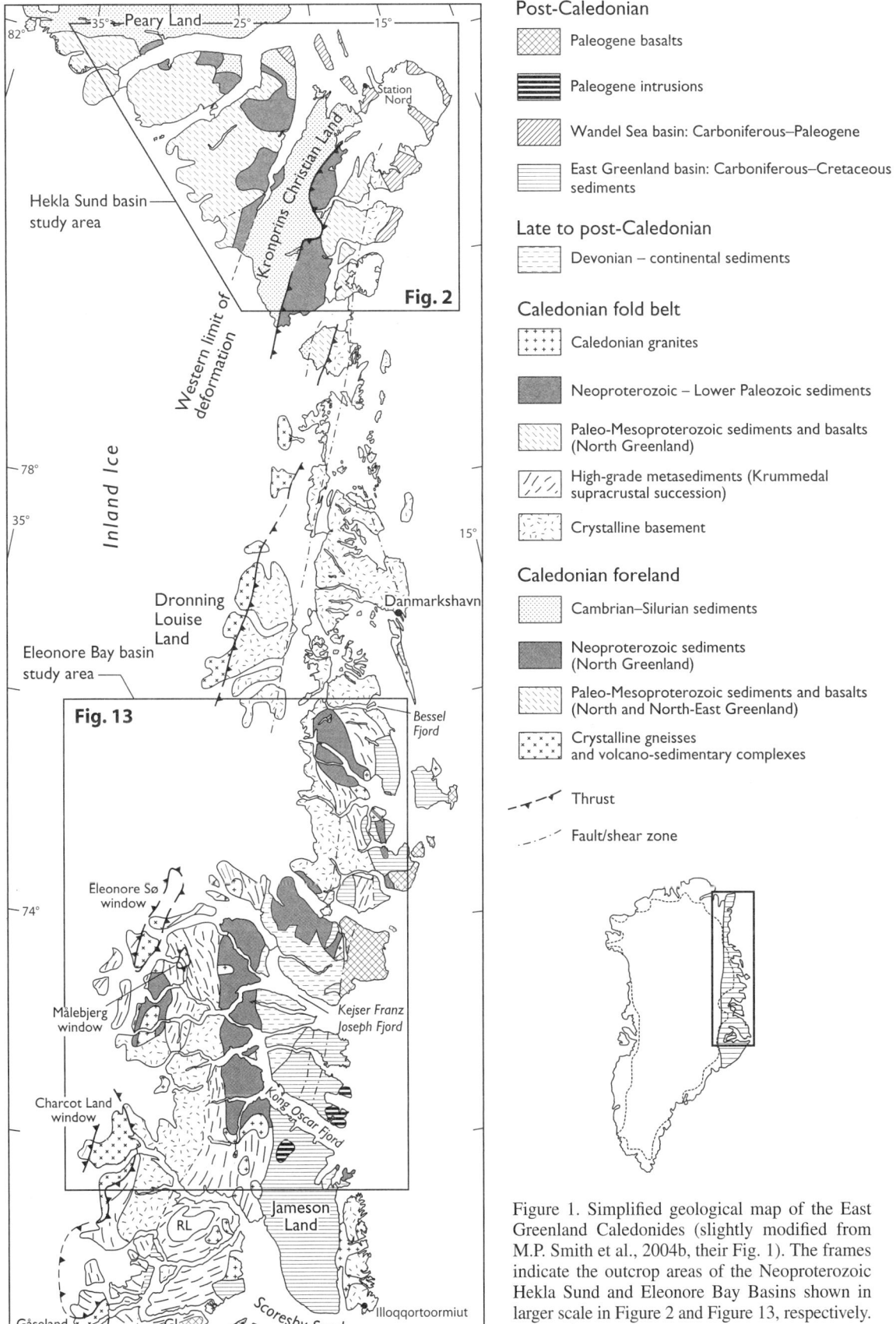

Figure 1. Simplified geological map of the East Greenland Caledonides (slightly modified from M.P. Smith et al., 2004b, their Fig. 1). The frames indicate the outcrop areas of the Neoproterozoic Hekla Sund and Eleonore Bay Basins shown in larger scale in Figure 2 and Figure 13, respectively. GL—Gåseland; RL—Renland.

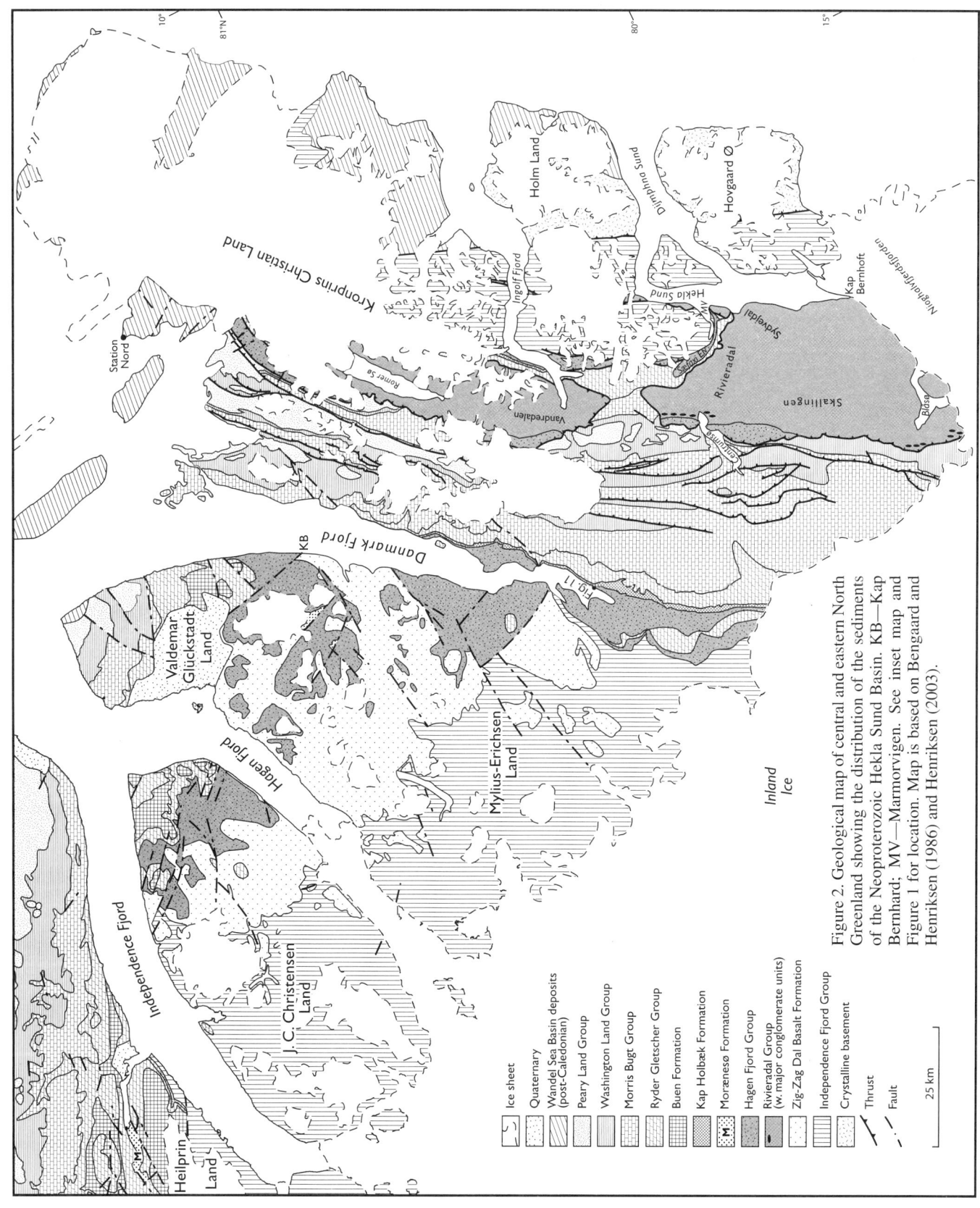

Figure 2. Geological map of central and eastern North Greenland showing the distribution of the sediments of the Neoproterozoic Hekla Sund Basin. KB—Kap Bernhard; MV—Marmorvigen. See inset map and Figure 1 for location. Map is based on Bengaard and Henriksen (1986) and Henriksen (2003).

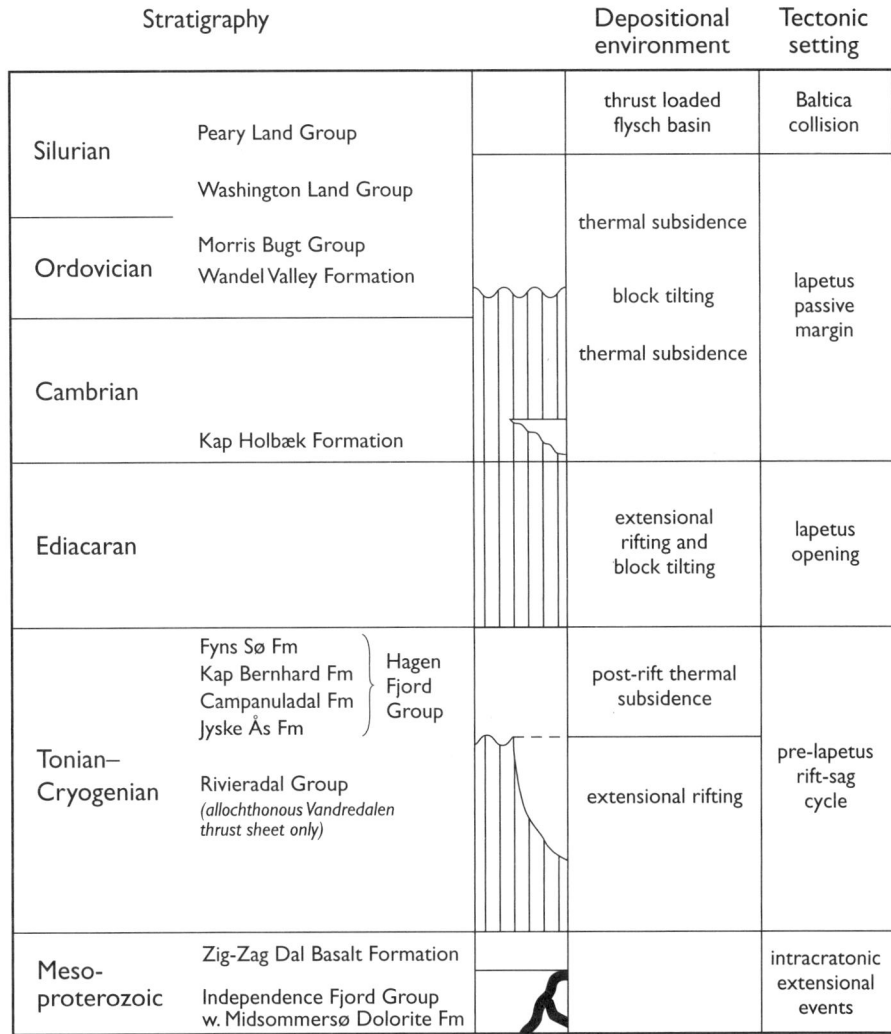

Figure 3. Summary stratigraphic scheme of Proterozoic and Paleozoic units shown in Figure 2 and their relationships to the opening of the Iapetus Ocean (modified from M.P. Smith et al., 1999, their Fig. 2). Vertical ruling depicts nondeposition or erosion. Used with permission from the Geological Society of London.

The Hagen Fjord Group unconformably overlies Mesoproterozoic strata that are intruded by 1380 Ma dolerite intrusions (Fig. 4; Upton et al., 2005), and it is composed of a lower, siliciclastic part (Jyske Ås, Campanuladal, and Catalinafjeld Formations) and an upper, carbonate-dominated part (Kap Bernhard and Fyns Sø Formations). Previously, the Kap Holbæk Formation was thought to constitute the top of the group (Clemmensen and Jepsen, 1992). However, a major unconformity has been demonstrated beneath the Kap Holbæk Formation, and the deep, decimeter-scale, burrows of the ichnogenus *Skolithos* that occur within it led M.P. Smith et al. (2004b) to remove the unit from the Hagen Fjord Group and to interpret it as a Lower Cambrian correlative of the Buen Formation of the Franklinian Basin.

The first systematic work on rocks of the Rivieradal Group was carried out by Fränkl (1954, 1955) in the area around Centrumsø in Kronprins Christian Land. He recognized that the Neoproterozoic succession could be divided into an autochthonous and an allochthonous succession separated by a major thrust, upon which his "main nappe" was transported to the west. The metasediments of the nappe were divided into a lower, more metamorphosed part (>1100 m) and an upper, less metamorphic part that also included the upper part of the Hagen Fjord Group succession (2000–3000 m; see Smith et al., 2004a).

Based on Geological Survey of Greenland reconnaissance field work, the allochthonous clastic succession underlying the Hagen Fjord Group was assigned to a single unit, referred to as the "Rivieradal sandstones" by Hurst and McKerrow (1980). Following the mapping of the region in 1994–1995, this unit was formally defined as the Rivieradal Group by Smith et al. (2004a).

Age constraints on the Rivieradal Group are poor and rely upon the ages of the underlying Midsommersø Dolerite Formation and the overlying Hagen Fjord Group, which together indicate that the Rivieradal Group was deposited in the later part of the interval between 1380 Ma and ~700 Ma (cf. Smith et al., 2004a). The age of the overlying Hagen Fjord Group is

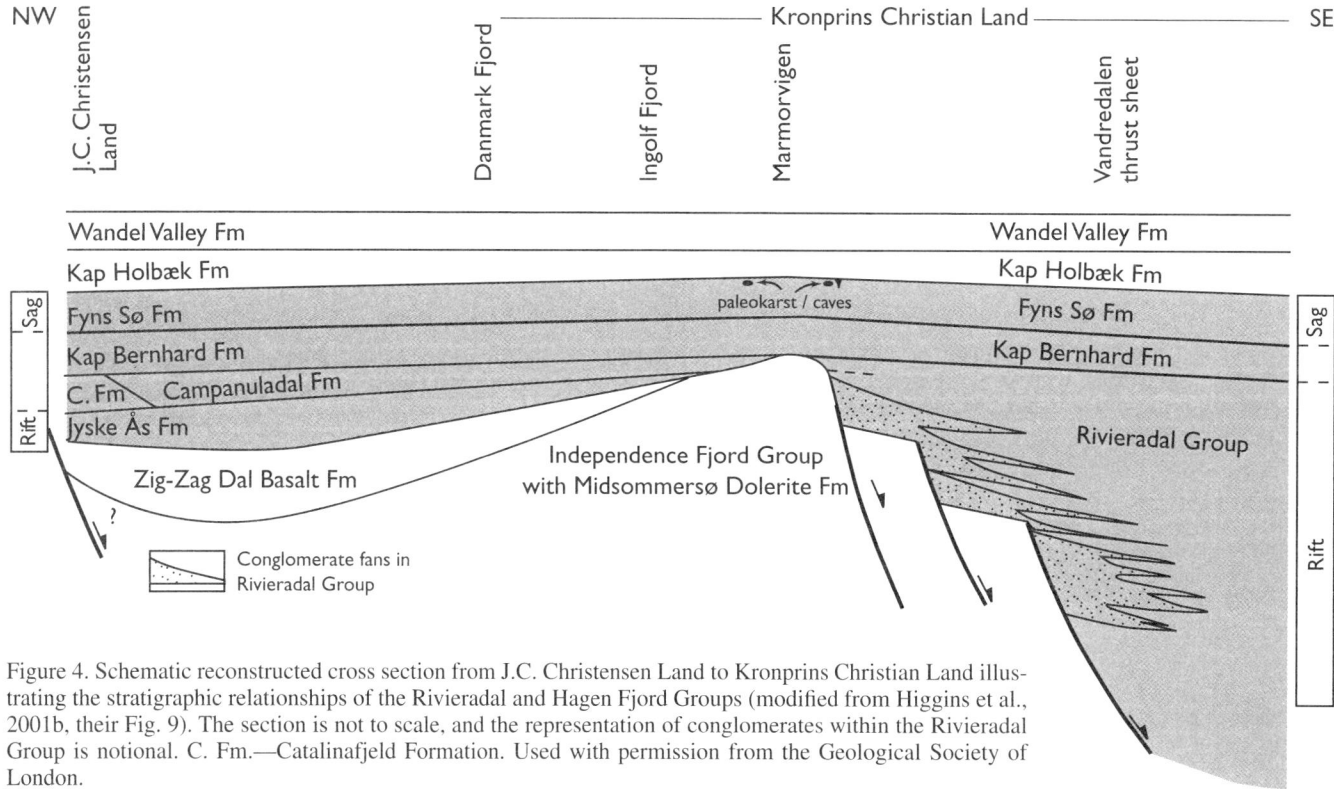

Figure 4. Schematic reconstructed cross section from J.C. Christensen Land to Kronprins Christian Land illustrating the stratigraphic relationships of the Rivieradal and Hagen Fjord Groups (modified from Higgins et al., 2001b, their Fig. 9). The section is not to scale, and the representation of conglomerates within the Rivieradal Group is notional. C. Fm.—Catalinafjeld Formation. Used with permission from the Geological Society of London.

also poorly constrained; it is based on a correlation between the Hagen Fjord Group and the Eleonore Bay Supergroup in eastern Greenland (see later section), which both show a transition from siliciclastic to carbonate deposition in the middle Neoproterozoic (Sønderholm and Jepsen, 1991; Sønderholm and Tirsgaard, 1993). The Eleonore Bay Supergroup is overlain by tillites of probable Marinoan age, so a pre–600 Ma age is thus suggested for the Hagen Fjord Group. Reworked glacial deposits of the Moræneso Formation are also present in the outcrop area of the Hagen Fjord Group (Fig. 2). They are of possible Marinoan age and occur in a depositional setting similar to the Tillit Nunatak Formation of the Gåseland, Charcot Land, and Målebjerg tectonic windows (see following; Collinson et al., 1989); however, their exact stratigraphic relationship with the Hagen Fjord Group is not clear.

Basin Evolution

Two major stages of basin evolution are recognized in the Hekla Sund Basin, along with a phase of initial rifting that created two major east-facing half-grabens. The two half-grabens show considerable differences in subsidence rates, since the deep-marine Rivieradal Group is only represented in the eastern half-graben, and the half-grabens have contrasting stratigraphies. The later postrift stage includes the deposits of the Hagen Fjord Group, which is represented in both the western and the eastern half-grabens (cf. Higgins et al., 2001b).

Stage 1: Rifting and Basin Initiation

Eastern half-graben. In the eastern half-graben of the Hekla Sund Basin, the initial stage of rifting is represented by the Rivieradal Group. The group is lithologically very variable and possesses a strong proximal to distal polarity in which the most distal sediments and stratigraphically lowest levels occur in the east, whereas proximal sediments and higher levels occur to the west (Figs. 4 and 5).

Structural studies in Rivieradal itself suggest that the maximum cumulative thickness of the various units within the entire Rivieradal Group is on the order of 7.5–10 km (Higgins et al., 2001b). However, restoration of the Vandredalen thrust suggests that the Rivieradal profile represents a >50-km-long E–W cross section of the half-graben (Smith et al., 2004a) and that several of the units thus may at least be partly lateral equivalents to each other. The actual stratigraphic thickness of the basin fill is therefore probably less than the 7.5–10 km thickness, and may be on the order of 5–6 km, comparable to the thickness of the Rivieradal Group in northern Vandredalen, where a total thickness of 4500 m has been measured (of which 3000 m are in a continuously exposed section). The basal 200 m of this section lie above a thrust contact with Ordovician carbonates and consist of strongly sheared conglomerates. The conglomerates are overlain by a 500-m-thick phyllite-dominated unit, and then by more than 2200 m of strata dominated by thick-bedded (30–250 cm), structureless to vaguely laminated sandstone turbidites (T_{a-c}) interbedded with dark pyritic mudstone, possibly deposited in a fan-delta

Figure 5. Lithological variation within the Rivieradal Group (from Higgins et al., 2001b, their Fig. 6). In general, the most distal sediments and stratigraphically deepest levels occur in the east, whereas proximal sediments and higher levels occur to the west along the leading edge of the Vandredalen thrust sheet. Broad arrows indicate point source of sediment input to the Hekla Sund Basin as determined by the location of coarse conglomerate units. BS—Brede Spærregletscher; DD—Dunkeldal; MV—Marmorvigen. Used with permission from the Geological Society of London.

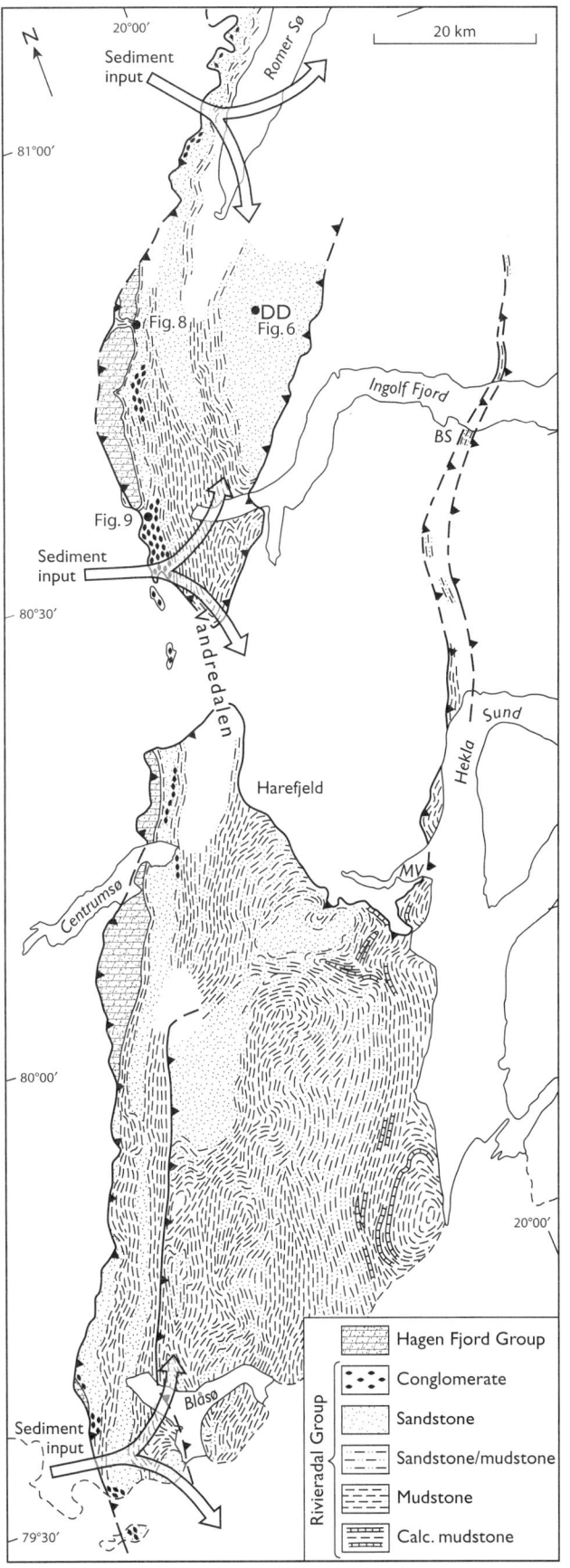

environment (Figs. 6 and 7A). Laterally and vertically, this succession grades into deep-marine, basin plain mudstone and equivalent phyllitic rocks that crop out over a large part of southeastern Kronprins Christian Land. On the west side of Vandredalen, to the northwest of innermost Ingolf Fjord, a 1000-m-thick succession consisting of storm- and tide-dominated shallow-marine deposits overlies the deep-marine succession. Current directions are generally toward the northeast parallel with the inferred paleocoastline (Figs. 5, 7A, and 8; Higgins et al., 2001b; Smith et al., 2004a).

The most proximal sediments are found along the leading edge of the Vandredalen thrust sheet. In this region, coarse-grained channelized, clast-supported conglomerate units occur along strike in three discrete areas (Figs. 5 and 9). The conglomerate clasts are well rounded and vary in size from a few decimeters to over a meter in the thicker beds, but boulders up to 3–4 m have been recorded. The clasts are derived from the Independence Fjord Group and the Midsommersø Dolerite Formation, which can be presumed to have been exposed to active erosion along the western margin of the Rivieradal Group basin. In a section west of Romer Sø, the thick conglomerate units can be traced eastward (basinward) over a distance of 1–2 km into coarsening- and thickening-upward sandstone units that sometimes grade into conglomerate. Farther eastward, these pass over a similar distance into coarsening- and thickening-upward mudstone–sandstone units. In the southern area around Blåsø, granite and quartz pebbles indicate an additional metamorphic basement source, suggesting that a deeper erosional level was reached here.

Toward the end of Rivieradal Group deposition, a general decrease in the rate of generated accommodation space is recorded by the presence of ~350 m of fluvial sediments that may be lateral equivalents to some of the conglomeratic units (Fig. 8; Sønderholm and Tirsgaard, 1998). Variations in fluvial style within this unit have been attributed to changing rates of generated accommodation space and a shift toward a more arid climate.

The more distal representatives of the Rivieradal Group are seen at the eastern end of the Rivieradal valley. Pelite with quartzite is exposed at the mouth of Rivieradal and is overlain to the west by pelite, semipelite, and calcareous pelite with prominent metacarbonate units (the "Stenørkenen Phyllites" and "Sydvejdal Marbles" of Fränkl, 1955). This unit is overlain by a thick succession of phyllite and quartzite (the "Taagefjeldene Greywackes" of Fränkl, 1955) similar to the turbidites seen in the section between Ingolf Fjord and Vandredalen (Higgins et al., 2001b).

Figure 6. (A) Thick succession of possible fan-delta turbidites exposed in the proximal part of the Rivieradal Group succession in "Dunkeldal," the valley connecting Vandredalen and Ingolf Fjord (for location, see Fig. 5). View is toward north; height of mountain is 600 m above lake level. The exposed section on the north side of the lake is more than 2000 m thick and youngs to the left (west). (B) Example of typical sandstone turbidite beds, structureless to vaguely laminated with common water-escape structures, interbedded with dark pyritic mudstone. Arrowed pencil = 15 cm.

Figure 7. (A) Conceptual diagram of the west side of the eastern half-graben of the Hekla Sund Basin, illustrating the fault-bounded control of the Rivieradal Group sedimentation and point-source input from rivers draining highland areas composed of Independence Fjord sandstones cut by dolerite of the Midsommersø Dolerite Formation (from Higgins et al., 2001b, their Fig. 8). (B) Late Hagen Fjord time. IF—Independence Fjord Group; RG—Rivieradal Group; HFG—Hagen Fjord Group. Used with permission from the Geological Society of London.

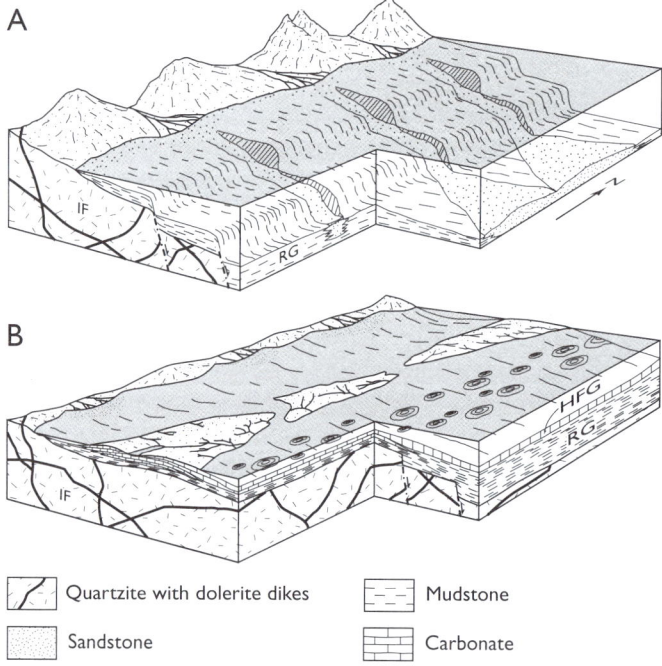

Overall, the Rivieradal Group basin is characterized by point sources of sediment input (Figs. 5 and 7A). The geometry of the conglomerate deposits, together with their discrete occurrences, suggests the presence of at least three discrete fan-delta systems that acted as major feeder distributary systems. The three conglomerate developments are all in the upper part of the Rivieradal Group succession, but not necessarily at the same stratigraphic level. The repeated cycles of coarsening-upward conglomerate deposition, recorded in the frontal parts of the thrust sheet, may have been controlled by seismic activity on the basin-margin fault systems. Substantial conglomerate fan-deltas accumulated in the immediate vicinity of the source areas and are associated with sandy, proximal turbidites. Between the fans and in the eastern distal part of the basin, sedimentation was dominated by mud and calcareous mud. As the basin filled, the depositional style switched from deep to shallow marine, and the upper part of the group is dominated by less localized and more laterally persistent, tidal- and storm-dominated shallow-marine and locally fluvial deposition (Sønderholm and Tirsgaard, 1998; Higgins et al., 2001b).

Western half-graben. During the initial phase of basin development, sedimentation in the western half-graben was probably very slow and restricted to fluvial and shallow-marine deposition represented by the Jyske Ås Formation, which forms the lower part of the Hagen Fjord Group.

Figure 8. Stacked, channelized clast-supported conglomerate units from upper part of Rivieradal Group in northern part of Vandredalen, Kronprins Christian Land (see Fig. 5 for location). Person for scale.

Figure 9. Cliff section in northern part of Vandredalen, Kronprins Christian Land (see Fig. 5 for location), showing basal part of shallow braided river deposits overlying marine inner-shelf deposits in the uppermost part of the Rivieradal Group. The thickness of the sandstone unit at the top of the mountain is ~200 m.

The Jyske Ås Formation is up to 500 m thick, and it records marine transgression following the long hiatus represented by the sub–Hagen Fjord Group unconformity. Although the basal, red part of the formation may include some fluvial sandstone, the main part consists of trough- or tabular, large-scale cross-bedded sandstone, which is interpreted to be of beach and shallow-tidal shelf origin (Clemmensen and Jepsen, 1992). The dominance of sediment-transport directions toward the northeast probably reflects ebb-tidal or storm-enhanced offshore-flowing tidal currents. The northwesternmost exposures of the formation appear, however, to be entirely of fluvial origin.

Direct lithostratigraphic correlation between the eastern and western half-grabens of the Hekla Sund Basin cannot be established during this phase of basin evolution since deposition in the two half-graben segments reflects very different sedimentary and structural environments. This suggests that the rift shoulder between the two half-graben segments—now seen in the area between Harefjeld and Marmorvigen—separated two subbasins that had significant differences in tectonic evolution, with subsidence rates varying by an order of magnitude (Fränkl, 1955; Higgins et al., 2001b).

Stage 2: Postrift Thermal Equilibration

The transition from the rift phase to the postrift thermal equilibration phase of the Hekla Sund Basin probably occurred during early Campanuladal Formation time, since this formation is recognized in both the western and the eastern half-graben segments, indicating that the rift shoulder of the eastern half-graben did not form a barrier at this time (Figs. 4 and 7B).

The Campanuladal Formation (and the correlative Catalinafjeld Formation in the extreme northwest) overlies the Jyske Ås Formation in the foreland region and the Rivieradal Group in the Vandredalen thrust sheet (Fränkl, 1955; Clemmensen and Jepsen, 1992; Smith et al., 2004a).

The Campanuladal Formation (110–175 m) consists mainly of a variegated succession of fine- to medium-grained sandstone and siltstone units, including a distinctive stromatolitic dolostone. The units are arranged in a characteristic sequence that is recognizable at most localities (Fig. 10). The fine-grained sandstone and siltstone are horizontally laminated and show small-scale wave- and current-formed cross-lamination. Desiccation cracks and gutter casts showing a preferred NE–SW orientation are common. The succession records a transition from inter- and supratidal deposition to more offshore conditions (Clemmensen and Jepsen, 1992). The stromatolitic dolostone horizon in the upper part of the formation probably developed in shallow subtidal environments as a response to reduced clastic influx. Current indications are similar to those observed in the Jyske Ås Formation, suggesting that the open sea was to the northeast (Clemmensen and Jepsen, 1992). In the allochthon, the Campanuladal Formation is represented by an ~80-m-thick, strongly tectonized, marly, variegated mudstone-dominated unit, which overlies the fluvial sandstone seen in the uppermost part of the Rivieradal Group (Sønderholm and Tirsgaard, 1998).

The Catalinafjeld Formation (260–350 m) mainly consists of laminated mudstone, including thin sandstone turbidite beds that indicate transport directions toward the east. The sediments are considered to represent flooding and the establishment of deeper-marine environments in the northwestern part of the region. Locally, coarsening- and thickening-upward successions record episodes of shoreline progradation (Sønderholm and Jepsen, 1991; Clemmensen and Jepsen, 1992).

The Campanuladal Formation is rather abruptly overlain by the Kap Bernhard Formation (150–215 m), which marks a change from siliciclastic, shallow-shelf deposition to incipient carbonate-platform deposition (Fig. 10). The Kap Bernhard Formation mainly consists of reddish-brown limestone. In the lower part of the formation, soft-sediment deformation structures are abundant, and intraformational breccias are locally conspicuous. Upward, the degree of soft-sediment deformation decreases, and intervals with edge-wise breccias and stromatolitic units occur (Clemmensen and Jepsen, 1992). The carbonate platform may have been initiated along the former rift shoulder and subsequently spread into the eastern and western segments of the basin. The rather sudden shift from siliciclastic shallow-marine deposition to incipient carbonate-platform deposition seen in both basin segments is probably a result of reduced siliciclastic influx related to a climatic change toward more arid conditions during a period of general retrogression. The thicker development of the Catalinafjeld Formation in the westernmost exposures compared to the underlying Jyske Ås Formation of stage 1 could suggest that siliciclastic deposition prevailed for a longer time along the western margin of the basin due to the presence of a larger catchment area in this region. The upper

Figure 10. Outcrop of the Hagen Fjord Group at Kap Bernhard, J.C. Christensen Land (see Fig. 2 for location) based on Clemmensen and Jepsen (1992, their Fig. 5). The Fyns Sø Formation is unconformably overlain by Cambrian deposits of the Portfjeld (PF) and Buen Formations. JÅ—Jyske Ås Formation; CD—Campanuladal Formation; KB—Kap Bernhard Formation; FS—Fyns Sø Formation. Height of cliff is ~600 m.

Figure 11. Top of Hagen Fjord Group (Fyns Sø Formation, FS) exposed on the east coast of innermost Danmark Fjord (see Fig. 2 for location). The top of the Fyns Sø Formation includes conspicuous microbialite mound structures and is overlain by the Lower Cambrian Kap Holbæk Formation (KH) with a major hiatus. The Kap Holbæk Formation is unconformably overlain by carbonate-platform deposits of the Lower Ordovician Wandel Valley Formation (WV). Thickness of Kap Holbæk Formation is approximately 150 m.

part of the Catalinafjeld Formation may thus correlate with the lower part of the Kap Bernhard Formation farther to the east.

Overlying the incipient platform deposits of the Kap Bernhard Formation, the Fyns Sø Formation (up to ~325 m) records the establishment of a well-developed prograding carbonate platform (Fig. 10). The formation consists of generally massive dolostone, which, in the upper part, is commonly interbedded with siltstone. Sedimentary structures and textures in the dolostone are locally preserved, including slump structures, intraformational breccias, and rare ripple marks. Stromatolitic horizons occur throughout the formation and are especially common in the uppermost part, where they locally form spectacular linked mounds with a relief of up to 2 m (Fig. 11; Sønderholm and Jepsen, 1991; Clemmensen and Jepsen, 1992). Conical columnar stromatolites ("conophyton") have also been reported from the formation. These stromatolites were probably restricted to subtidal environments, and they often form the only stromatolitic component in basinal and slope deposits (Donaldson, 1976; Hoffman, 1976; Grotzinger, 1989). The co-occurrence of "conophyton" and slumped horizons suggests subtidal deposition on the slope of a prograding carbonate platform (Sønderholm and Jepsen, 1991).

The Fyns Sø Formation, which forms the uppermost unit of the Hekla Sund Basin, is at most localities overlain by the Early Cambrian Kap Holbæk Formation, which represents early Iapetus passive-margin sedimentation (see Smith and Rasmussen, this volume). Although thought at one time to be conformable, it is now known that this boundary marks a significant hiatus, across which much of the Neoproterozoic is absent, and a substantial paleokarst development occurs in the upper parts of the Fyns Sø Formation (Figs. 7B and 12). Near the crest of

Figure 12. Paleokarst within the uppermost part of the Fyns Sø Formation at Marmorvigen, Kronprins Christian Land (see Figs. 2 and 5 for location). The fill of the phreatic tube is composed of quartzarenites of the Lower Cambrian Kap Holbæk Formation; the Fyns Sø Formation is unconformably overlain by Early Ordovician carbonates of the Wandel Valley Formation (in the background; see Smith and Rasmussen, this volume). The measuring pole is divided into 20 cm intervals.

the rift shoulder, seen at Sæfaxi Elv, the Kap Holbæk Formation is present only as the fill of paleokarst within the Fyns Sø Formation (Fig. 12), which, at this locality, is overlain by Ordovician carbonates (cf. M.P. Smith et al., 1999, 2004b).

ELEONORE BAY BASIN

Deposits of the Neoproterozoic Eleonore Bay Basin are assigned to the Eleonore Bay Supergroup, which is more than 14 km thick, and to the overlying 800–1300-m-thick Tillite Group, which occurs in an outcrop belt (the Franz Joseph allochthon) that can be followed for 500 km from N to S along the strike of the orogen (Fig. 13). Recent work on the tectonic architecture of this part of the orogen has demonstrated that, although the maximum E–W width between Canning Land and Scoresby Land in the south to Bessel Fjord in the north is currently a maximum of 100 km, the total westward displacement of the thrust sheets was around 200–400 km, with estimated shortening of 40%–60% (Higgins et al., 2004a).

The lower ~12 km of the Eleonore Bay Supergroup are made up mainly of shallow-marine siliciclastic sediments (Nathorst Land and Lyell Land Groups), whereas the upper 2 km are dominated by carbonate-platform deposits (Ymer Ø and Andrée Land Groups; Fig. 14; Sønderholm and Tirsgaard, 1993). The overlying Tillite Group includes five formations, the lower three of which are Cryogenian (mid-Neoproterozoic) in age and include diamictites and marine deposits. The upper two formations are of Ediacaran age, and they consist of dolomitic mudstone and sandstone of shallow-marine to supratidal origin (Figs. 14 and 15; Hambrey and Spencer, 1987). The Tillite Group is unconformably overlain by the Cambrian–Ordovician Kong Oscar Fjord Group, which represents early Iapetus passive-margin sedimentation (see Smith and Rasmussen, this volume).

The nature of the lower boundary of the Eleonore Bay Supergroup has been widely debated. The oldest sediments of the Nathorst Land Group lie in contact with metasediments of the Krummedal supracrustal succession (Figs. 13 and 16). This boundary has been variously interpreted as transitional (Higgins et al., 1981), an extensional shear zone (White and Hodges, 2002; White et al., 2002), and an unconformity (Friderichsen and Thrane, 1998), although most workers are now agreed that it is an extensional detachment with a significant increase in metamorphic grade between footwall and hanging wall (e.g., Watt et al., 2000). It is uncertain whether the Nathorst Land Group was once in unconformable contact with the underlying Krummedal succession, but it is probable. The relationship is further complicated by extensive anatexis and presence of Caledonian granites that yield dates of 435–425 Ma (Hartz et al., 2000; Watt et al., 2000; Kalsbeek et al., 2001; White and Hodges, 2002; Andresen et al., 2004; Higgins et al., 2004a). The Krummedal succession of the Hagar Bjerg thrust sheet, which forms the footwall of the extensional detachment that bounds the Nathorst Land Group, is also intruded by a suite of older leucogranites that have yielded emplacement ages of 950–920 Ma (Kalsbeek et al., 2000; Watt et al., 2000; Watt and Thrane, 2001; Higgins et al., 2004a). Metamorphic zircons from nonmigmatitic paragneisses of the Krummedal succession have also yielded dates of 955–938 Ma, albeit with larger error bars (Watt and Thrane, 2001). Farther south, in Renland, granites have yielded slightly younger protolith ages of 915–910 Ma that postdate a period of deformation and upper-amphibolite-facies metamorphism at 1000 Ma (Leslie and Nutman, 2000).

The relationship of the Nathorst Land Group to the underlying Krummedal succession and the 950–940 Ma granites is critical to an understanding of the timing of basin formation and the onset of deposition in Eleonore Bay Basin. The Krummedal succession contains detrital zircon populations that are dominated by grains of Paleoproterozoic and Mesoproterozoic age, clustering in the range 1700–1200 Ma, but the youngest grains are 1070 Ma (Kalsbeek et al., 2000; Watt et al., 2000; Higgins et al., 2004b). This provides a tightly constrained window for deposition of the Krummedal succession between 1070 Ma and the metamorphic event and intrusion of granites at 955 Ma onward. Watt et al. (2000) noted that detrital zircon suites in two samples from the base and top of the Nathorst Land Group did not contain a component derived from the 950–920 granites and concluded that the Krummedal sediments and the Nathorst Land Group were part of the same depositional sequence, rather than the latter being part of the Eleonore Bay Supergroup. We consider this to be highly improbable on a number of grounds: (1) the sedimentary facies, depositional style, and architecture of the Nathorst Land Group are strikingly similar to those of the overlying Lyell Land Group; (2) in terms of regional metamorphism, there is a significant contrast between the metasediments of the Krummedal succession and those of the Nathorst Land Group; and (3) the 950–920 suite of granites is widely present within the Krummedal succession of the Hagar Bjerg thrust sheet but is entirely absent from the Nathorst Land Group. We therefore consider the absence of 950–920 granites to reflect the absence of unroofing of the granites at the time of deposition of the Nathorst Land Group. Furthermore, additional data from detrital zircons from the Nathorst Land Group suggest a maximum age for the group of 987 ± 18 Ma (Dhuime et al., 2007). Thus, the upper age limit for the initiation of subsidence in Eleonore Bay Basin can be placed at ~900 Ma.

Figure 13. Simplified geological map of eastern Greenland showing the distribution of the sediments of the Neoproterozoic Eleonore Bay Basin (modified from Henriksen, 2003; Bengaard, 1991). See Figure 1 for location. The "central fjord zone" coincides with the Eleonore Bay Supergroup outcrop area between Alpefjord and Waltershausen Gletscher. AB—Albert Heim Bjerge; AF—Alpefjord; EB—Eleonore Bugt; ED—Eremitdal; EØ—Ella Ø; FF—Forsblad Fjord; GF—Geologfjord; HB—Hagar Bjerg thrust sheet; KW—Kap Weber; LBL—Louise Boyd Land; MB—Målebjerg; NG—Nordenskiöld Gletscher; PB—Petermann Bjerg; SD—Snestormdal; SG—Sorteelv Gletscher; SL—Strindberg Land; VS—Vibeke Sø; WFZ—western fault zone. A–B is position of section shown in Figure 16.

Neoproterozoic sedimentary basins with glacigenic deposits

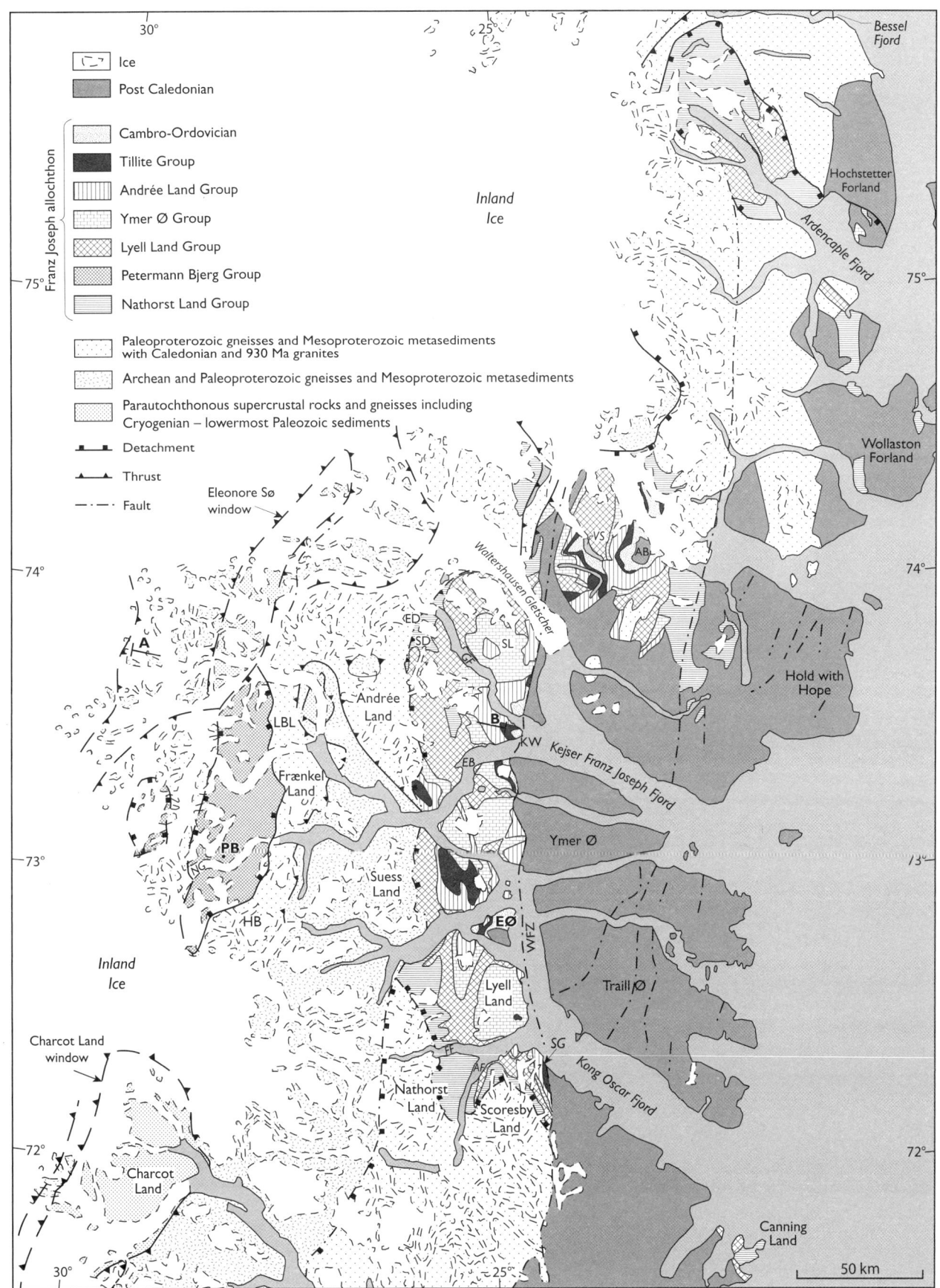

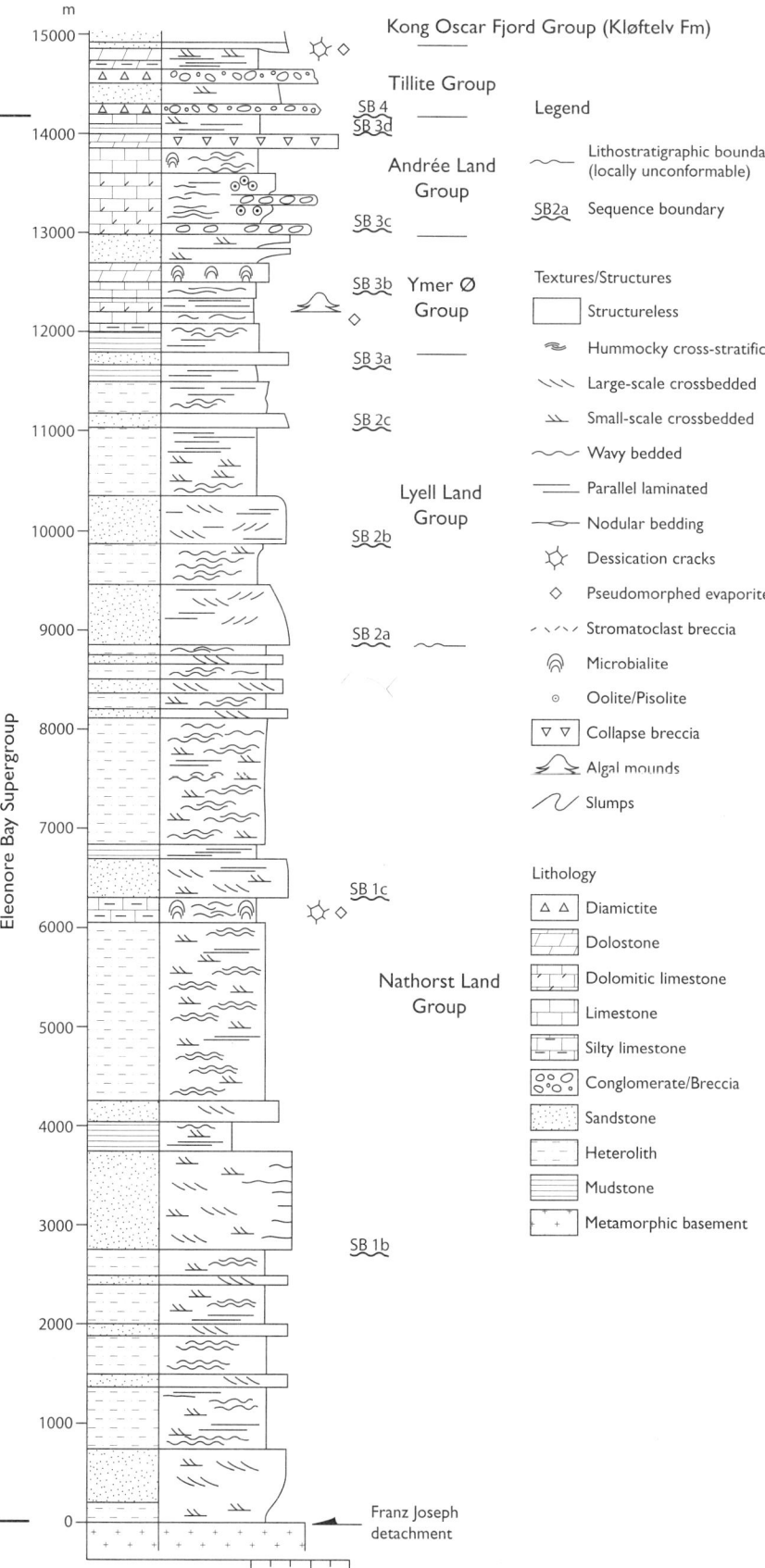

Figure 14. Generalized composite section of the Eleonore Bay Supergroup from the central fjord zone, based on Hambrey and Spencer (1987), Sønderholm and Tirsgaard (1993), Tirsgaard and Sønderholm (1997), and Smith and Robertson (1999a). SB—sequence boundary. Height in meters.

Figure 15. Lithostratigraphic subdivision of the Neoproterozoic Eleonore Bay Basin deposits (Franz Joseph allochthon, East Greenland Caledonides; modified after Sønderholm and Tirsgaard, 1993). Chronostratigraphic scale follows the International Commission on Stratigraphy (Gradstein et al., 2004).

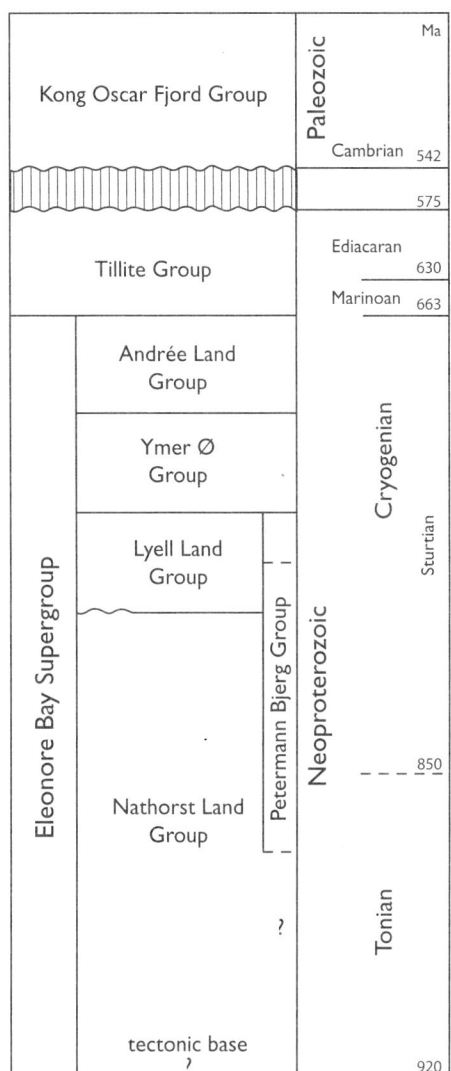

The upper age limit of the Eleonore Bay Supergroup is better defined since it is conformably overlain by the diamictites and associated sediments of the Tillite Group (Fig. 15). Although the correlation and age of this glacial episode have been the subject of some debate (see following discussion), there is now a considerable body of evidence in support of a Marinoan age (for reviews, see Fairchild and Hambrey, 1995; Halverson et al., 2004, 2005), corresponding to a 663 Ma age for the top of the Eleonore Bay Supergroup (Halverson et al., 2005). The younger of the two diamictites in the Tillite Group is overlain by a cap carbonate, which defines the base of the Ediacaran and which thus lies at the base of the Canyon Formation, corresponding to a date of 636 Ma (Hambrey and Spencer, 1987; Halverson et al., 2004, 2005). Up to 345 m of Ediacaran sediments overlie the youngest diamictite and are overlain with very low-angle unconformity by the Lower Cambrian Kløftelv Formation of the Kong Oscar Fjord Group (Fig. 15; Henriksen and Higgins, 1976; Hambrey, 1989; Hambrey et al., 1989; Smith et al., 2004b). However, the distinctive suite of fossils that encompasses the Ediacaran fauna has not been recovered from the upper Tillite Group, leading Halverson et al. (2004) to suggest a minimum age limit of 575 Ma for the group, the age of the oldest Ediacaran faunas. The unconformable break between the Tillite and Kong Oscar Fjord Groups would thus span an interval of at least 30 m.y.

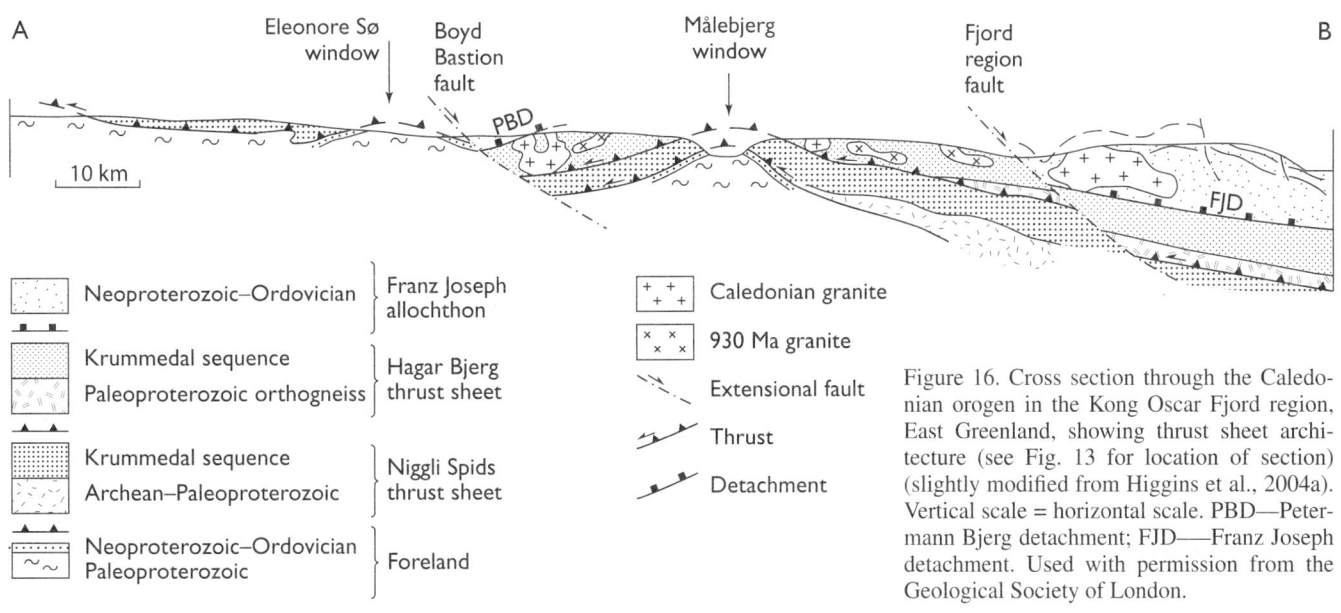

Figure 16. Cross section through the Caledonian orogen in the Kong Oscar Fjord region, East Greenland, showing thrust sheet architecture (see Fig. 13 for location of section) (slightly modified from Higgins et al., 2004a). Vertical scale = horizontal scale. PBD—Petermann Bjerg detachment; FJD—Franz Joseph detachment. Used with permission from the Geological Society of London.

Outcrops of the Eleonore Bay Basin are cut by the "western fault zone," a major normal fault zone that bounds the Devonian basin of North-East Greenland (Fig. 13; Larsen and Bengaard, 1991). Based on evidence from the Devonian succession, sinistral strike-slip movements in the order of 10–20 km along the western fault zone were estimated by Larsen and Bengaard (1991). Correlation of six stratigraphic markers in the Neoproterozoic Andrée Land and Tillite Groups east and west of the fault zone may, however, suggest a sinistral displacement of 80–90 km along the fault (Frederiksen, 2000b).

Previous Work

The spectacular Neoproterozoic succession of East Greenland (Fig. 17) has attracted interest since the earliest geological investigations in the 1870s, although the succession was not studied in detail until the period between the mid-1920s and the late 1950s, when systematic mapping efforts were undertaken on expeditions under the leadership of Lauge Koch (for accounts on previous studies, see Hambrey and Spencer, 1987; Sønderholm and Tirsgaard, 1993). During this time, many lithological and structural studies were carried out, and a lithostratigraphic scheme began to emerge.

During the mapping program of the Geological Survey of Greenland in 1968–1978, the lower part of the Eleonore Bay Supergroup in the Alpefjord region and the rather poorly known succession in Canning Land were investigated, resulting in the first sedimentologic studies of these deposits (cf. Bertrand-Sarfati and Caby, 1976). In the 1980s, several groups studied various aspects of the uppermost part of Eleonore Bay Supergroup and Tillite Group succession, resulting in some detailed sedimentologic work (Hambrey and Spencer, 1987; Herrington and Fairchild, 1989; Manby and Hambrey, 1989; Swett and Knoll, 1989; Moncrieff and Hambrey, 1990), descriptions of different microfossil assemblages (Knoll et al., 1986; Green et al., 1987, 1988, 1989), and a chemostratigraphic correlation of the Eleonore Bay Supergroup with equivalent strata on Svalbard (Knoll et al., 1986).

As part of mapping activities by the Survey in 1988–1990 and 1997–1998 (Higgins et al., 2004a; Henriksen and Higgins, this volume), sedimentologic and stratigraphic studies were carried out throughout the region (Sønderholm and Tirsgaard, 1993; Tirsgaard, 1993, 1996; Tirsgaard and Sønderholm, 1997; Henriksen, 1999; Smith and Robertson, 1999a; Frederiksen, 2000a, 2000b).

Lithostratigraphy

The Eleonore Bay Supergroup was formally defined and divided into five groups by Sønderholm and Tirsgaard (1993) using, in part, the more or less informal stratigraphic schemes presented by earlier workers (Fig. 15). The main outcrop area is the central fjord zone (72°N–74°30′N), where the Nathorst Land, Lyell Land, Ymer Ø, and Andrée Land Groups attain a

Figure 17. The spectacular character of the outcrops of the Eleonore Bay Supergroup along the coastal cliffs of Geologfjord between Strindberg Land and Andrée Land. LLG—Lyell Land Group; YØG—Ymer Ø Group; ALG—Andrée Land Group; BB—Berzelius Bjerg Formation; KA—Kap Alfred Formation; VS—Vibeke Sø Formation; SB—Skjoldungebræ Formation; TS—Teufelsschloss Formation; KP—Kap Petersens Formation; AS—Antarctic Sund Formation; TF—Tågefjeld Formation; RK—Rytterknægten Formation; SV—Skildvagten Formation; EB—Elisabeth Bjerg Formation. Height of cliff face is 1000 m.

combined thickness in excess of 14 km. However, in Andrée Land, 2.5–3 km of the Nathorst Land Group were eroded away at a well-exposed angular unconformity (Figs. 18 and 19) prior to the deposition of the Lyell Land Group (Henriksen, 1999; Smith and Robertson, 1999a), indicating the presence of at least one significant break in the succession. The Eleonore Bay Supergroup is also readily distinguished farther to the north in the Ardencaple Fjord–Bessel Fjord region (Nathorst Land Group and lower part of Lyell Land Group) and in southern Hochstetter Forland (Nathorst Land and Ymer Ø Groups; Sønderholm and Tirsgaard, 1993), although it is not known whether the unconformity seen at the base of the Lyell Land Group in Andrée Land is present in this area.

Approximately 50 km west of the central fjord zone, sediments crop out in an area centered on Petermann Bjerg (Fig. 13), and these sediments have long been correlated with the deposits of the Eleonore Bay Supergroup (cf. Wenk and Haller, 1953). Since a direct correlation with the main outcrop area could not be established with certainty, Sønderholm and Tirsgaard (1993) assigned these deposits to a separate unit—the Petermann Bjerg Group. More recent investigations (Smith and Robertson, 1999a) have suggested that the lower four units of the group can be correlated with the upper units of the Nathorst Land Group, and that the tentative correlation of the upper part of the Petermann Bjerg Group with the lower part of the Lyell Land Group suggested by Tirsgaard and Sønderholm (1997) is valid (Fig. 15).

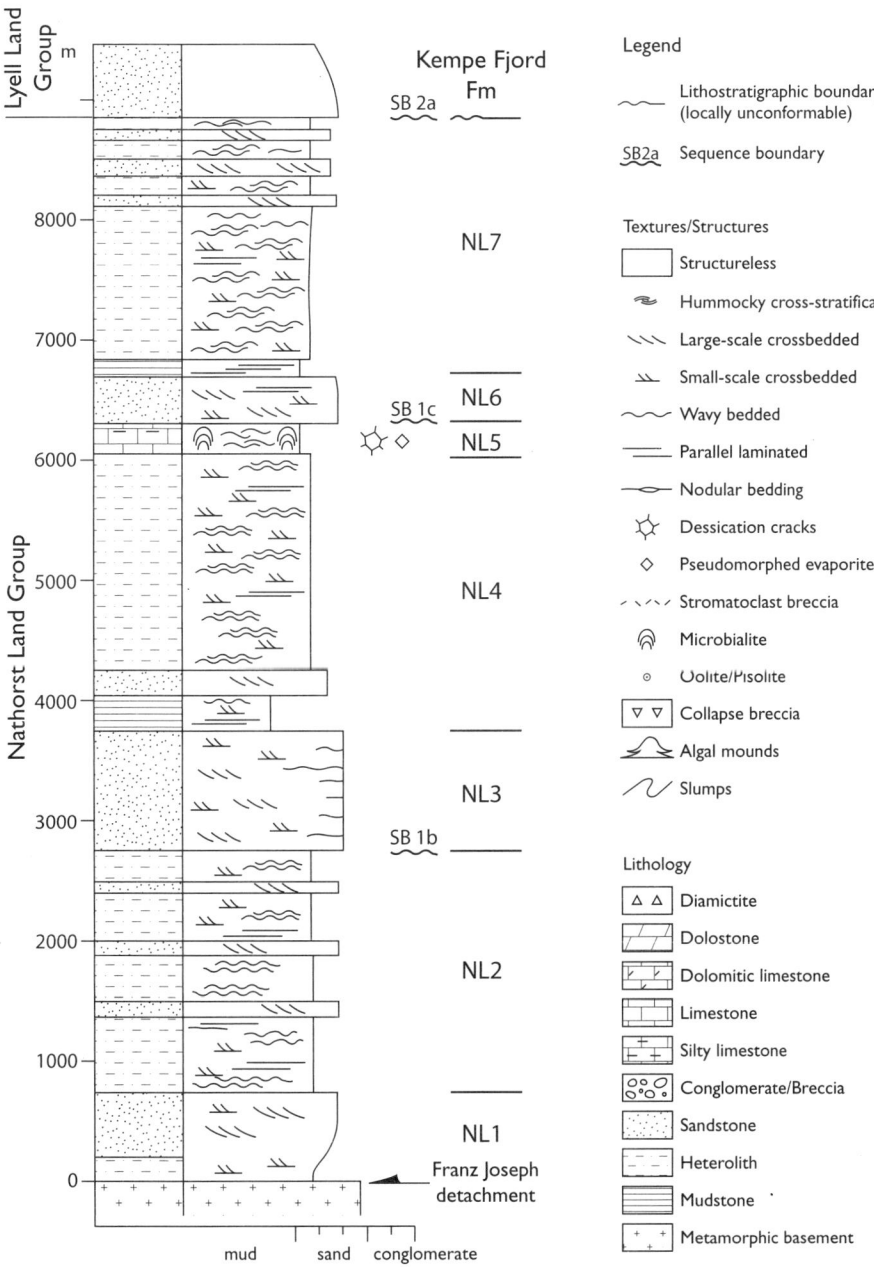

Figure 18. Composite sedimentary log of the Nathorst Land Group in the type area around Alpefjord and Forsblad Fjord; major sequence boundaries are shown. The base of the group is not seen because the oldest sediments are extensively intruded by granites. Division into formations follows Smith and Robertson (1999a). Height in meters.

Figure 19. Low-angle unconformity at the base of the Lyell Land Group on the south side of Eremitdal, opposite the junction with Snestormdal (see Fig. 13 for location). The pale sandstones of the basal Lyell Land Group can be seen to truncate brown-weathering marker horizons within formation NL4 of the Nathorst Land Group, which youngs eastwards. The cliff is ~800 m high.

The Tillite Group, originally defined by Haller (1971), includes five formations, two of which consist of glacigenic diamictites (Figs. 14, 32; Hambrey and Spencer, 1987). In the foreland to the west, correlatives of the Tillite Group directly overlie basement in the Målebjerg, Charcot Land, and Gåseland tectonic windows (Fig. 13; Higgins et al., 2001a, 2004a).

Sequence Stratigraphy

The development of depositional models for Proterozoic successions in order to elucidate basin evolution is strongly hampered by the lack of reliable biostratigraphic data. Correlations must therefore rely on lithostratigraphic principles. However, these may, if applied rigorously, often lead to false conclusions (van Wagoner et al., 1990), especially in correlations perpendicular to the general coastline trend.

Regional unconformities (i.e., sequence boundaries) may be difficult to recognize in Precambrian successions, which, in consequence, are often falsely regarded as thick, conformable successions (e.g., Harris and Eriksson, 1990); sequence stratigraphic principles can therefore be applied only with difficulty (Christie-Blick et al., 1988). However, the formation of sequence boundaries, particularly on gently dipping shelves and ramps, is often characterized by an associated significant basinward translation of facies, which results in major geographic reorganizations of the basin (van Wagoner et al., 1988; Posamentier and James, 1993). In some successions, flooding surfaces are readily defined, and, where these can be confidently correlated and placed in a sequence stratigraphic framework, they may provide a better basis for the subdivision of a succession and the subsequent definition of genetic stratigraphic packages (e.g., Galloway, 1989).

Neither continental deposits nor horizons of fluvial incision have been observed in the Eleonore Bay Supergroup. Furthermore, there is only one unequivocal semiregional unconformity present within the entire 14 km succession, and it is of tectonic origin (Figs. 14, 18, 19, and 20). However, it is apparent that major paleogeographic reorganization took place several times during the deposition of the succession. It seems highly improbable, therefore, that the 14 km of sediments comprising the Eleonore Bay Supergroup form an entirely conformable succession (cf. Tirsgaard and Sønderholm, 1997).

The widespread nature of the major lithostratigraphic units (formations) and their component facies within the rather narrow window of exposure in the central fjord zone suggest that depositional conditions, including sediment influx and subsidence, were uniform and parallel to basin strike for most of the time during deposition of (at least) the dominant siliciclastic part of the succession (Nathorst Land and Lyell Land Groups) and the lower part of the carbonate succession (Ymer Ø Group). During this time, the coastline was oriented approximately parallel to the window of exposure (present-day N–S; cf. Tirsgaard and Sønderholm, 1997). Therefore, units that show consistent stacking patterns within each sequence, and that are correlatable throughout the area, are assumed to have formed in response to regional changes in relative sea level and therefore have chronostratigraphic significance.

Within the Eleonore Bay Supergroup, large-scale cyclic sedimentary patterns have been recognized, and, in order to divide the succession into genetically related packages, second-order sequence boundaries have been placed where the main regional unconformities are considered to be located (Tirsgaard and Sønderholm, 1997). Since only one hiatus is demonstrably present within the basin (at the base of the Lyell Land Group; Figs. 14, 18, 19, and 20), sequence boundaries have been inferred from the succession of facies. On this basis, sequence boundaries are considered to be located where the main regional basinward translations of facies have occurred, and these can be recognized throughout the central fjord zone and in the northern outcrops. They also appear to be associated with laterally extensive erosion of the underlying deposits, and in carbonate deposits with regional dolomitization and formation of dissolution collapse breccias (Fig. 20). The location of the maximum flooding surface

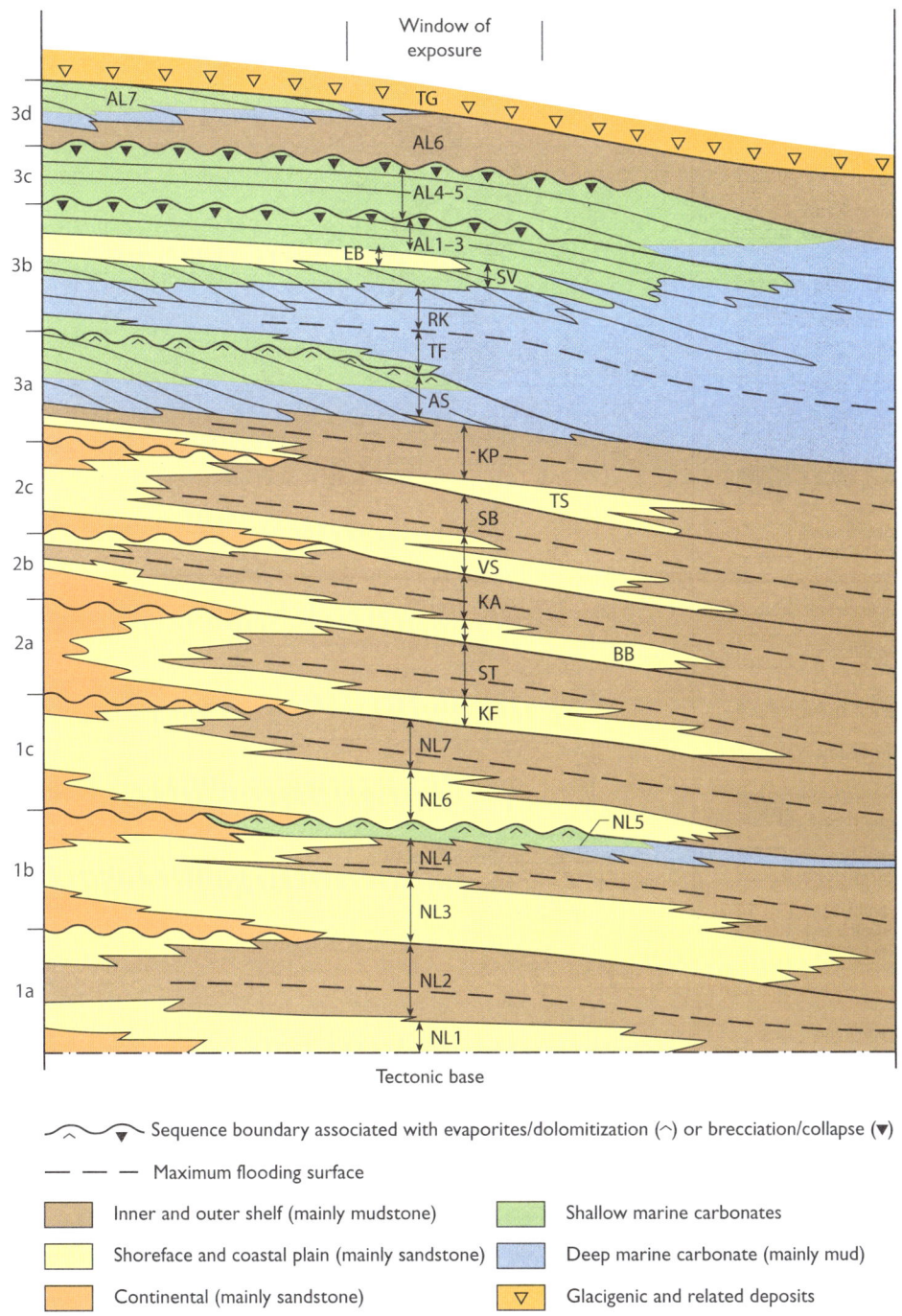

Figure 20. Tentative sequence stratigraphic cross section showing the characteristic stacking patterns of the lowstand, transgressive, and highstand systems tracts envisaged for the three stages of evolution of the Eleonore Bay Basin. Window of central fjord zone exposure is indicated. Stage 1 is represented by formations NL1–NL7 of the Nathorst Land Group. Stage 2 is represented by the Kempe Fjord (KF), Sandertop (ST), Berzelius Bjerg (BB), Kap Alfred (KA), Vibeke Sø (VS), Skjoldungebræ (SB), Teufelsschloss (TS), and Kap Peterséns (KP) Formations. Stage 3 is represented by the Antarctic Sund (AF), Tågefjeld (TF), Rytterknægten (RK), Skildvagten (SV), and Elisabeth Bjerg (EB) Formations, and formations AL1–AL7 of the Andrée Land Group. Glacigenic deposits are represented by the Tillite Group (TG). See text for further explanation.

is difficult to determine, but it is placed in the middle of the most fine-grained interval or at the shift from retrogradational to progradational deposition (Tirsgaard and Sønderholm, 1997).

Basin Evolution

In the Eleonore Bay Basin, four major stages of basin evolution may be distinguished based on major paleogeographic reorganizations of the shelf and a change from siliciclastic to carbonate deposition shown by the Eleonore Bay Supergroup, before glacigenic deposition eventually became dominant as recorded by the sediments of the Tillite Group.

Stage 1: Rapidly Subsiding Siliciclastic Shelf

Stage 1 of the evolution of the Eleonore Bay Basin is represented by the Nathorst Land Group, the type area of which is in the cliffs of Alpefjord and Forsblad Fjord (Fig. 21). Due to remoteness of outcrops, difficult access, and the vast thick-

Figure 21. The upper part of the Nathorst Land Group on the western side of Alpefjord, the type area of the group. Sequence boundary SB 1c lies at the base of formation NL6; see Figure 18 for a composite sedimentary log through the group. The summit of the highest peak is at 1800 m above the fjord in the foreground.

ness of the group, it has not yet been studied in the same detail as the overlying, but very similar, Lyell Land Group. The total thickness is 9 km, and it was divided into seven informal formations (NL1–NL7) by Smith and Robertson (1999a; Figs. 14, 18, and 20); these seven units correspond in turn to the three informal units (NL1–NL3) of Sønderholm and Tirsgaard (1993). Interpretation is further impeded by the tectonic overprint. Metamorphic grade increases rapidly toward the base of the group, attaining garnet zone commonly and sillimanite grade in places. Away from faults and more ductile detachments, deformation is, however, limited to the sporadic development of slaty cleavage, and sedimentologic and sequence stratigraphic studies are possible to some degree at most localities (Caby and Bertrand-Sarfati, 1988; Smith and Robertson, 1999a).

The nature of the lower boundary of the Eleonore Bay Supergroup has been debated for many years, since the contact of the original surface is obscured by abundant granite intrusions. The most recent studies (Watt and Thrane, 2001; White and Hodges, 2002; White et al., 2002; Higgins et al., 2004a) have suggested that the contact is associated with a major décollement surface rather than being a gradual transition to underlying amphibolitic gneisses. The apparent transitional nature of the contact zone resulted from movements on this surface during progressive Caledonian metamorphism (Higgins et al., 1981, 2004a; cf. Gilotti and McClelland, this volume).

Stage 1 may be divided into three second-order sequences that contain a similar range of lithofacies to the overlying stage 2 deposits, but they are distinctive in their exceptional thickness. Sequence 1a is a minimum of 2.4 km thick, 1b is 3.55 km thick, and 1c is 3.95 km thick. Lower-order sequences are not consistently recognizable, but they are clearly exhibited in the carbonate unit, NL5, and the overlying sandstone of NL6.

Sequence 1a. The lowest sequence of the Nathorst Land Group (Figs. 18 and 20) is the least amenable to sedimentologic and sequence stratigraphic study. The lowest lithostratigraphic unit, NL1 (750–950 m), is a quartzite unit that everywhere overlies the Franz Joseph detachment (Fig. 16; Higgins et al., 2004a). Toward the base, there is extensive anatexis and intrusion of granites, and the lowest exposed parts of the unit occur as xenoliths composed of quartzites and thin interbedded semipelitic gneissose rocks, both of which contain cordierite and sillimanite (Smith and Robertson, 1999a). However, toward the top of the unit, the metamorphic grade decreases, and the higher parts consist of a distinctive banded unit with medium-gray, ripple-laminated and cross-bedded sandstone together with very pale cross-bedded quartzarenites. Thinner, muddier, ripple-laminated beds occur throughout the unit and probably correspond to the gneissose semipelite in the lower part. NL1 is overlain by a mixed succession of heterolithic siltstone, fine-grained sandstone, and mudstone interbedded with quartzarenite and gray sandstone, the latter of which are large-scale cross-bedded or parallel laminated. This unit, assigned to NL2 by Smith and Robertson (1999a), is at least 2000 m thick.

Sequence 1b. The bipartite division seen in sequence 1a, which consists of sandstone (many of them of quartzarenite composition) overlain by heterolithic sediments, is maintained through sequences 1b and 1c, and indeed into the overlying Lyell Land Group. The progradational part of sequence 1b (Figs. 18 and 20) is NL3, a distinctive 1000-m-thick sandstone unit that consists of thick bedded sandstone with large-scale cross-bedding in the lower part overlain by quartzarenites in which strongly wedge-shaped units are present within massive gray sandstone and ripple-laminated sandstone. The high maturity, the marked absence of mud, and the presence of dunes and megaripples at a variety of scales are together interpreted as a shoreface facies association. The NL3–NL4 boundary is a prominent flooding surface, and there is evidence of a marked shift to outer-shelf, storm-dominated sediments consisting of normally graded, parallel- and ripple-laminated siltstone and fine-grained sandstone. This lower, 150-m-thick unit is overlain by 150 m of wavy bedded siltstone and 180 m of cross-bedded quartzarenites, but the remaining 1800 m is lithologically monotonous, composed of interbedded cross-bedded sandstone and wave-rippled, wavy bedded heterolithic siltstone/mudstone interpreted as a storm- and wave-dominated inner-shelf facies association. NL4 passes upward into the only carbonates that occur within the Nathorst Land Group (Figs. 18 and 20). NL5 is a 250 m carbonate unit composed of calcareous wavy-bedded heterolithic siltstone intensely veined by carbonates, which are probably pseudomorphed evaporites. In the upper part of the unit, low-relief stromatolites are preserved and shallowing-upward parasequences are common. Typical parasequences have calcareous heterolithic sediments at the base, which become increasingly veined upward, with a microbialite cap made up of stromatolites and crinkly laminated mats with desiccation cracks and tepee structures. NL5 is capped by the quartzarenites of NL6 (sequence 1c), and the total thickness of sequence 1b is 3.55 km.

Sequence 1c. The base of sequence 1c (Figs. 18 and 20) is marked by the sharp erosive base of quartzarenites that constitute NL6. The quartzarenites are white, fine-grained, and channelized in places; trough cross-bedding, parallel lamination, and wave ripples are present. In cliff sections, NL6 (400 m) clearly exhibits three parasequences made up of pale quartzarenite interbedded with darker, finer-grained units. The quartzarenites, like those at the base, are dominantly trough cross-bedded with wedge-shaped beds and rippled tops. The darker units are composed of heterolithic, ripple-laminated very fine sand with muddy drapes. An abrupt boundary to NL7 is overlain by 130 m of graded siltstone to mudstones, in packets a few centimeters thick; these are interpreted as outer-shelf, sub–storm-wave-base deposits, where the NL6–NL7 boundary represents a major flooding surface. NL7 is 2000 m thick in total and consists of wavy-bedded silts, sands, and muds with interbedded sharp-based, tabular, parallel-laminated and current- and wave-rippled sandstone; the current ripples are frequently bidirectional. The sediments of NL7 are interpreted as an association of storm- and wave-dominated sediments.

The construction of a full sequence stratigraphic model for the Nathorst Land Group is hampered by the narrow, linear outcrop belt within the Franz Joseph allochthon. However, some lateral variability is evident northward from the type area, 170 km along strike, and between the Nathorst Land Group and the isolated klippe of coeval sediments, assigned to the Petermann Bjerg Group, which lie 110 km northwest of the type area of Alpefjord.

The lateral variability along strike is the result of a pronounced angular unconformity that occurs at the base of the Lyell Land Group in Ermitdal, at the junction with Snestormdal in Andrée Land, where it rests on NL4 (Fig. 20). The absence of NL5–NL7 indicates that at least 2.5–3 km of the succession seen farther south are missing at this boundary and provides evidence for intra–Eleonore Bay Supergroup uplift and erosion.

The Petermann Bjerg Group (Fig. 22) crops out in a klippe of the Franz Joseph allochthon, centered on Louise Boyd Land and Frænkel Land, that measures 80 km N–S and 60 km E–W. Although it has long been clear that the group is correlative with the Eleonore Bay Supergroup (Wenk and Haller, 1953; Haller, 1971; Sønderholm and Tirsgaard, 1993), a detailed correlation has not been available until recently (Smith and Robertson, 1999a). The metamorphic grade makes detailed sedimentology impossible in all but the quartzites, but it is now clear that the Petermann Bjerg Group is an attenuated correlative of the upper Nathorst Land Group and lowermost Lyell Land Group (Fig. 23). More specifically, formations PB1 and PB2 and the lower part of formation PB3 correspond to sequence 1b (Fig. 23), and the upper part of formation PB3 and formation PB4 correspond to sequence 1c; formations PB5 and PB6 are correlative with the Kempe Fjord and Sandertop Formations, respectively, which form the basal part of the Lyell Land Group (Smith and Robertson, 1999a). This correlation gives critical three-dimensionality to interpretations of stage 1. Sequence 1b thins from 3.55 km in the Alpefjord–Forsblad Fjord area to around 2300 m in the Petermann Bjerg Group (Fig. 23), and it becomes correspondingly dominated by the quartzarenite shoreface association—in Alpefjord, this accounts for 28% of sequence 1b, whereas in the Petermann Bjerg Group, it amounts to around 44%. Similarly, sequence 1c thins from 2 km in the type area of the Nathorst Land Group to a little over 1 km in the Petermann Group, and quartzarenites again dominate the sequence in the latter case— the quartzarenites and associated sediments of the shoreface facies association comprise 20% of sequence 1c in the type area of the Nathorst Land Group, whereas they constitute 79% of the same sequence in the Petermann Bjerg Group (Fig. 23).

Stage 2: Stable Siliciclastic Shelf

Stage 2 makes up the major part of the Lyell Land Group, which has been studied in detail in the region between Canning Land and Vibeke Sø (Fig. 13; Tirsgaard, 1993; Tirsgaard and Sønderholm, 1997).

Stage 2 is composed of three second-order sequences (Figs. 20 and 24), all of which are characterized by thick retrogradational to progradational successions that show a similar trend in sedimentary evolution through time. In these sequences, the sequence boundary is placed where shoreface or coastal plain deposits sharply overlie outer-shelf or storm-dominated inner-shelf deposits. The vertical development of facies associations indicates that a major regional translation of facies is associated with the abrupt transition from shelf mudstone to shoreface or coastal plain sandstone. This translation is also interpreted to have been associated with regional erosion and appears to have formed in relation to a large-scale forced regression.

Sequence 2a. The lower sequence is represented by the Kempe Fjord and Sandertop Formations, which have a total thickness that reaches 600–1000 m in the central fjord zone (Fig. 24). It overlies a major angular unconformity that cuts out 2.5–3 km of the top of the Nathorst Land Group in Andrée Land (Fig. 20). The regional extent of this unconformity is unknown, but both 50 km to the south in Lyell Land (along apparent basinal strike) and 50 km to the WSW (approximately perpendicular to basinal strike), the Lyell Land Group rests without angular discordance

Figure 22. View of the upper part of the Petermann Bjerg Group northward across Nordenskiöld Gletscher toward Petermann Bjerg and Lille Petermann in Frænkel Land (see Fig. 13 for location). The upper part of formation PB3 (1400 m thick in total; "Layered Series" of Wenk and Haller 1953) is overlain by formation PB4, which is 350–470 m thick ("Shoulder Series" of Wenk and Haller 1953). The overlying formation PB5 is marked by a prominent pale-weathering sandstone that is considered to be correlative with the base of the Lyell Land Group (see Fig. 23). The vertical distance from Nordenskiöld Gletscher to the summit of Lille Petermann is ~900 m.

upon the uppermost formation of the Nathorst Land Group. The lower boundary is, however, marked by an abrupt shallowing from outer-shelf mudstone to shoreface and coastal plain sandstone at other localities. The initial shallowing is followed by a more gradual shallowing that is reflected by the lower 200 m of the Kempe Fjord Formation, where deposits of the storm- and wave-dominated shoreface association are overlain by subtidal sandstone. Toward the top, these grade into a 300-m-thick succession of predominantly intertidal deposits. The intertidal channel deposits are succeeded by ~100 m of sandstone of mainly subtidal origin, implying a subtle rise in relative sea level. This weak deepening trend observed in the upper part of the Kempe Fjord Formation becomes more pronounced in the Sandertop Formation (Fig. 24). The deepening of the shelf is reflected in a change in the lower 80 m of the Sandertop Formation, where storm-dominated inner-shelf deposits gradually give way to outer-shelf deposits that represent the maximum flooding of the basin. Above this unit, storm- and wave-dominated inner-shelf deposits gradually become more abundant, signifying renewed progradation (Fig. 20).

A poorly developed overall shallowing is signified by the overlying 320 m of sediment, which constitute the rest of the Sandertop Formation (Fig. 24). Sharp-based shoreface deposits occur within the storm-dominated inner-shelf deposits and reflect minor progradational events, possibly caused by forced regressions.

Sequence 2b. This sequence is represented by deposits of the Berzelius Bjerg and Kap Alfred Formations; it varies in thickness between 750 and 1150 m in the central fjord zone (Figs. 17 and 24).

The sequence boundary is located at the contact between the Sandertop Formation and the Berzelius Bjerg Formation (Fig. 24), and it marks an abrupt shift from heterolithic mudstone deposited in a storm- and wave-dominated inner-shelf environment to tidally dominated coastal plain sandstone. In contrast to sequence 2a, shoreface deposits are only a few meters thick, and coastal plain deposits are present almost directly above the sequence boundary. Similar to the development seen in the lower part of sequence 2a, the lower 400 m of sequence 2b reflect a stillstand in relative sea level, following the abrupt fall inferred at the base. A relative sea-level rise is indicated by the deepening trend recorded by the top of the Berzelius Bjerg Formation, where the coastal plain deposits pass up into 80 m of fine-grained shoreface sandstone. This trend continues into the Kap Alfred Formation, where the shoreface deposits pass into storm- and wave-dominated inner-shelf deposits that form a relatively uni-

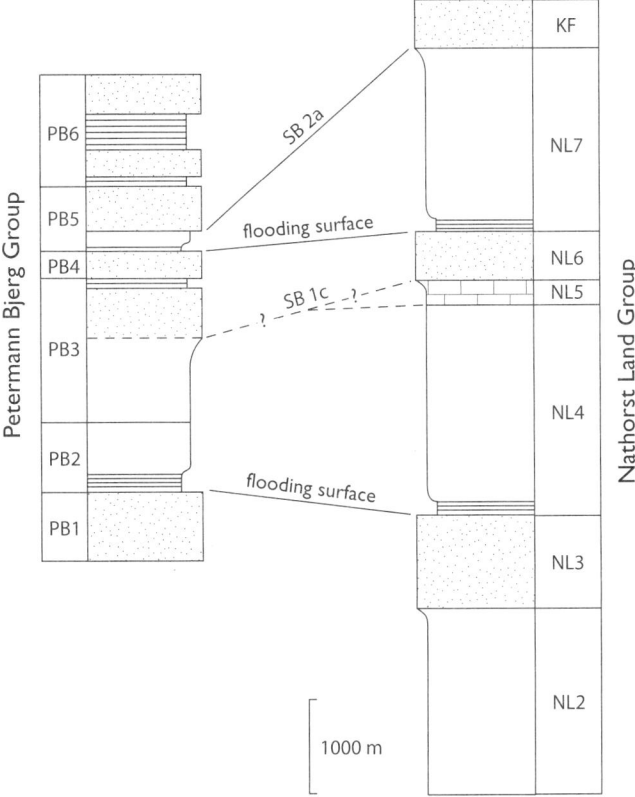

Figure 23. Correlation of the Petermann Bjerg Group (formations PB1–PB6) with the Nathorst Land and Lyell Land Groups. Stippled ornament indicates sand-dominated units, and horizontal ruling indicates mudstone- and siltstone-dominated units of outer-shelf origin; unornamented intervals indicate interbedded sandstones and finer-grained units, predominantly heterolithic siltstones or mudstones. Sequence boundaries (SB) discussed in the text are indicated (see also Figs. 14 and 20). KF—Kempe Fjord Formation (Lyell Land Group); ST—Sandertop Formation.

form 130-m-thick succession (Fig. 24). On top of this, there is a 400-m-thick succession that consists of interbedded tidally influenced shoreface and storm- and wave-dominated inner-shelf deposits, which mark a shift toward a more tidally dominated shelf. The uppermost 100 m of the sequence consists entirely of outer-shelf mudstone, representing the culmination of the overall deepening (Figs. 20 and 24).

Sequence 2c. Sequence 2c is made up of the sediments of the Vibeke Sø and the Skjoldungebræ Formations (Figs. 17 and 24); it reaches a thickness of 500–550 m in the central fjord zone and in Canning Land.

Sequence 2c starts with abrupt shift from outer-shelf mudstone to mature, structureless sandstone deposited in a storm- and wave-dominated shoreface environment. Numerous mudstone intraclasts and a highly irregular base indicate that the sequence boundary was erosive. Following the initial pronounced sea-level fall, the sequence records a sea-level stillstand followed by a gradual deepening. The lower 150 m of the sequence consist primarily of shoreface deposits. However, in the central fjord zone,

10–20 m of heterolithic storm- and wave-dominated inner-shelf deposits occur 20–25 m above the base, reflecting minor, high-frequency variations in relative sea level.

Above the shoreface deposits, there is ~150 m of sediments that show a recurrent interbedding of 1–5-m-thick shoreface deposits and 5–40-m-thick storm-dominated inner-shelf deposits, reflecting repeated episodes of regression followed by transgression. An overall deepening is manifested by a gradual upward thinning of the sandstone beds and a gradual thickening of the heterolithic deposits. In the uppermost 200 m of the sequence, storm-dominated inner-shelf deposits give way to outer-shelf mudstone interbedded with thin, sharp-based shoreface sandstone. A weak tendency toward upward coarsening is present in the uppermost 50 m, where a return to interbedded inner-shelf and shoreface deposits is seen (Fig. 24).

The repeated episodes of forced regression seen throughout most of this sequence reflect the superimposition of high-frequency relative-sea-level variations upon the overall relative sea-level rise. The high-frequency sea-level variations give rise to 20–50-m-thick regressive-transgressive cycles, similar to those observed in the sequences below. These cycles cannot convincingly be correlated between outcrops, and their lateral extent is unknown. Cyclic regressive-transgressive events on a scale of 70–120 m are also visible within the sequence and appear to be correlatable throughout the fjord zone and out into Canning Land. The cyclic pattern is produced by stacked successions of upward-thickening sandstone units.

Stage 3: Carbonate-Platform Development

Stage 3 is represented by the uppermost formation of the Lyell Land Group and the Ymer Ø and Andrée Land Groups, and it includes four second-order sequences with a total thickness of ~2700 m in the central fjord zone (Figs. 20 and 25). The lower sequence marks the change from siliciclastic to dominantly carbonate deposition and is characterized by a progradational evolution of the carbonate platform (sensu Read, 1982). The second sequence reflects a retrogradational to progradational and aggradational carbonate-platform succession. A shift toward a more humid climate in the later part is suggested by a marked increase in red siliciclastic mud that intermittently caused carbonate deposition to cease, at least on the inner parts of the platform. The third sequence shows an overall retrogradational pattern terminated by an abrupt fall in sea level that resulted in exposure of most of the carbonate platform (Figs. 20 and 25). The upper sequence, heralding the Marinoan glaciation represented by the deposits of the Tillite Group, consists of a strongly retrogradational succession that indicates drowning of the carbonate platform of the previous stage and deposition of deep-marine deposits, followed by a short period of carbonate-platform progradation before glacigenic deposition took over (Fig. 20).

Sequence 3a. Sequence 3a is made up of sediments of the Teufelsschloss, Kap Peterséns, and Antarctic Sund Formations, and it reaches a thickness of ~500 m in the central fjord zone (Figs. 17, 25, and 26). The sequence was induced by a major

Figure 24. Sedimentary log of the Lyell Land Group showing sequence boundaries of stage 2 in the development of the Eleonore Bay Basin (slightly modified from Tirsgaard and Sønderholm, 1997). For legend, see Figure 18. Height in meters.

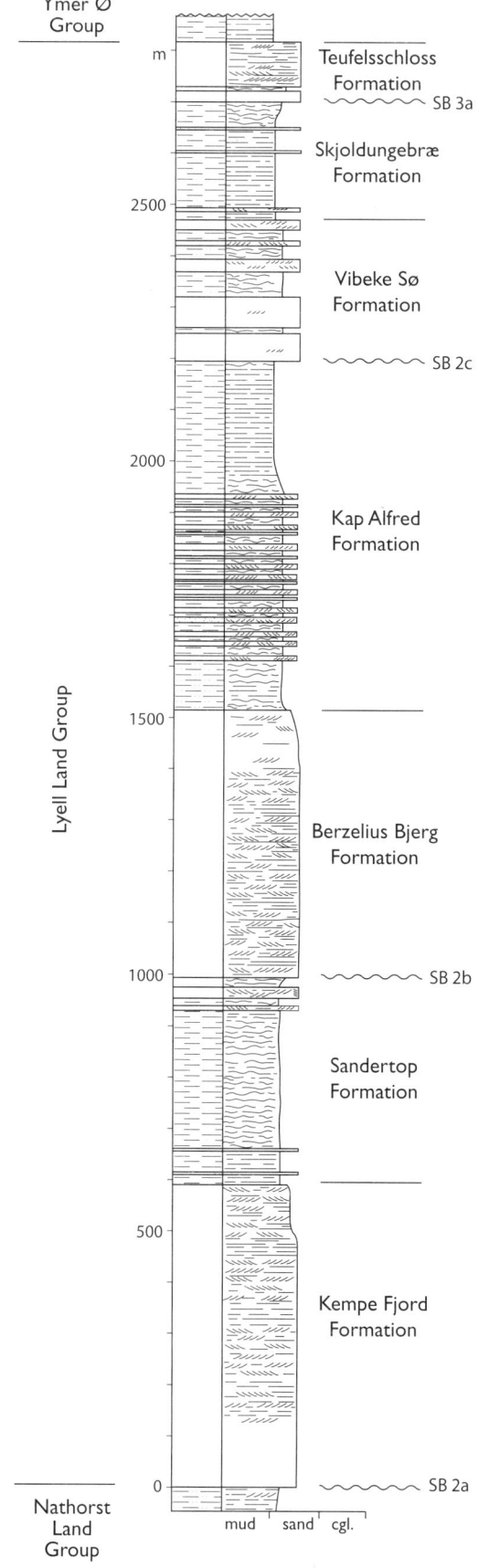

transgressive event, possibly accompanied by a shift to a more arid climate, and it records a gradual change from siliciclastic to carbonate deposition (Fig. 20).

Overlying the sequence boundary, initial lowstand forced regression deposits are represented by shoreface sandstone of the Teufelsschloss Formation (Fig. 26), which attains a thickness of around 130 m in most of the outcrop area. However, rapid thinning occurs in the southeastern part of the fjord zone and in Canning Land, where thicknesses of only 15–60 m are reported (Tirsgaard and Sønderholm, 1997). The shoreface deposits are succeeded by outer-shelf mudstone of the Kap Peterséns Formation (~280 m), and local large-scale channelized slump structures and progradational patterns become conspicuous upward. However, the shift to a more arid climate and ensuing carbonate deposition resulted in declining sedimentation rates and renewed generation of accommodation space. Initial progradation of a carbonate platform is indicated by a gradual increase in carbonate content in the mudstone of the Kap Peterséns Formation (Fig. 25). These are succeeded by cherty limestone of the Antarctic Sund Formation (115 m), in which common resedimented and slumped horizons reflect continued progradation of carbonate and slope and shelf sediments (Fig. 27). The top of the sequence is marked by a dolomitized, erosional unconformity.

Sequence 3b. Sequence 3b is made up of sediments of the Tågefjeld, Rytterknægten, Skildvagten, and Elisabeth Bjerg Formations of the Ymer Ø Group and formations AL1–AL3 of the Andrée Land Group (Figs. 17, 20, 25, and 27; cf. Sønderholm and Tirsgaard, 1993). The sequence reaches a thickness of ~1050 m in the central fjord zone. The formations of the Ymer Ø Group are relatively constant in thickness throughout the area of exposure, whereas the formations of the Andrée Land Group show marked local thickness variations in both S-N and W-E directions.

Although the facies development in the Andrée Land Group seems to record a deepening trend toward the north or northeast, there are no signs of a major tectonic reorganization of the overall approximate N–S basinal trend during this time. The change from a N–S coastline trend during the early part of the sequence to a more E–W trend reported by Frederiksen (2000a) therefore probably reflects the development of a more indented coastline resulting from varying growth rates of individual ramp segments along the general N–S coastline trend.

The basal part of sequence 3b is represented by the Tågefjeld Formation (~190 m) and is characterized by a retrogradational succession initiated by widespread succession of evaporitic, shallow-water carbonates and lagoonal mudstone (Fig. 27) grading upward into deeper-water limestone. These

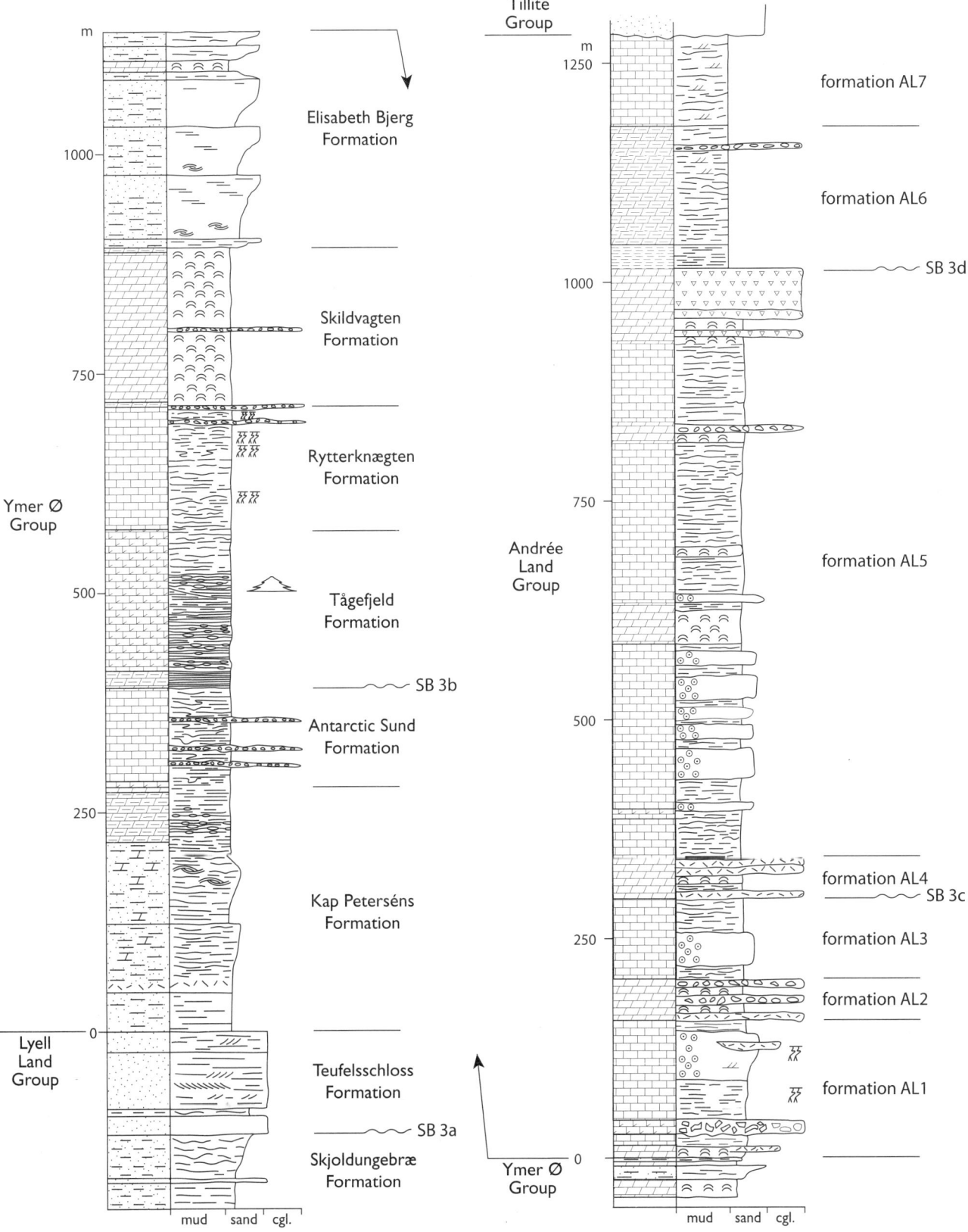

Figure 25. Sedimentary log of the top of the Lyell Land Group and the Ymer Ø and Andrée Land Groups, and sequence boundaries of stage 3 in the development of the Eleonore Bay Basin (slightly modified from Sønderholm and Tirsgaard, 1993; Frederiksen, 2000b). For legend, see Figure 24. Height in meters.

Figure 26. Major transgressive event shown by the significant basinward translation of facies recorded by the abrupt change from outer-shelf mudstones of the Skjoldungebræ Formation (SB) to pale shoreface sandstones that form the base of the Teufelsschloss Formation (TS; 125 m thick). The upper part of the Teufelsschloss Formation is composed of sandstones deposited in a coastal plain environment, and these are outer-shelf mudstones of the Kap Peterséns Formation (KP). A change to carbonate deposition in sequence 3a is recorded by the Antarctic Sund Formation (AS). Location shown is south coast of Ymer Ø (from Tirsgaard and Sønderholm, 1997).

are locally associated with complex algal mound structures up to 80 m thick and 250 m wide and possible large-scale slumped units, which suggest platform-margin deposition. This major transgressive phase resulted in drowning of the platform and eventually led to a change to basinal and slope deposition recorded by finely laminated, locally shaly limestone of the 140-m-thick Rytterknægten Formation.

Renewed platform progradation is indicated by an upward increase in abundance of slumped horizons in the Rytterknægten Formation (Fig. 20). These deposits are followed by a distinctive dolomite unit (Skildvagten Formation; 180 m) in which irregular stromatolite growth patterns reflect progradation of high-energy platform-margin environments. A shift toward a more humid climate is suggested by a marked increase in red siliciclastic mud, which eventually caused carbonate deposition to cease, at least on the inner parts of the platform, where a mixed siliciclastic-carbonate succession consisting of stacked inner- and outer-shelf deposits was laid down (Figs. 20 and 27; Elisabeth Bjerg Formation; 290 m). These are traceable throughout the entire outcrop area and probably reflect a response to third-order eustatic sea-level changes (Tirsgaard, 1996). An overall decrease in siliciclastic sediment supply within the Elisabeth

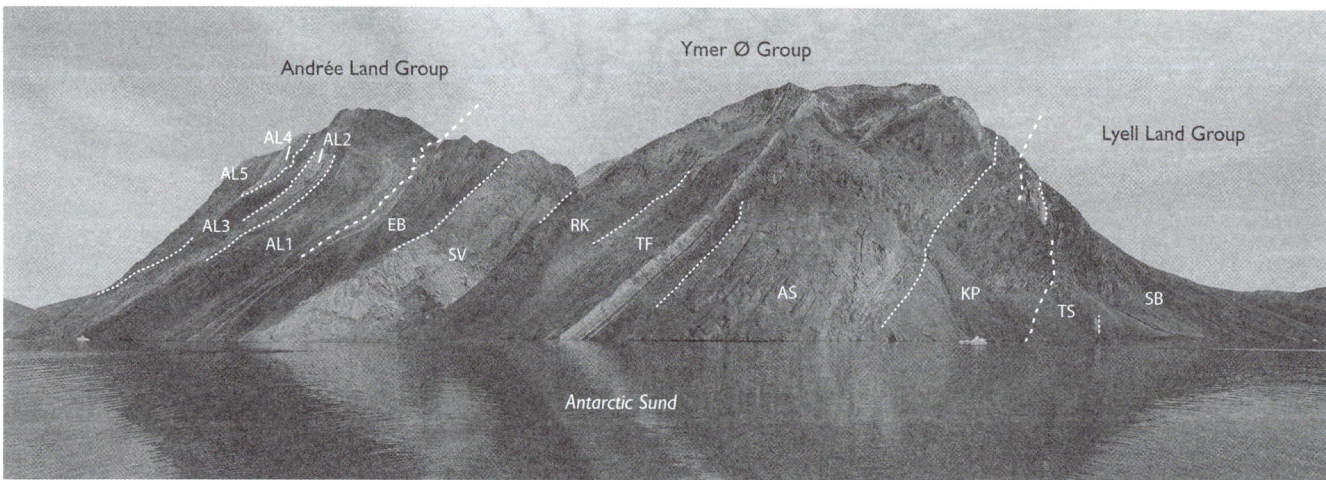

Figure 27. Type locality of the Ymer Ø Group on the south coast of Ymer Ø along Antarctic Sund, the strait between Ymer Ø and Suess Land (see Fig. 13 for location). Height of eastern mountain (to the right) is ~1250 m. The formations of the Ymer Ø Group are, from base to top, Kap Peterséns Formation (KP), Antarctic Sund Formation (AS), Tågefjeld Formation (TF), Rytterknægten Formation (RK), Skildvagten Formation (SV), and Elisabeth Bjerg Formation (EB). To the east is the top of the Lyell Land Group with the Skjoldungebræ Formation (SB) and Teufelsschloss Formation (TS). To the west, formations AL1–AL5 of the Andrée Land Group are well exposed. The pale unit in the lower part of the Tågefjeld Formation represents a widespread unit of evaporitic sulfate deposition (cf. Fig. 17).

Bjerg Formation heralds the return to inner-shelf, high-energy carbonate deposition and evaporite formation, recorded by the lower part of the Andrée Land Group.

The lower three formations of the Andrée Land Group reflect a complex relationship between storm-dominated, shallow-marine microbial and pisolitic ramp environments and mid- to outer-ramp environments (Fig. 25; Frederiksen et al., 1999; Frederiksen, 2000a, 2000b). There is no overall trend in stacking patterns, suggesting a general aggradation of the carbonate ramps during times when carbonate production, sea-level change, and subsidence were more or less in balance. The general tendency for very shallow environments and the exposures of parts of the inner ramps with subsequent evaporation, dissolution, and collapse suggest, however, that deposition occurred during an overall fall in relative sea level (Fig. 20).

Formation AL2 consists of dolomites that show a high proportion of polymict, stromatoclast, and dissolution collapse breccias. The polymict breccias, which show a high diversity of clast types, suggesting that the channel fills were probably derived from a large source area, are interpreted to have formed in an extensive, laterally migrating channel belt (Frederiksen, 2000b) that may have formed a lowstand, incised valley system.

Sequence 3c. Sequence 3c includes sediments of formations AL4 and AL5 of the Andrée Land Group (sensu Fränkl, 1953; Sønderholm and Tirsgaard, 1993). The thickness of this sequence is ~700 m in the central fjord zone (Figs. 25 and 27).

The lower sequence boundary is placed at the base of dolomitic polymict, stromatoclast, and dissolution collapse breccias that occur close to the base of formation AL4 (Figs. 20 and 25). The formation is 30–130 m thick and shows considerable thickness variation. During sea-level lowstand, the polymict and stromatoclast breccias were probably formed in channels and around storm-dominated microbial reef tracts that occurred between subaerially exposed highs experiencing dissolution collapse (Frederiksen, 2000b).

The subsequent overall transgression is documented by the lower ~500 m of formation AL5, in which inner- to mid-ramp environments such as subaerially exposed flats, lagoons, microbial reefs, pisoid shoals and flats, and channel flats are followed by outer-ramp environments.

A major sea-level fall resulted in exposure of most of the carbonate platform and ensuing extensive dolomitization and dissolution collapse. This is shown by a 20–130-m-thick dolomitic unit, which forms the top of formation AL5, that can be traced throughout most of the region (Frederiksen, 2000b). The upper sequence boundary is placed on top of this unit, since it records dissolution rather than erosion and redeposition (Fig. 25).

Sequence 3d. The uppermost sequence is a retrogradational to progradational succession made up of formations AL6 and AL7 (Figs. 20, 25, and 28). The combined thickness of this succession is ~250 m, but individual units show large regional variation in thickness (Figs. 25 and 28).

A major relative sea-level rise resulted in drowning of the partly exposed carbonate platform of sequence 3c (Fig. 20; Frederiksen, 2000b). A thin succession of possible shallow-water carbonates overlies the dissolution collapse breccias, followed by ~100 m of deep-water shaly carbonate turbidites and slump breccias of formation AL6.

In the northern part of the central fjord zone, shallow-water carbonate-platform sedimentation resumed (formation AL7) before glacigenic deposition abruptly took over as recorded by the sharp but transitional contact to the overlying Tillite Group (Fig. 28). Carbonate-platform conditions were not established further to the south, and it is therefore possible that the upper part of formation AL6 is time equivalent to formation AL7 (Fig. 20; Herrington and Fairchild, 1989).

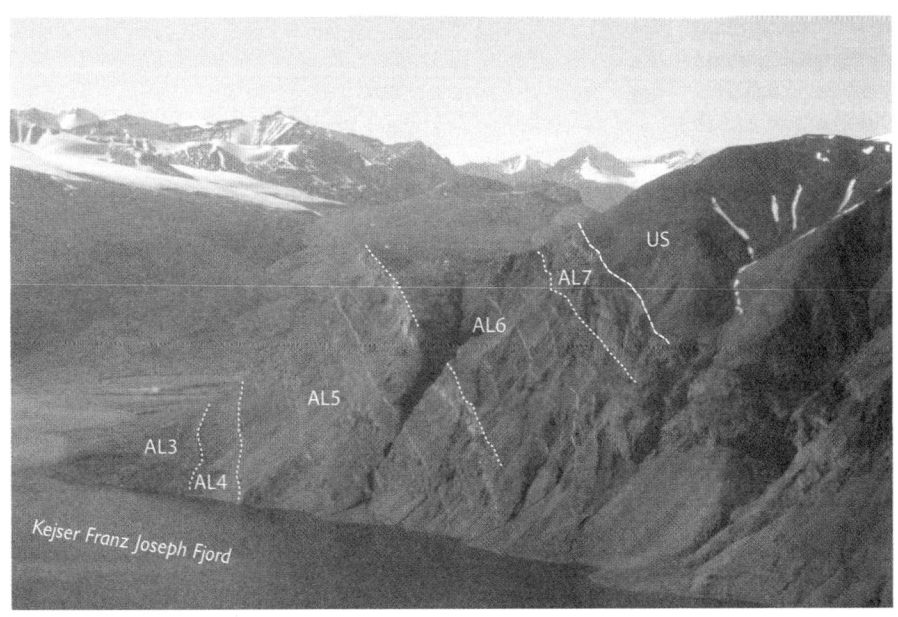

Figure 28. Upper part of Andrée Land Group (formations AL3–AL7) as exposed at Kap Weber, Andrée Land, showing boundary to the Ulvesø Formation (US), which forms the base of the Tillite Group. Note the conspicuous pale-weathering dissolution collapse breccia horizon that forms the top of formation AL5. The succession from sea level to the base of Ulvesø Formation is ~1100 m thick at this locality.

The abrupt increase in accommodation space, recorded by the cessation of stable shallow-water carbonate-platform environments and documented by the change from sequences 3a–3c to the mixed siliciclastic–carbonate deep-water sediments of sequence 3d, could reflect a change toward a more humid climate combined with a major relative sea-level rise. It has been suggested that the sea-level rise may have been driven by tectonically induced subsidence and tilting related to an initial rifting phase associated with the opening of the Iapetus Ocean (Herrington and Fairchild, 1989; Fairchild and Hambrey, 1995; Frederiksen, 2000b). However, direct evidence of synsedimentary faulting, apart from large variations in depositional environment and thickness of individual units over short distances, has not been observed.

Stage 4: Glacigenic Deposition

The Tillite Group overlies the Eleonore Bay Supergroup with no apparent major hiatus, and it is divided into five formations (Fig. 29; Hambrey and Spencer, 1987). The oldest unit, the Ulvesø Formation (100–318 m), is dominated by diamictite and is overlain by dolomitic shale, siltstone, and sandstone of the Arena Formation (223–320 m), which is in turn overlain by a second diamictite unit, the Storeelv Formation (60–223 m; Hambrey and Spencer, 1987). The Storeelv Formation is overlain by a cap carbonate, which marks the base of the Ediacaran (Knoll et al., 2004) and is assigned to the Canyon Formation (up to 300 m). The bulk of the Canyon Formation is made up of dolomitic shales, and stromatolitic dolostone is developed toward the top. The youngest unit in the Tillite Group, the Spiral Creek Formation (up to 45 m), is developed only in the central part of the basin and is overlain with slight angular unconformity by quartzarenites of the Lower Cambrian Kløftelv Formation.

The Ulvesø Formation contains a wide range of glacial and related lithofacies, documented by Hambrey and Spencer (1987) and Moncrieff and Hambrey (1988, 1990). These include waterlain tillite, debris-flow diamictite, glaciomarine deposits, and eolian sandstone together with indications of permafrost conditions (contraction wedges and load structures). The environment is interpreted as one in which a floating, debris-bearing ice shelf supplied sediment into a shallow-marine environment (Fig. 30). There are no close analogues of this situation at present—true ice shelves are restricted to Antarctica at the present but are debris-free unless freeze-on of saline ice allows sediment to be retained until calving takes place (M. Hambrey, 2007, personal commun.). Current activity was present beneath the floating ice in the Ulvesø Formation, and it produced lenses of sorted material in the waterlain tillites and glaciomarine facies (Moncrieff and Hambrey, 1990).

In most parts of the basin, massive sandstone assigned to the Arena Formation directly overlies the Ulvesø Formation, and there is no cap carbonate (Hambrey and Spencer, 1987). In general, the Arena Formation fines upward to shale or interbedded sandstone and shale, with ripple lamination and interference ripples present. At the snout of Sorteelv Gletscher, in northern Scoresby Land, two thin diamictite horizons similar to those in the Ulvesø Formation

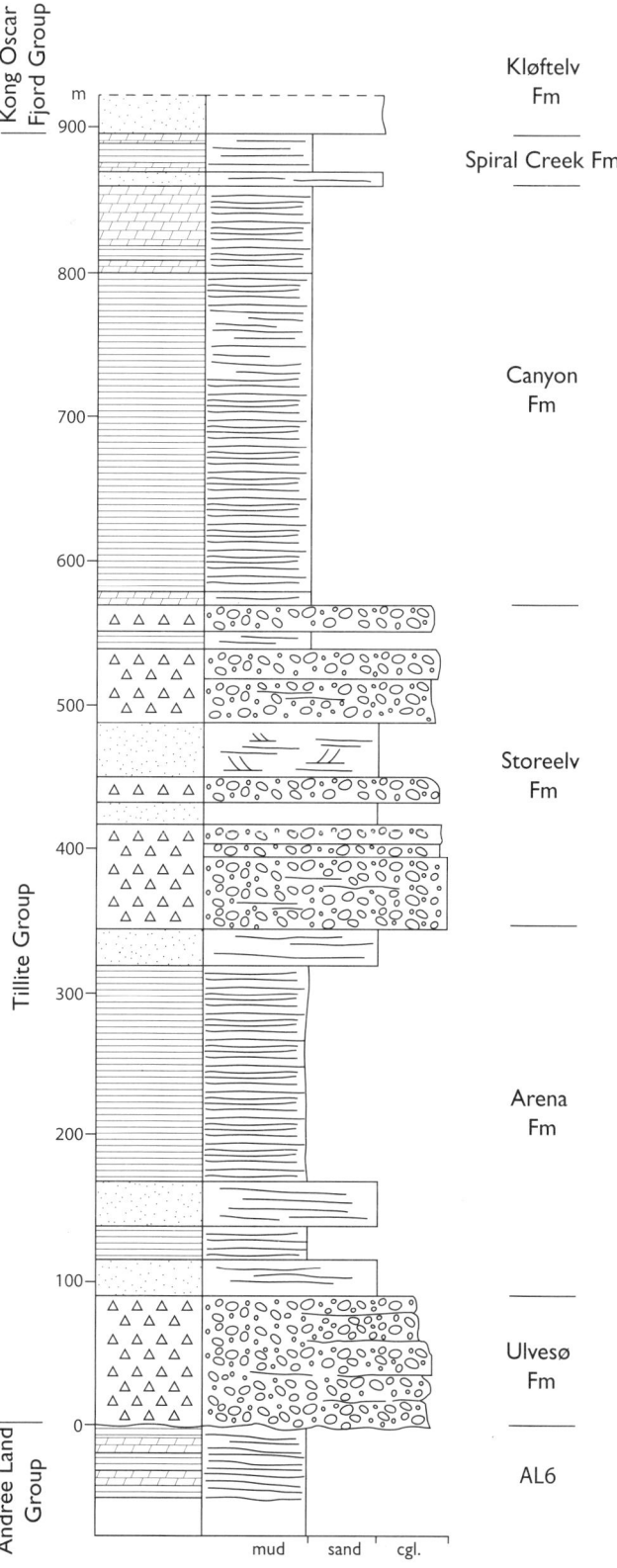

Figure 29. Sedimentary log through the Tillite Group at Ella Ø (modified from Hambrey and Spencer, 1987). Note that the carbonate-platform deposits of formation AL7 of the underlying Andrée Land Group are not present in this area. For legend, see Figure 24. Height in meters. Used with permission from Meddelelser om Grønland.

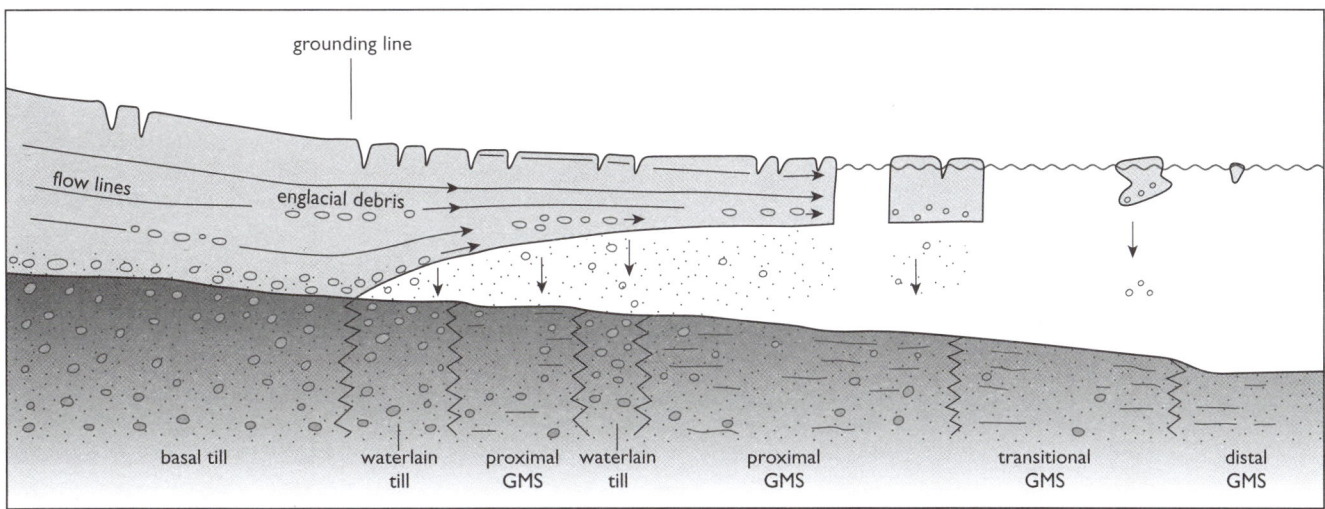

Figure 30. Depositional model for the glacial deposits of the Tillite Group showing the relationship in cross section among grounded basal till, waterlain till, and glaciomarine sediments (GMS; from Moncrieff and Hambrey, 1990). Used with permission from the Geological Society of London.

are present in the lower two-thirds of the Arena Formation, each of which is up to a few meters thick. The Arena Formation ranges in thickness from 100 to 360 m and was probably deposited in a marine environment (Hambrey and Spencer, 1987).

The younger diamictite unit, the Storeelv Formation (Figs. 29 and 31; 60–223 m), contains a suite of waterlain tillite and glaciomarine lithofacies similar to those in the Ulvesø Formation, but it also contains clast-rich muddy diamictite units with aligned pebble fabrics that have been interpreted as lodgment tills representing deposition beneath grounded ice. In places, this lithofacies is seen to overlie a striated boulder pavement (Moncrieff and Hambrey, 1990). Furthermore, at Kap Weber, a horizon of the lodgment till is seen to pass laterally into a proximal waterlain till (Moncrieff and Hambrey, 1990), which represents the point at which the ice detached from the substrate and began to float. Whereas the Ulvesø Formation is dominated by intrabasinal clasts, derived principally from the Eleonore Bay Supergroup, the Storeelv Formation contains a higher proportion of exotic clasts, including crystalline basement lithologies (Hambrey and Spencer, 1987; Hambrey et al., 1989).

In contrast to the older glacial deposits of the Ulvesø Formation, the Storeelv Formation is overlain by a characteristic 6–13-m-thick cap carbonate (Fig. 29; Hambrey and Spencer, 1987; Fairchild and Hambrey, 1995). The peritidal carbonates of this unit, the Canyon Formation, pass upward into dolomitic shales (25–35+ m), and the deepening trend continues into a 160–190-m-thick mudstone-siltstone unit, which is interpreted as having been deposited at outer-shelf depths. The Canyon Formation shows a marked shallowing trend through the uppermost 50 m, where the shale facies passes upward into peritidal dolostone with storm deposits (Fairchild and Herrington, 1989; Fairchild and Hambrey, 1995). This shallowing trend is capped by the Spiral Creek Formation (24–45 m), which was deposited

Figure 31. Marinoan glacial diamictites of the Storeelv Formation, Tillite Group, on Ella Ø (for location, see Fig. 13).

in playa-lake environments and which is composed of sandstone, siltstone, and mudstone with abundant halite pseudomorphs and gypsum together with mud cracks, ripple lamination, and silicified pebbles (Hambrey and Spencer, 1987; Fairchild and Hambrey, 1995). The formation is present only in the central part of the basin, on Ella Ø and in Andrée Land—to the north and south, the overlying Kløftelv Formation (Lower Cambrian) rests unconformably on the Canyon Formation.

Units of inferred glacial origin have also been identified in the parautochthonous foreland windows of the orogen. Although the Tillite Group, within the Franz Joseph allochthon, is now superimposed upon this foreland, it is separated from the foreland by both the Hagar Bjerg and Niggli Spids thrust sheets (Higgins et al., 2004a). Palinspastic reconstructions of the orogen suggest that the original separation between the foreland tillites and those in the Franz Joseph allochthon was in the order of 200–400 km (Higgins and Leslie, 2000; Higgins et al., 2004a).

Tillites have been identified in the tectonic windows in Gåseland (Wenk, 1961; Phillips et al., 1973), Charcot Land (Moncrieff, 1989), and at Målebjerg (Fig. 32; Smith and Robertson, 1999b). They differ significantly from those in the Franz Joseph allochthon because they directly overlie crystalline basement, indicating that the entire 13+ km of the Eleonore Bay Supergroup is absent. Moncrieff (1989) documented the tillite of the Charcot Land and Gåseland windows, where tillites sit in hollows on the eroded top of Paleoproterozoic gneisses and supracrustal rocks, beneath thrusts that transport Archean–Paleoproterozoic gneisses overlain by the Krummedal succession. Moncrieff (1989) erected different names for the tillite units in each of the two windows, the Tillit Nunatak Formation in the Charcot Land window and the Støvfanget Formation (misspelled as Stofvanget Formation) in the Gåseland window, but these are clearly lateral equivalents; the former name is selected here for the tillites that occur in this depositional and tectonic context in the Charcot Land, Gåseland, and Målebjerg windows. The lithofacies preserved in the two windows include tillite, sandstone, and laminated mudstone with dropstone. All of the clasts were derived locally and range up to 2 m. The maximum thickness is less than 200 m at Tillit Nunatak, but it is much less at most localities, rarely exceeding 25 m. Moncrieff (1989) interpreted the unit as consisting of lodgment tills near the base of the unit with diamictites deposited from floating ice (ice shelf and icebergs) and debris flows higher in the unit. The finer interbedded sediments were interpreted as being of predominantly turbiditic origin and, in Gåseland, proglacial outwash. Moncrieff (1989) correlated the Tillit Nunatak Formation with the Tillite Group of the Franz Joseph allochthon, and more specifically with the Storeelv Formation.

The glacial sediments assigned to the Tillit Nunatak Formation in the Målebjerg window (Smith and Robertson, 1999b) are directly comparable with those in windows farther to the south, but rather than being truncated by thrusts, they are unconformably overlain by the Lower Cambrian quartzites of the Slottet Formation (Smith et al., 2004b). The unit has a maximum thickness of 31 m and includes two diamictites separated by 21 m of

Figure 32. (A) View southward to Målebjerg (for location, see Fig. 13), where metasediments of the Niggli Spids thrust sheet are thrust over parautochthonous foreland. Within the foreland, in the foreground, glacial diamictites (beneath the hammer) unconformably overlie Paleoproterozoic orthogneisses and are unconformably overlain by Lower Cambrian quartzarenites of the Slottet Formation. The glacial deposits rest in an erosional hollow, and within a few hundred meters, the Slottet Formation rests directly on the gneisses. The maximum thickness of the glacial unit is 31 m. (B) Tillite unit of the Tillit Nunatak Formation. The largest clasts are of granitic composition, and other clast lithologies include vein quartz, metasandstones, and occasional metacarbonates. Lens cap is for scale.

phyllites, platy quartzites, and semipelites. Clasts in the upper diamictite range up to 6 m, and the bed is dominated by clasts of granitic lithology.

As with the glacial deposits of the Franz Joseph allochthon, those of the Tillit Nunatak Formation display a close interaction with adjacent marine environments. There is no evidence that these lay within "paleohighlands" as suggested by Halverson et al. (2004), since there is no evidence of significant relief, but there is evidence of deposition from floating ice (Moncrieff, 1989). These localities were well inboard of the Eleonore Bay Basin (Higgins et al., 2004a). They did not, however, constitute highlands, since they show evidence of marine environments.

Age of the Tillite Group

The age of the Tillite Group has been the subject of significant controversy. Early work correlated the unit with the Varanger tillites of northern Norway (Troelsen, 1956; Haller, 1971; Hambrey and Harland, 1981; Hambrey, 1983) and, more particularly, with the Neoproterozoic tillites of NE Spitsbergen and Nordaustlandet (Svalbard). Detailed work in Svalbard and eastern Greenland has demonstrated unit by unit, and to some extent bed by bed, correlation of the glacial horizons and associated units (cf. Fairchild and Hambrey, 1995), to the extent that it is clear that the two areas must have been part of the same basin prior to Caledonian collision and strike-slip dismemberment of the Laurentian margin. In considering the age of the glacial events in the Eleonore Bay Basin, it is therefore possible to utilize the combined data set from the two regions.

The development of interest in Neoproterozoic glaciation and the "Snowball Earth" hypothesis has led to a considerable focus on developing an understanding of the number, correlation, and dating of the events (for reviews, see Hambrey and Harland, 1985; Knoll, 2000; Hoffman and Schrag, 2002). There is now general consensus that there were two widespread, if not global, Neoproterozoic glaciations—the Sturtian and the Marinoan, which have been recognized and radiometrically dated in the same section of the Flinders Range in South Australia (Knoll et al., 2004)—together with an uncertain number of more localized and/or poorly known events. There has been uncertainty, however, over whether the events recorded in the Tillite Group and its Svalbard correlatives represent the Marinoan glacial event (Halverson et al., 2004) or the Sturtian (Brasier and Shields, 2000; Robb et al., 2004). The situation has been further confused by uncertainty over whether some of the glacial events contain paired tillites, or whether the two events recorded in Greenland could represent combined Sturtian + Marinoan events (Kennedy et al., 1998) or Marinoan + post-Marinoan (Gaskiers) events (Kaufman et al., 1997), with a condensed succession or hiatus between them. Thus, in the last ten years, virtually every possible permutation of correlation for the glacial events in the Eleonore Bay Basin has been advanced. Detailed work by Halverson et al. (2004, 2005) may point to a resolution of this debate. The detailed correlation of the East Greenland and Svalbard successions proposed by earlier authors was confirmed by Halverson et al. (2004), while Halverson et al. (2005) recognized the presence of the distinctive pre-Marinoan negative $\delta^{13}C$ anomaly (the "Trezona anomaly") beneath the lowest tillite in NE Svalbard and confirmed that the cap carbonate is of typical end-Marinoan type. Recently, this anomaly has also been identified in formation AL7, which underlies the Tillite Group (Kristiansen, 2007). This indicates that both tillite horizons in Svalbard and East Greenland are correlatives of the Marinoan glaciation. Furthermore, this suggests that the Tillite Group does not extend upward as far as the 580 Ma Gaskiers glacial event, which supports the conclusions based on the absence of an Ediacaran fauna, and it may be concluded that the Sturtian glaciation is not represented in the Eleonore Bay Basin. This may be because the Sturtian glaciation did not extend to this part of Laurentia (see Halverson et al. [2004] for discussion) or because the oldest basin infill postdates the Sturtian event.

COMPARISONS WITH NEOPROTEROZOIC BASINS ALONG THE LAURENTIAN MARGIN OF THE IAPETUS OCEAN

The late Neoproterozoic breakup of Rodinia and the development of the Iapetus margin of Laurentia involved a protracted and complex history. Two pulses of rifting and associated extrusive igneous activity at ~760–700 Ma and ~620–590 Ma have been recognized in the Appalachians; however, only the late phase proceeded to continental separation and opening of the Iapetus Ocean (Aleinikoff et al., 1995). In Newfoundland, this final pulse was a two-stage process involving the formation of the Iapetus Ocean between Laurentia and Gondwana at 570 Ma followed by continued rifting that generated one or more terranes at around 535 Ma (Fig. 33; Cawood et al., 2001).

Pre-Iapetus basins that are possibly coeval with those in the Greenland Caledonides have been recognized in Scotland and Ireland (Dalradian Supergroup basin, Moine Supergroup basin, and the Sleat, Stoer, and Torridon Group basins), and Svalbard (the basin or basins filled by the Hekla Hoek succession; Higgins et al., 2001b).

The successions that show the greatest similarity are those of the Eleonore Bay Basin and the Tonian–Cryogenian (early–middle Neoproterozoic) part of the Hekla Hoek succession of northeastern Svalbard. These two successions show much the same sedimentary evolution from siliciclastic to carbonate deposition capped by Marinoan glacigenic deposits, and they seem to lack major regional stratigraphic breaks and evidence of rift-related sedimentation. Furthermore, they are similar in attaining very substantial thicknesses (Harland and Gayer, 1972; Harland, 1985, 1997; Hambrey, 1989; Harland et al., 1992). The close relationship of these two Neoproterozoic successions is also suggested by chemostratigraphic analysis (Knoll et al., 1986) and microfossil assemblages (Vidal, 1985; Green et al., 1989; Swett and Knoll, 1989). It therefore seems reasonable to consider that the two successions formed within one or a series of connected ensialic basins and that eastern Svalbard was positioned very close to central East Greenland during the Neoproterozoic (Hambrey, 1989). It is evident that both successions reflect major, and probably long-lived, extension-related subsidence. Problems with this interpretation arise from the anticipated strain-strengthening of the lithosphere at slow strain rates, and these have been discussed by Soper (1994).

Sandelin et al. (2001) undertook a detailed study of the lower part of the Murchisonfjorden Supergroup in Nordaustlandet (NE Svalbard). Significantly, the lowest unit of the Murchisonfjorden Supergroup, the Galtedalen Group, has a well-defined base and well-developed basal conglomerates, and it rests with major unconformity on Mesoproterozoic sediments intruded by

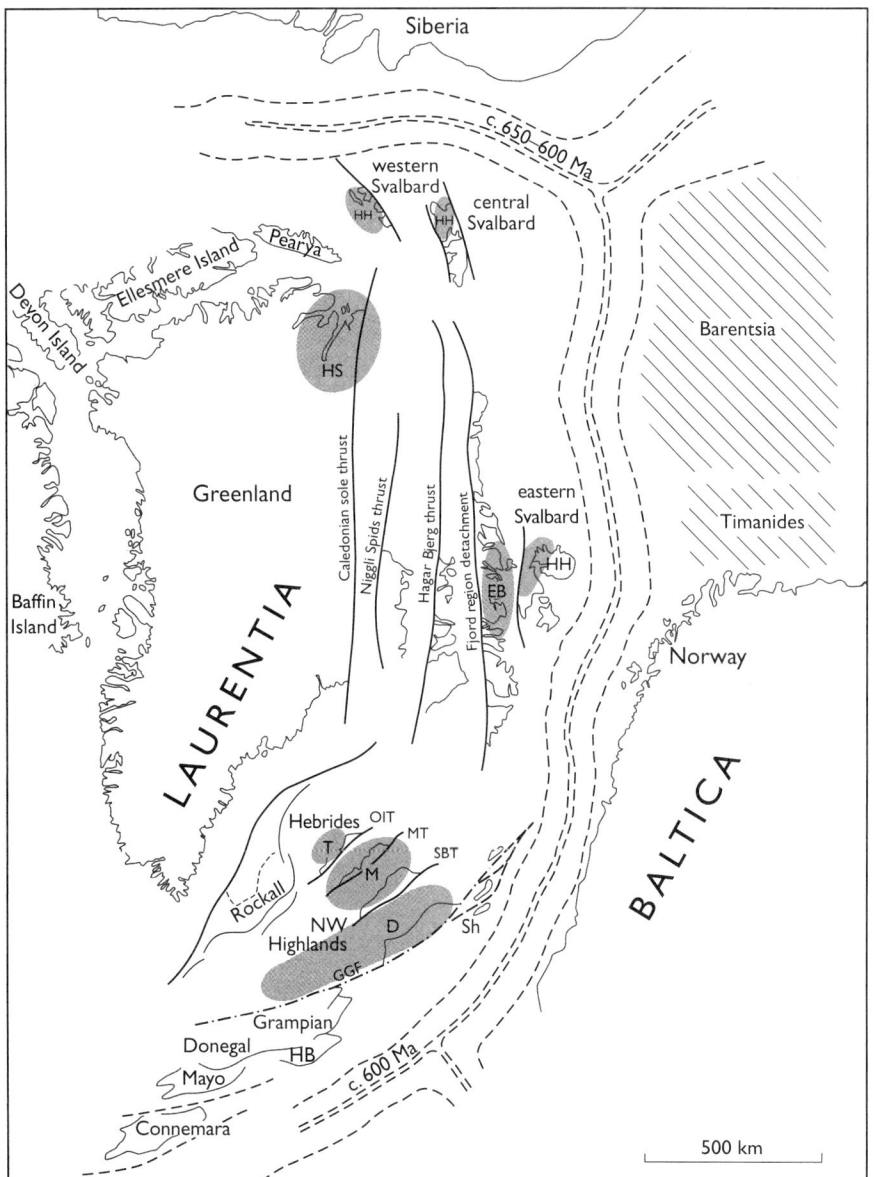

Figure 33. Palinspastic reconstruction of the northeastern margin of Laurentia showing the Vendian position of Greenland relative to components of Svalbard and the NW British Isles. Caledonian thrust sheets and detachment zones in East Greenland are indicated. It should be noted that the positions of western and central Svalbard are relatively poorly constrained relative to eastern Svalbard. Map is based on Higgins et al. (2001b, their Fig. 12) and Leslie et al. (this volume, their Fig. 1). Used with permission from the Geological Society of London. D—Dalradian Supergroup basin; EB—Eleonore Bay Basin; GGF—Great Glen fault; HB—Highland Border; HH—Hekla Hoek succession basins; HS—Hekla Sund Basin; M—Moine Supergroup basin; MT—Moine thrust; OIT—Outer Isles thrust; SBT—Sgurr Beag thrust; Sh—Shetland Islands; T—Torridonian basin.

950 Ma granites. The overlying part of the Galtedalen Group and the succeeding Franklinsundet Group make up alternating units of quartzite and shale with subordinate carbonates (Sandelin et al., 2001). This succession, which is equivalent to the lower part of the Lomfjorden Supergroup in Ny Friesland (Harland, 1997; Sandelin et al., 2001), is almost certainly a correlative of the Nathorst Land Group in which the basal relationships with Krummedal equivalents are preserved, although precise correlations are not yet possible.

In North Greenland, although the depositional evolution of the Hekla Sund Basin resembles the Eleonore Bay Basin, the tectonic setting is entirely different, since at least the early history of the Hekla Sund Basin is closely linked to extensional rifting (Rivieradal Group and lower part of Hagen Fjord Group). A correlation between the postrift carbonate-dominated succession (upper part of Hagen Fjord Group) of the Hekla Sund Basin and the upper part of the Eleonore Bay Basin (Ymer Ø and Andrée Land Groups) suggests the presence of a very extensive carbonate platform that testifies to a period of regional tectonic stability prior to the onset of Iapetus rifting during the late Neoproterozoic. It is possible that the extent of this carbonate platform was even larger than suggested by the Arctic localities, since it may have extended into the Caledonides of Scotland and Ireland, where a similar succession is present (Appin Group and Port Askaig tillite; see following).

Within the Scottish Highlands, the siliciclastic sediments of the Moine Supergroup are entirely allochthonous, and they have a depositional age that is well constrained between the youngest detrital zircon ages at 979 ± 77 Ma and 926 ± 68 Ma, and the intrusion of the West Highland granitic gneiss at 870 Ma

(Millar, 1999; Cawood et al., 2003). The Moine Supergroup was deposited in a marine environment, but it has often been correlated with the terrestrial, fluvially dominated, sediments of the Torridon Group, which is present largely in the foreland and in large-scale thrust sheets below the Moine thrust. The youngest detrital zircons in the Torridon Group have an age of 1060 ± 18 Ma (Rainbird et al., 2001), which is in agreement with the Rb-Sr whole-rock age of 977 ± 39 Ma that dates diagenetic mineral growth (Turnbull et al., 1996). A.D. Stewart has long favored a rift origin for the Torridon Group (see Stewart [2002] for review), although this has been countered by Prave (1999), who interpreted the Torridon Group as the molasse of the Grenville orogen and the Moine Supergroup as its distal marine equivalent, thus reviving the correlation of Peach et al. (1907). More recent work has noted, however, that although the Torridon Group and Moine Supergroup were deposited at broadly the same time, there are relatively low proportions of Archean and Ketilidian detrital grains in the Moine Supergroup compared to the Torridon Group (Cawood et al., 2003; Friend et al., 2003), and that they differ in other aspects of sediment provenance too, suggesting that the two units were deposited in coeval but spatially discrete basins. The age distribution of detrital zircon grains from the Moine nappe (Friend et al., 2003, their Fig. 3a) is strikingly similar to that of grains from the Krummedal succession (Watt et al., 2000, their Fig. 7a). Although, as noted already, the profile of detrital zircon ages from the Nathorst Land Group is also similar to that from the Krummedal succession and the Moine Supergroup (Watt et al., 2000; Cawood et al., 2001), we do not support a correlation on the grounds of the structural, metamorphic, and sedimentologic context of the Nathorst Land Group.

In the Grampian Highlands of Scotland, to the south of the Great Glen fault, the oldest part of the Dalradian Basin, the Grampian Group, is interpreted as the fill of several NE-SW–trending rift basins (Glover et al., 1995; Robertson and Smith, 1999), and this group rests on basement of enigmatic affinity, referred to as the Dava and Glen Banchor successions. The depositional age of the sub–Grampian Group basement is constrained by the youngest detrital grain at 900 Ma, metamorphism at 840 Ma, and ductile shearing at 806 Ma (Noble at al., 1996; Highton et al., 1999; Cawood et al., 2003). The depositional age of this basement is thus similar to the ages for the Moine and Krummedal successions (Cawood et al., 2003), and the initiation of Dalradian deposition must postdate ca. 800 Ma. Interestingly, however, the Grampian Group and the underlying basement share similar detrital zircon profiles (Cawood et al., 2003; Banks et al., 2007), despite an undoubted depositional break (M. Smith et al., 1999). This is very similar to the situation regarding the zircon profiles of the Krummedal succession and Nathorst Land Group and lends weight to the argument that detrital zircon distributions cannot be used to infer continuous deposition between the two Greenland units (contra Watt and Thrane, 2001).

The presence of the 800 Ma Knoydartian event in the Moine Supergroup (Rogers et al., 1998; Vance et al., 1998; Millar, 1999) and in sub–Grampian Group basement (Highton et al., 1999) and its absence elsewhere on the Laurentian margin introduce the possibility that these rocks represent an exotic terrane, and that Scandian thrusting was a final, minor event in comparison with earlier strike-slip juxtaposition, as suggested by Bluck et al. (1997). Although this hypothesis retains some currency (see Oliver [2002] for review), the detrital zircon data sets of Cawood et al. (2003, 2004), Friend et al. (2003), and Banks et al. (2007) suggest similar provenance for the Moine Supergroup, sub–Grampian Group basement, and Krummedal successions. This indicates that these basin fills were contiguous, if not actually part of the same basin, during the Neoproterozoic. Nevertheless, the Knoydartian event is not represented in East Greenland, and the localized nature of this event remains an enigma.

The Grampian Group of the Central Highlands of Scotland consists of ~8 km of metasediments, including fluvial fan-delta conglomerate and psammite, which pass upward into submarine-fan psammite-semipelite couplets, followed by shallow-marine arkosic sediments (Glover et al., 1995; Banks and Winchester, 2004). In places, there is a passage up into the Appin Group, but elsewhere, the Appin Group is transgressive across block-tilted Grampian Group strata. It is noteworthy that the transition between the Grampian and Appin Groups has been interpreted as onlap onto an active rift shoulder, such that a conformable relationship pertains away from the margins, but the Appin Group rests unconformably on older Grampian Group sediments on the rift shoulder (Robertson and Smith, 1999, their Fig. 5; M. Smith et al., 1999). This relationship has clear parallels with the observed relationships between the Nathorst Land and the Lyell Land Groups of the Eleonore Bay Supergroup. The Appin Group is a generally upward-shallowing sequence, characterized by phyllites and black shales with wedges of quartzite that pass up into subtidal carbonates and the Port Askaig tillites. As with the Tillite Group in East Greenland, the age of the tillites is clearly critical to determining the upper limit for the depositional age of the Grampian–Appin interval. However, there is currently a lack of consensus over the age and correlation of the Port Askaig tillite; some authors (Brasier and Shields, 2000; Condon and Prave, 2000) favor a Sturtian (~700 Ma) age, and others (Halverson et al., 2005) prefer a Marinoan correlation.

The thick siliciclastic succession overlying the Port Askaig tillite is assigned to the Argyll and Southern Highland Groups. In contrast to a maximum preserved thickness of ~340 m between the Ulvesø Formation and the base of the Cambrian Kløftelv Formation in East Greenland (Hambrey and Spencer, 1987), there are up to 9 km of Argyll Group sediments between the Port Askaig tillite and the 600 Ma Tayvallich volcanics, and a further 4 km of Southern Highland Group sediments. Regardless of arguments concerning the correlation of the upper parts of the Southern Highland Group (see Trewin [2002] for reviews), there is general consensus that the group reaches at least into the Lower Cambrian. This correlation points to a significant contrast in subsidence histories for the Dalradian and Eleonore Bay Basins during the later part of the Neoproterozoic.

SUMMARY

The Neoproterozoic successions in the Hekla Sund and Eleonore Bay Basins probably formed in response to an early pulse of Iapetan rifting at around 760–700 Ma. The deposits are well exposed in thrust sheets of the East Greenland Caledonides, and both basins attain thicknesses in excess of 10 km. Their lower parts show strong tectonic overprint and metamorphic recrystallization, which impede sedimentologic studies, but sedimentary structures are generally well preserved in the upper parts, making detailed sedimentologic interpretations possible.

The Hekla Sund Basin in eastern North Greenland reflects deposition during active rifting in two half-grabens. The rift phase is represented by the Rivieradal Group, which is only developed in the eastern half-graben, and the much thinner lower part of the Hagen Ford Group in the western half-graben. The lower part of the up to 10-km-thick Rivieradal Group is composed of a thick succession of phyllite, quartzite, and marble that is exposed toward the east. This succession may in part be equivalent to the upper, less-deformed western part that includes conglomerates and thick-bedded sandstone turbidites and mudstones deposited in a fan-delta and deep-marine basin plain setting. The uppermost part of the group reflects siliciclastic deposition in shallow-marine and fluvial environments. Major point sources of sediment input are indicated by the localized presence of thick units of channelized debris-flow deposits. In the western half-graben, the rift-phase deposits are only ~500 m thick, and they are represented by fluvial to shallow-marine deposits. The postrift, thermal subsidence phase of the Hekla Sund Basin is represented by the upper part of the Hagen Fjord Group (750 m), which is recognized in both half-graben segments of the basin. This phase is characterized by a change from shallow-marine siliciclastic to carbonate-platform deposition. The top of the Hagen Fjord Group is marked by a paleokarst development, reflecting a major hiatus across which much of the later Neoproterozoic is absent.

The deposits of the Eleonore Bay Basin of East Greenland attain a combined thickness of almost 15 km. Major regional stratigraphic breaks seem to be absent, as is evidence of rift-related sedimentation, and deposition within one or a series of connected ensialic basins is thus envisaged. Four stages of basin evolution have been recognized in the more than 14-km-thick Eleonore Bay Supergroup; these stages reflect major paleogeographic reorganizations of the shelf and, in the upper part of the succession, a change from shallow-marine siliciclastic to carbonate-platform deposition prior to the onset of Marinoan glacigenic deposition. Although neither continental deposits nor horizons of fluvial incision have been recognized in the Eleonore Bay Basin succession, at least 10 second-order sequence boundaries have been recognized. The widespread nature of these major lithostratigraphic units and their facies suggest that depositional conditions were uniform and parallel to basin strike for most of the duration of deposition.

Glacigenic deposition in the Eleonore Bay Basin is recorded by the Tillite Group, which is up to 750 m thick and overlies the Eleonore Bay Supergroup. The Tillite Group consist of two formations of diamictite separated by sediments deposited in a marine environment. A conspicuous negative $\delta^{13}C$ anomaly (the "Trezona anomaly") has been recognized beneath the lowest tillite in the Eleonore Bay Supergroup, indicating a Marinoan age for the tillites. The upper tillite is overlain by a cap carbonate that marks the base of the Ediacaran and consists of peritidal carbonate passing into outer-shelf mudstone. These deposits are locally overlain by playa lake sediments that form the top of the Tillite Group. An angular unconformity separates the Eleonore Bay Basin succession from overlying Cambrian deposits.

Pre-Iapetus basins along the Laurentian margin that are possibly coeval with those in the Greenland Caledonides have been recognized in Scotland and Ireland (represented by the Dalradian Supergroup, the Moine Supergroup, and the Sleat, Stoer, and Torridon Groups), and on Svalbard (represented by the Hekla Hoek succession). Of these, the Eleonore Bay Basin and the Hekla Hoek successions show the greatest similarity. Both successions show much the same sedimentary evolution from siliciclastic to carbonate deposition, capped by Marinoan tillite, and they probably reflect long-lived, extension-related subsidence.

The Grampian and Appin Groups of the Central Highlands of Scotland also record a change from siliciclastic to carbonate deposition that in many ways resembles the evolution observed within the Eleonore Bay Supergroup. They are overlain by the Port Askaig tillite, which has been attributed both a Sturtian (~700 Ma) and a Marinoan (~635 Ma) age, demonstrating that good dating of the glacigenic deposits is critical in correlating the successions of Scotland and East Greenland.

ACKNOWLEDGMENTS

The authors thank the reviewers Christopher Banks and Michael J. Hambrey for their valuable comments and suggestions, which greatly benefited this manuscript. The paper is dedicated to the memory of Steve Robertson, who died shortly after completing his work on the Nathorst Land Group.

This paper is published with the permission of the Geological Survey of Denmark and Greenland. Field work in North Greenland was partly funded by the Carlsberg Foundation (grant No. 93-0254/10) and Maersk Oil and Gas.

REFERENCES CITED

Adams, P.J., and Cowie, J.W., 1953, A geological reconnaissance of the region around the inner part of Danmarks Fjord, Northeast Greenland: Meddelelser om Grønland, v. 111, no. 7, 24 p.

Aleinikoff, J.N., Zartman, R.E., Walter, M., Rankin, D.W., Lyttle, P.T., and Burton, W.C., 1995, U-Pb ages of metarhyolites of the Catoctin and Mount Rogers Formations, central and southern Appalachians: Evidence for two pulses of Iapetan rifting: American Journal of Science, v. 295, p. 428–454.

Andresen, A., Rehnström, E.F., and Holte, M.K., 2004, Synchronous, but contrasting deformational regimes at different structural depths: Geochronology of syn-tectonic granites in the NE Greenland Caledonides: Geologiska Föreningens i Stockholm Förhandlingar, v. 126, p. 77.

Banks, C.J., and Winchester, J.A., 2004, Sedimentology and stratigraphic affinities of Neoproterozoic coarse clastic succession, Glenshirra

Group, Inverness-shire, Scotland: Scottish Journal of Geology, v. 40, p. 159–174.

Banks, C.J., Smith, M., Winchester, J.A., Horstwood, M.S.A., Noble, S.R., and Ottley, C.J., 2007, Provenance of intra-Rodinian basin-fills: The lower Dalradian Supergroup, Scotland: Precambrian Research, v. 153, p. 46–64, doi: 10.1016/j.precamres.2006.11.004.

Bengaard, H.-J., 1991, Upper Proterozoic (Eleonore Bay Supergroup) to Devonian Central Fjord Zone, East Greenland, Geological Map: Copenhagen, Geological Survey of Greenland, scale 1:250,000.

Bengaard, H.-J., and Henriksen, N., 1986, Geological map of Greenland, Sheet 8, Peary Land: Copenhagen, Geological Survey of Greenland, scale 1:500,000.

Bertrand-Sarfati, J., and Caby, R., 1976, Carbonates et stromatolites du sommet du Group d'Eleonore Bay (Précambrien terminal) au Canning Land (Groenland oriental): Bulletin Grønlands Geologiske Undersøgelse, v. 119, 51 p.

Bluck, B.J., Dempster, T.J., and Rogers, G., 1997, Allochthonous metamorphic blocks on the Hebridean passive margin, Scotland: Journal of the Geological Society of London, v. 154, p. 921–924, doi: 10.1144/gsjgs.154.6.0921.

Brasier, M.D., and Shields, G., 2000, Neoproterozoic chemostratigraphy and correlation of the Port Askaig glaciation, Dalradian Supergroup of Scotland: Journal of the Geological Society of London, v. 157, p. 909–914.

Caby, R., and Bertrand-Sarfati, J., 1988, The Eleonore Bay Group (central East Greenland), *in* Winchester, J.A., ed., Later Proterozoic Stratigraphy of the Northern Atlantic Regions: Glasgow, Blackie & Sons Ltd., p. 212–236.

Cawood, P.A., McCausland, P.J.A., and Dunning, G.R., 2001, Opening Iapetus: Constraints from the Laurentian margin in Newfoundland: Geological Society of America Bulletin, v. 113, p. 443–453, doi: 10.1130/0016-7606 (2001)113<0443:OICFTL>2.0.CO;2.

Cawood, P.A., Nemchin, A.A., Smith, M., and Loewy, S., 2003, Source of the Dalradian Supergroup constrained by U-Pb dating of detrital zircon and implications for the East Laurentian margin: Journal of the Geological Society of London, v. 160, p. 231–246.

Cawood, P.A., Nemchin, A.A., Strachan, P.S., Kinny, P.D., and Loewy, S., 2004, Laurentian provenance and an intracratonic setting for the Moine Supergroup, Scotland, constrained by detrital zircons from the Loch Eil and Glen Urquhart successions: Journal of the Geological Society of London, v. 161, p. 861–874, doi: 10.1144/16-764903-117.

Christie-Blick, N., Grotzinger, J.P., and von der Borch, C.C., 1988, Sequence stratigraphy in Proterozoic successions: Geology, v. 16, p. 100–104, doi: 10.1130/0091-7613(1988)016<0100:SSIPS>2.3.CO;2.

Clemmensen, L.B., and Jepsen, H.F., 1992, Lithostratigraphy and Geological Setting of Upper Proterozoic Shoreline-Shelf Deposits, Hagen Fjord Group, North Greenland: Rapport Grønlands Geologiske Undersøgelse, v. 157, 27 p.

Collinson, J.D., Bevins, R.E., and Clemmensen, L.B., 1989, Post-glacial mass flow and associated deposits preserved in palaeovalleys: The Late Precambrian Morænesø Formation: Meddelelser om Grønland, Geoscience, v. 21, 26 p.

Collinson, J.D., Kalsbeek, F., Jepsen, H.F., Pedersen, S.A.S., and Upton, B.G.J., 2008, this volume, Paleoproterozoic and Mesoproterozoic sedimentary and volcanic successions in the northern parts of the East Greenland Caledonian orogen and its foreland, *in* Higgins, A.K., Gilotti, J.A., and Smith, M.P., eds., The Greenland Caledonides: Evolution of the Northeast Margin of Laurentia: Geological Society of America Memoir 202, doi: 10.1130/2008.1202(04).

Condon, D.J., and Prave, A.R., 2000, Two from Donegal: Neoproterozoic glacial episodes on the northeast margin of Laurentia: Geology, v. 28, p. 951–954, doi: 10.1130/0091-7613(2000)28<951:TFDNGE>2.0.CO;2.

Dhuime, B., Bosch, D., Bruguier, O., Caby, R., and Pourtales, S., 2007, Age, provenance and post-deposition metamorphic overprint of detrital zircons from the Nathorst Land group (NE Greenland)—A LA-ICP-MS and SIMS study: Precambrian Research, vol. 155, p. 24–46, doi: 10.1016/j.precamres.2007.01.002.

Donaldson, J.A., 1976, Paleoecology of conophyton and associated stromatolites in the Precambrian Dismal Lakes and Rae Groups, Canada, *in* Walter, M.R., ed., Stromatolites: Amsterdam, Elsevier, p. 523–534.

Fairchild, I.J., and Hambrey, M.J., 1995, Vendian basin evolution in East Greenland and NE Svalbard: Precambrian Research, v. 73, p. 217–223, doi: 10.1016/0301-9268(94)00079-7.

Fairchild, I.J., and Herrington, P.M., 1989, A tempestite-stromatolite-evaporite association (late Vendian, East Greenland): A shoreface-lagoon model: Precambrian Research, v. 43, p. 101–127, doi: 10.1016/0301-9268(89)90007-7.

Fränkl, E., 1953, Geologische Untersuchungen in Ost-Andrées Land (NE-Grønland: Meddelelser om Grønland, v. 113, no. 4, 160 p.

Fränkl, E., 1954, Vorläufige Mitteilungen über die Geologie von Kronprins Christian Land (NE-Grönland): Meddelelser om Grønland, v. 116, no. 2, 85 p.

Fränkl, E., 1955, Weitere Beiträge zur Geologie von Kronprins Christian Land (NE-Grönland): Meddelelser om Grønland, v. 103, no. 7, 34 p.

Frederiksen, K.S., 2000a, Evolution of a late Proterozoic carbonate ramp (Ymer Ø and Andrée Land Groups, Eleonore Bay Supergroup, East Greenland): Response to relative sea-level rise: Polarforschung, v. 68, p. 125–130.

Frederiksen, K.S., 2000b, A Neoproterozoic carbonate ramp and base-of-slope succession, the Andrée Land Group, Eleonore Bay Supergroup, North-East Greenland: Sedimentary facies, stratigraphy and basin evolution [Ph.D. thesis]: Copenhagen, University of Copenhagen, Denmark, unpaginated.

Frederiksen, K.S., Craig, L.E., and Skipper, C.B., 1999, New observations of the stratigraphy and sedimentology of the Upper Proterozoic Andrée Land Group, East Greenland: Supporting evidence for a drowned carbonate ramp: Danmarks og Grønlands Geologiske Undersøgelse Rapport, v. 1999/19, p. 145–158.

Friderichsen, J.D., and Thrane, K., 1998, Caledonian and pre-Caledonian geology of the crystalline complexes of the Stauning Alper, Nathorst Land, and Charcot Land, East Greenland: Danmarks og Grønlands Geologiske Undersøgelse Rapport, v. 1998/28, p. 55–71.

Friend, C.R.L., Strachan, R.A., Kinny, P.D., and Watt, G.R., 2003, Provenance of the Moine Supergroup of NW Scotland; evidence from geochronology of detrital and inherited zircons from (meta)sedimentary rocks, granites and migmatites: Journal of the Geological Society of London, v. 160, p. 247–257.

Galloway, W.E., 1989, Genetic stratigraphic sequences in basin analysis. I: Architecture and genesis of flooding-surface bounded depositional units: American Association of Petroleum Geologists Bulletin, v. 73, p. 125–142.

Gilotti, J.A., and McClelland, W.C., 2008, this volume, Geometry, kinematics, and timing of extensional faulting in the Greenland Caledonides—A synthesis, *in* Higgins, A.K., Gilotti, J.A., and Smith, M.P., eds., The Greenland Caledonides: Evolution of the Northeast Margin of Laurentia: Geological Society of America Memoir 202, doi: 10.1130/2008.1202(10).

Glover, B.W., Key, R.M., May, F., Clark, G.C., Phillips, E.R., and Chacksfeld, B.C., 1995, Neoproterozoic multi-phase rift sequence: The Grampian and Appin Groups of the southwestern Monadhliath Mountains of Scotland: Journal of the Geological Society of London, v. 152, p. 391–406.

Gradstein, F.M., Ogg, J.G., and Smith, A.G., 2004, A Geologic Time Scale 2004: Cambridge, Cambridge University Press, 610 p.

Green, J.W., Knoll, A.H., Golubic, S., and Swett, K., 1987, Paleobiology of distinctive benthic microfossils from the Upper Proterozoic Limestone-Dolomite 'Series,' central East Greenland: American Journal of Botany, v. 74, no. 6, p. 928–940, doi: 10.2307/2443874.

Green, J.W., Knoll, A.H., and Swett, K., 1988, Microfossils from oolites and pisolites of the Upper Proterozoic Eleonore Bay Group, central East Greenland: Journal of Paleontology, v. 62, p. 835–852.

Green, J.W., Knoll, A.H., and Swett, K., 1989, Microfossils from silicified stromatolitic carbonates of the Upper Proterozoic Limestone-Dolomite 'Series,' central East Greenland: Geological Magazine, v. 126, p. 567–585.

Grotzinger, J.P., 1989, Facies and evolution of Precambrian carbonate depositional systems: Emergence of the modern platform archetype, *in* Crevello, P.D., Wilson, J.L., Sarg, J.F., and Read, J.F., eds., Controls on Carbonate Platform and Basin Development: Society of Economic Paleontologists and Mineralogists (SEPM) Special Publication 44, p. 79–106.

Haller, J., 1961, The Carolinides: An orogenic belt of Upper Precambrian age in Northeast Greenland, *in* Raash, G.O., ed., Geology of the Arctic: Toronto, Toronto University Press, p. 155–159.

Haller, J., 1971, Geology of the East Greenland Caledonides: New York, Interscience Publishers, 413 p.

Halverson, G.P., Maloof, A.C., and Hoffman, P.F., 2004, The Marinoan glaciation (Neoproterozoic) in northeast Svalbard: Basin Research, v. 16, p. 297–324, doi: 10.1111/j.1365-2117.2004.00234.x.

Halverson, G.P., Hoffman, P.F., Schrag, D.P., Maloof, A.C., and Rice, A.H.N., 2005, Toward a Neoproterozoic composite carbon-isotope record: Geological Society of America Bulletin, v. 117, p. 1181–1207, doi: 10.1130/B25630.1.

Hambrey, M.J., 1983, Correlation of Late Proterozoic tillites in the North Atlantic region and Europe: Geological Magazine, v. 120, p. 290–320.

Hambrey, M.J., 1989, The late Proterozoic sedimentary record of East Greenland: Its place in understanding the evolution of the Caledonide orogen, in Gayer, R.A., ed., The Caledonide Geology of Scandinavia: London, Graham & Trotman, p. 257–262.

Hambrey, M.J., and Harland, W.B., eds., 1981, Earth's Pre-Pleistocene Glacial Record: Cambridge, Cambridge University Press, 1004 p.

Hambrey, M.J., and Harland, W.B., 1985, The late Proterozoic glacial era: Palaeogeography, Palaeoclimatology, Palaeoecology, v. 51, p. 255–272, doi: 10.1016/0031-0182(85)90088-4.

Hambrey, M.J., and Spencer, A.M., 1987, Late Precambrian glaciation of Central East Greenland: Meddelelser om Grønland, Geoscience, v. 19, 50 p.

Hambrey, M.J., Peel, J.S., and Smith, M.P., 1989, Upper Proterozoic and Lower Palaeozoic in northern East Greenland: Rapport Grønlands Geologiske Undersøgelse, v. 145, p. 103–108.

Harland, W.B., 1985, Caledonide Svalbard, in Gee, D.G., and Sturt, B.A., eds., The Caledonide Orogen—Scandinavia and Related Areas: London, John Wiley, p. 999–1016.

Harland, W.B., 1997, The Geology of Svalbard: Geological Society of London Memoir 17, 521 p.

Harland, W.B., and Gayer, R.A., 1972, The arctic Caledonides and earlier oceans: Geological Magazine, v. 109, p. 289–384.

Harland, W.B., Scott, R.A., Auckland, K.A., and Snape, I., 1992, The Ny Friesland Orogen, Spitsbergen: Geological Magazine, v. 129, p. 679–708.

Harris, C.W., and Eriksson, K.A., 1990, Allogenic controls on the evolution of storm to tidal shelf sequences in the early Proterozoic Uncompahgre Group, southwestern Colorado, USA: Sedimentology, v. 37, p. 189–213, doi: 10.1111/j.1365-3091.1990.tb00955.x.

Hartz, E.H., Andresen, A., Martin, M.W., and Hodges, K.V., 2000, U-Pb and $^{40}Ar/^{39}Ar$ constraints on the fjord region detachment zone: A long-lived extensional fault in the central East Greenland Caledonides: Journal of the Geological Society of London, v. 157, p. 795–809.

Henriksen, N., 1999, Conclusion of the 1:500,000 mapping project in the Caledonian fold belt in North-East Greenland: Geology of Greenland Survey Bulletin, v. 183, p. 10–22.

Henriksen, N., 2003, Caledonian orogen East Greenland 70°–82°N, lithotectonic map: Copenhagen, Geological Survey of Denmark and Greenland, scale 1:1,000,000.

Henriksen, N., and Higgins, A.K., 1976, East Greenland Caledonian fold belt, in Escher, A., and Watt, W.S., eds., Geology of Greenland: Copenhagen, Geological Survey of Greenland, p. 182–246.

Henriksen, N., and Higgins, A.K., 2008, this volume, Geological research and mapping in the Caledonian orogen of East Greenland 70°N–82°N, in Higgins, A.K., Gilotti, J.A., and Smith, M.P., eds., The Greenland Caledonides: Evolution of the Northeast Margin of Laurentia: Geological Society of America Memoir 202, doi: 10.1130/2008.1202(01).

Herrington, P.M., and Fairchild, I.J., 1989, Carbonate shelf and slope facies evolution prior to Vendian glaciation, central East Greenland, in Gayer, R.A., ed., The Caledonide Geology of Scandinavia: London, Graham & Trotman, p. 263–273.

Higgins, A.K., and Leslie, A.G., 2000, Restoring thrusting in the East Greenland Caledonides: Geology, v. 28, p. 1019–1022, doi: 10.1130/0091-7613(2000)28<1019:RTITEG>2.0.CO;2.

Higgins, A.K., Friderichsen, J.D., and Thyrsted, T., 1981, Precambrian metamorphic complexes in the East Greenland Caledonides (72°–74°N)—Their relationships to the Eleonore Bay Group and Caledonian orogenesis: Rapport Grønlands Geologiske Undersøgelse, v. 104, p. 4–46.

Higgins, A.K., Leslie, A.G., and Smith, M.P., 2001a, Neoproterozoic–Lower Palaeozoic stratigraphical relationships in the marginal thin-skinned thrust belt of the East Greenland Caledonides: Comparisons with the foreland of Scotland: Geological Magazine, v. 138, p. 143–160, doi: 10.1017/S0016756801005076.

Higgins, A.K., Smith, M.P., Soper, N.J., Leslie, A.G., Rasmussen, J.A., and Sønderholm, M., 2001b, The Neoproterozoic Hekla Sund Basin, eastern North Greenland: A pre-Iapetan extensional sequence thrust across its rift shoulders during the Caledonian orogeny: Journal of the Geological Society of London, v. 158, p. 487–499.

Higgins, A.K., Elvevold, S., Escher, J.C., Frederiksen, K.S., Gilotti, J.A., Henriksen, N., Jepsen, H.F., Jones, K.A., Kalsbeek, F., Kinny, P.D., Leslie, A.G., Smith, M.P., Thrane, K., and Watt, G.R., 2004a, The foreland-propagating thrust sheet architecture of the East Greenland Caledonides 72°–75°N: Journal of the Geological Society of London, v. 161, p. 1009–1026, doi: 10.1144/0016-764903-141.

Higgins, A.K., Soper, N.J., Smith, M.P., and Rasmussen, J.A., 2004b, The Caledonian thin-skinned thrust belt of Kronprins Christian Land, eastern North Greenland: Geological Survey of Denmark and Greenland Bulletin, v. 6, p. 41–56.

Highton, A.J., Hyslop, E.K., and Noble, S.R., 1999, U-Pb zircon geochronology of migmatization in the northern central Highlands; evidence for pre-Caledonian (Neoproterozoic) tectonometamorphism in the Grampian block, Scotland: Journal of the Geological Society of London, v. 156, p. 1195–1204, doi: 10.1144/gsjgs.156.6.1195.

Hoffman, P., 1976, Environmental diversity of Middle Precambrian stromatolites, in Walter, M.R., ed., Stromatolites: Amsterdam, Elsevier, p. 599–611.

Hoffman, P.F., and Schrag, D.P., 2002, The snowball Earth hypothesis: Testing the limits of global change: Terra Nova, v. 14, p. 129–155, doi: 10.1046/j.1365-3121.2002.00408.x.

Hurst, J.M., and McKerrow, W.S., 1980, The Caledonian nappes of Kronprins Christian Land, eastern North Greenland: Rapport Grønlands Geologiske Undersøgelse, v. 106, p. 15–19.

Hurst, J.M., and McKerrow, W.S., 1981, The Caledonian nappes of eastern North Greenland: Nature, v. 290, p. 772–774, doi: 10.1038/290772a0.

Hurst, J.M., and McKerrow, W.S., 1985, The origin of the Caledonian nappes of eastern North Greenland, in Gee, D.G., and Sturt, B.A., eds., The Caledonide Orogen: Scandinavia and Related Areas: London, John Wiley, p. 1065–1069.

Hurst, J.M., Jepsen, H.F., Kalsbeek, F., McKerrow, W.S., and Peel, J.S., 1985, The geology of the northern extremity of the East Greenland Caledonides, in Gee, D.G., and Sturt, B.A., eds., The Caledonide Orogen: Scandinavia and Related Areas: London, John Wiley, p. 1047–1063.

Kalsbeek, F., Thrane, K., Nutman, A.P., and Jepsen, H.F., 2000, Late Mesoproterozoic to early Neoproterozoic history of the East Greenland Caledonides: Evidence for Grenvillian orogenesis?: Journal of the Geological Society of London, v. 157, p. 1215–1225.

Kalsbeek, F., Jepsen, H.F., and Nutman, A.P., 2001, From source migmatites to plutons: Tracking the origin of c. 435 Ma granites in the East Greenland Caledonian orogen: Lithos, v. 57, p. 1–21, doi: 10.1016/S0024-4937(00)00071-2.

Kaufman, A.J., Knoll, A.H., and Narbonne, G.M., 1997, Isotopes, ice ages, and terminal Proterozoic Earth history: Proceedings of the National Academy of Sciences of the United States of America, v. 94, p. 6600–6605, doi: 10.1073/pnas.94.13.6600.

Kennedy, M.J., Runnegar, B., Prave, A.R., Hoffman, K.H., and Arthur, M., 1998, Two or four Neoproterozoic glaciations?: Geology, v. 26, p. 1059–1063, doi: 10.1130/0091-7613(1998)026<1059:TOFNG>2.3.CO;2.

Knoll, A.H., 2000, Learning to tell Neoproterozoic time: Precambrian Research, v. 100, p. 3–20, doi: 10.1016/S0301-9268(99)00067-4.

Knoll, A.H., Hayes, J.M., Kaufman, A.J., Swett, K., and Lambert, I.B., 1986, Secular variation in carbon isotope ratios from Upper Proterozoic successions of Svalbard and East Greenland: Nature, v. 321, no. 6073, p. 832–838, doi: 10.1038/321832a0.

Knoll, A.H., Walter, M.R., Narbonne, G.M., and Christie-Blick, N., 2004, A new period for the geologic time scale: Science, v. 305, p. 621–622, doi: 10.1126/science.1098803.

Kristiansen, K.K., 2007, Den Neoproterozoiske Marinoan glaciation i Tillit Gruppen: et studie i $\delta^{13}C$ variationerne i øvre del af Eleonore Bay Supergruppen og Tillit Gruppen på Ella Ø, Nordøstgrønland [M.Sc. thesis]: Copenhagen, University of Copenhagen, Denmark, 94 p.

Larsen, P.H., and Bengaard, H.-J., 1991, Devonian basin initiation in East Greenland: A result of sinistral wrench faulting and Caledonian extensional collapse: Journal of the Geological Society of London, v. 148, p. 355–368, doi: 10.1144/gsjgs.148.2.0355.

Leslie, A.G., and Nutman, A.P., 2000, Episodic tectono-thermal activity in the southern part of the East Greenland Caledonides: Geology of Greenland Survey Bulletin, v. 186, p. 42–49.

Leslie, A.G., Smith, M., and Soper, N.J., 2008, this volume, Laurentian margin evolution and the Caledonian orogeny—A template for Scotland and East Greenland, in Higgins, A.K., Gilotti, J.A., and Smith, M.P., eds., The Greenland Caledonides: Evolution of the Northeast Margin of Laurentia: Geological Society of America Memoir 202, p. doi: 10.1130/2008.1202(13).

Manby, G.M., and Hambrey, M.J., 1989, The structural setting of the late Proterozoic tillites of East Greenland, *in* Gayer, R.A., ed., The Caledonide Geology of Scandinavia: London, Graham & Trotman, p. 299–312.

Millar, I., 1999, Neoproterozoic extensional basic magmatism associated with the West Highlands granite gneiss in the Moine Supergroup of NW Scotland: Journal of the Geological Society of London, v. 156, p. 1153–1162, doi: 10.1144/gsjgs.156.6.1153.

Moncrieff, A.C.M., 1989, The Tillite Group and related rocks of East Greenland: Implications for late Proterozoic palaeogeography, *in* Gayer, R.A., ed., The Caledonian Geology of Scandinavia: London, Graham and Trotman, p. 285–297.

Moncrieff, A.C.M., and Hambrey, M.J., 1988, Late Precambrian glacially-related grooved and striated surfaces in the Tillite Group of central East Greenland: Palaeogeography, Palaeoclimatology, Palaeoecology, v. 65, p. 183–200, doi: 10.1016/0031-0182(88)90023-5.

Moncrieff, A.C.M., and Hambrey, M.J., 1990, Marginal-marine glacial sedimentation in the late Precambrian succession of East Greenland, *in* Dowdeswell, J.A., and Scourse, J.D., eds., Glacimarine Environments: Processes and Sediments: Geological Society of London Special Publication 53, p. 397–410.

Noble, S.R., Highton, A.J., and Hyslop, E.K., 1996, High-precision U-Pb monazite geochronology of the c. 806 Ma Grampian shear zone and the implications for the evolution of the Central Highlands of Scotland: Journal of the Geological Society of London, v. 153, p. 511–514, doi: 10.1144/gsjgs.153.4.0511.

Oliver, G.J.H., 2002, Chronology and terrane assembly, new and old controversies, *in* Trewin, N.H., ed., The Geology of Scotland (4th edition): London, Geological Society, p. 201–211.

Peach, B.N., Horne, J., Gunn, W., Clough, C.T., Hinxman, L.W., and Teall, J.J.H., 1907, The Geological Structure of the Northwest Highlands of Scotland: Memoirs of the Geological Survey of Great Britain, 668 p.

Peel, J.S., and Sønderholm, M., eds., 1991, Sedimentary basins of North Greenland: Bulletin Grønlands Geologiske Undersøgelse, v. 161, 164 p.

Phillips, W.E.A., Stillman, C.J., Friderichsen, J.D., and Jemelin, L., 1973, Preliminary results of mapping in the western gneiss and schist zone around Vestfjord and inner Gåsefjord, south-west Scoresby Sund: Rapport Grønlands Geologiske Undersøgelse, v. 58, p. 17–32.

Piper, J.D.A., 1982, The Precambrian paleomagnetic record: The case for the Proterozoic supercontinent: Earth and Planetary Science Letters, v. 59, p. 61–89, doi: 10.1016/0012-821X(82)90118-2.

Posamentier, H.W., and James, D.P., 1993, An overview of sequence stratigraphic concepts: Uses and abuses, *in* Posamentier, H.W., Summerhayes, C.P., Haq, B.U., and Allen, G.P., eds., Sequence Stratigraphy and Facies Associations: Oxford, Blackwell Publishing, p. 3–18.

Prave, A.R., 1999, The Neoproterozoic Dalradian Supergroup of Scotland—An alternative hypothesis: Geological Magazine, v. 136, p. 609–617, doi: 10.1017/S0016756899003155.

Rainbird, R.H., Hamilton, M.A., and Young, G.M., 2001, Detrital zircon geochronology and provenance of the Torridonian, NW Scotland: Journal of the Geological Society of London, v. 158, p. 15–27.

Read, J.F., 1982, Carbonate platforms of passive (extensional) continental margins: Types, characteristics and evolution: Tectonophysics, v. 81, p. 195–212, doi: 10.1016/0040-1951(82)90129-9.

Robb, L.J., Knoll, A.H., Plumb, K.A., Shields, G.A., Strauss, H., and Veizer, J., 2004, The Precambrian: The Archaean and Proterozoic Eons, *in* Gradstein, F.M., Ogg, J.G., and Smith, A.G., eds., A Geologic Time Scale 2004: Cambridge, Cambridge University Press, p. 129–140.

Robertson, S., and Smith, M., 1999, The significance of the Geal Charn-Ossian Steep Belt in basin development in the Central Scottish Highlands: Journal of the Geological Society of London, v. 156, p. 1175–1182, doi: 10.1144/gsjgs.156.6.1175.

Rogers, G., Hyslop, E.K., Strachan, R.A., Peterson, B.A., and Holdsworth, R.E., 1998, The structural setting and U-Pb geochronology of Knoydartian pegmatites in W Inverness-shire; evidence for Neoproterozoic tectonothermal events in the Moine of NW Scotland: Journal of the Geological Society of London, v. 155, p. 685–696, doi: 10.1144/gsjgs.155.4.0685.

Sandelin, S., Tebenkov, A.M., and Gee, D.G., 2001, The stratigraphy of the lower part of the Neoproterozoic Murchisonfjorden Supergroup in Nordaustlandet, Svalbard: Geologiska Föreningens i Stockholm Förhandlingar, v. 123, p. 113–127.

Smith, M., Robertson, S., and Rollin, K.E., 1999, Rift basin architecture and stratigraphical implications for basement-cover relationships in the Neoproterozoic Grampian Group of the Scottish Caledonides: Journal of the Geological Society of London, v. 156, p. 1163–1173, doi: 10.1144/gsjgs.156.6.1163.

Smith, M.P., and Rasmussen, J.A., 2008, this volume, Cambrian–Silurian development of the Laurentian margin of the Iapetus Ocean in Greenland and related areas, *in* Higgins, A.K., Gilotti, J.A., and Smith, M.P., eds., The Greenland Caledonides: Evolution of the Northeast Margin of Laurentia: Geological Society of America Memoir 202, doi: 10.1130/2008.1202(06).

Smith, M.P., and Robertson, S., 1999a, The Nathorst Land Group (Neoproterozoic) of East Greenland—Lithostratigraphy, basin geometry and tectonic history, *in* Higgins, A.K., and Frederiksen, K.S., eds., Geology of East Greenland 72°–75°N, mainly Caledonian: Preliminary reports from the 1998 expedition: Danmarks og Grønlands Geologiske Undersøgelse Rapport, v. 1999/19, p. 127–144.

Smith, M.P., and Robertson, S., 1999b, Vendian–Lower Palaeozoic stratigraphy of the parautochthon in the Målebjerg and Eleonore Sø windows, East Greenland Caledonides: Danmarks og Grønlands Geologiske Undersøgelse Rapport, v. 1999/19, p. 169–182.

Smith, M.P., Soper, N.J., Higgins, A.K., Rasmussen, J.A., and Craig, L.E., 1999, Palaeokarst systems in the Neoproterozoic of eastern North Greenland in relation to extensional tectonics on the Laurentian margin: Journal of the Geological Society of London, v. 156, p. 113–124, doi: 10.1144/gsjgs.156.1.0113.

Smith, M.P., Higgins, A.K., Soper, N.J., and Sønderholm, M., 2004a, The Neoproterozoic Rivieradal Group of Kronprins Christian Land, eastern North Greenland: Geological Survey of Denmark and Greenland Bulletin, v. 6, p. 29–39.

Smith, M.P., Rasmussen, J.A., Robertson, S., Higgins, A.K., and Leslie, A.G., 2004b, Lower Palaeozoic stratigraphy of the East Greenland Caledonides: Geological Survey of Denmark and Greenland Bulletin, v. 6, p. 5–28.

Sønderholm, M., and Jepsen, H.F., 1991, Proterozoic basins of North Greenland, *in* Sønderholm, M., and Peel, J.S., eds., Sedimentary basins of North Greenland: Bulletin Grønlands Geologiske Undersøgelse, v. 160, p. 49–69.

Sønderholm, M., and Tirsgaard, H., 1993, Lithostratigraphic framework of the Upper Proterozoic Eleonore Bay Supergroup of East and North-East Greenland: Bulletin Grønlands Geologiske Undersøgelse, v. 167, 38 p.

Sønderholm, M., and Tirsgaard, H., 1998, Proterozoic fluvial styles: Response to changes in accommodation space (Rivieradal sandstone, eastern North Greenland): Sedimentary Geology, v. 120, p. 257–274, doi: 10.1016/S0037-0738(98)00035-9.

Soper, N.J., 1994, Neoproterozoic sedimentation on the NE margin of Laurentia and the opening of the Iapetus: Geological Magazine, v. 131, p. 291–299.

Stewart, A.D., 2002, The Later Proterozoic Torridonian Rocks of Scotland: Their Sedimentology, Geochemistry and Origin: Geological Society of London Memoir 24, 130 p.

Swett, K., and Knoll, A.H., 1989, Marine pisolites from Upper Proterozoic carbonates of East Greenland and Spitsbergen: Sedimentology, v. 36, p. 75–93, doi: 10.1111/j.1365-3091.1989.tb00821.x.

Tirsgaard, H., 1993, The architecture of Precambrian high energy tidal channels: An example from the Lyell Land Group (Eleonore Bay Supergroup): East Greenland: Sedimentary Geology, v. 88, p. 137–152, doi: 10.1016/0037-0738(93)90154-W.

Tirsgaard, H., 1996, Cyclic sedimentation of carbonate and siliciclastic deposits on a late Precambrian ramp: The Elisabeth Bjerg Formation (Eleonore Bay Supergroup): East Greenland: Journal of Sedimentary Research, v. 66B, p. 699–712.

Tirsgaard, H., and Sønderholm, M., 1997, Lithostratigraphy, sedimentary evolution and sequence stratigraphy of the Upper Proterozoic Lyell Land Group (Eleonore Bay Supergroup) of East and North-East Greenland: Geology of Greenland Survey Bulletin, v. 178, 60 p.

Trewin, N.H., ed., 2002, The Geology of Scotland (4th edition): London, Geological Society of London, 576 p.

Troelsen, J.C., 1949, Contributions to the geology of the area round Jørgen Brønlund Fjord, Peary Land, North Greenland: Meddelelser om Grønland, v. 149, no. 2, 29 p.

Troelsen, J.C., 1956, The Cambrian of North Greenland and Ellesmere Island, *in* Rodgers, J., ed., El Sistema Cámbrico, su paleogeografica y el problema de su base: 20th International Geological Congress, Mexico: International Commission on Stratigraphy Symposium, v. 1, p. 71–90.

Turnbull, M.J.M., Whitehouse, M.J., and Moorbath, S., 1996, New isotopic age determinations for the Torridonian, NW Scotland: Journal of the Geological Society of London, v. 153, p. 955–964, doi: 10.1144/gsjgs.153.6.0955.

Upton, B.G.J., Rämö, O.T., Hearnan, L.M., Blichert-Toft, J., Barry, T.L., Kalsbeek, F., and Jepsen, H.F., 2005, The Zig-Zag Dal Basalts and associated intrusions of eastern North Greenland: Progressive mantle plume–lithosphere interaction: Contributions to Mineralogy and Petrology, v. 149, p. 40–56, doi: 10.1007/s00410-004-0634-7.

Vance, D., Strachan, R.A., and Jones, K.A., 1998, Extensional versus compressional settings for metamorphism; garnet chronometry and pressure-temperature-time histories in the Moine Supergroup, northwest Scotland: Geology, v. 26, p. 927–930, doi: 10.1130/0091-7613(1998)026<0927:EVCSFM>2.3.CO;2.

van Wagoner, J.C., Posamentier, H.W., Mitchum, R.M., Vail, P.R., Sarg, J.F., Loutit, T.S., and Hardenbol, J., 1988, An overview of the fundamentals of sequence stratigraphy and key definitions, *in* Wilgus, C.K., Hastings, B.S., Kendall, C.G.S.C., Posamentier, H.W., Ross, H.W., and van Wagoner, J.C., eds., Sea-Level Changes—An Integrated Approach: Society of Paleontologists and Mineralogists Special Publication 42, p. 39–45.

van Wagoner, J.C., Mitchum, R.M., Campion, K.M., and Rahmanian, V.E., 1990, Siliciclastic sequence stratigraphy in well logs, core and outcrop: Concepts for high-resolution correlation of time and facies: American Association of Petroleum Geologists, Methods in Exploration Series, v. 7, 55 p.

Vidal, G., 1985, Biostratigraphic correlation of the Upper Proterozoic and Lower Cambrian of the Fennoscandian Shield and the Caledonides of East Greenland and Svalbard, *in* Gee, D.G., and Sturt, B.A., eds., The Caledonide Orogen—Scandinavia and Related Areas: London, John Wiley and Sons Ltd., p. 331–338.

Watt, G.R., and Thrane, K., 2001, Early Neoproterozoic events in East Greenland: Precambrian Research, v. 110, p. 165–184, doi: 10.1016/S0301-9268(01)00186-3.

Watt, G.R., Kinny, P.D., and Friderichsen, J.D., 2000, U-Pb geochronology of Neoproterozoic and Caledonian tectonothermal events in the East Greenland Caledonides: Journal of the Geological Society of London, v. 157, p. 1031–1048.

Wenk, E., 1961, On the crystalline basement and the basal part of the Precambrian Eleonore Bay Group in the southwestern part of Scoresby Sund: Meddelelser om Grønland, v. 168, no. 1, 54 p.

Wenk, E., and Haller, J., 1953, Geological explorations in the Petermann region, western part of Frænkels Land, East Greenland: Meddelelser om Grønland, v. 111, 48 p.

White, A.P., and Hodges, K.V., 2002, Multistage extensional evolution of the central East Greenland Caledonides: Tectonics, v. 21, no. 5, art. no. 1048, doi: 10.1029/2001TC001308,2002.

White, A.P., Hodges, K.V., Martin, M.W., and Andresen, A., 2002, Geological constraints on middle-crustal behavior during broadly synorogenic extension in the central East Greenland Caledonides: International Journal of Earth Sciences, v. 91, p. 187–208, doi: 10.1007/s005310100227.

Winchester, J.A., 1988, Late Proterozoic environments and tectonic evolution in the northern Atlantic lands, *in* Winchester, J.A., ed., Later Proterozoic Stratigraphy of the Northern Atlantic Regions: London, Blackie & Sons Ltd., p. 253–270.

MANUSCRIPT ACCEPTED BY THE SOCIETY 14 JANUARY 2008

The Geological Society of America
Memoir 202
2008

Cambrian–Silurian development of the Laurentian margin of the Iapetus Ocean in Greenland and related areas

M. Paul Smith*
Lapworth Museum of Geology, University of Birmingham, Edgbaston, Birmingham B15 2TT, UK

Jan Audun Rasmussen
Natural History Museum of Denmark (Geology), University of Copenhagen, Øster Voldgade 5-7, DK-1350 Copenhagen K, Denmark

ABSTRACT

The Iapetus margin of Laurentia is preserved, with varying degrees of deformation, along a belt that extends for 1300 km along the eastern coast of Greenland, from Scoresby Sund in the south to Kronprins Christian Land at the northernmost extent of the Caledonian–Appalachian orogen. Along the length of the Greenland Caledonides, deformation is restricted to a single orogenic phase, the Scandian, at around 425 Ma, which represents the continent-continent collision of Laurentia and Baltica. The Lower Paleozoic stratigraphy can be closely correlated with the palinspastically contiguous terranes of NE Spitsbergen, Bjørnøya, and NW Scotland, and, farther to the south, that of western Newfoundland. In Greenland itself, Lower Paleozoic sediments are present in the foreland, parautochthon, and the highest allochthonous sheet of the orogen, the Franz Joseph allochthon. In the Franklinian Basin of eastern North Greenland, unconformity-bounded Lower Cambrian sediments can be correlated with the Sauk I sequence of cratonic North America. These Cambrian sediments are separated from younger units by a significant hiatus, the sub–Wandel Valley unconformity, but above that surface, the succession extends without major breaks from the major flooding event at the base of Sauk IV (Early Ordovician) through to the early Wenlock. The carbonate platform in this region foundered from late Llandovery time onward due to loading by thrust sheets, and turbidite deposition replaced platform carbonate deposition. Caledonian thrusts truncate the youngest preserved sediments, which are of early Wenlock age. The punctuated, attenuated stratigraphy seen in Kronprins Christian Land continues southward along the length of the parautochthon, through Lambert Land, Nørreland, and Dronning Louise Land, to a series of tectonic windows in the southern part of the Greenland Caledonides. In contrast to the stratigraphy seen in the parautochthon, the Franz Joseph allochthon contains one of the thickest Cambrian–Middle Ordovician successions in Laurentia, including a complete succession from Sauk I to Tippecanoe II.

Keywords: Cambrian, Ordovician, Greenland, Svalbard, basins, tectonics.

*m.p.smith@bham.ac.uk

INTRODUCTION

Perspectives of the Iapetus margin in Greenland and its subsequent deformation during the Scandian phase of the Caledonian orogeny are very different in the two main areas of Lower Paleozoic outcrop, although, in both cases, the margin is incorporated within the Caledonian orogen. In the Caledonides of southern North-East Greenland (Figs. 1 and 2), the classic Cambrian-Ordovician sections of the fjord region (Cowie and Adams, 1957; Smith, 1991; Stouge et al., 2001, 2002) lie within an allochthonous sheet, the Franz Joseph allochthon, which is paradoxically little deformed and which has been subjected to maximum burial temperatures of only 140 °C. Understanding of the Cambrian-Ordovician development of the Iapetus margin in this area has been hampered historically by a lack of understanding of the tectonic architecture and the consequent inability to restore the margin palinspastically to its original configuration, although this has to a large extent been overcome by recent research (Higgins and Leslie, 2000, 2004; Higgins et al., 2004a). However, due to concealment beneath the Inland Ice, there is still very little evidence for the position of the western edge of the orogen and the nature of the foreland succession, although attenuated parautochthonous successions have recently been discovered in tectonic windows close to the ice margin (Leslie and Higgins, 1998; Higgins et al., 2004a; Smith et al., 2004). There is also some evidence available from erratics at the ice margin (Haller, 1971; Smith et al., 2004).

In distinct contrast, at the northernmost limit of the orogen (Figs. 1–3), there is complete exposure from the orogen to the parautochthon in Kronprins Christian Land and then for 1000 km across the foreland in what is one of the best-preserved and exposed margins of any ancient orogen. Nevertheless, reconstruction of the original Iapetus margin is hampered here because the preserved orogen itself is relatively narrow, and a substantial portion must be present on the shelf beneath the Greenland Sea to the east. There is good evidence that Caledonian structures extend eastward to the edge of the continental shelf and that the island of Bjørnøya, currently located between Spitsbergen and the northern tip of Norway, was a part of the North Greenland margin until the opening of the North Atlantic in the Paleogene (Smith, 2000). The nature of the Iapetus margin at this point is critical because Kronprins Christian Land lies at an original right-angled turn in the Laurentian margin (Fig. 1) that has been relatively little altered by post-Caledonian tectonics. To the south of Kronprins Christian Land, the original Laurentian margin extends as far as the southeastern United States and has been disrupted to greater or lesser degrees by successive Caledonian–Appalachian orogenic phases. To the west, the Laurentian margin extends into Arctic Canada as the Franklinian Basin, which has been successively deformed, not by Caledonian events, but by the Early Devonian–Carboniferous Ellesmerian orogeny and the Cretaceous–Paleogene Eurekan phase (Okulitch and Trettin, 1991; Soper and Higgins, 1991a, 1991b; Trettin et al., 1991).

Within the Greenland Caledonides, between the well-exposed areas of Lower Paleozoic geology in Kronprins Christian Land to the north and the fjord region of North-East Greenland to the south (Fig. 2), there is an area of extensive ice cover that contains nunataks and small land masses of various sizes that provide limited data concerning correlations and relationships between the two areas. However, these areas are very remote, and the Lower Paleozoic geology is not as well understood as it is in areas to the north and south. Nevertheless, some information is available from scattered outcrops of foreland and parautochthon in Dronning Louise Land, parautochthon in Nørreland, and thrust sheets containing Ordovician carbonates in Lambert Land (Figs. 2 and 3) (Smith et al., 2004).

The Greenland Caledonides offer a distinctive insight into the history and development of the Iapetus margin of Laurentia. In contrast to areas farther to the south, in the British, Irish, Canadian, and U.S. sectors, the subsequent deformation of the margin is relatively simple, with only one major orogenic phase, the Scandian. Immediately to the south of Greenland, the Scandian orogeny overprints earlier 470–465 Ma Grampian-Taconic deformation in NW Scotland (Leslie et al., this volume), and understanding of the pre-orogenic margin becomes correspondingly more difficult to extract. However, this phase of island-arc collision is absent to the north of the "Scottish Promontory" (Dalziel and Soper 2001), which marks a major inflection in the continental margin of Laurentia.

A Note on Lower Paleozoic Global Stratigraphy

Although internationally recognized and applied series and stages for the Silurian have been in existence for several decades, and those for the Ordovician have recently been ratified, the same is not true for the Cambrian. In the case of the Ordovician, a tripartite division into series has been ratified, as has the further division of the Lower, Middle, and Upper series into two, two, and three stages, respectively; global stratotype sections and points (GSSPs) have also been defined for each case (Cooper and Sadler, 2004). Reference is also made to Laurentian chronostratigraphy where appropriate. The Cambrian is in a state of considerable flux, and very few GSSPs for series or stage divisions have been ratified to date. The Cambrian Subcommission has recently voted to divide the Cambrian into four series with two, two, three, and three stages from oldest to youngest. Of these, only the youngest series, the Furongian, and its oldest stage, the Paibian, have been formalized (Shergold and Cooper, 2004), and the rest await full ratification. Pending ratification, the tripartite series divisions of Shergold and Cooper (2004) are used here, and, for clarity, reference to both the Laurentian and Siberian standards (cf. Shergold and Cooper, 2004, their Figure 11.1) is made wherever possible. Absolute ages follow Gradstein et al. (2004), and all chronostratigraphy has been recalibrated to the new time scale. Cited biozones are Laurentian conodont zones unless otherwise stated.

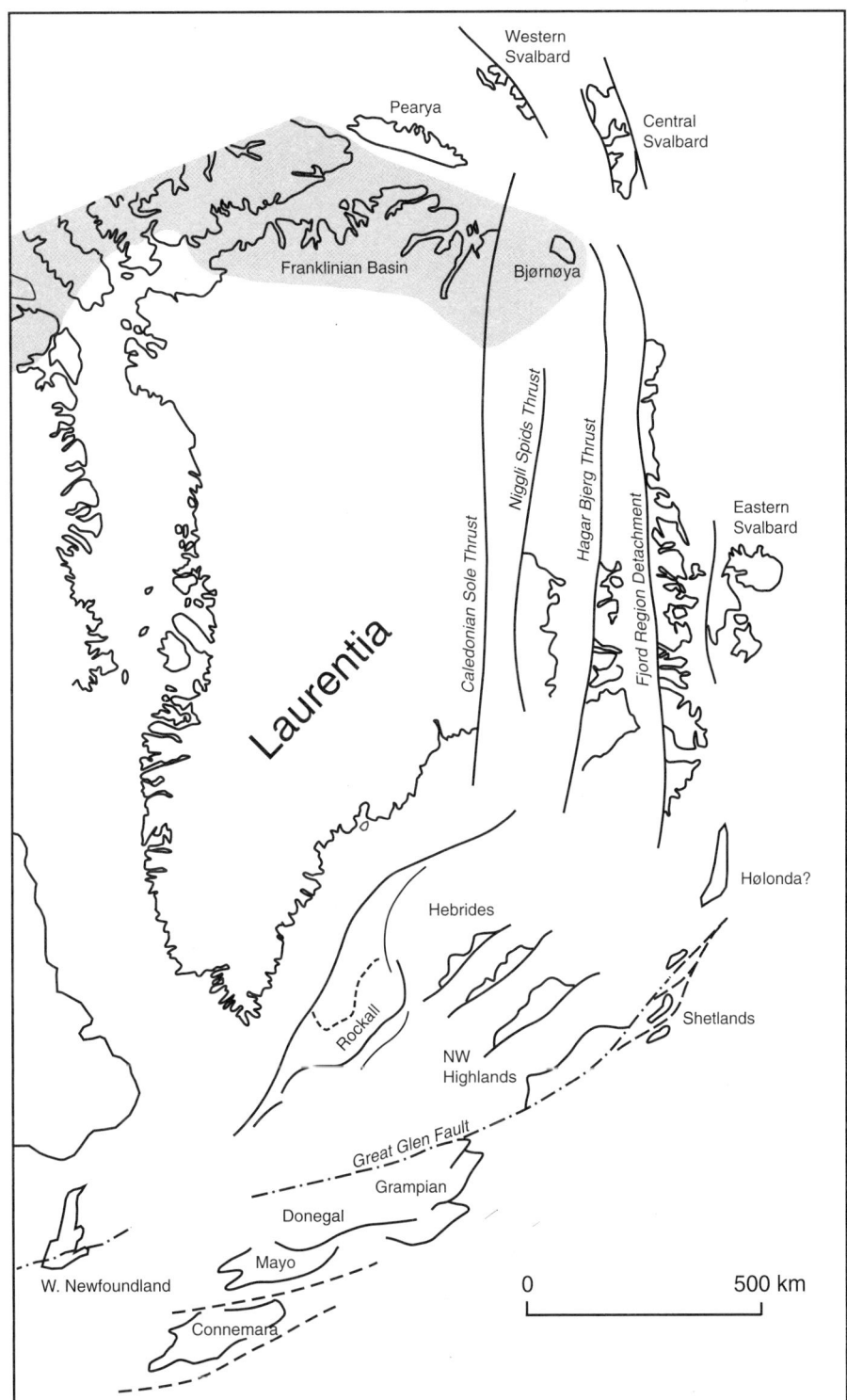

Figure 1. Palinspastic reconstruction of the northeastern margin of Laurentia, incorporating Greenland, northwestern Britain and Ireland, western Newfoundland, and parts of Svalbard, prior to Taconic-Grampian and Scandian collisions. Modern coastlines are provided for reference, and, for clarity, the Caledonian internal deformation within allochthonous blocks is not depicted. The pre-Caledonian positions of the Western and Central Provinces of Svalbard are poorly constrained relative to the Eastern Province and Bjørnøya. The Franklinian Basin of North Greenland and Canada is shown in gray. Modified in part after Higgins et al. (2001) and Leslie et al. (this volume), utilizing data from Soper (1994) and Smith (2000).

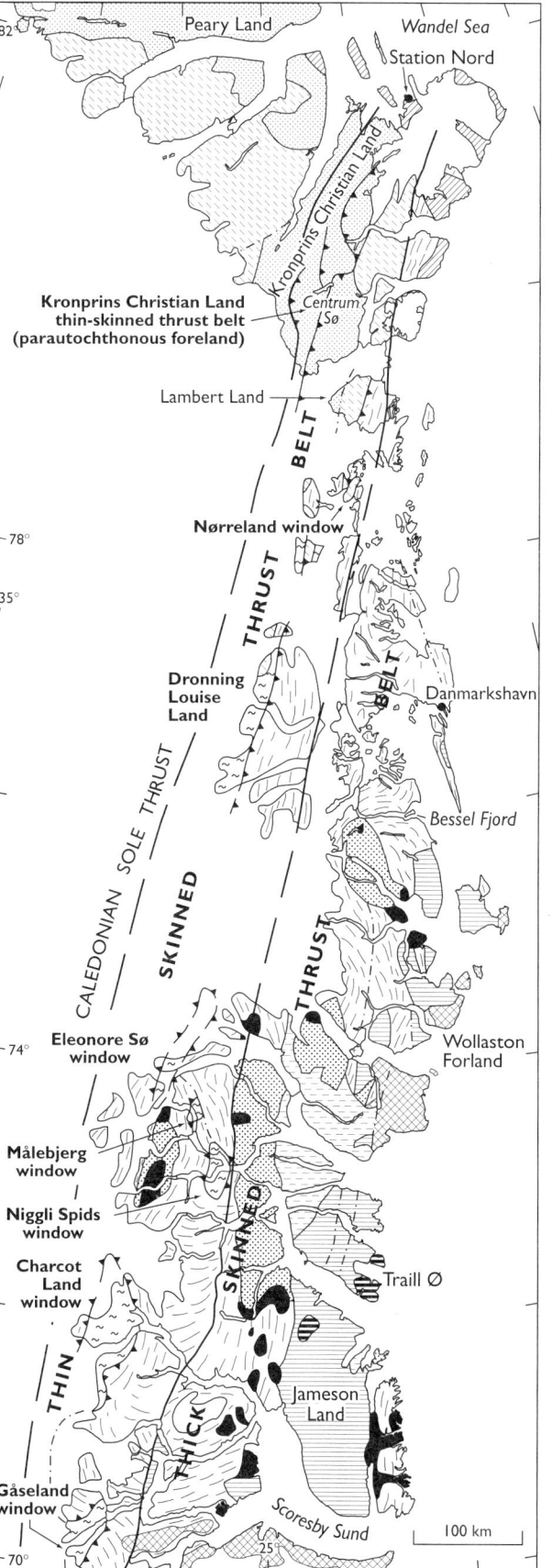

Figure 2. Map showing areas of Neoproterozoic and Lower Paleozoic outcrop in the Caledonides of North-East and North Greenland together with major structural features (after Higgins and Leslie, 2000). Five main areas of outcrop are present: the thick succession of the Franz Joseph allochthon in the fjord region of southern North-East Greenland; the parautochthonous succession within tectonic windows; the remote outliers of Dronning Louise Land and Nørreland; the parautochthonous and allochthonous succession in Kronprins Christian Land; and the undeformed succession of the Franklinian Basin that constitutes the foreland of the orogen in North Greenland.

PALEOGEOGRAPHY AND MARGINAL TERRANES

There is considerable consensus that, paleogeographically, Laurentia lay in an equatorial position throughout the early Paleozoic and that it was largely contained between latitudes 30°N and 30°S (Scotese and McKerrow, 1990; Torsvik et al., 1996; Harper et al., 1996; MacNiocaill et al., 1997; Cocks and Torsvik, 2002; Fortey and Cocks, 2003). Paleomagnetic and zoogeographic data are largely in agreement with regards to the location of the main craton of Laurentia, and Fortey and Cocks (2003) noted that it appears to have been the least mobile and most stable of early Paleozoic plates. There is less certainty about the positional history of terranes incorporated within marginal orogenic belts because, predictably, structural overprinting during collision has obscured primary data. Terranes of Laurentian affinity that are relevant to the Greenland sector of the Caledonides include: West, Central, and East Spitsbergen, Bjørnøya, the Upper and Uppermost Allochthons of the Scandinavian Caledonides, NW Scotland, and western Newfoundland (Fig. 1).

Displacement within Thrust Sheets in the East Greenland Caledonides

In addition to terrane separation and emplacement, any palinspastic restoration of the Iapetus margin in Greenland must of course take into account shortening caused by thrusting within the orogen. In Kronprins Christian Land, section balancing has produced an estimate of 17.6 km of shortening within the parautochthon (Higgins et al., 2004b), and there are no Lower Paleozoic sediments preserved within the allochthon in this area. However, immediately to the south in Lambert Land (northernmost North-East Greenland; Figs. 2 and 3), Lower Ordovician carbonates are present in the westernmost thrust sheets of the allochthon, where they unconformably overlie sandstones of the Independence Fjord Group (Smith et al., 2004; Collinson et al., this volume). Displacement on these thrust sheets is estimated to be on the order of 40–50 km (Higgins et al., 2004b).

Cambrian-Ordovician sediments in Nørreland, Dronning Louise Land, and the tectonic windows of Eleonore Sø, Målebjerg, and Gåseland (Higgins et al., 2001; Smith et al., 2004) (Figs. 2 and 4) lie either in the foreland or have been subject to minimal amounts of displacement within the parautochthon. In distinct contrast, the Cambrian-Ordovician sediments of the fjord region of North-East Greenland have been subject to considerable displacement. The tectonic architecture of the orogen between 72°N and 75°N was elucidated by Higgins et al. (2004a), who recognized three major elements: the Niggli Spids and Hagar Bjerge thrust sheets, plus the little-deformed and structurally highest Franz Joseph allochthon (Fig. 4). The thick Cambrian-Ordovician successions of the Kong Oscar Fjord Group lie within the Franz Joseph allochthon, which has a minimum documented transport distance of 200–400 km following 40%–60% shortening within the orogen (Higgins et al., 2004a). Palinspastically, there was thus a distance of several hundred kilometers between the Cambrian sediments of the parautochthon and those of the Franz Joseph allochthon.

Spitsbergen Terranes

As early as 1965, Harland had discussed the possible presence of large-scale strike-slip faulting between Svalbard and Greenland (Harland, 1965, 1969). A transpressional model for Svalbard terranes was developed through a series of papers (see Harland [1997] for a review). Three principal terranes, or "provinces," were recognized, all of which have Laurentian affinity. Of these, the Eastern Province, corresponding to NE Spitsbergen and Nordaustlandet, is the best constrained in terms of its position, and a number of authors have documented the close correlations between it and the Franz Joseph allochthon of North-East Greenland. This is particularly true of the Neoproterozoic tillite successions (Hambrey, 1983, 1988; Fairchild and Hambrey, 1995; Halverson et al., 2004) and the stratigraphy of the Cambrian-Ordovician interval (Swett, 1981; Harland et al., 1988). There is now reasonable certainty that the Eastern Province of Svalbard, encompassing Ny Friesland and Nordaustlandet (Figs. 1 and 5), was outboard of the Franz Joseph allochthon as an integral part of the Iapetus margin (Fig. 1), which extends the distance from the parautochthon to the continental margin even farther. The very close similarity of the tillite successions is a significant obstacle to the alternative terrane hypothesis of Gee and Teben'kov (2004), which suggested that the Svalbard terranes represent the direct northern continuation of the Greenland Caledonides.

The Western Province of Svalbard has been compared with North Greenland (Harland, 1985; Harland et al., 1988), but, in this case, the correlation is less secure and lacks stratigraphic detail. The most poorly constrained terrane is the Central Province, which has a thick Cambrian-Ordovician succession with Laurentian conodont faunas, including NE Laurentian endemics (Szaniawski, 1994), but is otherwise extremely poorly known. The terrane has not yet been documented in any detail—even to the extent that the age of the youngest pre-Caledonian sediments in the terrane, the Arkfjellet "series" (Major and Winsnes, 1955; Birkenmajer, 1978; Harland et al., 1988), remains completely unknown. The sedimentary facies, the thickness of the succession, and its present-day location to the west of the Eastern Province suggest that, relative to the Laurentian margin, it was positioned farther inboard than the Eastern Province (Fig. 1), but more precise correlations await much needed further work.

Bjørnøya

Bjørnøya, to the south of Spitsbergen, was also considered to be a Laurentian terrane by Harland (1985), who described it as a constituent part of the Central Province. Evidence for its Laurentian affinity includes the presence of carbonates of Laurentian

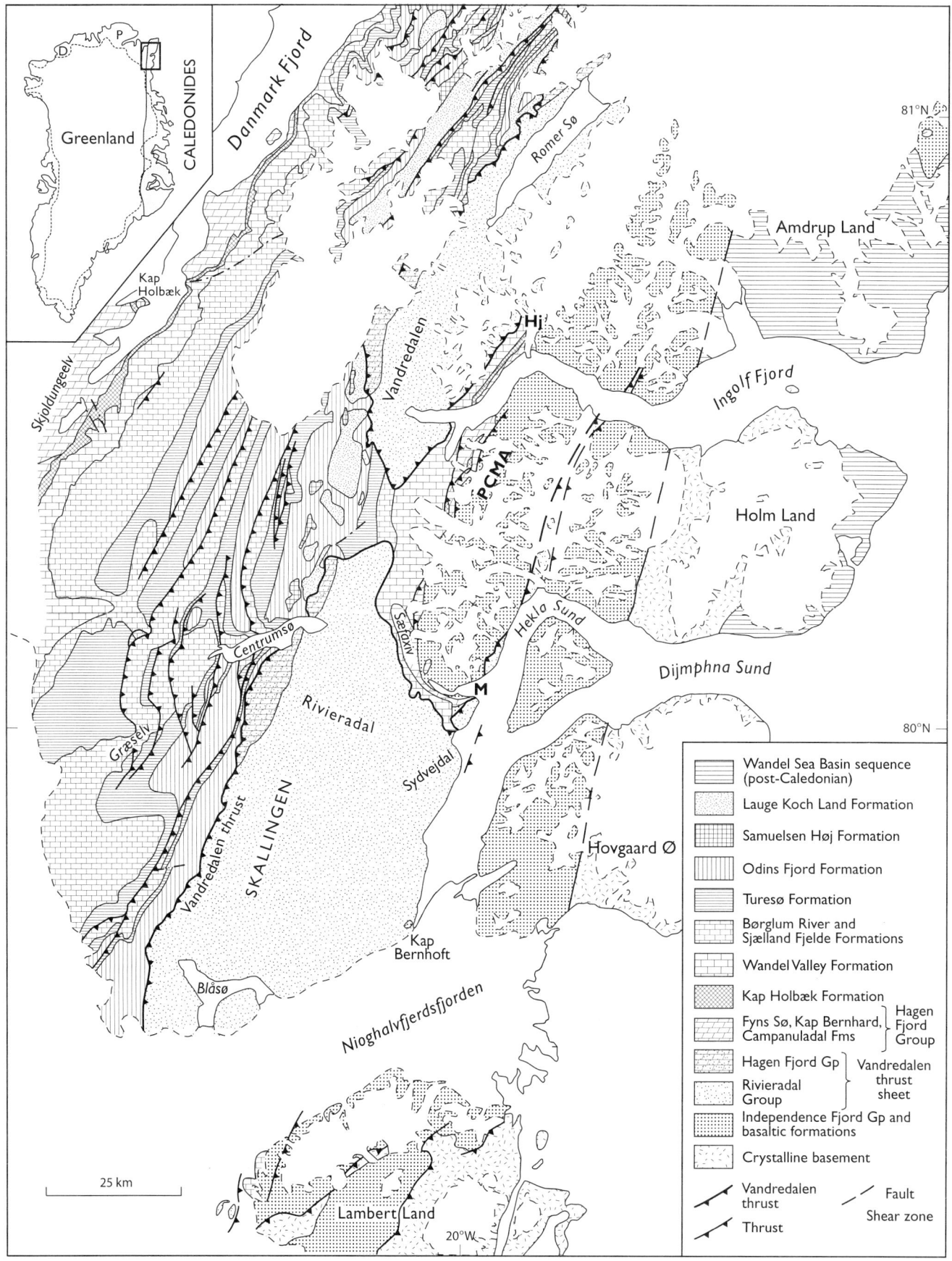

type (Smith, 2000), conodonts and gastropods of Midcontinent biogeographic affinity (Smith, 2000), and the presence of WNW-directed thrusting (Braathen et al., 1999; Smith, 2000). Detailed biostratigraphic correlation and sequence stratigraphy led Smith (2000) to conclude that the unconformity between the base of the Ordovician succession (Fig. 5B) and the Neoproterozoic diamictites of the Sørhamna Formation was exactly equivalent to the distinctive sub-Ibexian unconformity in eastern North Greenland (see following). Bjørnøya is now considered to have been an integral part of the North Greenland carbonate platform during the early Paleozoic (Fig. 1), and it was only detached as a terrane during the Paleogene as the Greenland Sea opened (Smith, 2000). The presence of the distinctive and areally restricted sub-Wandel Valley unconformity precludes significant strike-slip displacement during Caledonian deformation, but the presence of Caledonian thrusting means that the pre-Paleogene position of Bjørnøya provides important constraints on the total width of the Caledonian orogen at its northernmost extremity—the minimum width was 350 km, of which 250 km are currently submerged on the continental shelf.

Upper and Uppermost Allochthons of the Scandinavian Caledonides

The presence of faunas of Laurentian affinity in the upper levels of the Scandinavian Caledonides is now well documented (for reviews, see Bruton and Harper, 1988; Neuman and Harper, 1992; Harper et al., 1996; Bergström, 1990). More specifically, the Hølonda and Smøla macrofaunas of the Upper Allochthon are characteristic of the Toquima–Table Head brachiopod province and are interpreted as having originated on the peri-Laurentian Taconic island arc that collided with Laurentia during the early mid-Ordovician (Dapingian) to produce the Taconic-Grampian phase of the Caledonian orogeny (Harper et al., 1996; MacNiocaill et al., 1997). In addition, mid-Ordovician conodont faunas from the Upper Allochthon of the Trondheim region display clear Laurentian affinities (Bergström, 1979, 1997). Furthermore, when the oceanic conodont taxa are excluded from the data set, it becomes evident that the Hølonda Limestone shares more characteristics with the marginal Laurentian deposits of the Table Point Formation of western Newfoundland (Stouge, 1984) than with the more proximal parts of the Laurentian shelf (Rasmussen, 1998). Components of the northern part of the Taconic terrane were subsequently thrust eastward as part of the Upper Allochthon during the Scandian phase, incorporating a diverse assemblage of oceanic, magmatic arc, and marginal basin associations (Roberts, 2003).

Diagnostic Laurentian faunas have not yet been recovered from the Uppermost Allochthon, which occupies the coastal area of central north Norway. However, there is growing evidence from structural, sedimentologic, and stable isotopic–geochemical data for the incorporation of Laurentian shelf and slope-rise successions in this allochthon (Roberts, 2003). Thick metacarbonates are present along strike for hundreds of kilometers and have provided apparent depositional ages of Proterozoic–late Early Cambrian (Melezhik et al., 2000, 2002; Roberts et al., 2002). The Fauske Conglomerate in the lower part of the Uppermost Allochthon contains carbonate debris flows and turbidites that are consistent with a shelf-edge position (Melezhik et al., 2000; Roberts et al., 2002). Farther north, thick metalimestone and metadolostone units have yielded Neoproterozoic–Cambrian, and possible Silurian, ages on the basis of stable isotope paleoseawater chronology (Melezhik et al., 2002; Roberts et al., 2002).

In addition, the earliest deformation in the Uppermost Allochthon is recorded by NW-vergent and SE-dipping thrust ramps and flattening of clasts with a NW-SE stretching direction (Roberts et al., 2002). These are transected by later SE-verging structures that are characteristic of the Scandian phase of deformation. The earliest, NW-vergent phase of deformation is interpreted as representing Taconic-Grampian deformation (Roberts et al., 2001, 2002; Roberts, 2003). This is critical in any palinspastic restoration of the Laurentian margin, since NW-vergent Grampian structures are present as far north in the Caledonian orogen as Scotland but are absent in the East Greenland Caledonides. The only evidence of events of Taconic age in the East Greenland Caledonides is a single I-type granodiorite intrusion in east Milne Land that has yielded a SHRIMP age of 466 Ma (F. Kalsbeek, 2005, personal commun.), although this awaits confirmation from other similar intrusions in this area in the southernmost East Greenland Caledonides. If the carbonates of the Uppermost Allochthon in the Scandinavian Caledonides formed part of the Laurentian continental margin, then it is probable that it lay to the south of the Franz Joseph allochthon and the Eastern Province of Spitsbergen (Fig. 1).

Parautochthon of NW Scotland and Autochthon of Western Newfoundland

It has long been recognized that the Caledonian "foreland," actually parautochthon, of NW Scotland was part of the Laurentian margin of Iapetus, and that there are strong stratigraphic similarities with successions in western Newfoundland to the south (in palinspastic terms) and Greenland and NE Spitsbergen to the north (Swett and Smit, 1972; Swett, 1981). More recent work has begun to make more detailed comparisons of the Cambrian part of the succession with that in the Labrador and Port

Figure 3. Geological map of Kronprins Christian Land and Lambert Land (after Smith et al., 2004). The region includes autochthonous foreland in the west around Kap Holbæk and Danmark Fjord and passes eastward through a thin-skinned fold-and-thrust belt to a major allochthonous thrust sheet, the Vandredalen thrust sheet, which transports Neoproterozoic sedimentary rocks of the Independence Fjord, Rivieradal, and Hagen Fjord Groups. In Lambert Land, a very thin cover of Ordovician carbonates of the Wandel Valley Formation overlies the Independence Fjord Group within the westernmost thrust sheet. Hj—Hjørnegletscher; M—Marmorvigen; PCMA—Prinsesse Caroline-Mathilde Alper. On the inset map of Greenland: D—Daugaard-Jensen Land; P—Peary Land.

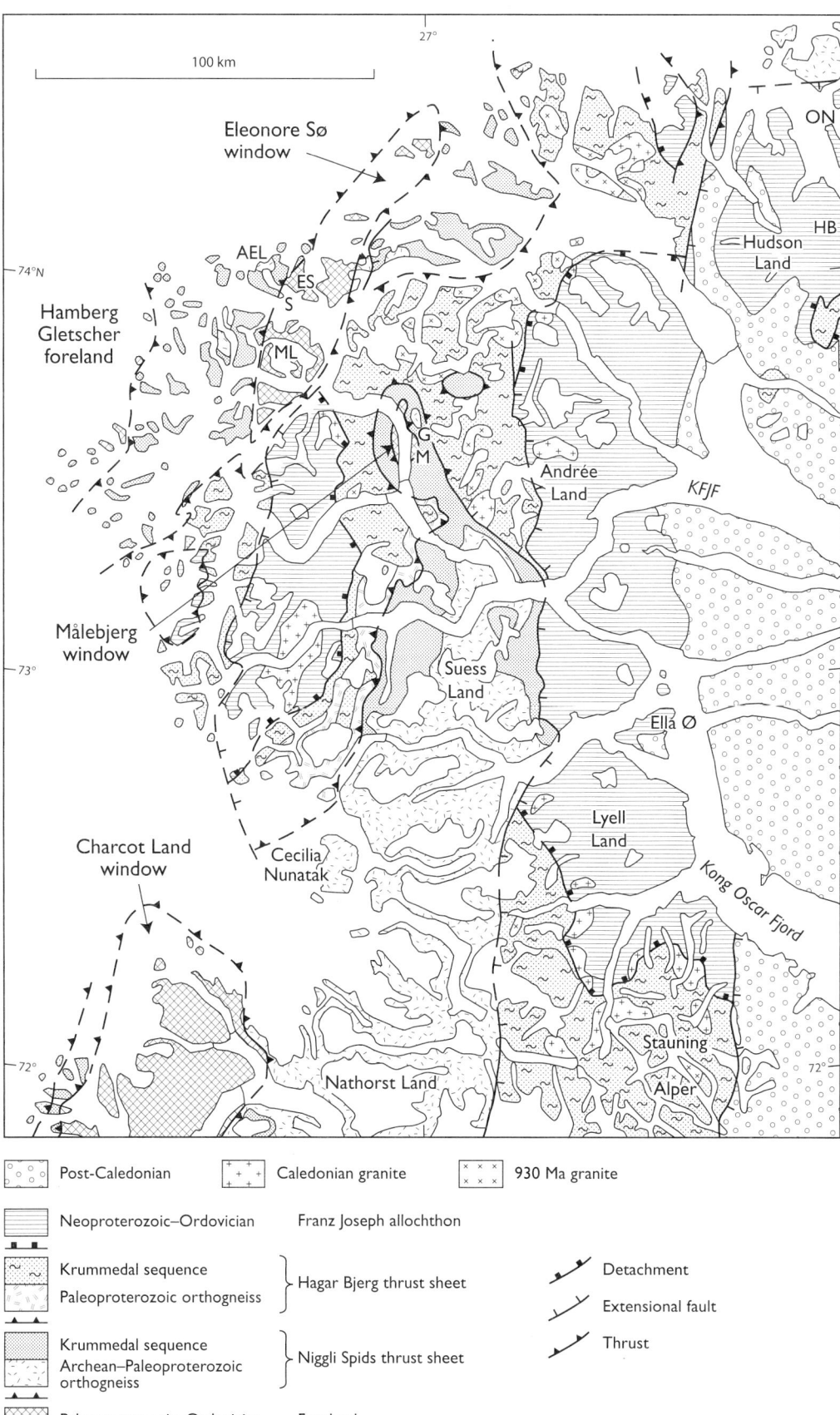

Figure 4. Geological map of southern North-East Greenland from 71°50′N to 74°30′N, showing the distribution of Cambrian-Ordovician sediments within the foreland windows and the Franz Joseph allochthon. AEL—Arnold Escher Land; ES—Eleonore Sø; G—Gemmedal; HB—Albert Heim Bjerge; KFJF—Kejser Franz Joseph Fjord; M—Målebjerg; ML—J.L. Mowinckel Land; ON—C.H. Ostenfeld Nunatak; S—Slottet.

au Port Groups of western Newfoundland (Wright and Knight, 1995; Park et al., 2002), and detailed work on the Ordovician part of the succession in NW Scotland is currently under way.

STRATIGRAPHY OF THE EAST GREENLAND CALEDONIDES: KRONPRINS CHRISTIAN LAND AND LAMBERT LAND

Context

Kronprins Christian Land is unique because it preserves an original and relatively undeformed promontory of the Laurentian margin, with an ~90° change in orientation of the continental margin (from N-S to E-W in modern coordinates) (Fig. 1). The area constitutes the easternmost part of the Lower Cambrian–Silurian Franklinian Basin, which continues westward for ~2000 km into the Canadian Arctic Islands (Trettin et al., 1991); it is made up of a deep-water clastic trough to the north and a carbonate-dominated shallow-water shelf and slope to the south (Higgins et al., 1991) (Fig. 3). The Lower Paleozoic stratigraphy of Kronprins Christian Land (Figs. 5A, 6, and 7) differs significantly from the fjord region farther to the south in the East Greenland Caledonides, but it also differs from the typical shelf stratigraphy of the Franklinian Basin because of the presence of uplift and erosion in this part of the basin from the Early Cambrian to the mid–Early Ordovician, which produced the distinctive stratigraphic signature of the sub–Wandel Valley unconformity (Fig. 8). This structure can be recognized from Peary Land eastward to Kronprins Christian Land and southward to Lambert Land (Higgins et al., 1991; Smith et al., 1999), but it is also present on Bjørnøya (Fig. 5B), where it is a key piece of evidence for the interpretation of the island as a terrane of the Franklinian Basin (Smith, 2000).

Kronprins Christian Land and Lambert Land: The Early Cambrian Siliciclastic Shelf

The oldest Paleozoic sediments in Kronprins Christian Land are the Lower Cambrian sandstones of the Kap Holbæk Formation. In the foreland of the Caledonides, this unit overlies stromatolitic dolostones of the Neoproterozoic Fyns Sø Formation (Hagen Fjord Group). The age and stratigraphic relationships of the Kap Holbæk Formation have been the subject of some debate (see Smith et al. [2004] for review), but the formation contains deep *Skolithos* burrows, which are indicative of an age no older than Tommotian (mid–Early Cambrian). In some parts of the parautochthon, for example the northern part of Vandredalen and Ingolf Fjord, the same stratigraphic relationships of the Kap Holbæk Formation are retained. However, in the Sæfaxi Elv and Marmorvigen areas, the overlying Wandel Valley Formation rests directly on Neoproterozoic carbonates of the Fyns Sø Formation, and the Kap Holbæk Formation is present only as the fill of cave systems developed within the latter (Smith et al., 1999, 2004).

The Kap Holbæk Formation is up to 150 m thick and is composed of fine- to coarse-grained sandstones with interbedded mudstones. Many of the sandstones are massive, but large-scale cross-bedding and wave-ripple lamination is present in places (Clemmensen and Jepsen, 1992). The *Skolithos* burrows have diameters up to 1 cm, and they are up to 0.5 m long. The formation belongs to a phase of siliciclastic shelf development, with associated turbidites in the deep-water trough to the north, present across the whole of North Greenland. This phase was initiated by an increase in subsidence and/or a sea-level rise, and most of the shelf sandstones are assigned to the Buen Formation, a lateral equivalent of the Kap Holbæk Formation. The Lower Cambrian Buen Formation is composed of tidally influenced and storm-derived sandstones that pass upward into lower-energy, deeper-water mudstones. Northward, toward the slope, the formation passes into more ubiquitous mudstone facies, and, at the shelf-slope break, black and green outer-shelf mudstones give way to the variegated mudstones of the slope. This slope was bypassed by sand transported from the shelf that supplied the sandy turbidites of the Polkoridorren Group, which were deposited in the deep-water trough (Higgins et al., 1991).

Kronprins Christian Land and Lambert Land: The Early Ordovician to Early Silurian Carbonate Shelf

Deposition on the shelf in central and western North Greenland was continuous from the Early Cambrian to the Early Ordovician (Higgins et al., 1991), but in eastern North Greenland, uplift produced a high that was not inundated until the major sea-level rise at the base of the Floian (late Early Ordovician) (Figs. 5A and 6). The onset of uplift may have begun as early as the late Early Cambrian (Higgins et al., 2001), but there is abundant stratigraphic and structural evidence for active uplift at the Cambrian–Ordovician boundary, when a wedge of quartz-rich siliciclastic sediments (the Perssuaq Gletscher Formation) prograded northward into the trough, feeding turbidite systems of the Vølvedal Group (Bryant and Smith, 1990, their Fig. 13); these sediments were probably derived from the unroofing of the Proterozoic Independence Fjord Group in the eastern part of the uplifted area (see Collinson et al., this volume). At the same time, a series of eustatic lowstands resulted in quartzarenites being worked westward across the shelf area to produce the Permin Land Formation and its distal equivalent, the Kap Coppinger Member (Bryant and Smith, 1990). Folding at the western limit of uplift deformed sediments of the *manitouensis* biozone (486 Ma; early Tremadocian–earliest Ordovician), and sediments that overlie this erosionally truncated fold belong to the same conodont biozone (Bryant and Smith, 1990). There is no evidence for active uplift after this time, and subsidence must have recommenced by the base of the Floian, when the Wandel Valley Formation began to accumulate across the uplifted area.

The late Early Ordovician sea-level rise that inundated the sub–Wandel Valley Formation unconformity surface (Figs. 6 and 8) marks the origin of a carbonate-dominated shelf that was maintained until the late Llandovery (Early Silurian).

				Kronprins Christian Land				North-East Greenland					
Syst.	Series	Stage	Sequence	Danmark Fjord	Centrumsø	Lambert Land	Nørreland	Dronning Louise Land	Målebjerg	Ella Ø	Albert Heim Bjerge	C.H. Ostenfeld Nunatak	
				foreland	parautochthon	allochthon	foreland	foreland	parautochthon	Franz Joseph allochthon			
Silurian	Prd		T III		Profilfjeldet Member (SH)								
	Lud		Tippencanoe II		Odins Fjord								
	Wen	Tel											
	Lly	Aer	Tippencanoe I	Børglum River	Turesø								
		Rhu											
		Hir			Børglum River								
Ordovician	U	Kat		Sjælland Fjelde	Sjælland Fjelde					Heimbjerge	Heimbjerge		
		San	Sauk IV	Alexandrine Bjerge Member	Alexandrine Bjerge Member	Amdrup Member				Narwale Sound	Narwhale Sound	Narwhale Sound	
	M	Dar		Amdrup Member	Amdrup Member				Målebjerg	Cape Weber	Cape Weber	Cape Weber	
		Dap		DFM	DFM	DFM	DFM	'Zebra Series 2'		Antiklinalbugt	Antiklinalbugt	Antiklinalbugt	
	L	Flo	Sauk III							Dolomite Point	Dolomite Point	Dolomite Point	
		Tre						'Zebra Series 1'	Slottet				
Cambrian	Fur	Toy	Sauk II							Hyolithus Creek	Hyolithus Creek	Hyolithus Creek	
		Bot								Ella Island	Ella Island		
	M	Atb	Sauk I							Bastion	Bastion	Bastion	
	L	Tom								Kløftelv	Kløftelv	Kløftelv	
		Nkd								Kap Holbæk			
Npt	Edi												

Ma: 420, 430, 440, 450, 460, 470, 480, 490, 500, 510, 520, 530, 540, 550

A

Figure 5 (*on this and previous page*). Correlation chart of Cambrian–Silurian units in North-East and eastern North Greenland (A), together with adjacent areas on the Laurentian margin (B). The left-hand columns follow the IUGS/ICS (International Union of Geological Science/International Commission on Stratigraphy) standard (Gradstein et al., 2004); boundaries of megasequences and supersequences are from Palmer (1981) and Golonka and Kiessling (2002). Data for the Greenland sections were compiled from Smith (1985, 1991), Tull (1988), Smith and Bjerreskov (1994), and Huselbee (1998). NE Spitsbergen data are from Fortey and Bruton (1973) and Fortey and Barnes (1977); Bjørnøya data are from Smith (2000); Newfoundland data were compiled from sources mentioned in the text. Edi—Ediacaran; L—Lower; M—Middle; U—Upper; Fur—Furongian; Lly—Llandovery; Wen—Wenlock; Lud—Ludlow; Prd—Pridoli; Nkd—Nemakit-Daldynian; Tom—Tommotian; Atb—Atdabanian; Bot—Botomian; Toy—Toyonian; Tre—Tremadocian; Flo—Floian; Dap—Dapingian; Dar—Darriwilian; San—Sandbian; Kat—Katian; Hir—Hirnantian; Rhu—Rhuddanian; Aer—Aeronian; Tel—Telychian; DFM—Danmarks Fjord Member; SH—Samuelsen Høj Formation; BCM—Barbace Cove Member.

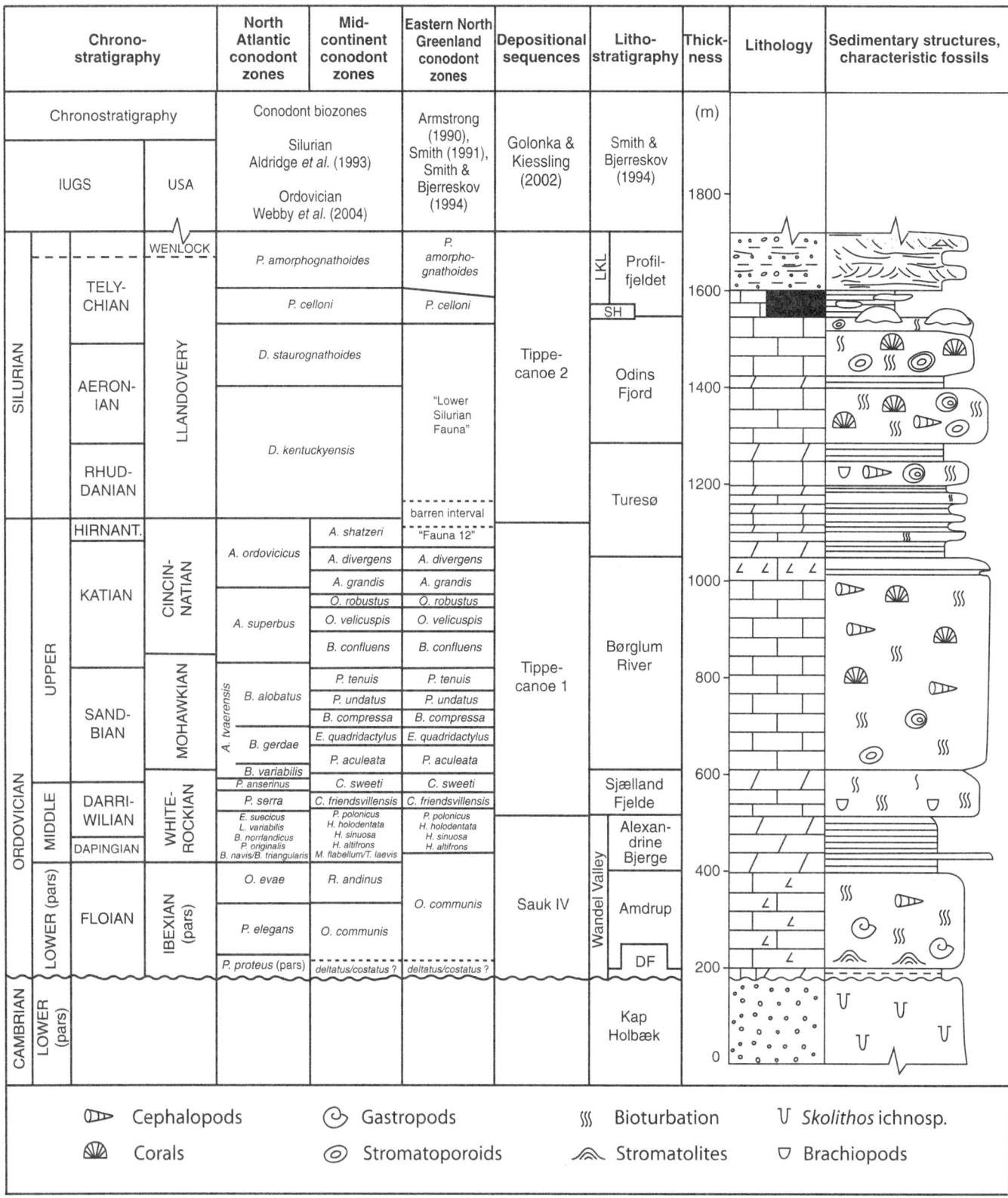

Figure 6. Composite lithological section for Kronprins Christian Land, eastern North Greenland, with correlation to international and Laurentian chronostratigraphic schemes, North Atlantic and Midcontinent conodont zonations, and depositional megasequences. DF—Danmarks Fjord Member; LKL—Lauge Koch Land Formation; SH—Samuelsen Høj Formation.

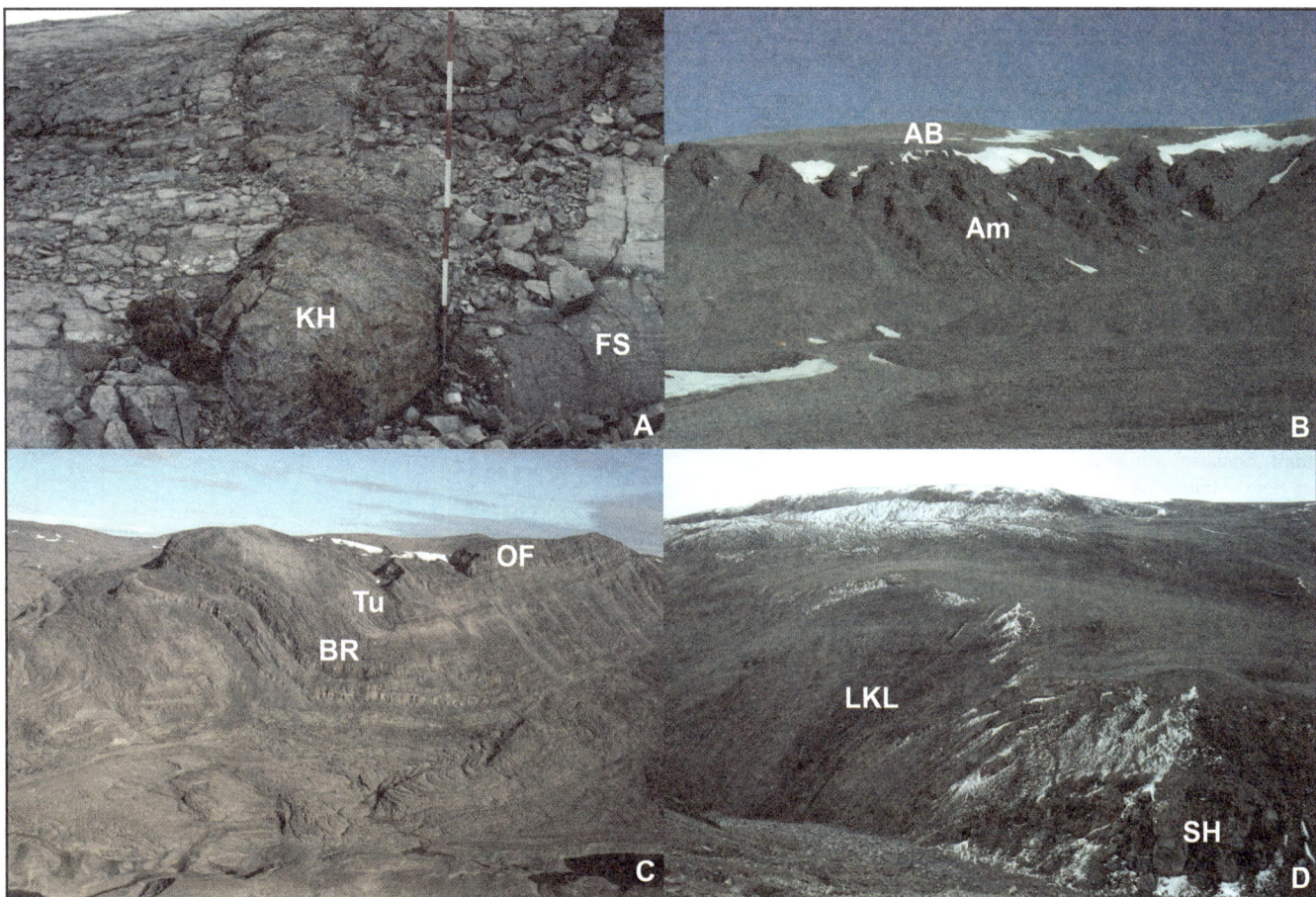

Figure 7. Cambrian-Ordovician stratigraphy of Kronprins Christian Land. (A) Lower Cambrian sandstones of the Kap Holbæk Formation (KH) filling phreatic tubes developed within the Proterozoic Fyns Sø Formation (FS) at Marmorvigen. Graduations on the staff are 20 cm. (B) Pale dolostones of the Middle Ordovician Alexandrine Bjerge Member (AB, Wandel Valley Formation) conformably overlying dark, highly strained burrow mottled limestones of the Lower Ordovician Amdrup Member (Am, Wandel Valley Formation) on Harefjeld, Vandredalen. There is ~400 m of relief in the photograph. (C) Folded Morris Bugt Group to the west of Centrumsø, Kronprins Christian Land. Cliff-forming Upper Ordovician limestones of the Børglum River Formation (BR) are conformably overlain by striped alternations of limestone and dolostones assigned to the Turesø Formation (Tu), which is in turn overlain by the Llandovery Odins Fjord Formation (OF). The Ordovician-Silurian boundary lies within the Turesø Formation. The summit of the hill is 500 m above the lake in the bottom right corner. (D) Black mudstones and interbedded carbonates of the Profilfjeldet Member (LKL—Lauge Koch Land Formation) draped over a late Llandovery reef of the Samuelsen Høj Formation (SH). The black mudstones are overlain by sandstone turbidites, also assigned to the Lauge Koch Land Formation, which occupy the middle distance. The reef is 50 m across and 20 m high.

The succession can be divided into three low-order sequences—late Early Ordovician to Middle Ordovician; Late Ordovician; and Early Silurian—that are traceable across the Franklinian Basin from Kronprins Christian Land to the Canadian Arctic Islands (Trettin et al., 1991). The Ordovician succession consists mainly of altering units of burrow mottled limestones and laminated dolostones, which may be distinguished by both their fossil evidence and chemostratigraphic characteristics, such as the content of manganese and the strontium/calcium ratio in whole-rock samples (Rasmussen and Smith, 1996).

The oldest sequence in Kronprins Christian Land, which rests on the unconformity surface, commences with peritidal dolostones and cyclic peritidal-subtidal carbonates of the Danmarks Fjord Member (Wandel Valley Formation), which represent the transgressive systems tract (TST). The base of the member is sandy at most localities, representing the reworking of terrestrial siliciclastic sediments on the unconformity surface, and it passes upward into current-laminated dolostones with evaporite collapse breccias (Peel and Smith, 1988; Smith et al., 2004). There is an abrupt and regionally synchronous shift to subtidal, burrow-mottled wackestones (typically 60%–80% $CaCO_3$) of the Amdrup Member, Wandel Valley Formation, that occurs at or close to the base of the *communis* biozone (478 Ma; lower Floian) (Smith, 1991; Smith and Bjerreskov, 1994). This monotonous lithofacies persists throughout the remainder of the Lower Ordovician with no expression of internal parasequences. An abrupt shift to recessive, peritidal dolostones of the Alexandrine Bjerge Member (Wandel Valley Formation) occurs in the uppermost part of the

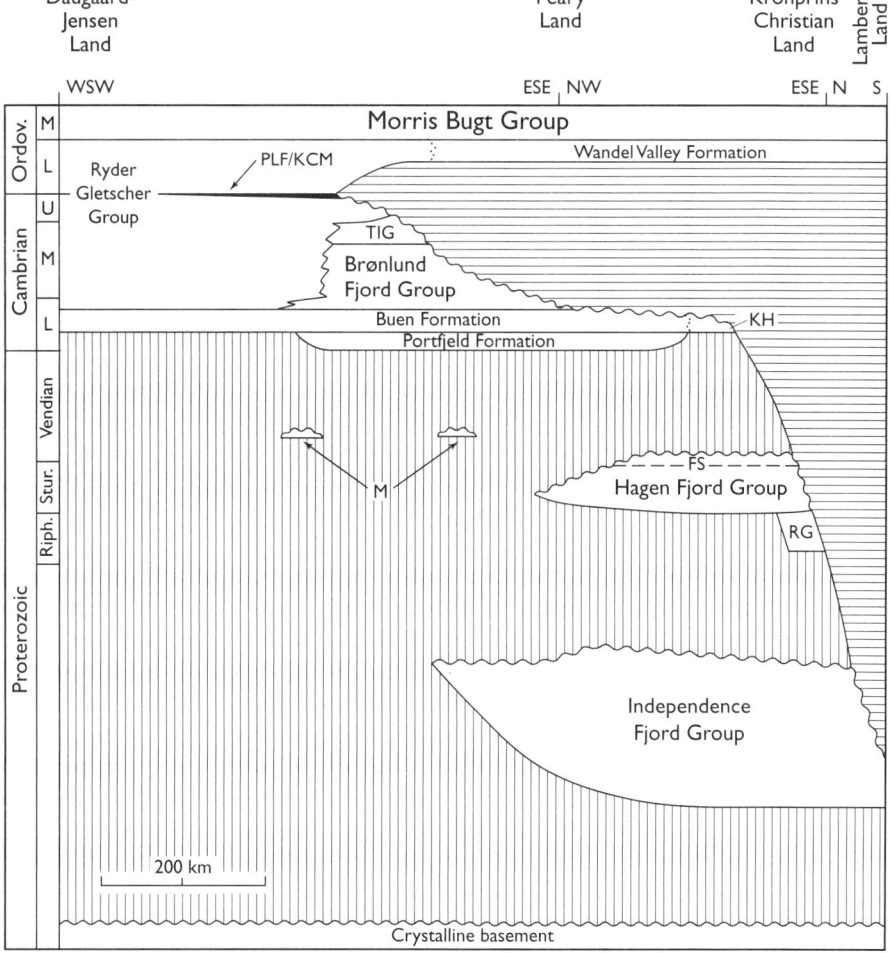

Figure 8. Proterozoic–Late Ordovician stratigraphic relationships on the platform area of North and North-East Greenland, showing the extent and magnitude of the sub–Wandel Valley unconformity (modified after Smith et al., 2004). Maximum uplift and associated erosion occur within the allochthon of Lambert Land, at the extreme right of the diagram, showing the increase in uplift eastward. FS—Fyns Sø Formation (Hagen Fjord Group); KH—Kap Holbæk Formation; M—glacial sediments of the Moræneso Formation; PLF/KCM—quartzarenite sandstone sheet assigned to the Permin Land Formation and the Kap Coppinger Member (Cass Fjord Formation); RG—Rivieradal Group; TIG—Tavsens Iskappe Group.

andinus biozone (473 Ma), and this marks the shift to a falling-stage systems tract (FSST), and consequent forced regression and basinward progradation of peritidal facies is observed throughout Laurentia (Ross et al., 1982). The Alexandrine Bjerge Member is ~200 m thick and consists entirely of pale-weathering laminated dolostones (60%–70% dolomite) with occasional stromatolites, flat pebble conglomerates, and shaly partings (Peel and Smith, 1988). The Ibexian-Whiterockian boundary occurs around 60 m above the base of the member, and the youngest parts extend into the *polonicus* conodont biozone (465 Ma).

In contrast to areas farther to the west, the older Ordovician megasequence in Kronprins Christian Land has a lower-order sequence developed in the uppermost part, a distinctive, 100 m thick, transgressive-regressive couplet assigned to the Sjælland Fjelde Formation. The unit spans the interval from the *polonicus* biozone to the Sandbian (early Late Ordovician; 457 Ma), and the lower part consists of burrow-mottled limestone facies (70%–90% $CaCO_3$) similar to the Amdrup Member, while the upper part is lithologically similar to the peritidal sediments of the Alexandrine Bjerge Formation (Ineson et al., 1986; Smith and Bjerreskov, 1994).

The base of the second Ordovician megasequence in the easternmost part of the Franklinian Basin is marked by a major flooding surface and the abrupt switch to subtidal carbonates of the Børglum River Formation (Fig. 6), corresponding to the Black Riveran transgression of mainland North America. The formation is composed of lithologically monotonous, burrow-mottled lime mudstones and wackestones (65%–90% $CaCO_3$) that are highly fossiliferous in the upper part. In Kronprins Christian Land, there is little expression of higher-order sequences, but these are evident farther to the east in Peary Land (Smith et al., 1989, 2004). Commencing with a prominent dolostone marker bed that occurs 20 m below the top of the formation in Kronprins Christian Land, higher-order parasequences become prominently expressed across the boundary into the overlying Turesø Formation. The Børglum River–Turesø Formation boundary lies within the upper Katian (Upper Ordovician), and the Ordovician-Silurian boundary is present ~50–80 m above the base, although the position of the latter is rather poorly constrained at present (Armstrong, 1990; Smith and Bjerreskov, 1994; Smith et al., 2004). However, beds containing pentamerid brachiopod coquinas do appear close to the first occurrence of Silurian conodonts (species of *Ozarkodina*) in southern Kronprins Christian Land, giving rough macroscopic evidence for the position of the Ordovician-Silurian boundary.

The Turesø Formation in Kronprins Christian Land (up to 320 m) is a distinctively banded unit containing parasequences composed of subtidal, burrow-mottled limestones (40%–90% $CaCO_3$) that are capped by peritidal dolostones (60%–75% dolomite). However, a 90 m interval in the lower to middle part of the unit, probably within the Ordovician part of the formation or straddling the boundary, lacks clear cyclicity and is dominated by burrow-mottled limestones (Smith et al., 2004). The Turesø Formation is bounded above by a flooding surface across which the well-developed parasequences give way to relatively undifferentiated limestones of the Odins Fjord Formation. In Kronprins Christian Land, the boundary is marked by a change from a dark-gray dolostone-dominated succession to golden brown–weathering, highly fossiliferous limestones (30%–90% $CaCO_3$) of the Odins Fjord Formation (Smith et al., 2004), which is of probable Aeronian (mid-Llandovery) age (436–439 Ma) (Hurst, 1984; Armstrong, 1990). Above the lower 100 m, there is a distinctive dolostone, up to 35 m thick, containing probable pseudomorphed evaporites, which is most likely a correlative of a peritidal division, the Melville Land Member, recognized to the northwest in Peary Land (Hurst, 1984; Smith et al., 2004). The remainder of the formation, which is around 220 m thick in Kronprins Christian Land, has similar lithology to the lower 100 m, with no clear expression of parasequences. The Odins Fjord Formation is capped across its outcrop area by reefs of the Samuelsen Høj Formation. The timing of the initiation of reef growth is well-constrained by conodonts to the *celloni* biozone (mid-Telychian, upper Llandovery), and it is probable that reef growth was restricted to this zone (430–435 Ma) (Armstrong, 1990).

In Kronprins Christian Land, the reef bodies of the Samuelsen Høj Formation are up to 300 m in maximum thickness, and they have diameters of up to 5 km, though most are significantly smaller than this (Hurst, 1984). The pale limestones of the reef cores are composed of massive lime mudstones, often with clotted textures suggestive of a microbialitic origin; stromatactis-like cavities are also present. The reef flanks consist of bioclastic rudstones with a wackestone, packstone, or grainstone matrix.

Within the wider context of the Franklinian Basin, Higgins et al. (1991) attributed the lower two megasequences seen in Kronprins Christian Land to a phase of aggradational platform growth with slow, continuous subsidence. This extended across North Greenland into the Canadian part of the basin, and units may be traced, with name changes but with little change in facies, for over 2000 km laterally across the platform. The aggradational platform growth was matched in the deep-water trough by sediment starvation and the deposition of black and green mudstones and radiolarian cherts, with some turbidites and base-of-slope conglomerates, all of which are assigned to the Lower Ordovician–lower Silurian Amundsen Land Group (Higgins et al., 2001). Although the deposition of the carbonate conglomerates was linked by Higgins et al. (2001) to the uplift associated with the sub–Wandel Valley unconformity, this must represent an oversimplification, since deposition of the conglomerates continued from the early Ibexian through until the mid-Whiterockian, when subsidence and sediment accumulation resumed in eastern North Greenland. It is more likely that the conglomerates, which are assigned to the Kap Mjølner Formation (Higgins et al., 1991), record a composite tectonic-eustatic signal analogous to the Cow Head Group of western Newfoundland (James and Stevens, 1986). Deposition of the Amundsen Land Group continued until the early Silurian, when a dramatic depositional change occurred in the deep-water trough.

Kronprins Christian Land: Silurian Foundering and Turbidite Deposition

The final phase in the development of the eastern Franklinian Basin occurred through Llandovery time into the Wenlock, when the slow depositional rates of the Amundsen Land Group were replaced by the deposition of a major longitudinal, sand-rich turbidite system (Hurst and Surlyk, 1982; Higgins et al., 1991). The timing of this major depositional switch is constrained by graptolites, and it appears to have occurred diachronously E-W across the deep-water trough—the first evidence of change is recorded by the deposition of fine-grained turbidites assigned to the Merqujôq Formation in the eastern part of the area close to the Ordovician-Silurian boundary (Hurst and Surlyk, 1982). However, a major influx of up to 2.8 km of sandy turbidites, also assigned to the Merqujôq Formation, occurred in the *turriculatus–spiralis* graptolite biozones and developed synchronously across the basin (mid- to late Telychian; late Llandovery; 435–430 Ma) (Larsen and Escher, 1985) at the available level of biostratigraphic resolution. The turbidite system is highly elongate, parallel to the shelf-trough boundary, and the sandstone turbidites record dominant westward flow (Hurst and Surlyk, 1982; Higgins et al., 1991); with its easterly source, the turbidite system almost certainly records the onset of Scandian deformation and uplift at 435 Ma.

The sandy turbidites of the Merqujôq Formation, derived from erosion of the rising Caledonian mountains, are coeval with the shelf carbonates of the Odins Fjord and Samuelsen Høj Formations; this timing points to the operation of an efficient bypass mechanism during Llandovery time. However, the onset of reef development in the Samuelsen Høj Formation on the shelf during the late Telychian probably reflects increased subsidence and relative sea-level rise associated with loading of the turbidite trough to the north and the Scandian thrust sheets to the east (Higgins et al., 1991). This process accelerated in latest Llandovery time, and the reefs became progressively drowned across the region. In Kronprins Christian Land, the Samuelsen Høj reefs are abruptly overlain by black mudstones and bituminous carbonates of the Profilfjeldet Member (Lauge Koch Land Formation). The youngest faunas recorded from the reefs are conodonts of the *celloni* biozone (late but not latest Llandovery) (Armstrong, 1990). Limited graptolite evidence suggests that the drowning of the reefs youngs from east to west across the platform, in support of a progressive loading model; the youngest sediments to overlie the reef are of the *griestoniensis* biozone (mid-Telychian; 432 Ma) in eastern Kronprins Christian Land, *spiralis* biozone (late Telychian; 429–431 Ma) in western Kronprins Christian Land, and *sakmaricus-laqueus* biozone (latest Llandovery; 429 Ma) in Valdemar Glückstadt Land and Peary Land (Hurst and Surlyk, 1982; Bjerreskov, 1989). This is supported by conodont faunas from the black limestones of the reefs in Kronprins Christian Land, where the *celloni– amorphognathoides* biozone boundary occurs a short distance above the reefs (Armstrong, 1990, p. 16). There may also be diachroneity toward the south, since Lane (1972) recorded an anomalously young, mid-Wenlock, graptolite fauna 50 m above the base of the Profilfjeldet Member, near Centrumsø.

In the parautochthon of eastern Kronprins Christian Land, the black mudstones and carbonates attain a thickness of around 50 m, and they are overlain by up to 150 m of $T_{a-c,e}$ and T_{b-e} sandstone turbidites that are truncated by thrusts within the parautochthon (Smith et al., 2004). Farther to the northwest, in Peary Land, black mudstones of the Thors Fjord Member (Lauge Koch Land Formation), equivalent to the Profilfjeldet Member, are also overlain by sandstone turbidites (Hurst and Surlyk, 1982; Higgins et al., 1991). The initiation of turbidite deposition on what was formerly a carbonate platform represents the rapid westward progradation of a submarine-fan system following foundering and the backstepping of the southern margin of the trough (Higgins et al., 1991). A system of stacked fan valleys in the east passes westward to a braided mid-fan environment and, west of Peary Land, into outer-fan, fan-fringe and basin plain environments (Hurst and Surlyk, 1982; Surlyk and Hurst, 1984; Higgins et al., 1991).

In eastern and central North Greenland, the final preserved phase of sedimentation in the Franklinian Basin is a major phase of conglomerate deposition, represented by the Nordkronen Formation. The formation is up to 700 m thick, though it is more usually truncated by the modern erosion surface at lesser values. Lithologically, it is composed of conglomerates, pebbly sandstones, and sandstones. The conglomerates are nongraded, weakly graded, or inverse graded and have a subhorizontal or weakly imbricated clast fabric. The clasts are of medium pebble to cobble grade and are dominated lithologically by black and green cherts, together with some quartzite and basement clasts (Hurst and Surlyk, 1982). The conglomerates represent deposition from high-density turbidity currents traveling westward from the Caledonian uplands (Higgins et al., 1991). Some of the cherts contain Ordovician radiolaria (Hurst and Surlyk, 1982), and Higgins et al. (1991) speculated that the most likely sources were the chert-rich basinal mudstones of the Vølvedal and Amundsen Land Groups, since no other units contain sufficient volumes of chert. An additional possibility, however, is that the incorporation of Ordovician platform carbonates within thrust sheets during the Wenlock, and their subsequent erosion, would have yielded very significant quantities of black cherts, which were then reworked in the shoreface before being transported westward in turbidity currents.

Lambert Land

Until recently, it was not known that Lower Paleozoic rocks cropped out in the area between southern Kronprins Christian Land and northern Dronning Louise Land, but survey mapping has revealed that a small outlier of Ordovician carbonates occurs in westernmost Lambert Land (79°20′N). Unlike any other locality in the Franklinian Basin, they rest unconformably on the Independence Fjord Group (Smith et al., 2004), reflecting continued unroofing to the east below the sub–Wandel Valley unconformity. A lower member, which consists of 25 m of strained peritidal dolostones, wavy and flaser bedding, and probable pseudomorphed evaporites, has yielded highly fragmentary Floian (late Early Ordovician) coniform conodont elements, and it was assigned to the Danmarks Fjord Member (Wandel Valley Formation) by Smith et al. (2004). The lower peritidal interval is conformably overlain by dark, highly strained, burrow-mottled and wavy laminated carbonates (80%–90% $CaCO_3$), dolomitized to varying degrees, which are assigned to the Amdrup Member of the Wandel Valley Formation (Figs. 5A and 6). The lithostratigraphic correlations are supported by chemostratigraphic evidence (Rasmussen and Smith, 1996).

The Wandel Valley Formation in Lambert Land occurs within a thrust-bounded horse that lies directly beneath the Vandredalen thrust, which transported the Neoproterozoic sediments of the Rivieradal Group westward (see Sønderholm et al., this volume). The sediments of the Independence Fjord Group that the Wandel Valley Formation unconformably overlies are part of a large-scale thrust stack that is composed of the Independence Fjord Group interleaved with Paleoproterozoic crystalline complexes and that lies tectonically beneath the Vandredalen thrust sheet. The Wandel Valley Formation is thus allochthonous, and it is the only part of the platform of the Franklinian Basin that is truly allochthonous within the Caledonides. Displacement of this Lower Paleozoic succession is constrained by estimates of ~17.5 km of shortening within the parautochthon to the north (Higgins et al., 2004b) and shortening of 17–32 km on the Vandredalen thrust and within the overlying thrust sheet (Higgins et al., 2001). The depositional location of the Lambert Land Lower Paleozoic succession was thus 35–50 km farther to the east, a relatively insignificant distance within the context of the size of the Franklinian platform.

STRATIGRAPHY OF THE EAST GREENLAND CALEDONIDES: NØRRELAND AND DRONNING LOUISE LAND

The least known Lower Paleozoic successions within the East Greenland Caledonides lie within the very highly strained tectonic window in Nørreland (~78°N) and the parautochthon in Dronning Louise Land (~78°N; Figs. 2 and 5).

In Nørreland, an anticlinal culmination exposes a 20-m-long, 10-km-wide window of quartzitic rocks beneath a thrust sheet of basement orthogneisses and amphibolites (Higgins et al., 2001).

The quartzites are lithologically comparable with the Proterozoic Independence Fjord Group, and, characteristically for that unit, they are cut by a network of basic intrusions that can be correlated with the Mesoproterozoic Midsommersø Dolerites. While major mylonitic fabrics are not present, a blue-gray platy metacarbonate crops out in the contact zone on the western side of the culmination. Despite the high strain levels, samples of the platy carbonate have yielded Ordovician conodonts that have conodont alteration index (CAI) values of 5–6, corresponding to an approximate overburden of 10–12.5 km (Rasmussen and Smith, 2001).

Dronning Louise Land is divided by a N-S–trending imbricate zone into a foreland area in the west and parautochthonous to allochthonous Paleoproterozoic gneiss complexes in the east. The foreland succession was first documented by members of the British North Greenland expedition (Peacock, 1956, 1958), who recognized a succession of sandstones intruded by dolerites (the Trekant series) unconformably overlain by the Zebra series, a succession of quartzites, mudstones, iron-rich sandstones, and carbonates. The Zebra series occurs in the foreland of the orogen, but Peacock (1956, 1958) also described a succession of metacarbonates and sandstones from within the deformed area that was assigned to the newly erected Britannia Sø Group and correlated with the Zebra series of the foreland. In contrast, Haller (1971) correlated the Britannia Sø Group with the older Trekant series. The area has been visited few times since the original description, and the Zebra series is generally considered to be of late Precambrian age (e.g., Henriksen and Higgins, 1976), largely in the absence of positive evidence.

The most detailed work carried out to date is that of Friderichsen et al. (1990) and Strachan et al. (1994), who demonstrated that the Zebra series must be no older than Cambrian by recovering *Skolithos* from the upper part of the sandstone sequence (Strachan et al., 1994, their Fig. 2). The documented succession consists of 3–10 m of pebbly conglomerates overlain by 10–15 m of purple-white striped quartzites that pass up into yellow to white, medium- to coarse-grained quartzites (5–30 m) and interbedded sandstones, siltstones, and mudstones (10 m) (Friderichsen et al., 1990). The top of the section is marked by 10 m of gray-black carbonates. The fining-upward trend from pebbly conglomerates to interbedded sandstone and mudstones, capped by carbonates, is compellingly similar to the Slottet and Målebjerg Formations of the foreland windows farther to the south (see following), although the thicknesses are different. There is little doubt that these units are both coeval and record the feather edge of deposition along this sector of the Laurentian margin.

In the lower structural levels of the imbricate zone, Friderichsen et al. (1990) recorded a similar succession to that in the foreland; it is made up of 2 m of pebbly sandstones overlain by 45 m of rusty, tabular cross-bedded sandstones and then 50 m of heterogeneous siltstones, mudstones, sandstones, and limestones. The base of the heterogeneous unit contains *Cruziana*, indicating the presence of Lower Cambrian or younger siliciclastic sediments in at least the lower thrust sheets, where they are again overlain by carbonates, with a minimum of 120 m preserved (Friderichsen et al., 1990).

Structurally higher within the imbricate zone, Strachan et al. (1994) confirmed that the Britannia Sø Group of Peacock (1956, 1958) is correlative with both the Trekant and Zebra series and therefore recommended abandonment of the term.

STRATIGRAPHY OF THE EAST GREENLAND CALEDONIDES: NUNATAK REGION, 71°50′N–74°30′N

Haller (1971, his Fig. 48) speculated that a source of quartzarenite clasts containing *Skolithos* must underlie the margins of the Inland Ice in North-East Greenland on the basis of the distribution of *Skolithos*-bearing erratic boulders, but it was not until the late 1990s that the source was discovered at outcrop in a series of tectonic windows (Fig. 4). The Slottet Formation (Smith et al., 2004) is present in the parautochthon within the Målebjerg and Eleonore Sø windows, where it rests unconformably on Marinoan tillites (see Sønderholm et al., this volume) and gneisses of probable Paleoproterozoic age (Figs. 5A, 9, and 10A). The "Slottet Quartzite" was first described by Katz (1952) but was originally considered to be part of the Eleonore Bay Supergroup.

The Slottet Formation ranges from 143 to 350 m in thickness and is composed of cross-bedded or structureless fine to very coarse quartzarenites. Above 79 m in the type section at Målebjerg, there is an abrupt change to a rust-weathering alternation of quartzarenites and sandy shales, and the unit is capped by a 14-m-thick massive arenite bed (Fig. 9). *Skolithos* burrows (Fig. 10C), measuring several tens of centimeters in length, first appear at the base of the rust-weathering unit, at 79 m, but there are more equivocal examples as low as 45 m.

The Slottet Formation is overlain by the Målebjerg Formation (Figs. 9, 10B, and 10D), a thin highly tectonized carbonate unit made up of sandy dolostones overlain by alternating dark-gray–weathering medium-gray dolostones and pale-gray–weathering pale-gray dolostones. In low strain areas, these are observed to be current-laminated and burrow-mottled, respectively, and they occur as typical, Laurentian shallowing-upward parasequences that are a few meters thick (Smith et al., 2004). The maximum recorded thickness of the unit is 45 m, but it is everywhere truncated by the Niggli Spids thrust (Fig. 10A). Although the precise age remains uncertain, the stratigraphic and tectonic context indicates a Cambrian–Early Ordovician age (Smith et al., 2004); the overall stratigraphic context is very similar to that seen in Kronprins Christian Land and the carbonates may be equivalent to the Danmarks Fjord Member of the Wandel Valley Formation.

The presence of the Slottet Formation within the parautochthon of the Målebjerg and Eleonore Sø windows, in conjunction with the estimates of shortening within the orogen in this region (Higgins et al., 2004a; see previous), provides constraints on the minimum width of the Early Cambrian shelf. Prior to Scandian shortening, the relatively attenuated, *Skolithos*-bearing Slottet Formation lay 200–400 km inboard of the thick, coeval sandstones of the Kløftelv and Bastion Formations that lie within the Franz Joseph allochthon (see following). Furthermore, the Slottet Formation must continue farther westward underneath the margin

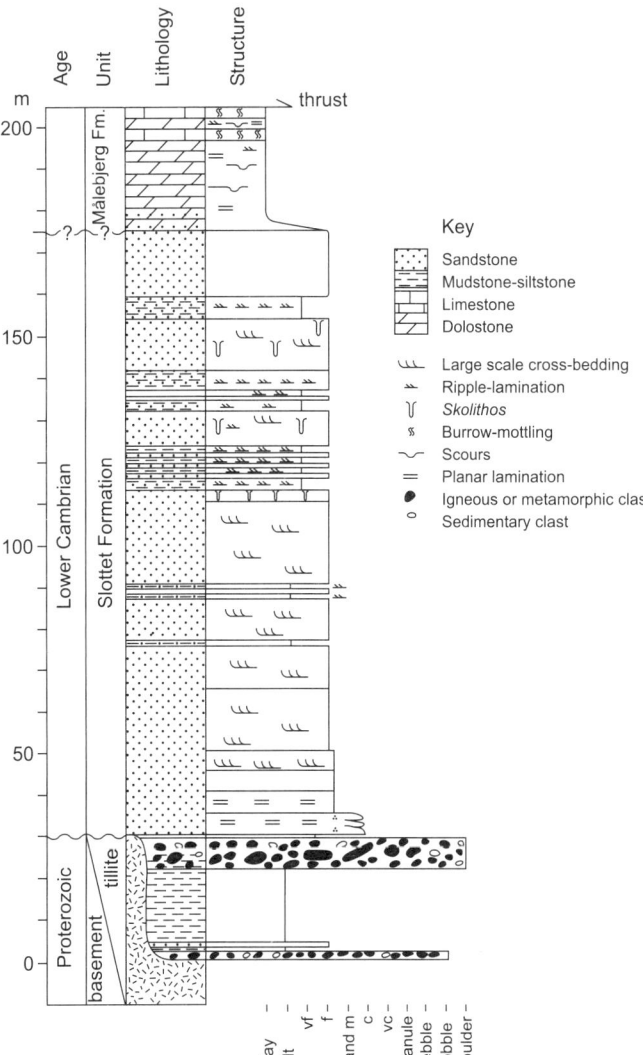

Figure 9. Composite log of the Slottet and Målebjerg Formations at the type locality in the Målebjerg window (after Smith et al., 2004) (see Figs. 2 and 4). The clastic succession beneath the Slottet Formation, with two diamictites, occupies an erosional depression in the gneissic basement and is interpreted as a Marinoan tillite (cf. Sønderholm et al., this volume). The age of the Slottet Formation is constrained by the presence of the ichnogenus *Skolithos*, but that of the Målebjerg Formation is less well constrained (see text).

of the Inland Ice, as evidenced by the distribution of *Skolithos*-bearing erratics (Haller, 1971, his Fig. 48), giving a minimum shelf width of 300–500 km over which the deposition of Lower Cambrian quartzarenite sheets took place.

The most southerly of the Caledonian tectonic windows, the Gåseland window, lies in western Gåseland and around the head of Scoresby Sund (70°10′N–70°40′N). Although there is no equivalent of the Slottet Formation, carbonates overlie well-developed glacial diamictites. The carbonates are highly deformed, strongly laminated cream-colored marbles, and they lie immediately beneath a major thrust, but it is possible that they are of early Paleozoic age and represent part of the Cambrian-Ordovician foreland succession (Phillips et al., 1973; Higgins et al., 2001) and are thus correlatives of the Målebjerg Formation.

STRATIGRAPHY OF THE EAST GREENLAND CALEDONIDES: FJORD REGION OF SOUTHERN NORTH-EAST GREENLAND, 71°36′N–74°17′N

By far the thickest of the Lower Paleozoic successions within the East Greenland Caledonides is that of the Franz Joseph allochthon—the classical Cambrian-Ordovician succession of the fjord region of North-East Greenland. This succession (Figs. 5A, 11, and 12) is significantly thicker than most coeval Laurentian margin intervals and must reflect a distinctively different subsidence history. The earliest sediments are of Early Cambrian age, and the youngest sediments preserved are of late Darriwilian or early Sandbian age (minimum 460 Ma); all are assigned to the Kong Oscar Fjord Group (Smith et al., 2004). The oldest unit within the group is the Kløftelv Formation, a 70–75-m-thick quartz-rich sandstone that contains some bedding-parallel burrows but which lacks *Skolithos*. Little sedimentological work has been carried out to date, but the extensively planar tabular cross-bedded unit is generally attributed to a tidally influenced environment (Swett and Smit, 1972). Tirsgaard (*in* Pickerill and Peel, 1990) suggested that three sequences were identifiable within the unit. The sandstones of the Kløftelv Formation are overlain by heterolithic sandstones-siltstones with interbedded thin ripple-laminated sandstones of the Bastion Formation (Tirsgaard *in* Pickerill and Peel, 1990). Three sequences were identified in the latter unit by Tirsgaard. The lowest is 20 m thick and has a conglomerate that contains glauconitic and phosphatic nodules at the base. A second sequence, ~30 m thick, begins at around 20 m and has a similar conglomerate at the base. The lower two sequences are interpreted as having been deposited beneath fair-weather wave base, where the sandstones represent storm events. The uppermost sequence (87–99 m) has the Lower Shell Limestone at the base, a 50 cm glauconitic limestone that is made up almost entirely of shelf fragments but that also contains phosphatic pebbles (Henriksen and Higgins, 1976). The shelly material makes up the oldest body fossils within the succession. The overlying sediments mark a shift to less siliciclastic, more carbonate-dominated sedimentation characterized by an alternation of shales with thin beds of limestone and mudstone. Tirsgaard (*in* Pickerill and Peel, 1990) interpreted the depositional environment as being well below wave base, where most deposition occurred out of suspension. Pickerill and Peel (1990) recorded a diverse ichnofauna from the Lower Bastion Formation, beneath the Lower Shell Bed, and a sparser assemblage from the upper sequence above the shell bed.

Above the Bastion Formation, the Kong Oscar Group is carbonate-dominated, but the succeeding three units have been little studied and are poorly understood. The Ella Island Formation is 80–140 m thick (Cowie and Adams, 1957; Hambrey et al., 1989) and is composed of thin, interbedded laminated

Figure 10. (A) Lower Cambrian quartzites of the Slottet Formation (SF) on Målebjerg unconformably overlying Paleoproterozoic gneisses. The thin pale stripe above the Slottet Formation is composed of carbonates of the Målebjerg Formation. The latter are truncated by the Niggli Spids thrust, which transports the Krummedal sequence. There is ~1500 m of relief in the photograph. (B) The Slottet and Målebjerg Formations in their type section on the north side of the ice-dammed lake, north of Målebjerg. The unconformity at the base of the Slottet Formation is on the left, the base of the Målebjerg Formation is at the left edge of the scree-covered interval on the right of the photograph, and the Niggli Spids thrust is present to the right of the scree. The Slottet Formation is 143 m thick at this, the type, locality. (C) Strained *Skolithos* burrows in quartzite from the Slottet Formation in its type section. Distortion of the upper part of the burrows, along bedding planes, is a consequence of westward displacement of the overriding Caledonian thrust sheets. (D) Strained, parallel-laminated dolostones and burrow-mottled limestones of the Målebjerg Formation in the type section. The lens cap in C and D is 65 mm in diameter.

siltstones and mudstones in the lower part of the unit overlain by thin-bedded lime mudstones in 1–3-m-thick sequences interbedded with 30–70 cm grainstones and flake conglomerates (Stouge et al., 2001). At the northernmost exposures, on C.H. Ostenfeld Nunatak (74°20′N), slump structures are common and ripple and parallel lamination is pervasive. The *Salterella* and trilobite-dominated macrofauna (Cowie and Adams, 1957) indicates a *Bonnia–Olenellus* zone (late Early Cambrian–early Middle Cambrian) age.

The overlying Hyolithus Creek Formation is dominated by dolostone rather than limestone and is 145–215 m thick (Cowie and Adams, 1957; Stouge et al., 2001). The dark-gray to black dolostones contain abundant intraformational conglomerates (Cowie and Adams, 1957), and Stouge et al. (2001) recorded mound and laminated facies, with vestigial limestones, close to the top of the unit on Ella Ø. These limestones contain *Salterella*, which is found sporadically throughout the unit (Hambrey et al., 1989; Stouge et al., 2001), indicating that the Hyolithus Creek Formation is also restricted to the *Bonnia–Olenellus* zone.

The Dolomite Point Formation (260–421 m; Cowie and Adams, 1957; Stouge et al., 2001) marks a change to pale-weathering microcrystalline dolostones and interbedded muddy dolostones and green-gray shales (Stouge et al., 2001). The upper part of the formation on Ella Ø consists of buff-weathering, pale-gray, current-laminated dolostones with parallel lamination, some ripple lamination, and scours. Thin siliciclastic mudstones, up to 0.5 cm thick, mark some bed tops. Toward the boundary with the overlying Antiklinalbugt Formation, shallowing-upward cycles containing burrow-mottled dolostones, current-laminated dolostones, and bacterial mats become evident. Although stromatolites are present, the macrofauna is very scarce. Nevertheless, the unit must span all of the Middle and Upper Cambrian, and

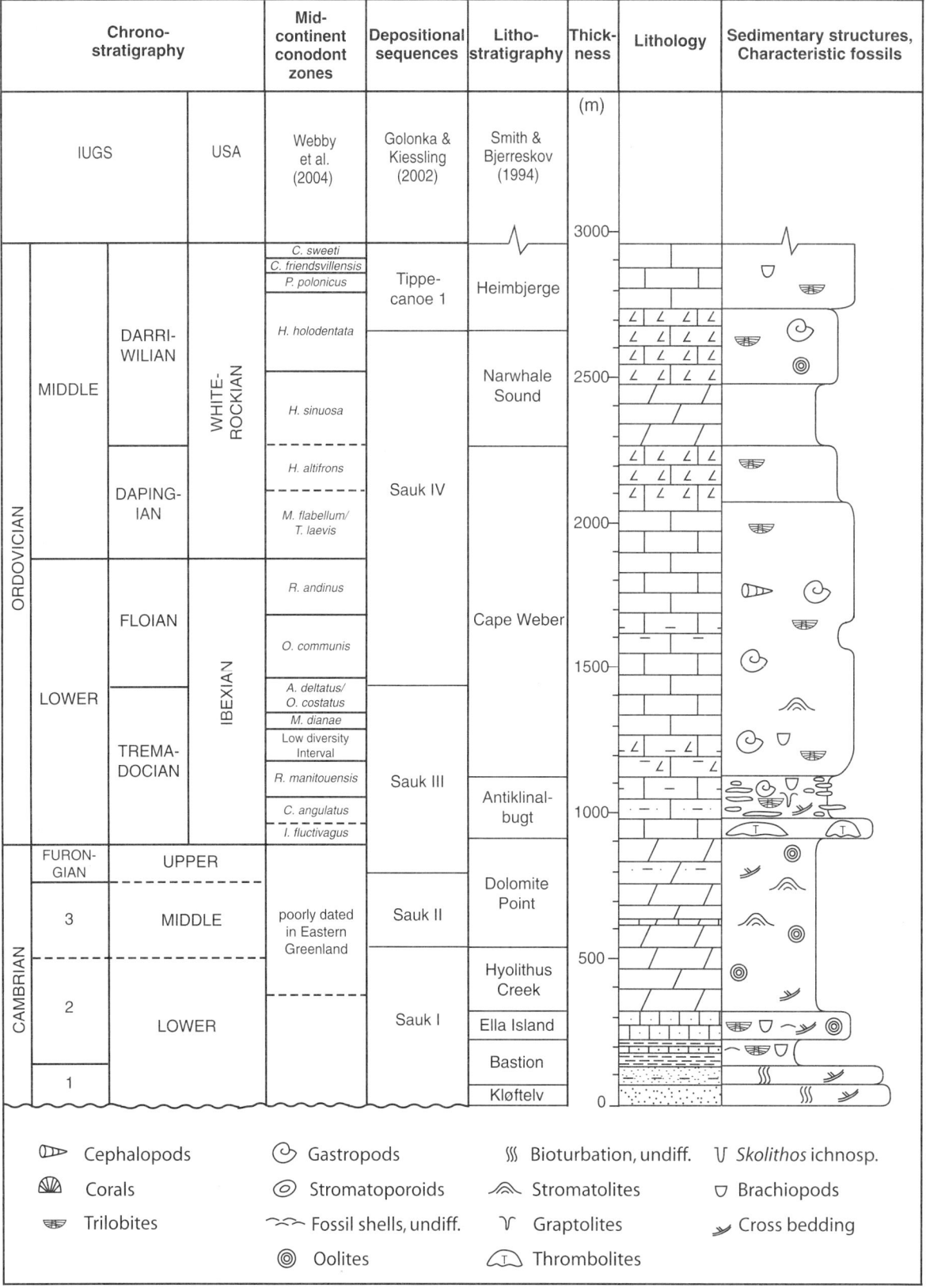

Figure 11. Composite lithologic section of the Kong Oscar Fjord Group of the Franz Joseph allochthon and correlation to international and Laurentian chronostratigraphic schemes, North Atlantic and Midcontinent conodont zonations, and depositional megasequences.

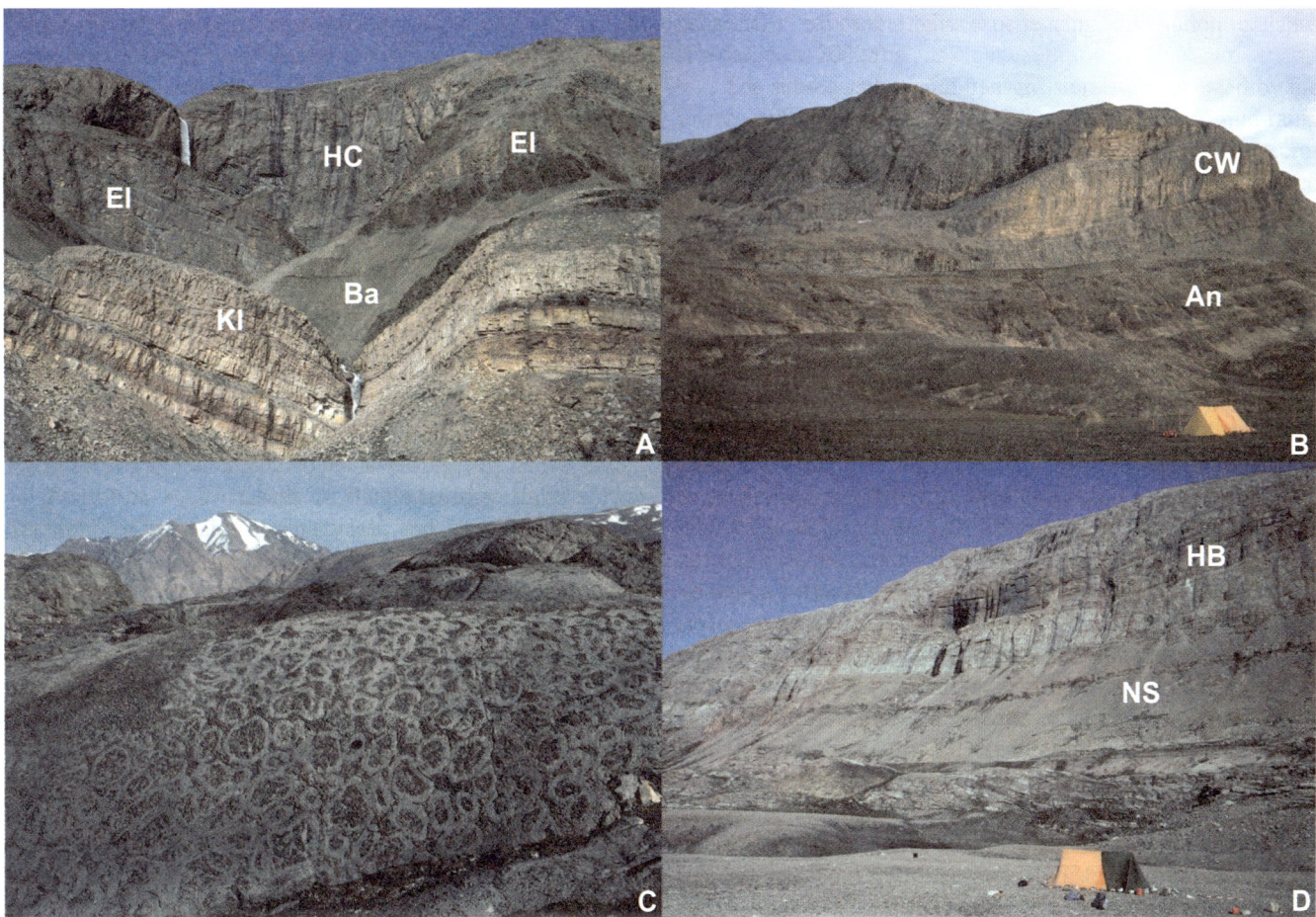

Figure 12. Cambrian-Ordovician stratigraphy of the lower Kong Oscar Fjord Group of the Franz Joseph allochthon. (A) The Kløftelv (Kl), Bastion (Ba), Ella Island (EI), and Hyolithus Creek (HC) Formations on Albert Heim Bjerge. The profile is 600 m high. (B) Early Ordovician carbonates of the Kong Oscar Fjord Group on Ella Ø. The Antiklinalbugt Formation (An) overlies the Dolomite Point Formation, upon which the tent is pitched. The Antiklinalbugt Formation is conformably overlain by the Cape Weber Formation (CW), and the formation boundary is located at the base of the massive interval in the cliff. The highest summit is 500 m above the tent in the foreground. (C) The upper surface of a composite thrombolite-stromatolite reef in the Cape Weber Formation on Albert Heim Bjerge. The lens cap in the center of the photograph is 65 mm in diameter, and the thrombolites are typically around 25–30 cm in diameter. (D) Massive cliff-forming limestones of the Heimbjerge Formation (HB) overlying recessive dolostones of the Narwhale Sound Formation (NS) on Albert Heim Bjerge. The cliff is 450 m high from tent to skyline.

Huselbee (1998) recorded a number of latest Cambrian conodonts from the top of the formation on Ella Ø, Albert Heim Bjerge, and C.H. Ostenfeld Nunatak. On Albert Heim Bjerge (74°10′N), these included *Clavohamulus hintzei* at 17 m below the top of the unit, a species restricted to the upper part of the uppermost Cambrian *Cordylodus intermedius* conodont biozone (Sweet and Tolbert, 1997).

The Cambrian-Ordovician boundary interval in the Kong Oscar Fjord Group has poor yields, but Huselbee (1998) demonstrated that the base of the *Cordylodus angulatus* biozone lies no higher than 39 m above the base of the Antiklinalbugt Formation on Albert Heim Bjerge, indicating that the base of the Ordovician lies beneath this but almost certainly within the Antiklinalbugt Formation. The macrofauna, dominated by gastropods, brachiopods, and trilobites, supports an early Tremadocian (Early Ordovician) age for all except the base of the unit (Stouge et al., 2001). Several authors (Miller and Kurtz, 1979; Smith and Bjerreskov, 1994; Huselbee, 1998) have postulated a degree of diachronism for the lower boundary, but there is no evidence that this is anything other than an artifact due to poor conodont recovery.

The base of the Antiklinalbugt Formation (210–270 m; Cowie and Adams, 1957) is readily identified by the abrupt change to pale-gray–weathering limestones. On Albert Heim Bjerge, the lower part of the formation is dominated by wavy bedded limestones, with occasional intraformational conglomerates. Above ~70 m, the character changes, and the remainder of the formation is made up of rubbly limestones with green mud partings and/or structureless green calcareous mudstones alternating with burrow-mottled limestones. Occasional intraformational conglomerates and columnar stromatolites are present throughout. The general pattern is similar on Ella Ø, with the exception of a major stromatolite-thrombolite reef complex that lies 25–40 m above the base of the formation. The reef consists of "organ pipes" up to 70 cm in diameter with laminated stromatolitic walls and thrombolitic cores; in places, the columns pass laterally into bacterial mats and low domal stromatolites.

The Cape Weber Formation (1000–1200 m) is a thick, superficially monotonous unit. The lower part of the unit is composed of meter-thick carbonate beds, with low stromatolitic and thrombolitic mounds, overlain by an interval in which cyclicity is more clearly developed with repeated bioturbated limestones, thrombolitic mounds, and laminated dolomitic limestones (Stouge et al., 2002). The middle part of the formation (Unit C of Stouge et al., 2002) is made up of 330 m of thick-bedded peloidal and bioturbated packstones and grainstones associated with mounds and abundant stratiform chert. The upper 460 m of the formation (Unit D, Stouge et al., 2002) are composed of lime mudstone, peloidal limestones, bituminous mudstones, and thickly bedded, bioturbated carbonates with minor dolomitic horizons.

Smith (1991) recorded *manitouensis* biozone conodonts from the base of the Cape Weber Formation and *andinus* biozone conodonts from the top, indicating that the unit spans the late Skullrockian to Blackhillsian stages of the Ibexian (Early Ordovician) in the North American standard. Stouge et al. (2001, 2002) invoked a major disconformity between the Antiklinalbugt and Cape Weber Formations, but conodont faunas are continuous through these two units and show no significant hiatuses (Smith, 1985, 1991; Huselbee, 1998).

The Narwhale Sound Formation (270–460 m) records the major sea-level lowstand at the Ibexian-Whiterockian boundary in the North American standard; on Ella Ø, nearly the whole unit is made up of microcrystalline or sucrosic dolostones with some finer-grained dolostones and occasional limestones. Small thrombolitic mounds are present (Stouge et al., 2002), and cherts are abundant. On Ella Ø, the formation is truncated by the Devonian erosion surface, but on Albert Heim Bjerge, the upper part of the formation is present. The lower boundary is gradational with the underlying Cape Weber Formation and is mainly marked by a change to more recessive weathering (Hambrey et al., 1989).

The youngest unit of the Kong Oscar Fjord Group, the Heimbjerge Formation, is present only in the northernmost part of the outcrop belt within the Franz Joseph allochthon, on Albert Heim Bjerge and C.H. Ostenfeld Nunatak; elsewhere, the sub-Devonian unconformity truncates older units. On Albert Heim Bjerge, the formation is 300 m thick, but to the north on C. H. Ostenfeld Nunatak, Frykman (1979) recorded 1200 m. The base of the formation contains *holodentata* biozone conodonts of mid-Darriwilian age, and the youngest part on C.H. Ostenfeld Nunatak is of *anserinus* biozone (late Darriwilian–early Sandbian) age (Smith, 1985; Smith and Bjerreskov, 1994). The reconnaissance-level conodont sampling (Smith, 1985; Smith and Bjerreskov, 1994) did not reveal any thrust repetitions; Frykman's thickness therefore appears to be valid and, furthermore, is supported by backstripping calculations (see following). Lithologically, the Heimbjerge Formation is of rather uniform appearance for the most part; it consists of predominantly pale-gray–weathering limestones, but there is cryptic variation within the unit. Stouge et al. (2002) noted that, on Albert Heim Bjerge, the basal 10–15 m consist of brown and dark red laminated lime mudstones with bird's-eye fenestrae, stromatolites, and bioturbated limestones; these are succeeded by 50 m of laminated lime mudstones with intraclast conglomerates overlain by platy lime mudstones, and a further 10 m of heterogeneous massive, platy, and bioturbated limestones with stromatoporoid mounds. There is then a transition to the typical lithofacies of the Heimbjerge Formation, with pale-gray, bioturbated wackestones to grainstones containing abundant chert nodules interbedded with stromatoporoid mounds. Above around 200 m in the type section, the Heimbjerge Formation remains unstudied sedimentologically, and the only documentation is that of Cowie and Adams (1957). Despite the very substantial thickness of 1200 m on C.H. Ostenfeld Nunatak, which remains almost completely unstudied in sedimentologic terms, the conodont biostratigraphy indicates that the Heimbjerge Formation is restricted to a narrow time interval of around 6 m.y. within the Middle to earliest Late Ordovician, illustrating the very high subsidence rates during this interval.

SUBSIDENCE HISTORY

The temporal resolution of the new stratigraphic framework for North-East and North Greenland permits a detailed comparison of the subsidence histories along the Laurentian margin from western Newfoundland to the NE tip of Greenland, a distance of over 3000 km. Attempts to determine the subsidence history of the Laurentian margins were initiated by Bond et al. (1984), who examined a number of localities on the western and southeastern margins of Laurentia, although they did not consider the Scotland–Greenland sector in detail. They did, however, conclude that on the eastern Laurentian margin, in SW Newfoundland and Virginia, the initial rifting occurred at ca. 600 Ma, with the transition to postrift thermal subsidence in the latest Neoproterozoic. Cawood et al. (2001) carried out a more detailed study of the margin in Newfoundland, utilizing new radiometric dates and the revised geological time scale for the Neoproterozoic–Cambrian. The initial separation of Laurentia from western Gondwana was estimated to have occurred ca. 570 Ma, and a wide Iapetus Ocean was established by 550 Ma (latest Neoproterozoic). The rift-drift transition in Newfoundland was estimated at 540–535 Ma, although it was recognized that the pattern of rifting along the margin was complex and produced a number of terranes of Laurentian affinity, such as the Argentinian Precordillera.

For this study, the sections illustrated in Figure 13 were first decompacted and then backstripped using the algorithms of Sclater and Christie (1980). Tectonic subsidence curves were then calculated using the model of McKenzie (1978). The latter calculations were carried out both with and without incorporating estimates of depositional depth and water loading—the domi-

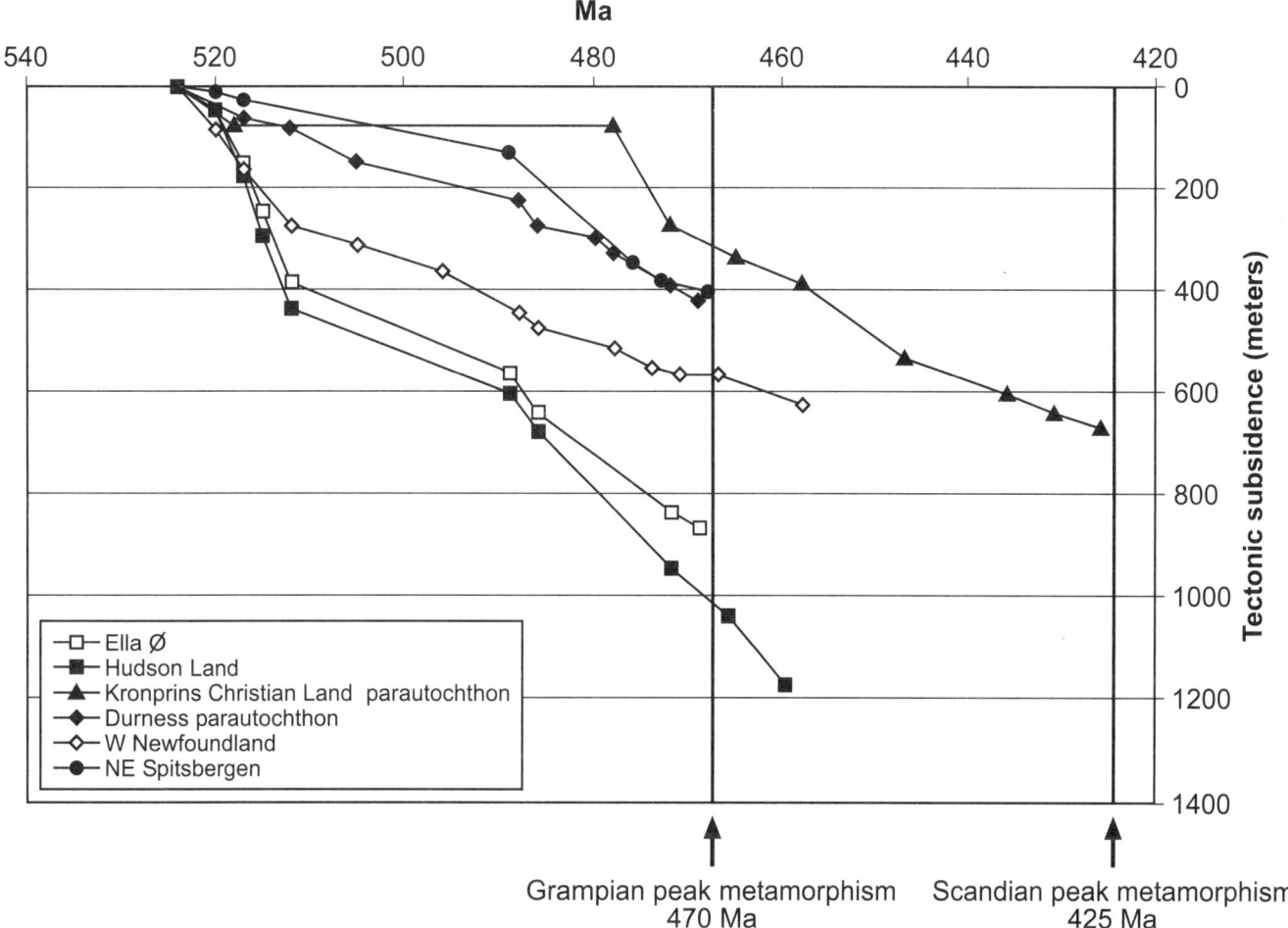

Figure 13. Tectonic subsidence on the eastern margin of Laurentia (paleo- and modern coordinates) in the sector that stretches from western Newfoundland to Kronprins Christian Land, a distance of around 2500 km. The curves for Hudson Land and Ella Ø represent localities in the Franz Joseph allochthon of the fjord region of North-East Greenland; the Durness parautochthon refers to the parautochthonous foreland of NW Scotland. The time scale and stratigraphic data are from Figure 5. The sections were decompacted using the algorithms of Sclater and Christie (1980) and backstripped following the method of McKenzie (1978). Other details are given in the text.

nance of shallow-water deposition means that this parameter had relatively little significance, and the results obtained with and without a bathymetric correction were very similar. Corrections for eustasy were not attempted in the current absence of widely accepted eustatic curves for this stratigraphic interval, let alone reliable estimates of magnitudes.

Western Newfoundland was included as a control, since the tectonic history is well understood (cf. Williams, 1995; Cawood et al., 2001, for reviews). The tectonic subsidence curve for western Newfoundland (Fig. 13) shows the onset of thermal subsidence ca. 525 Ma, which continues until 471 Ma. At this point, a period of uplift and nondeposition at the base of the Table Head Group occurs, which has been attributed to flexure associated with the development of a peripheral bulge during the Taconic orogeny (Stenzel et al., 1990; Knight et al., 1991, 1995). This was followed by foundering of the platform and the development of a classic foreland basin, represented by the Table Head Group (James et al., 1989; Knight et al., 1995). The parautochthonous succession in NW Scotland (Durness) exhibits a very similar pattern, although in the latter case, there is no preserved development of a foreland basin—the shallow-water carbonate succession is truncated by a major thrust that juxtaposes Neoproterozoic metasediments of the Moine Supergroup against the Cambrian-Ordovician Durness Group in the footwall.

Kronprins Christian Land

At the northernmost limit of the East Greenland Caledonides, the subsidence curve for the Lower Paleozoic succession in Kronprins Christian Land (Fig. 13) contrasts markedly with those for western Newfoundland and NW Scotland. Although there is an initial phase of Early Cambrian subsidence, and associated deposition of quartzarenites, this is followed by a long interval of uplift and nondeposition, corresponding to the sub–Wandel Valley unconformity. This tectonic phase unroofed many kilometers of Proterozoic–Cambrian carbonate and siliciclastic sediments with very little associated deformation. The amount of uplift increases from west to east, where, in the thrust sheets of Lambert Land, the Wandel Valley Formation rests unconformably on Proterozoic sandstones of the Independence Fjord Group (Fig. 8). The sub–Wandel Valley unconformity is also present on Bjørnøya, where the Lower Ordovician Younger Dolomite rests unconformably on Neoproterozoic diamictites, demonstrating that Bjørnøya was an integral part of the Laurentian margin in North-East Greenland until Paleogene rifting (Smith, 2000).

The progressive uplift of the carbonate platform from late Early Cambrian through to Early Ordovician time in central and eastern North Greenland was accompanied by the deposition of a stratigraphically complex series of carbonate and siliciclastic sediments (the Brønlund Fjord, Tavsens Iskappe, and lower Ryder Gletscher Groups) within platform-interior, platform-margin, and carbonate slope-apron to deep-shelf settings (Ineson and Peel, 1997). Uplift culminated in the earliest Ordovician with the development of a thin quartzarenite sheet, the Permin Land Formation, which prograded westward across the strike of the shelf for over 400 km as part of a forced regression during a eustatic lowstand (Bryant and Smith, 1990; Higgins et al., 1991). Deposition of the Permin Land Formation was coincident with the onset of turbidite deposition in the deep-water basin to the north after a period of relative starvation. The 600–700-m-thick Vølvedal Group is composed of dark cherts and shales with quartzose turbidites and base-of-slope conglomerates deposited in a series of small borderland fans (Surlyk, 1991; Higgins et al., 1991). The abrupt incoming of abundant quartzose sediment at the Cambrian-Ordovician boundary on both the platform and in the deep-water basin was probably due to the eventual unroofing of the Independence Fjord Group in eastern North Greenland. From earliest Ordovician (basal Tremadocian) time, there is no evidence for continued uplift, but the easternmost part of the Franklinian platform in North Greenland remained emergent until it was flooded by the basal Floian (late Early Ordovician; ca. 477 Ma) sea-level rise, leading to deposition of the Wandel Valley Formation, abandonment of the Vølvedal Group fans, and the initiation of starved deep-water basin conditions represented by the Amundsen Land Group.

Although the stratigraphic pattern and areal extent of Cambrian to Early Ordovician uplift are now well constrained, the tectonic processes responsible remain far from clear. It has been suggested (Surlyk and Hurst, 1984; Surlyk, 1991) that the uplift may have been a consequence of an early collision on the Iapetus margin with an arc or microplate, leading to the development of a peripheral bulge. This is here considered to be unlikely since the uplift spans the interval from late Early Cambrian to mid–Early Ordovician, a period of ~35 m.y. during which lithostatic loading and flexure would have to have been maintained, and during which there is no evidence of bulge migration.

From 480 to 425 Ma, subsidence in Kronprins Christian Land continued at an approximately even rate, with an average subsidence rate of 10.9 m m.y.$^{-1}$ (Fig. 13). During this interval from the mid–Early Ordovician (478 Ma) to the late Llandovery (431 Ma), sedimentation was dominated by carbonates in which the principal depositional control was eustasy (see following). The late Llandovery deepening event associated with the transition from the peritidal-subtidal cycles of the Turesø Formation to the predominantly deep subtidal sediments of the Odins Fjord Formation (Fig. 6) was attributed by Higgins et al. (1991) to a downflexing of the platform. However, this is not evident in the subsidence modeling (Fig. 13), and it is more likely that this transition was due to the progressive middle to late Llandovery second-order eustatic sea-level rise documented by Johnson et al. (1991), Johnson (1996), and Loydell (1998) (see following).

A major change in carbonate depositional patterns that has been more widely attributed to foundering of the carbonate platform occurs at the top of the Odins Fjord Formation in Kronprins Christian Land, where there is an abrupt transition to the reefs of the Samuelsen Høj Formation. The reef and mound belt is traceable from Kronprins Christian Land westward for ~850 km to Washington Land and is up to 100 km wide (Sønderholm and Harland, 1989; Higgins et al., 1991). Although mound formation ended in the latest Llandovery in Kronprins Christian Land, there is some evidence that it may have extended into the Přídolí on the western part of the shelf (Higgins et al., 1991). The foundering of the platform was at first thought to be entirely attributable to loading by Caledonian thrust sheets (Hurst et al., 1983), but the scale of the foundering, across the entire 900 km of the Franklinian platform in North Greenland, renders this unlikely (Surlyk, 1991), and the subsidence curve does not have the characteristic convex-up geometry of a typical foreland basin.

The first evidence of sediment input from the rising Caledonian orogen is close to the Ordovician-Silurian boundary in eastern North Greenland, and, through the Llandovery, this developed into an elongate fan system that was parallel to the Franklinian platform margin and that extended from easternmost North Greenland for 1500–2000 km into the Canadian Arctic Islands. The succession is up to 5 km thick and has an estimated volume of 2×10^6 km^3 (Hurst and Surlyk, 1982; Surlyk and Hurst, 1984; Higgins et al., 1991; Surlyk, 1991). By the end of the Llandovery, the turbidites had filled the deep-water basin to the edge of the platform, and a combination of sediment loading in the basin, a major eustatic sea-level rise, and loading by Caledonian thrust sheets led to downflexing of the platform in North Greenland and the deposition of hemipelagic mudstones and siltstone-sandstone turbidites across the platform area, rather than carbonates (Surlyk, 1991; Higgins et al., 1991). In Kronprins Christian Land, reef growth was terminated by the deposition of condensed, anoxic deep-water siltstones of the Profilfjeldet Member (Lauge Koch Land Formation), which pass upward within the member to sandstone turbidites. These are the youngest sediments preserved in the vicinity of the orogen itself, where they lie in the footwall to thrusts in the parautochthon of Kronprins Christian Land (Hurst and Surlyk, 1982; Smith et al., 2004).

Fjord Region of Southern North-East Greenland

The Lower Paleozoic successions in the Franz Joseph allochthon of southern North-East Greenland exhibit perhaps the most distinctive subsidence curves in this study. It has long been known (cf. Cowie and Adams, 1957) that the Cambrian-Ordovician successions are particularly thick in this region, particularly in the northern area of Hudson Land, where on Albert Heim Bjerge and C.H. Ostenfeld Nunatak, there is a more complete succession preserved underneath the sub-Devonian unconformity than on Ella Ø farther to the south. Modeling of the subsidence history confirms the distinctive, if enigmatic, character of this part of the Iapetus margin. The plots for Ella Ø and Hudson Land show very similar trajectories, and the principal difference is the termination of the Ella Ø curve at 471 Ma, which is the erosion surface beneath the Devonian Vilddal Group (Fig. 13).

In the Lower Cambrian of western Newfoundland, NW Scotland (Durness), Ella Ø, and Hudson Land, the onset of sedimentation marks the transition from rifting to thermal subsidence at around 525 Ma. In western Newfoundland, rift-related deposits are preserved beneath the Bradore Formation (Cawood et al., 2001), but in NW Scotland and East Greenland, the Lower Cambrian strata rest on prerift sediments and/or crystalline basement, reflecting the steer's head geometry of these relatively inboard successions. The concave-up inflection at 512–514 Ma is probably artifactual and relates to the shift from clastic to carbonate deposition. After ca. 510 Ma, both the Ella Ø and Hudson Land sections show an approximately constant rate of subsidence through to the end of the preserved sedimentary record. The average rate for this interval on Ella Ø is 11 m m.y.$^{-1}$, whereas in Hudson Land, there is a slightly higher rate of 14 m m.y.$^{-1}$. These rates are comparable with that obtained through the same interval in Kronprins Christian Land, but they are significantly higher than either NW Scotland or western Newfoundland.

Both the Franz Joseph allochthon and NW Scotland subsidence curves show a marked convex-up inflection in the latest Cambrian at 488–489 Ma. These inflections are again considered most likely to be artifactual; there is lithological uniformity and very little, if any, biostratigraphic control through the Middle and Upper Cambrian at these localities, resulting in a long undivided interval between the base of the Middle Cambrian and the *angulatus* biozone eustatic flooding event in the earliest Tremadocian. In this context, it is noteworthy that western Newfoundland, where there is good biostratigraphic control through the Cambrian (Chow and James, 1987, their Fig. 2) does not show this pattern.

The subsidence curves help to inform the debate over the presence (Stouge et al., 2001, 2002) or absence (Smith and Bjerreskov, 1994; Smith et al., 2004) of a significant hiatus in the Lower Ordovician Cape Weber Formation of the Franz Joseph allochthon. The rate of subsidence in the Ibexian of the Franz Joseph allochthon is 19 m m.y.$^{-1}$ in comparison with a figure of 7 m m.y.$^{-1}$ for western Newfoundland. If Stouge et al. (2002) are correct in inferring the presence of a disconformity spanning 5–6 m.y. (Stouge et al., 2002, their Fig. 2; Webby et al., 2004), then average subsidence rate for the Ibexian increases to 26.5–29 m m.y.$^{-1}$, an improbably high value that is around four times the rate seen in western Newfoundland and NW Scotland. In addition, the presence of subsidence rates in the region of 19 m m.y.$^{-1}$ renders the presence of significant disconformities unlikely, as accommodation space would have been constantly available.

In western Newfoundland, the subsidence curve shows a marked convex-up deflection at 467 Ma, which corresponds to the onset of foreland basin formation as a consequence of loading by Taconic thrust sheets. The migration of a peripheral bulge, which generated the St. George unconformity, followed by foundering of the platform and a switch to deep-water deposition have been well-documented (Knight et al., 1991, 1995). As noted already, there is no Taconic-Grampian deformation in the Greenland Caledonides, on the far side of the Scottish promontory, and the subsidence curve indicates the same rate across this event until it terminates at the modern erosion surface. There are no available constraints on when subsidence and sedimentation ended in this part of the orogen.

EUSTATIC CONTROLS AND SEQUENCE STRATIGRAPHY

Laurentian successions that lie between an Early Cambrian transgressive surface and the late mid-Ordovician have classically been referred to as the Sauk sequence or megasequence, the oldest of a series of sequences defined by Sloss (1963) that represent first-order tectono-eustatic sea-level fluctuations (Golonka and Kiessling, 2002). A combination of trilobite biomeres and major sea-level events has subsequently been used to divide the Sauk interval into three, and more recently four, second-order supersequences, generally termed Sauk I to Sauk IV (Palmer, 1981; Sloss, 1988; Golonka and Kiessling, 2002), although some authors have disputed the applicability of sequence stratigraphic concepts to these megasequences and their subdivisions (e.g., Sloss, 1993). The base of Sauk I is the transgressive Early Cambrian surface, and the Sauk I-II boundary coincides with the Lower–Middle Cambrian boundary (Sloss, 1988; Golonka and Kiessling, 2002). Sauk II spans the Middle and the lower part of the Upper Cambrian, and the upper boundary is at the Dresbachian–Franconian boundary (mid–Late Cambrian; mid-Steptoean sensu Ludvigsen and Westrop, 1985; Palmer, 1998). The Sauk II-III boundary is also marked by a major $\delta^{13}C$ excursion (the Steptoean Positive Isotopic Carbon Excursion [SPICE] event; Saltzman et al., 2000, 2004). Sauk IV was named more recently (Golonka and Kiessling, 2002), and it extends up to the unconformity at the traditional Tremadoc-Arenig boundary (Fortey, 1984), corresponding to a mid-Tulean age in the Ibexian standard (Webby et al., 2004). This level corresponds to the megacycle I-II boundary of Knight and James (1987) in western Newfoundland, but it is significantly above the level of the Sauk III-IV boundary postulated by Holmer et al. (2005).

The base of the succeeding Tippecanoe megasequence is placed at a major sea-level rise that was initiated in the mid-Darriwilian (Golonka and Kiessling, 2002). The Tippecanoe I-II boundary is coincident with the major drawdown of sea level that resulted from end-Ordovician glaciation (Sloss, 1988; Golonka and Kiessling, 2002). Tippecanoe II spans the Llandovery, and Tippecanoe III extends from the early Wenlock to the early Přídolí (Sloss, 1988; Golonka and Kiessling, 2002).

Prominent large-scale cycles, termed grand cycles, within the Sauk and Tippecanoe supersequences were first recognized in the Canadian Rockies by Aitken (1966). Until recently, there has been a lack of connection between the concept of "grand cycles," recognized principally using sedimentological criteria, and the Sauk supersequences, but there is an increasing determination to produce unified "whole evidence" models (e.g., Lavoie et al., 2003; Saltzman et al., 2004), and the grand cycles are now generally recognized as constituting third-order sequences (Cowan and James, 1993; Mount and Bergk, 1998; Spencer and Demicco, 2002). Each cycle is typically made up of a lower unit of fine-grained mudstones, limestones, and/or sandstones that is gradationally overlain by an upper unit of limestone and/or dolostone. Cycle boundaries are sharp and often coincide with trilobite biomeres (Palmer, 1965). Most studies of grand cycles and their sequence stratigraphy have been undertaken in central and western Laurentia, but work has been carried out on the Cambrian–Ordovician of Newfoundland and adjacent areas of Québec (Chow and James, 1987; Knight and James, 1987; Cowan and James, 1993; Lavoie et al., 2003) and in NW Scotland (Wright and Knight, 1995), providing the basis for correlation along the Laurentian margin to Greenland.

In the Greenland Caledonides (Figs. 5A, 6, and 11), Sauk I is represented by the unconformity-bounded Slottet and Kap Holbæk Formations of the parautochthon and the Kløftelv–Hyolithus Creek interval of the Franz Joseph allochthon. Sauk II and III are absent in the more inboard localities of the parautochthon along the entire length of the Greenland Caledonides, but they are present in the Franz Joseph allochthon and the palinspastically adjacent terrane of NE Spitsbergen. Within the Cambrian succession of the Franz Joseph allochthon, Sauk II lies entirely within the relatively unstudied Dolomite Point Formation, and it is not currently possible to derive any information relating to component third-order sequences, as recognized by Chow and James (1987) in western Newfoundland. In the Franz Joseph allochthon, Sauk III spans the upper Dolomite Point, Antiklinalbugt, and lower Cape Weber Formations, with major sea-level rise at the base of the Tremadocian.

The base of the Sauk IV supersequence marks the initiation of a major flooding event along the Greenland margin of Iapetus and is also clearly identifiable in NW Scotland, Newfoundland (where it corresponds to the base of megacycle 2 of Knight and James, 1987), and southward into mainland North America. In North Greenland, the Danmarks Fjord Member (Wandel Valley Formation) is the transgressive systems tract (TST) of Sauk IV, and the maximum flooding surface lies within the overlying Amdrup Member, where the Alexandrine Bjerge Member represents the upper part of the sequence.

The sea-level rise associated with Tippecanoe I began in strata of Darriwilian (late Whiterockian) age. The highstand systems tract (HST) is probably represented by the switch to subtidal deposition that is represented by the lower Sjælland Fjelde Formation, but this was short-lived, and there was a rapid return to peritidal deposition. A second megacycle is represented by the more long-lived change to subtidal deposition at the base of the Morris Bugt Group across the Franklinian Basin, represented in eastern North Greenland by the base of the Børglum River Formation (Smith et al., 1989). In the Franz Joseph allochthon, it is significant that the Heimbjerge Formation marks an approximately synchronous change to subtidally dominated deposition (see following) at the base of the Tippecanoe megasequence. The youngest units in the Caledonides of North Greenland are coeval with the Tippecanoe II–III supercycles, but by this time, the tectonic signal caused by loading was swamping the eustatic control, and the correlation with Sloss sequences breaks down.

THE DEMISE OF THE LAURENTIAN MARGIN

The Greenland sector of the Iapetus margin of Laurentia has a distinctive collisional history in comparison with areas to the south. From Scotland southward, the Caledonian-Appalachian orogen records successive arc-continent collisions (Penobscotian and Taconic-Grampian phases) followed by the Acadian continent-continent collision of Gondwana with Laurentia (McKerrow et al., 2000). In contrast, the Greenland margin records a much simpler history, with just a single event—the Scandian phase of Baltica-Laurentia collision (Higgins et al., 2001, 2004a). In Newfoundland and Scotland, Taconic-Grampian deformation resulting from extensive island-arc collision was responsible for the bulk of Dalradian deformation (Dewey and Shackleton, 1984; McKerrow, 1988; Cawood and Suhr, 1992; Cawood et al., 1995; Soper et al., 1999; Leslie et al., this volume), but it does not extend into Greenland. This may be, in part, because of the 120° inflection in the Laurentian margin between Canada and Greenland, which reflects the orientation of pre-Iapetus rifting, where Scotland sits on the intervening promontory (Soper, 1994; Dalziel and Soper, 2001).

In the northernmost part of the Greenland Caledonides, in Kronprins Christian Land and Lambert Land, sedimentation is continuous from the Lower Ordovician to the lower Wenlock, and no direct or indirect evidence of deformation and orogenesis exists during this interval (Smith et al., 2004). Deformation and metamorphism are restricted to a single orogenic phase, the Scandian, during the Silurian, and the earliest thrusting emplaced Proterozoic sediments and extrusive igneous rocks over Lower Paleozoic foreland, where conodont color alteration indices show that the amount of overburden in the footwall of the Vandredalen thrust ramp reached a maximum of 12.5 km (Higgins et al., 2001, 2004b; Rasmussen and Smith, 2001). The Lower Paleozoic foreland was dominated by shallow-water carbonate deposition until the late Llandovery, when it was replaced first by black mudstones and then by turbidites. The foundering of the shelf is not attributable to the development of a foreland basin, in the strict sense, because it extends too far from east to west across the Franklinian Basin (Surlyk, 1991). Rather, it reflects the position of eastern North Greenland on an original inflection in the Laurentian margin. Erosion products from the Scandian thrust pile initially bypassed the platform and accumulated as the 3-km-thick turbidite succession of the Llandovery Merqujôq Formation. Combined loading by the turbidite pile to the north and Scandian thrust sheets to the east then led to the extensive foundering of the platform in latest Llandovery time (Surlyk, 1991; Higgins et al., 1991).

Although no sediments younger than earliest Late Ordovician are preserved in the Franz Joseph allochthon of southern North-East Greenland, it is clear from recent work that there was no pre-Scandian deformation of thrust sheets in the southern part of the Greenland Caledonides (Higgins et al., 2004a). In the Franz Joseph allochthon, the youngest pre-orogenic sediments are overlain unconformably by Eifelian (early mid-Devonian) coarse fluvial sediments of the Solstrand Formation (Vilddal Group) (Larsen et al., this volume). The sub-Devonian erosion surface cuts down strongly from north to south such that the youngest sediments preserved beneath the unconformity on Ella Ø (72°50′N) are of early mid-Ordovician age, whereas 180 km farther to the north on Albert Heim Bjerge and C.H. Ostenfeld Nunatak (74°23′N), they are of late mid-Ordovician to earliest Late Ordovician age. On C.H. Ostenfeld Nunatak (74°17′N), the youngest unit preserved beneath the unconformity is the Heimbjerge Formation. The youngest age-diagnostic conodont samples recovered to date are from 90 m below the top of the 1200-m-thick formation (see previous), and they contain *Belodella robusta* Ethington and Clark, *Belodina monitorensis* (Ethington and Schumacher), *Dapsilodus*? *nevadensis* (Ethington and Schumacher), *Drepanoistodus suberectus* (Branson and Mehl), *Erraticodon balticus* Dzik, *Panderodus* sp., and *Pygodus anserinus* Lamont and Lindström. This is indicative of an *anserinus* biozone age (Smith, 1985; Smith and Bjerreskov, 1994), which is of latest Whiterockian age in terms of North American chronostratigraphy and Darriwilian–Sandbian in the global standard (Middle–Late Ordovician boundary), equivalent to a date of 461.5–459.5 Ma (Webby et al., 2004). Higher conodont samples are productive but yield only long-ranging taxa, including *Amorphognathus* sp. and *Appalachignathus delicatulus* Bergström et al., the latter of which indicates an age no younger than late Sandbian (>457 Ma) (Sweet, 1984).

In NW Scotland, ~700 km to the south of the Greenland Caledonides in palinspastic terms, it has been postulated that the termination of Durness Group sedimentation in late Dapingian (earliest Whiterockian) time was a consequence of uplift, or more specifically, the migration of a peripheral bulge produced as a flexural response to Grampian thrust loading (Dewey, 1971,

1982; Soper et al., 1999). This effect is better documented and constrained in western Newfoundland, ~1500 km palinspastically to the southwest of Scotland, where the boundary between the platform carbonates of the St. George Group and the carbonates and shales of the Table Head Group is an unconformity that both youngs and decreases in duration toward the craton (Knight et al., 1991). Deposition at the base of the Table Head Group occurred in peritidal environments across a series of downfaulted blocks, but this gave way to complete inundation and the rapid deposition of subtidal carbonates across the block-faulted terrane, and deposition was unable to keep pace with subsidence (James et al., 1989; Knight et al., 1995). Continued foundering of the platform and the further development of a foreland basin led to the deposition of deep-water clastic and carbonate sediments, including flysch (Knight et al., 1995).

The youngest preserved sediments on the Laurentian margin in the southern Greenland Caledonides are thus significantly younger than those in NW Scotland, by 7.5–10.5 m.y., and they postdate Grampian deformation and metamorphism (Oliver, 2001). It is thus clear that sedimentation in the southern part of the Greenland Caledonides was not terminated by Grampian orogenesis, and that the deformation is entirely Scandian in origin. Thus, sedimentation in the Franz Joseph allochthon is more closely related to that in the Franklinian Basin of North Greenland than either one is to the Iapetus margin farther south, reflecting the different history of collisions in this sector of the Laurentian margin.

ACKNOWLEDGMENTS

The Carlsberg Foundation is thanked for financial support to Rasmussen. Peter Cawood and David Harper are thanked for insightful and helpful reviews, and Tony Higgins for much assistance. Niels Henriksen (Oscar) is thanked for organizing the many expeditions during which data was retrieved for this paper.

REFERENCES CITED

Aitken, J.D., 1966, Middle Cambrian to Middle Ordovician cyclic sedimentation, southern Rocky Mountains of Alberta: Bulletin of Canadian Petroleum Geology, v. 14, p. 405–441.

Aldridge, R.J., Jeppsson, L., and Dorning, K.J., 1993, Early Silurian oceanic episodes and events: Journal of the Geological Society of London, v. 150, p. 501–513, doi: 10.1144/gsjgs.150.3.0501.

Armstrong, H.A., 1990, Conodonts from the Upper Ordovician–lower Silurian Carbonate Platform of North Greenland: Bulletin Grønlands Geologiske Undersøgelse, v. 159, 151 p.

Bergström, S.M., 1979, Whiterockian (Ordovician) conodonts from the Hølonda Limestone of the Trondheim Region, Norwegian Caledonides: Norsk Geologisk Tidsskrift, v. 59, p. 295–307.

Bergström, S.M., 1990, Relations between conodont provincialism and the changing palaeogeography during the early Palaeozoic, in McKerrow, W.S., and Scotese, C.R., eds., Palaeozoic Palaeogeography and Biogeography: Geological Society of London Memoir 12, p. 105–121.

Bergström, S.M., 1997, Conodonts of Laurentian faunal affinities from the Middle Ordovician Svartsætra Limestone in the Trondheim Region, central Norwegian Caledonides: Norges Geologiske Undersøkelse Bulletin, v. 432, p. 59–69.

Birkenmajer, K., 1978, Ordovician succession in south Spitsbergen: Studia Geologica Polonica, v. 59, p. 47–82.

Bjerreskov, M., 1989, Ordovician graptolite biostratigraphy in North Greenland: Rapport Grønlands Geologiske Undersøgelse, v. 144, p. 17–33.

Bond, G.C., Nickerson, P.A., and Kominz, M.A., 1984, Breakup of a supercontinent between 625 and 555 Ma: New evidence and implications for continental histories: Earth and Planetary Science Letters, v. 70, p. 325–345, doi: 10.1016/0012-821X(84)90017-7.

Braathen, A., Maher, H.D., Haabet, T.E., Kristensen, S.E., Tørudbakken, B.O., and Worsley, D., 1999, Caledonian thrusting on Bjørnøya: Implications for Palaeozoic and Mesozoic tectonism of the western Barents Shelf: Norsk Geologisk Tidsskrift, v. 79, p. 57–68, doi: 10.1080/002919699433915.

Bruton, D.L., and Harper, D.A.T., 1988, Arenig-Llandovery stratigraphy and faunas across the Scandinavian Caledonides, in Harris, A.L., and Fettes, D.J., eds., The Caledonian–Appalachian Orogen: Geological Society of London Special Publication 38, p. 477–498.

Bryant, I.D., and Smith, M.P., 1990, A composite tectonic-eustatic origin for shelf sandstones deposited at the Cambrian-Ordovician boundary in North Greenland: Journal of the Geological Society of London, v. 147, p. 795–809, doi: 10.1144/gsjgs.147.5.0795.

Cawood, P.A., and Suhr, G., 1992, Generation and obduction of ophiolites; constraints for the Bay of Islands Complex, western Newfoundland: Tectonics, v. 11, p. 884–897, doi: 10.1029/92TC00471.

Cawood, P.A., van Gool, J.A.M., and Dunning, G.R., 1995, Collisional tectonics along the Laurentian margin of the Newfoundland Appalachians, in Hibbard, J.P., van Staal, C.R., and Cawood, P.A., eds., Current Perspectives in the Appalachian–Caledonian Orogen: Geological Association of Canada Special Paper 41, p. 283–301.

Cawood, P.A., McCausland, P.J.A., and Dunning, G.R., 2001, Opening Iapetus: Constraints from the Laurentian margin in Newfoundland: Geological Society of America Bulletin, v. 113, p. 443–453, doi: 10.1130/0016-7606 (2001)113<0443:OICFTL>2.0.CO;2.

Chow, N., and James, N.P., 1987, Cambrian grand cycles—A northern Appalachian perspective: Geological Society of America Bulletin, v. 98, p. 418–429, doi: 10.1130/0016-7606(1987)98<418:CGCANA>2.0.CO;2.

Clemmensen, L.B., and Jepsen, H.F., 1992, Lithostratigraphy and geological setting of Upper Proterozoic shoreline-shelf deposits, Hagen Fjord Group, North Greenland: Rapport Grønlands Geologiske Undersøgelse, v. 157, 27 p.

Cocks, L.R.M., and Torsvik, T.H., 2002, Earth geography from 500 to 400 million years ago: A faunal and palaeomagnetic review: Journal of the Geological Society of London, v. 159, p. 631–644.

Cooper, R.A., and Sadler, P.M., 2004, The Ordovician Period, in Gradstein, F.M., Ogg, J.G., and Smith, A.G., eds., A Geologic Time Scale, 2004: Cambridge, Cambridge University Press, p. 165–187.

Cowan, C.A., and James, N.P., 1993, The interactions of sea-level change, terrigenous-sediment influx, and carbonate productivity as controls on Upper Cambrian grand cycles of western Newfoundland, Canada: Geological Society of America Bulletin, v. 105, p. 1576–1590, doi: 10.1130/0016-7606 (1993)105<1576:TIOSLC>2.3.CO;2.

Cowie, J.W., and Adams, P.J., 1957, The Geology of the Cambro-Ordovician Rocks of East Greenland: Meddelelser om Grønland, v. 153, no. 1, 193 p.

Dalziel, I.W.D., and Soper, N.J., 2001, Neoproterozoic extension on the Scottish Promontory of Laurentia: Paleogeographic and tectonic implications: The Journal of Geology, v. 109, p. 299–317, doi: 10.1086/319974.

Dewey, J.F., 1971, A model for the Lower Palaeozoic evolution of the southern margin of the early Caledonides of Scotland and Ireland: Scottish Journal of Geology, v. 7, p. 219–240.

Dewey, J.F., 1982, Plate-tectonics and the evolution of the British Isles: Journal of the Geological Society of London, v. 139, p. 371–412.

Dewey, J.F., and Shackleton, R.M., 1984, A model for the evolution of the Grampian tract in the early Caledonides and Appalachians: Nature, v. 312, p. 115–121, doi: 10.1038/312115a0.

Fairchild, I.J., and Hambrey, M.J., 1995, Vendian basin evolution in East Greenland and NE Svalbard: Precambrian Research, v. 73, p. 217–233, doi: 10.1016/0301-9268(94)00079-7.

Fortey, R.A., 1984, Global earlier Ordovician transgressions and regressions and their biological implications, in Bruton, D.L., ed., Aspects of the Ordovician System: Palaeontological Contributions of the University of Oslo, v. 295, p. 37–50.

Fortey, R.A., and Barnes, C.R., 1977, Early Ordovician conodont and trilobite communities of Spitsbergen: Influence on biogeography: Alcheringa, v. 1, p. 297–309.

Fortey, R.A., and Bruton, D.L., 1973, Cambrian-Ordovician rocks adjacent to Hinlopenstretet, north Ny Friesland, Spitsbergen: Geological Society of America Bulletin, v. 84, p. 2227–2242, doi: 10.1130/0016-7606(1973)84 <2227:CRATHN>2.0.CO;2.

Fortey, R.A., and Cocks, L.R.M., 2003, Palaeontological evidence bearing on global Ordovician-Silurian continental reconstructions: Earth-Science Reviews, v. 61, p. 245–307, doi: 10.1016/S0012-8252(02)00115-0.

Friderichsen, J.D., Holdsworth, R.E., Jepsen, H.F., and Strachan, R.A., 1990, Caledonian and pre-Caledonian geology of Dronning Louise Land, North-East Greenland: Rapport Grønlands Geologiske Undersøgelse, v. 148, p. 133–141.

Frykman, P., 1979, Cambro-Ordovician rocks of C.H. Ostenfeld Nunatak, northern East Greenland: Rapport Grønlands Geologiske Undersøgelse, v. 91, p. 125–132.

Gee, D.G., and Teben'kov, A.M., 2004, Svalbard: A fragment of the Laurentian margin, in Gee, D.G., and Pease, V.L., eds., The Neoproterozoic Timanide Orogen of Eastern Baltica: Geological Society of London Memoir 30, p. 191–206.

Golonka, J., and Kiessling, W., 2002, Phanerozoic time scale and definition of time slices, in Kiessling, W., Flügel, A., and Golonka, J., eds., Phanerozoic Reef Patterns: SEPM (Society for Sedimentary Geology) Special Publication 72, p. 11–20.

Gradstein, F.M., Ogg, J.G., and Smith, A.G., eds., 2004, A Geologic Time Scale, 2004: Cambridge, Cambridge University Press, 598 p.

Haller, J., 1971, Geology of the East Greenland Caledonides: New York, Interscience Publishers, 413 p.

Halverson, G.P., Maloof, A.C., and Hoffman, P.F., 2004, The Marinoan glaciation (Neoproterozoic) in northeast Svalbard: Basin Research, v. 16, p. 297–324, doi: 10.1111/j.1365-2117.2004.00234.x.

Hambrey, M.J., 1983, Correlation of late Proterozoic tillites in the North Atlantic region and Europe: Geological Magazine, v. 120, p. 290–320.

Hambrey, M.J., 1988, Late Proterozoic stratigraphy of the Barents Shelf, in Harland, W.B., and Dowdeswell, E.K., eds., Geological Evolution of the Barents Shelf Region: London, Graham and Trotman, p. 49–72.

Hambrey, M.J., Peel, J.S., and Smith, M.P., 1989, Upper Proterozoic and Lower Palaeozoic strata in northern East Greenland: Rapport Grønlands Geologiske Undersøgelse, v. 145, p. 103–108.

Harland, W.B., 1965, The tectonic evolution of the Arctic–North Atlantic region: Philosophical Transactions of the Royal Society of London, Series A, v. 258, p. 59–75.

Harland, W.B., 1969, Contribution of Spitsbergen to understanding of tectonic evolution of North Atlantic region, in Kay, M., ed., North Atlantic Geology and Continental Drift: American Association of Petroleum Geologists Memoir 12, p. 817–855.

Harland, W.B., 1985, Caledonide Svalbard, in Gee, D.G., and Sturt, B.A., eds., The Caledonide Orogen: Scandinavia and Related Areas: London, Wiley and Sons, p. 999–1016.

Harland, W.B., 1997, The Geology of Svalbard: Geological Society of London Memoir 17, 521 p.

Harland, W.B., Perkins, P.J., and Smith, M.P., 1988, Cambrian through Devonian stratigraphy and tectonic development of the western Barents Shelf, in Harland, W.B., and Dowdeswell, E.K., eds., Geological Evolution of the Barents Shelf Region: London, Graham and Trotman, p. 73–88.

Harper, D.A.T., MacNiocaill, C., and Williams, S.H., 1996, The palaeogeography of Early Ordovician Iapetus terranes: An integration of faunal and palaeomagnetic constraints: Palaeogeography, Palaeoclimatology, Palaeoecology, v. 121, p. 297–312, doi: 10.1016/0031-0182(95)00079-8.

Henriksen, N., and Higgins, A.K., 1976, East Greenland Caledonian fold belt, in Escher, A., and Watt, W.S., eds., Geology of Greenland: Copenhagen, Geological Survey of Greenland, p. 182–246.

Higgins, A.K., and Leslie, A.G., 2000, Restoring thrusting in the East Greenland Caledonides: Geology, v. 28, p. 1019–1022, doi: 10.1130/0091-7613 (2000)28<1019:RTITEG>2.0.CO;2.

Higgins, A.K., and Leslie, A.G., 2004, The Eleonore Sø and Målebjerg foreland windows, East Greenland Caledonides, and the demise of the "stockwerke" concept: Geological Survey of Denmark and Greenland Bulletin, v. 6, p. 77–93.

Higgins, A.K., Ineson, J.R., Peel, J.S., Surlyk, F., and Sønderholm, M., 1991, The Franklinian Basin in North Greenland: Bulletin Grønlands Geologiske Undersøgelse, v. 160, p. 71–139.

Higgins, A.K., Leslie, A.G., and Smith, M.P., 2001, Neoproterozoic–Lower Palaeozoic stratigraphical relationships in the marginal thin-skinned thrust belt of the East Greenland Caledonides: Comparisons with the foreland in Scotland: Geological Magazine, v. 138, p. 143–160, doi: 10.1017/S0016756801005076.

Higgins, A.K., Elvevold, S., Escher, J.C., Frederiksen, K.S., Gilotti, J., Henriksen, N., Jepsen, H.F., Jones, K.A., Kalsbeek, F., Kinny, P.D., Leslie, A.G., Smith, M.P., Thrane, K., and Watt, G., 2004a, The foreland-propagating thrust architecture of the East Greenland Caledonides, 72°–75°N: Journal of the Geological Society of London, v. 161, p. 1009–1026, doi: 10.1144/0016-764903-141.

Higgins, A.K., Soper, N.J., Smith, M.P., and Rasmussen, J.A., 2004b, The Caledonian thin-skinned thrust belt of Kronprins Christian Land, eastern North Greenland: Geological Survey of Denmark and Greenland Bulletin, v. 6, p. 41–56.

Holmer, L.E., Popov, L.E., Streng, M., and Miller, J.F., 2005, Lower Ordovician (Tremadocian) lingulate brachiopods from the House and Fillmore Formations, Ibex area, western Utah, USA: Journal of Paleontology, v. 79, p. 884–906, doi: 10.1666/0022-3360(2005)079[0884:LOTLBF]2.0.CO;2.

Hurst, J.M., 1984, Upper Ordovician and Silurian Carbonate Shelf Stratigraphy, Facies and Evolution, Eastern North Greenland: Bulletin Grønlands Geologiske Undersøgelse, v. 148, 73 p.

Hurst, J.M., and Surlyk, F., 1982, Stratigraphy of the Silurian Turbidite Sequence of North Greenland: Bulletin Grønlands Geologiske Undersøgelse, v. 145, 121 p.

Hurst, J.M., McKerrow, W.S., Soper, N.J., and Surlyk, F., 1983, The relationship between Caledonian nappe tectonics and Silurian turbidite deposition in North Greenland: Journal of the Geological Society of London, v. 140, p. 123–131.

Huselbee, M.Y., 1998, Late Cambrian to Earliest Ordovician (Ibexian) Conodont Evolution and Biogeography of Greenland and Northwest Scotland [Ph.D. thesis]: Birmingham, University of Birmingham, 296 p.

Ineson, J.R., and Peel, J.S., 1997, Cambrian Shelf Stratigraphy of North Greenland: Geology of Greenland Survey Bulletin, v. 173, 120 p.

Ineson, J.R., Peel, J.S., and Smith, M.P., 1986, The Sjælland Fjelde Formation: A new Ordovician formation from eastern North Greenland: Rapport Grønlands Geologiske Undersøgelse, v. 132, p. 27–37.

James, N.P., and Stevens, R.K., 1986, Stratigraphy and Correlation of the Cambro-Ordovician Cow Head Group, Western Newfoundland: Geological Survey of Canada Bulletin, v. 366, 143 p.

James, N.P., Stevens, R.K., Barnes, C.R., and Knight, I., 1989, Evolution of a Lower Paleozoic continental-margin carbonate platform, northern Canadian Appalachians, in Crevello, J.L., Wilson, J.L., Sarg, J.F., and Read, J.F., eds., Controls on Carbonate Platform and Basin Development: SEPM (Society of Economic Paleontologists and Mineralogists) Special Publication 44, p. 123–146.

Johnson, M.E., 1996, Stable cratonic sequences and a standard for Silurian eustasy, in Witzke, B.J., Ludvigson, G.A., and Day, J., eds., Paleozoic Sequence Stratigraphy: Views from the North American Craton: Geological Society of America Special Paper 306, p. 203–211.

Johnson, M.E., Kaljo, D., and Rong, J.-Y., 1991, Silurian eustasy, in Bassett, M.G., Lane P.D., and Edwards, D., eds., The Murchison Symposium, Proceedings of an International Conference on the Silurian System: Special Papers in Palaeontology, v. 44, p. 145–163.

Katz, H.R., 1952, Ein Querschnitt durch die Nunatakzone Ostgrönlands (ca. 74°N). Ergebnisse einer Reisen vom Inlandseis (in Zusammenarbeit mit den Expéditions olaires Françaises von P. E. Victor) ostwarts bis in die Fjord-region, ausgeführt im Sommer, 1951: Meddelelser om Grønland, v. 144, no. 8, 65 p.

Knight, I., and James, N.P., 1987, Stratigraphy of the St. George Group (Lower Ordovician) western Newfoundland: The interaction between eustasy and tectonics: Canadian Journal of Earth Sciences, v. 24, p. 1927–1952.

Knight, I., James, N.P., and Lane, T.E., 1991, The Ordovician St. George unconformity, northern Appalachians—The relationship of plate convergence at the St. Lawrence promontory to the Sauk–Tippecanoe sequence boundary: Geological Society of America Bulletin, v. 103, p. 1200–1225, doi: 10.1130/0016-7606(1991)103<1200:TOSGUN>2.3.CO;2.

Knight, I., James, N.P., and Williams, H., 1995, Cambrian–Ordovician carbonate sequence (Humber zone), in Williams, H., ed., Geology of the Appalachian–Caledonian Orogen in Canada and Greenland: Geological Survey of Canada, Geology of Canada, v. 6, p. 67–87.

Lane, P.D., 1972, New trilobites from the Silurian of northeast Greenland: Palaeontology, v. 15, p. 336–364.

Larsen, P.-H., and Escher, J.C., 1985, The Silurian turbidite sequence of the Peary Land Group between Newman Bugt and Victoria Fjord, western North Greenland: Rapport Grønlands Geologiske Undersøgelse, v. 126, p. 47–67.

Lavoie, D., Burden, E., and Lebel, D., 2003, Stratigraphic framework for the Cambrian–Ordovician rift and passive margin successions from southern Québec to western Newfoundland: Canadian Journal of Earth Sciences, v. 40, p. 177–205, doi: 10.1139/e02-078.

Leslie, A.G., and Higgins, A.K., 1998, On the Caledonian geology of Andrée Land, Eleonore Sø and adjacent nunataks (73°30′–74°N), East Greenland: Danmarks og Grønlands Geologiske Undersøgelse Rapport, v. 1998/28, p. 11–27.

Leslie, A.G., Smith, M., and Soper, N.J., 2008, this volume, Laurentian margin evolution and the Caledonian orogeny—A template for Scotland and East Greenland, in Higgins, A.K., Gilotti, J.A., and Smith, M.P., eds., The Greenland Caledonides: Evolution of the Northeast Margin of Laurentia: Geological Society of America Memoir 202, p. doi: 10.1130/2008.1202(13).

Loydell, D.K., 1998, Early Silurian sea-level changes: Geological Magazine, v. 135, p. 447–471, doi: 10.1017/S0016756898008917.

Ludvigsen, R., and Westrop, S.R., 1985, Three new Upper Cambrian stages for North America: Geology, v. 13, p. 139–143, doi: 10.1130/0091-7613(1985)13<139:TNUCSF>2.0.CO;2.

MacNiocaill, C., van der Pluijm, B.A., and Van der Voo, R., 1997, Ordovician paleogeography and the evolution of the Iapetus Ocean: Geology, v. 25, p. 159–162, doi: 10.1130/0091-7613(1997)025<0159:OPATEO>2.3.CO;2.

Major, H., and Winsnes, T.S., 1955, Cambrian and Ordovician Fossils from Sørkapp Land, Spitsbergen: Norsk Polarinstitutt Skrifter, v. 106, 47 p.

McKenzie, D.P., 1978, Some remarks on the development of sedimentary basins: Earth and Planetary Science Letters, v. 40, p. 25–32, doi: 10.1016/0012-821X(78)90071-7.

McKerrow, W.S., 1988, The development of the Iapetus Ocean from the Arenig to the Wenlock, in Harris, A.L., and Fettes, D.J., eds., The Caledonian–Appalachian Orogen: Geological Society of London Special Publication 38, p. 405–412.

McKerrow, W.S., MacNiocaill, C., and Dewey, J.F., 2000, The Caledonian orogeny redefined: Journal of the Geological Society of London, v. 157, p. 1149–1154.

Melezhik, V.A., Heldal, T., Roberts, D., Gorokhov, I.M., and Fallick, A.E., 2000, Depositional environment and apparent age of the Fauske carbonate conglomerate, north Norwegian Caledonides: Norges Geologiske Undersøkelse Bulletin, v. 436, p. 147–168.

Melezhik, V.A., Roberts, D., Gorokhov, I.M., Fallick, A.E., Zwaan, K.B., Kuznetsov, A.B., and Pokrovsky, B.G., 2002, Isotopic evidence for a complex Neoproterozoic to Silurian rock assemblage in the north-central Norwegian Caledonides: Precambrian Research, v. 114, p. 55–86, doi: 10.1016/S0301-9268(01)00218-2.

Miller, J.F., and Kurtz, V.E., 1979, Reassignment of the Dolomite Point Formation of East Greenland from the Middle Cambrian(?) to the Lower Ordovician: Geological Society of America Abstracts with Programs, v. 11, no. 7, p. 480.

Mount, J.F., and Bergk, K.J., 1998, Depositional sequence stratigraphy of Lower Cambrian grand cycles, southern Great Basin, U.S.A., in Ernst, W.G., and Nelson, C.A., eds., Integrated Earth and Environmental Evolution of the Southwestern United States: Columbia, Maryland, Bellwether Publishing, p. 180–202.

Neuman, R.B., and Harper, D.A.T., 1992, Paleogeographic significance of Arenig–Llanvirn Toquima–Table Head and Celtic brachiopod assemblages, in Webby, B.D., and Laurie, J.R., eds., Global Perspectives in Ordovician Geology: Rotterdam, Balkema, p. 241–254.

Okulitch, A.V., and Trettin, H.P., 1991, Late Cretaceous–early Tertiary deformation, Arctic Islands, in Trettin, H.P., ed., Geology of the Innuitian Orogen and Arctic Platform of Canada and Greenland: Geological Survey of Canada, Geology of Canada, v. 3, p. 469–489.

Oliver, G.J.H., 2001, Reconstruction of the Grampian episode in Scotland: Its place in the Caledonian orogeny: Tectonophysics, v. 332, p. 23–49, doi: 10.1016/S0040-1951(00)00248-1.

Palmer, A.R., 1965, Biomere—A new kind of biostratigraphic unit: Journal of Paleontology, v. 39, p. 149–153.

Palmer, A.R., 1981, Subdivision of the Sauk Sequence, in Taylor, M.E., ed., Short Papers for the Second International Symposium on the Cambrian System: U.S. Geological Survey Open-File Report 81-743, p. 160–162.

Palmer, A.R., 1998, A proposed nomenclature for stages and series for the Cambrian of Laurentia: Canadian Journal of Earth Sciences, v. 35, p. 323–328, doi: 10.1139/cjes-35-4-323.

Park, R.G., Stewart, A.D., and Wright, D.T., 2002, The Hebridean terrane, in Trewin, N.H., ed., The Geology of Scotland (4th edition): London, Geological Society of London, p. 45–80.

Peacock, J.D., 1956, The Geology of Dronning Louise Land, N.E. Greenland: Meddelelser om Grønland, v. 137, no. 7, 38 p.

Peacock, J.D., 1958, Some Investigations into the Geology and Petrography of Dronning Louise Land, N.E. Greenland: Meddelelser om Grønland, v. 157, no. 4, 139 p.

Peel, J.S., and Smith, M.P., 1988, The Wandel Valley Formation (Early–Middle Ordovician) of North Greenland and its correlatives: Rapport Grønlands Geologiske Undersøgelse, v. 137, p. 61–92.

Phillips, W.E.A., Stillman, C.J., Friderichsen, J.D., and Jemelin, L., 1973, Preliminary results of mapping in the western gneiss and schist zone around Vestfjord and inner Gåsefjord, south-west Scoresby Sund: Rapport Grønlands Geologiske Undersøgelse, v. 58, p. 17–32.

Pickerill, R.K., and Peel, J.S., 1990, Trace fossils from the Lower Cambrian Bastion Formation of North-East Greenland: Bulletin Grønlands Geologiske Undersøgelse, v. 147, p. 5–43.

Rasmussen, J.A., 1998, A reinterpretation of the conodont Atlantic realm in the late Early Ordovician (early Llanvirn), in Szaniawski, H., ed., Proceedings of the VI European Conodont Symposium ECOS VI: Palaeontologica Polonica, v. 58, p. 67–77.

Rasmussen, J.A., and Smith, M.P., 1996, Lower Palaeozoic carbonates in eastern North Greenland, and the demise of the "Sæfaxi Elv nappe": Bulletin Grønlands Geologiske Undersøgelse, v. 172, p. 49–54.

Rasmussen, J.A., and Smith, M.P., 2001, Conodont geothermometry and tectonic overburden in the northernmost East Greenland Caledonides: Geological Magazine, v. 138, p. 687–698.

Roberts, D., 2003, The Scandinavian Caledonides: Event chronology, palaeogeographic settings and likely modern analogues: Tectonophysics, v. 365, p. 283–299, doi: 10.1016/S0040-1951(03)00026-X.

Roberts, D., Heldal, T., and Melezhik, V.M., 2001, Tectonic structural features of the Fauske conglomerates in the Løvgavlen quarry, Nordland, Norwegian Caledonides, and regional implications: Norsk Geologisk Tidsskrift, v. 81, p. 245–256.

Roberts, D., Melezhik, V.M., and Heldal, T., 2002, Carbonate formations and NW-directed thrusting in the highest allochthons of the Norwegian Caledonides: Evidence of a Laurentian ancestry: Journal of the Geological Society of London, v. 159, p. 117–120.

Ross, R.J., Adler, F.J., Amsden, T.W., Bergstrom, D., Bergström, S.M., Carter, C., Churkin, M., Cressman, E.A., Derby, J.R., Dutro, J.T., Ethington, R.L., Finney, S.C., Fisher, D.W., Fisher, J.H., Harris, A.G., Hintze, L.F., Ketner, K.B., Kolata, D.L., Landing, E., Neuman, R.B., Sweet, W.C., Pojeta, J., Potter, A.W., Rader, E.K., Repetski, J.E., Shaver, R.H., Thompson, T.L., and Webers, G.F., 1982, The Ordovician System in the United States of America: Correlation Chart and Explanatory Notes: International Union of Geological Sciences Publication 12, 73 p.

Saltzman, M.R., Ripperdan, R.L., Brasier, M.D., Lohmann, K.C., Robison, R.A., Chang, W.T., Peng, S.C., Ergaliev, E.K., and Runnegar, B., 2000, A global carbon isotope excursion (SPICE) during the Late Cambrian: Relation to trilobite extinctions, organic-matter burial and sea level: Palaeogeography, Palaeoclimatology, Palaeoecology, v. 162, p. 211–223, doi: 10.1016/S0031-0182(00)00128-0.

Saltzman, M.R., Cowan, C.A., Runkel, A.C., Runnegar, B., Stewart, M.C., and Palmer, A.R., 2004, The Late Cambrian SPICE ($\delta^{13}C$) event and the Sauk II–Sauk III regression: New evidence from Laurentian basins in Utah, Iowa, and Newfoundland: Journal of Sedimentary Research, v. 74, p. 366–377, doi: 10.1306/120203740366.

Sclater, J.G., and Christie, P.A.F., 1980, Continental stretching: An explanation of the post–mid-Cretaceous subsidence of the central North Sea basin: Journal of Geophysical Research, v. 85, p. 3711–3739, doi: 10.1029/JB085iB07p03711.

Scotese, C.R., and McKerrow, W.S., 1990, Revised world maps and introduction, in McKerrow, W.S., and Scotese, C.R., eds., Palaeozoic Palaeogeography and Biogeography: Geological Society of London Memoir 12, p. 1–21.

Shergold, J.H., and Cooper, R.A., 2004, The Cambrian Period, in Gradstein, F.M., Ogg, J.G., and Smith, A.G., eds., A Geologic Time Scale, 2004: Cambridge, Cambridge University Press, p. 147–164.

Sloss, L.L., 1963, Sequences in the cratonic interior of North America: Geological Society of America Bulletin, v. 74, p. 93–113, doi: 10.1130/0016-7606 (1963)74[93:SITCIO]2.0.CO;2.

Sloss, L.L., 1988, Forty years of sequence stratigraphy: Geological Society of America Bulletin, v. 100, p. 1661–1665, doi: 10.1130/0016-7606(1988) 100<1661:FYOSS>2.3.CO;2.

Sloss, L.L., 1993, Sequence stratigraphy on the craton: Caveat emptor, in Witzke, B.J., Ludvigson, G.A., and Day, J., eds., Paleozoic Sequence Stratigraphy: Views from the North American Craton: Geological Society of America Special Paper 306, p. 425–434.

Smith, M.P., 1985, Ibexian–Whiterockian (Ordovician) Conodont Palaeontology of East and Eastern North Greenland [Ph.D. thesis]: Nottingham, UK, University of Nottingham, 372 p.

Smith, M.P., 1991, Early Ordovician Conodonts of East and North Greenland: Meddelelser om Grønland Geoscience, v. 26, 81 p.

Smith, M.P., 2000, Cambro-Ordovician stratigraphy of Bjørnøya and North Greenland: Constraints on tectonic models for the Arctic Caledonides and the Tertiary opening of the Greenland Sea: Journal of the Geological Society of London, v. 157, p. 459–470.

Smith, M.P., and Bjerreskov, M., 1994, The Ordovician System in Greenland: Correlation Chart and Stratigraphic Lexicon: International Union of Geological Sciences Special Publication 29A, 46 p.

Smith, M.P., Sønderholm, M., and Tull, S.J., 1989, The Morris Bugt Group (Middle Ordovician–Silurian) of North Greenland and its correlatives: Rapport Grønlands Geologiske Undersøgelse, v. 143, p. 5–20.

Smith, M.P., Soper, N.J., Higgins, A.K., and Rasmussen, J.A., 1999, Palaeokarst systems in the Late Proterozoic of eastern North Greenland and their stratigraphic and tectonic significance: Journal of the Geological Society of London, v. 156, p. 113–124.

Smith, M.P., Rasmussen, J.A., Higgins, A.K., and Leslie, A.G., 2004, Lower Palaeozoic stratigraphy of the East Greenland Caledonides: Geological Survey of Denmark and Greenland Bulletin, v. 6, p. 5–28.

Sønderholm, M., and Harland, T.L., 1989, Franklinian reef belt, Silurian, North Greenland, in Geldsetzer, H.J.J., James, N.P., and Tebbutt, G.E., eds., Reefs—Canada and Adjacent Areas: Canadian Society of Petroleum Geologists Memoir 13, p. 256–366.

Sønderholm, M., Frederiksen, K.S., Smith, M.P., and Tirsgaard, H., 2008, this volume, Neoproterozoic sedimentary basins with glacigenic deposits of the East Greenland Caledonides, in Higgins, A.K., Gilotti, J.A., and Smith, M.P., eds., The Greenland Caledonides: Evolution of the Northeast Margin of Laurentia: Geological Society of America Memoir 202, doi: 10.1130/2008.1202(05).

Soper, N.J., 1994, Was Scotland a Vendian RRR junction?: Journal of the Geological Society of London, v. 151, p. 579–582, doi: 10.1144/gsjgs.151.4.0579.

Soper, N.J., and Higgins, A.K., 1991a, Devonian–Early Carboniferous deformation and metamorphism, North Greenland, in Trettin, H.P., ed., Geology of the Innuitian Orogen and Arctic Platform of Canada and Greenland: Geological Survey of Canada, Geology of Canada, v. 3, p. 283–288.

Soper, N.J., and Higgins, A.K., 1991b, Late Cretaceous–early Tertiary deformation, North Greenland, in Trettin, H.P., ed., Geology of the Innuitian Orogen and Arctic Platform of Canada and Greenland: Geological Survey of Canada, Geology of Canada, v. 3, p. 461–465.

Soper, N.J., Ryan, P.D., and Dewey, J.F., 1999, Age of the Grampian orogeny in Scotland and Ireland: Journal of the Geological Society of London, v. 156, p. 1231–1236, doi: 10.1144/gsjgs.156.6.1231.

Spencer, R.J., and Demicco, R.V., 2002, Facies and sequence stratigraphy of two Cambrian grand cycles: Implications for Cambrian sea level and origin of grand cycles: Bulletin of Canadian Petroleum Geology, v. 50, p. 478–491, doi: 10.2113/50.4.478.

Stenzel, S.R., Knight, I., and James, N.P., 1990, Carbonate platform to foreland basin: Revised stratigraphy of the Table Head Group (Middle Ordovician), western Newfoundland: Canadian Journal of Earth Sciences, v. 27, p. 14–26.

Stouge, S., 1984, Conodonts of the Middle Ordovician Table Head Formation, Western Newfoundland: Fossils and Strata, v. 16, 145 p.

Stouge, S., Boyce, D.W., Christiansen, J., Harper, D.A.T., and Knight, I., 2001, Vendian–Lower Ordovician stratigraphy of Ella Ø, North-East Greenland: New investigations: Geology of Greenland Survey Bulletin, v. 189, p. 107–114.

Stouge, S., Boyce, D.W., Christiansen, J.L., Harper, D.A.T., and Knight, I., 2002, Lower–Middle Ordovician stratigraphy of North-East Greenland: Geology of Greenland Survey Bulletin, v. 189, p. 117–125.

Strachan, R.A., Friderichsen, J.D., Holdsworth, R.E., and Jepsen, H.F., 1994, Regional geology and Caledonian structure, Dronning Louise Land, North-East Greenland, in Higgins, A.K., ed., Geology of North-East Greenland: Rapport Grønlands Geologiske Undersøgelse, v. 162, p. 71–76.

Surlyk, F., 1991, Tectonostratigraphy of North Greenland: Bulletin Grønlands Geologiske Undersøgelse, v. 160, p. 25–47.

Surlyk, F., and Hurst, J.M., 1984, The evolution of the early Paleozoic deep-water basin of North Greenland: Geological Society of America Bulletin, v. 95, p. 131–154, doi: 10.1130/0016-7606(1984)95<131:TEOTEP>2.0.CO;2.

Sweet, W.C., 1984, Graphic correlation of upper Middle and Upper Ordovician rocks, North American Midcontinent Province, U.S.A., in Bruton, D.L., ed., Aspects of the Ordovician System: Palaeontological Contributions of the University of Oslo, v. 295, p. 23–35.

Sweet, W.C., and Tolbert, C.M., 1997, An Ibexian (Lower Ordovician) reference section in the southern Egan Range, Nevada, for a conodont-based chronostratigraphy: U.S. Geological Survey Professional Paper 1579, p. 53–84.

Swett, K., 1981, Cambro-Ordovician strata in Ny Friesland, Spitsbergen and their palaeotectonic significance: Geological Magazine, v. 118, p. 225–250.

Swett, K., and Smit, D.E., 1972, Paleogeography and depositional environments of the Cambro-Ordovician shallow marine facies of the North Atlantic: Geological Society of America Bulletin, v. 83, p. 3223–3248, doi: 10.1130/0016-7606(1972)83[3223:PADEOT]2.0.CO;2.

Szaniawski, H., 1994, Ordovician conodonts from the Hornsund Region, southern Spitsbergen, in Zalewski, S.M., ed., XXI Polar Symposium: 60 years of Polish Research of Spitsbergen: Warsaw, Institute of Geophysics of Polish Academy of Sciences, p. 39–44.

Torsvik, T.H., Smethurst, M.A., Meert, J.G., van der Voo, R., McKerrow, W.S., Brasier, M.D., Sturt, B.A., and Walderhaug, H.J., 1996, Continental break-up and collision in the Neoproterozoic and Palaeozoic—A tale of Baltica and Laurentia: Earth-Science Reviews, v. 40, p. 229–258, doi: 10.1016/0012-8252(96)00008-6.

Trettin, H.P., Okulitch, A.V., Packard, J.J., Smith, G.P., Harrison, J.C., Fox, F.G., and Brent, T.A., 1991, Silurian–Early Carboniferous deformational phases and associated metamorphism and plutonism, Arctic Islands, in Trettin, H.P., ed., Geology of the Innuitian Orogen and Arctic Platform of Canada and Greenland: Geological Survey of Canada, Geology of Canada, v. 3, p. 295–341.

Tull, S.J., 1988, Conodont Micropalaeontology of the Morris Bugt Group (Middle Ordovician–Early Silurian) [Ph.D. thesis]: Nottingham, UK, University of Nottingham, 366 p.

Webby, B.D., Cooper, R.A., Bergström, S.M., and Paris, F., 2004, Stratigraphic framework and time slices, in Webby, B.D., Paris, F., Droser, M.L., and Percival, I.G., eds., The Great Ordovician Biodiversification Event: New York, Columbia University Press, p. 41–47.

Williams, H., ed., 1995, Geology of the Appalachian–Caledonian Orogen in Canada and Greenland: Geological Survey of Canada, Geology of Canada, v. 6, 944 p.

Wright, D.T., and Knight, I., 1995, A revised chronostratigraphy for the lower Durness Group: Scottish Journal of Geology, v. 31, p. 11–22.

MANUSCRIPT ACCEPTED BY THE SOCIETY 14 JANUARY 2008

The Geological Society of America
Memoir 202
2008

Foreland-propagating Caledonian thrust systems in East Greenland

A. Graham Leslie*
British Geological Survey, Murchison House, Edinburgh EH9 3LA, UK

A.K. Higgins
Geological Survey of Denmark and Greenland, Øster Voldgade 10, DK-1350 Copenhagen K, Denmark

ABSTRACT

The 1300-km-long, up to 300-km-wide onshore segment of the East Greenland Caledonian orogen is divided into distinct structurally bound geological domains that originally evolved as major westward-displaced thrust units during collision with Baltica. The thrust systems accommodated contraction of an already complex Laurentian assembly of Archean to Neoproterozoic and Cambrian to Silurian lithostratigraphic units and are a consequence of the convergence, and final collision, of Baltica with Laurentia in the mid- to late Silurian Scandian orogeny. The transition from undisturbed foreland to orogen is perfectly preserved in the extreme north of the East Greenland Caledonides, where a younger lower (Vandredalen) thrust sheet carrying older thrust sheets (Western thrust belt) is displaced westward across a thin-skinned fold-and-thrust belt. In the southern half of the orogen, a pile of far-traveled thrust sheets (from youngest to oldest, Gemmedal, Niggli Spids, Hagar Bjerg thrust sheets) is displaced WNW across parautochthonous foreland windows, and the intact foreland is only intermittently exposed at the margin of the Inland Ice in the far west. These westward- and foreland-propagating systems are distinct from the Nørreland thrust sheet, the coastal region between 76°N and 79°N, in which Paleoproterozoic basement gneiss lithologies host enclaves of Devonian and Carboniferous eclogite-facies rocks. These rocks must have been exhumed from the roots of the collisional orogen, and their age suggests that the Nørreland thrust may be out of sequence relative to the main WNW foreland-propagating systems.

Keywords: Caledonian, thrust units, foreland propagating, WNW transport.

INTRODUCTION

The Caledonian orogen of East Greenland (Fig. 1) is composed of a sequence of westward-propagating thrust sheets that were created during the collision of a broad sector of Baltica with the northeastern margin of Laurentia. The west margin of the Caledonian orogen is largely obscured by the central continental ice cap (Inland Ice), but areas interpreted as Caledonian foreland and foreland windows have been recognized throughout the length of the orogen. The existence of foreland areas structurally underlying major thrust sheets was first demonstrated in Kronprins Christian Land in the extreme north of the orogen by Fränkl (1954, 1955), in Dronning Louise Land by the 1952–1954 British North Greenland Expedition (Peacock,

*agle@bgs.ac.uk

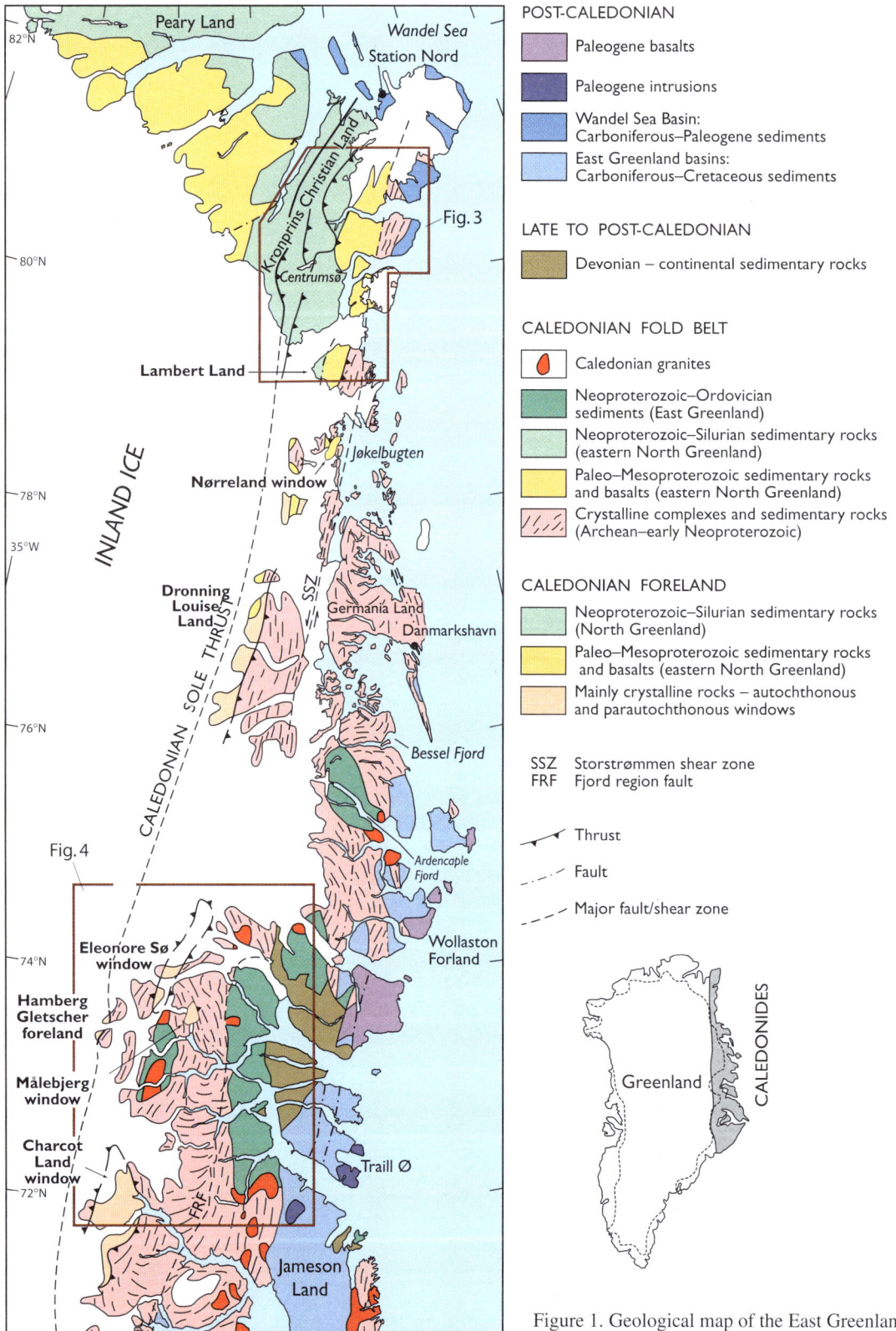

Figure 1. Geological map of the East Greenland Caledonide orogen, showing location of the main foreland windows, the approximate line of the Caledonian sole thrust, and some of the major faults and shear zones (Elvevold et al., 2000). Boxes outline regions shown in more detail in Figures 3 and 4.

1956, 1958), and in the extreme south of the orogen in Gåseland by Wenk (1961). Since 1968, the Geological Survey of Greenland (GGU) and its successor, the Geological Survey of Denmark and Greenland (GEUS), have carried out a systematic geological mapping program that has revealed evidence for several additional foreland windows (Fig. 1). Displacements on the order of hundreds of kilometers were convincingly demonstrated for Caledonian thrusts in central parts of East Greenland during the 1997–1998 mapping of the Kong Oscar Fjord region (72°N–74°N; Higgins and Leslie, 2000).

The orogen can be broadly divided into a number of structural domains (Fig. 2), which are outlined in more detail by Higgins and Leslie (this volume). In Kronprins Christian Land, in the northernmost sector of the orogen, the transition from intact to imbricated foreland is perfectly preserved in a foreland-propagating succession of major thrust sheets affecting Paleoproterozoic to Neoproterozoic and Cambrian to Silurian rocks. Farther south, Paleoproterozoic to Mesoproterozoic sedimentary and basic igneous rocks are present in significant amounts in thrust sheets exposed in Lambert Land and in the western nunatak region. Present-day exposure in the coastal region between eastern Kronprins Christian Land and Bessel Fjord (76°N) is dominated by Paleoproterozoic gneiss complexes, characterized over large areas by the presence of eclogitic enclaves (Gilotti, 1993; Brueckner et al., 1998). The southern half of the orogen (south of Bessel Fjord, 76°N) is dominated by two major foreland-propagating thrust sheets (Niggli Spids thrust sheet and Hagar Bjerg thrust sheet), with the uppermost structural level of the latter consisting of the thick Neoproterozoic–Ordovician succession of the Eleonore Bay Supergroup, Tillite Group, and Kong Oscar Fjord Group. These two major thrust sheets have traveled hundreds of kilometers across the foreland, as revealed in anticlinal structural windows. A thin and discontinuous lowermost thrust unit has also been identified between these windows and the base of the thick overriding thrust sheets.

This paper describes W-E cross sections constructed from field observations of the thrust systems preserved in the northern and southern segments of the East Greenland Caledonian orogen. These cross sections provide a record of the structures that resulted from the collision of Laurentia and Baltica during the mid- to late Silurian Scandian events. The completely exposed transition from foreland to orogen in the northern orogen is examined in two cross sections through Kronprins Christian Land (~80°30′N; Fig. 1); the architecture of the far-traveled thrust sheets in the southern orogen is described in profiles that include the recently discovered foreland windows at Eleonore Sø and Målebjerg (~73°40′N; Fig. 1). The stratigraphic framework and structural background for the orogen are summarized briefly next (Higgins and Leslie [this volume] provide additional detail). The high-grade Nørreland thrust sheet is only represented at the eastern end of one of the northern cross sections, and we do not attempt to deal here with details of the generation and exhumation of the high-pressure lithologies preserved in that thrust sheet. Caledonian metamorphic patterns are described by Gilotti et al. (this volume), and aspects of extensional faulting superimposed on the thrust systems are dealt with by Gilotti and McClelland (this volume).

LITHOSTRATIGRAPHIC AND TECTONIC FRAMEWORK

This section outlines the lithostratigraphic elements recognized in the foreland and in the principal parautochthonous to allochthonous structural domains (see also Higgins and Leslie, this volume, their Table 1). Particular emphasis is placed on the domains represented in the cross sections through the northern and southern segments of the orogen that form the subject of the detailed descriptions in later sections (Figs. 2, 3, and 4).

Foreland (70°N–82°N; Figs. 1 and 2)

In the autochthonous foreland region west of Kronprins Christian Land, the oldest strata exposed are the sandstones of the ~2000-m-thick Independence Fjord Group. The base of the succession is not exposed anywhere in this area, and the minimum age of the succession is constrained only by the oldest dated 1382 Ma dolerite dikes that cut them (Upton et al., 2005). As noted elsewhere (e.g., Collinson et al., this volume), isotopic evidence from eastern Kronprins Christian Land suggests that deposition of the Independence Fjord Group may have begun in the later part of the Paleoproterozoic (see also Kalsbeek et al., 1999).

The Independence Fjord Group is overlain by a succession of tholeiitic flood basalts (Zig-Zag Dal Basalt Formation; Kalsbeek and Jepsen, 1984; Upton et al., 2005), which are up to 1350 m thick. The abundant doleritic dike swarms (Midsommersø Dolerites) that cut the sandstones of the Independence Fjord Group are broadly correlated with the overlying basalts.

The 1000-m-thick Neoproterozoic Hagen Fjord Group onlaps the Independence Fjord Group and Zig-Zag Dal Formation with a very low-angle unconformity (Sønderholm and Jepsen, 1991). The lower siliciclastic formations are overlain by orange- to red-weathering limestones (Kap Bernhard Formation), and yellow-weathering dolomites with abundant developments of stromatolites (Fyns Sø Formation). The Early Cambrian Kap Holbæk Formation rests unconformably upon the Hagen Fjord Group (Smith et al., 2004a). Regional uplift of eastern North Greenland during the Cambrian and Early Ordovician culminated in the west to east overstepping of older strata by the Early–Middle Ordovician to Silurian carbonate succession.

In westernmost Kronprins Christian Land, the west margin of the Caledonian orogen corresponds to the westernmost limit of deformation in the Thin-skinned fold-and-thrust belt (Fig. 3). The west margin of the orogen can be traced throughout Kronprins Christian Land, but, southward, it disappears beneath the Inland Ice; its southern continuation has been portrayed in various ways, and a link with the imbricate thrust zone that divides Dronning Louise Land into western and eastern parts is the structure favored in older descriptions (Haller, 1971; Henriksen and Higgins, 1976).

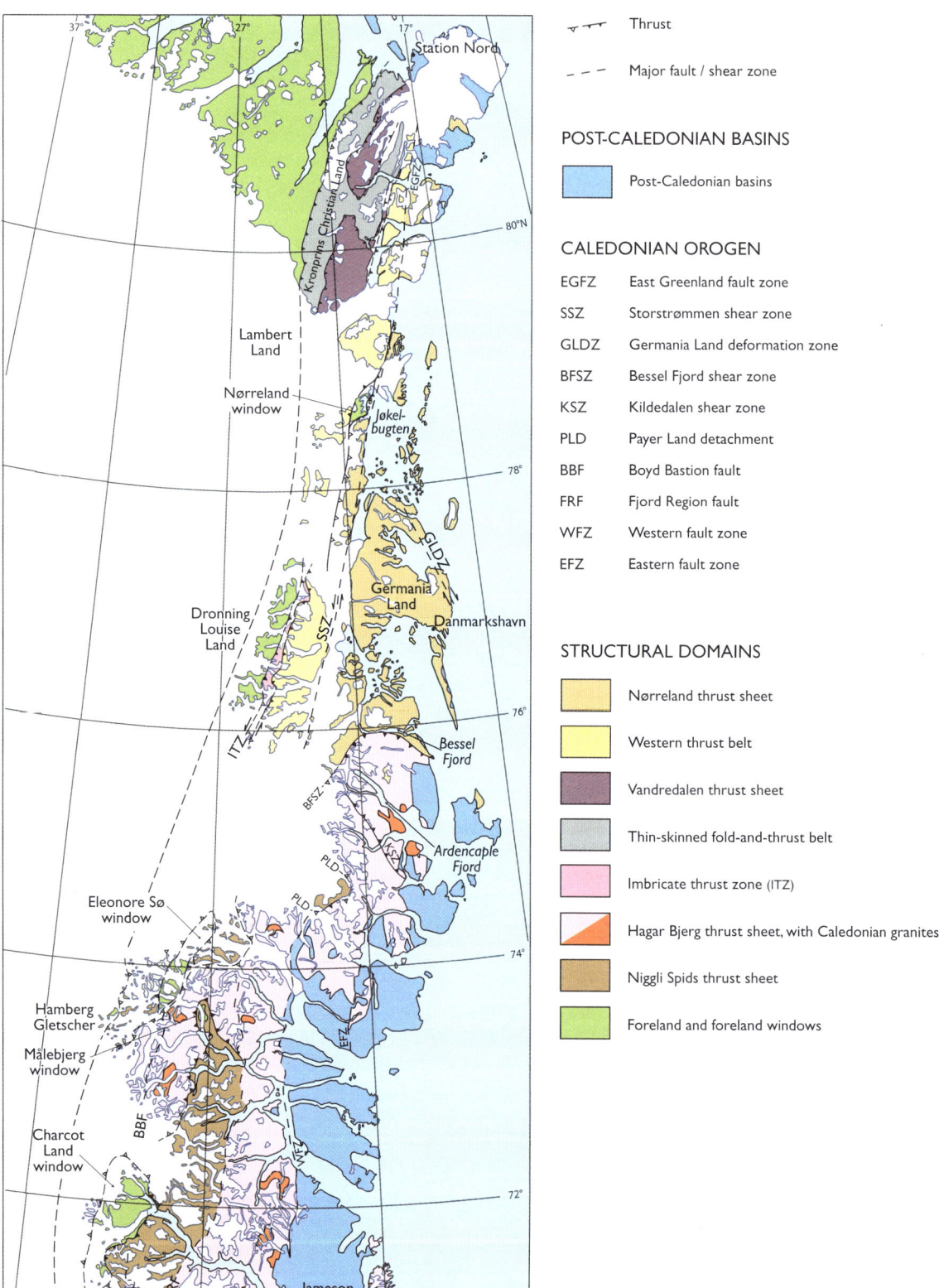

Figure 2. Principal geological domains of the East Greenland Caledonian orogen (after Higgins and Leslie, this volume). The locally developed Gemmedal thrust sheet, beneath the Niggli Spids thrust sheet, is too thin to be depicted here, but it is shown on the map of Figure 4. The main shear zones and faults are also indicated.

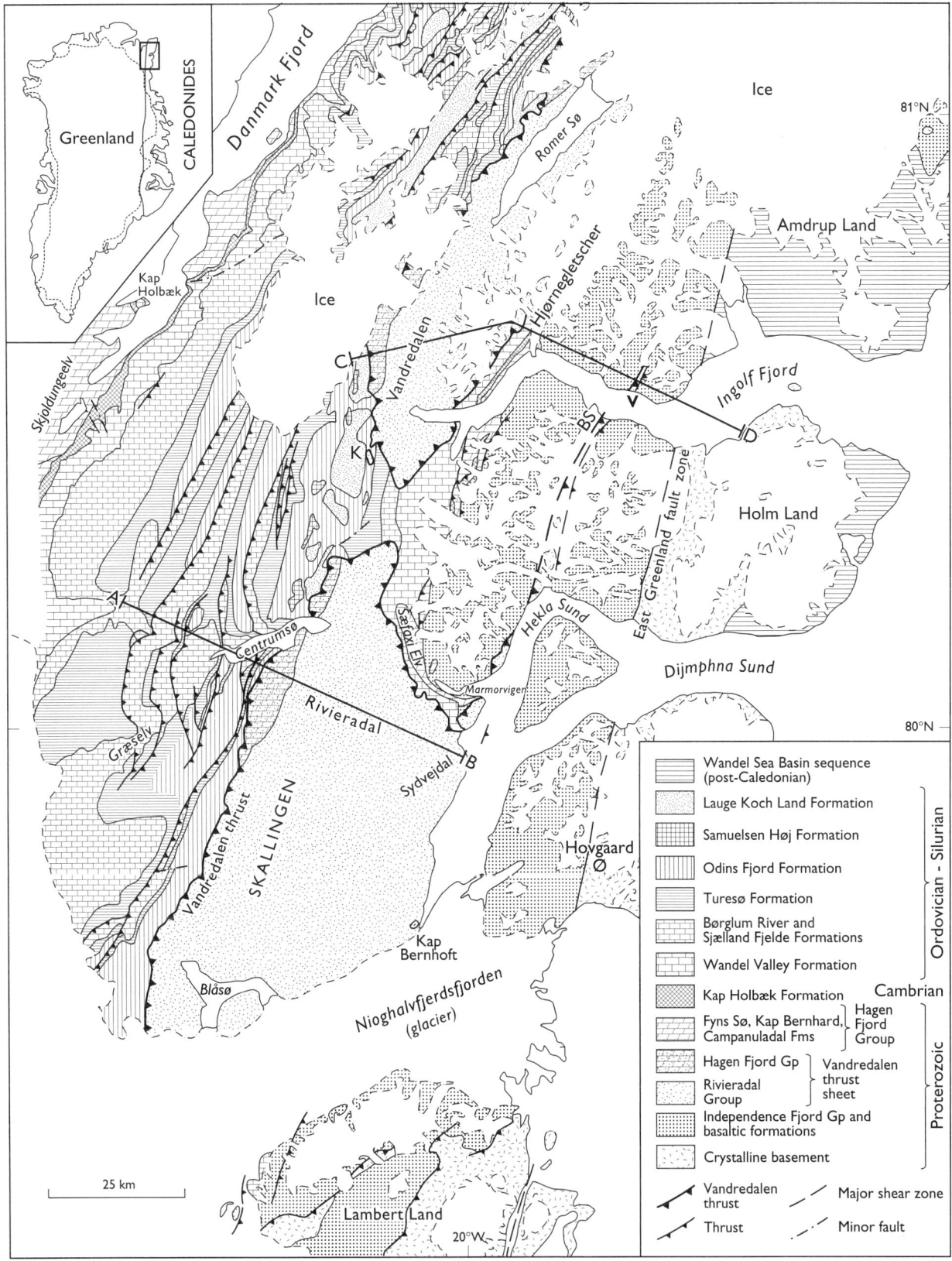

Figure 3. Geological map of Kronprins Christian Land and Lambert Land in the northern part of the Caledonian orogen (modified after Rasmussen and Smith, 2001). A–B and C–D are cross-section lines illustrated in Figures 8 and 10. BS—Brede Spærregletscher; K—Keglen; V—Vardedalen.

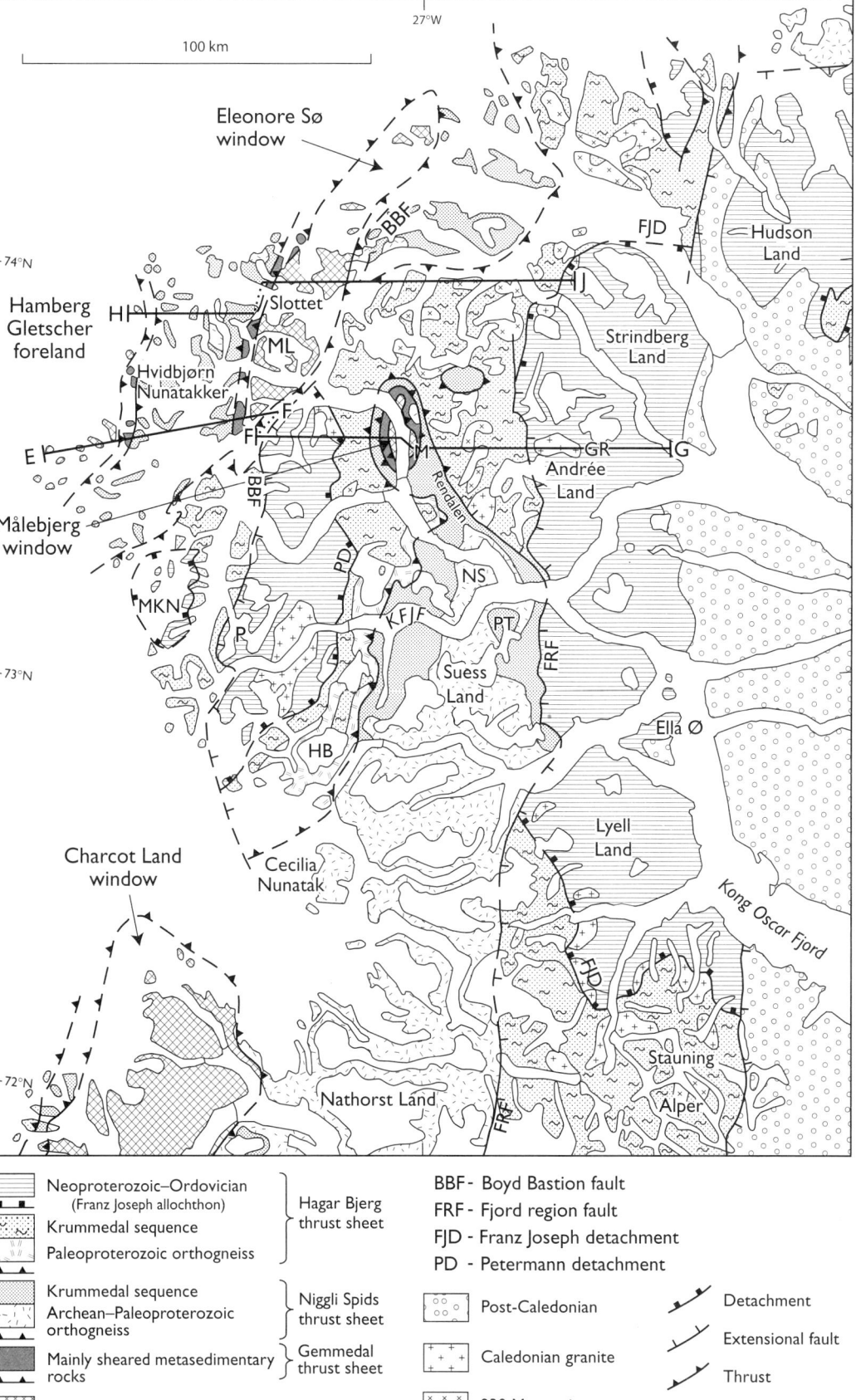

Figure 4. Geological map of the East Greenland Caledonian orogen from 72°N to 74°30′N, showing location of the Charcot Land, Eleonore Sø, and Målebjerg windows, and the Hamberg Gletscher foreland. E–F–G and H–J are the lines of the cross sections shown in Figures 13 and 16. A klippe of Eleonore Bay Supergroup sedimentary rocks in the Martin Knudsen Nunatakker (MKN) forms the westernmost exposure known in the region. GR—Grejsdalen; HB—Hagar Bjerg; KFJF—Kejser Franz Joseph Fjord; M—Målebjerg; ML—J.L. Mowinckel Land; NS—Niggli Spids; P—Petermann Bjerg; PT—Payer Tinde. Modified from Smith et al. (2004a).

However, the western half of Dronning Louise Land shows extensive Caledonian deformation (Strachan et al., 1992, their Fig. 7) and is interpreted here as parautochthonous rather than autochthonous foreland, where the Caledonian orogenic margin is located below the west side of Dronning Louise Land, probably above a blind thrust that branches off the imbricate thrust zone.

The parautochthonous foreland of western Dronning Louise Land is dominated by Paleoproterozoic gneisses. These are overlain in a few areas by two distinct sequences of quartzitic rocks, a lower sequence cut by dolerite dikes known as the Trekant "series" and an upper sequence lacking dikes known as the Zebra "series" (Strachan et al., 1994). The Trekant "series" is considered to be a southern correlative of the Independence Fjord Group that in Dronning Louise Land preserves the basal unconformity not known farther north. The quartzitic units of the overlying Zebra "series" locally preserve *Skolithos* ichnofossils and have been correlated with the Early Cambrian Kap Holbæk Formation in the foreland west of Kronprins Christian Land and the Slottet Formation of the Eleonore Sø and Målebjerg windows to the south.

Farther south, autochthonous foreland is exposed in the Hamberg Gletscher area in the far west and the Gåseland window in the south, and parautochthonous foreland is exposed in the Eleonore Sø, Målebjerg, and Charcot Land windows (Fig. 4). Paleoproterozoic gneisses are exposed in all windows, where they are overlain by a variety of sedimentary and volcanic rocks. The similarities and correlations between the different foreland areas have been presented elsewhere by Higgins et al. (2001a, 2004a). Rift-related pre–1900 Ma volcanic and sedimentary rocks are preserved in the Hamberg Gletscher foreland and the Charcot Land and Eleonore Sø foreland windows, but they are not represented in the structurally overlying thrust sheets. Conversely, the widespread Mesoproterozoic to early Neoproterozoic Krummedal metasedimentary rocks of the overlying thrust sheets are absent in the foreland. Diamictites correlated with Marinoan tillites of the Tillite Group (Sønderholm et al., this volume) have been recognized in the Gåseland, Charcot Land, and Målebjerg foreland windows, where they occupy depressions in the peneplaned gneissic basement. In the Målebjerg window, the diamictites are overlain by a thin, <400-m-thick sequence of Lower Paleozoic sedimentary rocks (Slottet and Målebjerg Formations). The Slottet Formation quartzites preserve abundant *Skolithos* ichnofossils, and this formation is clearly the source of the abundant glacial erratics of *Skolithos*-bearing quartzites known from central East Greenland (Haller, 1971).

Thin-Skinned Fold-and-Thrust Belt (79°30′N–81°30′N; Figs. 2 and 3)

Throughout western Kronprins Christian Land, the undeformed Ordovician–Silurian carbonates of the Iapetus passive margin preserved in the foreland pass eastward into a thin-skinned fold-and-thrust belt developed in the same strata. A series of NNE-SSW–striking and E-dipping Caledonian thrusts characterizes this belt, and there are typically a few kilometers of displacement on each thrust. Eastward, structurally below the Vandredalen thrust sheet, the thin-skinned thrusts penetrate to deeper levels, and thrust duplexes and west-vergent recumbent folds are developed in the Hagen Fjord Group and Independence Fjord Group sandstones (see later description of section C–D, Kronprins Christian Land). These folds and thrusts raise the Vandredalen thrust sheet above exposure level, and well-exposed cliff sections occur in the western parts of central Ingolf Fjord.

A line-and-area balance along a well-exposed WNW-ESE section through the western part of the fold-and-thrust belt along Centrumsø (Higgins et al., 2004b; see later discussion of section A–B, Kronprins Christian Land) demonstrates ~17 km total displacement, equivalent to 45% shortening in the line of section. On the basis of the alteration colors of conodonts in the carbonate succession, used for biostratigraphic control, Rasmussen and Smith (2001) estimated the burial temperatures and thus the thickness of overlying thrust sheets. Overburden was calculated to range from 6 km to ~12.5 km from west to east across the fold-and-thrust belt, but as the thickness of the Vandredalen thrust sheet alone was insufficient to yield the temperatures attained, higher thrust sheets (parts of the Western thrust belt) must have extended across the region.

The southward continuation of the thin-skinned fold-and-thrust belt is hidden beneath the Inland Ice south of 79°30′N, and no further exposures are known.

Vandredalen Thrust Sheet (79°20′N–81°15′N; Figs. 2 and 3)

The Vandredalen thrust sheet structurally overlies the Thin-skinned fold-and-thrust belt, and it is made up of rocks belonging to the Neoproterozoic Rivieradal and Hagen Fjord Groups. This coherent package of distinctive rock types has been thrust westward some 35–50 km (Smith et al., 2004b), and its well-exposed NNE-SSW–trending frontal ramp can be followed for 200 km through central parts of Kronprins Christian Land. Matching cutoffs of the Fyns Sø Formation (Hagen Fjord Group) in the hanging wall and footwall of the Vandredalen thrust provide the estimates of displacement. The Rivieradal Group was deposited in an east-facing half-graben (Hekla Sund Basin; Higgins et al., 2001b), and the remnants of the west margin of this rift, and the root zone of the Vandredalen thrust sheet, can be traced as a narrow strip of metasedimentary rocks from Marmorvigen northward along the west side of Hekla Sund, and along Store Bredegletscher to Ingolf Fjord and Vardedalen (Figs. 3 and 5).

The Rivieradal Group is widely exposed in Skallingen and northern parts of Vandredalen (Fig. 3; see sections A–B and C–D discussed later). The northernmost exposures are around Romer Sø, and the southernmost exposures occur on small nunataks west of Lambert Land. The present-day extent of the Rivieradal Group within the Vandredalen thrust sheet indicates that the half-graben Hekla Sund Basin in which it was deposited was at least 50 km wide and more than 200 km long. The

Figure 5. Looking northward from Marmorvigen to the west side of Hekla Sund. At left, Independence Fjord Group quartzites cut by doleritic sills and dikes lie in the footwall of the Vandredalen thrust, which dips steeply eastward (to the right) in the center of the photograph. At right, in the hanging wall, steeply dipping, dark-colored shaley metasedimentary rocks of the Rivieradal Group can be seen. Profile height is ~1200 m.

westernmost outcrops of the Rivieradal Group, preserved along the frontal thrust ramp, including coarse conglomerates derived from a source region to the west and the white quartzite and dark dolerite clasts, can be matched with the Independence Fjord Group and the Midsommersø Dolerites. The Independence Fjord Group and the Midsommersø Dolerites also make up the west margin of the original rift basin exposed along Hekla Sund and at Marmorvigen. During the Caledonian displacement of the fill of the Hekla Sund Basin across its western rift shoulder (Higgins et al., 2001b), small and large quartzitic segments were excised from the rift shoulder and carried westward along the Vandredalen thrust plane, where they formed isolated horses characterized by anomalous younger-on-older-on-younger relationships between the Rivieradal Group sedimentary rocks in the hanging wall and Ordovician–Silurian carbonates in the footwall (see also section C–D discussed later).

Western Thrust Belt (76°N–81°N; Fig. 2)

The Western thrust belt is composed of a group of poorly defined thrust sheets that structurally overlie the Vandredalen thrust sheet. In Kronprins Christian Land, the northern part of the Western thrust belt is dominated by white quartzites of the Independence Fjord Group, which are cut by numerous black doleritic diks and sills (Midsommersø Dolerites), interbedded with a number of volcanic units (Hekla Sund Formation; Pedersen et al., 2002; Collinson et al., this volume). The Western thrust belt can be traced from east-central Ingolf Fjord southward to Hekla Sund and through western Hovgaard Ø (Figs. 2 and 3). Farther south, the thrust belt continues into Lambert Land, where a pile of both north- and west-directed thrust units has been recognized (K.A. Jones and J.C. Escher, 1995, personal commun.). The thrust sheets in Lambert Land are dominated by Independence Fjord Group quartzites and cross-cutting dikes and sills, but several thrust sheets also incorporate slices of basement gneisses.

Farther south, the Western thrust belt continues through the large nunataks along the Inland Ice margin west of Jøkelbugten (Fig. 2), where the thrust sheets are composed of both gneisses and quartzitic rocks; however, the structure of this region is relatively poorly known due to the reconnaissance scale of the original mapping. The southward continuation of the thrust belt extends into the eastern half of Dronning Louise Land, which is dominated by amphibolite-facies gneisses and only few exposures of overlying quartzitic rocks (Trekant "series" and Zebra "series"; Strachan et al., 1994).

Nørreland Thrust Sheet (76°N–81°N; Fig. 2)

The Nørreland thrust sheet is made up of a broad zone of Paleoproterozoic gneiss complexes (Kalsbeek et al., 1993; Kalsbeek, 1995) that contain abundant Caledonian eclogitic enclaves over large areas (Brueckner et al., 1998; Gilotti and Ravna, 2002; Gilotti et al., this volume). These high-grade rocks dominate the coastal region between the major south-dipping shear zone along Bessel Fjord (76°N; Fig. 2), which marks its southern limit, and the northernmost outcrops in Hovgaard Ø and Holm Land (easternmost Kronprins Christian Land; Figs. 2 and 3). The presence of Devonian eclogites (Gilotti et al., 2004) testifies to late Caledonian exhumation from depths in excess of 50 km.

The western limit of this high-grade domain is placed where gneiss complexes containing eclogitic enclaves structurally overlie lower-grade gneiss complexes imbricated with epidote-amphibolite–facies Paleoproterozoic quartzites; the best known outcrops are in eastern Lambert Land and west of Jøkelbugten around the Nørreland window (Fig. 1), from which the Nørreland thrust sheet takes its name. Gilotti et al. (this volume) state that the extensive high-grade region making up the Nørreland thrust sheet is unlikely to represent a single continuous crustal slab, but that no internal thrust divisions have been recognized to date. In view of the Devonian age of the eclogites, the Nørreland thrust is likely to be out of sequence relative to the earlier foreland-propagating systems.

In the cross sections described in this chapter, the high-grade Nørreland thrust sheet is only represented at the eastern end of section C–D in Holm Land, Kronprins Christian Land (Fig. 3; see also section C–D detailed later). The boundary between the eclogite-bearing gneisses and the lower-grade rocks to the west that make up the Western thrust belt is a prominent lineament known as the East Greenland fault zone (Fig. 3); this structure is believed to have controlled deposition in a Carboniferous basin that lay to the east of the lineament (Stemmerik and Håkansson, 1991). In present-day exposures, Carboniferous sedimentary rocks unconformably overlie the eclogite-bearing gneisses in eastern Holm Land. Workers looking for the northward continuation of the prominent NNW-SSE–trending Storstrømmen shear zone, which can be traced continuously from the eastern side of Dronning Louise Land to eastern Lambert Land (Figs. 1 and 2), have speculated that it might coincide with the East Greenland fault zone (Larsen and Bengaard, 1991; Holdsworth and Strachan, 1991; Jepsen et al., 1994); this implies that the East Greenland fault zone might have a pre-Carboniferous tectonic history (see also description of section C–D).

Gemmedal Thrust Sheet (73°N–74°N; Fig. 4)

A relatively thin assemblage of diverse lithologies at the base of the Niggli Spids thrust sheet is identified as the Gemmedal thrust sheet. The main exposures of the Gemmedal thrust sheet are shown on the geological map of the southern orogen (Fig. 4). The lithologies represented include low-grade black shales in upper parts of the Hamberg Gletscher area, dark mylonitic metasedimentary units along the west side of the Eleonore Sø window, and mica schists associated with quartzites and carbonates around the Målebjerg window. There appear to be no equivalents of the Gemmedal thrust sheet around the Charcot Land window. However, analogous occurrences may be present in the Gåseland window in the extreme south of the orogen, where a several-hundred-meter-thick sequence of strongly sheared carbonates and quartzites occurs above the basal thrust (Phillips et al., 1973).

Niggli Spids Thrust Sheet (70°N–76°30′N; Figs. 2 and 4)

The Niggli Spids thrust sheet is the lower of the two major thrust sheets recognized throughout the southern half of the Caledonian orogen. In most of the region, the lower part of the thrust sheet is made up of Archean or Paleoproterozoic basement gneiss complexes, and the overlying, upper part is made up of a thick succession of Mesoproterozoic to lower Neoproterozoic metasedimentary rocks known as the Krummedal supracrustal sequence (Higgins, 1988). The Niggli Spids thrust sheet is separated from the underlying foreland in some areas by the low-grade metasedimentary rocks distinguished as the Gemmedal thrust sheet.

The basement gneiss complexes that form the lower part of the thrust sheet are widely exposed in the western part of the Scoresby Sund region (70°N–72°N), and they form spectacular cliffs in the inner fjord system in the Kong Oscar Fjord region (72°N–74°N; Fig. 6). The basement gneiss complexes overlie the eastern margins of both the Gåseland and Charcot Land windows, but they wedge out westward such that Krummedal sequence metasedimentary rocks of the upper part of the Niggli Spids thrust sheet overlie the western parts of both windows. The basement gneiss complexes also wedge out locally northward, and, around the Målebjerg window, they are only a few hundred meters thick.

The Krummedal sequence metasedimentary rocks include dominantly semipelitic and psammitic lithologies, and the sequence varies in thickness from 2 to 4 km over much of the region. The overall dark brown–weathering color contrasts with the underlying light-colored gneisses (Fig. 6), and the contact with the gneisses is thought to have once been an unconformity. However, original relationships are nearly everywhere marked by strong Caledonian shearing. Three or four Caledonian deformation phases are recognized (Escher and Jones, 1998, 1999; Leslie and Higgins, 1998, 1999), and metamorphic grade is medium to high amphibolite facies (Gilotti et al., this volume), with widespread Caledonian high-pressure relics. In contrast to the Krummedal sequence in the overlying Hagar Bjerg thrust sheet, production of Caledonian granitic melt was insignificant, and granite veins and sheets are only locally abundant, for example, in Payer Land and Gletscherland (Gilotti et al., this volume).

Displacement indicators throughout the thick metasedimentary pile are top-to-the-WNW (Fig. 7). The WNW-ESE distance from the east side of the Hamberg Gletscher foreland

Figure 6. View eastward along Kejser Franz Joseph Fjord, with Payer Tinde (2320 m), the prominent peak, at right. The steep cliffs at left (1200–1400 m high) are eroded in the Niggli Spids thrust sheet. Light-colored Paleoproterozoic orthogneisses form the cliffs and are overlain by a thin unit of light-colored carbonate and a thicker succession of dark-weathering sandstones and mica schists (Krummedal supracrustal sequence).

Figure 7. Shear sense criteria in the Gemmedal, Niggli Spids, and Hagar Bjerg thrust sheets observed across the area transected by the cross sections in Figures 13 and 16. All field photographs and photomicrographs are oriented such that the observer is looking approximately north. The sense of shear is top-to-the WNW in A–E and G, but top-to-the ESE in H. (A–C) In the footwall to the Franz Joseph detachment, shear sense is defined by lenticular bodies of quartzofeldspathic granitic neosome developed at a variety of scales within garnet-kyanite–grade semipelitic and pelitic gneisses of the Hagar Bjerg thrust sheet, locality k on Figure 16. In A the person is 1.8 m tall, in B the lens cap is 52 mm across, and in C the hammer head is ~20 cm long. (D) Shear sense is defined by sigma-type quartzofeldspathic augen within semipelitic and pelitic schists of the Gemmedal thrust sheet. Location is above the Eleonore Sø window in J.L. Mowinckel Land, near locality b in Figure 13. Hammer head is ~20 cm long. (E) Shear sense is defined by elongated quartz-rich lenticles in garnet-biotite–grade semipelitic and psammitic schists of the Niggli Spids thrust sheet, east of Målebjerg, near locality f on Figure 13. Pencil is 18 cm long. (F) A lower-hemisphere stereographic projection of all measured lineations associated with the top-to-the-WNW shear fabrics observed throughout the area traversed by the constructed cross sections of Figures 13 and 16. ESE-plunging lineations are dominant, and WNW-plunging data are observed on the western flanks of the antiformal windows. (G) Shear sense is defined by lenticular aggregates of muscovite and biotite (+ garnet and quartz). An earlier kyanite-garnet assemblage predates shearing and is represented by the garnet porphyroclast in the top right of the view. Field of view is 1 mm. Sample is from pelitic gneiss at the base of the Hagar Bjerg thrust sheet. Locality: Hagar Bjerg thrust in Rendalen, just south of section line in panel 4, Figure 13. (H) Top-to-the-ESE shearing is indicated by lenticular feldspar porphyroclasts within a mylonitic quartz-rich foliated matrix. Field of view is 1 mm. Locality is same as G. This shear zone (H) transects the mylonitic rocks shown in part G (dipping more steeply to the ESE) and illustrates the polyphase nature of these major thrust sheet–bounding structures.

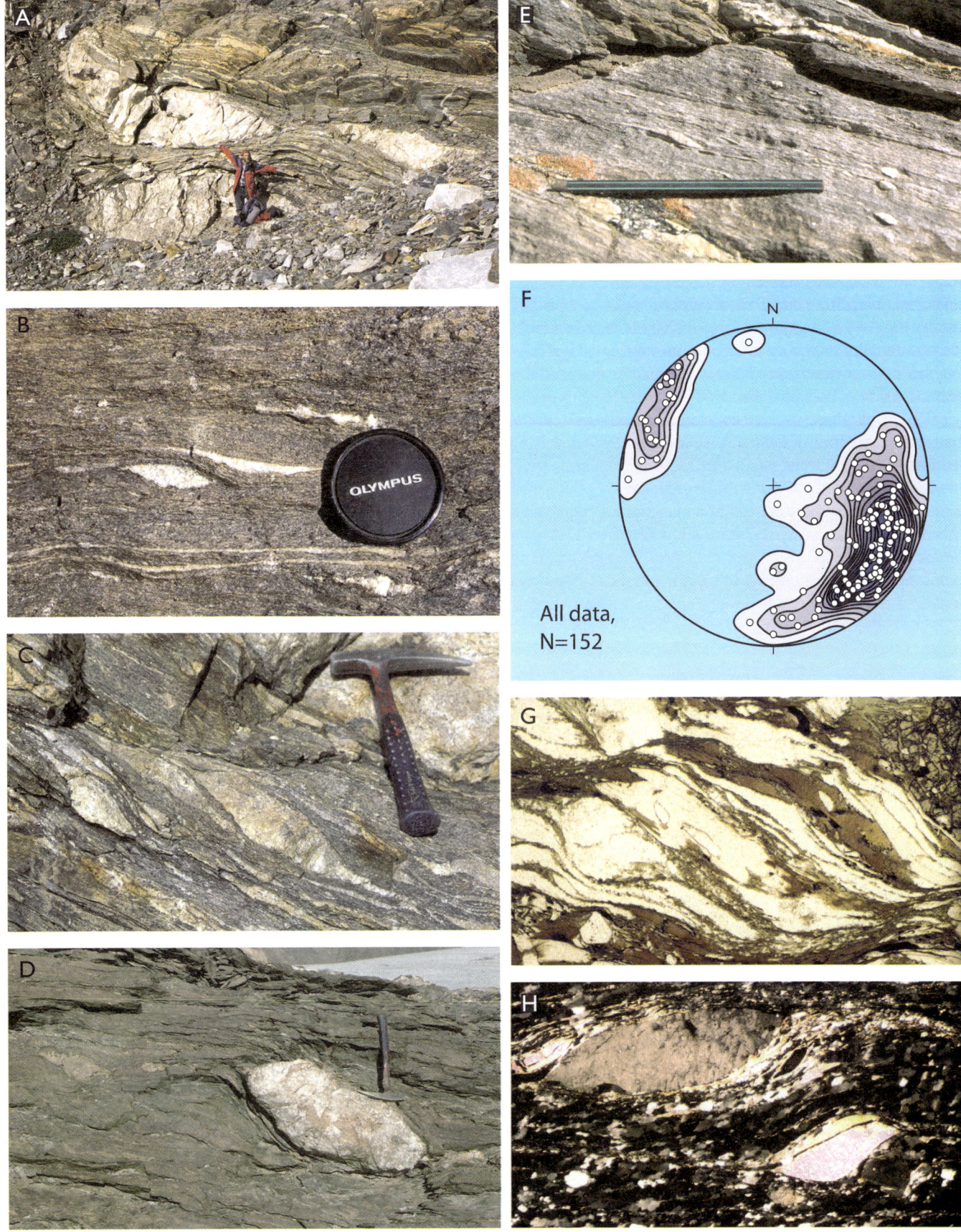

to the east side of the Målebjerg window provides a minimum estimate of displacement for the Niggli Spids thrust sheet of 90 km (carrying the older Hagar Bjerg thrust sheet piggyback above it), but outcrops attributed to the Niggli Spids thrust sheet as far east as the Fjord Region fault imply that WNW-ESE displacements relative to the underlying foreland windows may have been much greater.

Hagar Bjerg Thrust Sheet and Franz Joseph Allochthon (70°N–74°30′N; Figs. 2 and 4)

The Hagar Bjerg thrust sheet everywhere overlies the Niggli Spids thrust sheet along the Hagar Bjerg thrust. This thrust sheet contains a few large crystalline gneiss units (e.g., the "Hagar Bjerg migmatite sheet" of Haller, 1971), but it is largely composed of a considerable thickness of high-grade, migmatitic, metasedimentary rocks assigned to the late Mesoproterozoic Krummedal supracrustal sequence. The uppermost segment of the Hagar Bjerg thrust sheet consists of the distinctive Neoproterozoic–Ordovician succession distinguished as the Franz Joseph allochthon.

The Krummedal sequence metasedimentary rocks in the Hagar Bjerg thrust sheet are everywhere extensively migmatized, and over a S-N distance of at least 400 km (70°N–74°N), they are intruded by two suites of granites (see Kalsbeek et al., this volume). In western Andrée Land, the Hagar Bjerg thrust crops out as a distinct NW-SE–trending contact, where high-grade Krummedal sequence semipelitic and kyanite-bearing pelitic rocks of the Hagar Bjerg thrust sheet, containing abundant granites, occur in the hanging wall, and garnet-bearing pelitic and semipelitic Krummedal rocks of the Niggli Spids thrust sheet, lacking granites, occur in the footwall.

Caledonian regional metamorphic grade in the Krummedal sequence metasedimentary rocks locally reaches high-temperature, low-pressure amphibolite-facies conditions, and in a few areas, high-temperature granulite-facies conditions (Jones and Escher, 2002; Leslie and Nutman, 2003). Caledonian high-grade metamorphism also led to new zircon growth on detrital grains in the form of rare partial rims (Kalsbeek et al., 2000).

The older of the two suites of granites in the Krummedal sequence is composed of thick sheets of leucogranite, often characterized by conspicuously large feldspar augen. U-Pb sensitive high-resolution ion microprobe (SHRIMP) zircon analyses have established emplacement ages of 910–940 Ma for many different granite bodies between northern Andrée Land and the Stauning Alper (Jepsen and Kalsbeek, 1998; Kalsbeek et al., 2000, this volume; Watt and Thrane, 2001). Locally, the granites postdate a phase of isoclinal folding that affected the Krummedal metasedimentary rocks (Leslie and Nutman, 2003), an indication that orogenic deformation was associated with granite emplacement. The younger granite suite is composed of Caledonian leucogranites (415–435 Ma; Rex and Gledhill, 1981; Hartz et al., 2000; Kalsbeek et al., this volume, and references therein).

The up to 18-km-thick Neoproterozoic–Ordovician sedimentary succession distinguished as the Franz Joseph allochthon forms the uppermost segment of the Hagar Bjerg thrust sheet. It includes the Eleonore Bay Supergroup, the Tillite Group, and the Cambrian-Ordovician Kong Oscar Fjord Group (Sønderholm et al., this volume; Smith and Rasmussen, this volume). The lower contact of the Eleonore Bay Supergroup against the underlying migmatitic Krummedal metasedimentary rocks is a bedding-parallel shear zone (Petermann Bjerg detachment in the west, Franz Joseph detachment in the east) that exhibits evidence of *both* contractional and extensional displacements. Contractional, top-to-the-WNW displacement indicators are dominant in the footwall of the Franz Joseph detachment (Figs. 7A–7C). However, age relationships are younger-on-older, and the simple pattern of deformation in the rock units that make up the Franz Joseph allochthon suggests that displacement along the detachment is perhaps no more than a few tens of kilometers. If this is the case, the Neoproterozoic–Ordovician sedimentary rocks that make up the Franz Joseph allochthon can be envisaged as having been transported more or less passively westward as the uppermost part of the Hagar Bjerg thrust sheet (Higgins et al., 2004a).

Conodonts extracted from the nonmetamorphic Ordovician carbonates at the highest levels of the Franz Joseph allochthon have very low color alteration indices (Smith, 1991), which demonstrate the succession was never overlain by higher, thick thrust units (nor any significant thickness of post-Caledonian sedimentary rocks); therefore, it is considered to be the highest structural level of the Caledonian orogen. Metamorphic grade increases gradually downward through the Eleonore Bay Supergroup, and Caledonian granitic melt generated in the underlying Krummedal sequence of the Hagar Bjerg thrust sheet has migrated upward across the detachment at the base of the Franz Joseph allochthon to form substantial granite plutons in the lower levels of the Eleonore Bay Supergroup.

The Hagar Bjerg thrust sheet, with the thick sedimentary package of the Franz Joseph allochthon at the top, is very widely distributed in the central fjord region of East Greenland (Figs. 2 and 4). Since it everywhere overlies the Niggli Spids thrust sheet, it can be argued that displacement on the Hagar Bjerg thrust relative to the underlying Niggli Spids thrust sheet is of a similar order of magnitude (>90 km, and possibly considerably more; Higgins and Leslie, 2000). However, the most convincing evidence for very substantial displacements for the Niggli Spids and Hagar Bjerg thrust sheets is the fact that the up to 18-km-thick Neoproterozoic–Ordovician succession found in the Franz Joseph allochthon structurally overlies an equivalent, <400-m-thick succession preserved in the Målebjerg and Eleonore Sø windows (see Smith and Rasmussen, this volume).

THE CROSS SECTIONS

The foreland-propagating Caledonian thrust systems of East Greenland are illustrated here by two pairs of parallel cross sections, a northern pair across Kronprins Christian Land in the northern part of the orogen (section lines A–B and C–D on Fig. 3), and a southern pair across the Eleonore Sø and Målebjerg windows in the southern part of the orogen (section lines E–F–G and H–J on Fig. 4).

In Kronprins Christian Land, the southern of the two cross sections (A–B on Fig. 3) extends from undisturbed Ordovician–Silurian shelf carbonates, through the Thin-skinned fold-and-thrust belt in the same strata, and into the structurally overlying Vandredalen thrust sheet along the valley of Rivieradal. The cross section ends near Marmorvigen, at the western rift margin of Hekla Sund Basin, in which the Rivieradal Group was deposited. The northern cross section (C–D on Fig. 3) extends from the Vandredalen thrust sheet front on the west side of Vandredalen eastward along the north side of Ingolf Fjord. The section here passes through an elevated stack of west-directed folds and thrusts developed in Paleoproterozoic to Mesoproterozoic quartzites and dolerites of the eastern part of the Thin-skinned fold-and-thrust belt and crosses the narrow root zone of the Vandredalen thrust sheet. The structurally higher thrust imbricates of the Western thrust belt occupy a broad zone of east-central Ingolf Fjord, and the section ends in the high-grade gneisses of the Nørreland thrust sheet.

The structure of the central fjord zone in the southern half of the orogen is also illustrated by two parallel west-to-east cross sections. The southern cross section (E–F–G on Fig. 4) extends from the Hamberg Gletscher foreland in the west, across the Eleonore Sø and Målebjerg windows, and through the entire thrust pile (Gemmedal thrust sheet, Niggli Spids thrust sheet, Hagar Bjerg thrust sheet) to the highest structural levels of the Franz Joseph allochthon in the east. The northern of the two sections (H–J on Fig. 4) does not cross the Målebjerg window but otherwise illustrates the same structural elements.

Kronprins Christian Land: Section A–B (Figs. 3 and 8)

Centrumsø (Fig. 8, Panels 1 and 2)

Section A–B (Fig. 8) begins in the extreme west, in the flat-lying Ordovician to Silurian carbonates of the foreland, with the Sjælland Fjelde, Børglum River, Tureso, and Odins Fjord Formations exposed mainly in valley walls (Fig. 3; see also Higgins and Leslie, this volume, their Table 1). Eastward, these same stratigraphic units are involved in a series of minor thrusts and associated belts of folding. Displacements on individual thrusts are generally small, from a few hundred meters to several kilometers. A thin-skinned deformation style for this region was suggested in the earliest studies around Centrumsø by Fränkl (1954, 1955), and it was confirmed by Peel (1980) in an investigation of good valley sections northwest of Romer Sø. A line-and-area restoration of the thin-skinned part of the cross section east and west of Centrumsø (Higgins et al., 2004b) indicates duplication of parts of the succession on long flats (a on Fig. 8) and suggests that all the thrusts west of Centrumsø root into a sole thrust near the base of the Wandel Valley Formation, resulting in a total measured displacement of ~9 km. The several higher thrusts east of Centrumsø are all shown to root into a flat thrust at the base of the Børglum River Formation that has a total calculated displacement of ~8 km (b on Fig. 8), and this thrust appears to merge eastward with the Vandredalen thrust. The floor thrust at the base of the Wandel Valley Formation continues eastward beneath the Vandredalen thrust sheet (see following).

Rivieradal to Marmorvigen (Fig. 8, Panels 2 and 3)

The Vandredalen thrust climbs a moderately steep ramp on the west side of Vandredalen, cutting through the Ordovician to Silurian succession from the base of the Børglum River Formation upward (Fig. 8) and eventually overlying the Silurian turbiditic sedimentary rocks of the Lauge Koch Land Formation (Fig. 3). This NNE-SSW–striking thrust ramp can be traced for 200 km through central Kronprins Christian Land. In the central part of Vandredalen, the base of the Vandredalen thrust sheet is locally elevated above present-day exposure level, but it can be traced for ~30 km on the southwest side of Sæfaxi Elv (Fig. 3), where it follows a long flat in the upper Alexandrine Bjerge Member of the Wandel Valley Formation. Westward, the thrust changes level upward to follow a flat in the upper dolostone unit of the Sjælland Fjelde Formation. As indicated already, the lower thrust at the base of the Wandel Valley Formation has a displacement of ~9 km near the Vandredalen thrust ramp (b on Fig. 8). The eastward continuation of this lower thrust beneath the Vandredalen thrust is uncertain, but displacement may dissipate along several bedding-parallel shear zones in the lower Wandel Valley Formation, reported along Sæfaxi Elv by M.P. Smith (1995, personal commun.), each of which could have been the site of significant displacement. Just north of Sæfaxi Elv (Fig. 3), the thrusts in the Thin-skinned fold-and-thrust belt penetrate to deeper levels, involving the Independence Fjord Group; some displacement here may be taken up on a side-wall ramp.

The sedimentologic and stratigraphic variations within the Rivieradal Group are well exposed in the 35-km-long section along Rivieradal between the thrust sheet front at the east end of Centrumsø and its trailing edge near Marmorvigen. The frontal region of the thrust sheet is characterized by simple, large-scale upright folds developed in the Fyns Sø, Kap Bernhard, and Campanuladal Formations of the Hagen Fjord Group that conformably overlie the Rivieradal Group (see also Higgins and Leslie, this volume, their Table 1). A large-scale hanging-wall anticline is outlined in the steep to vertical dips observed in the well-bedded sandstones in the upper levels of the Rivieradal Group. North and south of the line of section, conglomeratic successions are prominent, and they contain well-rounded boulders of quartzite and dolerite in upward-coarsening units (Fig. 9). Thick successions of mudstones and siltstones, with rare sandstone units, exhibit occasional upright folds. An abrupt change to tightly developed folds occurs in central Rivieradal, most likely in the hanging wall of a subsidiary thrust (c on Fig. 8). These tight folds become progressively more intense eastward, and the inclination of the axial surfaces decreases from moderate eastward dip to almost horizontal. In the eastern part of the section (Fig. 8, panel 3), deformation

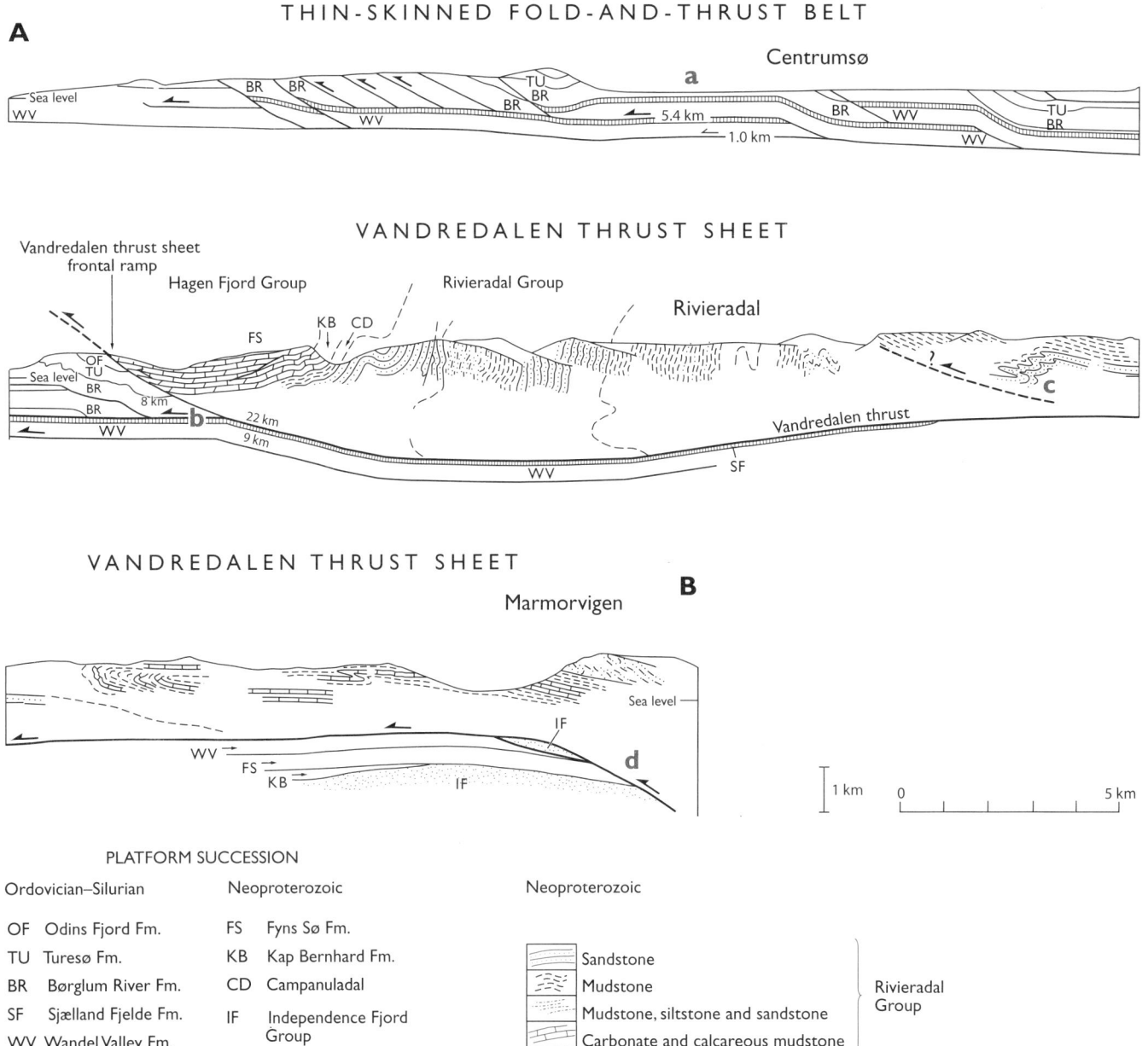

Figure 8. Cross section A–B extends from undisturbed foreland through the Thin-skinned fold-and-thrust belt and into the Vandredalen thrust sheet (modified after Higgins et al., 2001b, 2004b). The Vandredalen thrust sheet roots downward at the east end of the section near Marmorvigen; a–d indicate localities referred to in the text. See Figure 3 for section line.

is intense with complete erasure of sedimentary way-up indicators in long-limbed isoclinal folds; the mudstone and calcareous mudstone units exposed in this part of the thrust sheet correspond to the most distal parts of the Rivieradal Group.

The Vandredalen thrust sheet roots to depth into what is believed to be the sheared remnant of the original Hekla Sund Basin (d in Fig. 8). At Marmorvigen, the steeply dipping thrust plane cuts upward across the Fyns Sø and Wandel Valley Formations before following a flat in the upper part of the Wandel Valley Formation. Correlations of the cutoff of the Fyns Sø Formation in the hanging wall east of Centrumsø, and in the footwall at Marmorvigen, provide an estimate of displacement of 35 km on the Vandredalen thrust. A large horse of Independence Fjord Group rocks at Marmorvigen is interpreted as a fragment detached from the rift margin of the Hekla Sund Basin during thrusting (Higgins et al., 2001b). Where the Vandredalen thrust cuts into the Independence Fjord Group, which makes up the uplifted west margin of the Hekla Sund Basin (in which the Rivieradal Group was deposited), relationships are "younger-on-older" with the Rivieradal Group in the hanging wall. East of the

Figure 9. Klippe of the Vandredalen thrust sheet at Keglen on the west side of Vandredalen, looking west. At this locality, the Proterozoic Rivieradal Group consists of numerous upward-coarsening packets of conglomerate in the hanging wall of the Vandredalen thrust (VT), which dips moderately to steeply toward the observer. The Ordovician–Silurian platform carbonates of the Turesø Formation (TU) and Odins Fjord Formation (OF) in the footwall of the thrust contain conodonts, the alteration colors of which are indicative of a former overburden ~10.7 km in thickness (Rasmussen and Smith, 2001). Summit of Keglen is 850 m above the valley floor in the foreground.

line of section, the waters of the 6-km-wide Hekla Sund conceal the contact with structurally higher thrust sheets to the east. Independence Fjord Group quartzites and interbedded basalts are exposed on western Hovgaard Ø, and crystalline gneisses crop out on eastern Hovgaard Ø (see also section C–D next).

Kronprins Christian Land: Section C–D (Figs. 3 and 10)

Vandredalen to Hjørnegletscher (Fig. 10, Panels 1 and 2)

The northern cross section (Fig. 10) begins on the west side of Vandredalen, where the Vandredalen thrust cuts up through the Lower Paleozoic carbonate succession on an east-dipping ramp. Orange brown and yellow carbonates of the Kap Bernhard and Fyns Sø Formations are deformed into major anticlines and synclines in the hanging wall of the thrust, and there are well-exposed spectacular cutoffs against the thrust plane. A conglomerate unit, up to 130 m thick, occurs within the upper sandstone unit of the Rivieradal Group in Vandredalen, but it thins to zero within 5 km to both north and south. About 20 km south of the line of section, a 460-m-thick succession of boulder conglomerates includes up to 34 upward-coarsening conglomerate beds; these boulder conglomerates are believed to represent one of the point sources that fed sedimentary rocks into the Hekla Sund Basin.

Coarsening-upward sequences of parallel-laminated, lenticular, and wavy-bedded mudstones overlain by trough and planar cross-bedded sandstones are exposed beneath the conglomerates, and these correspond, in part, to the original "Rivieradal sandstones" of Fränkl (1955). The sandstones often show spectacular tight to isoclinal westward-overturned folding. In the line of section, the east side of Vandredalen is dominated by a succession of over 2200 m of sandstone turbidites (e in Fig. 10) interbedded with dark pyritic mudstones ("Taagefjeldene Greywackes" of Fränkl, 1955; Smith et al., 2004b).

The Vandredalen thrust reemerges on the west side of Hjørnegletscher (f in Fig. 10) and dips at moderate angles to the west (Fig. 11A); strongly deformed Sjælland Fjelde Formation gray carbonates are exposed in the footwall, and the thrust plane follows the same long flat in this formation as seen along Sæfaxi Elv, 70 km to the south. In the hanging wall, a few tens of meters above the thrust, thin conglomerate units within the "Taagefjeldene Greywackes" have stretched clasts oriented almost E-W (277°–097°). As at Marmorvigen, conspicuous horses of Paleoproterozoic to Mesoproterozoic Independence Fjord Group rocks occur locally along the thrust plane (Fig. 11B). The horses are structurally overlain by younger Neoproterozoic Rivieradal Group sedimentary rocks, and they overlie Ordovician platform carbonate rocks. Such younger-on-older-on-younger relationships developed locally within the thrust system are readily attributed to the prethrust architecture, such that the Vandredalen thrust, which transported the Rivieradal Group out of its depositional rift basin, ramped across the faulted western rift margin and sampled the Independence Fjord Group quartzites from the elevated rift shoulder in the footwall.

The Neoproterozoic Hagen Fjord Group, the Cambrian Kap Holbæk Formation, and Ordovician shelf carbonates exposed in the footwall are an eastward extension of the parautochthonous, thin-skinned fold-and-thrust belt of western Kronprins Christian Land. These units are locally strongly deformed by upright to

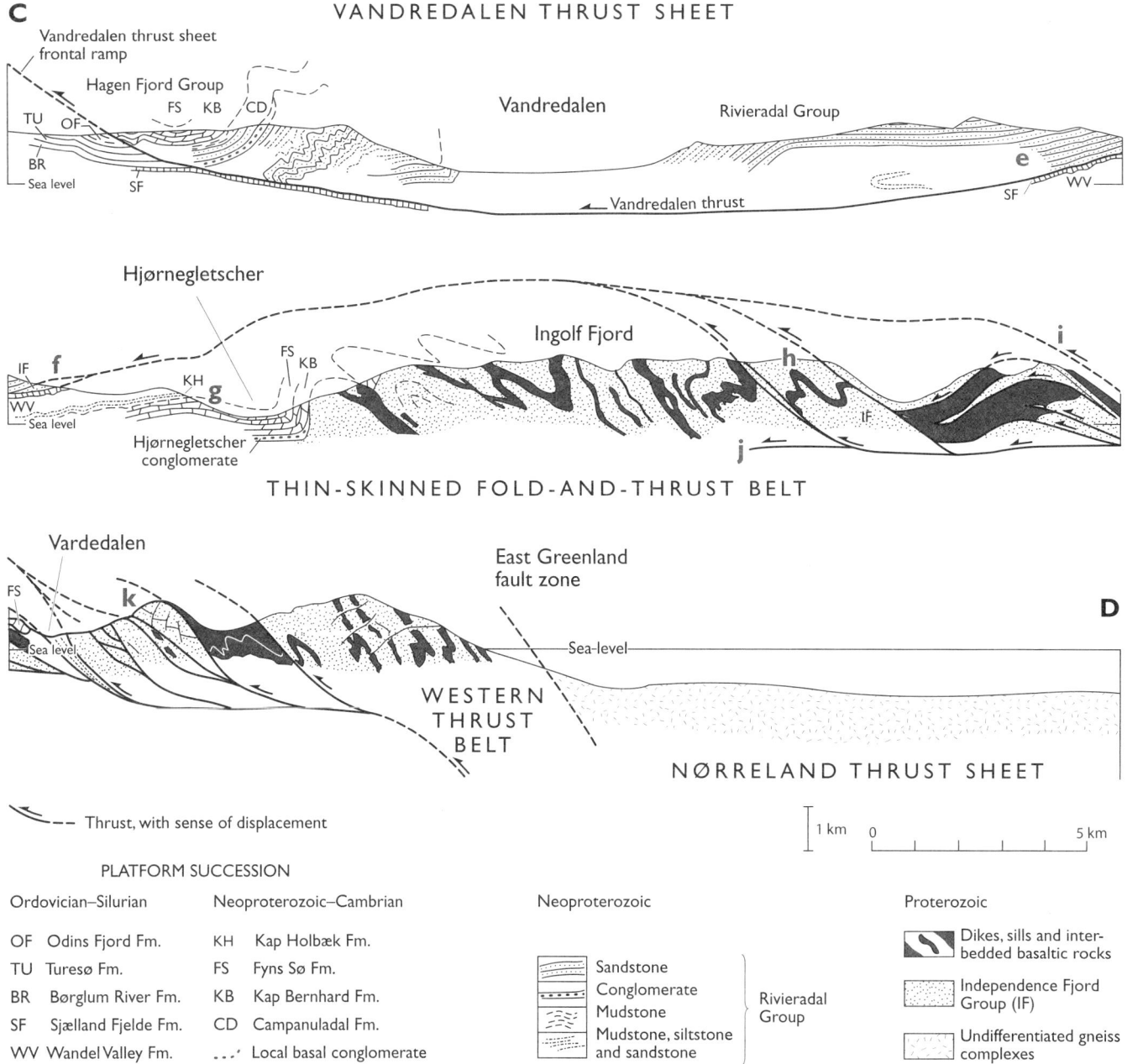

Figure 10. Cross section C–D begins at the east-dipping thrust ramp of the Vandredalen thrust sheet on the west side of Vandredalen. See Figure 3 for section line. At Hjørnegletscher, the now west-dipping Vandredalen thrust rises above exposure level beneath an imbricate fold-and-thrust pile, part of the Thin-skinned fold-and-thrust belt. The Vandredalen thrust sheet roots to depth at Vardedalen, shown in detail in Figure 12. East of Vardedalen, the Western thrust belt is composed of an imbricate thrust sequence of the Independence Fjord Group and associated rocks. The section ends in the high-grade rocks of the Nørreland thrust sheet; e–k indicate localities referred to in the text.

westerly overturned fold structures and, east of Hjørnegletscher, are elevated above exposure level by a large-scale fold-and-thrust duplex developed in the Independence Fjord Group that is well exposed along west-central Ingolf Fjord. The base of the Hagen Fjord Group is marked locally and in the line of section by a basal clastic sequence, originally described as the "Hjørnegletscher conglomerate" (Fig. 10; see Jepsen and Kalsbeek, 1981).

The sandstones and siltstones of the Kap Holbæk Formation that overlie the Hagen Fjord Group (g in Fig. 10) infill channels and cavities eroded into the Fyns Sø Formation and represent one of the oldest examples of paleokarst known (Smith et al., 1999, 2004a). The overlying Ordovician carbonate succession is dominated by the Wandel Valley Formation with the Sjælland Fjelde Formation at the top.

Figure 11. (A) The west-dipping Vandredalen thrust (VT) in the upper reaches of Hjørnegletscher, looking northward. The well-bedded sequence in the hanging wall is part of the >2000 m succession of sandy turbidites in the Proterozoic Rivieradal Group described by Smith et al. (2004b). The footwall succession of Lower Paleozoic carbonates of the Wandel Valley Formation (WV) and Sjælland Fjelde Formation (SF) is deformed by slightly westerly overturned fold structures. IF is a thrust-bounded enclave of Independence Fjord Group quartzite (see part B). Profile height is ~750 m. Photo is by Jakob Lautrup. (B) Close-up of the Vandredalen thrust in upper Hjørnegletscher seen in A, showing a horse of Independence Fjord Group quartzite (IF, ~15 m thick) along the thrust plane between the Rivieradal Group in the hanging wall and the Ordovician Sjælland Fjelde Formation in the footwall.

Hjørnegletscher to Vardedalen (Fig. 10, Panels 2 and 3)

Between Hjørnegletscher and Vardedalen, Independence Fjord Group sandstones and dolerite sills and dikes are arranged, with unconformable Hagen Fjord Formation rocks in the west, in a series of NE-SW– to NNE-SSW–trending tight folds, and thrusts and/or attenuation of the overturned common fold limbs increase in significance eastward. Whereas the floor thrust in the parallel cross section to the south (section A–B; Fig. 8) has remained in the Ordovician carbonates as far as the rift shoulder at Marmorvigen, in the alpine region north of Sæfaxi Elv and Marmorvigen and east of Hjørnegletscher, thrusting has evidently penetrated to deeper levels, with consequent elevation of the deformed Independence Fjord Group to exposure level, and the Vandredalen thrust above exposure level.

This part of the cross section is constructed from the spectacular 1350-m-high cliff sections in western-central Ingolf Fjord, east of Hjørnegletscher, that illustrate the characteristic contrast between the white-weathering Independence Fjord Group quartzites and black-weathering dikes and sills (cf. Fig. 5). The section immediately east of Hjørnegletscher is deformed by a series of kilometer-scale anticlines and synclines that can be traced without discontinuity along Ingolf Fjord for at least 5 km. Individual closures and axial surfaces may be readily correlated from north to south across the fjord. These folds also deform the unconformity with the Hagen Fjord Group rocks on the eastern flanks of Hjørnegletscher.

Passing eastward, the effects of fold-related thrusting become more significant; thin dolerite dikes may serve to preferentially locate the position of developing fold limbs. Several dikes lie along axial planes of large-scale folds and clearly accommodate thrusting as they rotate into easy slip orientations, constraining the structural framework. Note that while dikes are clearly observed to cut sills, they are in fact, typically more strongly deformed than the concordant intrusions due to their intimate association with thrust propagation.

A strong axial planar cleavage is seen in both limbs of these folds, although it is typically more intense in the inverted limb. Bedding-cleavage relationships are consistent with the major folds, which generally plunge at a small angle to the SSW. In the higher strain zones, a foliation, which always has a consistent, top-to-the-NW sense of shear, is seen to develop conspicuous shear bands around more resistant elements of the lithology. This fabric, and the related axial planar cleavage in less strongly deformed areas, dips at ~45°–50° toward ~140°. A prominent mineral or rodding lineation is commonly developed and is nearly always downdip. A second transecting shear fabric (in s/c relationship) has been observed and also shows top-to-the-NW sense of shear on planes dipping 25°–30° to the SE. A weak mineral lineation may be developed and is again essentially downdip.

The first prominent thrust observed in this section of Ingolf Fjord is steep on the northern wall of the fjord, but it has a shallow dip on the southern wall of the fjord. Farther eastward, in the hanging wall of this thrust, a large-scale antiform-synform fold pair is picked out by a prominent dolerite sill (h on Fig. 10).

Eastward again, the structures are dominated by the effects of thrusting. An antiformal stack is evident, and a series of imbricates repeats a thick dolerite sill. The stack pinches out to the southwest beneath Brede Spærregletscher on the south side of Ingolf Fjord. The Vandredalen thrust makes up the eastern roof thrust to this duplex (i on Fig. 10) and must therefore have been elevated above the fold-and-thrust complex in western Ingolf Fjord. The antiformal stack formed by footwall imbrication beneath and in front of the root zone to the Vandredalen thrust sheet, where the higher-level older thrusts became folded and elevated in the growing stack as thrusting propagated westward.

Minimum shortening between Hjørnegletscher and Vardedalen is estimated to be on the order of 50%–55%; a nominal area balance suggests that décollement may have occurred at only shallow depths (~1 km) below this section. A blind floor thrust is thus depicted propagating westward beneath the highly deformed part of the section (j on Fig. 10), and it supports the interpretation that, in this section, the Thin-skinned fold-and-thrust belt extends eastward as far as the root zone of the Vandredalen thrust sheet at Vardedalen (V on Fig. 3).

Root Zone of the Vandredalen Thrust Sheet, Vardedalen
(Fig. 10, Panel 3; Fig. 12)

The tectonic slice of Rivieradal Group sedimentary rocks at Vardedalen constitutes the root zone of the Vandredalen thrust sheet; the root zone has a maximum structural thickness of only 250 m at this location. Relationships at Vardedalen are summarized in Figure 12. Two principal high strain zones can be recognized in the root zone of the thrust sheet in a section where simple shear deformation is pervasive.

The Vandredalen thrust (A on Fig. 12) places ~200 m of phyllonitic Rivieradal Group siltstones and sandstones over carbonates of the Fyns Sø Formation, and the latter rest unconformably on Independence Fjord Group quartzites and Midsommersø Dolerite dikes. The thrust is marked by ~20 m of black micaceous phyllonite with distinctive oyster-shell texture developed in places. In the hanging wall of the thrust, the black phyllonite grades upward into gray to green phyllonites, where silty laminae and thin layers of pale-gray quartz sandstone begin to appear within low-strain lenses. Sense of shear indicators in these phyllonitic rocks are strongly top-to-the-NW, but the original sedimentary lamination is locally relatively well preserved. Beds that are 5–10 cm thick are commonly preserved within a fabric that makes up the anastomosing shear bands, and graded bedding is preserved in some low-strain augen.

A footwall syncline is preserved in the carbonates immediately below the Vandredalen thrust. The stratigraphically lowest of the Fyns Sø Formation carbonates in the syncline consists of a conglomerate (unconformable on Independence Fjord Group quartzite) with quartzite clasts in a matrix of dark-brown impure marble. As at Marmorvigen, the underlying Kap Bernhard Formation is missing at the rift margin. Poorly preserved stromatolites occur in the upper part of this lower unit of the Fyns Sø Formation. The brown marbles pass upward into yellow- to cream-colored marbles, which, at lower strains, show NW-vergent asymmetrical folds on a tens of centimeters scale. Above that, a penetrative L-S fabric is developed in gray, cream, or brown calcite mylonite. These

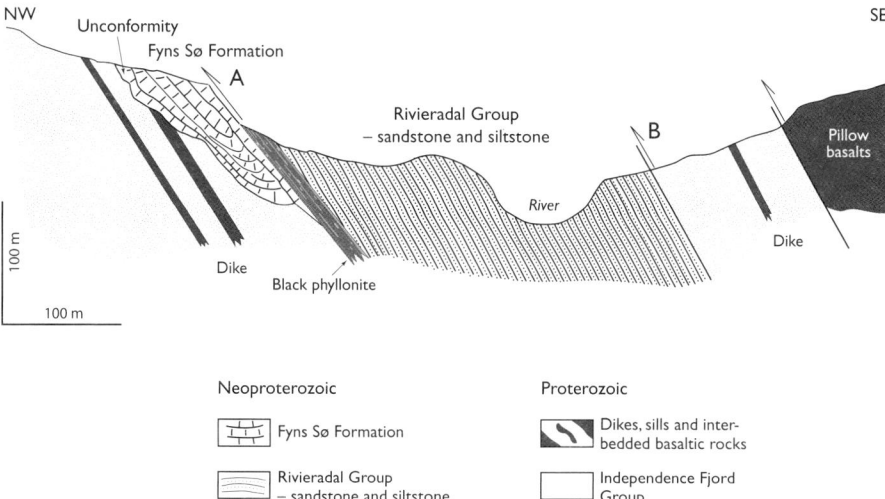

Figure 12. Detailed section at Vardedalen on the north side of Ingolf Fjord, showing the root zone of the Vandredalen thrust sheet. A strip of Rivieradal Group phyllonitic sandstone and siltstone, ~300 m wide, is bounded on both sides by steeply dipping thrusts (A, B). The lower thrust cuts through a folded remnant of Fyns Sø Formation dolomites, resting unconformably on the Independence Fjord Group.

relationships are identical to those that occur 70 km to the south at Marmorvigen, where the Vandredalen thrust cuts up through the unconformity between the Fyns Sø Formation and Independence Fjord Group (see Fig. 8). Correlation of the footwall cutoff for the Fyns Sø carbonates on the Vandredalen thrust at Vardedalen with the hanging wall cutoff west of Vandredalen suggests ~45–50 km of movement on the Vandredalen thrust in the line of section C–D. The trace of the Vandredalen thrust, and thus the root zone of the Vandredalen thrust sheet, extends southward from Vardedalen and Ingolf Fjord along Brede Spærregletscher (BS on Fig. 3) and along Hekla Sund to Marmorvigen (Fig. 5).

The higher roof thrust (B on Fig. 12) is marked by ~20 m of strikingly planar quartz mylonite with Independence Fjord Group rocks thrust over the phyllonitic Rivieradal Group rocks. The contact is sharp. Gray phyllonites become increasingly planar upward in the footwall and carry a prominent downdip rodding lineation. The quartz mylonites show a very faint L-S fabric but grade upward into Independence Fjord Group quartzites from which they are clearly derived.

East-Central Ingolf Fjord (Fig. 10, Panel 3)

This part of the section is viewed as the northward extension of the Western thrust belt (Fig. 2). In Kronprins Christian Land, where the Thin-skinned fold-and-thrust belt and Vandredalen thrust sheet occupy wide areas, the term "Western" may seem incongruous, but farther south in Lambert Land and the nunatak region west of Jøkelbugten, where the two former domains are absent or covered by ice, the Western thrust belt is the westernmost exposed part of the orogen. Immediately east of Vardedalen in the line of section (Fig. 10), a large-scale thrust duplex repeats a unit of pillow basalts and conglomeratic quartzites. The section clearly shows evidence of preserved extensional fault systems, which predate the thrusting and were subsequently reactivated with the opposite sense of slip during Caledonian deformation (S.A.S. Pedersen, 1995, personal commun.). Rotated segments of a dolerite sill with prominent dark margins clearly show the effect of this early extensional faulting (k on Fig. 10). These extensional systems are probably linked to the original rifting of the Hekla Sund Basin, in which the Rivieradal Group sedimentary rocks were deposited.

Farther east, across a prominent thrust ramp, a complex pattern of phyllonitic horizons and tight to isoclinal anticlinal folds is developed in a volcanic and volcaniclastic sequence. This intensely deformed section is bounded to the east by another prominent high-strain zone. Structures at outcrop scale are more intensely deformed than equivalent structures seen in west-central Ingolf Fjord; thrust-related cleavages (two sets) and lineations are parallel with those observed to the west.

Toward the eastern end of the alpine region, the outcrop is dominated by thin sills within white quartzites with preserved cross-bedding. A subhorizontal intersection lineation is evident in the quartzites, but there is only a slight grain-shape lineation parallel to the regional stretching direction. In contrast, the sills show very obvious hornblende-plagioclase mineral lineations and are typically highly schistose. All show a NW-directed sense of shear. The most easterly exposures observed are rather more strongly sheared, but, even here, bedding is still evident in the sandstones, and doleritic textures survive in the intrusions.

Outer Ingolf Fjord (Fig. 10, Panel 3)

The east boundary of the alpine part of the section is a prominent NNE-SSW–trending lineament known as the East Greenland fault zone; this structure is considered to have controlled deposition in a Carboniferous basin (Stemmerik and Håkansson, 1991). However, the lineament has also been viewed as a northward continuation of the Storstrømmen shear zone (Jepsen et al., 1994), a major NNE-SSW lineament traceable from eastern Dronning Louise Land to at least Lambert Land (Figs. 1 and 2). The present-day exposures of western Holm Land are high-grade gneisses with eclogitic enclaves that have been identified as part of the Nørreland thrust sheet, and these are unconformably overlain in eastern Holm Land by Lower Carboniferous (Visean) sedimentary rocks, the lowest level of the Carboniferous–Paleogene succession known as the Wandel Sea Basin (Fig. 1).

Higgins et al. (2001b) speculated that the gneiss areas of easternmost Kronprins Christian Land might represent the roots of deep-seated thrust sheets that formerly projected westward above the Western thrust zone, and that may have had displacements of as much as 100 km (see also Higgins and Leslie, this volume). In those parts of the Western thrust zone southward from Lambert Land, thrust units of Precambrian gneisses are interleaved with the Independence Fjord Group and associated dikes and sills, and similar gneiss-dominated thrust units were probably once present above the Western thrust zone along Ingolf Fjord. Since the high-grade gneisses exposed in the line of section are now considered part of the Nørreland thrust sheet, the present-day East Greenland fault zone probably obscures the original thrust relationship.

The East Greenland fault zone itself is not exposed in the line of section in Holm Land. The nearest possible exposure is found on the south side of Hovgaard Ø, where a several-kilometer-wide, steeply east-dipping to near-vertical shear zone exposes mylonitic Independence Fjord Group quartzites to the west in contact with mylonitic granitic gneisses containing sheared basic sheets to the east (K.A. Jones and J.C. Escher, 1995, personal commun.). The nature of this shear zone is perhaps more reminiscent of other exposures of the Storstrømmen shear zone rather than the East Greenland fault zone, except that shear sense is here dominantly vertical, east-side-up, rather than sinistral strike-slip; Jones and Escher (1995, personal commun.) reported that the mylonitic fabric developed under amphibolite- to epidote-amphibolite–facies metamorphism and suggested that the shear zone may have played a role in the exhumation of the high-grade rocks of the Nørreland thrust sheet to the east. The shear sense supports the idea that the high-grade thrust sheet (Nørreland thrust sheet) may have been displaced upward above the lower-grade rocks making up the Western thrust zone. The East Greenland fault zone may be viewed as a normal fault that here follows the line of the Storstrømmen shear zone.

Central Fjord Zone: Section E–F–G (Figs. 4 and 13)

Hamberg Gletscher to Boyd Bastion Fault
(Fig. 13, Panels 1–3)

The thrust architecture of the southern part of the orogen (Fig. 4) is illustrated by two profiles (E–F–G and H–J) that traverse the Eleonore Sø and Målebjerg foreland windows. The western part of the southern section line (E–F–G, Fig. 13) extends from the isolated nunataks of the Hamberg Gletscher foreland at the margin of the Inland Ice, across the southern part of the Eleonore Sø window, to the Boyd Bastion fault. The age of the volcano-sedimentary units in this southern part of the Hamberg Gletscher foreland is currently unknown. An extensive gabbroic complex intrudes a volcano-sedimentary sequence; parts of the volcanic sequence locally retain pillow structures and are thus likely to have been lavas erupted in a marine setting. None of the rocks preserves a tectonic fabric, and they have not been subject to penetrative ductile deformation; they are thus likely to be a part of the intact foreland to the Caledonian orogen but younger than the ca. 1900 Ma granitic rocks exposed 40–50 km to the north, at the west end of section H–J (see following section).

The easternmost localities visited in the gabbroic units of the Hamberg Gletscher foreland are strongly sheared mylonitic rocks that preserve top-to-the-WNW shear fabrics and prominent transport lineations on the gently ESE-dipping mylonitic fabric (i.e., typical Caledonian; a in Fig. 13). There are no exposures of the Slottet Formation quartzites in the foreland at this location. To the east of the shear zone and still structurally beneath the Niggli Spids thrust sheet, the western Hvidbjørn Nunatakker consists of low-grade amphibolite-facies metasedimentary rocks with a penetrative Caledonian schistosity, and they are assigned to the Gemmedal thrust sheet. Well-preserved sedimentary structures occur in laminated to finely bedded metasiltstones and metasandstones that preserve upward-younging and facing textures. Pelitic lithologies are preserved as dark-gray phyllites with pinhead garnet ± chloritoid; occasional calcareous layers and nodules preserve calc-silicate mineralogies. There is no evidence that these low-grade rocks have been subject to contact metamorphism despite geophysical evidence that the gabbroic rocks continue eastward for some distance at depth (Schlindwein, 1998). While only mapped in a few nunataks in the west, the presence of more widespread developments of such low-grade rocks at depth is suggested by abundant sedimentary xenoliths in Paleogene volcanic plugs. Further outcrops of low-grade rocks, with shear sense defined by sigma-type quartzofeldspathic augen (Fig. 7D), occur on the flanks of the Eleonore Sø and Målebjerg windows and point to the presence of a thin but more-or-less continuous Gemmedal thrust sheet (<2 km thick; b on Fig. 13) underlying the Niggli Spids thrust sheet.

In the western nunataks around Hamberg Gletscher, the Niggli Spids thrust is for the most part concealed beneath ice. The eastern Hvidbjørn Nunatakker consists of amphibolite-facies Krummedal sequence rocks, typically brown-weathering, coarse-grained, garnet–mica schists alternating with psammitic beds. These medium-grade Krummedal metasedimentary rocks host only sparse developments of granites, seen as white granite stringers on some nunatak cliffs. The garnet-grade, often coarse-grained, schists have a more complex structural architecture than the dark-colored, finer-grained low-grade metasedimentary rocks of the Gemmedal thrust sheet structurally beneath and to the west, further indicating a tectonic contact.

The sheet dip in the Niggli Spids thrust sheet, although gently undulating across open upright folds, is generally flat-lying across these nunataks. Several cliff sections do display large recumbent tight to isoclinal folds within that sheet dip (Fig. 14), and these folds are refolded at decameter-scale by discrete zones of tight reclined west-verging asymmetrical folds. The tight asymmetrical folds are consistent with the foreland-propagating top-to-the-WNW Caledonian thrust stack that now dominates the tectonostratigraphy of this region, and there is the possibility that the large-scale isoclinal folds are of similar age to those generated during crustal thickening immediately prior to augen granite generation at ca. 910 Ma farther south in East Greenland (71°30′N; Leslie and Nutman, 2003). Given that the low-grade rocks of the Gemmedal thrust sheet are exposed farther east in the immediate hanging wall of the thrust bounding the Eleonore Sø window (b in Fig. 13), then the Gemmedal thrust sheet should lie at no great depth below the present exposure level.

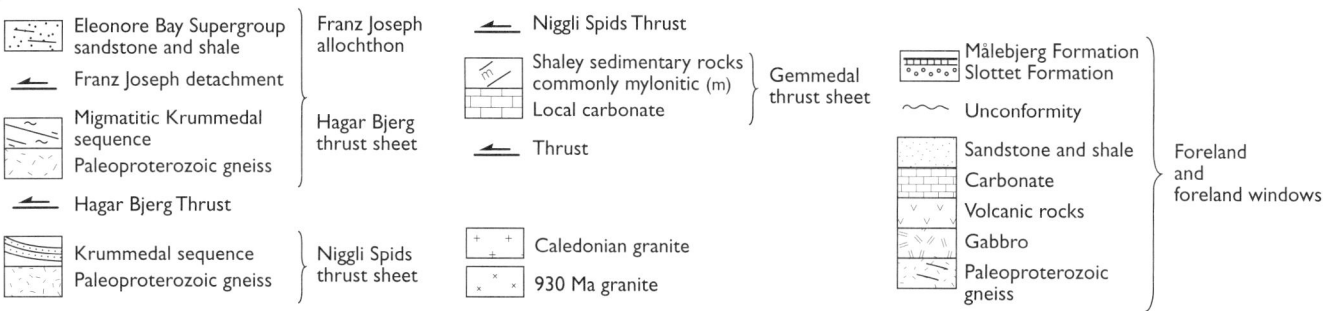

Figure 13 (*on this and following page*). Cross section E–F–G, which extends from the Hamberg Gletscher foreland in the west, across the Eleonore Sø and Målebjerg windows, to the Eleonore Bay Supergroup, making up the Franz Joseph allochthon at Grejsdalen in Andrée Land. See Figure 4 for section line; a–f are localities referred to in the text.

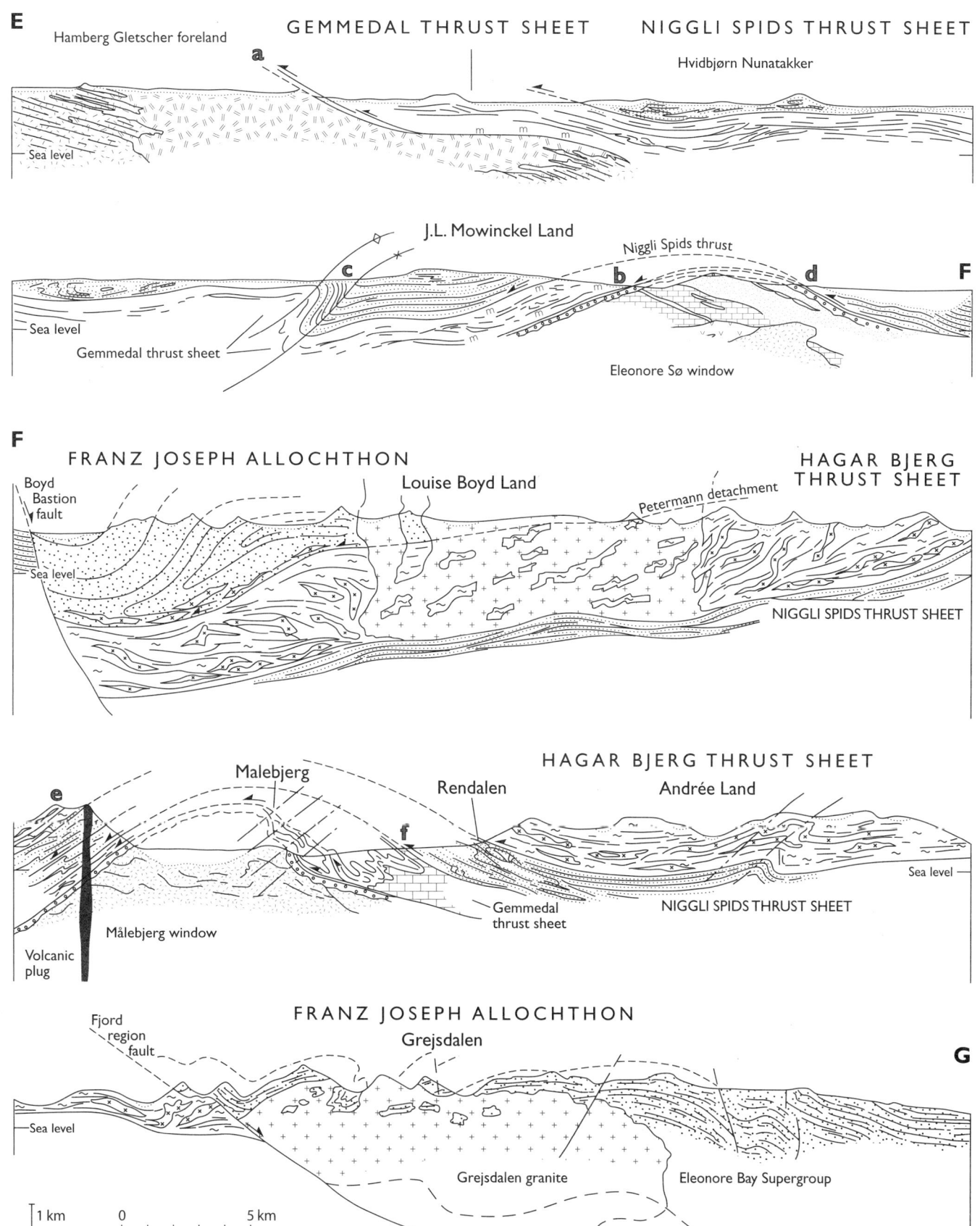

Figure 14. Interbedded metamorphosed sandstones and shales of the Krummedal sequence, with recumbent tight to isoclinal folds. Locality is in the upper part of the Niggli Spids thrust sheet north of the Hvidbjørn Nunatakker. Profile height is ~400 m.

Eastward, the section continues in Krummedal metasedimentary rocks into the southern part of J.L. Mowinckel Land (see Fig. 13, panel 2). The sheet dip remains broadly flat-lying in the west and has the same structural elements of early isoclinal folds and tight asymmetrical folds followed by more open and upright folds. The central part of panel 2 is dominated by a large-scale reclined east-verging asymmetrical fold (c in Fig. 13) that is also seen in the northern section in central J.L. Mowinckel Land. A panel of steep dips and numerous west-verging close folds coincides with intense crenulation of the regional schistosity. This large-scale fold probably refolds the entire thrust pile and may thus be a back-fold that formed within a pop-up that developed above deeper thrusts in the foreland-propagating system (see also section H–J).

The Gemmedal thrust sheet reemerges farther east in the section, and this unit is more or less continuously exposed along the western margin of the Eleonore Sø window as dark, commonly phyllitic, metasedimentary rocks of generally uniform appearance. A few enclaves of white- to yellow-weathering carbonates occur in the vicinity of the section line.

The *Skolithos*-bearing quartzites of the Slottet Formation define the dome-like structure of the southern part of the Eleonore Sø window. To the east, they are overlain by Målebjerg Formation carbonates immediately beneath a thrust contact with overriding coarse-gained Krummedal metasedimentary rocks (Niggli Spids thrust sheet; d in Fig. 13), indicating that the Gemmedal thrust sheet must locally be absent along the east side of the window. Abundant shear sense indicators in the Niggli Spids thrust sheet rocks confirm top-to-the-WNW transport with respect to the Lower Paleozoic foreland successions in the eastern part of the window. Slottet Formation quartzites are horizontal on the mountain tops in the central culmination of the window, and they crop out locally to the west, in continuous section with the Målebjerg Formation carbonates (Smith and Robertson, 1999), as a west-dipping sequence that includes spectacular calcareous mylonites immediately below the roof thrust. The persistent occurrence of carbonates beneath the roof thrust suggests that the deformation exploited the more incompetent rheology of the Målebjerg Formation carbonates, and the thrust followed a regionally extensive flat in this lithology.

Boyd Bastion Fault to Rendalen (Fig. 13, Panels 3 and 4)

The central part of section E–F–G (Fig. 13, panels 3 and 4) begins at the Boyd Bastion fault, where the Franz Joseph allochthon, made up here of the Eleonore Bay Supergroup, is dropped down ~10–15 km on the east side of a major late Caledonian extensional fault (Escher and Jones, 1998, 1999). The Neoproterozoic rocks are commonly steeply inclined in spectacular cliff sections, but eastward, they pass into flat-lying structures with large-scale isoclines locally observed in cliff sections. A very large west-facing recumbent isocline in the northern part of the outcrop (Escher and Jones, 1999) is consistent with the top-to-the-WNW emplacement of the Franz Joseph allochthon along the Petermann detachment at its base (Escher and Jones, 1998, 1999). The Franz Joseph allochthon is, along the line of section, underlain by kyanite-grade Krummedal metasedimentary rocks and sheeted granites of the Hagar Bjerg thrust sheet. Both elements of the thrust sheet are intruded by a large Caledonian granite characterized by abundant metasedimentary rafts; these relationships are spectacularly displayed in cliff exposures in Louise Boyd Land (Wenk and Haller, 1953; see also Kalsbeek et al., this volume). The granites were likely generated, in keeping with other Caledonian plutons such as the Grejsdalen granite in Andrée Land (Fig. 13, panel 5), prior to displacement on the major bounding thrusts (Kalsbeek et al., this volume).

Sheet dips increase toward the west margin of the Målebjerg window. The Niggli Spids thrust sheet, here almost entirely made up of Krummedal sequence metasedimentary rocks, is well exposed in cliff sections on the west side of the window. The western cliff section of the Målebjerg window reveals the full succession in the hanging wall of the roof thrust defining the window (Fig. 13, panel 4). Low-grade calcareous schists and dark chlorite-biotite–grade phyllites (Gemmedal thrust sheet) lie above the thrust and are bounded upward by a thin quartz mylonite, which marks the junction with overlying migmatite-free Krummedal metasedimentary rocks (Niggli Spids thrust sheet). The highest levels that form the summit ridge are composed of kyanite-grade Krummedal metasedimentary rocks and granites of the Hagar Bjerg thrust sheet (e on Fig. 13). A prominent Paleogene volcanic plug cuts through the west margin of the Målebjerg window; this plug is circular, ~200 m in diameter, and forms a summit ~1600 m high.

The Målebjerg window is the easternmost of the known parautochthonous foreland windows, and it places minimum constraints on the scale of displacement of the overriding thrust sheets relative to the foreland (Higgins and Leslie, 2000). The key horizon is the recognition, on both sides of the window, of early Cambrian *Skolithos*-bearing Slottet Formation quartzites that rest unconformably on basement orthogneisses and that are overlain by a thin unit of Lower Paleozoic carbonates assigned to the Målebjerg Formation (Leslie and Higgins, 1999; Smith et al., 2004a). The carbonates provide an easy-glide horizon, as in the Eleonore Sø window, and so they lie immediately beneath the arched roof thrust. The farthest west-traveled part of the Niggli Spids thrust sheet yet identified (Fig. 13) lies 90–100 km west of the east margin of the Målebjerg foreland window, and it provides a minimum estimate of thrust displacement. With respect to the Hagar Bjerg thrust sheet, which everywhere overlies the Niggli Spids thrust sheet, a minimum of 90–100 km can also be argued. Total displacement for the two major thrust sheets would thus be ~200 km, but may be substantially more.

Low-grade metasedimentary rocks of the Gemmedal thrust sheet are well exposed above the thrust on the eastern flank of the Målebjerg window (Fig. 13, panel 4). Marbles, calcareous schists, and chlorite-biotite phyllites lie in a complex series of fold closures and intensely attenuated fold limbs and subsidiary thrusts. Fold vergence and movement is always top-to-the-WNW on these structures. A large horse of granitic augen gneiss is also present (not in section line) interleaved with the low-grade rocks; this horse may have been excised during thrusting from a structural (footwall?) high in the basement below the original depocenter for the low-grade metasedimentary succession now present in the Gemmedal thrust sheet. East of Målebjerg, the Niggli Spids thrust is once again marked by a distinctive white quartzite mylonite (f on Fig. 13); this persistent marker band repeated across the glacier from the west may be derived from a sheared equivalent of the Slottet Formation quartzites that originally extended across the depocenter for the metasedimentary rocks now preserved in the Gemmedal thrust sheet. Garnet-biotite–grade semipelitic and psammitic schists assigned to the Niggli Spids thrust sheet contain quartz-rich lenticles that define top-to-the-WNW shear sense (Fig. 7E).

East-verging greenschist-facies ductile crenulation fabrics and decimeter-scale folding have been identified at numerous locations along the eastern edge of the Målebjerg window in the low-grade phyllitic metasedimentary rocks of the Gemmedal thrust sheet. These crenulations deform the NW-verging thrust-related fabric and folding but are broadly coaxial and coplanar with the earlier structure, and they suggest that thickening associated with thrusting was partially restored by movement down to the east, probably partially accommodated on the newly established thrust architecture.

Rendalen to Grejsdalen (Fig. 13, Panels 4 and 5)

The Hagar Bjerg thrust is represented in southern Andrée Land by the so-called "Rendalen thrust zone," a complex and conspicuous east-dipping structure. In northern Rendalen, this structure is a bedding-parallel thrust with high-grade migmatitic Krummedal metasedimentary rocks in the hanging wall, which contrast with migmatite-free Krummedal metasedimentary rocks of the Niggli Spids thrust sheet in the footwall (Fig. 15). Rocks in the thrust zone at the junction between the two thrust sheets confirm the regional top-to-the-WNW contractional movements demonstrated for all levels of the thrust pile (Fig. 7F). However, in addition to the low-angle top-to-the-WNW contractional movement (Fig. 7G), there is a superimposed slightly steeply SE-dipping top-to-the-ESE extensional movement (Fig. 7H). Farther southeast along Rendalen, the Hagar Bjerg thrust merges with extensional splays of the Fjord region fault, and the Hagar Bjerg thrust sheet wedges out in southernmost Andrée Land (see Fig. 4). The migmatitic Krummedal metasedimentary rocks are well exposed in the valley walls of central Andrée Land, where they host large and small concordant to highly discordant granite bodies. Sheet dips are generally gently inclined, occasionally disturbed by large-scale open to tight folds.

Traced eastward in the line of section, the outcrop of the Hagar Bjerg thrust sheet terminates against a moderately to steeply inclined extensional fault (Fjord region fault), which carries outcrops of the Eleonore Bay Supergroup (Franz Joseph allochthon) in the hanging wall that have been deformed into large-scale simple anticlines and synclines. The walls of Grejsdalen exhibit spectacular sections through the Grejsdalen Caledonian granite body, which encloses large xenoliths of Eleonore Bay Supergroup lithologies.

Central Fjord Zone: Section H–J (Figs. 4 and 16)

Hamberg Gletscher Foreland to Boyd Bastion Fault (Fig. 16, Panels 1 and 2)

Section H–J starts in the banded gneisses of isolated nunataks along the margin of the Inland Ice, in the northernmost representatives of the Hamberg Gletscher foreland. A granitic gneiss

Figure 15. The moderately dipping Hagar Bjerg thrust (HBT) at the east side of Rendalen, Andrée Land. Migmatitic metasedimentary rocks with abundant granite sheets and veins (Hagar Bjerg thrust sheet) in the hanging wall overlie metasedimentary rocks lacking granite sheets and veins in the footwall (Niggli Spids thrust sheet). Profile height is ~900 m.

sample from this locality has yielded a SHRIMP zircon age of ca. 1900 Ma (F. Kalsbeek, 1999, personal commun.). The gneisses are pierced on one of these nunataks by a spectacular Paleogene volcanic plug containing cognate dunitic xenoliths. The plug is interpreted, along with others discovered in the region, as a feeder to the highly alkaline Paleogene ultrabasic lavas (maymechites) that occur as scattered outliers on the peaks of several nunataks in the region (Bernstein et al., 2000). The rocks of the Hamberg Gletscher foreland are overthrust by garnet-grade Krummedal metasedimentary rocks of the Niggli Spids thrust sheet, which locally incorporate a very few minor occurrences of granite stringers and veins. The Niggli Spids thrust is here hidden beneath the ice. However, another basalt plug south of the line of section (g on Fig. 16) contains abundant hornfelsed metasedimentary xenoliths, the lithologies of which suggest that, here as well, the gently east-dipping garnet–mica schists of the Niggli Spids thrust sheet are underlain by the same discrete package of lower-grade micaceous phyllites identified in the Gemmedal thrust sheet, and perhaps also by Lower Cambrian Slottet Formation quartzites of the underlying foreland.

East-vergent major folds deform both the Niggli Spids and Hagar Bjerg thrust sheets and, as noted previously, may be back-folds generated by movement on the thrust that floors the Målebjerg and Eleonore Sø windows (e.g., h on Fig. 16). The asymmetrical syncline core is occupied by a granite pluton that has a sheeted contact with the Hagar Bjerg thrust sheet rocks (Fig. 16, panel 1). Although undated, this granite shares some similarities (including enclosed amphibolite sheets) with the ca. 930 Ma augen granites characteristic of the Hagar Bjerg thrust sheet farther east in Andrée Land (Kalsbeek et al., this volume). Pelites carry intense west-verging crenulations of the regional schistosity on the steep short limb of the asymmetrical fold pair.

On the west side of the Eleonore Sø window, medium-grade Krummedal sequence metasedimentary rocks (kyanite–garnet–mica schists) of the Niggli Spids thrust sheet dip moderately to steeply westward beneath the higher-grade Krummedal metasedimentary rocks and granites of the Hagar Bjerg thrust sheet. The central part of the section (Fig. 16, panel 2) illustrates that an ~2-km-thick unit of intensely deformed, highly schistose, gray to black chlorite-biotite–grade phyllites of the Gemmedal thrust sheet rests on the westward-dipping roof thrust of the window and dips west beneath the Niggli Spids thrust sheet.

The spectacular Eleonore Sø window occurs beneath an arched roof thrust (Fig. 16, panel 2) that encloses an antiformal stack developed in the Paleoproterozoic metasedimentary and metavolcanic rocks of the Eleonore Sø supracrustal succession. Sheared quartzites of the Slottet Formation mark the location of some of the thrust surfaces within the window. Development of the antiformal stack implies that a floor thrust (h in Fig. 16) underlies at least part of the window. At the eastern exposed boundary of the window, the Slottet Formation quartzites are overlain by limestones of the Målebjerg Formation (Smith et al., 2004a), and the roof thrust appears to follow a long flat in this formation, as shown by its occurrence in analogous structural settings in other parautochthonous foreland windows (Higgins et al., 2001a). The Eleonore Sø window best demonstrates the spectacular unconformity at the base of the Early Cambrian

Figure 16. Cross section H–J extends from the Hamberg Gletscher foreland across the Eleonore Sø window to the Eleonore Bay Supergroup, which makes up the Franz Joseph allochthon in northern Strindberg Land; g–k are localities referred to in the text. See Figure 4 for section line.

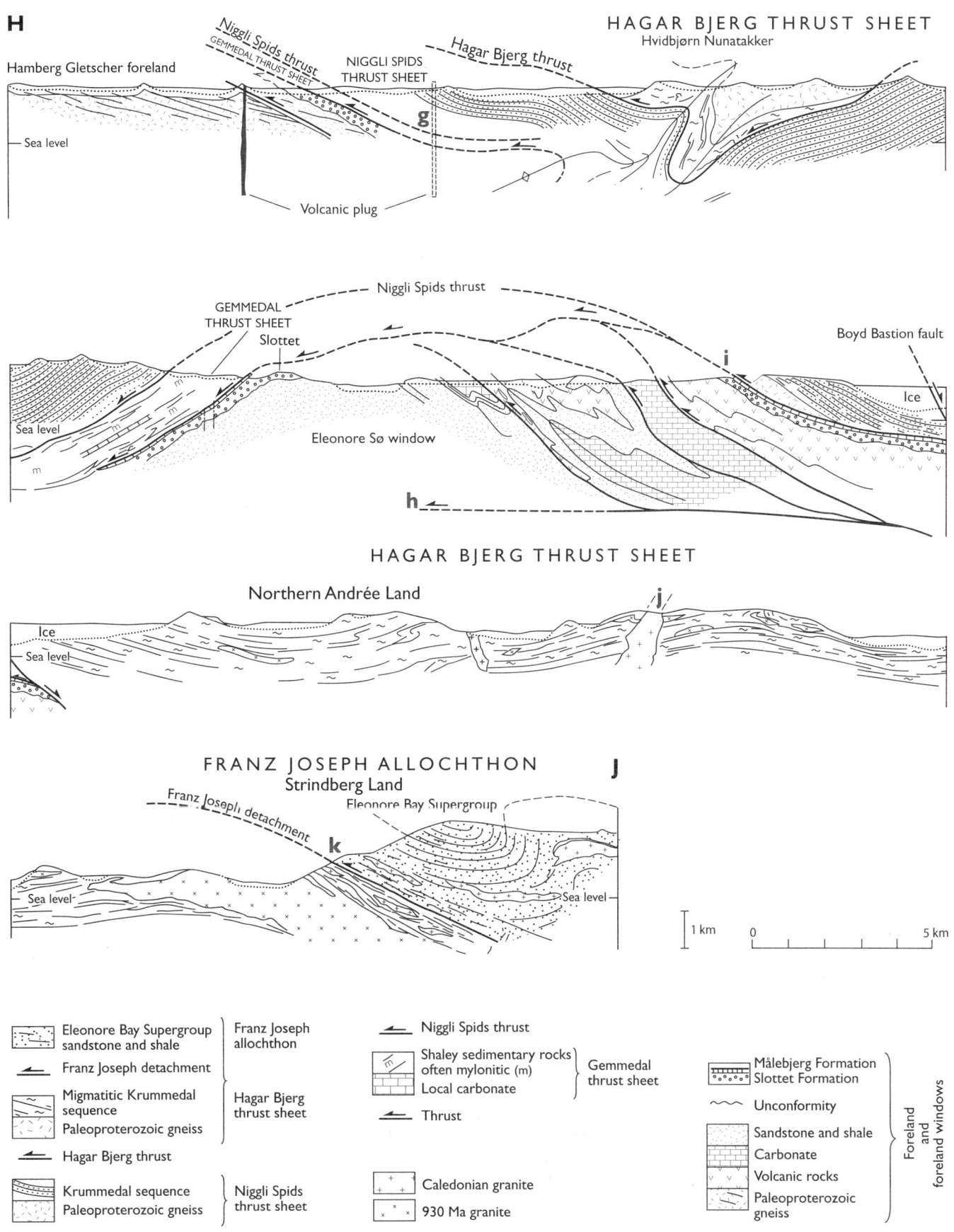

Slottet Formation (Smith et al., 2004a). White Slottet Formation quartzites rest on dark psammitic metasedimentary rocks of the Eleonore Sø supracrustal sequence (Fig. 17); gravels and pebble-grade, often black, sandstones are present at the unconformity surface. The fabrics deforming the Eleonore Sø supracrustal succession pass without break across the unconformity and testify to the Caledonian age of these fabrics and the associated thrusting. The roof thrust is hidden by ice west of the mountain Slottet, but it is exposed elsewhere on the western side of the window.

Metasedimentary rocks of the Niggli Spids thrust sheet overlie the east margin of the Eleonore Sø window. The occurrence of high-grade metasedimentary and meta-igneous rocks of crystalline basement affinities in the immediate hanging wall of the roof thrust beneath the Krummedal succession rocks (i on Fig. 16) points to unresolved heterogeneities within the overriding Niggli Spids thrust sheet. The 2-km-thick sheet of low-grade phyllitic rocks recognized all along the western edge of the window does not occur here in the east; this further reinforces the asymmetrical tectonostratigraphy across these windows, where greater thicknesses of basement gneisses often occur in the east.

Boyd Bastion Fault to Strindberg Land (Fig. 16, Panels 3 and 4)

Although hidden below ice in the line of section, the Boyd Bastion fault has a significant downthrow of some 10–15 km, and the nunataks in northern Andrée Land are composed of garnet-kyanite–grade Krummedal metasedimentary rocks and granite sheets of the Hagar Bjerg thrust sheet. Sheet dips are generally gently inclined and deformed by broad, open antiformal and synformal structures. Belts of tight to isoclinal intrafolial folds are locally observed. D2 fold vergence in decameter-scale folds at outcrop suggests the presence of a large westward-closing recumbent fold, similar to other kilometer-scale examples that occur in the hanging walls of other major thrusts elsewhere in the region.

In addition to sheets of coarse-grained, ca. 930 Ma augen granites, the garnet-kyanite–grade Krummedal metasedimentary rocks also host foliated white or pink Caledonian granite sheets (ca. 425 Ma; Kalsbeek et al., this volume) that are broadly concordant with the prevailing structure. These Caledonian granites are syn- to post-tectonic with respect to the pervasive top-to-the-WNW shear fabric associated with emplacement of the thrust sheets, and thus they show a variety of geometries. Many granites are intensely deformed in noncoaxial simple shear (Fig. 7A), while other sheets escape this attenuation but are folded in Caledonian D2 fold hinges. Still other granite sheets are unfoliated and thus must be slightly younger than the deformation; these sharply crosscut the regional structure (j in Fig. 16).

Although too small to be visible on the cross sections, it is worth noting here that the highest summits both north and south of the line of section in this remote nunatak region preserve outliers of flat-lying Paleogene maymechite lava flows (Bernstein et al., 2000).

The eastern part of the section H–J (Fig. 16, panel 4) exposes the uppermost levels of the Hagar Bjerg thrust sheet in northern Strindberg Land. Conspicuous tabular sheets of S-type garnetiferous augen granite (Fig. 18) emplaced ca. 910–940 Ma can be shown to derive from the Krummedal sequence metasedimentary

Figure 17. Lower Cambrian quartzites of the Slottet Formation at the west margin of the Eleonore Sø window, resting with marked unconformity on dark-colored Paleoproterozoic sandstones. The summit ridge of Slottet (skyline at right center) is 600 m above the glacier surface in the right foreground.

rocks that host them (Kalsbeek et al., 2000, this volume; Leslie and Nutman, 2003).

The Franz Joseph detachment is a bedding-parallel shear zone, up to 1 km thick, that separates the main part of the Hagar Bjerg thrust sheet from its uppermost segment, known as the Franz Joseph allochthon. In this paper (see also Higgins and Leslie, this volume), the Franz Joseph detachment is considered to be distinct from the extensional, down-to-the-east Fjord region fault, which cuts steeply across the bedding in the Eleonore Bay Supergroup; Gilotti and McClelland (this volume) interpret both structures as parts of their "Fjord region detachment system." The original detachment contact at the base of the Franz Joseph allochthon is well exposed in northern Strindberg Land (k in Fig. 16) and northern Andrée Land (Fig. 18); we consider it likely that this detachment (or shear zone), which exhibits evidence for both low-angle top-to-the-WNW ductile contractional shearing and superimposed, more steeply SE-dipping, top-to-the-ESE brittle-ductile extensional movement, may have developed at the site of an original unconformity at the base of the Eleonore Bay Supergroup. Lenticular or pod-like Caledonian granite sheets in garnet-kyanite–grade Krummedal metasedimentary rocks are geometrically consistent with the top-to-the-WNW shearing fabrics in the footwall of the detachment (Figs. 7A–7C). Lithological units in the lower Eleonore Bay Supergroup in the hanging wall, and sporadic granite sheets, are coplanar with the shear zone. The Eleonore Bay Supergroup is deformed by a series of N-S–trending major upright folds (see, e.g., Haller, 1971), some of which have a "box-fold" style.

DISCUSSION AND CONCLUSIONS

The Caledonian orogeny in East Greenland was the result of Silurian collision of Baltica with the margin of Laurentia. The structural record and architecture of that part of the orogen preserved onshore show that the collision produced a system of WNW-directed, i.e., foreland-propagating, thrust sheets that were derived from the Laurentian margin and that translated westward across the orogenic foreland (Higgins and Leslie, 2000; Higgins et al., 2004a). Orogen-scale restoration of the mapped and recorded thrusting illustrated here indicates that the original site of collision was probably several hundred kilometers east of the present-day part of the orogen preserved onshore.

Mid-Silurian Scandian ductile thrusting in the Scottish sector of the Caledonides was followed by sinistral strike-slip displacements along an array of structures that developed prior to, and during, the oblique collision of Avalonia and Baltica with Laurentia throughout the Late Silurian to Early Devonian (Soper et al., 1992) as the thickened orogenic welt began to dissect along its axis. In Scotland, the most prominent structures are the Great Glen–Walls Boundary and Highland Boundary faults, along which hundreds of kilometers of displacement may have occurred (Dewey and Strachan, 2003). Blastomylonitic rocks preserved in the core of the Great Glen fault zone may reflect the presence of an exhumed positive flower structure formed during sinistral transpression along the same zone of weakness (Stewart et al., 1999). Relationships among fault zone structures, dated igneous intrusions, and postorogenic sedimentary rocks constrain

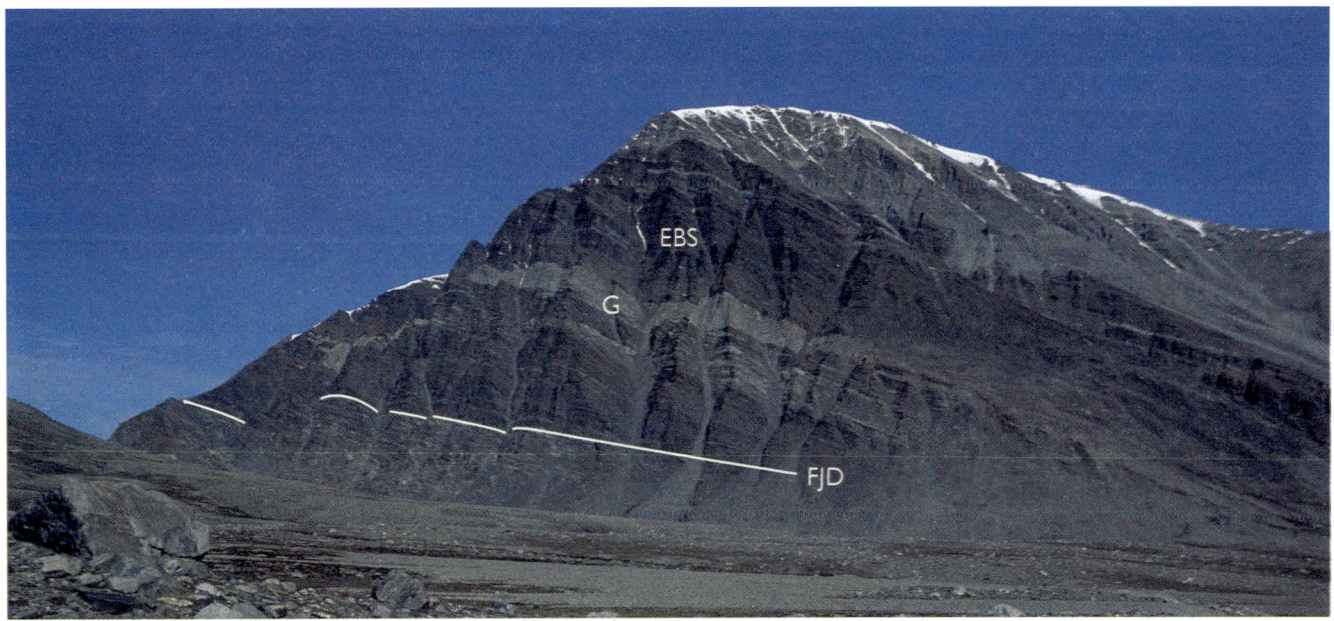

Figure 18. The Franz Joseph detachment (FJD) in northern Andrée Land is a bedding-parallel shear zone between the Nathorst Land Group of the Eleonore Bay Supergroup (EBS) in the hanging wall and migmatitic Krummedal sequence metasedimentary rocks in the footwall. Granite sheets in the shear zone exhibit top-to-the-WNW shear fabrics. The hanging wall sedimentary rocks are intruded by concordant light-colored sheets of Caledonian granite (G). The pale-colored rocks forming the summit ridge at right are sandstones of the upper Nathorst Land Group. View looks northward; profile height is ~1500 m. Modified from Elvevold et al. (2000).

the main sinistral displacement along the Great Glen fault to the period between ca. 428 Ma and ca. 400 Ma (Stewart et al., 1999). In East Greenland, while such strike-slip motion may have initiated at about this time, it clearly continued after exhumation of eclogitic rocks in the mid-Devonian (Gilotti et al., 2004).

The cross sections presented here in effect examine the consequences of Laurentia-Baltica collision in East Greenland, up to the point that wrench-dominated orogen-parallel tectonics become dominant. A number of comparable sections are also provided on the new 1:1,000,000 scale geological map of East Greenland that accompanies this volume. The thrust architecture was transected and displaced by a regional system of major late extensional faults and shear zones (Gilotti and McClelland, this volume, and references therein), but a number of key observations emerge, many of which underpin the foreland-propagating model presented here:

Kronprins Christian Land exposes a complete section through the Caledonian orogenic elements from autochthonous foreland in the west, through a transition into deeper-level, and thus farther-traveled, crystalline thrust sheets in the east. In this structural transition, the westernmost element is a thin-skinned fold-and-thrust belt developed in the Lower Paleozoic platform succession. Recorded thrust displacements in the fold-and-thrust belt are on the order of tens of kilometers.

The Vandredalen thrust sheet structurally overlies the fold-and-thrust belt, displaced some 35–50 km toward the WNW on the Vandredalen thrust. The floor thrust is well exposed where it follows a ramp that cuts up through the Lower Paleozoic rocks. Hanging wall cutoffs in the Vandredalen thrust sheet were juxtaposed with this steep ramp in the footwall succession, after which deformation propagated forward into the foreland.

Thrusts in the fold-and-thrust belt have propagated westward from deeper stratigraphic levels penetrated farther east. WNW-vergent, WNW-facing overturned folds are developed in a fold-and-thrust duplex of Independence Fjord Group rocks and doleritic dikes and sills in the alpine region around west-central Ingolf Fjord. Thrusting becomes more dominant eastward and structurally upward toward the Vandredalen thrust, and the Vandredalen thrust sheet has become elevated above exposure level.

The Vandredalen thrust sheet roots along a steeply dipping zone traceable from Marmorvigen northward along the western side of Hekla Sund to Ingolf Fjord. At Vardedalen in central Ingolf Fjord, both floor and roof thrusts are preserved, and an antiformal stack exhibits footwall imbrication to the west beneath the emerging Vandredalen thrust sheet.

Color alteration indices of conodonts in the Lower Paleozoic dolomites and limestones indicate that higher thrust sheets formerly extended westward above both the Vandredalen thrust sheet and the Thin-skinned fold-and-thrust belt. The overlying thrust pile is estimated to have reached a total thickness of ~10.7 km at the frontal ramp of the Vandredalen thrust sheet (Rasmussen and Smith, 2001).

Sediment shed from the eroding, emergent Caledonian thrust stack was transported as turbidity currents northward (Hurst et al., 1983), and in western Kronprins Christian Land, turbiditic sandstones and shales of the Lauge Koch Land Formation brought platform carbonate sedimentation to a close in the Wenlockian.

The most intense Caledonian deformation in Kronprins Christian Land is seen in the cross section along eastern-central Ingolf Fjord (east of Vardedalen), where the imbricate thrust pile (Western thrust belt) has a moderate to steep eastward dip. Toward the eastern end of the section, NW-directed thrusts appear to be localized along thin doleritic sills.

East-dipping extensional normal faults that postdate sill emplacement in the Mesoproterozoic but predate Caledonian thrusting are seen at a number of places along the Ingolf Fjord section.

The marked topographic lineament of the East Greenland fault zone, which follows the boundary with the basement gneisses in the outer part of Ingolf Fjord, is not exposed in the line of section; the highly strained quartzites and dolerites west of the boundary display a pronounced horizontal intersection lineation but no sign of a downdip stretching lineation. As noted already, mylonitic rocks exposed in southern Hovgaard Ø have been interpreted as a continuation of the Storstrømmen shear zone, but they display a top-up-to-the-east rather than a sinistral shear sense. The East Greenland fault zone may be interpreted as a normal fault that here follows the approximate line of the Storstrømmen shear zone.

The Vandredalen thrust ramps westward very close to the east-dipping footwall high of the extensional fault system that defines the Hekla Sund Basin in which the Rivieradal Group was deposited. The thrust excised segments of the Independence Fjord Group rocks that make up the footwall high and transported them westward as thrust-bounded horses. The Ingolf Fjord profile demonstrates footwall imbrication beneath this ramp structure and development of an antiformal stack in the Independence Fjord Group rocks.

The southern half of the orogen (70°N–76°N) exhibits a distinctive tectonostratigraphy with two major far-traveled thrust sheets, the Hagar Bjerg and Niggli Spids thrust sheets, which can be traced over a wide region between Bessel Fjord and Scoresby Sund, and a minor lowermost (Gemmedal) thrust sheet that is periodically present.

The intact foreland is largely concealed beneath the Inland Ice in the southern part of the orogen, and only the Hamberg Gletscher foreland and the rock units beneath the flat-lying basal thrust of the Gåseland window are interpreted as undisturbed autochthonous foreland.

The parautochthonous Caledonian foreland is exposed in several anticlinal windows exposed through overriding thrust sheets. The stratigraphy in these windows includes Paleoproterozoic igneous and volcanic elements that are not represented in the overlying thrust sheets. A major hiatus covering the Mesoproterozoic and most of the Neoproterozoic seems to have prevailed on the foreland craton at the same time as the adjacent marginal areas of Laurentia were major centers of deposition.

The lowest exposed thrust sheet is the Gemmedal thrust sheet, which has variable thickness and is made up of low-grade metasedimentary and phyllonitic rocks. The Lower Paleozoic Slottet and Målebjerg Formations on the western side of the Eleonore Sø window and on both western and eastern sides of the Målebjerg window are overlain by variably sheared phyllitic to phyllonitic metasedimentary rocks.

The Niggli Spids thrust sheet is composed of a lower basement gneiss unit and an upper metasedimentary unit correlated with the Krummedal supracrustal sequence. Unlike the Krummedal sequence metasediments in the older, overlying Hagar Bjerg thrust sheet, the Krummedal sequence of the Niggli Spids thrust sheet did not generate significant granitic melt during Caledonian orogenesis.

The Krummedal supracrustal sequence of the Hagar Bjerg thrust sheet is characterized by amphibolite-facies migmatitic metasedimentary rocks that host both ca. 930 Ma and Caledonian granites. The upper part of this thrust sheet is made up of a thick Neoproterozoic–Ordovician sedimentary succession (Franz Joseph allochthon) that hosts Caledonian (ca. 425 Ma) granites in its lowest levels.

Crustal shortening led to emplacement of the Hagar Bjerg thrust sheet above what was to become the Niggli Spids thrust sheet and, with continued shortening, emplacement of both thrust sheets above the discontinuous Gemmedal thrust sheet or the foreland. Westward displacement of the two major thrust sheets equaled a minimum of 90–100 km for each, from more outboard locations in the collision zone onto the Laurentian foreland. The strongly contrasting characteristics of the different thrust levels imply that thrust translation for each thrust sheet may have been significantly more than the minimum 90–100 km.

The thrusts sheets and the underlying foreland are deformed into long-wavelength, low-amplitude folds such that the foreland windows occupy elongate anticlinal fold cores. The asymmetrical distribution of tectonostratigraphy across the Eleonore Sø and Målebjerg foreland windows from east to west suggests that the defining thrusts are ramping through preferred stratigraphic levels as imbricate faults. The thrust sheets show a general pattern of stratal thinning (with loss of the gneissose lower element) from east to west across the windows.

The foreland windows in the southern part of the orogen are generally aligned perpendicular to the thrust transport direction. These antiformal culminations are most likely located above blind thrusts that developed as forward propagation of the eastward-thickening thrust sheets began to lock up. New thrust systems would develop structurally below, and in front of, the juxtaposition of any large-scale thickness changes in the hanging wall against ramps or other irregularities in the footwall.

Major extensional faults (Boyd Bastion fault, Fjord region fault) postdate thrust emplacement and juxtapose markedly different structural levels.

The present distribution of Caledonian domains across the northern Atlantic region is largely a consequence of the transtensional shearing that resulted from relative lateral motion of two large plates (Laurentia and Baltica) after their respective continental margins were brought into contact and subduction of oceanic crust had ceased (Soper et al., 1992; Dewey and Strachan 2003). After collision, Greenland became part of the Southern Hemisphere Devonian continent of Euramerica, consisting of most of North America, Greenland, and northern Europe. Marine conditions lay to the south in the Rhenohercynian Basin and in the Variscan orogen, which owed its existence to the relative northward movement of North Gondwana (Africa–South America plate). Tectonic syntheses by Coward (1993) and Maynard et al. (1997) suggest the progressive eastward expulsion of a triangular European-Baltica block that was squeezed between Greenland toward the northwest and the Variscides to the south. Left-lateral wrench-dominated tectonics thus continued to affect the northwestern margin of the European-Baltica block, including the East Greenland Caledonides, throughout the Devonian until the Early Carboniferous. The change to right-lateral strike-slip occurred when the European-Baltica block was pushed back to the west with orogeny in the Uralides. The "late Caledonian" record of deformation in East Greenland seems to reflect intraplate jostling of major blocks of continental crust along preexisting lines of weakness. The extensive pattern of open upright folding affecting rocks as young as the Late Devonian and that recognized in the coastal region of East Greenland north of Kong Oscar Fjord are geometrically consistent with right-lateral deformation in the Carboniferous. The Caledonian orogen was only dismantled and dispersed by Paleogene opening of the North Atlantic Ocean, so that widely separated parts now occur in East Greenland, Svalbard, Scandinavia, and the NW British Isles.

ACKNOWLEDGMENTS

The interpretations presented here have evolved over the course of a number of field seasons in East Greenland and draw on discussions and debate with numerous colleagues during that time, many of whom are the authors of other chapters in this volume. The authors recognize especially the input of N.J. Soper, M.P. Smith, J.A. Rasmussen, and S.A.S. Pedersen to the work in the Centrumsø, Rivieradal, Vandredalen, and Ingolf Fjord areas. The perceptive comments of the two reviewers, Allen Dennis and Joseph Hull, and editor Jane A. Gilotti have helped greatly to shape the final version of the manuscript.

REFERENCES CITED

Bernstein, S., Leslie, A.G., Higgins, A.K., and Brooks, C.K., 2000, Tertiary alkaline volcanics in the nunatak region, North-East Greenland: New observations and comparison with Siberian maymechites: Lithos, v. 53, p. 1–20, doi: 10.1016/S0024-4937(00)00012-8.

Brueckner, H.K., Gilotti, J.A., and Nutman, A.P., 1998, Caledonian eclogite-facies metamorphism of Early Proterozoic protoliths from the North-East Greenland eclogite province: Contributions to Mineralogy and Petrology, v. 130, p. 103–120, doi: 10.1007/s004100050353.

Collinson, J.D., Kalsbeek, F., Jepsen, H.F., Pedersen, S.A.S., and Upton, B.G.J., 2008, this volume, Paleoproterozoic and Mesoproterozoic sedimentary and volcanic successions in the northern parts of the East Greenland Caledonian orogen and its foreland, in Higgins, A.K., Gilotti, J.A., and Smith, M.P., eds., The Greenland Caledonides: Evolution of the Northeast

Margin of Laurentia: Geological Society of America Memoir 202, doi: 10.1130/2008.1202(04).
Coward, M.P., 1993, The effect of late Caledonian and Variscan continental escape tectonics on basement structure, Palaeozoic basin kinematics and subsequent Mesozoic basin development in NW Europe, *in* Parker, J.R., ed., Petroleum Geology of Northwest Europe, Proceedings of the 4th Conference: London, Geological Society of London, p. 1095–1108.
Dewey, J.F., and Strachan, R.A., 2003, Changing Silurian-Devonian relative plate motion in the Caledonides: Sinistral transpression to sinistral transtension: Journal of the Geological Society of London, v. 160, p. 219–229.
Elvevold, S., Escher, J.C., Frederiksen, K.S., Friderichsen, J.D., Gilotti, J.A., Henriksen, N., Higgins, A.K., Jepsen, H.F., Jones, K.A., Kalsbeek, F., Kinny, P.D., Leslie, A.G., Robertson, R., Smith, M.P., Thrane, K., and Watt, G.R., 2000, Tectonic architecture of the East Greenland Caledonides 72°–74°30′ N: Danmarks og Grønlands Geologiske Undersøgelse Rapport, 2000/88, 34 p.
Escher, J.C., and Jones, K.A., 1998, Caledonian thrusting and extension in Frænkel Land, East Greenland (72°30′–73°N): Preliminary results, *in* Higgins, A.K., and Frederiksen, K.S., eds., Caledonian Geology of East Greenland 72°–74°N: Preliminary Reports from the 1997 Expedition: Danmarks og Grønlands Geologiske Undersøgelse Rapport, v. 1998/28, p. 29–42.
Escher, J.C., and Jones, K.A., 1999, Caledonian geology of Frænkel Land and adjacent areas (73°00′–73°30′N), East Greenland, *in* Higgins, A.K., and Frederiksen, K.S., eds., Geology of East Greenland 72°–75°N, Mainly Caledonian: Preliminary Reports from the 1998 Expedition: Danmarks og Grønlands Geologiske Undersøgelse Rapport, v. 1999/19, p. 27–36.
Fränkl, E., 1954, Vorläufige Mitteilung über die Geologie von Kronprins Christian Land (NE-Grönland): Meddelelser om Grønland, v. 116, no. 2, 85 p.
Fränkl, E., 1955, Weitere Beiträge zur Geologie von Kronprins Christian Land (NE-Grönland): Meddelelser om Grønland, v. 103, no. 7, 35 p.
Gilotti, J.A., 1993, Discovery of a medium-temperature eclogite province in the Caledonides of North-East Greenland: Geology, v. 21, p. 523–526, doi: 10.1130/0091-7613(1993)021<0523:DOAMTE>2.3.CO;2.
Gilotti, J.A., and McClelland, W.C., 2008, this volume, Geometry, kinematics, and timing of extensional faulting in the Greenland Caledonides—A synthesis, *in* Higgins, A.K., Gilotti, J.A., and Smith, M.P., eds., The Greenland Caledonides: Evolution of the Northeast Margin of Laurentia: Geological Society of America Memoir 202, doi: 10.1130/2008.1202(10).
Gilotti, J.A., and Ravna, E.J.K., 2002, First evidence for ultrahigh-pressure metamorphism in the North-East Greenland Caledonides: Geology, v. 30, p. 551–554, doi: 10.1130/0091-7613(2002)030<0551:FEFUPM>2.0.CO;2.
Gilotti, J.A., Nutman, A.P., and Brueckner, H.K., 2004, Devonian to Carboniferous collision in the Greenland Caledonides: U-Pb zircon and Sm-Nd ages of high-pressure and ultrahigh-pressure metamorphism: Contributions to Mineralogy and Petrology, v. 148, p. 216–235, doi: 10.1007/s00410-004-0600-4.
Gilotti, J.A., Jones, K.A., and Elvevold, S., 2008, this volume, Caledonian metamorphic patterns in Greenland, *in* Higgins, A.K., Gilotti, J.A., and Smith, M.P., eds., The Greenland Caledonides: Evolution of the Northeast Margin of Laurentia: Geological Society of America Memoir 202, doi: 10.1130/2008.1202(08).
Haller, J., 1971, Geology of the East Greenland Caledonides: London, Interscience, 413 p.
Hartz, E.H., Andresen, A., Martin, M.W., and Hodges, K.V., 2000, U-Pb and $^{40}Ar/^{39}Ar$ constraints on the Fjord region detachment zone: A long-lived extensional fault in the central East Greenland Caledonides: Journal of the Geological Society of London, v. 157, p. 795–809.
Henriksen, N., and Higgins, A.K., 1976, East Greenland Caledonides, *in* Escher, A., and Watt, W.S., eds., Geology of Greenland: Copenhagen, Grønlands Geologiske Undersøgelse, p. 182–246.
Higgins, A.K., 1988, The Krummedal supracrustal sequence in East Greenland, *in* Winchester, J.A., ed., Later Proterozoic Stratigraphy of the Northern Atlantic Regions: Glasgow and London, Blackie and Son Ltd., p. 86–96.
Higgins, A.K., and Leslie, A.G., 2000, Restoring thrusting in the East Greenland Caledonides: Geology, v. 28, p. 1019–1022, doi: 10.1130/0091-7613(2000)28<1019:RTITEG>2.0.CO;2.
Higgins, A.K., and Leslie, A.G., 2008, this volume, Architecture and evolution of the East Greenland Caledonides—An introduction, *in* Higgins, A.K., Gilotti, J.A., and Smith, M.P., eds., The Greenland Caledonides: Evolution of the Northeast Margin of Laurentia: Geological Society of America Memoir 202, doi: 10.1130/2008.1202(02).
Higgins, A.K., Leslie, A.G., and Smith, M.P., 2001a, Neoproterozoic–Lower Palaeozoic stratigraphical relationships in the marginal thin-skinned thrust belt of the East Greenland Caledonides: Comparisons with the foreland of Scotland: Geological Magazine, v. 138, p. 143–160, doi: 10.1017/S0016756801005076.
Higgins, A.K., Smith, M.P., Soper, N.J., Leslie, A.G., Rasmussen, J.A., and Sønderholm, M., 2001b, The Neoproterozoic Hekla Sund Basin, eastern North Greenland: A pre-Iapetan extensional sequence thrust across its rift shoulders during the Caledonian orogeny: Journal of the Geological Society of London, v. 158, p. 487–499.
Higgins, A.K., Elvevold, S., Escher, J.C., Frederiksen, K.S., Gilotti, J.A., Henriksen, N., Jepsen, H.F., Jones, K.A., Kalsbeek, F., Kinny, P.D., Leslie, A.G., Smith, M.P., Thrane, K., and Watt, G.R., 2004a, The foreland-propagating thrust architecture of the East Greenland Caledonides 72°–75°N: Journal of the Geological Society of London, v. 161, p. 1009–1026, doi: 10.1144/0016-764903-141.
Higgins, A.K., Soper, N.J., Smith, M.P., and Rasmussen, J.A., 2004b, The Caledonian parautochthonous fold and thrust belt of Kronprins Christian Land, eastern North Greenland: Geological Survey of Denmark and Greenland Bulletin, v. 6, p. 41–56.
Holdsworth, R.E., and Strachan, R.A., 1991, Interlinked system of ductile strike-slip and thrusting formed by Caledonian sinistral transpression in northeastern Greenland: Geology, v. 19, p. 510–513, doi: 10.1130/0091-7613(1991)019<0510:ISODSS>2.3.CO;2.
Hurst, J.M., McKerrow, W.S., Soper, N.J., and Surlyk, F., 1983, The relationship between Caledonian nappe tectonics and Silurian turbidite deposition in North Greenland: Journal of the Geological Society of London, v. 140, p. 123–132, doi: 10.1144/gsjgs.140.1.0123.
Jepsen, H.F., and Kalsbeek, F., 1981, Non-existence of the Carolinidian orogeny in the Prinsess Caroline–Mathilde Alper of Kronprins Christian Land, eastern North Greenland: Rapport Grønlands Undersøgelse, v. 106, p. 7–14.
Jepsen, H.F., and Kalsbeek, F., 1998, Granites in the Caledonian fold belt of East Greenland, *in* Higgins, A.K., and Frederiksen, K.S., eds., Caledonian Geology of East Greenland 72°–74°N: Preliminary Reports from the 1997 Expedition: Danmarks og Grønlands Geologiske Undersøgelse Rapport, v. 1998/28, p. 73–82.
Jepsen, H.F., Escher, J.C., Friderichsen, J.D., and Higgins, A.K., 1994, The geology of the north-eastern corner of Greenland—Photogeological studies and 1993 field work: Rapport Grønlands Geologiske Undersøgelse, v. 161, p. 21–33.
Jones, K.A., and Escher, J.C., 2002, Near-isothermal decompression within a clockwise P–T evolution recorded in migmatitic mafic granulites from Clavering Ø, NE Greenland: Implications for the evolution of the Caledonides: Journal of Metamorphic Geology, v. 20, p. 365–378, doi: 10.1046/j.1525-1314.2002.00375.x.
Kalsbeek, F., 1995, Geochemistry, tectonic setting, and poly-orogenic history of Palaeoproterozoic basement rocks from the Caledonian fold belt of North-East Greenland: Precambrian Research, v. 72, p. 301–315, doi: 10.1016/0301-9268(94)00097-B.
Kalsbeek, F., and Jepsen, H.F., 1984, The late Proterozoic Zig-Zag Dal Basalt Formation of eastern North Greenland: Journal of Petrology, v. 25, p. 644–664.
Kalsbeek, F., Nutman, A.P., and Taylor, P.N., 1993, Palaeoproterozoic basement province in the Caledonian fold belt of North-East Greenland: Precambrian Research, v. 63, p. 163–178, doi: 10.1016/0301-9268(93)90010-Y.
Kalsbeek, F., Nutman, A.P., Escher, J.C., Friderichsen, J.D., Hull, J.M., Jones, K.A., and Pedersen, S.A.S., 1999, Geochronology of granitic and supracrustal rocks from the northern part of the East Greenland Caledonides: Ion microprobe U-Pb zircon ages: Geology of Greenland Survey Bulletin, v. 184, p. 31–48.
Kalsbeek, F., Thrane, K., Nutman, A.P., and Jepsen, H.F., 2000, Late Mesoproterozoic to early Neoproterozoic history of the East Greenland Caledonides: Evidence for Grenvillian orogenesis?: Journal of the Geological Society of London, v. 157, p. 1215–1225.
Kalsbeek, F., Higgins, A.K., Jepsen, H.F., Frei, R., and Nutman, A.P., 2008, this volume, Granites and granites in the East Greenland Caledonides, *in* Higgins, A.K., Gilotti, J.A., and Smith, M.P., eds., The Greenland Caledonides: Evolution of the Northeast Margin of Laurentia: Geological Society of America Memoir 202, doi: 10.1130/2008.1202(09).

Larsen, H.C., and Bengaard, H.-J., 1991, Devonian basin initiation in East Greenland: A result of sinistral wrench faulting and Caledonian extensional collapse: Journal of the Geological Society of London, v. 148, p. 355–368, doi: 10.1144/gsjgs.148.2.0355.

Leslie, A.G., and Higgins, A.K., 1998, On the Caledonian geology of Andrée Land, Eleonore Sø and adjacent nunataks (73°30′–74°N), East Greenland, in Higgins, A.K., and Frederiksen, K.S., eds., Caledonian Geology of East Greenland 72°–74°N: Preliminary Reports from the 1997 Expedition: Danmarks og Grønlands Geologiske Undersøgelse Rapport, v. 1998/28, p. 11–27.

Leslie, A.G., and Higgins, A.K., 1999, On the Caledonian (and Grenvillian) geology of Bartholin Land, Ole Rømer Land, and adjacent nunataks, East Greenland, in Higgins, A.K., and Frederiksen, K.S., eds., Caledonian Geology of East Greenland 72°–74°N: Preliminary Reports from the 1997 Expedition: Danmarks og Grønlands Geologiske Undersøgelse Rapport, v. 1999/19, p. 11–26.

Leslie, A.G., and Nutman, A.P., 2003, Evidence for Neoproterozoic orogenesis and early high temperature Scandian deformation events in the southern East Greenland Caledonides: Geological Magazine, v. 140, p. 309–333, doi: 10.1017/S0016756803007593.

Maynard, J.R., Hofman, W., Dunay, R.E., Bentham, P.N., Dean, K.P., and Watson, I., 1997, The Carboniferous system of Europe: The development of a petroleum system: Petroleum Geoscience, v. 3, p. 97–115.

Peacock, J.D., 1956, The Geology of Dronning Louise Land, N.E. Greenland: Meddelelser om Grønland, v. 137, no. 7, 38 p.

Peacock, J.D., 1958, Some Investigations into the Geology and Petrography of Dronning Louise Land, N.E. Greenland: Meddelelser om Grønland, v. 157, no. 4, 139 p.

Pedersen, S.A.S., Craig, L.E., Upton, B.G.J., Rämö, O.T., Jepsen, H.F., and Kalsbeek, F., 2002, Palaeoproterozoic (1740 Ma) rift-related volcanism in the Hekla Sund region, eastern North Greenland: Field occurrence, geochemistry and tectonic setting: Precambrian Research, v. 114, p. 327–346, doi: 10.1016/S0301-9268(01)00234-0.

Peel, J.S., 1980, Geological reconnaissance in the Caledonian foreland of eastern North Greenland with comments on the Centrum Limestone: Rapport Grønlands Geologiske Undersøgelse, v. 99, p. 61–72.

Phillips, W.E.A., Stillman, C.J., Friderichsen, J.D., and Jemelin, L., 1973, Preliminary results of mapping in the western gneiss and schist zone around Vestfjord and inner Gåsefjord, south-west Scoresby Sund: Rapport Grønlands Geologiske Undersøgelse, v. 58, p. 17–32.

Rasmussen, J.A., and Smith, M.P., 2001, Conodont geothermometry and tectonic overburden in the northernmost East Greenland Caledonides: Geological Magazine, v. 138, p. 687–698.

Rex, D.C., and Gledhill, A.R., 1981, Isotopic studies in the East Greenland Caledonides (72°–74°N)—Precambrian and Caledonian ages: Rapport Grønlands Geologiske Undersøgelse, v. 104, p. 47–72.

Schlindwein, V., 1998, Architecture and Evolution of the Continental Crust of East Greenland from Integrated Geophysical Studies [Ph.D. dissertation]: Bremen, Universität Bremen, 144 p.

Smith, M.P., 1991, Early Ordovician conodonts of East and North Greenland: Meddelelser om Grønland: Geoscience, v. 28, 81 p.

Smith, M.P., and Rasmussen, J.A., 2008, this volume, Cambrian–Silurian development of the Laurentian margin of the Iapetus Ocean in Greenland and related areas, in Higgins, A.K., Gilotti, J.A., and Smith, M.P., eds., The Greenland Caledonides: Evolution of the Northeast Margin of Laurentia: Geological Society of America Memoir 202, doi: 10.1130/2008.1202(06).

Smith, M.P., and Robertson, S., 1999, The Nathorst Land Group (Neoproterozoic) of East Greenland—Lithostratigraphy, basin geometry and tectonic history: Danmarks og Grønlands Geologiske Undersøgelse Rapport, v. 1999/19, p. 127–143.

Smith, M.P., Soper, N.J., Higgins, A.K., and Rasmussen, J.A., 1999, Palaeokarst systems in the Late Proterozoic of eastern North Greenland and their stratigraphic and tectonic significance: Journal of the Geological Society of London, v. 156, p. 113–124, doi: 10.1144/gsjgs.156.1.0113.

Smith, M.P., Rasmussen, J.A., Robertson, S., Higgins, A.K., and Leslie, A.G., 2004a, Lower Palaeozoic stratigraphy of the East Greenland Caledonides: Geology Survey of Denmark and Greenland Bulletin, v. 6, p. 5–28.

Smith, M.P., Higgins, A.K., Soper, N.J., and Sønderholm, M., 2004b, The Neoproterozoic Rivieradal Group of Kronprins Christian Land, eastern North Greenland: Geological Survey of Denmark and Greenland Bulletin, v. 6, p. 29–39.

Sønderholm, M., and Jepsen, H.F., 1991, Proterozoic basins of North Greenland: Bulletin Grønlands Geologiske Undersøgelse, v. 160, p. 49–69.

Sønderholm, M., Frederiksen, K.S., Smith, M.P., and Tirsgaard, H., 2008, this volume, Neoproterozoic sedimentary basins with glacigenic deposits of the East Greenland Caledonides, in Higgins, A.K., Gilotti, J.A., and Smith, M.P., eds., The Greenland Caledonides: Evolution of the Northeast Margin of Laurentia: Geological Society of America Memoir 202, doi: 10.1130/2008.1202(05).

Soper, N.J., Strachan, R.A., Holdsworth, R.E., Gayer, R.A., and Greiling, R.O., 1992, Sinistral transpression and the Silurian closure of Iapetus: Journal of the Geological Society of London, v. 149, p. 871–880, doi: 10.1144/gsjgs.149.6.0871.

Stemmerik, L., and Håkansson, E., 1991, Carboniferous and Permian history of the Wandel Sea Basin, North Greenland, in Peel, J.S., and Sønderholm, M., eds., Sedimentary Basins of North Greenland: Bulletin Grønlands Geologiske Undersøgelse, v. 160, p. 141–151.

Stewart, M., Strachan, R.A., and Holdsworth, R.E., 1999, Structure and early kinematic history of the Great Glen fault zone, Scotland: Tectonics, v. 18, p. 326–342, doi: 10.1029/1998TC900033.

Strachan, R.A., Holdsworth, R.E., Friderichsen, J.D., and Jepsen, H.F., 1992, Regional Caledonian structure within an oblique convergence zone, Dronning Louise Land, NE Greenland: Journal of the Geological Society of London, v. 149, p. 359–371, doi: 10.1144/gsjgs.149.3.0359.

Strachan, R.A., Friderichsen, J.D., Holdsworth, R.E., and Jepsen, H.F., 1994, Regional geology and Caledonian structure, Dronning Louise Land, North-East Greenland, in Higgins, A.K., ed., Geology of North-East Greenland: Rapport Grønlands Geologiske Undersøgelse, v. 162, p. 71–76.

Upton, B.G.J., Rämö, O.T., Heaman, L.M., Blichert-Toft, J., Barry, T.L., Kalsbeek, F., and Jepsen, H.F., 2005, The Zig-Zag Dal Basalts and associated intrusions of eastern North Greenland: Progressive mantle plume–lithosphere interaction: Contributions to Mineralogy and Petrology, v. 149, p. 40–56, doi: 10.1007/s00410-004-0634-7.

Watt, G.R., and Thrane, K., 2001, Early Neoproterozoic events in East Greenland: Precambrian Research, v. 110, p. 165–184, doi: 10.1016/S0301-9268(01)00186-3.

Wenk, E., 1961, On the crystalline basement and the basal part of the pre-Cambrian Eleonore Bay Group in the southwestern part of Scoresby Sund: Meddelelser om Grønland, v. 168, no. 1, 54 p.

Wenk, E., and Haller, J., 1953, Geological explorations in the Petermann Bjerg region, western part of Frænkels Land, East Greenland: Meddelelser om Grønland, v. 111, no. 3, 48 p.

MANUSCRIPT ACCEPTED BY THE SOCIETY 14 JANUARY 2008

Caledonian metamorphic patterns in Greenland

Jane A. Gilotti*
Department of Geoscience, University of Iowa, Iowa City, Iowa 52242, USA

Kevin A. Jones
Hills Road Sixth Form College, Hills Road, Cambridge, CB2 8PE, UK

Synnøve Elvevold
Norwegian Polar Institute, Polar Environmental Center, N-9296 Tromsø, Norway

ABSTRACT

The Greenland Caledonides have a tectonic architecture built of Laurentian-margin Precambrian crystalline complexes and younger sedimentary successions that were metamorphosed during the Paleozoic collision with Baltica. Caledonian metamorphic patterns correspond to the gross structural levels of the orogen. The patterns are superimposed on earlier metamorphic histories in the Precambrian crystalline complexes, but they account for the sole metamorphism of Neoproterozoic to early Paleozoic sedimentary units. We describe the Caledonian metamorphism by dividing the orogen into two parts, a northern and a southern segment separated at 76°N by Bessel Fjord. North of Bessel Fjord, metamorphic grade increases eastward toward the hinterland in progressively higher thrust sheets, where it ultimately reaches ultrahigh-pressure conditions. The metamorphic pattern in the southern segment is complicated by regional extensional detachment faults. Very low-grade sedimentary rocks of the foreland are overlain by the deepest structural level, the Niggli Spids thrust sheet, which contains widespread relicts of high-pressure metamorphism. The overlying Hagar Bjerg thrust sheet is composed of a midcrustal-level migmatite complex that records high temperatures in the amphibolite to granulite facies. The Neoproterozoic to Ordovician sedimentary rocks of the uppermost unit, the Franz Joseph allochthon, reached greenschist- and locally amphibolite-facies conditions (garnet + staurolite) at their base. The Devonian and younger sedimentary basins are not significantly metamorphosed. Each of the main structural levels has a characteristic pressure-temperature path. Three main periods of metamorphism are currently recognized. The oldest, ca. 440–415 Ma, relates to the formation of migmatites and granites at midcrustal levels. This was followed by widespread high-pressure granulite- and eclogite-facies metamorphism from 410 to 390 Ma. A very late pulse of ultrahigh-pressure metamorphism occurred at 360–350 Ma and marked the end of the Caledonian collision.

Keywords: Greenland, Caledonides, metamorphic facies, pressure-temperature path, eclogites.

*jane-gilotti@uiowa.edu

INTRODUCTION

Caledonian metamorphic patterns in East Greenland correspond to different structural levels that developed mainly during the contractional phase of continent-continent collision of Laurentia with Baltica. Peak assemblages are commonly overprinted by retrograde reactions that occurred during exhumation. The Caledonian patterns of metamorphism are complicated to discern because of an earlier metamorphic history in the Precambrian rocks (Rex and Gledhill, 1981; Higgins, 1995); however, recent advances in U-Pb geochronology have allowed for the unequivocal identification of Caledonian metamorphism. We use the term Caledonian in the broad sense of the early Paleozoic collision between Baltica and Laurentia (McKerrow et al., 2000), and attempt to delineate metamorphic and deformational events within this broad framework.

Caledonian metamorphism ranges from very low grade in the Neoproterozoic to Silurian sedimentary sequences to ultrahigh pressure and high temperature in the crystalline rocks in the core of the orogen. The aim of this paper is to provide an overview of metamorphism along the length of the Greenland Caledonides and to highlight the significant discoveries of the recent 30 year mapping campaign by the Geological Survey of Denmark and Greenland. During these years, the approach to understanding the physical conditions of metamorphism has become more quantitative and directed at determining the entire pressure-temperature-time-deformation (P-T-t-d) history of rocks (e.g., Spear, 1993). Such quantitative studies are still in their infancy in the Greenland Caledonides, especially when compared to other orogens, such as the Alps and the New England portion of the Appalachians. This paper is intended to provide a basic framework for future studies of Caledonian metamorphism in East Greenland. It is not an exhaustive review but rather a synthesis of current knowledge. The reader should see Haller (1971) for references to the older literature.

The entire Caledonian belt in Greenland is part of the Laurentian plate margin. Archean and Paleoproterozoic crystalline complexes are continuous from northern Canada across to West Greenland and beneath the ice sheet to East Greenland (Kalsbeek et al., 1993; Thrane, 2002). Mesoproterozoic and Neoproterozoic to early Paleozoic sedimentary successions were deposited on this Laurentian basement substrate. Many of the exotic arc and ophiolite terranes (e.g., Stephens and Gee, 1989; Grenne et al., 1999) were stranded in the Scandinavian Caledonides by the opening of the North Atlantic; no exotic terranes have been identified in East Greenland. Laurentia is thought to have been the overriding plate in the collision with Baltica (Gee, 1975; Hossack and Cooper, 1986; Gilotti and McClelland, 2007), analogous to the Eurasian plate in the Himalayan orogen. Studies of Caledonian metamorphic patterns exhumed along the east coast of Greenland provide insight into the processes that result from the considerable tectonic thickening and subsequent thinning of the upper plate in a continent-continent collision setting.

This paper presents a regional map of metamorphic facies in the Greenland Caledonides, a summary of the physical conditions of metamorphism, and a brief synthesis of the tectonometamorphic evolution. We divide the Greenland Caledonides into a northern half and a southern half, split at Bessel Fjord (76°N). The northern half of the orogen is a relatively simple stack of thrust sheets that have an increasing metamorphic grade structurally upward (i.e., eastward) in the pile to eclogite-facies conditions toward the hinterland. South of Bessel Fjord, the structural complexity increases, in part due to extensional faulting, which essentially shuffled the metamorphic patterns. The structural levels in the south are delineated here on the basis of the major thrust sheets defined by Higgins et al. (2004a), rather than on their extensional geometry (e.g., Gilotti and McClelland, this volume). Metamorphism in each structural level is described geographically from north to south, emphasizing the most representative or best-known areas. For each structural level (north and south), we summarize the important lithologies, mineral assemblages, and metamorphic facies. We present pressure-temperature (P-T) estimates, P-T paths, and geochronology when possible. We restrict our review of geochronology to metamorphic or deformational events that have been directly dated. Rb-Sr and Ar-Ar ages are not reviewed here because their relationship to metamorphism in most cases is not well documented. The age of the Caledonian granites and migmatites is discussed by Kalsbeek et al. (this volume). Mineral abbreviations follow Spear (1993), unless otherwise stated.

CALEDONIDES NORTH OF BESSEL FJORD (76°N)

The Caledonides north of Bessel Fjord (76°N) have the geometry of a classic thrust belt, where progressively higher-grade rocks are situated in each higher thrust sheet from the foreland in the west to the hinterland in the east (Fig. 1). The metamorphism in three major thrust packages exhibits distinct metamorphic conditions. Each successively higher thrust package represents an older part of the Laurentian margin. They are: (1) the unmetamorphosed to low-grade Neoproterozoic to Silurian sedimentary rocks in the foreland; (2) an imbricate zone of medium-grade Paleoproterozoic to Mesoproterozoic metasedimentary successions with subordinate slices of orthogneiss; and (3) a high-grade, eclogite-bearing, Paleoproterozoic orthogneiss complex with subordinate paragneiss forming the hinterland.

Figure 1. Regional metamorphic facies map showing peak Caledonian conditions for the Greenland Caledonides north of Bessel Fjord (76°N), as well as major structures. Units for the same map base are shown on the synoptic tectonic map of Henriksen (2003, this volume). Conodont color alteration indices (CAI) are shown in red. Abbreviations: CF—Chatham Elv fault, D—Danmarkshavn, GLDZ—Germania Land deformation zone, N—Nordmarken, S—Sanddal, SSZ—Storstrømmen shear zone, VTS—Vandredalen thrust sheet, W—Weinschenck Ø. Exposures on Île de France are Quaternary ground moraine. UHP—ultrahigh pressure.

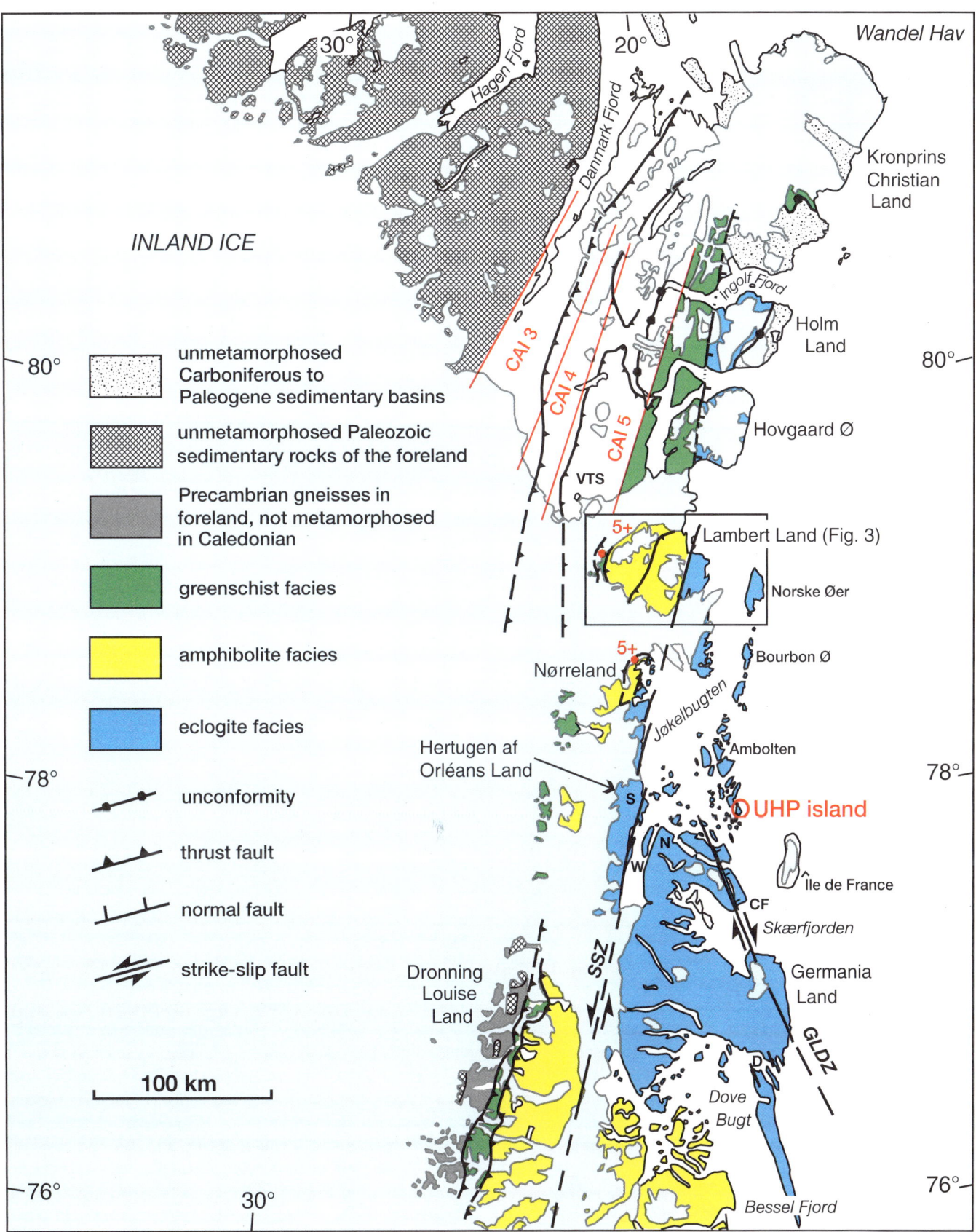

Neoproterozoic to Silurian Sedimentary Rocks in the Marginal Thrust Belt

A complete section from the autochthonous foreland to the frontal fold-and-thrust belt is preserved only in Kronprins Christian Land (79°30′N–81°30′N), the northernmost segment of the Greenland Caledonides (Higgins et al., 2004b). Elsewhere along the mountain front, large parts of the foreland and thrust belt are covered by the Inland Ice. The foreland stratigraphy is composed of rift-related Neoproterozoic sedimentary rocks unconformably overlain by the *Skolithus*-bearing Cambrian Kap Holbæk Formation (Smith et al., 2004b). Cambrian clastic sequences and Lower Ordovician to Middle Silurian carbonates form the Laurentian passive margin (Higgins et al., 2001a). Silurian flysch of the Lauge Koch Land Formation marks the destruction of the passive margin in Llandoverian to Wenlockian times, and it is interpreted as marking the onset of collision with Baltica (Hurst et al., 1983). The middle Wenlockian Profilfjældet Member of the Lauge Koch Land Formation is also the youngest unit cut by thrusts (Hurst and Surlyk, 1982); therefore, it gives the oldest possible age of thrusting.

The metamorphic patterns of the Cambrian to Silurian imbricate thrust zone are delineated by color alteration indices (CAI) of conodonts extracted from carbonate units (Rasmussen and Smith, 2001). CAI values increase steadily across Kronprins Christian Land, beginning at 2–3 in the west to >5 in the east, and they form thrust-parallel, NNE-striking contours (Fig. 1). CAI values of 5–5.5 generally correspond to greenschist-facies, chlorite-zone conditions (Rejebian et al., 1987) and maximum temperatures of 300–400 °C. The continuous, undisrupted pattern of the CAI isopleths is due to thrust loading and not to the original stratigraphic overburden (Rasmussen and Smith, 2001). The highest CAI values (5–6) come from slices of Ordovician Wandel Valley Formation in the thrust stack at Lambert Land and beneath the eclogite-bearing Nørreland thrust sheet.

A thick package of Neoproterozoic rift-related sedimentary rocks, the Rivieradal Group, is thrust westward over the Paleozoic sequences in a large, coherent mass known as the Vandredalen thrust sheet (Higgins et al., 2001b; Smith et al., 2004a). The mudstones and siltstones in this sequence record increasingly intense fabric development from west to east (A.G. Leslie, 2005, personal commun.). A sporadic solution cleavage is developed in the westernmost exposures above the Vandredalen thrust. This cleavage grades into a more widespread slaty cleavage defined by a preferred orientation of neoblastic white mica and chlorite that transect bedding. Farther east, a second crenulation cleavage overprints the first fabric and, in the easternmost part of the area, becomes the pervasive, dominant fabric. Phyllites, marbles, and quartzites are common in the eastern part of the Rivieradal Group beneath the next thrust sheet to the east. The highest metamorphic grade attained is greenschist facies.

South of Kronprins Christian Land, the foreland part of the thrust belt is limited to nunatak exposures. There, very low-grade metasedimentary rocks of the Neoproterozoic Hagen Fjord Group, which stratigraphically overlies the Rivieradal Group, are composed mainly of detrital chlorite, white mica, quartz, and feldspar, and there is only limited development of neomorphic white mica and chlorite. In the slaty mudstone-sandstone units, bedding and other sedimentary structures are preserved. S_1 is a bedding-parallel fabric defined by aligned white mica and chlorite. The bedding and S_1 fabrics are folded about asymmetric west-vergent folds that exhibit a nonpervasive, axial-planar fabric (S_2). The S_2 fabric is defined by aligned metamorphic white mica and chlorite and by solution cleavage in sandstone layers (Fig. 2A).

Imbricates and Windows of Paleoproterozoic Metasedimentary and Metavolcanic Rocks

An imbricate stack of thrust sheets, composed of metasandstones intruded by Paleoproterozoic volcanic rocks, both sills and dikes, overlies the Neoproterozoic to Paleozoic rocks in the foreland thrust belt (Fig. 1). Higgins and Leslie (this volume) refer to these units as the Western thrust belt. Metarhyolites and metaporphyries dated at ca. 1740–1730 Ma (Kalsbeek et al., 1999) provide an upper limit for the age of the Independence Fjord Group metasedimentary rocks. Metamorphic grade increases southward from greenschist facies in Kronprins Christian Land to epidote-amphibolite facies in Lambert Land and Nørreland. The Independence Fjord Group is dominated by quartzites and meta-arkoses, but it also contains minor pelitic units that are best studied together with the mafic intrusive rocks to assess metamorphic grade.

Pedersen et al. (2002) reported that the volcanic rocks in the Independence Fjord Group of Kronprins Christian Land were extensively metamorphosed in the greenschist facies to the point where most of the igneous textures have been obliterated. Epidote and clay minerals have replaced abundant plagioclase phenocrysts, whereas chlorite replaced the original ferromagnesian minerals. Amygdules contain epidote, chlorite, quartz, and calcite.

A west-to-east transition from greenschist to amphibolite facies is recorded in mafic intrusive rocks within the metasandstones across Lambert Land (Fig. 3). In low-strain areas, peak mineral assemblages randomly overgrow magmatic fabrics and/or are synchronous with early fabrics. In the west, the interiors of the metabasic intrusions have greenschist-facies assemblages (Act + Ab + Chl + Qtz ± Cal), which overprint relict magmatic fabrics, whereas peak epidote-amphibolite–facies assemblages (Hbl + Ab + Ep + Qtz + Ilm ± Spl) occur near the margins of the intrusions. Blue-green hornblende contains cores of actinolite and actinolitic hornblende (Fig. 2B). Thermobarometric calculations indicate peak metamorphic conditions of $P = 0.48$–0.6 GPa at $T = 460$–490 °C. The appearance of garnet coincides with a major decrease in modal chlorite and a progressive decrease in epidote (Fig. 2C). The peak assemblage Grt + Hbl + Ep + Ab + Qtz ± Ilm ± Rt (Fig. 2D) is developed in the east. The garnets across this zone preserve growth zoning. Temperatures estimated from Grt-Hbl thermometry are 530–640 °C, consistent with 0.86 (above the Grt-Ab-Ep ± Hbl bathograd of Begin, 1992) $< P < 1.2$ GPa. The main foliation on the margins of these bodies

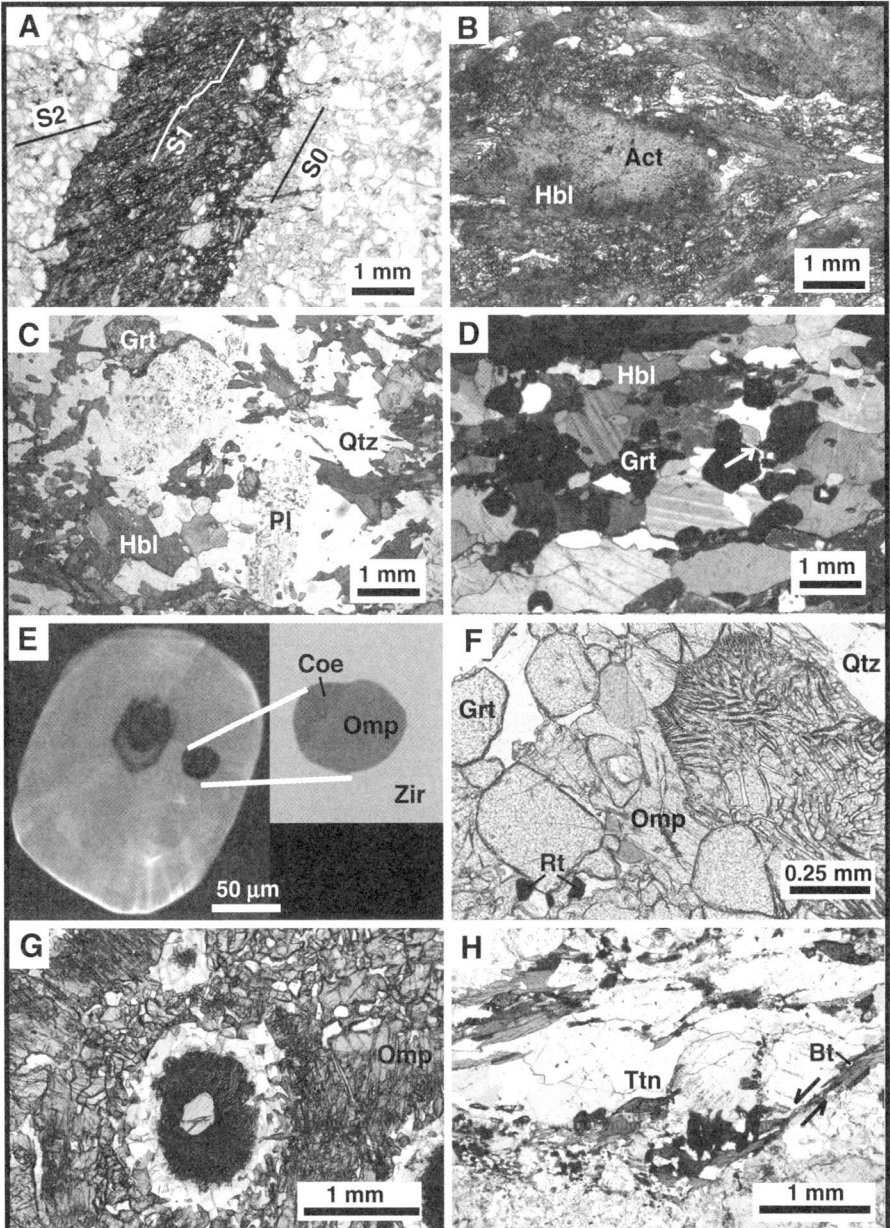

Figure 2. Representative metamorphic assemblages in the area north of Bessel Fjord. (A) Low-grade cleavage formation in the Hagen Fjord Group, western nunataks. Relict compositional layering (S_0) and detrital feldspar and quartz grains are overprinted by a bedding-parallel cleavage (S_1) and an oblique solution cleavage (S_2), which is axial planar to map-scale folds. Plane light. (B) Epidote-amphibolite facies metabasite from western Lambert Land. The zoned amphibole has a pale core of actinolite rimmed by hornblende; the matrix is Hbl + Ep + Ab + Qtz. Plane light. (C) Porphyroblasts of albite, outlined by epidote and amphibole inclusions, in an amphibolite-facies matrix of Qtz + Hbl + Ep + Grt. Plane light. (D) Peak amphibolite-facies assemblage in central Lambert Land of Grt + Ab + Hbl + Qtz; note polygonal, equilibrium texture. A thin reaction rim of albite (at arrow), separating garnet and hornblende, formed during decompression. (E) Cathodoluminescence (CL) image of zircon with a small, inherited core and newly grown metamorphic zircon (gray). Inset is a backscattered electron image of a composite coesite (Coe) and omphacite (Omp) inclusion. The coesite has been verified by Raman spectroscopy (McClelland et al., 2006). (F) Typical quartz eclogite with rutile from Sanddal. The omphacite on the right has exsolved to a symplectite of Di + Pl. Plane light. (G) Decompression textures in kyanite eclogite include sieved omphacite and Crn + Pl ± Spr symplectites around kyanite with an outer halo of Pl + Amp, Danmarkshavn. Plane light. (H) Titanite (Ttn) fish and biotite (Bt) shear bands give a sinistral sense of shear, mylonite, Storstrømmen shear zone, Sanddal. Plane light.

is synchronous with postpeak mineral growth, and it is linked to shearing that intensifies toward the major thrusts. An eastward transition from greenschist to amphibolite facies also occurs in the imbricate thrust zone on Dronning Louise Land (Strachan et al., 1992). Because the precise location of the transition is undocumented, it is shown as greenschist facies on Figure 1.

The Independence Fjord Group crops out in upright folds within the Nørreland window, ~30 km south of Lambert Land. Mafic rocks are in the epidote-amphibolite facies and consist of Amp + Olg + Ep + Qtz. The amphibole is tschermakitic hornblende. Rare, centimeter-scale rusty pelitic layers intercalated with the muscovite-bearing quartzites are composed of Grt + Ms + Qtz ± Bt + Ilm; tourmaline is a distinctive accessory phase. Euhedral garnets have prograde zoning with high-Ca cores and high-Mg rims. Both lithologies contain a strong foliation that is commonly crenulated.

Paleoproterozoic Gneiss Complexes of the Allochthonous Laurentian Margin

Crystalline basement rocks are emplaced as thrust sheets above Precambrian sedimentary sequences along the length of the orogen north of Bessel Fjord (Fig. 1). The main protolith of the gneiss complex is a granodioritic to tonalitic orthogneiss derived from 2.0 to 1.8 Ga calc-alkaline batholiths (Kalsbeek et al., 1993) that were deformed in the Paleoproterozoic prior to the

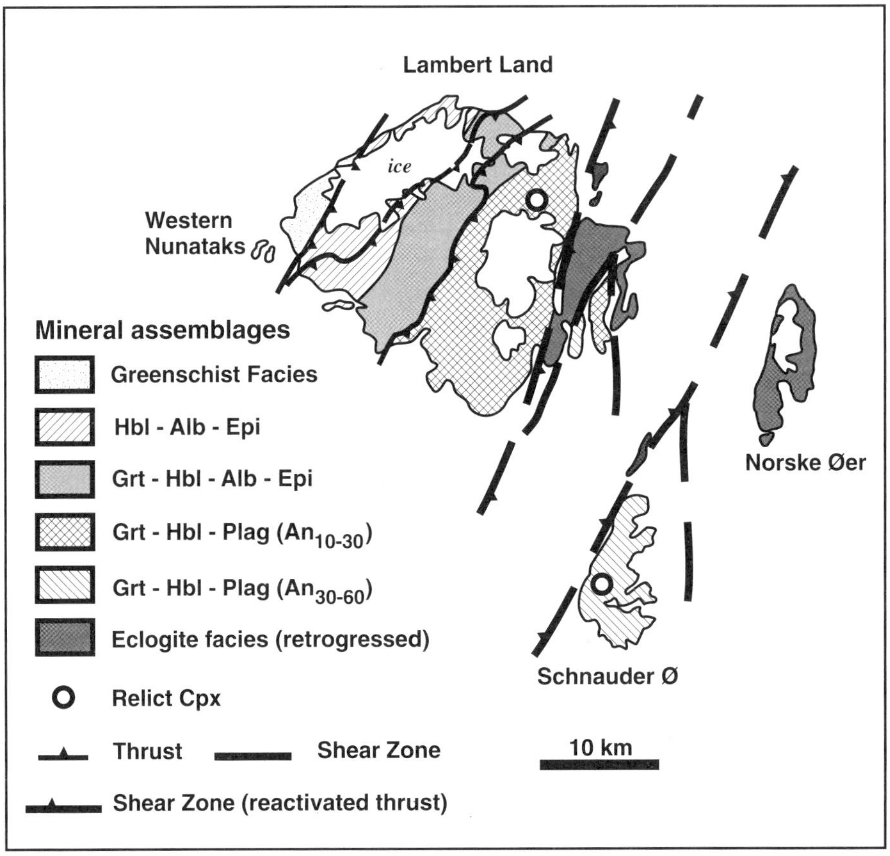

Figure 3. Metamorphic map of the Lambert Land region (79°N–79°30′N). Metamorphic zones indicate synthermal peak metamorphic assemblages, with the exception of retrogressed eclogite-facies rocks of the Nørreland thrust sheet exposed on eastern Lambert Land and Norske Øer. Locations where clinopyroxene is preserved are shown.

intrusion of 1.75 Ga granitoids (Hull et al., 1994; Kalsbeek, 1995). At least one suite of mafic dikes of unknown age intruded both the gneiss complex and the granitoids before the Caledonian collision. The fact that the mafic dikes and the 1.75 Ga granitoids preserve a single deformation fabric is taken as evidence that no deformation affected the gneiss complex between the Paleoproterozoic and the Paleozoic (Hull et al., 1994). These basement rocks are part of the Laurentian margin and must have been the substrate to the Precambrian sedimentary successions described previously here.

In general, the eastern parts of the basement thrust sheets are in the amphibolite facies (Fig. 1), and the metamorphic grade was determined by studying mafic rocks encased in quartzofeldspathic gneisses. An imbricate stack of interleaved basement and metasedimentary rocks in central Lambert Land (Fig. 3) is the best-studied representative of this structural level. Figure 1 shows the gneisses in Dronning Louise Land as amphibolite facies, equivalent to those in central Lambert Land; however, these rocks are on strike with the eclogite-bearing basement of Hertugen af Orléans Land and may contain unrecognized eclogites. The metamorphic history of the gneisses south of Dove Bugt and in eastern Dronning Louise Land is poorly known.

On Lambert Land, the cores of metabasic bodies preserve prograde Grt + Hbl (blue-green) + Ab + Ep ± Spl ± Rt assemblages. Many of the foliated rims lack prograde plagioclase and contain Hbl + Grt + Qtz ± Rt, possibly indicating a prograde transition to eclogite facies. Across Lambert Land, relicts of Ab + Ep are preserved in garnet cores and in the cores of zoned matrix feldspars. Garnets show prograde zoning. The early postpeak stage is characterized by the formation of coronae of feldspar around garnet, which indicate static decompression. In strongly foliated rocks, numerous plagioclase grains form porphyroblasts. Plagioclase composition in the reaction textures and the matrix changes systematically with increasing grade from NW to SE. In the NW, rare albite is preserved, whereas toward the SW, the anorthite content increases from oligoclase (An_{10-16}) to andesine and labradorite (An_{16-60}). Thermobarometry performed on these assemblages gives a range of P-T determinations consistent with a systematic increase in temperature and decrease in pressure (K.A. Jones, unpublished data). They are ~600–700 °C at 1.2–1.4 GPa (NW) to 600–800 °C at 0.9–1.2 GPa (SE) and are coincident with the change in feldspar composition. The earlier assemblage Grt + Hbl + Ab + Qtz ± Cpx ± Rt is replaced by Hbl + Pl (An_{16-60}). A composite P-T path for Lambert Land is given in Figure 4A.

North-East Greenland Eclogite Province

The North-East Greenland eclogite province (Gilotti, 1993) occupies the uppermost, highest-grade thrust sheet, covering an area of at least 100 × 400 km (Fig. 1), from Holm Land in the north to Dove Bugt (Chadwick et al., 1990) in the south.

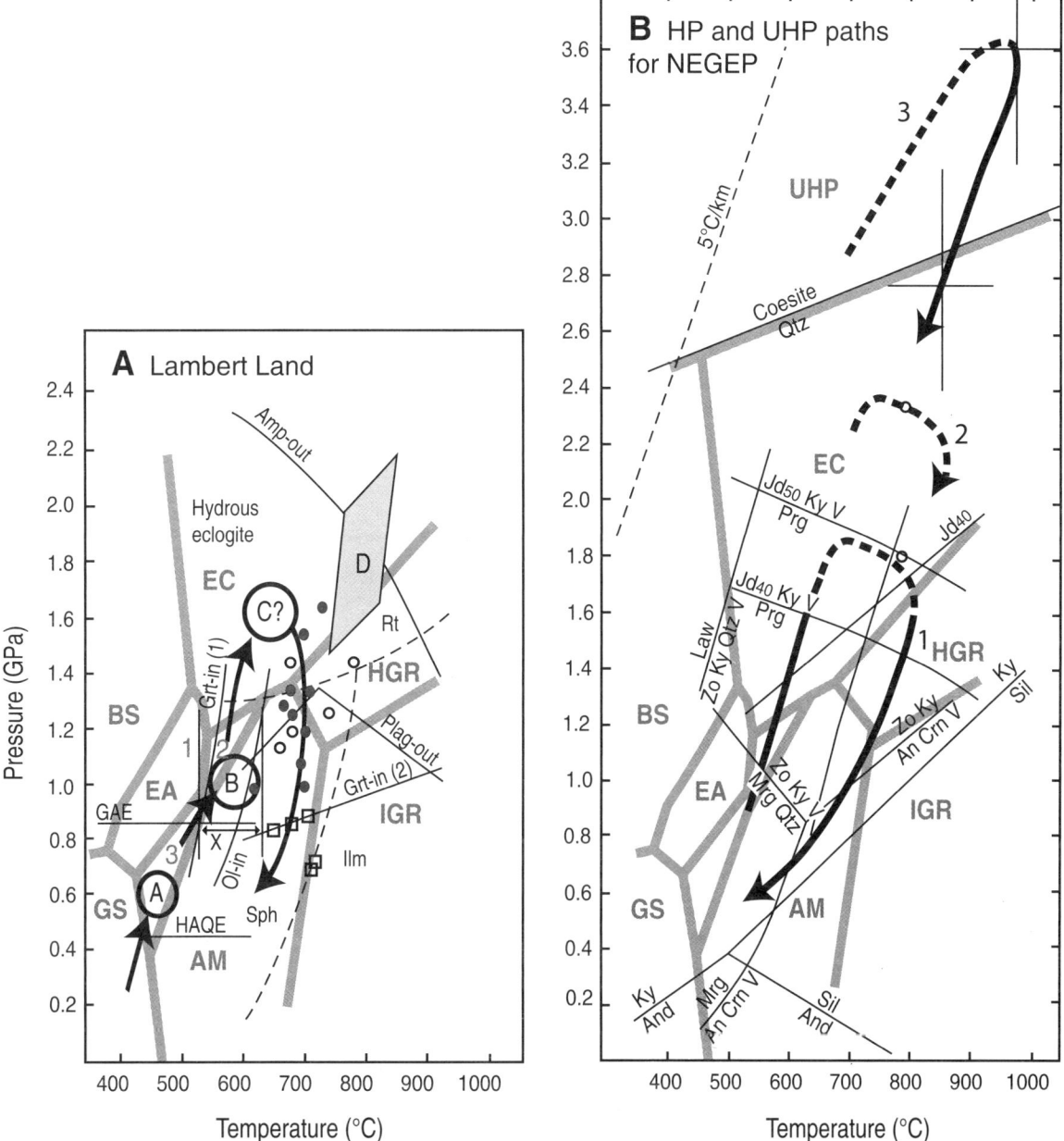

Figure 4. Pressure-temperature (*P-T*) paths for rocks north of Bessel Fjord. Facies grid is after Oh and Liou (1998) and Liou et al. (1998). Abbreviations: GS—greenschist; BS—blueschist; EC—eclogite; AM—amphibolite; HGR—high-pressure granulite; IGR—intermediate-pressure granulite; EA—epidote-amphibolite; UHP—ultrahigh pressure. (A) Lambert Land. Subfacies 1 (Bar-Ep-Ab), 2 (Hbl, Grt, Ab, Ep), and 3 (Hbl, Ep, Ab, Chl, Act) are shown for the EA facies. Reactions: Grt-in (1), Grt-in (2), Amp-out, Plag-out, and stability field of Rt, Spn, and Ilm (dashed lines) are after Ernst and Liu (1998). GAE (Grt-Ab-Ep bathograd) and HAQE (Hbl-Ab-Qtz-Ep bathograd) are after Begin (1992); Ol-in is after Maruyama et al. (1983). Symbols indicate *P-T* estimates from synthermal peak equilibrium textures and decompression textures: filled circles are from amphibolites of central Lambert Land; open circles are for eclogites of the North-East Greenland eclogite province from the Nørreland thrust (Norske Øer); and open squares are from amphibolites of Schnauder Ø. Open ellipses indicate peak pressure conditions along *P-T* trajectories indicated for: A—western Lambert Land, B—central Lambert Land (west), and C—central Lambert Land (east). Shaded box labeled D indicates *P-T* conditions for the North-East Greenland eclogite province from Norske Øer. X is the range of temperature estimates from Grt-Hbl thermometry for central Lambert Land (Grt-Hbl and Grt-Hbl-Ab zones). (B) High-pressure (HP) and ultrahigh-pressure (UHP) rocks of the North-East Greenland eclogite province (NEGEP). Path 1 is derived from retrogressed kyanite-eclogites on Weinschenck Ø (Elvevold and Gilotti, 2000) where only minimum *P* could be obtained. Path 2 is inferred from Danmarkshavn; circles are *P-T* conditions for garnet websterites from Brueckner et al. (1998). Path 3 is inferred for the ultrahigh-pressure terrane using peak and retrograde conditions calculated from the same kyanite-eclogite sample (Gilotti and Ravna, 2002).

The most abundant lithology is a quartzofeldspathic orthogneiss with subordinate paragneiss and mafic enclaves. Although most of the rocks are amphibolite-facies gneisses, many of the mafic rocks retain a Caledonian, eclogite-facies mineralogy (Brueckner et al., 1998). The eclogite province is sometimes referred to as the Nørreland thrust sheet because the basal thrust can be seen in Nørreland through a window that exposes a sliver of Ordovician carbonates and the underlying Paleoproterozoic metasedimentary section. The basal thrust is also exposed in Lambert Land. The North-East Greenland eclogite province is cut by two strike-slip zones, the sinistral NNE-striking Storstrømmen shear zone (Holdsworth and Strachan, 1991) and the dextral NNW-striking Germania Land deformation zone (Hull and Gilotti, 1994), which divide it into western, central, and eastern blocks. The absolute timing and interplay among foreland thrusts, the Nørreland thrust, and the strike-slip faults is unknown. Ultrahigh-pressure (UHP) eclogites have been found on one island in the eastern block (Gilotti and Ravna, 2002; McClelland et al., 2006), so it is unlikely that the Nørreland thrust sheet is a continuous crustal slab despite the lack of mapped internal thrusts.

The entire crustal slab of the North-East Greenland eclogite province experienced high-pressure or ultrahigh-pressure conditions despite a general lack of preservation of eclogite-facies assemblages in the quartzofeldspathic gneisses. The eclogitic rocks are typically found as meter-scale mafic lenses and pods enveloped by strongly deformed, amphibolite-facies, quartzofeldspathic orthogneiss. The protoliths of the high- and ultrahigh-pressure rocks were mafic xenoliths in the calc-alkaline batholiths, layered gabbroic plutons, and pre-Caledonian mafic dikes, all of which were an integral part of the continental crust prior to the onset of high-pressure metamorphism. Eclogite-facies assemblages have been found as inclusions in garnet in the host gneisses (Gilotti and Ravna, 2002), which also harbor coesite-bearing zircon (Fig. 2E; McClelland et al., 2006). The preservation of the high- and ultrahigh-pressure phases in the mafic pods and generally not in the quartzofeldspathic gneisses is attributed to the lack of fluid infiltration and/or strain within the pods. The lack of fluids is partly due to the continued stability of biotite and white mica even at ultrahigh pressure (Rubie, 1986; Hermann and Green, 2001).

The North-East Greenland eclogite province is similar in many ways to the Western Gneiss Region of Norway, which constitutes a high- and ultrahigh-pressure part of the margin of Baltica (Carswell and Cuthbert, 2003, and references therein). However, there are two significant differences. One difference is the general lack of mantle rocks intercalated with crust in the North-East Greenland eclogite province. Only a single dunite pod (Friderichsen et al., 1991) and two Ol + Spl ± Opx pods (Wyllie, 1957) have been observed compared to the plethora of mantle peridotites known from Norway (Krogh and Carswell, 1995). The other difference is in the age of the gneissic protolith, which is generally less than 1700 Ga in the Western Gneiss Region (Tucker et al., 1990; Corfu and Andersen, 2002).

Ultrahigh-Pressure Metamorphism

Ultrahigh-pressure metamorphism is known from one island in Jøkelbugten located at 78°00′N, 18°04′W (Fig. 1). Blocks of layered kyanite and bimineralic eclogite up to 400 m × 50 m are strung out within the steep, NNW-striking gneissosity of the enveloping quartzofeldspathic host rocks. Ultrahigh-pressure metamorphism was first recognized on the basis of polycrystalline quartz inclusions with palisade texture and radial fractures in garnet, omphacite, and kyanite (Gilotti and Ravna, 2002). Coesite inclusions in zircon (Fig. 2E) have now been positively identified in kyanite eclogites and their host gneisses using laser Raman spectroscopy (McClelland et al., 2006). Given the recent spate of discoveries of ultrahigh-pressure metamorphism around the globe (Carswell and Compagnoni, 2003), we predict that the "Rabbit Ears Island" localities are only the tip of the iceberg in terms of defining a sizeable ultrahigh-pressure terrane within the North-East Greenland eclogite province.

The ultrahigh-pressure assemblage in the kyanite eclogites is Grt + Omp + Ky + Coe + Phn + Rt. Omphacite compositions range from Jd_{49} to Jd_{32}. Geothermobarometry, using the method described by Ravna and Terry (2004), returns peak temperatures of 972 ± 93 °C and pressures of 3.6 ± 0.41 GPa (Gilotti and Ravna, 2002); however, the phengite inclusion used for this estimate was altered, and thus these numbers slightly overestimate temperature and underestimate pressure. Nearby metapelites record partial melting of phengite while the rocks were still in the coesite stability field (Lang and Gilotti, 2007). The coesite-bearing host gneiss is composed of the disequilibrium assemblage Qtz + Pl + Grt + Cpx ± Ky + Amp + Bt + Ep/Czo + zircon, but the same eclogite-facies assemblage described previously is preserved as inclusions in garnet. Similar geothermobarometric calculations return $P = 2.54 ± 0.41$ GPa and $T = 826 ± 92$ °C for the host gneiss, but these results record retrograde conditions because pressure is below the coesite stability field in rocks (Gilotti and Ravna, 2002). A tentative P-T path for ultrahigh-pressure rocks is given in Figure 4B.

The timing of ultrahigh-pressure metamorphism has been established at ca. 365–350 Ma using U-Pb sensitive high-resolution ion microprobe (SHRIMP) geochronology on zircons from the kyanite eclogites (Gilotti et al., 2004; McClelland et al., 2006). The dated zircons are typical of metamorphic ultrahigh-pressure zircons (Hoskin and Schaltegger, 2003), since they are small spheres (≈100 μm), with low U, Th, and Th/U values, no Eu anomaly, and abundant inclusions of high- and ultrahigh-pressure phases. The bright cores in cathodoluminescence (CL) images give a weighted mean $^{206}Pb/^{238}U$ age of 360 ± 5 Ma, which Gilotti et al. (2004) interpreted as the age of ultrahigh-pressure metamorphism. McClelland et al. (2006) confirmed the presence of coesite in the bright CL domains of zircon in a coesite-bearing eclogite and a host gneiss (Fig. 2E). The bright zircon domains growing around coesite in these samples give $^{206}Pb/^{238}Pb$ SHRIMP ages of 364 ± 8 Ma for the host gneiss to 350 ± 4 Ma for kyanite eclogite, which are in agreement with the earlier 360 Ma age for ultrahigh-pressure metamorphism. Younger, dark rim domains give ages from 357 ± 4 Ma for the host gneiss down to

342 ± 3 Ma for the eclogites. This age is the same as the garnet–omphacite–whole-rock Sm-Nd mineral isochron of 342 ± 3 Ma from a ultrahigh-pressure kyanite eclogite (Gilotti et al., 2004). Apparently 10 m.y. was required for the ultrahigh-pressure terrane to cool through the closure temperature for the Sm-Nd system. Zircons from two boudin neck pegmatites and a hornblende-bearing leucosome date the amphibolite-facies metamorphism at ca. 330 Ma (Gilotti and McClelland, 2007).

High-Pressure Metamorphism

The North-East Greenland eclogite province is dominated by medium-temperature eclogites (Carswell, 1990). The most common eclogite-facies lithologies are eclogites sensu stricto, garnet clinopyroxenites, garnet websterites, websterites, and coronitic metagabbros (Gilotti, 1993, 1994). The high-pressure assemblages are variably preserved, massive to strongly foliated, and display a wide variety of textures and grain size. The most common assemblage is Omp + Grt ± Rt ± Qtz (Fig. 2F). Other important eclogite-facies minerals include kyanite, zoisite/clinozoisite, K-feldspar, phlogopite, and rare phengite. In rocks with low bulk Na, such as the garnet clinopyroxenites, the clinopyroxene is diopside. The garnet websterites are Grt + Cpx + Opx ± Phl ± Rt with diopside ($X_{Jd} < 10\%$) and Mg-rich orthopyroxene (En_{83-63}), similar to the so-called orthopyroxene eclogites (Carswell et al., 1985) of western Norway.

Partially eclogitized, igneous bodies composed of gabbro, leucogabbro, anorthosite, and crosscutting diabase dikes are found across the eclogite province, from Sanddal (Lang and Gilotti, 2001) in the west to Ambolten (Gilotti and Elvevold, 1998) and other islands in Jøkelbugten (e.g., Sartini-Rideout et al., 2007). Plagioclase is rarely completely replaced in these rocks; rather, coronas of garnet, omphacite, diopside, orthopyroxene, and amphibole form discrete growth layers between plagioclase and igneous olivine and pyroxenes. Recrystallized plagioclase is more albitic than the original igneous composition, and is the nucleation site for kyanite, zoisite, corundum, and garnet. The unusual replacement of plagioclase by omphacite-spinel symplectites prior to garnet nucleation was observed by Lang and Gilotti (2001). The abundance of partially eclogitized gabbroic localities on the islands in Jøkelbugten does not preclude the existence of ultrahigh-pressure metamorphism in the eastern block, because coesite-bearing eclogites have been found adjacent to coronitic gabbros in the Sulu ultrahigh-pressure terrane in China (Zhang and Liou, 1997).

Estimates of pressure and temperature have been made around the North-East Greenland eclogite province, but a systematic spatial variation in these parameters has yet to be established. One difficulty is that only minimum pressures can be determined for the common bimineralic eclogites, and errors are large. For example, Gilotti (1993) used the Jd = Ab + Qtz barometer (Holland, 1980) to calculate minimum pressures between 0.9 and 1.4 GPa for widespread Grt + Omp eclogites. Temperatures for these eclogites, determined using Fe-Mg exchange thermometry in garnet and clinopyroxene (Ellis and Green, 1979), fall in the range of 600–850 °C, which is typical of medium-temperature eclogite-facies metamorphism (Carswell, 1990). The calculated range in temperature in part reflects the assumption that all Fe is ferrous. When possible, lower variance assemblages have been used for thermobarometry. Application of the Brey and Köhler (1990) thermobarometer to garnet websterites from Danmarkshavn yields pressures of 1.8–2.3 GPa at $T = 700$–800 °C (Brueckner et al., 1998).

The composite P-T path for the North-East Greenland eclogitic rocks presented here (Fig. 4) builds on a path obtained for retrogressed kyanite eclogites on Weinschenck Ø (Elvevold and Gilotti, 2000). The prograde part of the path is indicated by inclusion sequences and prograde zoning patterns in garnet. Garnet began to grow in the epidote-amphibolite facies, based on the inclusion assemblage Amp + Pl + Ep found in cores. Garnet-amphibole-plagioclase thermobarometry gives $T = 550$–630 °C and $P = 1.0$–1.2 GPa. Paragonite inclusions reveal that the reaction Prg = Jd + Ky + V was crossed. In Lambert Land and Norske Øer, K.A. Jones (unpublished data) has observed the inclusion suite Ep + Qtz + Pl (Ab–Olg) + Rt ± Hbl in garnet cores, with Omp (Jd_{22-28}) + Rt inclusions in their rims. These inclusion sequences are typical of the transition from epidote-amphibolite and amphibolite facies into eclogite facies during prograde garnet growth (Krogh, 1982). As discussed already, more robust barometry is needed to calculate peak pressures around the North-East Greenland eclogite province; the peak estimate from the Danmarkshavn garnet websterites is used in Figure 4B. Some of the eclogites show evidence of a high-pressure granulite-facies history, which suggests that on the clockwise P-T path, peak temperature was encountered after peak pressure. The evidence for this includes the ubiquitous diopside-plagioclase symplectites after omphacite (Fig. 2F), as well as the replacement of kyanite by sapphirine-plagioclase, spinel-plagioclase, and corundum-plagioclase symplectites (Fig. 2G; Elvevold and Gilotti, 2000). Temperatures of 620–750 °C at 1.0–1.3 GPa have been calculated for the Cpx (Jd_{8-14}) + Pl (An_{11-18}) ± Hbl symplectites for the Lambert Land and Norske Øer localities based on Grt + Cpx Fe-Mg exchange thermometry and Grt-Cpx-Pl-Qtz barometry (K.A. Jones, unpublished data).

The age of high-pressure metamorphism has been determined using Sm-Nd analysis of eclogite-facies minerals and U-Pb SHRIMP geochronology of metamorphic zircon. Brueckner et al. (1998) reported a spread of metamorphic ages from 440 to 370 Ma, based on these methods, but their more recent work (Gilotti et al., 2004) shows that high-pressure metamorphism is Devonian in the western and central blocks of the eclogite province. Grt + Omp ± Amp ± whole-rock Sm-Nd mineral isochrons from a quartz eclogite, a Grt + Omp + Rt eclogite, and a zoisite eclogite give ages of 401 ± 2 Ma, 402 ± 9 Ma, and 414 ± 18 Ma, respectively. Corresponding SHRIMP $^{206}Pb/^{238}U$ zircon ages from the same samples are 401 ± 7 Ma, 414 ± 13 Ma, and 393 ± 10 Ma. Metamorphic zircon domains were identified on the basis of morphology, CL imaging of rims, and low U and very low Th and Th/U ratios. Additional evidence that zircon grew in the eclogite facies is the lack of an Eu anomaly in the trace-element data measured in zircon for all the samples. In contrast to the ultrahigh-pressure eclogite, the Sm-Nd system in these high-pressure rocks closed at the same time zircon grew near peak pressure, perhaps because the high-pressure eclogites formed at lower temperatures.

Amphibolite- to Greenschist-Facies Exhumation Stage

All the eclogites show signs of retrogression through the amphibolite facies. Common features include: compositional zoning in garnet and clinopyroxene, direct replacement of omphacite by amphibole, kelyphites of Pl + Amp around garnet, breakdown of rutile to titanite, epidote overgrowths on zoisite and clinozoisite, hornblende poikiloblasts, and matrix plagioclase. A final retrogression is marked by the replacement of Grt + Cpx + Hbl by Ep + Hbl + Pl ± Ilm ± Spl.

Quartzofeldspathic orthogneisses have the metamorphic assemblage Qtz + Pl + Kfs + Hbl + Bt ± Grt ± Cpx ± Ep, with titanite, zircon, apatite, rutile, and opaque oxides as typical accessory minerals. Although the rocks are overwhelmingly in the amphibolite facies, the sporadic preservation of clinopyroxene suggests a hidden, higher-grade history. The 1.75 Ga granitoids contain the assemblage Kfs + Pl + Qtz + Grt + Amp + Bt with sphene, zircon, apatite, monazite, and opaque oxides as accessory minerals. The metamorphic fabric consists of alternating layers of quartz-feldspar ribbons with aligned biotite and amphibole that wrap around garnet porphyroblasts. Thin layers of paragneiss are intercalated with orthogneisses throughout the eclogite province. Unambiguous lithologies are marbles, garnet-bearing quartzites, and aluminous schists. The general lack of sillimanite in paragneisses suggests that the P-T path remained in the kyanite field. However, sillimanite-bearing paragneisses with the assemblage Qtz + Kfs + Grt + Sil + Hb + Ep + Bt have been recognized near Danmarkshavn (Sartini-Rideout et al., 2006). No quantitative thermobarometry exists for these rocks.

All the rocks in the North-East Greenland eclogite province are deformed and overprinted by lower-grade assemblages along the Storstrømmen shear zone (Strachan and Tribe, 1994; Hallett, 2005) and the Germania Land deformation zone (Hull and Gilotti, 1994). In both cases, amphibolite- to greenschist-facies mylonites are characterized by pronounced grain-size reduction compared to the protolith gneisses and an increase in modal phyllosilicates (Hallett et al., 2005; Sartini-Rideout et al., 2006). The mylonitic foliation of both shear zones is defined by quartz ribbons and a biotite–white mica–chlorite–hornblende grain shape preferred orientation. Amphibole crystallization within shear bands and subgrain formation in feldspar indicate that mylonitic deformation began at amphibolite-facies conditions. Ultramylonites and mylonites in the highest strain portions of the shear zones record deformation at lower-grade conditions by growth of biotite, chlorite, and epidote/clinozoisite. Elongate quartz, feldspar, and epidote/clinozoisite grains define a subhorizontal stretching lineation. Titanite, garnet, epidote/clinozoisite, and partially recrystallized feldspar occur as porphyroclasts in a micaceous matrix. Narrow zones of cataclasite mark a final phase of brittle deformation along these high strain zones.

The age of the amphibolite-facies metamorphism in the North-East Greenland eclogite province has been estimated from reconnaissance TIMS (thermal ionization mass spectrometry) U-Pb dating of titanite in the orthogneiss complex (Hallett et al., 2005). Precambrian titanite is preserved in the orthogneiss but is overgrown by new titanite probably formed during amphibolite-facies deformation of the orthogneisses (Hallett, 2005). Isotope-dilution (ID) TIMS analysis of titanite fractions from a biotite schist west of the Storstrømmen shear zone gives a U-Pb age of 384 ± 2 Ma, whereas titanite ages from orthogneisses east of the Storstrømmen shear zone on Weinschenck Ø and in Nordmarken are 392 ± 2 Ma, 381 ± 1 Ma, and 378 ± 2 Ma (Hallett et al., 2005). Mafic eclogite layers were commonly pulled apart during exhumation, and their margins were strongly foliated and retrogressed to amphibolite facies during this process. Granitic pegmatites that formed in the necks of these mafic boudins contain new igneous zircon. At Sanddal in the western block, oscillatory-zoned zircon from pegmatites in eclogite boudin necks yield weighted mean $^{206}Pb/^{238}U$ SHRIMP ages of 393 ± 3 Ma and 385 ± 5 Ma (McClelland and Gilotti, 2004). At Danmarkshavn in the central block, oscillatory-zoned igneous zircon from a pegmatite within the neck of a boudinaged garnet websterite gives a weighted mean $^{206}Pb/^{238}U$ age of 374 ± 3 Ma (Sartini-Rideout et al., 2006). The same garnet websterite yielded a Sm-Nd mineral isochron age of 370 ± 12 Ma (Brueckner et al., 1998), suggesting that the Sm-Nd system closed during the amphibolite-facies retrogression.

The two major strike-slip mylonite zones have different kinematics but are similar in metamorphic grade, microstructural development, and they are broadly similar in age. The sinistral Storstrømmen shear zone may have been active by 390 Ma at the start of amphibolite-facies conditions, whereas the occurrence of 350 Ma titanite porphyroclasts within the mylonites (Fig. 2H) suggests that deformation was younger than that (Hallett, 2005). Mylonite zones from Danmarkshavn have the same orientation and dextral kinematics as the main Germania Land deformation zone, and they probably started to localize in the amphibolite facies around 370 Ma (Sartini-Rideout et al., 2006). However, major displacement on the Germania Land deformation zone should be younger than the 360 Ma ultrahigh-pressure metamorphism on its eastern side. Ductile deformation ceased before ca. 340 Ma on the Danmarkshavn mylonites zones, bracketed by the emplacement age of a suite of crosscutting pegmatites dated using U-Pb SHRIMP on zircon (Sartini-Rideout et al., 2006). This may also mark the end of ductile deformation on the Germania Land structure. Later brittle deformation (with a dip-slip component) is localized along the Chatham Elv fault just east of the Germania Land deformation zone (Hull and Gilotti, 1994).

CALEDONIDES SOUTH OF BESSEL FJORD (76°N)

The Caledonides south of 76°N are divided into major lithotectonic units consisting of foreland windows overlain by two major basement-involved thrust sheet assemblages, the Niggli Spids and Hagar Bjerg thrust sheets, and an uppermost unit of

Neoproterozoic to Ordovician sedimentary sequences known as the Franz Joseph allochthon (Higgins et al., 2004a). These rocks are overlain by thick sequences of unmetamorphosed Devonian to Cretaceous sedimentary basins. We describe the metamorphic characteristics of this tectonic architecture next (excluding the Devonian and younger basins), from north to south within each unit, starting with the structurally lowest rocks. Peak metamorphic facies that developed during the Caledonian orogeny are depicted on a regional map (Fig. 5). The base for Figure 5 is the *Synoptic Tectonic Map 70°N –76°N* (Henriksen, 2003 and Henriksen and Higgins, this volume, Chapter 14, their Plate 1), which shows the individual lithotectonic units. Regional extensional detachment faults (see Gilotti and McClelland, this volume) considerably modify the map pattern of the thrust geometry and lithotectonic units of Higgins (2004a), as well as overprint the peak metamorphism and contractional fabrics, but they are treated cursorily here.

Foreland Windows

The foreland windows are characterized by a thin (220–400 m) Neoproterozoic to Lower Paleozoic succession that unconformably overlies the Precambrian basement (Higgins et al., 2004a). The Caledonian foreland is for the most part hidden by the Inland Ice, and outcrops are restricted to a small area of the westernmost nunataks (Hamberg Gletscher foreland) and three anticlinal windows through the thrust stack (Fig. 5). The rocks exposed in the windows show effects of Caledonian deformation and are considered parautochthonous. The overall metamorphic condition of the metasedimentary section within the windows is low grade (chlorite to biotite zone) (Leslie and Higgins, 1999). The Caledonian overprint on the basement rocks is assumed to be negligible. Detailed metamorphic studies are lacking.

Niggli Spids Thrust Sheet

The Niggli Spids thrust sheet structurally overlies the foreland windows and is composed of two main rock units: Precambrian orthogneiss complexes and Mesoproterozoic metasedimentary rocks of the Krummedal sequence (Higgins, 1974, 1988). Thrane (2002) described a diffuse boundary between the protolith ages of the orthogneiss complexes at 72°30′N. Paleoproterozoic (ca. 1900 Ma) gneisses dominate to the north, whereas the gneisses are Archean (2800–2700 Ma) farther south. The crystalline complex contains several different types of orthogneiss and lesser amounts of amphibolite, paragneiss, schist, marble, and ultramafic rock. The Niggli Spids sheet locally preserves evidence of high-pressure granulite- and eclogite-facies metamorphism regionally overprinted by amphibolite-facies assemblages.

Payer Land

A regional high-pressure granulite-facies gneiss complex was first recognized in northwest Payer Land by Larsen (1980) during a search for an explanation for a local, positive magnetic anomaly centered over the area. Larsen attributed the anomaly to exposures of magnetite- and ilmenite-bearing, granulite-facies rocks. The Payer Land gneiss complex consists of granitic and tonalitic orthogneisses, metabasites, and rare ultramafic rocks, which are tectonically interleaved with pelitic to psammitic paragneisses that are lithologically similar to the Krummedal unit. Metabasites are present as mappable units and as smaller pods and dikes that represent deformed dikes and sills within the ortho- and paragneiss. The Payer Land gneisses form a metamorphic core complex that is separated from the overlying low-grade sedimentary rocks of the Neoproterozoic Eleonore Bay Supergroup by an extensional fault, the Payer Land detachment (Gilotti and Elvevold, 2002; Gilotti and McClelland, 2005).

High-pressure granulite-facies mineral assemblages are well preserved in all lithologies (Gilotti and Elvevold, 2002; Elvevold et al., 2003). Orthopyroxene is common in the granitic orthogneisses. High-pressure, high-temperature metamorphism resulted in the formation of the peak assemblage Grt + Cpx + Amp + Qtz + Rt ± Pl in mafic rocks, and Grt + Ol + Cpx + Opx + Spl in ultramafic pods. The paragneisses are anatectic metapelites that are interlayered with semipelitic and psammitic gneisses and marbles. Metapelites and metapsammites contain the peak assemblage Grt + Ky + Kfs + antiperthite + Qtz ± Bt ± Rt. The metapelites contain centimeter-scale stringers and lenses of Ky + Kfs ± Grt + Pl + Qtz leucosomes, indicating melting at high-pressure conditions. Peak metamorphic assemblages formed around 800–850 °C at pressures of 1.4–1.7 GPa.

Abundant reaction textures, including the replacement of garnet in mafic rocks by symplectites of Opx + Spl + Pl, indicate that the high-pressure event was followed by decompression, while the granulites remained at elevated temperatures. A 1500-m-thick mylonite zone is found mostly within the paragneisses beneath the Payer Land detachment fault. The retrograde assemblage preserved in the mylonitic matrix is Qtz + Pl + Kfs + Bt + Sil, while garnet, kyanite, and the high-pressure leucocratic melt pods are preserved as porphyroclasts. The absence of cordierite suggests the *P-T* path shown in Figure 6A.

Caledonian metamorphic ages are recorded in the high-pressure granulites from Payer Land. U-Pb SHRIMP analysis of metamorphic rims on Precambrian zircons from a metapsammite yields a $^{206}Pb/^{238}U$ age of 403 ± 5 Ma, whereas, newly grown oscillatory-zoned zircons from a high-pressure kyanite-bearing melt pod derived from metapelite is 404 ± 4 Ma (McClelland and Gilotti, 2003). In contrast, zircons from the Payer Land orthogneisses do not record Caledonian metamorphism (Elvevold et al., 2003).

Isfjord Area

The Krummedal sequence in the Isfjord area is dominated by metapsammites, quartzites, calcareous schists, and marbles that provide few constraints on the metamorphic evolution of the metasedimentary rocks. Pelitic schists are rare but do occur within the lower part of this unit. Metapelites from Rendalen and Kap Madelaine contain Grt + St + Ky (±Sil) + Bt + Ms +

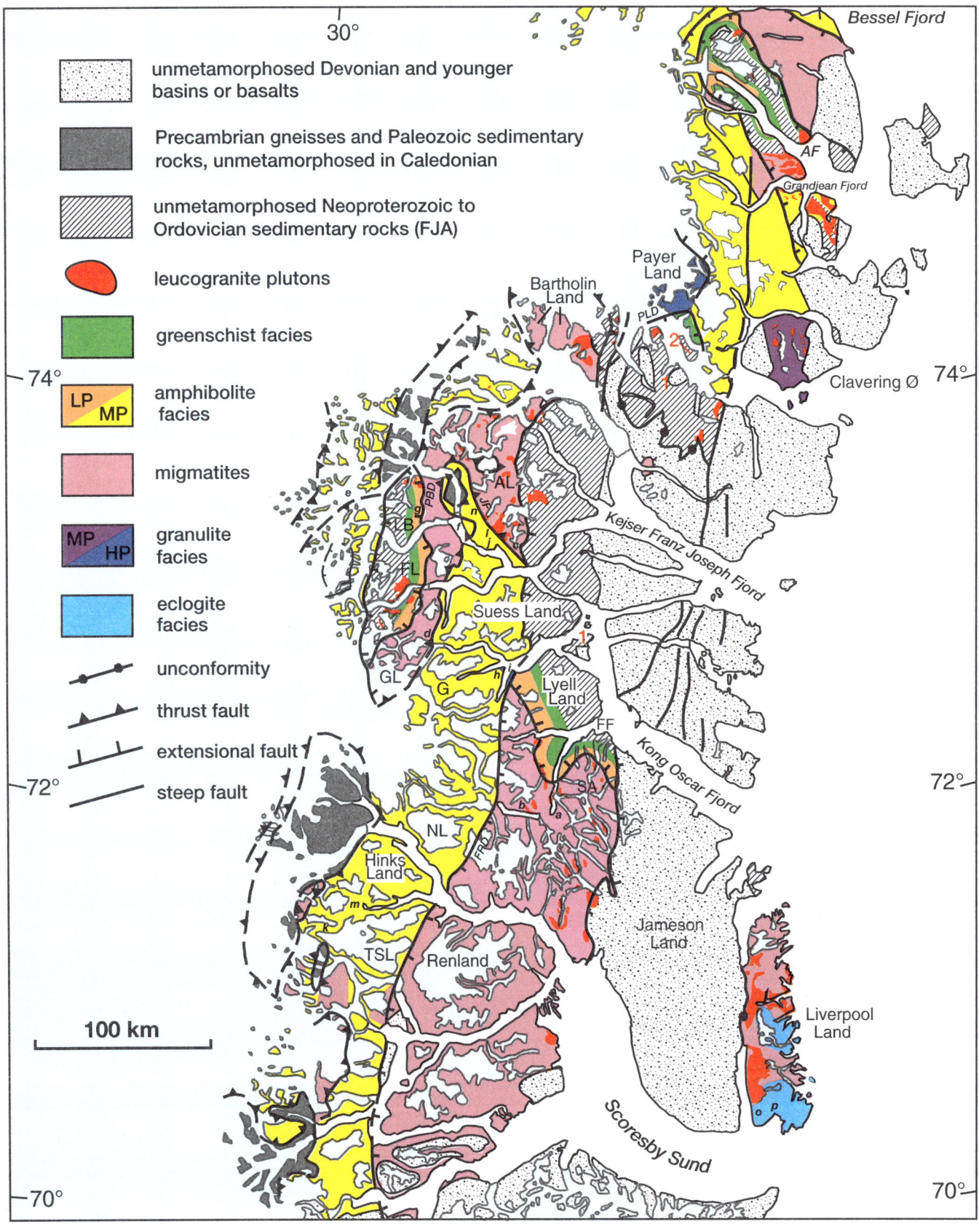

Figure 5. Regional metamorphic facies map showing peak Caledonian conditions for the Greenland Caledonides south of Bessel Fjord (70°N–76°N). Red numbers are conodont color alteration indices (CAI; Smith, 1991). Abbreviations are: AL—Andrée Land, AF—Ardencaple Fjord, FF—Forsblad Fjord, FJA—Franz Joseph allochthon, FL—Frænkel Land, FRD—Ford Region detachments, G—Gletscherland, GL—Goodenough Land, JF—Junctiondal fault, LB—Louise Boyd Land, NL—Nathorst Land, PLD—Payer Land detachment, PBD—Petermann Bjerg detachment, SA—Stauning Alper, TSL—Th. Sørensen Land. Additional abbreviations for individual localities are: a—Dammen, b—Furesø, c—Gauli Gletscher, d—Hagar Bjerg, e—Hamberg Gletscher, f—Isfjord, g—Jættedalen, h—Kap Bayard, i—Kap Hedlund, j—Kap Madelaine, k—Krummedal, l—Lacroix Bjerg, m—Rencontre Dal, n—Rendalen, o—Rendeelven, p—Tværdalen. LP, MP, HP—low, middle, and high pressure, respectively.

Qtz + Pl ± Rt ± Ilm. Garnet porphyroblasts contain abundant inclusions that form straight to curved trails and complex spirals. Early garnet is preserved in the cores of staurolite and kyanite porphyroblasts (Fig. 7A). Kyanite occurs in textural equilibrium with staurolite and includes early resorbed garnets and biotite. The coexistence of staurolite and kyanite indicates conditions of 600–700 °C at 0.6–0.7 GPa. The peak assemblages in this region are given by the occurrence of pegmatites containing Ky + Kfs as a result of crossing the reaction Ms + Qtz = Kfs + Ky + H_2O or melt, which indicates $P > 0.8$ GPa at temperatures of 700 °C.

In several higher-strain samples, sillimanite (fibrolite) occurs. Sillimanite may have developed on crossing the Ky-Sil curve; alternatively, the presence of mats of sillimanite and biotite surrounding staurolite may indicate that the St-out reaction (above) progressed into the sillimanite stability field.

Further constraints on the metamorphic peak are provided by rare Crd-orthoamphibole-Ky-Bt assemblages in samples from Gauli Gletscher (Fig. 7B). In these rocks, early staurolite occurs as inclusions and relicts in garnet, in an assemblage of Grt + Ky + orthoamphibole. The latter assemblage indicates

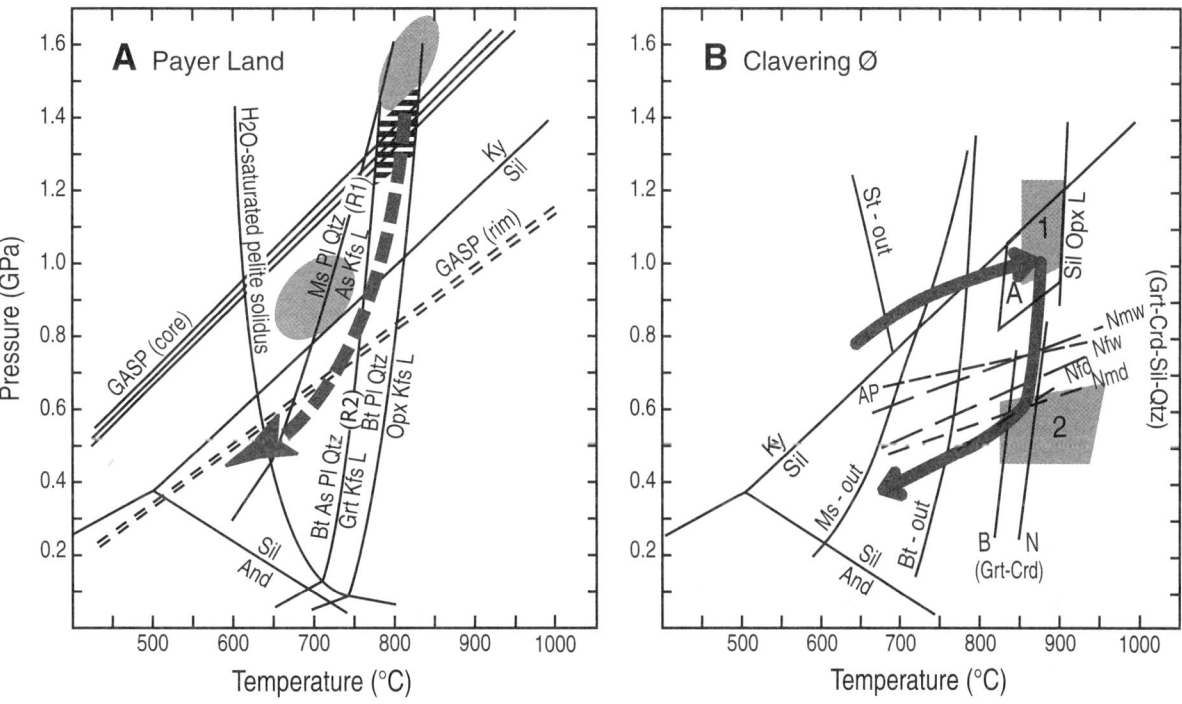

Figure 6. (A) Composite pressure-temperature (P-T) path for Payer Land high-pressure granulites showing the anatectic metapelites (Gilotti and Elvevold, 2002) in black and the mafic rocks (Elvevold et al., 2003) in gray. Dehydration melting reactions and H_2O-saturated pelite solidus are from Le Breton and Thompson (1988). GASP (garnet-kyanite-quartz-plagioclase) barometry on core and rim compositions is after Hodges and Crowley (1985). (B) Composite P-T path for granulite-facies metapelites (paragneisses) and metabasites from Clavering Ø. Box A indicates conditions of the metamorphic peak based on Grt-Bt thermometry (Ferry and Spear, 1978) and Grt-Als-Pl-Qtz barometry (Koziol and Newton, 1988). The P-T conditions of the decompression segment of the path were obtained by intersection of Grt-Crd thermometry (B—Bhattacharaya et al., 1988; N—Nichols et al., 1992) and Grt-Crd-Sil-Qtz barometry (AP—Aranovich and Podleskii, 1983; N—Nichols et al., 1992; where f and m relate to the Fe-Al-Si and Mg-Al-Si systems, and w and d indicate wet and dry conditions). Shaded boxes 1 and 2 indicate the peak and retrograde P-T conditions for the metabasites (Jones and Escher, 2002). Reaction abbreviations: Muscovite and biotite dehydration melting reactions are after Le Breton and Thompson (1988); St-out is from Bickle and Archibald (1984). Al_2SiO_5 stability is after Salje (1986). Lower-temperature limit of Sil-Opx-Melt is after Carrington and Harley (1995).

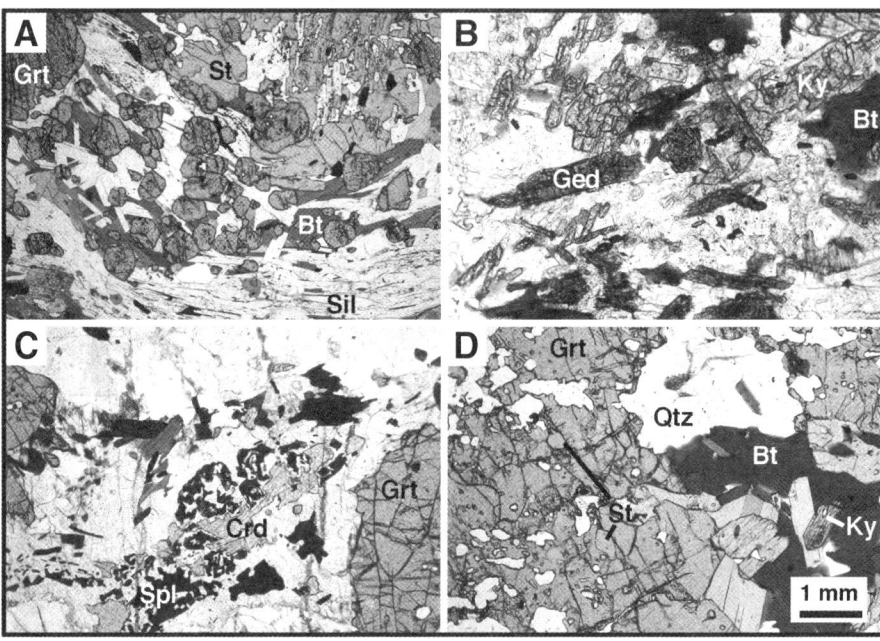

Figure 7. Plane-light photomicrographs of key metamorphic assemblages in the Niggli Spids and Hagar Bjerg thrust sheets. Scale is the same in each image. (A) Grt-St-Sil assemblage with early, oblique fabric preserved in staurolite (upper right). Garnet occurs as inclusions in staurolite as well as in the matrix. Rendalen. (B) Kyanite and gedrite in a pelite from Gauli Gletscher, Isfjord. (C) Grt + Sil + Crd + Spl (hercynite) assemblage in metapelites used to estimate high-temperature, low-pressure part of P-T path for Clavering Ø. (D) Staurolite inclusions in garnet; Ky + Bt in the matrix shows the reaction St + Ms + Qtz = Ky + Grt + Bt + V. Louise Boyd Land.

conditions in excess of ~700 °C at 0.8 GPa. Peak phases are replaced by cordierite, which indicates a decrease in P to <0.6 GPa and 600 °C (Spear, 1993) during the later part of the evolution. Decompression is also indicated by replacement of garnet by Hbl + Pl symplectites in dikes within the basement orthogneisses.

P-T estimates were calculated using Grt-Bt thermometry and Grt-Ky-Qtz-Pl (GASP) barometry on Grt + Ky + Bt + Pl (±St) assemblages from a sample in Rendalen. Recorded P-T conditions are 600–700 °C and 0.7–1.0 GPa (K.A. Jones, unpublished data).

Gletscherland and Goodenough Land

Gletscherland and adjacent areas are dominated by a lithologically varied orthogneiss terrane. The orthogneisses are tectonically interleaved with amphibolites, garnet-amphibolites, calc-silicate gneisses, and psammitic and pelitic paragneisses. The vast majority of orthogneisses and amphibolites contain assemblages (biotite, amphibole, K-feldspar, plagioclase, and quartz) typical of medium-pressure, amphibolite-facies conditions.

The earliest metamorphic mineral phases are found in rare, isolated lenses of mafic rocks in the easternmost part of the region at Kap Hedlund. A fine-grained mafic rock contains the well-preserved, high-pressure granulite-facies assemblage Grt + Cpx + Hbl + Pl + Qtz + Ilm + Bt (minor amounts). Geothermobarometry performed on the matrix minerals indicates that the high-pressure assemblage equilibrated at 700–750 °C and 1.1–1.3 GPa. Later reworking in the amphibolite facies has obscured evidence for the earlier high-pressure textures and minerals in the host orthogneisses.

Pelitic paragneiss from Kap Hedlund and Kap Bayard contains leucocratic layers and lenses interpreted as anatectic melts. The peak granulite-facies assemblage is Grt + Bt + Ky + Kfs + Pl + Qtz + Rt. The coexistence of Ky + Kfs + Qtz in the peak assemblage and the anatectic pods and veins suggest that partial melting of the pelites took place in the kyanite stability field. Phase-equilibria constraints (Spear et al., 1999) and geothermobarometry yield conditions of 700–750 °C and 0.8–1.0 GPa for the thermal peak. Rutile is included within garnet, whereas, ilmenite is the stable matrix Ti-phase. This suggests that the rock passed through the Rt + Grt stability field before falling pressures placed it within the ilmenite stability field. Postpeak textures are associated with an early stage of retrogression and then an extensional shearing. Replacement of garnet by the assemblage Sil + Bt + Pl is commonly observed.

Relicts of early high-pressure metamorphism are also preserved at the base of Hagar Bjerg, in the southeastern part of Goodenough Land. A recrystallized fine-grained mafic dike contains the high-pressure assemblage Grt + Cpx + Qtz + Hbl + Pl, and thermobarometric calculations yield temperatures of 650–680 °C at pressures of 1.0–1.2 GPa (S. Elvevold, unpublished data). Paragneisses from Hagar Bjerg are composed of Grt + Bt + Ky + Pl + Qtz + Rt. Retrograde muscovite mimics and cuts across the foliation and commonly includes kyanite. The late muscovite suggests the reaction Ky + (hydrous) melt ± Kfs = Ms + Qtz ± Pl. Back reaction between melt and restite also resulted in consumption of garnet to produce secondary biotite, kyanite, and plagioclase. The initial cooling thus took place in the stability field of kyanite, and the retrograde P-T path crossed above the Al_2SiO_5 triple point.

Garnets from the base of Hagar Bjerg display prograde growth zoning patterns, whereas garnets from eastern Gletscherland are characterized by broad homogeneous interiors, suggesting homogenization at high temperatures. The mineralogy, as well as the garnet-zoning trends, suggests that the metamorphic temperatures were higher in the eastern part of Gletscherland than to the west in Goodenough Land. Both areas exhibit evidence of an earlier high-pressure metamorphism, similar to Payer Land, but not as well preserved.

Nathorst Land and Hinks Land to Th. Sørensen Land

The crystalline orthogneiss complex in Nathorst Land consists of banded gray gneisses with amphibolite layers, abundant ultramafic lenses, and crosscutting amphibolitic mafic dikes (Higgins, 1982; Henriksen, 1986). Metasedimentary layers are present as relatively thin (≤300 m wide) micaceous and psammitic schist, marble, and quartzite. Typical Bt + Amp + Qtz + Pl + Kfs ± Grt parageneses in the orthogneisses are indicative of amphibolite-facies metamorphism.

The Krummedal supracrustal unit crops out in a broad strip from Hinks Land southward to eastern Rencontre Dal and the mouth of Krummedal. The typical paragenesis in metapelites is Grt + Ky + Bt + Ms + Qtz + feldspar (Higgins, 1974). Higgins (1974) also reported the presence of quartzofeldspathic sweats in the metapelites that may represent partial melts. Detailed metamorphic petrology is lacking for these areas.

Liverpool Land

The Archean and Paleoproterozoic orthogneisses from central and southern Liverpool Land, respectively, are described here with the Niggli Spids sheet because the mafic rocks preserve high-pressure metamorphic assemblages, despite the fact that Liverpool Land is an isolated area of crystalline rocks east of the Devonian and younger basins. Early workers recognized eclogite-like garnet-clinopyroxene rocks in Liverpool Land (Krank, 1935; Sahlstein, 1935), and these rocks were cited by Backlund (1936) to support his theory on the metamorphic origin of eclogites. Subsequent work (Coe and Cheeney, 1972; Cheeney, 1985) did not reveal any omphacite; the highest jadeite content they reported is Jd_{10}. However, an inferred supersilicic clinopyroxene in a sample from Rendeelven (Smith and Cheeney, 1980) and a chromite-garnet lherzolite from Tværdalen (Smith and Cheeney, 1981) suggest ultrahigh-pressure metamorphism. Omphacite-bearing rocks that record $P > 2.5$ GPa and $T = 800$ °C have recently been found; their age of metamorphism has been interpreted as ca. 400 Ma based on ID-TIMS U-Pb analysis of zircon (Hartz et al., 2005).

Summary of Metamorphic Characteristics of the Niggli Spids Thrust Sheet

Evidence of Caledonian high-pressure metamorphism is widespread throughout the Niggli Spids thrust sheet. The best preservation of high-pressure granulites and eclogites is found in Payer Land and Liverpool Land, respectively; vestiges of the high-pressure granulites are recognized in many places in between. These occurrences form the deepest level of crust exposed in this southern segment of the orogen and attest to significant crustal thickening at ca. 405–400 Ma, similar in age to the eclogite-facies metamorphism in the North-East Greenland eclogite province. The high-pressure rocks in Payer Land exhibit a clockwise P-T path and a high-temperature decompression segment (Fig. 6A).

The vast majority of orthogneisses and mafic rocks contain parageneses typical of medium-pressure, amphibolite-facies conditions. Younger greenschist-facies retrogression and hydrothermal veins are widespread. The orthogneisses generally lack garnet and provide few constraints on the metamorphic evolution. However, paragneisses and garnet-bearing mafic lithologies preserve a number of diagnostic mineral assemblages that record various stages of the P-T evolution.

Hagar Bjerg Thrust Sheet

The Hagar Bjerg thrust sheet (Higgins et al., 2004a) overlies the Niggli Spids sheet, but in many places, it is separated from it by an extensional detachment fault. Similar to the Niggli Spids thrust sheet, the Hagar Bjerg thrust sheet consists of two major units: Paleoproterozoic orthogneiss and the Mesoproterozoic Krummedal metasedimentary succession. In contrast, however, the metasedimentary rocks are much more abundant in the Hagar Bjerg unit. The Hagar Bjerg thrust sheet is distinguished from the Niggli Spids sheet by extensive midcrustal Caledonian migmatite complexes and S-type granites derived from partial melting of metapelites. The migmatite complexes are complicated by the presence of Neoproterozoic (ca. 930 Ma) granites (Kalsbeek et al., 2000, 2002, this volume).

Bessel Fjord to Grandjean Fjord

A Mesoproterozoic supracrustal unit, the Smallefjord succession, occurs in an arcuate belt running from the inner part of Grandjean Fjord (75°N) northward to inner Bessel Fjord (76°N). Based on similarities in structural setting, lithology, and age, the Smallefjord rocks are correlated with the Krummedal supracrustal unit (Friderichsen et al., 1994). The Smallefjord metasedimentary rocks are pervasively migmatized and dominated by metatexite fabrics, including discontinuous layers and augen of granitic leucosomes.

Metapelitic lithologies include metatexites and diatexites with the assemblage Grt + Ky/Sil + Bt + Pl + Qtz ± Kfs ± Ms ± Rt ± Ilm. Anatectic pelites in the Smallefjord unit record a clockwise P-T path (Jones and Strachan, 2000). The prograde path followed a near-isobaric slope constrained by mineral assemblages and growth zoning in garnet. Garnet cores contain straight to sigmoidal inclusion trails thought to record synthickening deformation. The outer parts of garnet contain Ky + Bt + Pl + Rt as included phases, interpreted to have resulted from St-out reactions and indicating that the heating path passed above the Al_2SiO_5 triple point. Melting began in the kyanite stability field by muscovite dehydration. Further melting continued within the sillimanite field with Grt + Kfs + Sil + melt as products of biotite dehydration melting. P-T estimates of the peak conditions were constrained by experimentally determined melting reactions and by Grt-Bt geothermometry and barometry (GASP, Grt-Als-Pl-Qtz and GRIPS, Grt-Rt-Ilm-Pl-Qtz) to 0.9–1.0 GPa at 790–850 °C. Prograde metamorphism was interrupted by the onset of decompression and retrograde extensional shear fabrics. Foliations and shear fabrics developed synkinematically with crystallization of melts during recrossing of biotite dehydration melting reactions at ~760 °C and 0.8 GPa, generating Bt + Sil–rich foliations. Continued cooling resulted in penetrative muscovite-rich foliations on recrossing of the muscovite dehydration reactions at 650–700 °C and <0.6 GPa.

A mafic dike within the Smallefjord sequence provides additional data on the metamorphic conditions (K.A. Jones, unpublished data). The dike contains fine-grained assemblages of Grt + Cpx + Pl + Qtz + Hbl ± Opx ± Rt ± Ilm. Limited evidence of an earlier assemblage is indicated by armored relics of orthopyroxene in matrix clinopyroxene and as inclusions in garnet. The matrix has symplectitic intergrowths of Cpx + Pl that replace coarse clinopyroxene grains, textures typical of decompression. Several samples show dynamic recrystallization of the earlier assemblage of Grt + Cpx + Pl + Qtz ± Hbl, which are interpreted as annealed mylonites. These textures indicate synkinematic recrystallization under transitional high-pressure granulite- to upper amphibolite–facies conditions. P-T estimates obtained from Grt-Cpx thermometry and Grt-Cpx-Pl-Qtz barometry on these recrystallized textures give 0.8–1.1 GPa and 650–750 °C and are thought to record the metamorphic conditions during decompression, comparable to those obtained on the Smallefjord metapelites. No constraints are available on the earlier Opx-Cpx–bearing assemblages.

A U-Pb SHRIMP age of 445 ± 10 Ma from two zircons in a garnet-kyanite schist within the Smallefjord sequence has been interpreted to date the time of Caledonian metamorphism (Strachan et al., 1995). However, this age is older than all the 430–415 Ma granite ages (Kalsbeek et al., this volume), and its significance is in doubt. The sample is complicated because it also contains a suite of Neoproterozoic (980–890 Ma) zircons.

Clavering Ø

The rocks of Clavering Ø represent the easternmost exposures of granulite-facies rocks in East Greenland (Jones and Escher, 2002). Granulite-facies metamorphism affected both basement orthogneiss complexes and metapelites disposed in imbricate thrust sheets.

Mafic granulites, occurring as sheet-like bodies within metapelitic paragneisses and orthogneisses, experienced a clockwise P-T history (Fig. 6B) that involved a significant component of near-isothermal decompression (Jones and Escher, 2002). The mafic granulites consist of the assemblage Grt + Cpx + Opx + Hbl + Pl + Rt + Qtz ± Ilm ± Spl. The earliest coarse-grained assemblages are Grt + Cpx/Opx + Pl + Qtz + melt. Inclusions of hornblende in garnet, clinopyroxene, orthopyroxene, and plagioclase suggest that these assemblages developed from the prograde dehydration melting of amphibole. In several samples, preserved magmatic textures indicate the presence of a silicate melt. P-T conditions of >0.8–1.1 GPa and 850–915 °C for this assemblage have been estimated from experimentally determined melting reactions in conjunction with thermometry (Grt-Cpx, Grt-Opx, Al in Opx) and barometry (Grt-Opx-Pl-Qtz and Grt-Cpx-Pl-Qtz) (Jones and Escher, 2002). These early coarse-grained assemblages have been partly to completely replaced by finer-grained intergrowths of Opx + Pl ± Spl ± Ilm. Along the margins of mafic bodies, the coarse-grained assemblages are deformed, and amphibole is the main fabric-forming mineral. These assemblages are also overprinted by Opx-Pl symplectites, suggesting that high-temperature ductile shearing was followed by (or synkinematic with) recrystallization. P-T estimates of the secondary assemblages obtained from thermometry (Grt-Opx, Al in orthopyroxene) and barometry (Grt-Opx-Plag-Qtz) give 850–915 °C and 0.5–0.65 GPa. These data suggest that the mafic granulites record a phase of near-isothermal decompression at 0.3–0.45 GPa as part of the clockwise P-T path.

Additional analysis of the metapelitic paragneisses (metatexites and diatexites) on Clavering Ø provides further constraints on the clockwise nature of the P-T path (K.A. Jones, unpublished data). A near-isobaric prograde path is inferred from mineral relicts and possible melting reactions (Fig. 6B). The initial prograde path is given by growth of Ky + Grt + Bt assemblages from St + Ms breakdown at 650–700 °C and <0.8 GPa. The initial melting occurred by muscovite dehydration, which led to the formation of Ky + Kfs + melt. Further biotite dehydration melting at higher temperatures produced peak assemblages of Sil + Kfs + melt. Peak assemblages are bracketed by experimentally determined equilibria to 790–915 °C at 0.9–1.0 GPa. Thermobarometry on postpeak reaction textures involving Grt + Sil ± Bt breakdown to Crd ± Spl give $P < 0.6$–0.7 GPa at $T = 830$–860 °C (Fig. 7C). Cooling is indicated by recrossing of biotite and muscovite melting reactions, and it commenced at 750 °C and 0.4–0.6 GPa. The high temperatures and near-isothermal decompression are similar to the history recorded by the mafic granulites.

Chronological constraint on the age of migmatization is limited to a U-Pb SHRIMP age of 429 ± 10 Ma (G. Watt and P.D. Kinney, 1999, personal commun.) obtained on zircons from a migmatitic leucosome and interpreted to date partial melting.

Bartholin Land, Andrée Land, Louise Boyd Land, and Frænkel Land

The Mesoproterozoic metasedimentary rocks of the Hagar Bjerg thrust sheet occupy extensive areas in Bartholin Land, Andrée Land, Louise Boyd Land, and Frænkel Land. The succession is dominated by sandstone and mudstone lithologies, and the metamorphism is medium to high grade, locally reaching anatexis in the granulite facies.

In Andrée Land and Louise Boyd Land, the orthogneiss unit of the Hagar Bjerg thrust sheet is missing, such that the high-grade metasedimentary rocks with abundant granites (Hagar Bjerg thrust sheet) are thrust over clastic metasediments belonging to the Niggli Spids thrust sheet that lack granites. The metasedimentary and metabasic lithologies of the Krummedal succession and basic sheets within the orthogneiss complex have been subjected to a medium- to high-pressure event. In the well-exposed sequences of Louise Boyd Land and Frænkel Land, the nature of the Caledonian metamorphism is identified in the metasedimentary rocks in the footwall of the Petermann Bjerg detachment. The metamorphic grade in these sections increases downward toward the basement orthogneisses.

The prograde segment of the P-T-d path has been established from inclusion sequences within porphyroblasts (K.A. Jones, unpublished data). Garnet porphyroblasts preserve prograde zoning. The garnet cores contain Qtz ± Ilm, whereas the outer cores contain rounded inclusions of Pl (strongly zoned) + Qtz + Rt ± St ± Bt (Fig. 7D). The outer parts of these garnets contain coarse idioblastic Ky + Rt + Bt + Qtz + Pl and xenoblastic relics of St + Qtz. Staurolite also occurs as relics in matrix kyanite porphyroblasts, with rutile and quartz, and as xenoblastic grains in matrix plagioclase. The textures suggest that the assemblage Grt + Ky + Bt resulted from staurolite breakdown by the reaction St + Qtz (±Ms) = Ky + Bt + Qtz + H_2O.

The onset of in situ partial melting is indicated by the presence of isolated leucocratic lenses and stromatic migmatites with leucosomes containing Ky + Pl + Qtz ± Kfs (microcline) surrounded by biotite-rich leucosomes with abundant kyanite porphyroblasts, indicating that partial melting occurred via muscovite dehydration melting within the kyanite stability field. In deeper parts of the section, the amount of quenched melt increases with the localized development of diatexite patches and pods. Mafic schlieren and enclaves of metatexite incorporated within the diatexites contain Ky + Bt + Grt assemblages. At these levels, melt migration along shear zones and coalescence into crosscutting diatexite sheets are observed.

Constraints on the prograde P-T-d path are provided by experimentally determined reactions. Kyanite indicates pressures above the Al_2SiO_5 triple point during prograde metamorphism. The prograde crossing of the St-out → Grt + Ky + Bt-in reaction must have occurred at $P > 0.78$ GPa and $T > 700$ °C. The crossing of GRAIL (Grt + Rt + Als = Ilm + Qtz) points to a pressure increase along the prograde path. Anatexis by muscovite dehydration melting is constrained to $P > 0.8$ GPa and $T > 720$ °C. GASP and Grt-Bt thermobarometry, as well as constraints from mineral equilibria to assemblages containing Grt + Ky + Pl + Bt + Qtz ± Rt ± Kfs + melt, give $T = 750-790$ °C at 1.0–1.2 GPa (Louise Boyd Land and Frænkel Land), with slightly higher pressures of 1.3–1.4 GPa recorded farther east (Lacroix Bjerg). The upper temperature limit is less than biotite dehydration melting (<790–800 °C). These P-T conditions are further supported by mineral assemblages and thermobarometric determinations obtained on metabasic intrusions that cut the Krummedal succession and from slightly discordant sheets within the basement orthogneiss complexes. The peak assemblage Grt + Cpx + Pl + Qtz ± Rt ± Hbl has given $T = 700-800$ °C and $P = 1.15-1.5$ GPa from Grt-Cpx thermometry and Grt-Cpx-Pl-Qtz barometry.

The retrograde part of the P-T path is indicated by the growth of sillimanite either as fasiculate bundles nucleated on kyanite and garnet, or as radiating needles grown along grain boundaries between plagioclase and quartz. Rutile is partially replaced by ilmenite. Lower-temperature deformation is associated with the transformation of Grt + Bt ± Ky/Sil + Kfs + Pl + Qtz + Rt/Ilm assemblages to Ms + Qtz + Ilm ± Pl. Application of thermobarometry to these assemblages has proven difficult as a result of intense retrogression and resetting of mineral compositions by late Fe-Mg exchange. Mineral equilibria for the assemblage Grt + Sil + Pl + Bt + Ilm + Qtz in the absence of muscovite indicates conditions of $T > 700$ °C at $P = 0.4-0.8$ GPa, whereas the assemblage Grt + Ms + Qtz + Plag + Ilm indicates $P = 0.4-0.65$ GPa at $T < 700$ °C.

Constraints on the timing of the metamorphic peak and anatexis are provided by a U-Pb SHRIMP age of ca. 428 ± 10 Ma obtained on zircon rims from a migmatitic garnet-kyanite gneiss from Jættedalen (G. Watt and P.D. Kinney, 1999, personal commun.), which confirms a Caledonian age for this metamorphism, despite the abundance of Neoproterozoic granites in some of the areas.

Lyell Land, Forsblad Fjord, and Stauning Alper Migmatite Complexes

A triangular-shaped belt of partially melted metasedimentary rocks and granites occurs in Lyell Land. The migmatite belt extends southward across inner Forsblad Fjord to Nathorst Land and the Stauning Alper. The Krummedal metasediments in Lyell Land are neosome-rich with abundant restitic inclusions of quartzite and biotite schist. The neosome is marked by ghost layering defined by biotite-rich stringers. Leucogranite forms sheets that are both parallel to and cut across a relict metasedimentary layering. Peak temperatures increase southward in this area.

Metapelites in Lyell Land contain the muscovite-absent assemblage Grt + Bt + Sil/(Ky) + Kfs + Pl + Qtz + Ilm. The coexistence of Sil + Kfs in the peak assemblage indicates temperature conditions above the muscovite dehydration melting curve. The biotite melting solidus was not crossed; peak temperatures are thus constrained to 650–750 °C. Myrmekite textures are common. Feldspar commonly contains rounded inclusions of quartz, which may have been individual droplets of melt. Retrograde consumption of garnet resulted in secondary Bt + Pl + Sil assemblages.

Garnet has preserved compositional growth zoning in some samples. Zoning patterns related to retrograde processes are present along the margin and around resorption embayments. Garnet-biotite thermometry and GASP performed on one sample yielded $T = 600-700$ °C and $P = 0.57-0.73$ GPa (S. Elvevold, unpublished data).

Krummedal metasedimentary rocks in the Forsblad Fjord region are typically banded with a stromatic layering. Anatexites are scattered throughout the sequence and increase in abundance up-section. The metasedimentary units have been studied by White et al. (2002) and White and Hodges (2003). Typical assemblages in metapelitic gneisses are Bt + Grt + Pl + Qtz ± Kfs ± Ms ± Ky/Sil ± Crd. Concordant and crosscutting leucosomes contain Kfs + Pl + Qtz ± Sil ± Bt. Peak conditions are estimated as ~1.05 GPa at 785 °C. The metapelites followed a clockwise *P-T* loop related to crustal thickening followed by tectonic denudation (White and Hodges, 2003). The decompression segment of the *P-T* path is attributed to 18 km of tectonostratigraphic throw along the Fjord region detachment. In situ U-Th-Pb monazite electron microprobe dating of the Krummedal metasedimentary rocks yields Caledonian ages (White and Hodges, 2003).

The migmatite zone continues southward to Furesø, Dammen, and the Stauning Alper region, where it covers an extensive area. All degrees of migmatization are present, from small-scale patchy leucosomes in metasediments to schlieren-rich granitic migmatites. Initial stromatic migmatites were formed in the kyanite stability field under water-saturated conditions. As temperature increased, vapor-absent melting reactions began to dominate. Diatexites and gneissic migmatites have the characteristic assemblage Grt + Crd + Sil + Spl + Bt, which is common in many granulite-facies migmatite terranes (Watt et al., 2000).

Renland

Southwest Renland is dominated by paragneisses correlated with the Krummedal sequence that are intruded by late Neoproterozoic augen granites and Caledonian granitoids. Upper amphibolite- to granulite-facies metamorphism produced garnetiferous melt segregations in the augen granite and in semipelitic components of the paragneisses, whereas orthopyroxene formed in rare mafic lithologies (Leslie and Nutman, 2003). Mineral assemblages and geothermobarometric estimations have not been published. The high-grade metamorphism in Krummedal rocks has been dated at ca. 430 Ma with U-Pb SHRIMP on zircon (Leslie and Nutman, 2003).

Summary of Metamorphic Characteristics of the Hagar Bjerg Thrust Sheet

Whereas the Krummedal metasedimentary rocks in the Niggli Spids thrust sheet show high-pressure metamorphism overprinted by amphibolite-facies assemblages, the equivalent succession in the Hagar Bjerg thrust sheet is characterized by widespread midcrustal anatexis and migmatization. The presence of Neoproterozoic granitoids complicates the Caledonian metamorphic pattern; however, gradients in temperature can be recognized locally. For example, the temperature of migmatization increases southward from Lyell Land to the Stauning Alper, and eastward toward Clavering Ø at the latitude of 74°N. Caledonian anatexis occurred at ca. 430–415 Ma, prior to high-pressure metamorphism of the underlying Niggli Spids units at 405–400 Ma.

Franz Joseph Allochthon

The Franz Joseph allochthon (Higgins et al., 2004a) is the structurally highest metamorphic unit in the Greenland Caledonides (Fig. 5). In many places, the lower boundary is marked by extensional detachment faults that place it in contact with either the Hagar Bjerg or the Niggli Spids thrust sheets. The Franz Joseph allochthon consists of Neoproterozoic to Ordovician sedimentary successions intruded by discrete, kilometer-scale leucogranitic plutons. Our description of metamorphism focuses on outliers of the Neoproterozoic Eleonore Bay Supergroup in the Ardencaple–Bessel Fjord region, Andrée Land, and Suess Land, and its western equivalent, the Petermann Bjerg Group, in Frænkel Land and Louise Boyd Land.

Ardencaple–Bessel Fjord Region

Metamorphism in this region is of low-pressure type, and metamorphic grade increases structurally downward toward the base of the Eleonore Bay Supergroup. Four metamorphic zones are identified (see Fig. 2 *in* Soper and Higgins, 1993): (1) rocks that retain essentially detrital mineralogy and textures (white mica + Qtz + Chl ± Pl ± Kfs ± Cc); (2) a recrystallized zone (Qtz + white mica + Chl); (3) a biotite zone (Bt + Ms + Chl + Qtz); and (4) a garnet + staurolite zone (Grt ± St ± And ± Sil + Bt + Ms + Pl + Qtz). The time of peak metamorphic mineral growth throughout the sequence was synchronous with, and outlasted, the formation of a pervasive, bedding-parallel, differentiated crenulation cleavage (S_2) (Fig. 8A). Earlier S_1 fabrics are preserved in low-strain zones, in microlithons of S_2 crenulations (Fig. 8A), and as inclusion sequences in porphyroblasts. Post-peak retrograde metamorphism accompanied the development of intense layer-parallel foliations (S_3), and weak fabrics (kink bands and crenulations) are associated with late upright folding.

Estimates for the conditions of peak metamorphism are based on published petrogenetic grids (e.g., Spear and Cheney, 1989), since mineral compositional data are absent. Application of simple grids implies temperatures >560 °C for the high-grade assemblages Grt + And + Bt, Grt + St + Bt, and And + St + Bt (+ Ms + Qtz ± Pl) (Fig. 8B). Calculated phase equilibria for similar assemblages (e.g., Dymoke and Sandiford, 1992; Reche et al., 1998) indicate a restricted stability range of 0.3–0.45 GPa at 550–620 °C. The presence of sillimanite (fibrolite) at the highest grade indicates *P-T* conditions of >600 °C at ~0.4–0.6 GPa.

Frænkel Land and Louise Boyd Land

Metamorphism of the Petermann Bjerg Group increases in grade structurally downward and eastward. Incipient metamorphism of pelitic lithologies is indicated by neomorphic growth of Chl + Ms overprinting detrital grains (Chl + Pl + Qtz ± Kfs). Progressive increases in grade are recorded first by the appearance of biotite (Bt + Ms + Chl + Qtz ± Pl ± Ep ± Ab) and second by garnet-bearing assemblages (Grt + Bt + Chl + Ms + Qtz ± Alb ± Olg). The highest-grade assemblage is Grt + Bt + Ms + Olg + Qtz.

Microstructures indicate that the timing of the peak thermal mineral assemblage is post-D_2 crenulation cleavage formation (Fig. 8C). Retrograde mineral growth is associated with formation of mylonitic and phyllonitic zones (D_3), the intensity of which increases toward the base of the Petermann Bjerg Group. Near-vertical D_4 fabrics are associated with upright folds.

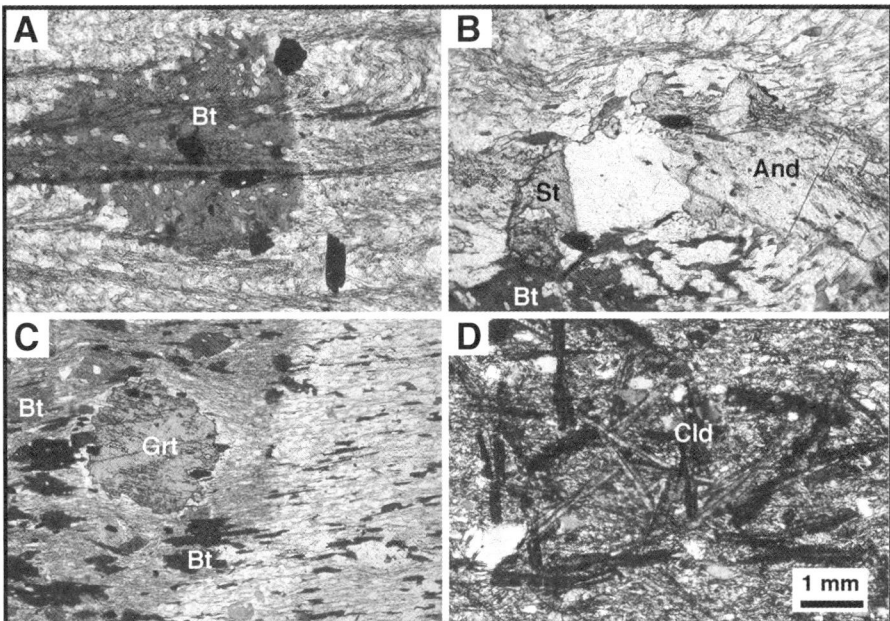

Figure 8. Plane-light photomicrographs showing metamorphic assemblages at the base of the Franz Joseph allochthon. Scale is the same in each image. (A) Biotite overgrowing deformation fabrics, Petermann Bjerg Group, Mysteriedalen. (B) Key assemblage of St + And, Eleonore Bay Supergroup, Bessel Fjord. (C) Typical garnet zone assemblage (Grt + Bt + Ms + Qtz), Petermann Bjerg Group, Frænkel Land. (D) Postkinematic chloritoid from low-grade pelites, Petermann Bjerg Group, Andrée Land.

P-T estimates obtained from Grt-Bt thermometry and Grt-Bt-(Ms)-Pl-Qtz barometry performed on a sample of pelitic schist containing the assemblage Grt + Bt + Chl + Pl + Qtz from Louise Boyd Land are $P = 0.4$–0.55 GPa and $T = 570$–650 °C (K.A. Jones, unpublished data). Garnet preserves normal prograde growth zoning. Additional P-T constraints are provided by the rare occurrence of pseudomorphs after andalusite(?) in a Grt + And + Ms + Chl + Qtz schist from Frænkel Land. The coexistence of And + Grt in Fe-rich lithologies implies conditions of 570–610 °C at 0.4 GPa (e.g., Dymoke and Sandiford, 1992).

The timing of peak metamorphism is constrained by field relationships with dated granitic intrusions. Major sheeted leucogranite complexes within the Petermann Bjerg Group cut the early upright folds (D_4) and yet are intruded along the axial planes of these folds, indicating that they are synkinematic. U-Pb SHRIMP dating of zircon from one of these granites has given an intrusive age of ca. 435 Ma (Kalsbeek et al., 2001a), which therefore puts the time of peak metamorphism before 435 Ma.

Andrée Land and Suess Land

A similar sequence of microstructures and metamorphic zones is preserved in rocks of the Eleonore Bay Supergroup above the Junctiondal fault in southeast Andrée Land and in northeast Suess Land. Rocks with detrital grains contain biotite- and garnet-grade assemblages. The notable development in this region is chloritoid in fine-grained slaty rocks (Fig. 8D). The presence of chloritoid in the assemblage Cld + Chl + white mica + Qtz and the absence of biotite are consistent with metamorphism of highly aluminous rocks under conditions equivalent to the biotite zone, i.e., $T < 400$ °C and $P < 0.3$–0.5 GPa (e.g., Spear and Cheney, 1989; Spear, 1993).

Summary of Metamorphism in the Franz Joseph Allochthon

Metamorphism in the Franz Joseph allochthon is low-pressure, subgreenschist, greenschist, and lower-amphibolite facies. Microstructural studies reveal a consistent pattern of mineral growth in which the peak thermal mineral assemblages outlasted regionally developed differentiated S_2 crenulation cleavage. The kinematics of this fabric have not been determined owing to the intensity of successive retrograde overprinting, and its significance in regional terms is currently difficult to assess. P-T constraints from published petrogenetic grids and preliminary thermobarometry indicate that the rocks have been subjected to a near-isobaric heating path at <400–660 °C and ~0.3–0.6 GPa. Field relationships suggest that metamorphism was concurrent with granite formation in the Hagar Bjerg thrust sheet at 435 Ma.

DISCUSSION

Caledonian metamorphism along the Laurentian margin in East Greenland is generally considered to have evolved in the overriding plate of a continent-continent collision with Baltica (Gee, 1975; Hossack and Cooper, 1986; Gilotti and McClelland, 2007). Metamorphic patterns reflect episodes of crustal thickening and melting that are spatially associated with structural levels. These structural levels are characterized by generally similar P-T paths. Low-grade metamorphism occurred in the sedimentary sequences of the Laurentian margin both in the foreland and in the upper structural level known as the Franz Joseph allochthon. Where fertile source rocks are present, midcrustal levels of the Hagar Bjerg thrust sheet melted to produce peraluminous granites and migmatites. The deepest parts of the orogen must have been formed by extreme crustal thickening implied by the widespread high- and ultrahigh-pressure metamorphism in the North-East Greenland eclogite province and the Niggli Spids thrust sheet. The main metamorphic units are separated by major tectonic breaks, which make it difficult to construct a simple orogenic model. In addition, the timing of metamorphism varies among the different units and, in the case of the low-grade rocks, has not been directly dated. Although we tend to think of the units in terms of adjacent structural levels, it is important to keep in mind that we lack the data to reconstruct their position relative to one another during each metamorphic event. Regional extensional faults have shuffled the original thrust geometries and juxtaposed rocks exhumed from various tectonic levels.

The oldest metamorphism recorded during the Caledonian collision is the melting of the Mesoproterozoic Krummedal succession, which produced the widespread migmatite and leucogranite complexes south of Bessel Fjord. Melting took place over an ~20 m.y. period from ca. 435 to 415 Ma (Kalsbeek et al., this volume) and was restricted to the southern half of the orogen, where the fertile, pelitic source rocks of the Krummedal unit were present. The suggestion that melting was triggered by decompression due to regional extensional faults (Watt et al., 2000) is not supported by P-T paths for these rocks (e.g., Fig. 6B; Jones and Strachan, 2000; White and Hodges, 2003). The P-T paths show that prograde, dehydration melting of micas at medium pressure produced the migmatites. We infer that anatexis was a result of crustal thickening. Some of the rocks reached granulite facies (e.g., Clavering Ø), requiring anomalously high temperatures for Barrovian metamorphism. Better documentation of metamorphic spatial variations within the Hagar Bjerg sheet might help unravel these processes; however, the heat source for the high-temperature metamorphism remains problematic (e.g., Jamieson et al., 1998).

The age of metamorphism of the Neoproterozoic metasedimentary rocks in the Franz Joseph allochthon has not been directly determined. It has been suggested (Peucat et al., 1985) that pelitic rocks in the Nathorst Land Group, which forms the basal 7 km of the Eleonore Bay Supergroup, have also been migmatized. If this is true, the metamorphism may have been similar in age to the Hagar Bjerg leucogranites and migmatites. However, the Nathorst Land Group and the Krummedal units are very difficult to distinguish when strongly deformed at high grades. Future provenance studies using detrital zircon may help to confirm melting of the Neoproterozoic rocks. Leucogranite plutons within the Franz Joseph allochthon show that it was coupled with the Hagar Bjerg unit at 435–415 Ma, assuming the plutons originated in the migmatite complexes (Kalsbeek et al., 2001a, 2001b). The leucogranites in the allochthon lack contact-metamorphosed aureoles and are unlikely to have been able to carry the heat necessary to produce the regional low-pressure isobaric heating of the basal Franz Joseph allochthon.

The cause of the low-pressure metamorphism in the Franz Joseph allochthon remains enigmatic. Haller's (1971) idea of a melted mobile infrastructure heating the overlying suprastructure was the basis of his "stockwerke" tectonic model. Although the stockwerke model has been invalidated by thrust tectonics (Higgins et al., 2004a), it does emphasize the crux of the problem: how to heat the allochthon from the bottom up? If the ~18-km-thick Franz Joseph allochthon were thrust over the Hagar Bjerg sheet, this would have provided the overburden needed to thicken the crust and contribute to heat production in the underlying Krummedal sequence. Static growth of peak phases at low pressure (e.g., garnet and staurolite) could mark the advection of heat into the orogenic lid. Thus, the low-pressure metamorphism may or may not be linked to granite formation. The question of whether the granites formed in a midcrustal channel (e.g., Beaumont et al., 2001) or during extension also must fit the metamorphic puzzle.

The widespread nature of high- and ultrahigh-pressure metamorphism in East Greenland does not conform to current models for eclogite-facies metamorphism via continental subduction (Chemenda et al., 1996; Ernst and Liou, 2000; Ernst, 2006), because the Laurentian margin formed the overriding plate in the collision with Baltic. High-pressure metamorphism of the North-East Greenland eclogite province and the Niggli Spids unit occurred from ca. 410 to 390 Ma, the same time as continental subduction produced ultrahigh-pressure metamorphism at the margin of Baltica in western Norway (Carswell et al., 2003; Root et al., 2004). Eclogite-facies metamorphism in the Greenland Caledonides is more likely to have been the result of crustal thickening of the overriding plate (Ryan, 2001). The Himalaya may provide the best analogy for the Caledonides, where Laurentia would be comparable to Eurasia as the upper plate in the collisional system. Conditions under which eclogite could form late in an orogenic cycle may exist today at the base of Tibet, where the thickest continental crust is known to be 90 km (Wittlinger et al., 2004). Future exhumation of deep Tibetan crust would expose high-pressure rocks that are younger than the suite of 20–15 Ma peraluminous granites of the High Himalayas (e.g., Searle et al., 2003), similar to the East Greenland situation where the leucogranites are older than the eclogites.

Localized ultrahigh-pressure metamorphism of the Laurentian margin at 365–350 Ma (Gilotti et al., 2004; McClelland et al., 2006) marked the end of Caledonian continental convergence. Pressures within the coesite stability field require that this part of the crust reached a depth in excess of 120 km—much greater than presently observed continental thicknesses, and outside the range of thicknesses predicted by Ryan's (2001) model. The high temperatures calculated for the Greenland ultrahigh-pressure rocks (>950 °C; Gilotti and Ravna, 2002) are consistent with the steep geotherms that might be encountered at the base of the crust, as opposed to those in a cold subduction zone. Gilotti and McClelland (2007) have proposed that intracratonic subduction occurred in the overriding plate of Laurentia at the waning stage of plate convergence, bringing the already overthickened crust of the North-East Greenland eclogite province into the coesite stability field.

Unmetamorphosed to low-grade sedimentary rocks of the foreland crop out along the length of the orogen but are best seen, together with their autochthonous equivalents, in Kronprins Christian Land. The foreland forms a classic fold-and-thrust belt with metamorphic grade, defined by CAI values, steadily increasing to the east (Rasmussen and Smith, 2001). The parallel pattern of like CAI numbers indicates that this metamorphism was related to burial by overlying thrust sheets. The timing of thrusting should, therefore, slightly precede the timing of metamorphism. Thrusting was thought to be Silurian because this was when the passive margin foundered, and it is also the age of the youngest rocks cut by thrusts (Hurst et al., 1983). However, slivers of Ordovician limestone with CAI values of 5+ lie beneath an eclogite-bearing thrust sheet in Lambert Land and Nørreland (Rasmussen and Smith, 2001), implying that metamorphism of the foreland must be younger than ca. 400 Ma, the age of the eclogites. Furthermore, the eclogite thrust sheet must have been considerably thinned before emplacement in order to keep the overburden below ~12 km (CAI = 5+). A similar situation exists in the southern part of the orogen where the Niggli Spids sheet, with its 405 Ma high-pressure relics (McClelland and Gilotti, 2003), is thrust over low-grade, foreland sedimentary rocks. Thrusts may have propagated toward the foreland (Higgins and Leslie, 2000; Leslie and Higgins, this volume), but the foreland thrusts developed after ca. 400 Ma. Models that invoke Silurian transpression for the closure of Iapetus (Holdsworth and Strachan, 1991; Soper et al., 1992; Dewey and Strachan, 2003) are inconsistent with Devonian to Carboniferous high- and ultrahigh-pressure metamorphism of the Laurentian basement in the Greenland Caledonides. Perhaps transpression occurred later in the collision; this should be tested by directly dating the foreland thrusting and the strike-slip shear zones.

An understanding of the metamorphic patterns acquired during the Caledonian collision is now beginning to emerge. Isograds have only been mapped in a few places, and future work should concentrate on establishing trends within individual units. For example, it would be enlightening to know how pressure and temperature increase spatially within the migmatite complexes, eclogite terranes, and lower-grade rocks in order to better understand metamorphic processes. Establishing the timing of metamorphism must also be a goal of future work. The Greenland Caledonides offer tremendous exposures of the deeper levels of an exhumed upper plate from a collisional orogen—the challenge is to piece the puzzle back together such that the metamorphic evolution can be understood spatially and chronologically.

CONCLUSIONS

1. The tectonic architecture of the Greenland Caledonides has characteristic metamorphic patterns, more or less related to structural level, that differ north and south of Bessel Fjord (76°N). North of Bessel Fjord, there is a stack of thrust sheets with metamorphic grade increasing toward the hinterland from unmetamorphosed to ultrahigh-pressure (UHP) in the east. South of Bessel Fjord, the metamorphic pattern is more complicated. Very low-grade foreland sedimentary rocks are overlain by the deepest structural level of the Niggli Spids thrust sheet, which contains high-pressure (HP) relics. A midcrustal migmatite complex occupies the overlying Hagar Bjerg thrust sheet; this dominantly sedimentary unit reached high temperatures from the amphibolite into the granulite facies. The uppermost metamorphic rocks, the Franz Joseph allochthon, are a thick package of Neoproterozoic to Ordovician sedimentary successions that reach greenschist- and local amphibolite-facies conditions at their base.

2. Each main structural level records a similar P-T path across large areas. For example, the deepest structural level shows a clockwise P-T loop and a steep prograde path reaching peak pressures of 1.5–2.3 GPa, followed by a steep, isothermal decompression segment. The midcrustal migmatite terranes were formed by prograde dehydration melting at much lower pressures ($P < 1$ GPa) along a shallow path, reaching peak temperatures up to 850 °C before following a near-isothermal decompression segment. Rocks in the Franz Joseph allochthon followed a low-pressure isobaric heating path.

3. The Laurentian margin of East Greenland experienced widespread high-pressure metamorphism, extending from the orthogneiss complex at 80°N to gneisses in Liverpool Land at 70°N. A small but significant area of ultrahigh-pressure metamorphism also occurs in the easternmost hinterland. This high- and ultrahigh-pressure metamorphism happened in the upper plate of the collision with Baltica, late in the collision, probably due to crustal thickening. The ultrahigh-pressure terrane may represent an exhumed portion of the overthickened welt that was taken into the coesite stability field in the footwall of an intracratonic subduction zone (Gilotti and McClelland, 2007).

4. Three main periods of Caledonian metamorphism are currently recognized. The oldest is the formation of granites and migmatites in the middle crust at 440–415 Ma, while high-pressure metamorphism was Devonian (410–390), and ultrahigh-pressure metamorphism occurred at 365–350 Ma near the end of Caledonian plate convergence.

ACKNOWLEDGMENTS

The authors thank Niels Henriksen for his outstanding leadership and organization of the many expeditions in which we participated. We are grateful to all of our colleagues with whom we worked in Greenland for their insight and camaraderie but extend special thanks to our field partners at various times: Hannes Brueckner, Jan Escher, the late J.D. Friderichsen, A.K. Higgins, Joseph M. Hull, and Bill McClelland. In addition to funding from the Danish Ministry of the Environment, we acknowledge the support of the Carlsberg Foundation (to Synnøve Elvevold), the Danish Natural Science Research Council (to Synnøve Elvevold), the National Geographic Society (6853-00 to Gilotti), and the U.S. National Science Foundation (EAR-9508218 and EAR-0208236 to Gilotti). We thank Bradley Hacker and David Roberts for their helpful reviews of the manuscript, and A.K. Higgins for his careful editorial work.

REFERENCES CITED

Aranovich, L.Y., and Podleskii, K.K., 1983, The cordierite-garnet-sillimanite-quartz equilibrium: Experiments and applications, *in* Saxena, S.K., ed., Kinetics and Equilibrium Mineral Reactions: New York, Springer-Verlag, p. 173–198.

Backlund, H.G., 1936, Zur genetischen deutunt der eklogite: Geologische Rundscahu, v. 27, p. 47–61.

Beaumont, C., Jamieson, R.A., Nguyen, M.H., and Lee, B., 2001, Himalayan tectonics explained by extrusion of a low viscosity crustal channel coupled to focused surface denudation: Nature, v. 414, p. 738–742, doi: 10.1038/414738a.

Begin, N.J., 1992, Contrasting mineral isograd sequences in metabasites of the Cape Smith Belt, northern Quebec, Canada: Three new bathograds for mafic rocks: Journal of Metamorphic Geology, v. 10, p. 685–704, doi: 10.1111/j.1525-1314.1992.tb00115.x.

Bhattacharaya, A., Mazumdar, A.C., and Sen, S.K., 1988, Fe-Mg mixing in cordierite: Constraints from natural data and implications for cordierite-garnet geothermometry in granulites: The American Mineralogist, v. 73, p. 338–344.

Bickle, M.J., and Archibald, N.J., 1984, Chloritoid and staurolite stability: Implications for metamorphism in the Archean Yilgarn block, Western Australia: Journal of Metamorphic Geology, v. 2, p. 179–203, doi: 10.1111/j.1525-1314.1984.tb00295.x.

Brey, G.P., and Köhler, T., 1990, Geothermobarometry in four-phase lherzolite: II. New thermobarometers, and practical assessment of existing thermobarometers: Journal of Petrology, v. 31, p. 1353–1378.

Brueckner, H.K., Gilotti, J.A., and Nutman, A., 1998, Caledonian eclogite facies metamorphism of early Proterozoic protoliths from the North-East Greenland eclogite province: Contributions to Mineralogy and Petrology, v. 130, p. 103–120, doi: 10.1007/s004100050353.

Carrington, D., and Harley, S.L., 1995, Partial melting and phase relation in high-grade metapelites: An experimental petrogenetic grid in the KFMASH system: Contributions to Mineralogy and Petrology, v. 20, p. 270–291.

Carswell, D.A., 1990, Eclogites and the eclogite facies: Definitions and classifications, *in* Carswell, D.A., ed., Eclogite Facies Rocks: New York, Chapman and Hall, p. 1–13.

Carswell, D.A., and Compagnoni, R., 2003, Introduction with review of the definition, distribution and geotectonic significance of ultrahigh pressure metamorphism, *in* Carswell, D.A., and Compagnoni, R., eds., Notes in Mineralogy, Volume 5: Budapest, European Mineralogical Union, p. 3–9.

Carswell, D.A., and Cuthbert, S.J., 2003, Ultrahigh-pressure metamorphism in the Western Gneiss Region of Norway, *in* Carswell, D.A., and Compagnoni, R., eds., Notes in Mineralogy, Volume 5: Budapest, European Mineralogical Union, p. 51–73.

Carswell, D.A., Krogh, E.J., and Griffin, W.L., 1985, Norwegian orthopyroxene eclogites: Calculated equilibrium conditions and petrogenetic implications, *in* Gee, D.G., and Sturt, B.A., eds., The Caledonide Orogen—Scandinavia and Related Areas: London, John Wiley and Sons, p. 823–841.

Carswell, D.A., Tucker, R.D., O'Brien, P.J., and Krogh, T.E., 2003, Coesite micro-inclusions and the U/Pb age of zircons from the Hareidland eclogite in the Western Gneiss Region of Norway: Lithos, v. 67, p. 181–190, doi: 10.1016/S0024-4937(03)00014-8.

Chadwick, B., Friend, C.R.L., and Higgins, A.K., 1990, The crystalline rocks of western and southern Dove Bugt, North-East Greenland: Rapport Grønlands Geologiske Undersøgelse, v. 148, p. 127–132.

Cheeney, R.F., 1985, The Plutonic Igneous and High-Grade Metamorphic Rocks of Southern Liverpool Land, Central East Greenland, Part of a Supposed Caledonian and Precambrian Complex: Rapport Grønlands Geologiske Undersøgelse, v. 123, 39 p.

Chemenda, A.I., Mattauer, M., and Bokun, A.N., 1996, Continental subduction and a mechanism for exhumation of high-pressure metamorphic rocks: New modeling and field data from Oman: Earth and Planetary Science Letters, v. 143, p. 173–182, doi: 10.1016/0012-821X(96)00123-9.

Coe, K., and Cheeney, R.F., 1972, Preliminary results of mapping in Liverpool Land, East Greenland: Rapport Grønlands Geologiske Undersøgelse, v. 48, p. 7–20.

Corfu, F., and Andersen, T.B., 2002, U-Pb ages of the Dalsfjord complex, SW Norway, and their bearing on the correlation of allochthonous crystalline segments of the Scandinavian Caledonides: International Journal of Earth Sciences, v. 91, p. 955–963, doi: 10.1007/s00531-002-0298-3.

Dewey, J.F., and Strachan, R.A., 2003, Changing Silurian-Devonian plate motion in the Caledonides: Sinistral transpression to sinistral transtension: Journal of the Geological Society of London, v. 160, p. 219–229.

Dymoke, P., and Sandiford, M., 1992, Phase relations in Buchan facies series pelitic assemblages: Calculations with applications to andalusite–staurolite paragenesis in Mount Lofty Ranges, South Australia: Contributions to Mineralogy and Petrology, v. 110, p. 121–132, doi: 10.1007/BF00310886.

Ellis, D.J., and Green, D.H., 1979, An experimental study of the effect of Ca upon garnet-clinopyroxene Fe-Mg exchange equilibrium: Contributions to Mineralogy and Petrology, v. 71, p. 13–22, doi: 10.1007/BF00371878.

Elvevold, S., and Gilotti, J.A., 2000, Pressure-temperature evolution of retrogressed kyanite eclogites, Weinschenck Island, North-East Greenland Caledonides: Lithos, v. 53, p. 127–147, doi: 10.1016/S0024-4937(00)00014-1.

Elvevold, S., Thrane, K., and Gilotti, J.A., 2003, Metamorphic history of high-pressure granulites in Payer Land, Greenland Caledonides: Journal of Metamorphic Geology, v. 21, p. 49–63, doi: 10.1046/j.1525-1314.2003.00419.x.

Ernst, W.G., 2006, Preservation/exhumation of ultrahigh-pressure subduction complexes: Lithos, v. 92, p. 321–335, doi: 10.1016/j.lithos.2006.03.049.

Ernst, W.G., and Liou, J.G., 2000, Overview of UHP metamorphism and tectonics in well-studied collisional orogens, *in* Ernst, W.G., and Liou, J.G., eds., Ultrahigh-Pressure Metamorphism and Geodynamics in Collision-Type Orogenic Belts: Final Report of the Task Group III-6 of the International Lithosphere Project: Columbia, Maryland, Bellwether Publishing, p. 3–19.

Ernst, W.G., and Liu, J., 1998, Experimental phase-equilibrium study of Al- and Ti-contents of calcic amphibole in MORB—A semiquantitative thermobarometer: The American Mineralogist, v. 83, p. 952–969.

Ferry, J.M., and Spear, F.S., 1978, Experimental calibration of the partitioning of Fe and Mg between biotite and garnet: Contributions to Mineralogy and Petrology, v. 66, p. 113–117, doi: 10.1007/BF00372150.

Friderichsen, J.D., Gilotti, J.A., Henriksen, N., Higgins, A.K., Hull, J.M., Jepsen, H.F., and Kalsbeek, F., 1991, The crystalline rocks of Germania Land, Nordmarken and adjacent areas, North-East Greenland: Rapport Grønlands Geologiske Undersøgelse, v. 152, p. 85–94.

Friderichsen, J.D., Henriksen, N.H., and Strachan, R.A., 1994, Basement-cover relationships and regional structure in the Grandjean Fjord–Bessel Fjord region (75°–76°N), North-East Greenland: Rapport Grønlands Geologiske Undersøgelse, v. 162, p. 17–33.

Gee, D.G., 1975, A tectonic model for the central part of the Scandinavian Caledonides: American Journal of Science, v. 275A, p. 468–515.

Gilotti, J.A., 1993, Discovery of a medium-temperature eclogite province in the Caledonides of North-East Greenland: Geology, v. 21, p. 523–526, doi: 10.1130/0091-7613(1993)021<0523:DOAMTE>2.3.CO;2.

Gilotti, J.A., 1994, Eclogites and related high-pressure rocks from North-East Greenland: Rapport Grønlands Geologiske Undersøgelse, v. 162, p. 77–90.

Gilotti, J.A., and Elvevold, S., 1998, Partial eclogitization of the Ambolten gabbro-norite, North-East Greenland Caledonides: Schweizerische Mineralogische und Petrographische Mitteilungen, v. 78, p. 273–292.

Gilotti, J.A., and Elvevold, S., 2002, Extensional exhumation of a high-pressure granulite terrane in Payer Land, Greenland Caledonides: Structural, petrologic and geochronologic evidence from metapelites: Canadian Journal of Earth Sciences, v. 39, p. 1169–1187, doi: 10.1139/e02-019.

Gilotti, J.A., and McClelland, W.C., 2005, Leucogranites and the time of extension in the East Greenland Caledonides: The Journal of Geology, v. 113, p. 399–417, doi: 10.1086/430240.

Gilotti, J.A., and McClelland, W.C., 2007, Characteristics of, and a tectonic model for, ultrahigh-pressure metamorphism in the overriding plate of the Caledonian orogen: International Geology Review, v. 49 p. 777–797.

Gilotti, J.A., and McClelland, W.C., 2008, this volume, Geometry, kinematics, and timing of extensional faulting in the Greenland Caledonides—A synthesis, *in* Higgins, A.K., Gilotti, J.A., and Smith, M.P., eds., The Greenland Caledonides: Evolution of the Northeast Margin of Laurentia: Geological Society of America Memoir 202, doi: 10.1130/2008.1202(10).

Gilotti, J.A., and Ravna, E.J.K., 2002, First evidence for ultrahigh-pressure metamorphism in the North-East Greenland Caledonides: Geology,

v. 30, p. 551–554, doi: 10.1130/0091-7613(2002)030<0551:FEFUPM>2.0.CO;2.
Gilotti, J.A., Nutman, A.P., and Brueckner, H.K., 2004, Devonian to Carboniferous collision in the Greenland Caledonides: U-Pb zircon and Sm-Nd ages of high-pressure and ultrahigh-pressure metamorphism: Contributions to Mineralogy and Petrology, v. 148, p. 216–235, doi: 10.1007/s00410-004-0600-4.
Grenne, T., Ihlen, P.M., and Vokes, F.M., 1999, Scandinavian Caledonide metallogeny in a plate tectonic perspective: Mineralium Deposita, v. 34, p. 422–471, doi: 10.1007/s001260050215.
Haller, J., 1971, The Geology of the East Greenland Caledonides: New York, Interscience Publishers, 413 p.
Hallett, B., 2005, U-Pb SHIRMP zircon and titanite age constraints on deformation associated with the Storstrømmen shear zone, North-East Greenland Caledonides [M.S. thesis]: Moscow, University of Idaho, 58 p.
Hallett, B., McClelland, W.C., Gilotti, J.A., Power, S.E., and Tucker, R.D., 2005, Timing of displacement on the Storstrømmen shear zone, North-East Greenland Caledonides: Geological Association of Canada–Mineralogical Association of Canada Abstracts, v. 30, p. 78.
Hartz, E.H., Condon, D., Austrheim, H., and Erambert, M., 2005, Rediscovery of the Liverpool Land eclogites (Central East Greenland): A post- and supra-subduction UHP province: Mitteilungen der Österreichischen Mineralogischen Gesellschaft, v. 150, p. 50.
Henriksen, N., 1986, Geological Map of Greenland Sheet 12, Scoresby Sund: Copenhagen, Geological Survey of Greenland, scale 1:500,000.
Henriksen, N., 2003, Geological Map of Caledonian Orogen, East Greenland 70–82°N: Copenhagen, Geological Survey of Denmark and Greenland, scale 1:1,000,000.
Henriksen, N., and Higgins, A.K., 2008, this volume, Caledonian orogen of East Greenland 70°N–82°N: Geological map at 1:1,000,000 scale—Concepts and principles of compilation, in Higgins, A.K., Gilotti, J.A., and Smith, M.P., eds., The Greenland Caledonides: Evolution of the Northeast Margin of Laurentia: Geological Society of America Memoir 202, doi: 10.1130/2008.1202(14).
Hermann, J., and Green, D.H., 2001, Experimental constraints on high pressure melting in subducted crust: Earth and Planetary Science Letters, v. 188, p. 149–168, doi: 10.1016/S0012-821X(01)00321-1.
Higgins, A.K., 1974, The Krummedal Supracrustal Sequence around Inner Nordvestfjord, Scoresby Sund, East Greenland: Rapport Grønlands Geologiske Undersøgelse, v. 67, 34 p.
Higgins, A.K., 1982, Geological Map of Greenland, Charcot Land 71 Ø.4 Nord, Krummedal 71 Ø.4 Syd, Descriptive Text: Copenhagen, Geological Survey of Greenland, scale 1:100,000.
Higgins, A.K., 1988, The Krummedal supracrustal sequence in East Greenland, in Winchester, J.A., ed., Later Proterozoic Stratigraphy of the Northern Atlantic Regions: Glasgow, Blackie and Son, p. 86–96.
Higgins, A.K., 1995, Caledonides of East Greenland, in Williams, H., ed., Geology of the Appalachian-Caledonian Orogen in Canada and Greenland: Boulder, Colorado, Geological Society of America, Geology of North America, v. F-1, p. 891–921.
Higgins, A.K., and Leslie, A.G., 2000, Restoring thrusting in the East Greenland Caledonides: Geology, v. 28, p. 1019–1022, doi: 10.1130/0091-7613(2000)28<1019:RTITEG>2.0.CO;2.
Higgins, A.K., and Leslie, A.G., 2008, this volume, Architecture and evolution of the East Greenland Caledonides—An introduction, in Higgins, A.K., Gilotti, J.A., and Smith, M.P., eds., The Greenland Caledonides: Evolution of the Northeast Margin of Laurentia: Geological Society of America Memoir 202, doi: 10.1130/2008.1202(02).
Higgins, A.K., and Soper, N.J., 1993, Basement-cover relationships in the East Greenland Caledonides: Evidence from the Eleonore Bay Supergroup at Ardencaple Fjord: Transactions of the Royal Society of Edinburgh, v. 84, p. 105–115.
Higgins, A.K., Leslie, A.G., and Smith, M.P., 2001a, Neoproterozoic–Lower Palaeozoic stratigraphical relationships in the marginal thin-skinned thrust belt of the East Greenland Caledonides: Comparisons with the foreland in Scotland: Geological Magazine, v. 138, p. 143–160, doi: 10.1017/S0016756801005076.
Higgins, A.K., Smith, M.P., Soper, N.J., Leslie, A.G., Rasmussen, J.A., and Sønderholm, M., 2001b, The Neoproterozoic Hekla Sund Basin, eastern North Greenland: A pre-Iapetan extensional sequence thrust across its rift shoulders during the Caledonian orogeny: Journal of the Geological Society of London, v. 158, p. 487–499.

Higgins, A.K., Elvevold, S., Escher, J.C., Frederiksen, K.S., Gilotti, J.A., Henriksen, N., Jepsen, H.F., Jones, K.A., Kalsbeek, F., Kinny, P.D., Leslie, A.G., Smith, M.P., Thrane, K., and Watt, G.R., 2004a, The foreland-propagating thrust architecture of the East Greenland Caledonides 72°–75°N: Journal of the Geological Society of London, v. 161, p. 1009–1026, doi: 10.1144/0016-764903-141.
Higgins, A.K., Soper, N.J., Smith, M.P., and Rasmussen, J.A., 2004b, The Caledonian thin-skinned thrust belt of Kronprins Christian Land, eastern North Greenland: Geological Survey of Denmark and Greenland Bulletin, v. 6, p. 41–56.
Hodges, K.V., and Crowley, P.D., 1985, Error estimation and empirical geothermobarometry for pelitic systems: The American Mineralogist, v. 70, p. 702–709.
Holdsworth, R.E., and Strachan, R.A., 1991, Interlinked system of ductile strike slip and thrusting formed by Caledonian sinistral transpression in northeastern Greenland: Geology, v. 18, p. 423–439.
Holland, T.J.B., 1980, The reaction albite = jadeite + quartz determined experimentally in the range 600–1200 °C: The American Mineralogist, v. 65, p. 129–134.
Hoskin, P.W.O., and Schaltegger, U., 2003, The composition of zircon and igneous and metamorphic petrogenesis, in Hanchar, J.M., and Hoskin, P.W.O., eds., Zircon: Reviews in Mineralogy and Geochemistry, v. 53, p. 27–62.
Hossack, J.R., and Cooper, M.A., 1986, Collision tectonics in the Scandinavian Caledonides, in Coward, M.P., and Ries, A.C., eds., Collision Tectonics: Geological Society of London Special Publication 19, p. 287–304.
Hull, J.M., and Gilotti, J.A., 1994, The Germania Land deformation zone and related structures, North-East Greenland: Rapport Grønlands Geologiske Undersøgelse, v. 162, p. 113–127.
Hull, J.M., Friderichsen, J.D., Gilotti, J.A., Henriksen, N., Higgins, A.K., and Kalsbeek, F., 1994, Gneiss complex of the Skærfjorden region, 76°–78°N, North-East Greenland: Rapport Grønlands Geologiske Undersøgelse, v. 162, p. 35–51.
Hurst, J.M., and Surlyk, F., 1982, Stratigraphy of the Silurian Turbidite Sequence of North Greenland: Bulletin Grønlands Geologiske Undersøgelse, v. 145, 121 p.
Hurst, J.M., McKerrow, W.S., Soper, N.J., and Surlyck, F., 1983, The relationship between Caledonian nappe tectonics and Silurian turbidite deposition in North Greenland: Journal of the Geological Society of London, v. 140, p. 123–131, doi: 10.1144/gsjgs.140.1.0123.
Jamieson, R.A., Beaumont, C., Fullsack, P., and Lee, B., 1998, Barrovian regional metamorphism: Where's the heat?, in Treloar, P.J., and O'Brien, P.J., eds., What Drives Metamorphism and Metamorphic Reactions?: Geological Society of London Special Publication 138, p. 23–51.
Jones, K.A., and Escher, J.C., 2002, Near-isothermal decompression within a clockwise P-T evolution recorded in migmatitic mafic granulites from Clavering Ø, NE Greenland: Implications for the evolution of the Caledonides: Journal of Metamorphic Geology, v. 20, p. 365–378, doi: 10.1046/j.1525-1314.2002.00375.x.
Jones, K.A., and Strachan, R.A., 2000, Crustal thickening and ductile extension in the NE Greenland Caledonides: A metamorphic record from anatectic pelites: Journal of Metamorphic Geology, v. 18, p. 719–735, doi: 10.1046/j.1525-1314.2000.00282.x.
Kalsbeek, F., 1995, Geochemistry, tectonic setting, and poly-orogenic history of Palaeoproterozoic basement rocks from the Caledonian fold belt of North-East Greenland: Precambrian Research, v. 72, p. 301–315, doi: 10.1016/0301-9268(94)00097-B.
Kalsbeek, F., Nutman, A.P., and Taylor, P.N., 1993, Palaeoproterozoic basement province in the Caledonian fold belt of North-East Greenland: Precambrian Research, v. 63, p. 163–178, doi: 10.1016/0301-9268(93)90010-Y.
Kalsbeek, F., Nutman, A.P., Escher, J.C., Friderichsen, J.D., Hull, J.M., Jones, K.A., and Pedersen, S.A.S., 1999, Geochronology of granitic and supracrustal rocks from the northern part of the East Greenland Caledonides: Ion microprobe U-Pb zircon ages: Geology of Greenland Survey Bulletin, v. 184, p. 31–48.
Kalsbeek, F., Thrane, K., Nutman, A.P., and Jepsen, H.F., 2000, Late Mesoproterozoic to early Neoproterozoic history of the East Greenland Caledonides: Evidence for Grenville orogenesis?: Journal of the Geological Society of London, v. 157, p. 1215–1225.
Kalsbeek, F., Jepsen, H.F., and Nutman, A.P., 2001a, From source migmatites to plutons: Tracking the origin of the c. 435 Ma S-type granites in the East Greenland Caledonian orogen: Lithos, v. 57, p. 1–21, doi: 10.1016/S0024-4937(00)00071-2.

Kalsbeek, F., Jepsen, H.F., and Jones, K.A., 2001b, Geochemistry and petrogenesis of S-type granites in the East Greenland Caledonides: Lithos, v. 57, p. 91–109, doi: 10.1016/S0024-4937(01)00038-X.

Kalsbeek, F.K., Thrane, K., Nutman, A.P., and Jepsen, H.F., 2002, Late Mesoproterozoic to early Neoproterozoic history of the East Greenland Caledonides: Evidence for Grenvillian orogenesis?: Journal of the Geological Society of London, v. 157, p. 1215–1225.

Kalsbeek, F., Higgins, A.K., Jepsen, H.F., Frei, R., and Nutman, A.P., 2008, this volume, Granites and granites in the East Greenland Caledonides, in Higgins, A.K., Gilotti, J.A., and Smith, M.P., eds., The Greenland Caledonides: Evolution of the Northeast Margin of Laurentia: Geological Society of America Memoir 202, doi: 10.1130/2008.1202(09).

Koziol, A.M., and Newton, R.C., 1988, Redetermination of the anorthite breakdown reaction and improvement of the plagioclase-garnet-Al_2SiO_5-quartz barometer: The American Mineralogist, v. 73, p. 216–223.

Krank, E.H., 1935, On the Crystalline Complex of Liverpool Land: Meddelelser om Grønland, v. 95, no. 7, 122 p.

Krogh, E.J., 1982, Metamorphic evolution of Norwegian country-rock eclogites, as deduced from mineral inclusions and compositional zoning in garnets: Lithos, v. 15, p. 305–321, doi: 10.1016/0024-4937(82)90021-4.

Krogh, E.J., and Carswell, D.A., 1995, HP and UHP eclogites and garnet peridotites in the Scandinavian Caledonides, in Coleman, R.G., and Wang, X., eds., Ultrahigh Pressure Metamorphism: New York, Cambridge University Press, p. 244–298.

Lang, H.M., and Gilotti, J.A., 2001, Plagioclase replacement textures in partially eclogitised gabbros from the Sanddal mafic-ultramafic complex, Greenland Caledonides: Journal of Metamorphic Geology, v. 19, p. 497–515, doi: 10.1046/j.0263-4929.2001.00325.x.

Lang, H.M., and Gilotti, J.A., 2007, Partial melting of metapelites at ultrahigh-pressure conditions, Greenland Caledonides: Journal of Metamorphic Geology, v. 25, p. 129–147, doi: 10.1111/j.1525-1314.2006.00687.x.

Larsen, H.C., 1980, A high-pressure granulite facies complex in North-West Payers Land, East Greenland fold belt: Bulletin of the Geological Society of Denmark, v. 29, p. 161–174.

Le Breton, N., and Thompson, A.B., 1988, Fluid absent dehydration melting of biotite in metapelites in the early stages of crustal anatexis: Contributions to Mineralogy and Petrology, v. 99, p. 226–237, doi: 10.1007/BF00371463.

Leslie, A.G., and Higgins, A.K., 1999, On the Caledonian (and Grenvillian) geology of Bartholin Land, Ole Rømer Land, and adjacent nunataks, East Greenland: Danmarks og Grønlands Geologiske Undersøgelse Rapport, v. 1999/19, p. 11–26.

Leslie, A.G., and Higgins, A.K., 2008, this volume, Foreland-propagating Caledonian thrust systems in East Greenland, in Higgins, A.K., Gilotti, J.A., and Smith, M.P., eds., The Greenland Caledonides: Evolution of the Northeast Margin of Laurentia: Geological Society of America Memoir 202, doi: 10.1130/2008.1202(07).

Leslie, A.G., and Nutman, A.P., 2003, Evidence for Neoproterozoic orogenesis and early high temperature Scandian deformation events in the southern East Greenland Caledonides: Geological Magazine, v. 140, p. 309–333, doi: 10.1017/S0016756803007593.

Liou, J.G., Zhang, R.Y., Ernst, W.G., Rumble, D., III, and Maruyama, S., 1998, High-pressure minerals from deeply subducted metamorphic rocks, in Hemley, R.J., ed., Ultrahigh-Pressure Mineralogy: Physics and Chemistry of the Earth's Deep Interior: Reviews in Mineralogy, v. 37, p. 33–96.

Maruyama, S., Suzuki, K., and Liou, J.G., 1983, Greenschist-amphibolite transition equilibria at low pressures: Journal of Petrology, v. 24, p. 583–604.

McClelland, W.C., and Gilotti, J.A., 2003, Late-stage extensional exhumation of high-pressure granulites in the Greenland Caledonides: Geology, v. 31, p. 259–262, doi: 10.1130/0091-7613(2003)031<0259:LSEEOH>2.0.CO;2.

McClelland, W.C., and Gilotti, J.A., 2004, Limits on the timing of HP/UHP metamorphism in the NE Greenland eclogite province based on pegmatites in boudin necks, in 32nd International Geologic Congress Abstract Volume, Part 1: Florence, Italy, International Geologic Congress, abs. 63–3, p. 312.

McClelland, W.C., Power, S.E., Gilotti, J.A., Mazdab, F., and Wopenka, B., 2006, U-Pb SHRIMP geochronology and trace element geochemistry of coesite-bearing zircons, North-East Greenland Caledonides, in Hacker, B., McClelland, W.C., and Liou, J.G., eds., Ultrahigh-Pressure Metamorphism: Deep Continental Subduction: Geological Society of America Special Paper 403, p. 23–43.

McKerrow, W.S., Mac Niocaill, C., and Dewey, J.F., 2000, The Caledonian orogeny redefined: Journal of the Geological Society of London, v. 157, p. 1149–1154.

Nichols, G.T., Berry, R.F., and Green, D.H., 1992, Internally consistent gahnitic spinel-cordierite-garnet equilibria in the FMASH Zn system: Contributions to Mineralogy and Petrology, v. 111, p. 362–377, doi: 10.1007/BF00311197.

Oh, C.W., and Liou, J.G., 1998, A petrogenetic grid for the eclogite and related facies under high-pressure metamorphism: The Island Arc, v. 7, p. 36–51, doi: 10.1046/j.1440-1738.1998.00180.x.

Pedersen, S.A.S., Craig, L., Upton, B.G.J., Rämö, O.T., Jepsen, H.F., and Kalsbeek, F., 2002, Paleoproterozoic rift-related volcanism in the Hekla Sund region, eastern North Greenland: Field occurrence, geochemistry and tectonic setting: Precambrian Research, v. 114, p. 327–346, doi: 10.1016/S0301-9268(01)00234-0.

Peucat, J.J., Tisserant, D., Caby, R., and Clauer, N., 1985, Resistance of zircons to U-Pb resetting in a prograde metamorphic sequence of Caledonian age in East Greenland: Canadian Journal of Earth Sciences, v. 22, p. 330–338.

Rasmussen, J.A., and Smith, M.P., 2001, Conodont geothermometry and tectonic overburden in the northernmost East Greenland Caledonides: Geological Magazine, v. 138, p. 687–698.

Ravna, E.J.K., and Terry, M.P., 2004, Geothermobarometry of UHP and HP eclogites and schists—An evaluation of equilibria among garnet-clinopyroxene-kyanite-phengite-coesite/quartz: Journal of Metamorphic Geology, v. 22, p. 579–592, doi: 10.1111/j.1525-1314.2004.00534.x.

Reche, J., Martinez, F.J., and Arboleya, M.L., 1998, Low- to medium-pressure Variscan metamorphism in Galicia (NW Spain): Evolution of a kyanite-bearing synform and associated bounding antiformal domains, in Treloar, P.J., and O'Brien, P.J., eds., What Drives Metamorphism and Metamorphic Reactions?: Geological Society of London Special Publication 138, p. 61–79.

Rejebian, V.A., Harris, A.G., and Huebner, J.S., 1987, Conodont color and texture alteration: An index to regional metamorphism, contact metamorphism and hydrothermal alteration: Geological Society of America Bulletin, v. 99, p. 471–479, doi: 10.1130/0016-7606(1987)99<471:CCATAA>2.0.CO;2.

Rex, D.C., and Gledhill, A.R., 1981, Isotopic studies in the East Greenland Caledonides (72°–74°N)—Precambrian and Caledonian ages: Rapport Grønlands Geologiske Undersøgelse, v. 104, p. 47–72.

Root, D.B., Hacker, B.R., Mattinson, J.M., and Wooden, J.L., 2004, Zircon geochronology and c. 400 Ma exhumation of Norwegian ultrahigh-pressure rocks: An ion microprobe and chemical abrasion study: Earth and Planetary Science Letters, v. 228, p. 325–341, doi: 10.1016/j.epsl.2004.10.019.

Rubie, D.C., 1986, The catalysis of mineral reactions by water and restrictions on the presence of aqueous fluid during metamorphism: Mineralogical Magazine, v. 50, p. 399–415, doi: 10.1180/minmag.1986.050.357.05.

Ryan, P.D., 2001, The role of deep basement during continent-continent collision: A review, in Miller, J.A., Holdsworth, R.E., Buick, I.S., and Hand, M., eds., Continental Reactivation and Reworking: Geological Society of London Special Publication 184, p. 39–55.

Sahlstein, Th.G., 1935, Petrographie der Eklogiteinschlüsse in den Gneisen des Südwestlichen Liverpool-Landes in Ost-Grønland: Meddelelser om Grønland, v. 95, no. 5, 43 p.

Salje, E., 1986, Heat capacities and entropies of andalusite and sillimanite: Influence of fibrolitisation on the phase diagram of the Al_2SiO_5 polymorphs: The American Mineralogist, v. 7, p. 366–371.

Sartini-Rideout, C., Gilotti, J.A., and McClelland, W.C., 2006, Geology and timing of dextral strike-slip shear zones in Danmarkshavn, North-East Greenland Caledonides: Geological Magazine, v. 143, p. 431–446, doi: 10.1017/S0016756806001968.

Sartini-Rideout, C., Gilotti, J.A., and Foster, C.T., Jr., 2007, Forward modeling corona growth in a partially eclogitized leucogabbro, Bourbon Island, North-East Greenland: Lithos, v. 95, p. 279–297, doi: 10.1016/j.lithos.2006.08.001.

Searle, M.P., Simpson, R.L., Law, R.D., Parrish, R.R., and Waters, D.J., 2003, The structural geometry, metamorphic and magmatic evolution of the Everest Massif, High Himalaya of Nepal–South Tibet: Journal of the Geological Society of London, v. 160, p. 345–366.

Smith, D.C., and Cheeney, R.F., 1980, Oriented needles of quartz in clinopyroxene: Evidence of exsolution of SiO_2 from a non-stoichiometric supersilicic clinopyroxene, *in* 26th International Geological Congress: Paris, France, International Geological Congress, abstract, p. 125.

Smith, D.C., and Cheeney, R.F., 1981, A new occurrence of garnet-ultrabasite in the Caledonides: A Cr-rich chromite-garnet-lherzolite from Tværdalen, Liverpool Land: East Greenland: Terra Cognita, v. 1, no. 1, p. 74.

Smith, M.P., 1991, Early Ordovician Conodonts of East and North Greenland: Meddelelser om Grønland: Geoscience, v. 26, 81 p.

Smith, M.P., Higgins, A.K., Soper, N.J., and Sønderholm, M., 2004a, The Neoproterozoic Rivieradal Group of Kronprins Christian Land, eastern North Greenland: Geological Survey of Denmark and Greenland Bulletin, v. 6, p. 29–39.

Smith, M.P., Rasmussen, J.A., Robertson, S., Higgins, A.K., and Leslie, A.G., 2004b, Lower Palaeozoic stratigraphy of the East Greenland Caledonides: Geological Survey of Denmark and Greenland Bulletin, v. 6, p. 5–28.

Soper, N.J., and Higgins, A.K., 1993, Basement-cover relationships in the East Greenland Caledonides: Evidence from the Eleonore Bay Supergroup at Ardencaple Fjord: Transactions of the Royal Society of Edinburgh, Earth Sciences, v. 84, p. 103–115.

Soper, N.J., Strachan, R.A., Holdsworth, R.E., Gayer, R.A., and Greiling, R.O., 1992, Sinistral transpression and the Silurian closure of Iapetus: Journal of the Geological Society of London, v. 149, p. 871–880.

Spear, F.S., 1993, Metamorphic Phase Equilibria and Pressure-Temperature-Time Paths: Washington, D.C., Mineralogical Society of America Monograph I, 799 p.

Spear, F.S., and Cheney, J.T., 1989, A petrogenetic grid for pelitic schists in the system SiO_2-Al_2O_3-FeO-MgO-K_2O-H_2O: Contributions to Mineralogy and Petrology, v. 101, p. 149–164, doi: 10.1007/BF00375302.

Spear, F.S., Kohn, M.J., and Cheney, J.T., 1999, *P-T* paths from anatectic pelites: Contributions to Mineralogy and Petrology, v. 134, p. 17–32, doi: 10.1007/s004100050466.

Stephens, M.B., and Gee, D.G., 1989, Terranes and polyphase accretionary history in the Scandinavian Caledonides, *in* Dallmeyer, R.D., ed., Terranes of the Circum-Atlantic Paleozoic Orogens: Geological Society of America Special Paper 230, p. 17–30.

Strachan, R.A., and Tribe, I.R., 1994, Structure of the Storstrømmen shear zone, eastern Hertugen af Orléans Land: Rapport Grønlands Geologiske Undersøgelse, v. 162, p. 103–112.

Strachan, R.A., Holdsworth, R.E., Friderichsen, J.D., and Jepsen, H.F., 1992, Regional Caledonian structure within an oblique convergence zone, Dronning Louise Land, NE Greenland: Journal of the Geological Society of London, v. 149, p. 359–371, doi: 10.1144/gsjgs.149.3.0359.

Strachan, R.A., Nutman, A.P., and Friderichsen, J.D., 1995, SHRIMP U-Pb geochronology and metamorphic history of the Smallefjord sequence, NE Greenland Caledonides: Journal of the Geological Society of London, v. 152, p. 779–784, doi: 10.1144/gsjgs.152.5.0779.

Thrane, K., 2002, Relationships between Archaean and Palaeoproterozoic crystalline basement complexes in the southern part of the East Greenland Caledonides: An ion microprobe study: Precambrian Research, v. 113, p. 19–42, doi: 10.1016/S0301-9268(01)00198-X.

Tucker, R.D., Krogh, T.E., and Råheim, A., 1990, Proterozoic evolution and age-province boundaries in the central part of the Western Gneiss Region, Norway: Results of U-Pb dating of accessory minerals from Trondheimsfjord to Geiranger, *in* Gower, C.F., Rivers, T., and Ryan, B., eds., Mid-Proterozoic Geology of the Southern Margin of Proto-Laurentia-Baltica: Geological Association of Canada Special Paper 241, p. 33–50.

Watt, G., Kinny, P.D., and Friderichsen, J.D., 2000, U-Pb geochronology of Neoproterozoic and Caledonian tectonothermal events in the East Greenland Caledonides: Journal of the Geological Society of London, v. 157, p. 1031–1048.

White, A.P., and Hodges, K.V., 2003, Pressure-temperature-time evolution of the Central East Greenland Caledonides: Quantitative constraints on crustal thickening and synorogenic extension: Journal of Metamorphic Geology, v. 21, p. 875–897.

White, A.P., Hodges, K.V., Martin, M.W., and Andresen, A., 2002, Geologic constraints on middle-crustal behavior during broadly synorogenic extension in the central East Greenland Caledonides: International Journal of Earth Sciences, v. 91, p. 187–208, doi: 10.1007/s005310100227.

Wittlinger, G., Vergne, J., Tapponnier, P., Farra, V., Poupinet, P., Jiang, M., Su, H., Herquel, G., and Paul, A., 2004, Teleseismic imaging of subducting lithosphere and Moho offsets beneath western Tibet: Earth and Planetary Science Letters, v. 221, p. 117–130, doi: 10.1016/S0012-821X (03)00723-4.

Wyllie, P.J., 1957, A Geological Reconnaissance through South Germania Land, North-East Greenland, Lat. 77°N, Long. 18 1/2°W to 22°W: Meddelelser om Grønland v. 157, 66 p.

Zhang, R.Y., and Liou, J.G., 1997, Partial transformation of gabbro to coesite-bearing eclogite from Yankou, the Sulu terrane, eastern China: Journal of Metamorphic Geology, v. 15, p. 183–202, doi: 10.1111/j.1525-1314.1997.00012.x.

MANUSCRIPT ACCEPTED BY THE SOCIETY 14 JANUARY 2008

Granites and granites in the East Greenland Caledonides

Feiko Kalsbeek*
A.K. Higgins
Hans F. Jepsen
Geological Survey of Denmark and Greenland, Øster Voldgade 10, DK-1350 Copenhagen K, Denmark

Robert Frei
Geological Institute, University of Copenhagen, Øster Voldgade 10, DK-1350 Copenhagen K, Denmark

Allen P. Nutman[†]
Research School of Earth Sciences, Australian National University, Canberra, ACT 0200, Australia

ABSTRACT

Caledonian (435–425 Ma) and "Grenvillian" (950–900 Ma) S-type leucogranites and augen gneisses are prominent in the thrust units that form the southern half of the East Greenland Caledonian orogen, south of 76°N. Such rocks do not occur further north (76°N–81°N), where the bedrock is dominated by Paleoproterozoic orthogneisses and metagranitoid rocks (2000–1750 Ma). More mafic Caledonian granitoid rocks (quartz diorites, granodiorites, quartz monzonites, syenites, etc.) are found only in the southernmost parts of the orogen (~71°N), side by side with S-type leucogranites. The S-type granites were formed by partial fusion of "fertile" lithologies within the late Mesoproterozoic Krummedal supracrustal sequence prior to or during emplacement of the thrust units and subsequent collapse of the orogen. The lack of similar granites north of 76°N is probably related to the absence of major units of metasedimentary rocks in that area. Among the granitoid rocks in the southernmost area, an early quartz-dioritic to granodioritic intrusion was dated at 466 ± 9 Ma; this is ~35 m.y. older than most Caledonian S-type granites. Quartz-monzonitic, granitic, and syenitic intrusions have yielded ages of 444–432 Ma. These rocks are geochemically similar to Caledonian granites in Scotland and may be related to subduction of oceanic lithosphere underneath East Greenland.

The north-south variation in the occurrence of granites in the East Greenland Caledonides is the expression of an original (pre-thrusting) west-east zonation. It is envisaged that the orogen consists of a number of parallel belts, now telescoped by thrusting: a southeastern belt containing supracrustal rocks (Krummedal sequence) with leucogranites, with more mafic granitoids in the southeast, and a northwestern belt where these rocks do not occur. These belts are envisaged to run from Scotland

*fkalsbeek@gmail.com
[†]Present address: Beijing SHRIMP Center, Institute of Geology, Chinese Academy of Geological Sciences, 26 Baiwangzhuang Road, Beijing 100037, China.

Kalsbeek, F., Higgins, A.K., Jepsen, H.F., Frei, R., and Nutman, A.P., 2008, Granites and granites in the East Greenland Caledonides, *in* Higgins, A.K., Gilotti, J.A., and Smith, M.P., eds., The Greenland Caledonides: Evolution of the Northeast Margin of Laurentia: Geological Society of America Memoir 202, p. 227–249, doi: 10.1130/2008.1202(09). For permission to copy, contact editing@geosociety.org. ©2008 The Geological Society of America. All rights reserved.

over the southern parts of the East Greenland Caledonides and, obliquely to the Greenland coast, over the North-East Greenland shelf to Svalbard and Norway, where similar rock units also occur.

Keywords: Caledonian orogen, East Greenland, S-type granites, I-type granites, geochemistry, isotopes, ages.

INTRODUCTION

Granitic rocks are common in the thrust sheets that build up the East Greenland Caledonide orogen (Haller, 1971; Higgins et al., 2004; Fig. 1). Their distribution is uneven and shows a clear zonation from north to south. Three distinct regions can be differentiated: (1) In the northern part of the orogen, an area of ~500 km along strike north of 76°N, all granites have yielded Paleoproterozoic ages, and Caledonian granites appear to be absent (Kalsbeek et al., 1993, 1999). (2) South of 76°N, Caledonian (435–425 Ma) and "Grenvillian" (950–900 Ma; Kalsbeek et al., 2000, this volume) granites are common. Both occur in close association with metasedimentary rocks of the late Mesoproterozoic Krummedal supracrustal sequence (see Kalsbeek et al., this volume). Most of these are S-type leucogranites, formed by anatexis of schists and paragneisses of the Krummedal sequence. Paleoproterozoic and Archean rocks that form most of the crystalline basement underlying the Krummedal sequence seem not to have participated in the generation of these granites. (3) In the southernmost part of the orogen, 70°N–72°N, more mafic granitoid rocks (granodiorites, quartz diorites, mafic monzonites, etc.) occur side by side with S-type granites; similar mafic granitoid rocks have not been encountered further north.

Tectonostratigraphic studies in the Kong Oscar Fjord region (72°N–74°N) have demonstrated that two major thrust sheets overlie the Caledonian foreland, a lower Niggli Spids thrust sheet and an upper Hagar Bjerg thrust sheet, both of which consist of crystalline basement rocks unconformably overlain by the Krummedal sequence (Figs. 2 and 3; Higgins et al., 2004; Leslie and Higgins, this volume). All Caledonian and Grenvillian granites occur in the structurally higher Hagar Bjerg thrust sheet, whereas they are absent from the Niggli Spids thrust sheet and the structurally underlying foreland. Granites are also present in the lower part of the Neoproterozoic Eleonore Bay Supergroup, which lies in the upper part of the Hagar Bjerg thrust sheet and is separated from the underlying Krummedal sequence by major extensional faults (Figs. 2 and 3; Gilotti and McClelland, this volume).

The description of the S-type granites from the central region (72°N–76°N) that follows is a summary of already published data. The more mafic granitoid rocks from south of 72°N, for which little information has been published previously, are treated more comprehensively. Archean and Paleoproterozoic granites are described by Kalsbeek et al. (this volume) and are not considered here.

CALEDONIAN AND "GRENVILLIAN" GRANITES, 72°N–76°N

Early dating of granites in the southern part of the Caledonian orogen demonstrated that two age groups are present. Most samples yielded Caledonian ages, mainly 435–425 Ma (see below), but a minority proved to be around 1000 Ma (e.g., Steiger et al., 1979; Rex and Gledhill, 1981). These Grenvillian granites were later dated at 950–900 Ma (Kalsbeek et al., 2000; Watt et al., 2000; Leslie and Nutman, 2003). Granites of the two age groups are lithologically similar and cannot be differentiated in the field.

Most Grenvillian and Caledonian granites are muscovite-biotite leucogranites, which occur as leucosomes within migmatized metasedimentary rocks of the Krummedal sequence as well as in sheets of varying width, up to hundreds of meters across (Fig. 4). Leucogranites also occur as plutonic bodies emplaced into the lower part of the Neoproterozoic Eleonore Bay Supergroup. These commonly contain large angular blocks of country rock (Fig. 4C). There is no evidence of significant contact metamorphism.

More mafic, biotite-rich, granodioritic rocks are also present. Grenvillian granodioritic augen gneisses form major sheets, hundreds of meters across. Lithologically similar Caledonian granites are also present, but they do not occur as such large sheets.

Caledonian and Grenvillian granites can only be distinguished by radiometric age determination, but for granites intruded into the lower part of the Neoproterozoic Eleonore Bay Supergroup, a Caledonian age is certain for stratigraphic reasons. For samples with sufficiently high Rb/Sr ratios (>~1.5), Sr-isotope analysis alone can differentiate between Caledonian and Grenvillian granites (Fig. 5A; Kalsbeek et al., 2001a).

Petrogenesis

Since the age of many of the investigated samples is not known, and dated Caledonian and Grenvillian granites are lithologically, petrographically, and chemically similar, they are treated together in the following.

There is convincing evidence that the leucogranites were formed by anatexis of "fertile" lithologies within the Krummedal sequence (Kalsbeek et al., 2001a). First, in the field, all transitions can be observed from leucosomes within migmatized metasedimentary rocks to granite sheets with various proportions of biotite-rich schlieren, and then to homo-

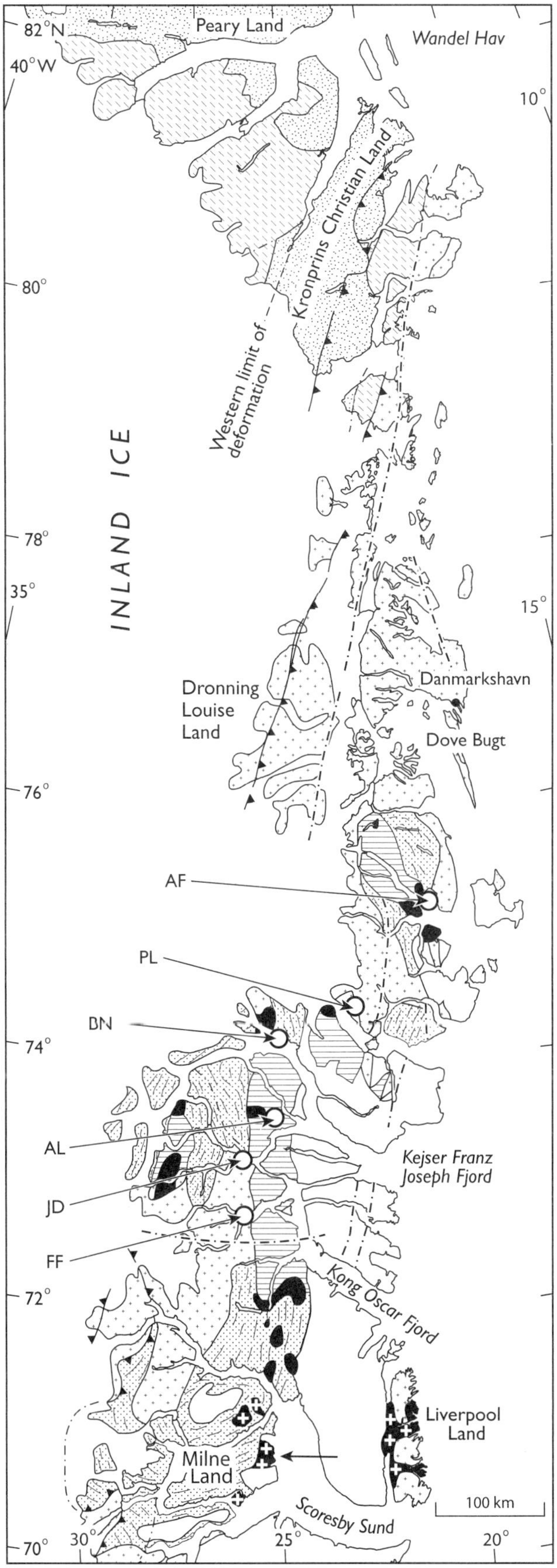

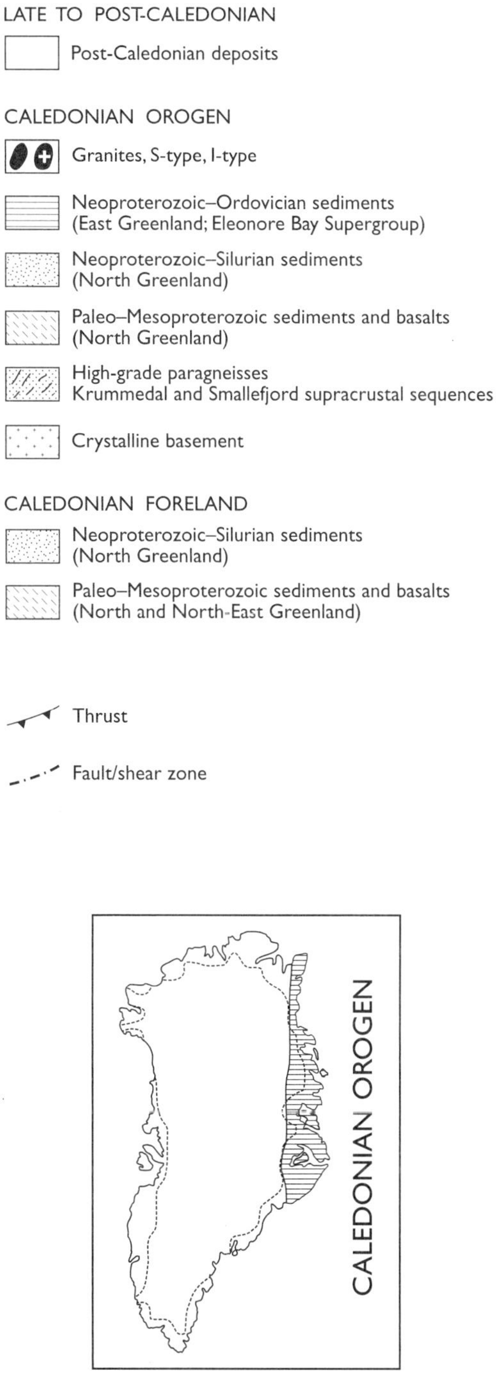

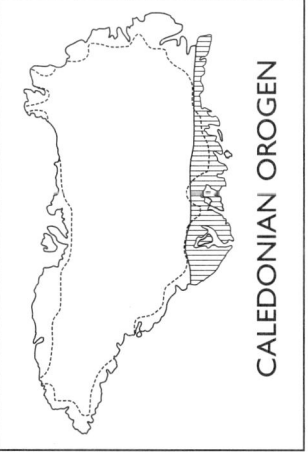

Figure 1. Geological sketch map of the East Greenland Caledonides (adapted from Henriksen, 2003) showing the distribution of Caledonian granites in the southern half of the orogen. Only the largest granite bodies are shown; a distinction is made between true granites (70°N–76°N) and more mafic granodiorites, monzonites, etc., between 70°N and 72°N. AF—Ardencaple Fjord; AL—Andrée Land; BN—Bartholin Nunatak; FF—Forsblad Fjord; JD—Junction Dal; PL—Payer Land. The arrow (~71°N) points to the East Milne Land plutonic complex. For the distribution of the different thrust sheets, see Figure 2.

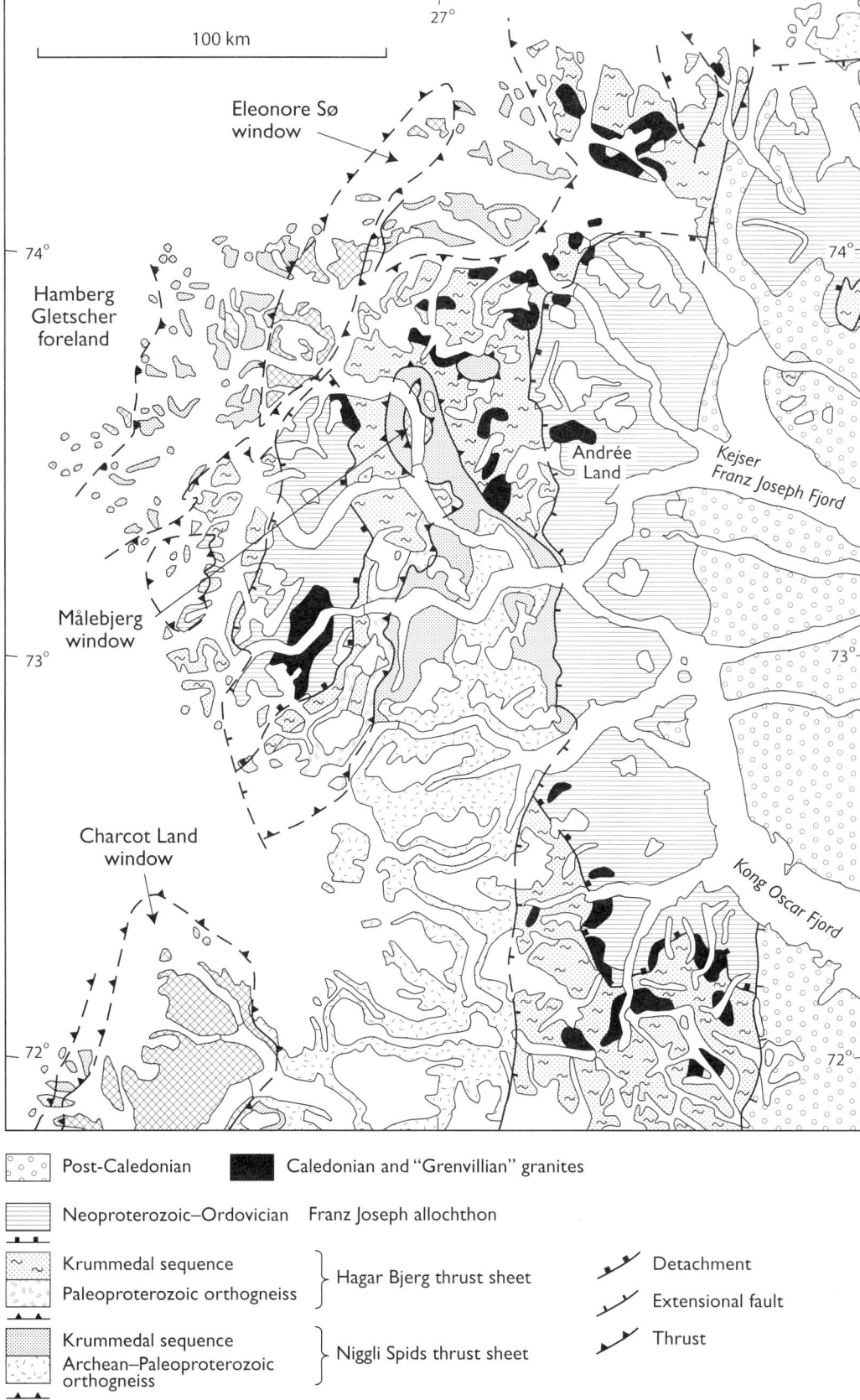

Figure 2. Simplified geological map of the Kong Oscar Fjord region, East Greenland Caledonides (adapted from Henriksen, 2003), showing the disposition of the two major thrust units. Caledonian and "Grenvillian" granites are only present in the upper Hagar Bjerg thrust sheet (see Fig. 3) and often occur near the base of the Eleonore Bay Supergroup.

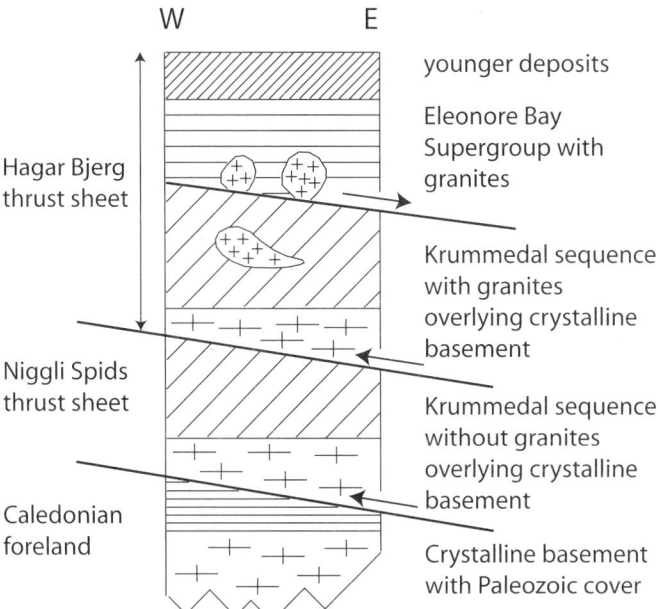

Figure 3. Schematic tectonostratigraphy of the Kong Oscar Fjord region (72°N–74°N). The Caledonian foreland is overlain by two major thrust units, the Niggli Spids thrust sheet and the Hagar Bjerg thrust sheet, of which only the latter contains granites. These granites were formed by anatexis of metasedimentary rocks of the late Mesoproterozoic Krummedal supracrustal sequence. The upper part of the Hagar Bjerg thrust sheet consists of the Neoproterozoic Eleonore Bay Supergroup and younger deposits, which have been juxtaposed with the high-grade Krummedal sequence along major extensional faults. Granites are prominent in the lower part of the Eleonore Bay Supergroup, close to the tectonic contact with the Krummedal sequence.

geneous leucogranites (Figs. 4 and 6). Second, despite a careful search, no similar granites have been found in the crystalline basement underlying the Krummedal sequence. Third, zircons separated from the leucogranites commonly contain inherited cores (Fig. 5B), the age distribution of which mimics that of detrital zircons from samples of the Krummedal sequence, and which is dissimilar to that of zircons from the crystalline basement (Fig. 5C).

Furthermore, Rb-Sr isotopic data (Fig. 5A) show that, at the time of their emplacement, both Grenvillian and Caledonian granites had $^{87}Sr/^{86}Sr$ ratios similar to those of Krummedal lithologies, consistent with a local origin of the granites. Chemical compositions of the leucogranites (see below) are also consistent with an origin of the leucogranites by anatexis of metasedimentary rocks of the Krummedal sequence. The biotite-rich granodiorites are chemically similar to Krummedal metasediments (Table 1), and it is envisaged that they represent partially melted pelitic rocks of the Krummedal sequence with large proportions of inherited biotite. A typical clockwise pressure-temperature (P-T) path for metamorphism and anatexis of metasedimentary rocks of the Krummedal sequence is given by Gilotti et al. (this volume).

Figure 4. Field occurrence of the leucogranites, showing the variation from (A) granitic veins in migmatized Krummedal metasediments, to (B) large granite sheets, up to tens of meters wide, and to (C) plutonic granite bodies within sediments of the Eleonore Bay Supergroup. The mountain in B is ~800 m high above the screes; the section in C is ~600 m. Note rotated blocks of country rock in C. Photographs are from Kalsbeek et al. (2001a, 2001b); used with permission from Elsevier.

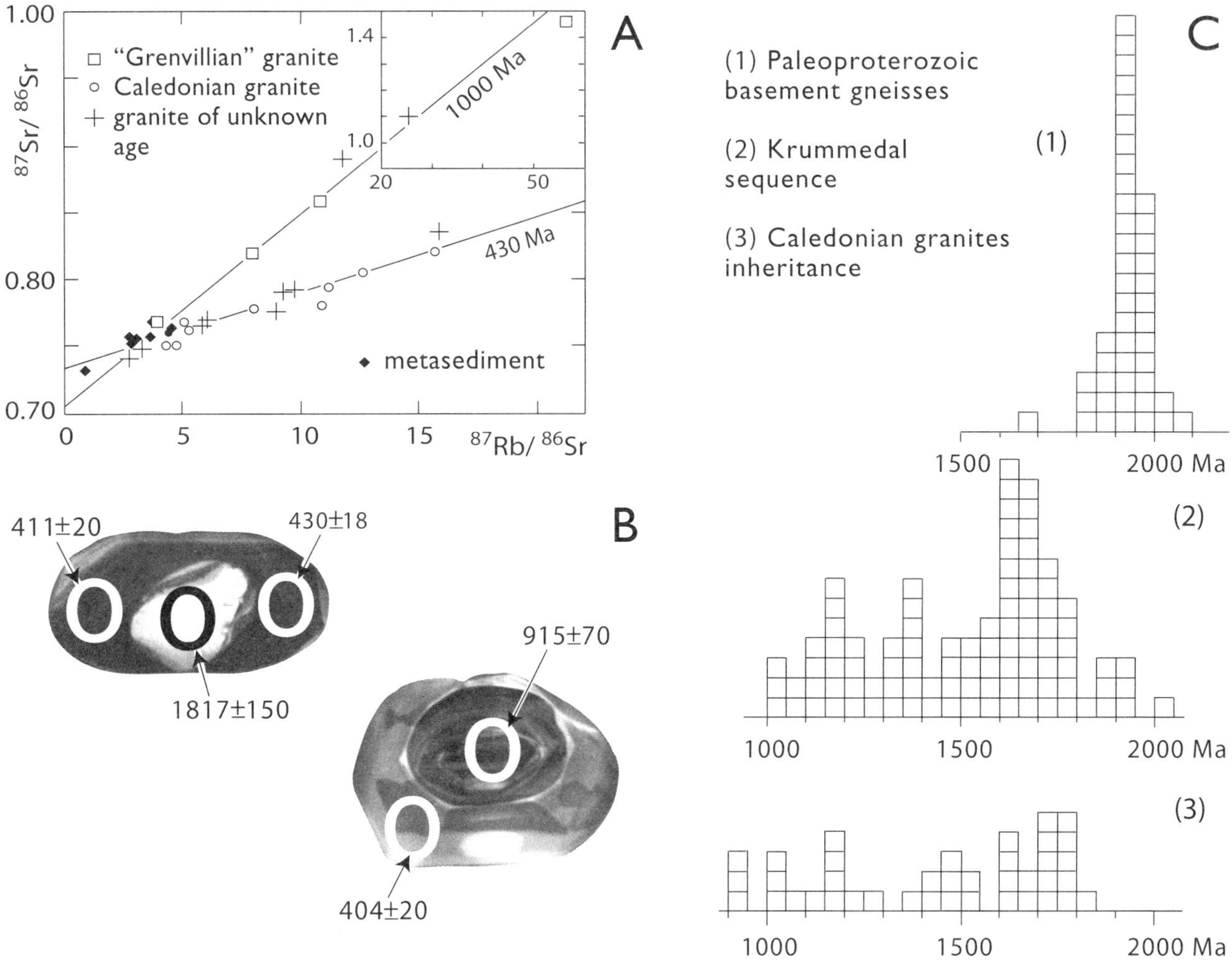

Figure 5. Age and petrogenesis of the leucogranites. (A) Rb-Sr isochron diagram showing the distinction between Caledonian and "Grenvillian" granites. (B) Typical Caledonian zircons with inherited cores. Ages in Ma. (C) Age distribution of inherited zircon (SHRIMP U-Pb data) in Caledonian granites (3) compared to that of detrital zircons in the Krummedal sequence (2) and in Paleoproterozoic basement gneisses (1). Each square represents one zircon age determination. Graphs in C were modified from similar diagrams in Kalsbeek et al. (2001a, 2001b), using additional data from Thrane (2002), Watt et al. (2000), and Leslie and Nutman (2003).

Figure 6. Photograph illustrating the inhomogeneous distribution of biotite as schlieren and patches in the leucogranites; the granite shown is of Grenvillian age. Photograph is reprinted from Kalsbeek et al. (2001b) with permission from Elsevier. Length of hammer is ~45 cm.

TABLE 1. REPRESENTATIVE ANALYSES OF SAMPLES FROM THE KRUMMEDAL SUPRACRUSTAL SEQUENCE AND S-TYPE GRANITES FROM THE EAST GREENLAND CALEDONIDES

Major/trace elements	Sample no.								
	427431	427463	Mean (7)	427410†	427491	427495	Mean (35)	427414†	427493
	Krummedal metasediments			Leucogranites				Dark granites	
SiO_2 (wt%)	69.43	61.81	65.6	76.59	72.37	73.92	73.1	67.36	65.08
TiO_2	0.83	1.09	0.9	0.10	0.27	0.24	0.2	0.77	1.07
Al_2O_3	13.44	18.86	15.8	12.99	14.70	14.25	14.8	15.02	15.69
Fe_2O_3	0.49	0.89	1.0	0.28	0.00	0.29	0.3	1.21	0.50
FeO	4.66	5.06	5.1	0.42	0.87	0.86	0.8	4.20	5.33
MnO	0.04	0.07	0.1	0.02	0.01	0.02	0.0	0.07	0.11
MgO	1.67	1.66	2.3	0.21	0.38	0.26	0.3	1.32	2.00
CaO	0.70	0.61	1.4	0.71	0.71	0.78	1.0	1.31	1.27
Na_2O	3.86	1.96	2.5	3.00	3.16	3.35	3.7	2.69	3.62
K_2O	3.04	5.00	3.5	4.89	7.00	5.42	4.9	4.69	3.07
P_2O_5	0.09	0.07	0.1	0.21	0.18	0.18	0.2	0.17	0.14
Volat.	1.39	2.66	1.7	0.58	0.50	0.68	0.7	1.28	1.95
Sum	99.64	99.76	100.0	99.98	100.14	100.25	99.8	100.09	99.82
A/CNK	1.23	1.94	1.51	1.13	1.04	1.11	1.13	1.26	1.35
Cs* (ppm)	7.8	18.5	10	2.0	6.1	10.5		17.5	6.1
Rb	132	209	153	130	239	359	242	260	144
Ba	527	633	626	255	1020	304	405	458	554
Pb	14	26	22	49	42	38	42	26	20
Sr	104	135	164	97	138	97	140	72	110
Y	29	39	31	17	4	16	15	81	45
Th*	10.7	17.4	12	1.3	4.3	12.0	9	19.6	11.9
Zr	333	284	241	31	34	90	84	319	275
Hf*	8.7	6.9	7	1.2	1.3	2.2		9.8	7.6
Nb	17	21	15	4	5	14	8	15	16
Ni	19	38	42	4	6	4	5	18	30
Sc	8	21	17	2	2	4	2	16	19
V	54	122	99	5	10	9	8	62	113
Cr	30	90	97	3	<3	4	6	43	72
Ga	15	23	20	14	16	25	22	25	22
La* (ppm)	53.5	60.9	46	5.4	13.1	30.9		62.8	50.9
Ce*	99.4	121	95	12	30	70		147	118
Pr**	9.93	13.1	10	1.26	2.84	6.97		14.8	14.7
Nd*	37.5	53.6	38	6	11	25		69	50
Sm*	5.82	9.13	7	1.78	2.14	4.25		13.4	9.27
Eu*	1.56	1.78	1.5	0.57	1.51	0.57		1.74	1.56
Gd**	5.02	7.57	6	2.22	1.31	2.54		12.4	9.16
Tb**	0.83	1.20	1.0	0.4	0.2	0.5		2.3	1.6
Dy**	4.91	6.62	6	2.85	0.75	2.48		13.4	8.76
Ho**	1.01	1.30	1.1	0.48	0.11	0.42		2.80	1.67
Er**	3.15	3.87	3.3	1.21	0.30	1.24		8.13	4.89
Tm**	0.45	0.54	0.5	0.17	0.04	0.20		1.30	0.78
Yb*	3.17	3.56	3.7	1.16	0.34	1.09		9.07	5.17
Lu*	0.48	0.54	0.6	0.16	0.05	0.16		1.42	0.78

Note: Sample numbers are according to the files of the Geological Survey of Denmark and Greenland. Major elements were analyzed at the Geological Survey of Denmark and Greenland by X-ray fluorescence spectrometry (XRF) on glass disks, Na_2O by atomic absorption spectrometry, and FeO by titration (Kystol and Larsen, 1999). Unmarked trace elements were analyzed by XRF on powder tablets at the Geological Institute, University of Copenhagen. Elements marked * were analyzed by instrumental neutron activation analysis, and elements marked ** were analyzed by inductively coupled plasma–mass spectrometry at Activation Laboratories Ltd., Ontario, Canada.
†427410 and 427414 are "Grenvillian" granites.

Geochemistry

A detailed description of the geochemistry of the granites has been presented by Kalsbeek et al. (2001b). In many leucogranites, biotite is patchily distributed (Fig. 6), and it is likely that at least some, and in some cases much, of the biotite was inherited from the source metasediments. Chemical analyses therefore do not necessarily represent the composition of a melt, which makes geochemical modeling difficult.

Major-element compositions of the leucogranites (Table 1) are similar to those of felsic S-type granites from other areas (e.g., Wickham, 1987; Inger and Harris, 1993; Chappell, 1999). In the normative Q-Ab-Or (quartz, albite, orthoclase) diagram (Fig. 7A), most leucogranites fall in the low-temperature

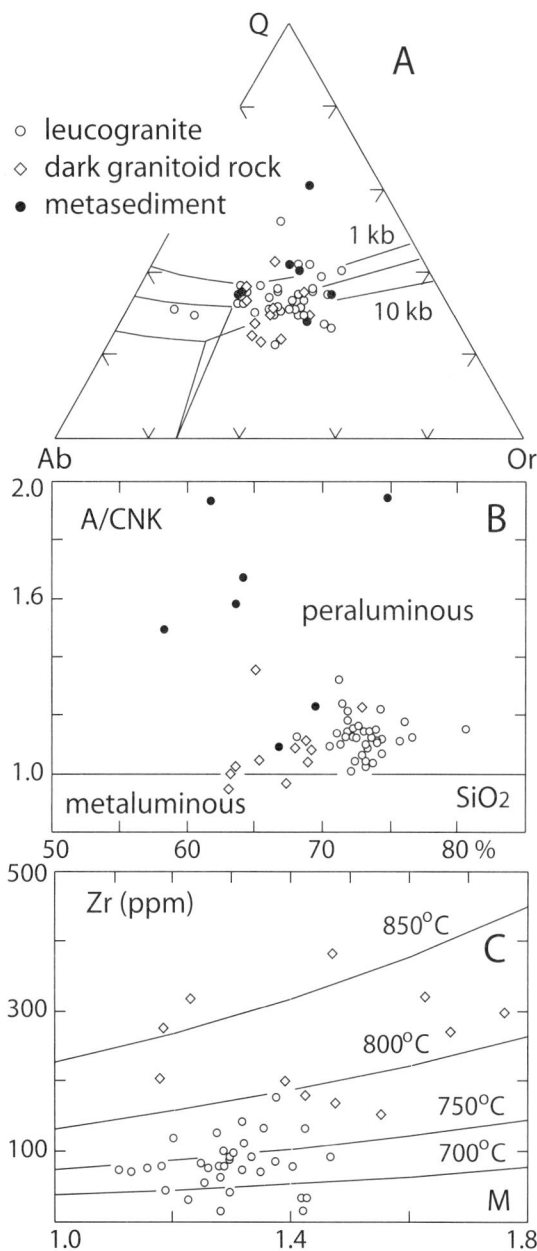

Figure 7. Chemical composition of granites from the Kong Oscar Fjord region (72°N–74°N). (A) Granite compositions plotted in the CIPW-normative Q-Ab-Or diagram with low-temperature melt compositions at 1, 5, and 10 kbar (Johannes and Holtz, 1996). (B) Peraluminous nature of the granites. A/CNK = molecular $Al_2O_3/(CaO + Na_2O + K_2O)$. (C) Zr concentrations plotted versus M (M = the cation ratio $[Na + K + 2Ca]/[Al \times Si]$) (Watson and Harrison, 1983). The lines marked 800 °C, etc., show the proportions of Zr that can be dissolved in granitic melts of different compositions at different temperatures (compositions expressed by the parameter M). Low concentrations of Zr in the leucogranites suggest low temperatures of formation. Samples shown as "dark granitoid rocks" represent the biotite-rich granodiorites, etc., mentioned in the text. Plots are modified from Kalsbeek et al. (2001a); used with permission.

field, between 1 and 10 kbar, and somewhat to the right of the cotectic-eutectic minima of the H_2O-saturated haplogranite system, as is commonly the case for natural low-temperature granites (Johannes and Holtz, 1996). Like other well-documented S-type granites, all leucogranites are peraluminous (Fig. 7B).

Figure 7C shows the concentrations of Zr in the granites compared to the proportions of Zr that can be dissolved in granitoid melts of different compositions at variable temperatures (Watson and Harrison, 1983). Granite compositions are expressed by the parameter M, the cation ratio $(Na + K + 2Ca)/(Al \times Si)$. The leucogranites have low Zr concentrations, in most cases <100 ppm. Since many zircons have inherited cores (Fig. 5B), Zr concentrations in the melt must have been even lower than those analyzed in the rock. These low Zr concentrations suggest that the leucogranites were formed at relatively low temperatures.

On the basis of trace-element data, Kalsbeek et al. (2001b) suggested that melting of fertile lithologies within the Krummedal sequence was aided by the presence of externally derived H_2O-rich fluids. Such fluids could have been transported along shear zones and fractures during tectonic activity. If this interpretation is correct, fluids would not have been dispersed throughout the rock, which would explain how enough "fertile" material was left in the Krummedal sequence after formation of "Grenvillian" granites at ca. 950 Ma for Caledonian granites to be formed, and how, after two periods of anatexis, fertile (muscovite-rich) metasedimentary rocks are still present.

Age Determinations

In a number of recent publications, attempts have been made to elucidate the structural evolution of the Caledonian orogen with the help of age determinations on leucogranites that have well-defined relationships to contractional and extensional structures (e.g., Hartz et al., 2000, 2001; Strachan et al., 2001; White et al., 2002; White and Hodges, 2002; Andresen et al., 2004; Gilotti and McClelland, 2005). In order to be successful in this respect, the granites must be dated precisely, which is very difficult.

Early Rb-Sr whole-rock isochron dating by Rex and Gledhill (1981) yielded ages between 560 and 377 Ma and should, according to these authors, be viewed with caution. Similarly, U-Pb dating of bulk zircon concentrates (e.g., Hansen et al., 1994) did not yield reliable ages because of the presence of zircon of variable ages in the concentrates (younger zircon overgrowing older cores; Fig. 5B). Although these early age determinations served to establish the Caledonian age of the granites, they lacked precision and are not further considered below.

Newer age determinations on the leucogranites have been carried out by U-Pb dating on selected grains of zircon, monazite, and xenotime. Two analytical methods have been employed, ID-TIMS (classical isotope dilution–thermal ionization mass spectrometry) and analysis by SHRIMP (sensitive high-resolution ion microprobe). Each of these methods has its inherent strengths and weaknesses.

ID-TIMS analyses can have a precision of ~1–2 m.y., but, since whole zircon or monazite crystals are analyzed, it is difficult entirely to avoid inherited or disturbed parts of the crystals if the latter are not visible by microscopic inspection. Although individual analyses commonly have a high precision, it may be difficult to judge which of the analyses should be used to define the age of emplacement of the rock. This is especially the case for monazite (e.g., Hartz et al., 2000, 2001; White et al., 2002).

Using SHRIMP, it is possible to select 20–50-μm-sized spots within individual zircon or monazite crystals so that cores and rims can be dated separately, thereby largely avoiding the problems that hamper ID-TIMS dating. This process is greatly aided by cathodoluminescence (CL) imaging, which can not only distinguish between cores and rims, but also identify recrystallized and metamict domains within individual zircons. However, SHRIMP dating is in general less precise, with precisions typically on the order of 5–10 m.y. Moreover, in dating these relatively young rocks by SHRIMP, $^{206}Pb/^{238}U$ ratios are used for the age calculation, which necessitates calibration of the $^{206}Pb/^{238}U$ ratios with the help of a standard. This may cause a systematic error of 1.5%–2% (e.g., Tucker and McKerrow, 1995).

Figure 8 and Table 2 present an overview of all published modern age determinations on Caledonian leucogranites from East Greenland north of 72°N. Ages older than 435 Ma, obtained by Kalsbeek et al. (2001a) by SHRIMP at the Australian National University, are probably the result of a slight calibration error. Two ages reported in that paper, 440 ± 8 Ma for a biotite-rich tonalitic sheet and 438 ± 5 Ma for metamorphic zircons in a metagabbroic rock, were checked by new SHRIMP analyses at the J.C. Roddick Ion Microprobe Laboratory in Ottawa by M.A. Hamilton (2000, personal commun., see Kalsbeek et al., 2001a), and they yielded ages of 436 ± 5 and 427 ± 4 Ma, respectively. The youngest zircon U-Pb ages (420 ± 8 Ma and 413 ± 3 Ma; Table 2) were also obtained by SHRIMP (Watt et al., 2000; Gilotti and McClelland, 2005).

Monazite ID-TIMS data have in a number of cases yielded a variety of ages between ca. 420 and 430 Ma for individual crystals (e.g., Hartz et al., 2000, 2001; Strachan et al., 2001; White et al., 2002). In such cases, interpretation of the data in terms of granite formation is not straightforward. Usually a selection of analyses (in some cases, one single analysis) is used to define the best estimate of the time of granite formation, where older grains are interpreted as inheritance and younger ages are interpreted as the result of later disturbance. All monazite ages younger than 425 Ma (and a few higher ages) were acquired in this manner, and, since this approach is not entirely objective, ages <425 Ma must be regarded with caution; in most cases, alternative interpretations of the data are possible. In some cases, monazite and xenotime data have yielded upper-intercept ages from well-fitted discordia plots; these age determinations may be more reliable.

Arguably, the most reliable ages obtained are from ID-TIMS zircon data (Strachan et al., 2001; Andresen et al., 2004). Most of these analyses have yielded well-fitted discordia plots that give ages between 432 and 425 Ma. A selection of analyses was used here as well, but in most cases, rejected analyses yield ages older than 1000 Ma and undoubtedly represent inherited zircon.

Hartz et al. (2000, 2001) and White and Hodges (2002) presented $^{40}Ar/^{39}Ar$ muscovite dates for a number of granites. With few exceptions, these are much younger than the age of the granites determined with the help of U-Pb analysis of zircon, monazite, and xenotime, and therefore they do not date emplacement of the granites.

Granite Formation and the Structural Evolution of the Orogen

The present distribution of rock units in the East Greenland Caledonian orogen is the result of two main structural events: (1) Large-scale west and northwestward transport of major

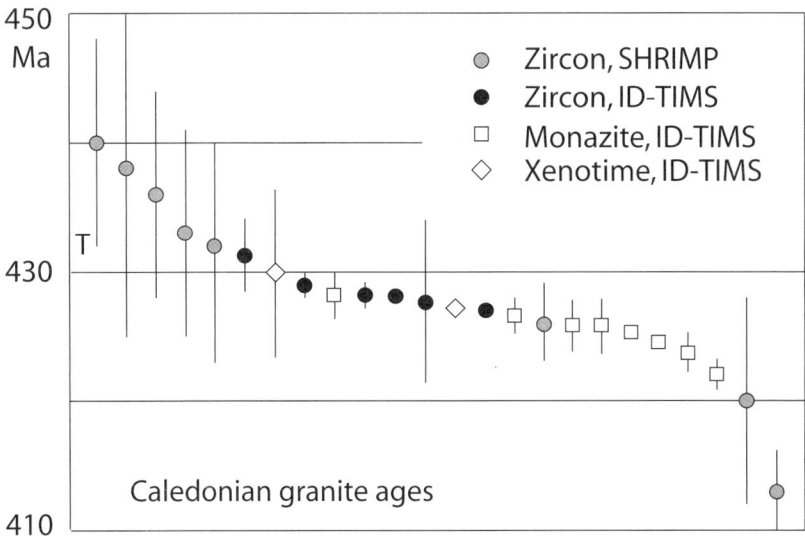

Figure 8. Diagrammatic representation of modern age determinations on granites from the East Greenland Caledonides, 72°N–76°N, arranged with decreasing ages from left to right. Data are from Table 2. Apart from the biotite-rich tonalitic rock marked T (440 Ma), all dated samples are leucogranites. SHRIMP—sensitive high-resolution ion microprobe; ID-TIMS—isotope dilution–thermal ionization mass spectrometry.

TABLE 2. MODERN AGE DETERMINATIONS OF CALEDONIAN LEUCOGRANITES IN THE EAST GREENLAND CALEDONIDES

	Author	Rock	Mineral, method	Age (Ma)	Comments
1	Watt et al. (2000)	Leucosome	Z, SHRIMP	420 ± 8	Anatectic rims; Wtd. mean $^{206}Pb/^{238}U$ age
2		Granite, pluton	Z, SHRIMP	433 ± 8	Wtd. mean $^{207}Pb/^{206}Pb$ age (?), method not stated
3	Hartz et al. (2000)	Granite mylonite	M, ID-TIMS	425.8 ± 2.0	U.I. age
4		Granite, dike	X, ID-TIMS	430 ± 6.5	U.I. age; monazite grains give younger ages
5	Hartz et al. (2001)	Granite	M, ID-TIMS	422.0 ± 1.2	Wtd. mean $^{207}Pb/^{235}U$ age of three selected analyses
6		Granite	M, ID-TIMS	426.6 ± 1.4	$^{207}Pb/^{235}U$ age, one selected analysis
7	Kalsbeek et al. (2001a)	Bi. tonalite, sheet	Z, SHRIMP	440 ± 8	Wtd. mean $^{206}Pb/^{238}U$ age. Check: 436 ± 5 Ma
8		Granite, sheet	Z, SHRIMP	432 ± 8	Wtd. mean $^{206}Pb/^{238}U$ age, selected analyses
9		Granite, pluton	Z, SHRIMP	436 ± 8	Wtd mean $^{207}Pb/^{206}Pb$ age, selected analyses
10		Granite, pluton	Z, SHRIMP	438 ± 14	Wtd. mean $^{206}Pb/^{238}U$ age, selected analyses
11	Strachan et al. (2001)	Granite, sheet	Z, ID-TIMS	428.2 ± 1.8	U.I. age, selected analyses
12		Granite, sheet	M, ID-TIMS	428.2 ± 1.0	Wtd. mean $^{207}Pb/^{235}U$ age, selected analyses
13		Granite, pluton	Z, ID-TIMS	427.7 ± 6.3	Wtd. mean $^{207}Pb/^{206}Pb$ age, selected analyses
14		Granite, pluton	Z, ID-TIMS	429 ± 1.0	U.I. age, selected analyses
15		Granite, pluton	Z, ID-TIMS	431.3 ± 2.8	U.I. age, selected analyses
16	White et al. (2002)	Leucosome	M, ID-TIMS	425.3 ± 0.3	Wtd. mean $^{207}Pb/^{235}U$ age, two selected analyses
17		Leucosome	M, ID-TIMS	425.8 ± 2.1	U.I. age
18		Granite, pluton	X, ID-TIMS	427.2 ± 0.8	U.I. age; X crystals interpreted as inherited grains
19		Same granite	M, ID-TIMS	423.8 ± 1.5	U.I. age, selected analyses from two samples
20		Granite, pluton	M, ID-TIMS	424.6 ± 0.5	$^{207}Pb/^{235}U$ age, one selected analysis
21	Andresen et al. (2004)	Granite, dike	Z, ID-TIMS	427	U.I. age, details not given
22		Granite, pluton	Z, ID-TIMS	428	U.I. age, details not given
23	Gilotti and McClelland (2005)	Granite, dike	Z, SHRIMP	426 ± 3	Tera-Wasserburg intercept, selected analyses
24		Granite, dike	Z, SHRIMP	413 ± 3	Tera-Wasserburg intercept, selected analyses

Note: Apart from no. 7, a biotite-rich tonalite, all dated rocks are anatectic leucogranites. Z—zircon, M—monazite, X—xenotime. SHRIMP—sensitive high-resolution ion microprobe; ID-TIMS—isotope dilution–thermal ionization mass spectrometry. Wtd. mean—weighted mean; U.I.—upper intercept.

thrust sheets over the western foreland (Higgins and Leslie, 2000; Higgins et al., 2004; Leslie and Higgins, this volume), and (2) extensional collapse of the orogen along major detachments (e.g., Strachan, 1994; Hartz and Andresen, 1995; Andresen et al., 1998; see Gilotti and McClelland, this volume).

First-order information on the relationship between granite formation and the various structures is provided by some general observations. (1) Caledonian migmatites and granites only occur in the structurally higher Hagar Bjerg thrust sheet and not in the underlying Niggli Spids thrust sheet (Higgins et al., 2004; Figs. 2, 3, and 9). Granite formation cannot, therefore, have significantly postdated emplacement of the Hagar Bjerg thrust sheet on the Niggli Spids thrust sheet. Either the granites were formed entirely before the main thrusting event, or else they were generated more or less synchronous with the thrusting. Gilotti et al. (this volume) argue that the leucogranites were formed on a prograde *P-T* path during crustal thickening. (2) The large extensional fault zones that bring low-grade rocks of the Eleonore Bay Supergroup into contact with high-grade rocks of the Krummedal sequence cut across the thrusts that separate the different thrust sheets (Henriksen, 2003) and must therefore postdate the main thrusting. (3) There is a concentration of granites near the tectonic contacts between the Eleonore Bay Supergroup and the Krummedal metasedimentary sequence (Figs. 1 and 2), but undeformed granite plutons have not been observed to cut the main extensional faults. Extension must therefore have continued after emplacement of these plutons.

Age of convergence. Hartz et al. (2001) reported monazite dates of 422.0 ± 1.2 Ma and 426.6 ± 1.4 Ma for granites believed to be contemporaneous with top-to-the-west thrusting on Bartholin Nunatak (BN, Fig. 1; 5 and 6 in Table 2). White et al. (2002) presented an age of 425.3 ± 0.3 Ma for monazites from a granitic leucosome, interpreted to be synkinematic with E-W contraction and simultaneous N-S extension at Forsblad Fjord (FF, Fig. 1), and an age of 425.8 ± 2.1 Ma for a granite postdating this event (16 and 17 in Table 2).

Age of extension. Hartz et al. (2000) studied localities in Junction Dal (JD, Fig. 1) and obtained a 425.8 ± 2.0 Ma monazite age for a mylonitic granite within a detachment zone, and an age of 430 ± 6.5 Ma for xenotime from a granite interpreted to be syntectonic with respect to extensional deformation (3 and 4 in Table 2). White et al. (2002) presented a monazite age of 423.8 ± 1.5 Ma for a granite from Forsblad Fjord, interpreted to have been emplaced during extensional deformation, and an age of 424.6 ± 0.5 Ma was obtained for a granite pluton in the lower part of the Eleonore Bay Supergroup, interpreted to predate major extension (19 and 20 in Table 2).

All of these age determinations are open to discussion—see comments in Table 2. However, taken together, they make a strong case for the initiation of extensional collapse immediately after, or in part synchronous with, emplacement of the thrust sheets around 425 Ma or a few million years earlier.

Strachan et al. (2001) obtained zircon and monazite U-Pb ages of 427.7 ± 6.3 Ma to 431.3 ± 2.8 Ma (all identical within error) on five leucogranite samples from the Ardencaple Fjord area (AF, Fig. 1; 11–15 in Table 2). Two of these were from granite sheets believed to be more or less synkinematic with extension; three were from granite plutons in the lower Eleonore

Figure 9. The Hagar Bjerg thrust sheet (HBTS) overlying the Niggli Spids thrust sheet (NSTS) in Andrée Land, looking north. At this locality, both thrust sheets are composed of metasedimentary rocks of the Krummedal supracrustal sequence; it is only in the Hagar Bjerg thrust sheet that the rocks are migmatized and contain granite sheets. Height of section above the screes is ~800 m.

Bay Supergroup. Andresen et al. (2004) reported a U-Pb zircon age of 428 Ma for a synextensional granite pluton in Andrée Land (AL, Fig. 1; 22 in Table 2).

Significantly different results were obtained by McClelland and Gilotti (2003). These authors reported SHRIMP zircon U-Pb dates of 403 ± 5 Ma for metamorphic rims on older zircon and 404 ± 4 Ma on zircons from felsic melt patches that formed during high-pressure granulite-facies metamorphism of metasedimentary rocks in Payer Land (PL, Fig. 1), where high-grade gneisses and low-grade metasedimentary rocks from the Eleonore Bay Supergroup are juxtaposed along the Payer Land detachment zone. These ages imply that exhumation of high-grade rocks there took place at least 20 m.y. later than extensional deformation elsewhere. Granitic dikes that predate deformation in the detachment zone yielded ages of 426 ± 3 Ma and 413 ± 3 Ma (Table 2, 23 and 24), whereas postdeformational pegmatites contain older zircon and monazite: monazites from one pegmatite yielded a U-Pb age of 434 ± 11 Ma (Gilotti and McClelland, 2005). These authors concluded that the older zircon and monazite represent inheritance from the pegmatite source, and, in contrast to the opinion expressed by all other authors quoted previously, they suggested that many of the previous results yielding earlier dates for extension may have involved inherited zircon and monazite. While they allow that syncontractional extension during emplacement of the older granite suite (ca. 425 Ma) may have occurred within upper structural levels of the orogen, Gilotti and McClelland (2005, this volume) conclude that exhumation of deep-level high-grade rocks was accommodated by displacements on regional detachment systems later than 405 Ma.

CALEDONIAN GRANITES SOUTH OF 72°N

In the southernmost part of the East Greenland Caledonides, south of 72°N, Caledonian granitoid rocks are present that are very different from the leucogranites further north (Fig. 1). These rocks (quartz diorites, granodiorites, quartz monzonites, granites, quartz syenites, etc.) were studied during Geological Survey of Greenland mapping projects in the period 1969–1972, and major-element analyses have been presented by Henriksen and Higgins (1988) and Coe (1975). Granitic rocks from Liverpool Land (Fig. 1) have been described by Coe (1975); in contrast to the leucogranites farther north, these granites were emplaced into crystalline basement rocks. However, no modern age determinations or trace-element data have been published for these rocks. One of the largest occurrences of these granitoids, the East Milne Land plutonic complex (Henriksen and Higgins, 1988; Fig. 10), is therefore described in more detail in the following section.

Most of Milne Land (Fig. 1) consists of migmatitic and leucogranitic rocks, similar to those found further north, formed by migmatization and anatexis of metasedimentary rocks of the Krummedal sequence (Henriksen and Higgins, 1988). Eastern Milne Land, however, is dominated by more mafic plutonic bodies, which occur together with granites and syenitic rocks. This area is separated from the area to the west by a major fault. Migmatites are rare, but high-grade metasedimentary rocks, regarded as parts of the Krummedal sequence, occur in the north (Fig. 10). The East Milne Land plutonic complex covers an area of ~30 km × 17 km, and several distinct plutonic bodies were differentiated during the mapping.

The largest intrusive unit in east Milne Land, the Korridoren granodiorite (1, Fig. 10), consists of granodioritic and quartz-dioritic rocks, and it is ~13 km in diameter. The western margin of this intrusion is formed by the fault that borders east Milne Land to the west, and the eastern margin is intruded by mafic rocks of the Bregnepynt quartz monzonite (4, Fig. 10; see following). The northern and southern contacts of the Korridoren intrusion dip at moderate to high angles to the NNE and are conformable with bordering marble bands, suggesting that the magma may have been emplaced as a thick sheet. The central part of the body consists of homogeneous massive hornblende-biotite granodiorite with numerous rounded mafic inclusions of quartz diorite and diorite (Fig. 11). Toward the margins of the pluton, the rocks become more mafic and grade into quartz diorite. The rocks in the marginal zones are strongly foliated, and the inclusions are strongly flattened.

Near the northern coast of east Milne Land, there is a small body of medium-grained hornblende-biotite quartz monzonite (2, Fig. 10). It has irregular intrusive contacts with the metasediments, and it is intruded by the leucocratic granites that border it to the south. This quartz monzonite has not been studied in detail because of the lack of suitable sample material.

Porphyritic biotite-muscovite leucogranites (~3% biotite and 2%–3% muscovite) form a pluton, ~8 km in diameter, in the northeast of the area (3, Fig. 10). The pluton cuts sharply across the metasediments and encloses large host-rock xenoliths. Toward the east, the leucogranites are intruded by the Bregnepynt quartz monzonite.

The Bregnepynt quartz monzonite (4, Fig. 10) in the eastern part of the area measures ~16 km from north to south. The quartz monzonite has intrusive contacts with the Korridoren granodiorite and is chilled against the leucocratic granite, and it is thus distinctly younger than both of these. The central part of the pluton consists of leucocratic quartz monzonites, dominated by centimeter-sized prismatic K-feldspar and plagioclase phenocrysts with dark-green clusters of hornblende and biotite. The chilled zone in the northwestern part of the intrusion is ~300 m wide and consists of mafic monzonitic to monzodioritic rocks. The pluton exhibits a roughly concentric zoning, truncated by the waters of Hall Bredning, suggesting that the greater part of the intrusion may lie offshore.

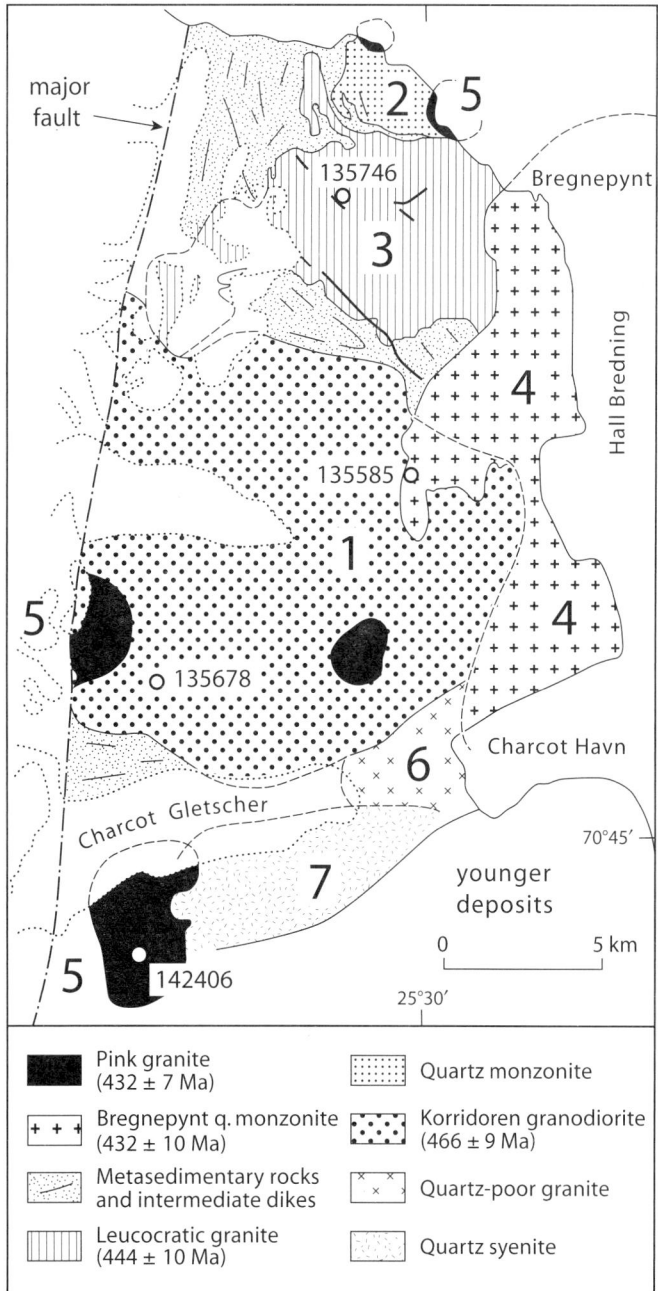

Figure 10. Sketch map showing the main plutonic bodies and the location of the dated samples in east Milne Land (modified from Henriksen and Higgins, 1988). For location, see Figure 1 (arrow).

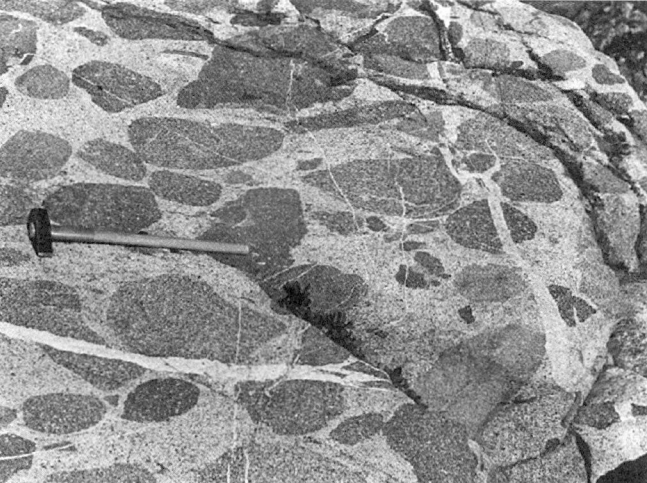

Figure 11. Rounded dioritic inclusions of variable texture and composition in leucocratic Korridoren granodiorite. Length of hammer is ~50 cm.

Pink and grayish pink, homogeneous, fine- to medium-grained granites occur throughout east Milne Land as small subcircular plutons and in bodies of more irregular shape (5, Fig. 10). Most of these appear to have formed late in the history of the area. Though distributed widely through east Milne Land, most of the pink granites are petrographically similar. The only significant mafic mineral is biotite (<5%); a little muscovite is sometimes present. Quartz, K-feldspar, and zoned oligoclase occur in roughly equal proportions.

An intrusive granitic body, a few kilometers across, in the southern part of east Milne Land (6, Fig. 10), has less quartz than most other pink granites (~20% vs. ~30% normally), and it has been referred to as the "Charcot Havn quartz-poor granite" (Henriksen and Higgins, 1988).

A pluton of syenitic rocks (7, Fig. 10), the "Charcot Gletscher quartz syenite," is exposed along the south side of Charcot Gletscher. Its contacts with other plutonic units are obscured by younger deposits, but in the west, the syenite is veined by pink granite. The main minerals in thin section are sodic plagioclase and microcline. Mafic minerals include garnet, hornblende, biotite, and sometimes a little augite. Quartz is usually present in small proportions. The syenite exhibits a large range in grain size and textures. Western outcrops are coarser grained (~1 cm) than those in the east and include leucocratic types with virtually no mafic minerals, as well as unusual mafic types rich in garnet and amphibole.

Relative age relationships, based on field observations, are as follows (Fig. 10): (1) The Korridoren granodiorite is probably the oldest intrusive unit in east Milne Land. (2) The age of the quartz monzonite, relative to that of the Korridoren granodiorite, is not known, but its lack of deformation suggests that it is younger. The quartz monzonite is followed, successively, by the leucocratic granite (3), the Bregnepynt quartz monzonite (4), and the pink granites (5), which appear to be the youngest. The relative ages of the quartz-poor granite (6) and the quartz syenite (7) in the south are poorly constrained.

Geochronology

Hansen and Tembusch (1979) obtained a Rb-Sr whole-rock isochron age of 453 ± 23 Ma (Sr_i 0.7071 ± 0.0002) for the Korridoren granodiorite, while one of the pink granites yielded an age of 373 ± 9 Ma (Sr_i 0.7164 ± 0.0009). Multigrain U-Pb zircon analyses yielded an age of 418 +5/–13 Ma for the Bregnepynt quartz monzonite, and an age of 353 +30/–15 Ma for the pink granite that had given the Rb-Sr isochron age of 373 ± 9 Ma (Hansen et al., 1987).

New zircon U-Pb age determinations were made by SHRIMP on samples from the Korridoren granodiorite, the porphyritic leucogranite, the Bregnepynt quartz monzonite, and one of the pink granites; for analytical procedures, see Williams (1998). Before analysis, the zircons were studied microscopically and by CL imaging to find the most suitable sites for analysis.

In order to detect possible small age differences between the different rocks, zircons were analyzed in a systematic manner: the first zircon from each of the four samples, followed by the second zircon from each of the samples, etc. In this way, minor instrumental drift during the analytical session could not give rise to spurious age differences. Analytical data are given in Table 3; results were as follows:

GGU 135678, Korridoren granodiorite. This sample yielded 200–300-μm-long, colorless, euhedral prismatic zircons with pronounced oscillatory zoning. Only a small minority of the grains had distinct cores. Eight analyses yielded a weighted mean $^{206}Pb/^{238}U$ age of 466 ± 9 Ma (Fig. 12A; mean square of weighted deviates [MSWD] = 0.87; all precisions are quoted at the 95% level of confidence). A core gave a $^{207}Pb/^{206}Pb$ age of 1265 Ma, indicating it represents inherited zircon.

GGU 135746, Leucogranite. Zircons from this sample are 150–300-μm-long, prismatic to bipyramidal colorless grains. Oscillatory zoning is visible, and many of the zircons show distinct cores. Eight analyses gave a weighted mean $^{206}Pb/^{238}U$ age of 444 ± 10 Ma (Fig. 12B; MSWD = 0.95). Two zircon cores yielded older $^{207}Pb/^{206}Pb$ ages, 1452 Ma and 842 Ma, and are interpreted to represent inheritance. One analysis yielded a $^{206}Pb/^{238}U$ age of ca. 400 Ma (Fig. 12B) and was excluded from the age calculation.

GGU 135585, Bregnepynt quartz monzonite. This sample yielded colorless, oscillatory-zoned, euhedral prisms, typically 150–300 μm long. No inherited cores were seen in the ~100 grains mounted and imaged. Nine analyses on zircons from the Bregnepynt quartz monzonite yielded a weighted mean $^{206}Pb/^{238}U$ age of 432 ± 10 Ma (Fig. 12C; MSWD = 1.06).

GGU 142406, Pink granite. The majority of the ~100 mounted grains are colorless, oscillatory-zoned euhedral prisms, typically 100–200 μm long, but a few are up to 400 μm; a few of the grains have distinct cores. Some grains have very thin (<15 μm; too thin to be analyzed) rims and pyramidal tips that appear dull in CL images; these rims are probably very rich in U and Th. Eight analyses on zircons from the pink granite yielded a weighted mean $^{206}Pb/^{238}U$ age of 432 ± 7 Ma (Fig. 12D; MSWD = 0.59).

These results are in agreement with the relative age relationships from field observations and the earlier age determinations described above. However, the age obtained for the pink granite (432 ± 7 Ma) is significantly different from that obtained by Hansen et al. (1987). Their multigrain U-Pb age of 352 +30/–15 Ma could have been influenced by U- and Th-rich rims like those observed in our sample GGU 142406, or else, pink granites of different ages may be present in east Milne Land.

The age data show that the Korridoren granodiorite is some 20–30 m.y. older than the other dated rocks. The distinction in age between the leucogranite and the Bregnepynt quartz monzonite, although in agreement with the field observations, is not statistically significant at the 95% level of confidence. On the other hand, the age difference between the leucogranite and the pink granite is significant.

TABLE 3. SHRIMP U-Pb DATA FOR GRANITOID ROCKS FROM EAST MILNE LAND

Spot	U (ppm)	Th/U	f_{206} (%)	$^{206}Pb/^{238}U$	$^{207}Pb/^{235}U$	Age (Ma)	Conc. (%)
GGU 135585, Bregnepynt quartz monzonite, 70°51′N, 25°31′W							
1.1	142	1.00	0.00	0.0730 ± 25	0.613 ± 38	454 ± 15	71
2.1	136	0.81	1.33	0.0736 ± 37	0.917 ± 228	458 ± 22	32
3.1	122	0.85	2.03	0.0656 ± 28	0.453 ± 90	409 ± 17	204
4.1	184	1.06	0.00	0.0703 ± 19	0.564 ± 26	438 ± 12	81
5.1	132	0.76	0.00	0.0714 ± 19	0.651 ± 55	445 ± 11	55
6.1	338	0.85	0.30	0.0680 ± 20	0.518 ± 29	424 ± 12	100
7.1	138	0.88	0.00	0.0678 ± 18	0.623 ± 33	423 ± 11	51
8.1	130	0.83	0.00	0.0689 ± 19	0.584 ± 25	430 ± 12	66
9.1	176	0.93	1.46	0.0677 ± 23	0.487 ± 48	422 ± 14	144
GGU 135678, Korridoren granodiorite, 70°48′N, 25°44′W							
1.1	295	0.59	0.45	0.0774 ± 18	0.656 ± 31	481 ± 11	73
2.1*	308	0.47	0.01	0.1153 ± 74	1.316 ± 117	1265 ± 109	56
3.1	288	0.54	1.14	0.0754 ± 20	0.542 ± 44	468 ± 12	161
4.1	212	0.51	0.13	0.0750 ± 21	0.664 ± 32	466 ± 13	63
4.2	173	0.54	0.70	0.0745 ± 18	0.567 ± 36	463 ± 11	111
5.1	320	0.58	0.44	0.0758 ± 18	0.613 ± 32	471 ± 11	85
6.1	320	0.56	0.45	0.0718 ± 23	0.575 ± 36	447 ± 14	84
7.1	151	0.61	1.86	0.0720 ± 21	0.472 ± 97	448 ± 13	590
8.1	183	0.64	0.00	0.0752 ± 22	0.660 ± 34	468 ± 13	64
GGU 135746, east Milne Land leucogranite, 70°56′N, 25°36′W							
c1.1*	388	0.31	0.16	0.2527 ± 82	3.139 ± 141	1452 ± 42	102
r1.2	1793	0.25	3.37	0.0634 ± 15	0.584 ± 81	396 ± 9	48
c2.1*	125	0.29	0.00	0.1396 ± 45	1.525 ± 81	842 ± 26	72
r2.2	2033	0.35	0.22	0.0734 ± 26	0.563 ± 23	456 ± 16	104
3.1	363	0.34	0.28	0.0721 ± 21	0.554 ± 22	449 ± 13	102
4.1	458	0.60	0.77	0.0710 ± 20	0.566 ± 29	442 ± 12	85
5.1	356	0.50	1.24	0.0709 ± 22	0.458 ± 31	441 ± 13	993
6.1	1382	0.26	0.26	0.0735 ± 16	0.613 ± 26	457 ± 10	74
7.1	1578	0.36	0.56	0.0695 ± 16	0.606 ± 26	433 ± 10	61
r8.1	752	0.36	0.09	0.0724 ± 24	0.597 ± 23	451 ± 14	76
9.1	686	0.37	0.44	0.0684 ± 19	0.520 ± 22	426 ± 11	102
GGU 142406, east Milne Land pink granite, 70°43′N, 25°46′W							
1.1	1005	0.48	0.23	0.0699 ± 17	0.526 ± 16	435 ± 10	110
2.1	407	0.46	0.10	0.0710 ± 19	0.544 ± 19	442 ± 11	101
3.1	443	0.27	0.00	0.0683 ± 17	0.513 ± 21	426 ± 10	108
4.1	103	0.41	1.15	0.0678 ± 21	0.492 ± 46	423 ± 12	136
5.1	702	0.44	0.05	0.0691 ± 18	0.536 ± 17	431 ± 11	94
6.1	587	0.32	0.51	0.0721 ± 19	0.549 ± 31	449 ± 12	106
7.1	160	0.29	0.46	0.0677 ± 23	0.518 ± 55	422 ± 14	98
8.1	402	0.16	0.65	0.0694 ± 30	0.543 ± 43	433 ± 18	90
9.1	564	0.44	0.51	0.0683 ± 17	0.521 ± 23	426 ± 10	100

Note: GGU sample numbers are given according to the files of the Geological Survey of Denmark and Greenland. f_{206} is the percentage of ^{206}Pb that is not radiogenic. Age is the $^{206}Pb/^{238}U$ age; for samples marked with * (inheritance) $^{207}Pb/^{206}Pb$ ages are given. Spots marked c and r represent cores and rims of grains, respectively. Conc. is the degree of concordance between $^{206}Pb/^{238}Pb$ and $^{207}Pb/^{206}Pb$ ages. Errors are at 1σ.

Major- and Trace-Element Geochemistry

Chemical analyses of representative samples from the East Milne Land plutonic complex are presented in Table 4, and Figure 13 displays aspects of their major-element chemistry. A plot of $Na_2O + K_2O$ versus SiO_2 (Fig. 13A; fields of various plutonic rocks are according to Wilson, 1989) illustrates the wide variation in composition of the different intrusions. Samples from the Korridoren complex plot in the quartz diorite and granodiorite fields, and the Bregnepynt quartz monzonites plot as syenodiorites and syenites. The curved line separates the fields of subalkaline and alkaline rocks (Miyashiro, 1978), and several of the east Milne Land intrusions, especially the Bregnepynt quartz monzonite and the Charcot Gletscher quartz syenite, are distinctly alkaline in composition. Most of the samples plot in the calc-alkaline field of the AFM diagram (Fig. 13B; $A = Na_2O + K_2O$; $F = FeO^{tot}$; $M = MgO$); however, the quartz syenites are characterized by much higher FeO^{tot}/MgO ratios than the other rocks. Most east Milne Land intrusions are rich in K_2O (Fig. 13C); Korridoren samples plot in the high-K field, and samples from the Bregnepynt quartz monzonite, the quartz syenites, and some of the pink granites plot in the shoshonitic field.

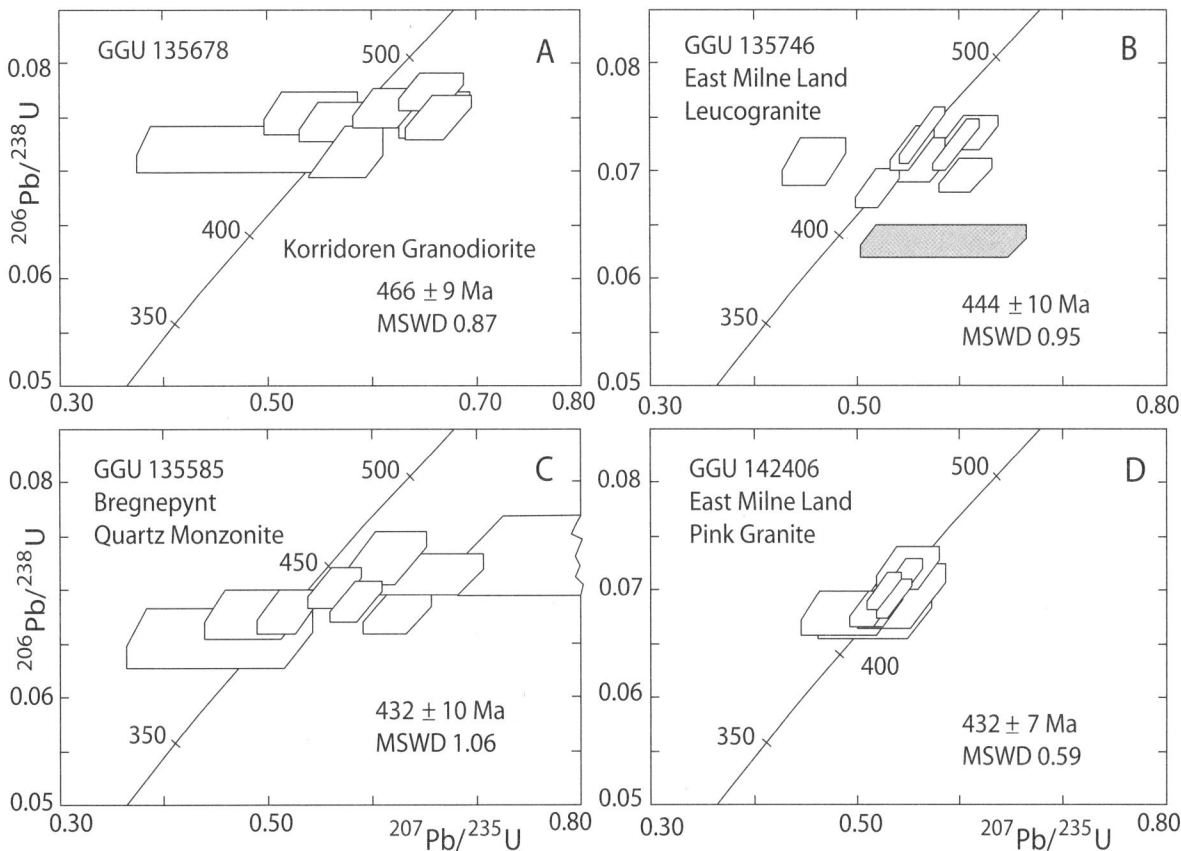

Figure 12. Concordia diagrams for samples from the East Milne Land plutonic complex. For description, see text. Error boxes represent 1σ errors. Errors of the ages of the different rocks are given at 2σ. GGU sample numbers are listed according to the files of the Geological Survey of Denmark and Greenland. MSWD—mean square of weighted deviates.

The compositions of the plutonic rocks from east Milne Land are very different from those of the granites of the Kong Oscar Fjord region (cf. Table 1). While the latter are peraluminous leucogranites, most east Milne Land rocks are more mafic and distinctly metaluminous (Table 4). The leucogranites from the Kong Oscar Fjord region have been interpreted as S-type granites that formed by fusion of metasedimentary rocks of the Krummedal sequence; this is clearly not the case for most plutonic rocks of east Milne Land.

Trace-element data (Table 4) are illustrated in Figure 14. In spider diagrams (Figs. 14A and 14C), all samples display the strong negative spikes for Nb and Ti that are typical for arc-related rocks; some samples also show negative spikes for Sr and P. Among the more mafic rocks, the Bregnepynt quartz monzonite is characterized by high concentrations of most incompatible elements: Ba and Sr up to ~2000 ppm, and a sum of rare earth elements (ΣREE) up to 600 ppm. Some of the quartz syenites and pink granites also have high contents of incompatible elements; a mafic quartz syenite (142415, Fig. 14A), for example, has ΣREE ~1400 ppm, and another syenite (142319, Fig. 14C) has ~250 ppm Th (Table 4). Zr concentrations (Fig. 14E) for most east Milne Land samples are much higher than those of the S-type leucogranites in the north (cf. Fig. 7C), up to >500 ppm for some samples of the Bregnepynt quartz monzonite. Only the leucogranites have comparable low concentrations of Zr. Concentrations of Nb and Y are relatively low, and most samples plot in the field of volcanic arc granites in the discrimination diagram of Pearce et al. (1984). The pink granites, however, plot in the field of syncollisional granites (Fig. 14F). It should be noted that there may be large differences between individual samples from the same rock types (e.g., pink granites 108724 and 135614, Table 4), suggesting differences in petrogenesis between superficially similar rocks.

REE data were acquired for 15 samples. They display variably fractionated chondrite-normalized spectra (Figs. 14B and 14D). La_N/Yb_N varies from 3 to 11 for samples from the Korridoren intrusion and from 15 to 27 for the Bregnepynt quartz monzonite. Some of the felsic rocks have much higher La_N/Yb_N, up to >400 for a quartz-poor granite (142397, Fig. 14D) with very low Yb (~0.03 ppm), probably related to equilibration with garnet in the source. Most samples display a small negative Eu anomaly (Eu/Eu* = 0.71–0.91); two samples have strong negative Eu anomalies, a pink granite and a quartz-poor granite (135614 and 142397, Fig. 14D), with Eu/Eu* 0.22 and 0.41, respectively.

TABLE 4. REPRESENTATIVE ANALYSES OF THE EAST MILNE LAND PLUTONIC COMPLEX

Major/trace elements	Sample no.									
	135598	135555	108719	142391	108724	135614	135757	135571	142415	142419
	Korridoren intrusion		Bregnepynt quartz monzonites		Pink granites		Leuco-granite	Quartz-poor granite	Quartz syenites	
SiO_2 (wt%)	50.74	65.19	53.20	54.97	69.53	72.92	72.86	64.96	48.16	65.53
TiO_2	0.73	0.49	0.92	1.52	0.72	0.21	0.15	0.75	0.87	0.22
Al_2O_3	15.15	16.13	14.06	16.35	14.66	14.33	15.13	15.30	13.12	17.40
Fe_2O_3	3.12	0.65	2.46	2.87	0.31	0.12	0.00	1.33	10.70	1.68
FeO	6.76	3.11	4.23	4.45	1.56	0.99	1.02	2.30	1.07	0.25
MnO	0.33	0.08	0.11	0.11	0.03	0.02	0.05	0.06	0.41	0.09
MgO	7.38	2.07	6.04	4.00	0.65	0.30	0.31	2.04	1.29	0.19
CaO	8.83	4.25	5.21	4.94	1.29	0.55	1.58	3.52	15.32	1.74
Na_2O	2.65	3.37	1.87	3.34	3.52	3.15	3.87	3.71	2.24	3.67
K_2O	1.71	2.97	7.22	4.62	5.88	5.53	3.54	4.09	3.75	7.91
P_2O_5	0.25	0.13	0.39	0.64	0.26	0.29	0.08	0.35	0.40	0.05
LOI	1.42	0.83	3.22	1.14	0.68	0.64	0.66	0.66	1.09	0.54
Sum	99.07	99.26	98.93	98.93	99.08	99.05	99.26	99.08	98.42	99.25
A/CNK	0.68	0.98	0.69	0.84	1.01	1.18	1.16	0.90	0.37	0.98
Cs* (ppm)	7.2	8.8	5.7	3.1	4.2	14.8	3.7	4.4	6.8	8.6
Rb	95	121	205	153	218	430	120	131	121	251
Ba	132	542	1390	1440	1360	138	577	1240	3530	2130
Pb	10	15	33	20	38	21	22	24	43	41
Sr	359	437	934	1210	680	43	236	1210	2020	1590
Y	42	19	30	36	12	14	10	19	50	19
Th*	6.3	12.1	7.8	6.1	39	17	7.3	24	158	249
Zr	125	123	197	534	380	98	85	196	785	169
Hf*	4	4.1	6.4	14	11	3.6	2.8	6.5	18.4	4.7
Nb	14	10	12	22	8.6	16	8.2	11	22	20
Ta*	0.7	1.7	0.7	1.1	1.8	2.5	3.2	n.d.	1.5	1.9
Ni	98	9	109	42	5	4	4	22	5	4
Sc	30	11	25	19	3	2	3	9	2	1
V	204	84	114	142	26	6	5	78	171	24
Cr	460	30	400	123	6	3	<3	27	5	<3
Ga	25	16	18	24	20	23	18	18	20	19
La* (ppm)	18.3	24.0	56.7	51.6	112	20.3	20.2	67.9	325	78.3
Ce*	44.6	48.2	102	114	222	48.9	35.6	142	605	140
Pr**	7.74	5.68	14.0	15.4	25.9	6.14	3.97	17.5	86.4	18.3
Nd*	34.5	20.5	53.7	64.7	87.9	22.0	13.3	66.0	314	59.3
Sm*	7.88	3.85	9.62	12.7	11.4	5.53	2.24	10.9	40.5	7.28
Eu*	1.92	1.01	2.60	3.29	2.10	0.36	0.60	2.64	8.18	1.33
Gd**	7.68	3.49	7.91	10.7	6.48	4.65	1.99	7.72	27.5	5.14
Tb**	1.20	0.58	1.09	1.43	0.57	0.73	0.32	0.89	2.34	0.63
Dy**	6.67	3.19	5.73	6.90	2.44	3.08	1.71	4.01	10.2	3.17
Ho**	1.34	0.63	1.05	1.21	0.34	0.43	0.32	0.67	1.53	0.53
Er**	4.05	1.89	2.85	3.20	0.69	0.94	0.92	1.67	3.63	1.50
Tm**	0.64	0.29	0.41	0.43	0.09	0.13	0.14	0.22	0.46	0.27
Yb*	4.49	2.06	2.64	2.68	0.46	0.64	0.98	1.41	2.64	2.07
Lu*	0.66	0.31	0.36	0.36	0.07	0.08	0.13	0.21	0.41	0.33

Note: Sample numbers refer to the files of the Geological Survey of Denmark and Greenland. Major elements were analyzed at the Geological Survey of Denmark and Greenland by X-ray fluorescence spectrometry (XRF) on glass disks, Na_2O was analyzed by atomic absorption spectrometry, and FeO was analyzed by titration (Kystol and Larsen, 1999). Unmarked trace elements were analyzed by XRF on powder tablets at the Geological Institute, University of Copenhagen. Elements marked * were analyzed by instrumental neutron activation analysis, and elements marked ** were analyzed by inductively coupled plasma–mass spectrometry at Activation Laboratories Ltd., Ontario, Canada. LOI—loss on ignition.

Rb-Sr and Sm-Nd Isotope Data

Rb-Sr and Sm-Nd isotope data were acquired for a few samples from each of the east Milne Land intrusions (Table 5). Most $^{87}Sr/^{86}Sr$ ratios calculated for the time of emplacement (Sr_i) scatter around 0.708; with few exceptions, they are much lower than those obtained for granites from the Kong Oscar Fjord region (Fig. 15A), further illustrating the fundamental difference in petrogenesis between the two kinds of granites.

Most ε_{Nd} values calculated for the time of emplacement of the intrusions ($\varepsilon_{Nd(t)}$) vary between –3 and –5 and form a well-defined cluster in the $\varepsilon_{Nd(t)}$ versus $\varepsilon_{Sr(t)}$ diagram (Fig. 15B). Four

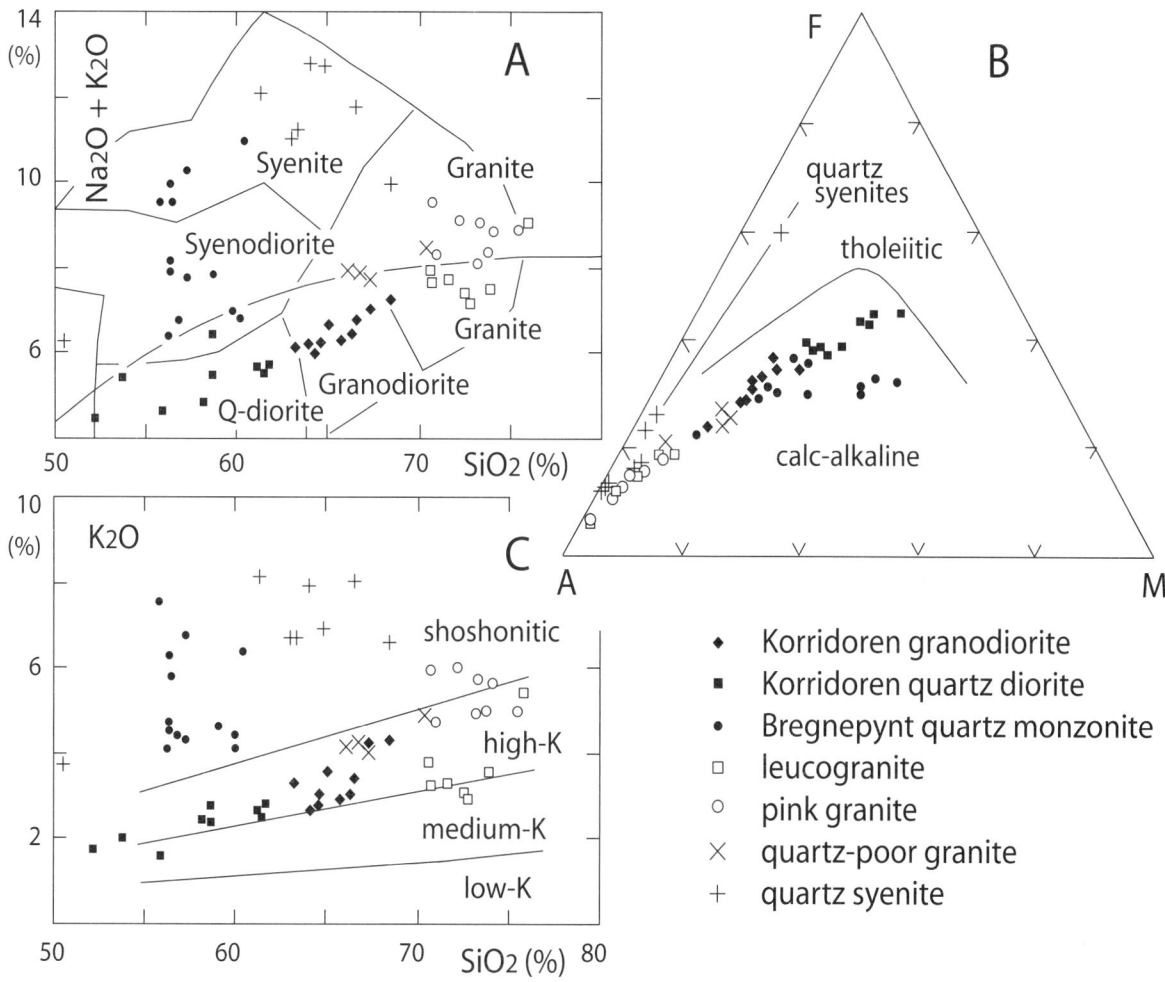

Figure 13. Aspects of the major-element chemistry of rocks from the East Milne Land plutonic complex. (A) $Na_2O + K_2O$ versus SiO_2; the compositional fields of common plutonic rocks are after Wilson (1989). (B) AFM diagram showing the calc-alkaline nature of the rocks (Irvine and Baragar, 1971), where $A = Na_2O + K_2O$; $F = FeO^{tot}$; $M = MgO$. (C) K_2O versus SiO_2, illustrating the potassic nature of most of the east Milne Land plutonic rocks. See text for discussion.

samples have more strongly negative $\varepsilon_{Nd(t)}$ values, –8 to –11. These samples, two pink granites, one leucogranite, and one quartz-poor granite, have high SiO_2, 69%–76% (Fig. 15C), and the combination of low $\varepsilon_{Nd(t)}$ and high SiO_2 suggests that they contain major contributions derived from crustal sources. For two of these four samples, full REE data sets are available, and they are the only samples that show strong negative Eu anomalies.

Interpretation

Korridoren granodiorite. Although individual age determinations for the different plutonic units of the East Milne Land complex are not very precise, there can be no doubt that the Korridoren granodiorite and associated quartz diorites are 20–30 m.y. older than the other intrusions and, indeed, Caledonian granites elsewhere in East Greenland. This, together with their calc-alkaline chemistry (Fig. 13B), suggests that they belong to a volcanic arc, related to subduction of Iapetus underneath this part of Laurentia. Strong deformation along its margins is consistent with a precollision origin of this intrusion. Emplacement of the Korridoren complex must have occurred shortly after deposition of Ordovician sediments along the passive margin of Iapetus ceased at ca. 460 Ma (Smith and Rasmussen, this volume).

Younger intrusions. Samples from the Bregnepynt quartz monzonite and several of the other late Caledonian intrusions in east Milne Land have high concentrations of Ba and Sr (and other incompatible trace elements, Table 4). In this and other respects (e.g., high K/Na), they are comparable to the coeval "Newer granites" and associated rocks of northern Scotland (e.g., Tarney and Jones, 1994; Fowler and Henney, 1996; Fowler et al., 2001; Strachan et al., 2002). These authors argued that the Scottish rocks were formed by fractionation of melts derived from an incompatible element–enriched mantle source ($\varepsilon_{Nd(t)}$ ~–3.6, $^{87}Sr/^{86}Sr_{(t)}$ ~0.706; Fowler and Henney, 1996), with limited contributions

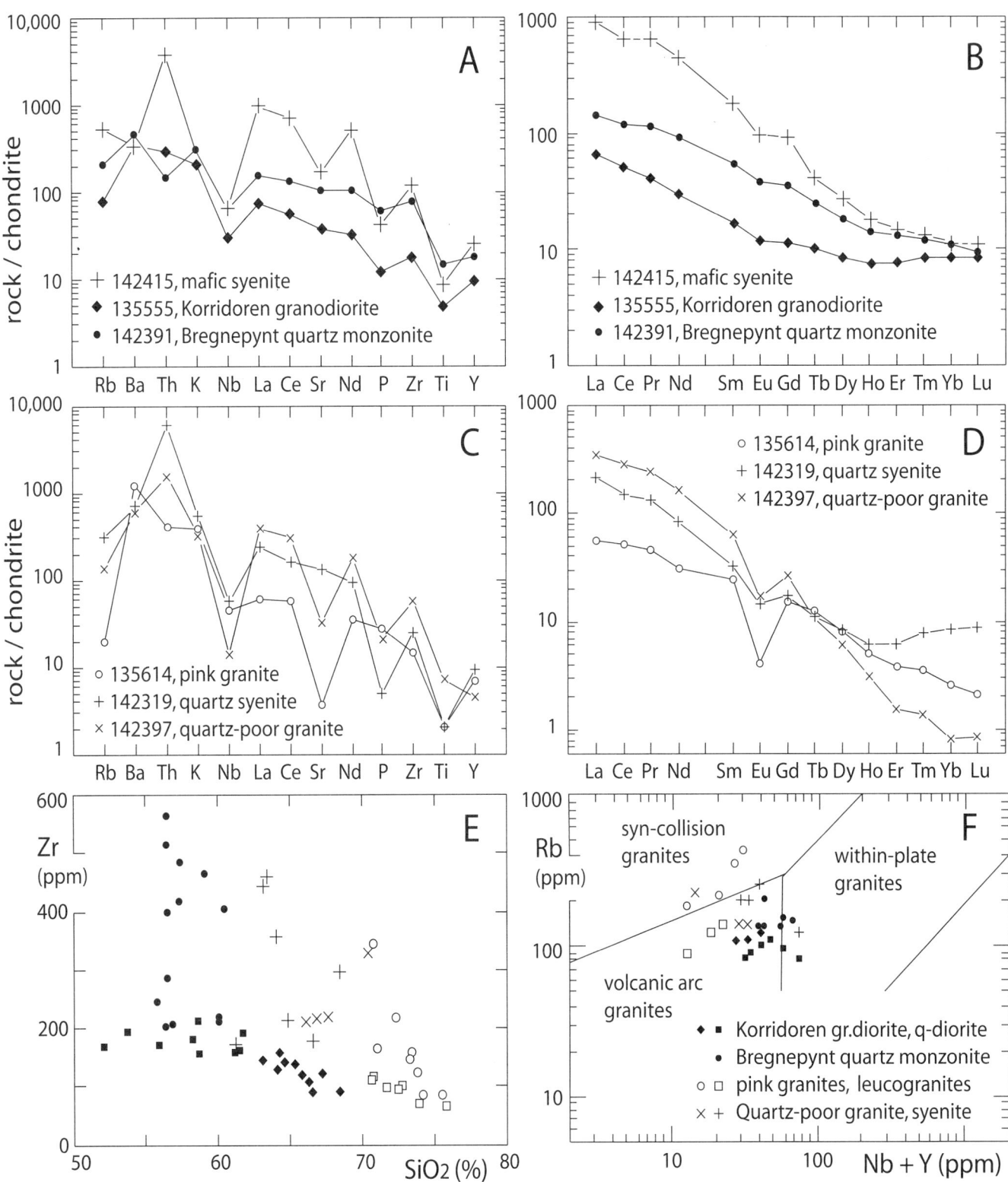

Figure 14. Trace-element compositions of rocks from the East Milne Land plutonic complex. Chondritic normalization values are from Thompson (1982) in the spider diagrams (A and C) and from Taylor and McLennan (1985) in the normalized rare earth element (REE) diagrams (B and D). See text for discussion. Legend in 14F also applies to 14E.

TABLE 5. Rb-Sr AND Sm-Nd ANALYSES FOR SAMPLES FROM THE EAST MILNE LAND PLUTONIC COMPLEX

GGU no.	Rb (ppm)	Sr (ppm)	^{87}Rb/^{86}Sr	^{87}Sr/^{86}Sr	Sr(i)	εSr$_{(i)}$	Sm (ppm)	Nd (ppm)	^{147}Sm/^{144}Nd	^{143}Nd/^{144}Nd	T_{DM} (Ga)	εNd$_{(i)}$
Korridoren intrusion												
135513	121	424	0.832	0.71280	0.7073	47.5	5.40	20.36	0.1521	0.512268	2.17	-4.56
135555	121	437	0.805	0.71244	0.7071	44.6	4.17	21.11	0.1131	0.512183	1.46	-3.90
135589	113	518	0.628	0.71172	0.7075	50.3	5.70	29.00	0.1127	0.512158	1.49	-4.37
135625	115	550	0.603	0.71132	0.7073	47.5	5.91	33.21	0.1020	0.512202	1.29	-3.86
135678	160	425	1.093	0.71420	0.7069	41.8	4.65	23.33	0.1142	0.512143	1.54	-4.74
Bregnepynt quartz monzonite												
108719	205	934	0.636	0.71201	0.7081	58.2	10.46	56.29	0.1065	0.512129	1.45	-4.95
135585	160	1283	0.315	0.71035	0.7084	62.5	15.33	85.19	0.1031	0.512121	1.42	-4.93
142391	153	1210	0.363	0.71063	0.7084	62.5	13.88	67.76	0.1174	0.512163	1.56	-4.89
142395	151	1980	0.219	0.70917	0.7078	54.0	22.91	132.89	0.0988	0.512148	1.33	-4.17
Pink granites												
135614	430	43	28.5	0.88002			5.62	22.70	0.1417	0.512079	2.26	-7.88
142406	188	285	1.906	0.73111	0.7194	218.8	8.04	44.47	0.1036	0.511800	1.86	-11.22
Leucogranite												
108685	136	400	0.988	0.71444	0.7082	59.9	4.35	25.85	0.0965	0.512169	1.27	-3.48
108699	234	85	7.99	0.75931	0.7088	68.4	2.75	13.58	0.1161	0.511901	1.94	-9.82
Quartz-poor granite												
135571	131	1210	0.311	0.70960	0.7077	52.6	11.42	66.83	0.0980	0.512177	1.28	-3.54
142397	227	376	1.760	0.72559	0.7148	153.4	15.51	114.18	0.0778	0.511757	1.56	-10.63
142400	125	1150	0.314	0.71028	0.7084	62.5	11.69	68.31	0.0980	0.512171	1.29	-3.66
Quartz syenite												
142405	201	679	0.859	0.71371	0.7084	62.5	4.48	32.51	0.0789	0.512091	1.20	-4.17
142415	121	2020	0.177	0.70939	0.7083	61.1	44.00	329.82	0.0764	0.512133	1.13	-3.21
142419	251	1590	0.456	0.71105	0.7082	59.7	7.90	61.80	0.0732	0.512103	1.14	-3.62

Note: Samples were analyzed at the Geological Institute, University of Copenhagen, Denmark. Rb and Sr were analyzed by X-ray fluorescence spectrometry; precision for ^{87}Rb/^{86}Sr was ~1% (1σ). Chemical separation of Sr from whole-rock powders was carried out on conventional cation-exchange columns, followed by a clean-up of the Sr fraction over miniature columns using Eichrom SrSpec resin. Sr isotopic analyses were performed on a VG Sector 54 IT mass spectrometer. A value of 0.1194 for the ^{88}Sr/^{86}Sr ratio was used for online mass fractionation correction, using the exponential bias law. The mean ^{87}Sr/^{86}Sr value of the NBS 987 Sr standard was 0.710248, with a 2σ external reproducibility of ±0.000011 (four measurements) during the analysis time of this project. For Sm-Nd analysis, the rock powders were spiked with a ^{149}Sm-^{150}Nd mixed spike, and the bulk rare earth elements (REEs) were separated over 15 mL glass stem columns charged with AG 50W cation resin. REEs were further separated over HDEHP-coated biobeads (BioRad®) loaded in 6 mL glass stem columns. Sm and Nd isotopic analyses were performed on a VG Sector 54 IT mass spectrometer, and both static and multidynamic routines were used for the collection of the isotopic ratios. The mean value for our internal JM Nd standard (referenced against La Jolla) during the period of measurement was 0.511115 for ^{143}Nd/^{144}Nd, with a 2σ external reproducibility of ±0.000013 (five measurements). T_{DM} is the Sm-Nd model age calculated according to the depleted mantle model of DePaolo et al. (1991).

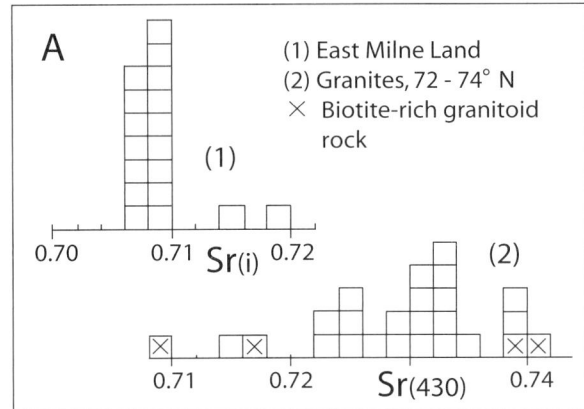

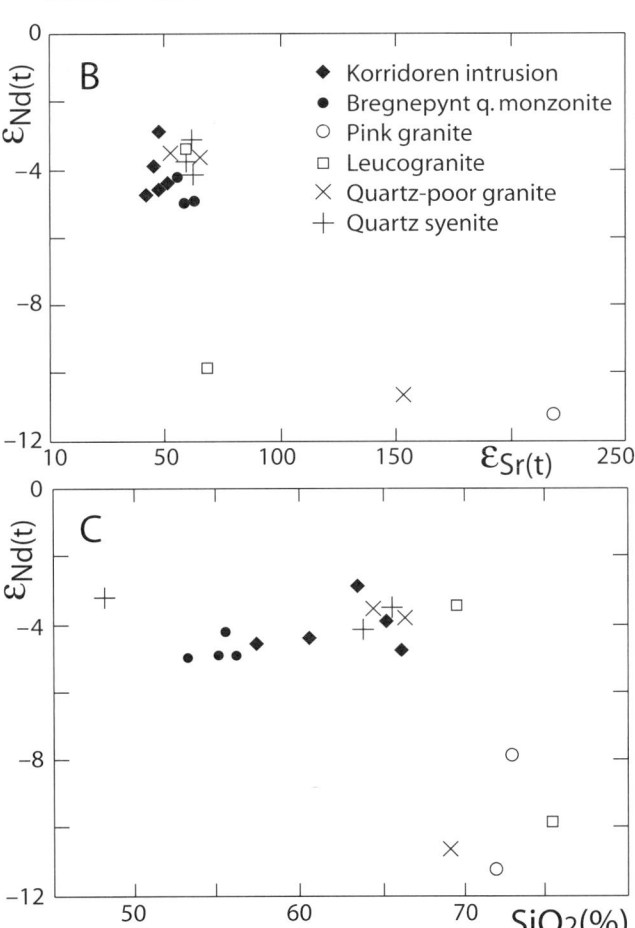

Figure 15. Sr and Nd isotopic compositions of rocks from the East Milne Land plutonic complex. (A) Comparison of initial $^{87}Sr/^{86}Sr$ ratios for samples from the east Milne Land plutonic complex (1) with those of S-type granites from the Kong Oscar Fjord region (2). Data from Kalsbeek et al. (2001a). (B–C) $\varepsilon_{Nd(t)}$ versus $\varepsilon_{Sr(t)}$ and $\varepsilon_{Nd(t)}$ versus SiO_2 diagrams, showing strongly negative $\varepsilon_{Nd(t)}$ values for some of the samples, and indicating that these samples were contaminated by components derived from continental crust.

from crustal sources. This model may also be applicable to the Ba- and Sr-rich east Milne Land rocks. The quartz monzonites, quartz syenites, and quartz-poor granites could represent different stages of fractional crystallization of similar K-, Ba-, and Sr-rich magmas. However, several samples of pink granites (135614, Table 4) and leucogranites (135757) have significantly lower Ba and Sr values and may have different origins.

Most east Milne Land samples have slightly lower $\varepsilon_{Nd(t)}$ values, and slightly higher $\varepsilon_{Sr(t)}$ values than those proposed for the Scottish lithospheric mantle (Fowler and Henney, 1996). This difference is compatible with the admixture of minor proportions of crustal components into the east Milne Land magmas. However, the tight clustering of isotopic data in the $\varepsilon_{Nd(t)}$ versus $\varepsilon_{Sr(t)}$ diagram and the lack of correlation between $\varepsilon_{Nd(t)}$ and SiO_2 (Figs. 15B and 15C) suggest that contamination with crustal material did not play a major role during later fractionation. Two pink granites, one leucogranite, and one quartz-poor granite for which Sr and Nd isotopic data are available (Table 5) have much more strongly negative $\varepsilon_{Nd(t)}$ values and apparently contain significantly larger proportions of crustal material. This is confirmed by the presence of inherited cores in zircons separated from leucogranite 135746 and pink granite 142406. However, different samples from the same intrusion may show large differences in $\varepsilon_{Nd(t)}$ (Figs. 15B and 15C), indicating the presence of highly variable proportions of inherited crustal components within individual intrusions.

Late Caledonian igneous activity in Scotland is interpreted to be the result of renewed subduction along the Iapetan margin (Brown, 1991; Strachan et al., 2002). A similar origin is proposed here for the younger east Milne Land intrusions. Emplacement of the Scottish granites was related to large-scale faulting, whereby pathways were created for the rising magmas (e.g., Strachan et al., 2002). The location of the East Milne Land plutonic complex adjacent to a major fault may suggest a similar mechanism of emplacement.

DISCUSSION: GRANITES AND LACK OF GRANITES IN THE EAST GREENLAND CALEDONIDES

There is a distinct north-to-south zonation in the East Greenland Caledonides with respect to the occurrence of granites. In the northern part of the orogen, north of 76°N, Caledonian granites appear to be absent; the area from 72°N to 76°N is characterized by the presence of leucogranites formed by anatexis of metasedimentary rocks of the Krummedal sequence, while more mafic arc-related granitoid rocks, such as those described here from east Milne Land, only occur in the southernmost part of the East Greenland Caledonian orogen, south of ~72°N.

In the Kong Oscar Fjord region (72°N–74°N), all Caledonian and "Grenvillian" granites occur within the Hagar Bjerg thrust sheet, whereas they are absent from the structurally underlying Niggli Spids thrust sheet and the parautochthonous and autochthonous foreland. The original location of the rocks forming the Niggli Spids thrust sheet was to the east of its present position, and the Hagar Bjerg thrust sheet was originally located even further east (Higgins and Leslie, 2000; Leslie and Higgins, this volume; Higgins et al., 2004). Apart from the north-to-south zonation in the distribution of the granites, there is thus a west-to-east zonation—Caledonian and Grenvillian granites are absent in the foreland and the Niggli Spids thrust sheet but are common in the originally easternmost, now overlying Hagar Bjerg thrust sheet. The southeastern belt also contains granitoid rocks related to arc magmatism in its southeasternmost parts; arguably, this belt may once have formed a more central part of the orogen. The different belts run at an angle to the present strike of the Caledonian orogen, and large-scale thrusting telescoped them into their present positions.

Since most granites in the southeastern belt were formed by fusion of Krummedal sequence lithologies, the lack of granites in the northwestern belt can plausibly be explained by the absence of the Krummedal sequence there. The absence of arc-type granites from most of the East Greenland Caledonides, however, is more enigmatic. It must be related to the disposition of the various subduction zones that were active during closure of Iapetus, the details of which are poorly known.

The Krummedal supracrustal sequence of North-East Greenland has been correlated with the Moine Supergroup of Scotland (e.g., Winchester, 1988) and with the Brennevinsfjorden Group and correlative strata of Svalbard (e.g., Gee and Tebenkov, 1996). Neoproterozoic S-type granites similar to the North-East Greenland Grenvillian granites occur both in Scotland (e.g., Friend et al., 1997) and on Svalbard (e.g., Johansson et al., 2000, 2004). Grenvillian granites have not been documented from northern Norway, but evidence of ca. 950 Ma metamorphism has been reported (Dallmeyer, 1992).

Caledonian granitoid intrusions comparable in age and character to the east Milne Land plutonic rocks are present in Scotland (the so-called "Newer granites"; see, e.g., Strachan et al., 2002). Caledonian granitoid rocks, both S-type granites and more mafic varieties, are also present on Svalbard (Gee et al., 1996; Harland, 1997; Johannson et al., 2004), and in central to northern Norway (the Bindal Batholith; Barnes et al., 1992; Birkeland et al., 1993; Nordgulen et al., 1993). In northern Norway, these rocks are restricted to the thrust sheets of the uppermost allochthon, for which a Laurentian origin is assumed (e.g., Roberts et al., 2002).

The distribution of Grenvillian and Caledonian granites described here suggests that they may once have formed a more-or-less continuous belt running from Scotland, over the southern part of the East Greenland Caledonides, obliquely to the coast onto what is now the North-East Greenland shelf. The shelf areas off North-East Greenland are up to 300 km wide, and the presence of a Caledonian basement underneath a thick sedimentary cover has recently been documented by seismic investigations (Hamann et al., 2005). A continuation of the granite-bearing belt to Svalbard and northern Norway is plausible, but, unfortunately, rapid relative movements of Svalbard and Baltica versus Laurentia during the later Paleozoic (Harland, 1997; Torsvik et al., 1996) hamper precise correlation of Greenland with these areas at the time of granite formation.

ACKNOWLEDGMENTS

The title of this paper was inspired by H.H. Read's (1948) opening paper, "Granites and Granites," at the 1947 annual meeting of the Geological Society of America, where the origin of granites was debated. At that time, the dispute between "plutonists" and "transformists" was very harsh; see Read (1957), "The Granite Controversy."

The investigations reported in this paper were carried out in relationship with regional mapping programs conducted by the Geological Survey of Denmark and Greenland (GEUS) in North-East Greenland under the inspiring direction of Niels Henriksen. We thank Calvin G. Barnes and Paul Tomascak for helpful reviews, Helle Zetterwall and Benny Schark for help with the figures, Elsevier BV (the Netherlands) for permission to reuse figures that were published earlier in *Lithos*, and GEUS for permission to publish this paper.

REFERENCES CITED

Andresen, A., Hartz, E.H., and Vold, J., 1998, A late orogenic extensional origin for the infracrustal gneiss domes of the East Greenland Caledonides (72–74°N): Tectonophysics, v. 285, p. 353–369, doi: 10.1016/S0040-1951(97)00278-3.

Andresen, A., Rehnström, E.F., and Holte, M.K., 2004, Synchronous, but contrasting deformational regimes at different crustal depths: Geochronology of syn-tectonic granites in the NE Greenland Caledonides: GFF, v. 126, p. 77.

Barnes, C.G., Prestvik, T., Nordgulen, Ø., and Barnes, M.A., 1992, Geology of three dioritic plutons in Velfjord, Nordland: Bulletin Norges Geologiske Undersøkelse, v. 423, p. 41–54.

Birkeland, A., Nordgulen, Ø., Cumming, G.L., and Bjørlykke, A., 1993, Pb-Nd-Sr isotopic constraints on the origin of the Caledonian Bindal Batholith, central Norway: Lithos, v. 29, p. 257–271, doi: 10.1016/0024-4937(93)90020-D.

Brown, P.E., 1991, Caledonian and earlier magmatism, *in* Craig, G.Y., ed., Geology of Scotland: London, Geological Society of London, p. 229–295.

Chappell, B.W., 1999, Aluminium saturation in I- and S-type granites and the characterization of fractionated haplogranites: Lithos, v. 46, p. 535–551, doi: 10.1016/S0024-4937(98)00086-3.

Coe, K., 1975, The Hurry Inlet Granite and Related Rocks of Liverpool Land, East Greenland: Bulletin Grønlands Geologiske Undersøgelse, v. 115, 34 p.

Dallmeyer, R.D., 1992, $^{40}Ar/^{39}Ar$ mineral ages within the Western Gneiss terrane, Troms, Norway: Evidence for polyphase Proterozoic tectonothermal activity (Svecokarelian and Sveconorwegian): Precambrian Research, v. 57, p. 195–206, doi: 10.1016/0301-9268(92)90002-6.

DePaolo, D.J., Lynn, A.M., and Schubert, G., 1991, The continental age distribution: Methods of determining mantle separation ages from Sm-Nd isotopic data and application to the southwestern United States: Journal of Geophysical Research, v. 96, p. 2071–2088, doi: 10.1029/90JB02219.

Fowler, M.B., and Henney, P.J., 1996, Mixed Caledonian appinite magmas: Implications for lamprophyre fractionation and high Ba-Sr granite genesis: Contributions to Mineralogy and Petrology, v. 126, p. 199–215, doi: 10.1007/s004100050244.

Fowler, M.B., Henney, P.J., Darbyshire, D.P.F., and Greenwood, P.B., 2001, Petrogenesis of high Ba-Sr granites: The Rogart pluton, Sutherland: Journal of the Geological Society of London, v. 158, p. 521–534.

Friend, C.R.L., Kinny, P.D., Rogers, G., Strachan, R.A., and Paterson, B.A., 1997, U-Pb zircon geochronological evidence for Neoproterozoic events in the Glenfinnan Group (Moine Supergroup): The formation of the Ardgour granite gneiss, north-west Scotland: Contributions to Mineralogy and Petrology, v. 128, p. 101–113, doi: 10.1007/s004100050297.

Gee, D.G., and Tebenkov, A.M., 1996, Two major unconformities beneath the Neoproterozoic Murchisonfjorden Supergroup in the Caledonides of central Nordaustlandet, Svalbard: Polar Research, v. 15, p. 81–91, doi: 10.1111/j.1751-8369.1996.tb00460.x.

Gee, D.G., Johansson, Å., Larionov, A.N., and Tebenkov, A.M., 1996 (reprinted in 1999), A Caledonian granitoid pluton at Djupkilsodden, Central Nordaustlandet, Svalbard: Age, magnetic signature and tectonic significance: Polarforschung, v. 66, p. 19–32.

Gilotti, J.A., and McClelland, W.C., 2005, Leucogranites and the time of extension in the East Greenland Caledonides: The Journal of Geology, v. 113, p. 399–417, doi: 10.1086/430240.

Gilotti, J.A., and McClelland, W.C., 2008, this volume, Geometry, kinematics, and timing of extensional faulting in the Greenland Caledonides—A synthesis, in Higgins, A.K., Gilotti, J.A., and Smith, M.P., eds., The Greenland Caledonides: Evolution of the Northeast Margin of Laurentia: Geological Society of America Memoir 202, doi: 10.1130/2008.1202(10).

Gilotti, J.A., Jones, K.A., and Elvevold, S., 2008, this volume, Caledonian metamorphic patterns in Greenland, in Higgins, A.K., Gilotti, J.A., and Smith, M.P., eds., The Greenland Caledonides: Evolution of the Northeast Margin of Laurentia: Geological Society of America Memoir 202, doi: 10.1130/2008.1202(08).

Haller, J., 1971, The Geology of the East Greenland Caledonides: New York, Interscience, 413 p.

Hamann, N.E., Whittaker, R.C., and Stemmerik, L., 2005, Structural and geological development of the North-East Greenland shelf, in Doré, A.G., and Vining, B.A., eds., Petroleum Geology: Northwest Europe and Global Perspectives: Proceedings of the 6th Petroleum Geology Conference: London, Geological Society of London, p. 887–902.

Hansen, B.T., and Tembusch, H., 1979, Rb-Sr isochron ages from east Milne Land, Scoresby Sund, East Greenland: Rapport Grønlands Geologiske Undersøgelse, v. 95, p. 96–101.

Hansen, B.T., Steiger, R.H., Henriksen, N., and Borchardt, B., 1987, U-Pb and Rb-Sr age determinations on Caledonian plutonic rocks in the central part of the Scoresby Sund region, East Greenland: Rapport Grønlands Geologiske Undersøgelse, v. 134, p. 5–18.

Hansen, B.T., Henriksen, N., and Kalsbeek, F., 1994, Age and origin of Caledonian granites in the Grandjean Fjord–Bessel Fjord region (75°–76°N), North-East Greenland: Rapport Grønlands Geologiske Undersøgelse, v. 162, p. 139–151.

Harland, B.W., 1997, The Geology of Svalbard: Geological Society of London Memoir 17, 521 p.

Hartz, E.[H.], and Andresen, A., 1995, Caledonian sole thrust of central East Greenland: A crustal-scale Devonian extensional detachment?: Geology, v. 23, p. 637–640.

Hartz, E.H., Andresen, A., Martin, M.W., and Hodges, K.V., 2000, U-Pb and $^{40}Ar/^{39}Ar$ constraints on the Fjord region detachment zone: A long-lived extensional fault in the central East Greenland Caledonides: Journal of the Geological Society of London, v. 157, p. 795–809.

Hartz, E.H., Andresen, A., Hodges, K.V., and Martin, M.W., 2001, Syncontractional extension and exhumation of deep crustal rocks in the East Greenland Caledonides: Tectonics, v. 20, p. 58–77, doi: 10.1029/2000TC900020.

Henriksen, N., 2003, Geological map of the Caledonian Orogen, East Greenland 70°–82°N: Copenhagen, Geological Survey of Denmark and Greenland, scale 1:1,000,000.

Henriksen, N., and Higgins, A.K., 1988, Descriptive Text to 1:100,000 Sheets Rødefjord 70 Ø.3 N and Kap Leslie 70 Ø.2 N: Copenhagen, Grønlands Geologiske Undersøgelse, 34 p.

Higgins, A.K., and Leslie, A.G., 2000, Restoring thrusting in the East Greenland Caledonides: Geology, v. 28, p. 1019–1022, doi: 10.1130/0091-7613 (2000)28<1019:RTITEG>2.0.CO;2.

Higgins, A.K., Elvevold, S., Escher, J.C., Frederiksen, K.S., Gilotti, J.A., Henriksen, N., Jepsen, H.F., Jones, K.A., Kalsbeek, F., Kinny, P.D., Leslie, A.G., Smith, M.P., Thrane, K., and Watt, G.R., 2004, The foreland-propagating thrust architecture of the East Greenland Caledonides, 72°–75°N: Journal of the Geological Society of London, v. 161, p. 1009–1026, doi: 10.1144/0016-764903-141.

Inger, S., and Harris, N., 1993, Geochemical constraints on leucogranite magmatism in the Langtang Valley, Nepal Himalaya: Journal of Petrology, v. 34, p. 345–368.

Irvine, T.N., and Baragar, W.R.A., 1971, A guide to the chemical classification of the common volcanic rocks: Canadian Journal of Earth Sciences, v. 8, p. 523–548.

Johannes, W., and Holtz, F., 1996, Petrogenesis and Experimental Geology of Granitic Rocks: Berlin, Springer, 335 p.

Johansson, Å., Larionov, A.N., Tebenkov, A.M., Gee, D.G., Whitehouse, M.J., and Vestin, J., 2000, Grenvillian magmatism of western and central Nordaustlandet, northeastern Svalbard: Transactions of the Royal Society of Edinburgh, Earth Sciences, v. 90, p. 221–254.

Johansson, Å., Larionov, A.N., Gee, D.G., Ohta, Y., Tebenkov, A.M., and Sandelin, S., 2004, Grenvillian and Caledonian tectono-magmatic activity in northeasternmost Svalbard, in Gee, D.G., and Pease, V., eds., The Neoproterozoic Timanide Orogen of Eastern Baltica: Geological Society of London Memoir 30, p. 207–232.

Kalsbeek, F., Nutman, A.P., and Taylor, P.N., 1993, Palaeoproterozoic basement province in the Caledonian fold belt of North-East Greenland: Precambrian Research, v. 63, p. 163–178, doi: 10.1016/0301-9268(93) 90010-Y.

Kalsbeek, F., Nutman, A.P., Escher, J.C., Friderichsen, J.D., Hull, J.M., Jones, K.A., and Pedersen, S.A.S., 1999, Geochronology of granitic and supracrustal rocks from the northern part of the East Greenland Caledonides: Ion microprobe U-Pb zircon ages: Geology of Greenland Survey Bulletin, v. 184, p. 31–48.

Kalsbeek, F., Thrane, K., Nutman, A.P., and Jepsen, H.F., 2000, Late Mesoproterozoic to early Neoproterozoic history of the East Greenland Caledonides: Evidence for Grenvillian orogenesis?: Journal of the Geological Society of London, v. 157, p. 1215–1225.

Kalsbeek, F., Jepsen, H.F., and Nutman, A.P., 2001a, From source migmatites to plutons: Tracking the origin of ca. 435 Ma S-type granites in the East Greenland Caledonian orogen: Lithos, v. 57, p. 1–21, doi: 10.1016/S0024-4937(00)00071-2.

Kalsbeek, F., Jepsen, H.F., and Jones, K.A., 2001b, Geochemistry and petrogenesis of S-type granites in the East Greenland Caledonides: Lithos, v. 57, p. 91–109, doi: 10.1016/S0024-4937(01)00038-X.

Kalsbeek, F., Thrane, K., Higgins, A.K., Jepsen, H.F., Leslie, A.G., Nutman, A.P., and Frei, R., 2008, this volume, Polyorogenic history of the East Greenland Caledonides, in Higgins, A.K., Gilotti, J.A., and Smith, M.P., eds., The Greenland Caledonides: Evolution of the Northeast Margin of Laurentia: Geological Society of America Memoir 202, doi: 10.1130/2008.1202(03).

Kystol, J., and Larsen, L.M., 1999, Analytical procedures in the rock geochemical laboratory of the Geological Survey of Denmark and Greenland: Geology of Greenland Survey Bulletin, v. 184, p. 59–62.

Leslie, A.G., and Higgins, A.K., 2008, this volume, Foreland-propagating Caledonian thrust systems in East Greenland, in Higgins, A.K., Gilotti, J.A., and Smith, M.P., eds., The Greenland Caledonides: Evolution of the Northeast Margin of Laurentia: Geological Society of America Memoir 202, doi: 10.1130/2008.1202(07).

Leslie, A.G., and Nutman, A.P., 2003, Evidence for Neoproterozoic orogenesis and early high temperature Scandian deformation events in the southern East Greenland Caledonides: Geological Magazine, v. 140, p. 309–333, doi: 10.1017/S0016756803007593.

McClelland, W.C., and Gilotti, J.A., 2003, Late-stage extensional exhumation of high-pressure granulites in the Greenland Caledonides: Geology, v. 31, p. 259–262, doi: 10.1130/0091-7613(2003)031<0259:LSEEOH> 2.0.CO;2.

Miyashiro, A., 1978, Nature of alkalic volcanic rock series: Contributions to Mineralogy and Petrology, v. 66, p. 91–104.

Nordgulen, Ø., Bickford, M.E., Nissen, A.L., and Wortman, G.L., 1993, U-Pb zircon ages from the Bindal Batholith, and the tectonic history of the Helgeland nappe complex, Scandinavian Caledonides: Journal of the Geological Society of London, v. 150, p. 771–783, doi: 10.1144/gsjgs.150.4.0771.

Pearce, J.A., Harris, N.B.W., and Tindle, A.G., 1984, Trace element discrimination diagrams for the tectonic interpretation of granitic rocks: Journal of Petrology, v. 25, p. 956–983.

Read, H.H., 1948, Granites and granites, in Gilluly, J., chairman, Origin of Granite: Geological Society of America Memoir 28, p. 1–19.

Read, H.H., 1957, The Granite Controversy: London, Thomas Murby & Co, 430 p.

Rex, D.C., and Gledhill, A.R., 1981, Isotopic studies in the East Greenland Caledonides (72°–74°N)—Precambrian and Caledonian ages: Rapport Grønlands Geologiske Undersøgelse, v. 104, p. 47–72.

Roberts, D., Melezhik, V.M., and Heldal, T., 2002, Carbonate formations and early NW-directed thrusting in the highest allochthons of the Norwegian Caledonides: Evidence of a Laurentian ancestry: Journal of the Geological Society of London, v. 159, p. 117–120.

Smith, M.P., and Rasmussen, J.A., 2008, this volume, Cambrian–Silurian development of the Laurentian margin of the Iapetus Ocean in Greenland and related areas, in Higgins, A.K., Gilotti, J.A., and Smith, M.P., eds., The Greenland Caledonides: Evolution of the Northeast Margin of Laurentia: Geological Society of America Memoir 202, doi: 10.1130/2008.1202(06).

Steiger, R.H., Hansen, B.T., Schuler, Ch., Bär, M.T., and Henriksen, N., 1979, Polyorogenic nature of the southern Caledonian fold belt in East Greenland: An isotopic age study: The Journal of Geology, v. 87, p. 475–495.

Strachan, R.A., 1994, Evidence in North-East Greenland for Late Silurian–Early Devonian regional extension during the Caledonian orogeny: Geology, v. 22, p. 913–916, doi: 10.1130/0091-7613(1994)022<0913:EINEGF> 2.3.CO;2.

Strachan, R.A., Martin, M.W., and Friderichsen, J.D., 2001, Evidence for contemporaneous yet contrasting styles of granite magmatism during extensional collapse of the northeast Greenland Caledonides: Tectonics, v. 20, p. 458–473, doi: 10.1029/2000TC001206.

Strachan, R.A., Smith, M., Harris, A.L., and Fettes, D.J., 2002, The Northern Highland and Grampian terranes, in Trewin, N.H., ed., The Geology of Scotland: London, Geological Society of London, p. 81–147.

Tarney, J., and Jones, C.E., 1994, Trace element geochemistry of orogenic igneous rocks and crustal growth models: Journal of the Geological Society of London, v. 151, p. 855–868, doi: 10.1144/gsjgs.151.5.0855.

Taylor, S.R., and McLennan, S.M., 1985, The Continental Crust: Its Composition and Evolution: Oxford, Blackwell, 312 p.

Thompson, R.N., 1982, British Tertiary volcanic province: Scottish Journal of Geology, v. 18, p. 49–107.

Thrane, K., 2002, Relationships between Archaean and Palaeoproterozoic crystalline basement complexes in the southern part of the East Greenland Caledonides: An ion microprobe study: Precambrian Research, v. 113, p. 19–42, doi: 10.1016/S0301-9268(01)00198-X.

Torsvik, T.H., Smethurst, M.A., Meert, J.G., Van der Voo, R., McKerrow, W.S., Brasier, M.D., Sturt, B.A., and Walderhaug, H.J., 1996, Continental break-up and collision in the Neoproterozoic and Palaeozoic—A tale of Baltica and Laurentia: Earth-Science Reviews, v. 40, p. 229–258, doi: 10.1016/0012-8252(96)00008-6.

Tucker, R.D., and McKerrow, W.S., 1995, Early Paleozoic chronology: A review in light of new U-Pb zircon ages from Newfoundland and Britain: Canadian Journal of Earth Sciences, v. 32, p. 368–379.

Watson, E.B., and Harrison, T.M., 1983, Zircon saturation revisited: Temperature and composition effects in a variety of crustal magma types: Earth and Planetary Science Letters, v. 64, p. 295–304, doi: 10.1016/0012-821X (83)90211-X.

Watt, G.R., Kinny, P.D., and Friderichsen, J.D., 2000, U-Pb geochronology of Neoproterozoic and Caledonian tectonothermal events in the East Greenland Caledonides: Journal of the Geological Society of London, v. 157, p. 1031–1048.

White, A.P., and Hodges, K.V., 2002, Multistage extensional evolution of the central East Greenland Caledonides: Tectonics, v. 21, no. 5, p. 1048, doi: 10.1029/2001TC001308.

White, A.P., Hodges, K.V., Martin, M.W., and Andresen, A., 2002, Geologic constraints on middle-crustal behavior during broadly synorogenic extension in the central East Greenland Caledonides: International Journal of Earth Sciences, v. 91, p. 187–208, doi: 10.1007/s005310100227.

Wickham, S.M., 1987, Crustal anatexis and granite petrogenesis during low-pressure regional metamorphism: The Trois Seigneurs Massif, Pyrenees, France: Journal of Petrology, v. 28, p. 127–169.

Williams, I.S., 1998, U-Th-Pb geochronology by ion microprobe, in McKibben, M.A., Shanks, W.C., III, and Ridley, W.I., eds., Applications of Microanalytical Techniques to Understanding Mineralizing Processes: Reviews in Economic Geology, v. 7, p. 1–35.

Wilson, M., 1989, Igneous Petrogenesis: London, Unwin Hyman, 466 p.

Winchester, J.A., 1988, Later Proterozoic environments and tectonic evolution in the northern Atlantic lands, in Winchester, J.A., ed., Later Proterozoic Stratigraphy of the Northern Atlantic Regions: Glasgow, Blackie and Son Ltd., p. 253–270.

MANUSCRIPT ACCEPTED BY THE SOCIETY 14 JANUARY 2008

Geometry, kinematics, and timing of extensional faulting in the Greenland Caledonides—A synthesis

Jane A. Gilotti*
Department of Geoscience, University of Iowa, Iowa City, Iowa 52242, USA

William C. McClelland†
Department of Geological Sciences, University of Idaho, Moscow, Idaho 83844, USA

ABSTRACT

The North-East Greenland Caledonides record a complex history of crustal thickening and extension during the Paleozoic collision of Baltica with Laurentia. We divide the southern portion of the orogen (70°N–76°N) into three plates separated by low-angle fault systems that are interpreted as extensional detachments superimposed on, and perhaps coeval with, the thrust geometry of the orogen. From structurally lowest to highest, the plates include amphibolite-facies Archean to Paleoproterozoic orthogneiss and lesser paragneiss that retain relics of Devonian high-pressure metamorphism, migmatitic Mesoproterozoic metasedimentary rocks with Silurian leucogranites and lesser orthogneiss at amphibolite-facies conditions, and low-grade Neoproterozoic to Ordovician sedimentary rocks. Individual detachments are characterized by superposition of cataclastic features on mylonitic fabrics, and they record progressive deformation that accommodated exhumation. The extensional faults define two detachment systems that evolved at different crustal levels during two episodes of movement. The upper detachment system, which separates the upper and middle plates, exhumed the midcrustal rocks after ca. 420 Ma. Extension was contemporaneous with crustal thickening and closely followed leucogranite emplacement. The structure may be analogous to the South Tibetan detachment system in the present-day Himalayas. Continental Old Red Sandstone deposition began in the Eifelian, closely following high-pressure metamorphism in the lower plate at ca. 405 Ma. The lower detachment was probably active at some depth below the evolving Devonian basins. The lower detachment system brought lower-plate metamorphic rocks to shallower crustal levels after 400 Ma, excising the overlying extensional system. This second period of extension was similar in timing and style to extension in the Scandinavian Caledonides. Displacement on the younger detachments, which exhumed lower-plate rocks, was broadly syncollisional, as indicated by the overlap in age with ultrahigh-pressure metamorphism

---

*jane-gilotti@uiowa.edu
†Present address: Department of Geoscience, University of Iowa, Iowa City, Iowa 52242, USA.

Gilotti, J.A., and McClelland, W.C., 2008, Geometry, kinematics, and timing of extensional faulting in the Greenland Caledonides—A synthesis, *in* Higgins, A.K., Gilotti, J.A., and Smith, M.P., eds., The Greenland Caledonides: Evolution of the Northeast Margin of Laurentia: Geological Society of America Memoir 202, p. 251–271, doi: 10.1130/2008.1202(10). For permission to copy, contact editing@geosociety.org. ©2008 The Geological Society of America. All rights reserved.

in the north at 365–350 Ma, and it may have been synchronous with young thrusts that emplaced high-pressure lower-plate rocks over the foreland and with strike-slip faults in the hinterland. Conversion to extension, accommodated by high-angle brittle faulting in the Carboniferous (after 345 Ma), may mark the final transition to plate divergence that ultimately led to continental rifting.

Keywords: detachment, excision, exhumation, extension, Greenland Caledonides.

INTRODUCTION

Regional extensional detachment faults have been recognized as important structures in most collisional orogens (e.g., Burchfiel and Royden, 1985; Dewey, 1988; Mezger et al., 1991; Burchfiel et al., 1992; Selverstone, 1988, 2005; Frotzheim et al., 1994; Andersen, 1998). Such regional detachment faults are commonly characterized by listric geometries and can be subhorizontal over long distances. Domed metamorphic core complex geometries are also common (Coney, 1980; Burg et al., 1984; Mattauer et al., 1988; Lister and Davis, 1989; Chen et al., 1990; Burg and Vanderhaeghe, 1993; Mancktelow and Pavlis, 1994; Wawrzyniec et al., 2001). Numerous tectonic processes have been postulated for the formation of extensional structures in orogenic belts, including synorogenic gravitational collapse of an overthickened wedge (Dewey, 1988), extrusion of migmatitic middle crust (Beaumont et al., 2001; Grujic et al., 2002), extension related to a change in plate motion at the end of collision (Fossen, 2000), or postorogenic collapse (Coney and Harms, 1984; Platt and Vissers, 1989). Our understanding of synorogenic extension, in particular, has been enhanced by studies of the ongoing collision between India and Asia to form the Himalayas and raise Tibet because the geometric and chronological relationships of extension to contraction and strike-slip are relatively clear (see review in Hodges, 2000). Similarly, the Alpine orogen provides modern analogues for late-stage extension in collisional orogens (Selverstone, 2005). The geometry and timing of extension in ancient, exhumed orogens are bound to be more complicated because superposition of the various mechanisms will have operated on different structural levels through time. For example, the tectonic significance of extensional detachments in the Paleozoic Caledonian orogen is an issue of some debate, in part due to the fact that the orogen began rifting to form the Atlantic Ocean in the Mesozoic. Resolution of the role of extension within ancient orogens, such as the Caledonides, will aid in understanding deep-level processes currently active in modern collisional settings.

Extensional fault systems are a prominent architectural element in the southern half of the Greenland Caledonides, from Bessel Fjord (76°N) to Scoresby Sund (70°N) (Fig. 1). Low-angle normal faults have long been recognized (e.g., Fränkl, 1953; Haller, 1971) and assigned regional significance (e.g., McClay et al., 1986). Nevertheless, an ongoing discussion within the Geological Survey of Denmark and Greenland (GEUS) mapping group and others concerns the significance of the extensional structures in terms of their regional extent, geometry, and displacement history. A minimalist treatment of extension is presented by Higgins et al. (2004), which emphasizes the thrust geometry associated with imbrication and juxtaposition of tectonostratigraphic units along low-angle faults. In this view, the extensional structures are interpreted as moderately steep, normal faults that offset the thrust stack by 10 km at most. This is the interpretation presented in the cross sections on the 1:1,000,000 map (Henriksen, 2003; see also map supplement with this volume). Others (e.g., Strachan, 1994; Hartz and Andresen, 1995; Watt et al., 2000; Gilotti and Elvevold, 2002; White et al., 2002) consider the extensional structures to be low-angle detachment surfaces that parallel the main tectonostratigraphic boundaries and are superimposed on the thrust architecture. The magnitudes of displacement are unknown, but judging from the pressure differences preserved on either side of the detachments, they are in excess of several tens of kilometers.

The aim of this paper is to review the geometry and kinematics of the map-scale extensional fault systems in the Greenland Caledonides. The timing of displacement on these structures is discussed in light of the available data. We then present a "maximalist" interpretation by assuming that the extensional geometry on documented detachments can be projected to lesser known areas. Our interpretation differs significantly from that shown on the 1:1,000,000 map provided with this volume because we distinguish between extensional listric detachment faults that parallel major tectonostratigraphic contacts and the moderately steep normal faults that simply displace the thrust packages or bound the Devonian and later sedimentary basins. The fundamental difference between the two interpretations is highlighted by a comparison of the relationships shown in cross

Figure 1. Geological map emphasizing extensional detachments in the Greenland Caledonides (70°N–76°N), modified from the 1:1,000,000 geological map (Henriksen, 2003; Henriksen and Higgins, Plate 1, Chapter 14, this volume). The bold arrows are displacement directions taken from the literature. Place name abbreviations: A—Alpefjord; AL—Andrée Land; AF—Ardencaple Fjord; FF—Forsblad Fjord; HL—Hudson Land; J—Junctiondal; K—Kempe Fjord; KB—Kap Bayard; KH—Kap Hedlund; M—Murgangsdal; N—Nordenskiöld Gletscher; S—Schaffhauserdalen; SA—Stauning Alper; SL—Strindberg Land; TTL—Th. Thomsen Land; W—Waltershausen Gletscher. Abbreviations of major structures: BSZ—Bessel Fjord shear zone; BBF—Boyd Bastion fault; EFZ—Eastern fault zone; FRD—Fjord region detachment; HD—Højedal detachment; KSZ—Kildedal shear zone; LSZ—Lerbugt shear zone; PLD—Payer Land detachment; RD—Rendalen detachment; TD—Tinderne detachment; PBD—Petermann Bjerg detachment; WFZ—Western fault zone.

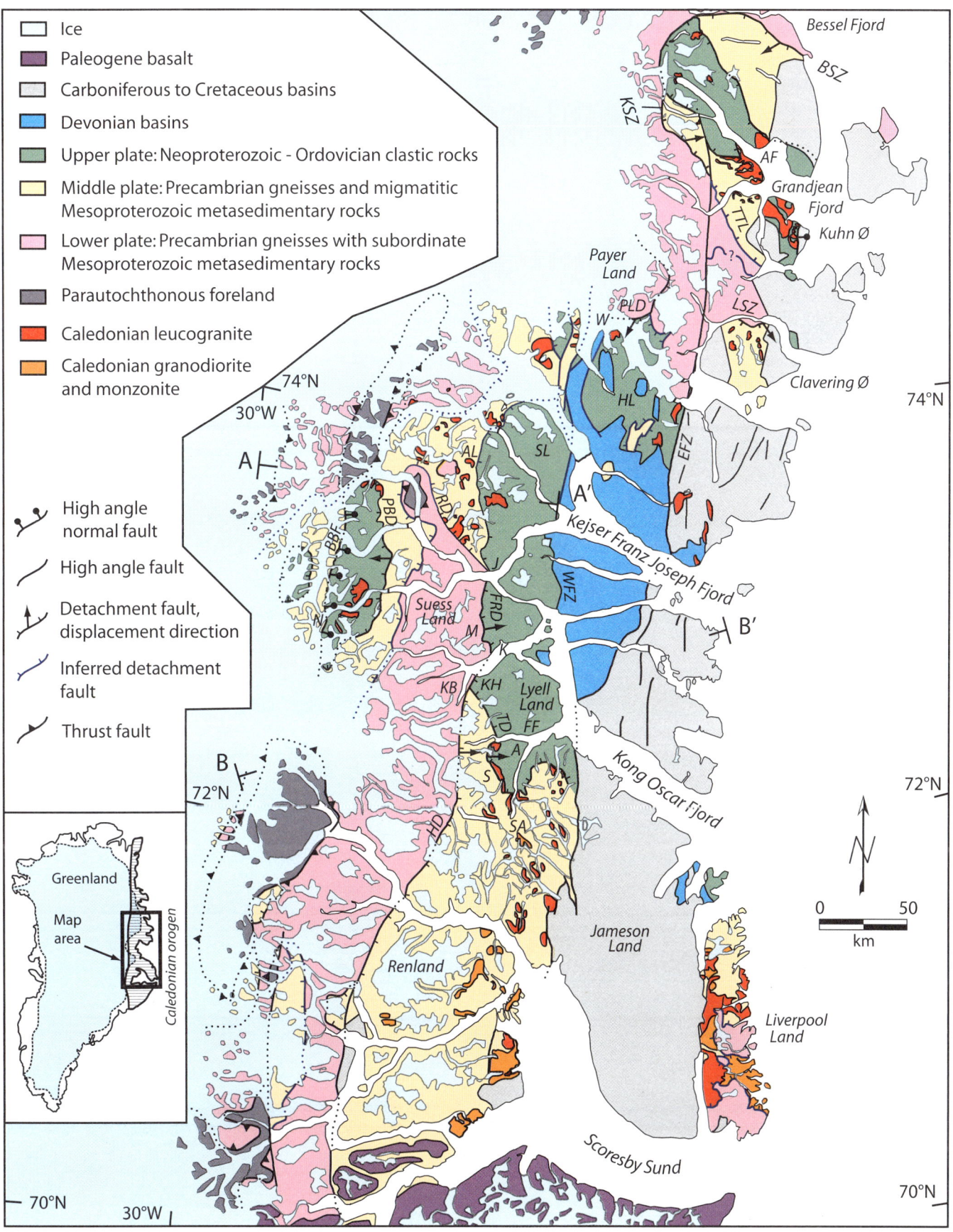

section A–A′ (Fig. 2) and the same cross section depicted in Higgins et al. (2004). Our ultimate goal is to assess the tectonic significance of regional extension in the Greenland Caledonides as it relates to the evolution of the orogen.

EXTENSIONAL ARCHITECTURE IN THE GREENLAND CALEDONIDES

The Caledonian orogen south of Bessel Fjord (76°N) consists of different tectonostratigraphic levels that are juxtaposed against one another by shear zones. The lowermost sheet consists of parautochthonous foreland sedimentary rocks best seen in windows through the overlying Niggli Spids thrust sheet, an Archean to Paleoproterozoic orthogneiss package with subordinate paragneiss. The overlying Hagar Bjerg thrust sheet contains the same Precambrian orthogneisses, but they are less abundant than the Mesoproterozoic Krummedal metasedimentary rocks. Neoproterozoic to Ordovician sedimentary rocks comprise the uppermost level in the stack and are known as the Franz Joseph allochthon. The faults framing the foreland windows are undisputed west-directed thrusts; they put older rocks on younger rocks and high-grade metamorphic rocks on low-grade metasedimentary rocks. The higher levels have also been described as a sequence of thrust sheets (Higgins and Leslie, 2000; Higgins et al., 2004; Leslie and Higgins, this volume). However, many of the contacts separating the upper tectonostratigraphic levels have been documented as extensional detachments, where retrograde mylonites capped by brittle faults have downdip kinematic indicators and place lower-grade units over high-grade rocks (e.g., Hartz and Andresen, 1995)—resulting in a significant difference of opinion as to whether some contacts have thrust or normal displacement, or both.

A synoptic map (Fig. 1) shows our interpretation of the maximum extent of the extensional province. Documented extensional structures are shown in black, and inferred segments are shown in blue to facilitate comparison with the 1:1,000,000 tectonic map (Henriksen, 2003). The maximum extension interpretation is based on extrapolating from the known extensional contacts to areas with a similar geometry and structural juxtaposition but undocumented kinematics. Brittle, steeper normal faults, including those that bound or cut the Devonian to Jurassic basins, are shown with different ornaments than the plastic to brittle extensional detachments.

We describe the extensional architecture of the orogen in terms of lower, middle, and upper plates, which roughly correspond to the Niggli Spids thrust sheet, the Hagar Bjerg thrust sheet, and the Franz Joseph allochthon (Higgins et al., 2004), respectively. Although relics of Precambrian structure and metamorphism exist, only Caledonian metamorphic patterns (Gilotti et al., this volume) are used to delineate extensional relationships. Devonian and younger basins situated in the eastern part of the Fjord region either unconformably or structurally overlie the older rocks.

The lower plate of the extensional system consists of Archean to Paleoproterozoic quartzofeldspathic orthogneiss complexes with subordinate mafic intrusive and metasedimentary rocks. Layers and lenses of marble, schist, and paragneiss are infolded with the orthogneisses and generally correlate with the Mesoproterozoic Krummedal sequence in the overlying structural level. Orthogneiss and paragneiss are cut by a variety of mafic dikes and sills of uncertain age. Structures in the lower plate are extremely complex making it difficult to unravel the Precambrian versus Caledonian deformation history. The majority of the rocks preserve amphibolite-facies metamorphic assemblages, but relics

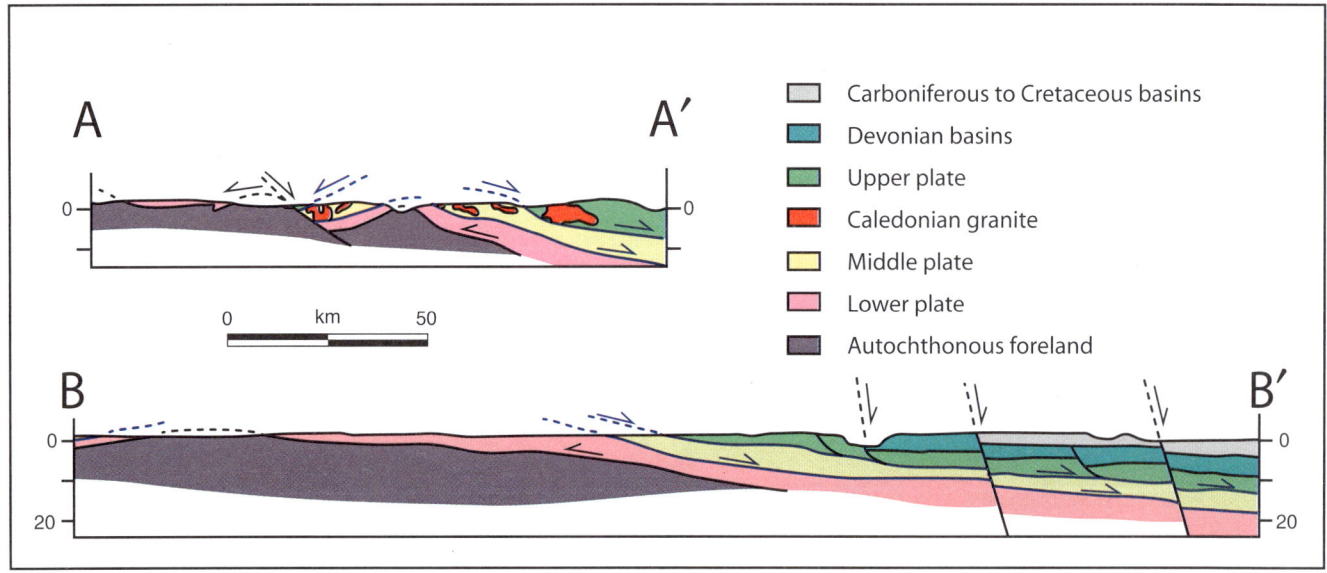

Figure 2. Regional cross sections (V=H) showing the extensional architecture of the Caledonides; see Figure 1 map for locations.

of a higher-pressure Caledonian history are locally preserved throughout the region (Gilotti et al., this volume). The lower plate, therefore, exposes the deepest levels of the orogen, even though most of the rocks are now in the amphibolite facies. Some authors (Andresen and Hartz, 1998; Andresen et al., 1998) take the view that all the amphibolite-facies, ductile fabrics are related to extensional flow in the middle crust. Others have argued for preservation of Caledonian if not older contractional structures within the lower plate (Higgins et al., 1981; Rex and Gledhill, 1981).

The middle plate is characterized by a thick Mesoproterozoic metasedimentary unit known as the Krummedal sequence (Higgins, 1988) with subordinate Precambrian orthogneiss. The supracrustal sequence consists of high-grade psammitic, semi-pelitic, and pelitic gneisses with subordinate amphibolite and marble interlayers. Metamorphic grade varies considerably but is consistent with this being a thick section subjected to midcrustal metamorphic conditions (Gilotti et al., this volume), even though pressures up to 1 GPa have been reported from Clavering Ø (Jones and Escher, 2002) and Forsblad Fjord (White and Hodges, 2003). Partial melting of the pelites in the Krummedal sequence produced abundant migmatites and leucogranites in the middle plate. Kalsbeek et al. (2000) showed that some of the granites have an age of 940 Ma and argued for local derivation of the granitic melts from the Krummedal sequence at that time. Caledonian leucogranites and migmatite complexes, formed by a second round of Silurian partial melting of the same unit (Kalsbeek et al., 2001a, 2001b, this volume), comprise vast areas of the middle plate in Stauning Alper, Lyell Land, Andrée Land, Kuhn Ø, and Ardencaple Fjord (Fig. 1).

The upper plate, corresponding to the Franz Joseph allochthon (Higgins et al., 2004; Higgins and Leslie, this volume), is composed of low-grade to unmetamorphosed, relatively undeformed Neoproterozoic to Ordovician sedimentary rocks. In the central Fjord region, the Neoproterozoic Eleonore Bay Supergroup consists of a >13-km-thick sequence of sedimentary rocks that is overlain by the Vendian Tillite Group (<1 km) and a 4-km-thick package of Cambrian to Ordovician clastic and carbonate sedimentary rocks (Sønderholm and Tirsgaard, 1993; Higgins et al., 2004). Kilometer-scale, Silurian leucogranitic plutons and steep dikes intrude the Eleonore Bay Supergroup throughout the Fjord region and are thought to have originated in the migmatite complexes of the middle plate (Kalsbeek et al., 2001a, 2001b). Metamorphic grade increases downward in the section and reaches greenschist or locally amphibolite facies near the base, particularly in the pelitic units of the lowermost Eleonore Bay Supergroup (Peucat et al., 1985; Soper and Higgins, 1993; Gilotti et al., this volume). The Franz Joseph allochthon is considered to have been carried passively in the upper part of the Hagar Bjerg thrust sheet (Higgins et al., 2004; Higgins and Leslie, this volume) prior to extension.

Devonian and younger sedimentary rocks both unconformably overlie the Neoproterozoic to Ordovician sequence and are in fault contact with the older rocks in the eastern part of the area (Fig. 1). Old Red Sandstone deposition in intermontane basins began in the Eifelian to Givetian, and it makes up two distinct sequences, the 3-km-thick Vilddal Supergroup and the 6.5-km-thick Gauss complex (Friend et al., 1983; Larsen et al., this volume). Deposition continued through a long period of rifting that began near the Devonian-Carboniferous boundary and continued through the Cretaceous (Hamann et al., 2005). True rifting associated with the opening of the Atlantic Ocean began in the mid-Jurassic and peaked near the Jurassic-Cretaceous boundary (Hamann et al., 2005). Geophysical models based on seismic refraction studies (Schlindwein and Jokat, 1999) depict rift-related crustal thinning toward the east. North of Kong Oscar Fjord, the Moho shows three steps, which Schlindwein and Jokat (1999) interpreted as a Caledonian, Devonian, and Mesozoic configuration. The Devonian step does not appear to the south where the Tertiary flood basalt province is exposed.

GEOMETRY AND KINEMATICS OF THE EXTENSIONAL DETACHMENTS

In the following sections, we describe the structural geometry and kinematics of studied extensional detachments from north to south. Some areas are documented in detail, but many areas have only reconnaissance information, and vast segments of inferred detachments are simply unknown. The point of this description is to outline the existing data in order to draw some general conclusions about the style of extensional deformation in East Greenland.

Ardencaple Fjord

Arcuate outcrops of the three structural levels are separated by extensional detachments in the area between 75°N and 76°N surrounding Ardencaple Fjord (Fig. 1; Soper and Higgins, 1993; Friderichsen et al., 1994; Strachan, 1994). Paleoproterozoic orthogneisses interpreted as belonging to the lower plate comprise the outer arc, while two wedges of the Smallefjord metasedimentary unit occur in the northeast and southwest parts of the area. The Smallefjord rocks are dominated by high-grade pelitic to psammitic schists and gneisses (Jones and Strachan, 2000) that are correlated with the Krummedal sequence (Friderichsen et al., 1994) and thus form the middle plate in our terminology. The Neoproterozoic Eleonore Bay Supergroup structurally overlies the middle plate and is in extensional contact with lower-plate gneisses in the northwest part of the area.

Two ductile, extensional detachments form the contact between Paleoproterozoic orthogneisses and the Mesoproterozoic Smallefjord sedimentary unit (Fig. 1). These are the Bessel Fjord shear zone in the northeast and the Kildedal shear zone in the southwest (Friderichsen et al., 1994). The western segment of the 70-km-long Bessel Fjord shear zone is an east-striking, moderately southward dipping, 150–200-m-thick zone of amphibolite-facies mylonites developed in both hanging wall schists and footwall orthogneisses. Stretching lineations in the mylonitic foliation plunge <30° toward the SW. Asymmetric porphyroclasts and shear bands give a top-to-the-SW sense of shear. In the mylonitic schists, sillimanite needles are parallel to the stretching lineation defined by quartz and feldspar aggregates, but fibrolite mats also overgrow the mylonitic foliation. Deformation is localized into three distinct shear zones in the eastern, SE-striking segment of the Bessel Fjord shear zone: one at the contact with the Niggli Spids orthogneiss and two within the lower part of the Smallefjord sequence. The mylonites dip moderately to steeply southwestward; we assume that the kinematics are similar to the western segment, even though the two segments are separated by a steep, brittle, NNE-striking dextral fault.

The Kildedal shear zone (Friderichsen et al., 1994) forms the SE-striking boundary of the Smallefjord sequence against the Paleoproterozoic orthogneisses in the southwest part of the area (Fig. 1). The orthogneisses beneath the contact are characterized by a 1.2–3.5-km-thick zone of strongly deformed, flaggy tectonites, while a 150 m section at the base of the Smallefjord rocks is mylonitic. A sharp brittle fault separates the two lithologies in Kildedalen, but it is not known if it persists along the entire contact. The amphibolite-facies mylonitic foliation is moderately east dipping, and stretching lineations plunge toward the east. Friderichsen (1999) mapped the southern extension of the Kildedal shear zone into Th. Thomsen Land, where it cuts upsection into a thick section of the Smallefjord sequence but retains a uniform 40°–50° eastward dip and east-plunging mineral and stretching lineations. Rare kinematic indicators, such as rotated porphyroclasts, shear bands, and offset veins, give a top-to-the-east sense of asymmetry along the length of the shear zone. This southern segment of the Kildedal shear zone does not everywhere follow the major tectonostratigraphic contact between Proterozoic orthogneiss and Smallefjord supracrustal rocks (i.e., the queried contact in Fig. 1), which we infer to be an older structure; instead it cuts the Smallefjord sequence. We assign the thick section of Smallefjord rocks in the footwall of the Kildedal shear zone to the lower plate but recognize that other interpretations are possible.

The inner arc of outcrop is composed of the Neoproterozoic Eleonore Bay Supergroup, which is the lower part of the Franz Joseph allochthon of Higgins et al. (2004) and the upper plate of the extensional architecture. The upper part of the section is unmetamorphosed, but metamorphic grade increases dramatically to garnet and staurolite zone at the base (Soper and Higgins, 1993). The entire outcrop area is bounded by at least three different fault segments (Fig. 1). (1) In the northwest part of the area, the Eleonore Bay Supergroup rests directly on lower-plate orthogneiss along the first fault segment, which connects the Kildedal and Bessel Fjord shear zones. The orthogneiss shows the same degree of deformation as in the Kildedal shear zone; however, the contact is marked by a brittle fault (Friderichsen et al., 1994). Kinematic information is not available, but the younger-on-older geometry and linkage to the Bessel Fjord and Kildedal shear zones suggest extensional displacement as well. (2) The second segment is a high strain zone bounding the west side of the Eleonore Bay Supergroup that shows evidence for top-to-the-east extensional shear (Friderichsen et al., 1994; Strachan, 1994). A 200–300-m-thick, shallow to moderately dipping, ductile mylonite zone is formed in both the Smallefjord sequence and in the low-grade sedimentary rocks of the Eleonore Bay Supergroup. In places, this contact is cut by brittle faults. (3) Finally, the northeastern boundary also juxtaposes the Neoproterozoic low-grade sedimentary rocks against Mesoproterozoic pelitic gneisses along a mylonite zone. Soper and Higgins (1993) recognized both top-to-the-east and top-to-the-west senses of shear along this west-dipping boundary, which they interpreted as Vendian extension overprinted by Caledonian thrusting on a paleorift shoulder. Based on present-day geometry, we depict this as a Caledonian extensional detachment in Figure 1, in keeping with our maximalist interpretation.

Melting of the Smallefjord sequence is thought to have been the source of the leucogranites, which form mostly concordant sheets within the foliated paragneisses and schists, and steep-sided plutons within the Eleonore Bay Supergroup. Conceptual models link melting to extension (Strachan et al., 2001), but direct evidence is weak. For example, no granites have been observed to cut the detachment zone mylonites, and the plutons are deformed by solid-state foliations and lineations (Friderichsen et al., 1994).

Clavering Ø

The Lerbugt shear zone (Jones and Escher, 1999, 2002) is an ESE-striking structure that clips the northern part of Clavering Ø and separates middle-plate (above) from lower-plate orthogneisses (below) (Figs. 1 and 3). The mylonitic foliation dips ~50°SE and contains a strong SE-plunging lineation. Movement sense from asymmetric pods and shear bands is top-down-to-the-SE. The footwall of the Lerbugt shear zone is composed of orthogneisses that retain a high-pressure granulite-facies signature in mafic lenses but are dominated by amphibolite-facies assemblages. The hanging wall is an interleaved sequence of subordinate orthogneiss slivers and Krummedal metasedimentary rocks in the high-temperature, lower-pressure granulite facies. The complex hanging wall contains a number of presumably older thrusts, and it is dissected by numerous, steep, NNE-striking, top-to-the-east extensional shear zones. Jones and Escher (1999) interpreted the structure as an extensional duplex floored by the Lerbugt shear zone (Fig. 3B). This style of deformation probably reflects extensional displacement at relatively deep levels in the orogen.

Payer Land Detachment

The Payer Land detachment (Fig. 1; Gilotti and Elvevold, 2002; Gilotti and McClelland, 2005) separates high-pressure granulite-facies orthogneisses and paragneisses in the lower plate from overlying klippe of greenschist-facies metasedimentary rocks of the Eleonore Bay Supergroup in the upper plate. The Payer Land detachment fault is defined as the brittle surface between penetratively deformed ultramylonite and cataclasite

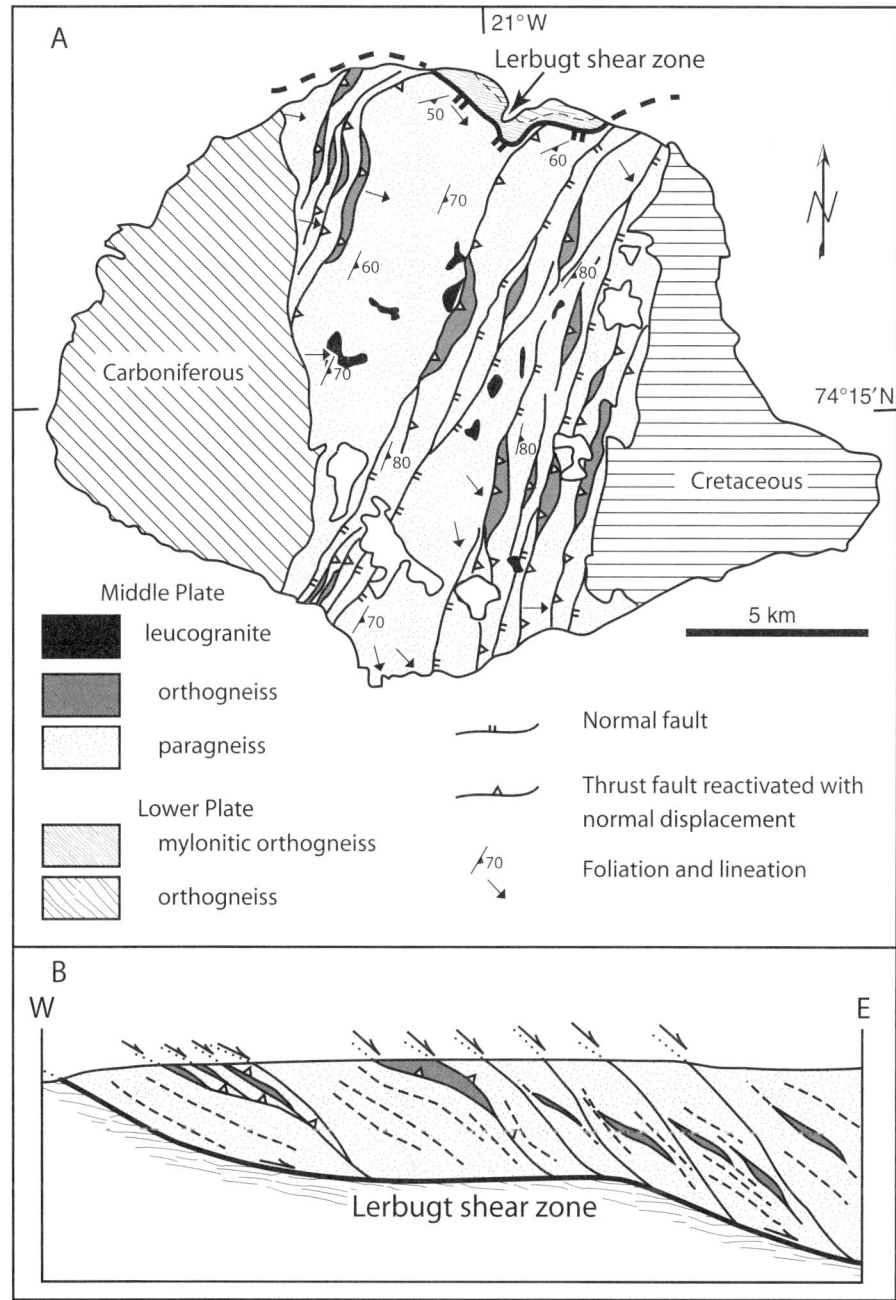

Figure 3. (A) Map and (B) cross section of an extensional duplex developed in the middle plate on Clavering Ø (modified from Jones and Escher, 1999).

of variable protolith affinity and the structurally overlying, less-deformed quartzites of the Eleonore Bay Supergroup. The mylonites beneath the brittle fault of the Payer Land detachment consist of a 1500-m-thick package of protomylonites, mylonites, and ultramylonites primarily derived from the high-grade paragneisses. The mylonitic foliation is defined by the <0.7 GPa retrograde assemblage of sillimanite + biotite + plagioclase + quartz ± muscovite, and it dips moderately to the SE. A sillimanite mineral lineation and a quartz-feldspar aggregate stretching lineation plunge shallowly toward the NE or SW. Displacement is top-to-the-SW based on abundant kinematic indicators, such as δ- and σ-porphyroclasts of kyanite, garnet, and feldspar, mica fish, shear bands, and offset markers on small shear zones.

The hanging wall of the Payer Land detachment fault is dominated by low-grade quartzite, metasandstone, and metasiltstone characteristic of the Neoproterozoic Eleonore Bay Supergroup. Only small klippe of the upper plate are preserved in Payer Land; therefore, it is unclear which part of the section is present. To the southeast in Hudson Land, the Eleonore Bay section is overlain by the Vendian Tillite Group and carbonate and clastic rocks of Paleozoic age (Fig. 1; Stouge et al., 2002), suggesting that the klippe lie at the base of what was originally a much larger expanse of the upper plate. Quasi-plastic mylonites are found in Eleonore Bay lithologies above the brittle Payer Land detachment fault. Chlorite ± biotite, associated with quartz microstructures, suggest a transition from plastic to brittle deformation under greenschist-facies conditions. Cumulative displacement along the Payer Land detachment fault ultimately juxtaposed the deep-level high-pressure–high-temperature granulites (maximum pressure [P] = 1.4–1.6 GPa) with the Eleonore Bay Supergroup at shallow crustal levels, and this represents the largest pressure difference across an extensional structure documented in East Greenland. The estimated throw on this detachment is 40–50 km, which requires displacements on the order of 80–100 km on a surface with a 30° dip.

Fjord Region Detachment System

The Fjord region detachment is the N-S striking, E-dipping, ductile extensional mylonite zone that separates lower-plate gneiss from overlying units in the central Caledonides from 72°N to 74°N (Hartz and Andresen, 1995). In many places, the ductile mylonites are truncated by a brittle normal fault, which is distinguished as the Fjord region detachment fault (Hartz et al., 2000). Branch lines at Kap Hedlund (Elvevold and Gilotti, 1998) and near Junctiondal (Escher and Jones, 1999) divide the Fjord region detachment into splays (Fig. 1). The lower splay separates lower-plate gneisses from migmatites in the middle plate, while the upper splay juxtaposes the middle plate and the upper plate. The entire brittle-ductile fault complex is referred to as the Fjord region detachment system.

Main Segment: Upper Plate on Lower Plate

The main segment of the Fjord region detachment separates the gneiss complexes of the lower plate from different stratigraphic levels of the Neoproterozoic Eleonore Bay Supergroup in the upper plate. The detachment is spectacularly exposed in the walls of Kejser Franz Joseph Fjord and Kempe Fjord, where it has received the most attention (Hartz and Andresen, 1995; Andresen et al., 1998; Hartz et al., 2000, 2001). A suite of common features characterizes the Fjord region detachment along its strike length. These are nicely illustrated by a completely exposed stream section in Murgangsdal, Suess Land (Fig. 4; Elvevold and Gilotti, 1998). The especially clean section was created in 1932 by a mudslide (i.e., *murgang*), which broke the ice dam of a large lake and filled the valley with a 1.5-km-wide swath of water 4 m deep in places (Koch, 1955). A cross section (Fig. 4B) illustrates the salient features of the Fjord region detachment.

The Fjord region detachment at Murgangsdal consists of an ~800-m-thick zone of plastic mylonites, derived from the footwall, capped by 80 m of dark-green cataclasites (Fig. 5). The cataclasites mark a brittle fault zone that is subparallel to the ductile detachment mylonites. The cataclasites contain feldspar clasts and fragments of mylonite that point to a gneissic protolith. The base of the upper plate begins in clean, bedded quartzites from the middle of the Eleonore Bay Supergroup. Unlike Payer Land, the quartzites in the hanging wall are not mylonitic, indicating that they have been truncated at shallow levels by the brittle fault.

The deformation zone is dominated by pinstriped, quartzofeldspathic protomylonitic orthogneiss, but it also contains metasedimentary protoliths, including marble, calc-silicate, and mylonitic schist. The mylonitic foliation dips shallowly to the east and contains an approximately downdip lineation (Fig. 4C). Lineations consist of stretched quartz-feldspar aggregates, aligned rod-shaped minerals, commonly amphibole and sillimanite, and pull-aparts of these rods (Fig. 6). The augen mylonites show progressive grain-size reduction approaching the cataclastic zone. Feldspar augen have poorly recrystallized tails and, thus, are not very useful kinematic indicators. A top-to-the-east shear sense is deduced from abundant shear bands, particularly in the mylonitic schists (Figs. 6A and 6B). Two sets of shear bands, a steep one and another at a low angle to the mylonitic foliation, are present in some cases. The preferred orientation of fine-grained biotite and chlorite defines the shear bands. Mesoscopic offsets of compositional layering also give a top-to-the-east sense of displacement. Deformation in the mylonite zone gradually decreases downward. Parts of the footwall beneath the mylonite zone contain gneisses with fabrics, dikes, and shear zones that are discordant to the Fjord region detachment (Fig. 5A), which we interpret as low-strain lenses that preserve the preexisting structure of the lower-plate gneisses.

The mylonitic schists contain garnet porphyroclasts and polyphase augen of white kyanite-bearing leucocratic melt pods. Good examples of these same garnet-kyanite schists with kyanite-bearing leucosomes are found 30 km farther south at Kap Bayard (Fig. 7A), and mylonitic versions are exposed at Kap Hedlund. These deformed quartz + K-feldspar + plagioclase + kyanite ± garnet leucosomes were formed by melting of pelitic schists under conditions grossly similar to the anatectic metapelites from Payer Land (800–850 °C and 1.4–1.5 GPa). The kyanite-bearing leucocratic augen are converted to stretched sillimanite-rich clumps in the footwall of the Fjord region detachment at Murgangsdal (Fig. 7B) and throughout the region. A comparison of the greenschist-facies (maximum) quartzites in the hanging wall at Murgangsdal with the higher-pressure gneisses in the footwall shows that the Fjord region detachment separates rocks with a minimum pressure difference of 0.8–0.9 GPa, which represents ~30 km of missing crust. Vold (1997) estimated peak conditions of 700 °C and 0.7 GPa from garnet-biotite thermometry and garnet-aluminosilicate-plagioclase (GASP) barometry of the schists at Kap Hedlund but recognized that this was a minimum estimate. Garnet + clinopyroxene + amphibole + plagioclase + quartz in mafic rocks at Kap Hedlund, similar to mafic granulites in Payer Land (Elvevold et al., 2003), also point to higher pressures in the footwall. The pressure difference is somewhat less at Kap Hedlund than Murgangsdal because the upper-plate rocks are in the garnet zone.

Deformation of the sedimentary rocks in the hanging wall of the Fjord region detachment is variable. Immediately above the detachment fault at Murgangsdal, the bedded quartzites are folded into a train of upright, small-amplitude décollement folds

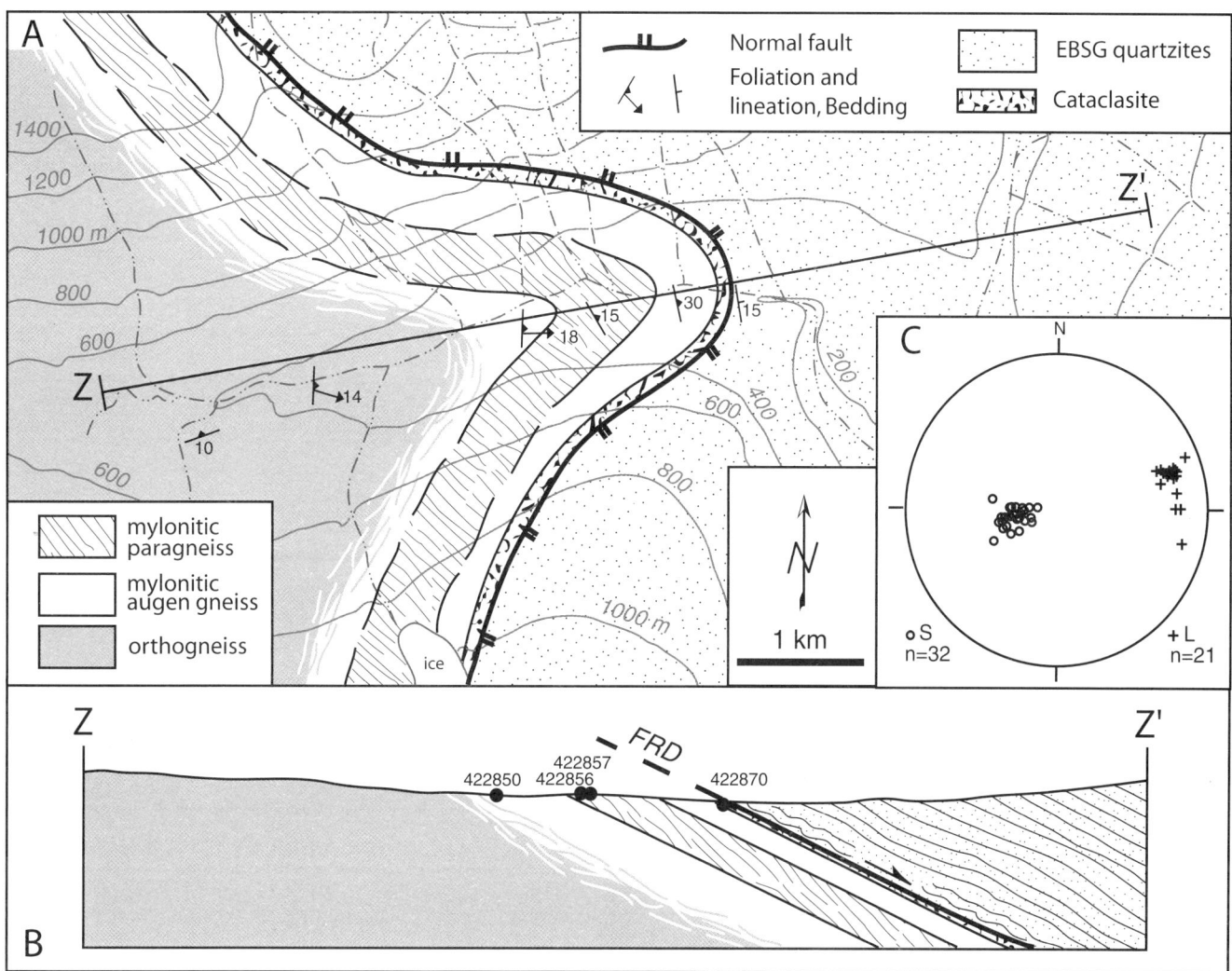

Figure 4. (A) Map and (B) profile of the Fjord region detachment (FRD) at Murgangsdal, Suess Land. The detachment is marked by a ductile, top-to-the-east, extensional deformation zone developed in lower-plate gneisses capped by a brittle, normal fault with low-grade sedimentary rocks in the hanging wall. Six-digit numbers, keyed to a GEUS database, represent samples shown in Figure 6. (C) Lower hemisphere, stereographic projection of lineations (crosses) and poles to mylonitic foliations (circles) (from Elvevold and Gilotti, 1998). EBSG—Eleonore Bay Supergroup.

that die out within meters above the fault (Fig. 5D). In other places, the Neoproterozoic to Ordovician section is folded by regional-scale, upright, N-S–striking folds (Henriksen, 2003). The section is cut by both antithetic and synthetic normal faults (e.g., Fig. 8), which do not appear to cut the main detachment, and thus this geometry supports Hartz and Andresen's (1995) hanging-wall collapse model. Leucogranitic plutons and dikes intrude the upper-plate sedimentary sequences but are everywhere truncated by the detachment.

Southern Segment: Middle Plate on Lower Plate

At Kap Hedlund, the Fjord region detachment splits into two strands; the branch line is marked by intense cataclasis (Fig. 1; Elvevold and Gilotti, 1998). The lower splay, known locally as the Højedal detachment in Forsblad Fjord (White et al., 2002), separates lower-plate orthogneiss from the migmatitic Krummedal units in the middle plate. Footwall orthogneisses south of Kap Hedlund are at the same structural level as those along the main segment but are dominated by Archean rather than Paleoproterozoic protoliths (Thrane, 2002). The geometry and kinematics are similar to the footwall along the northern segment of the Fjord region detachment, i.e., a 1–2-km-thick high-strain zone displays top-to-the-east shear sense and is capped by a subparallel, <5-m-thick, cataclastic zone marking a brittle fault (Rasmussen and Andresen, 1998; White et al., 2002). The mylonite zone associated with the Fjord region detachment appears to be cut out by brittle normal faults south of Forsblad Fjord (Friderichsen and Thrane, 1998).

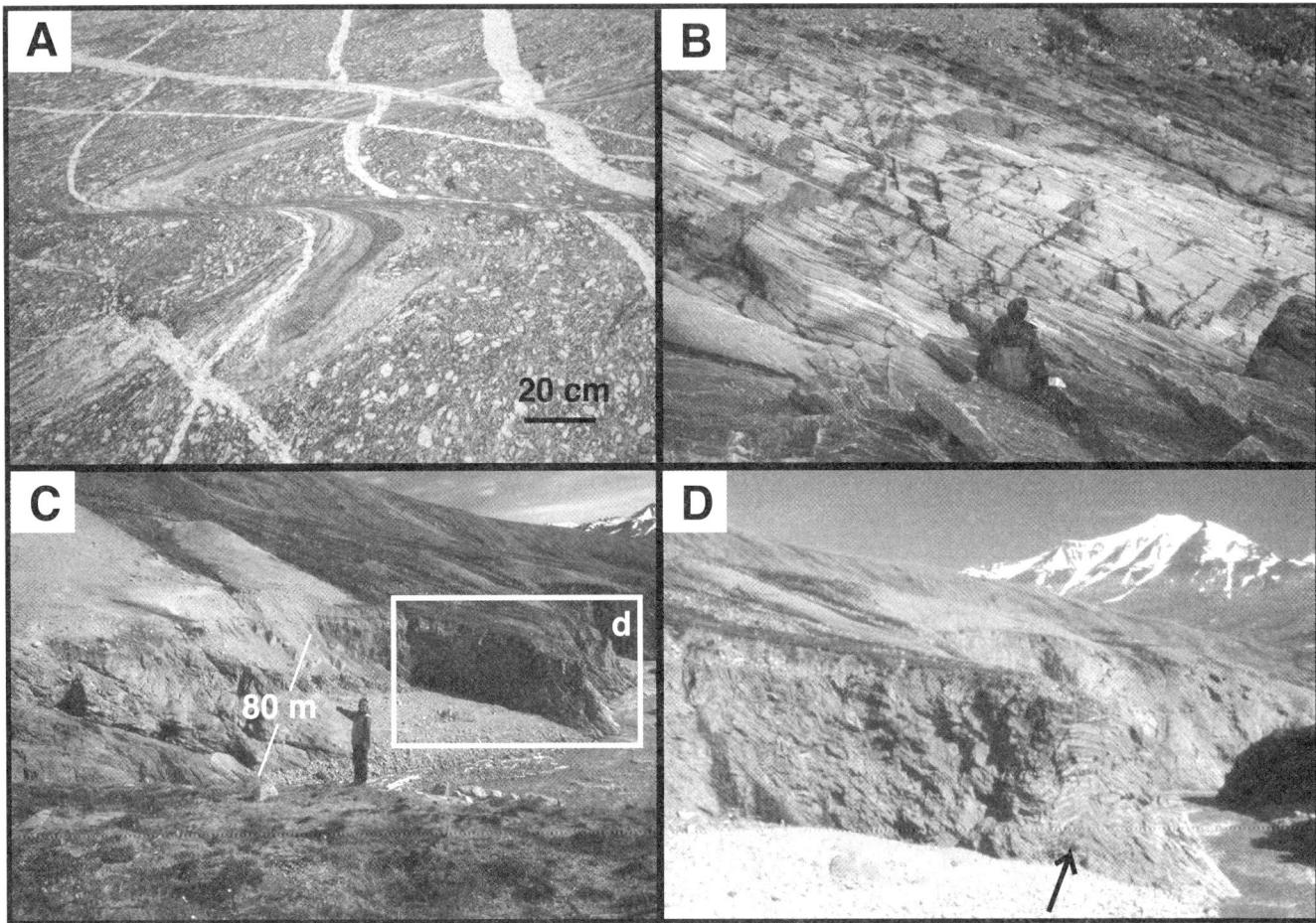

Figure 5. Outcrop photographs of relationships along the Fjord region detachment at Murgangsdal. (A) Low-strain lens of augen gneiss with crosscutting dikes and shear zones preserved in lower plate ~4.5 km upstream from the detachment fault. (B) Pinstriped orthogneiss within the detachment zone dip east; these are the deformed equivalents of the rocks in A, where all the older fabrics are highly strained and subparallel. Note the strong downdip stretching lineation and the scattered augen on the foliation surface where the person is sitting. (C) White lines delineate the 80-m-thick zone of green cataclasites that mark the brittle detachment fault. (D) Bedding in quartzites of the Eleonore Bay Supergroup is folded in the hanging wall of the Fjord region detachment. The arrow points to décollement-style folds die out upsection (~20 m) above the fault. View is toward the ENE in C and D.

The wedge of migmatitic Krummedal metasedimentary rocks that forms the middle plate (Fig. 1) is 14 km thick along the Forsblad Fjord section (White et al., 2002). Peak conditions for the assemblage garnet + biotite + plagioclase + quartz + kyanite/sillimanite are ~785 °C at 1.05 GPa, and they developed along a clockwise pressure-temperature (P-T) path (White and Hodges, 2003). Leucocratic melts were produced along this path. Assuming that the high pressures of 1.4–1.6 GPa inferred for the lower plate from Kap Hedlund carried southward, then the Højedal detachment delineates a pressure difference of ~0.4 GPa or ~15 km of missing section. Metamorphism of the middle plate reached higher temperatures further south in Stauning Alper, where garnet + spinel + sillimanite migmatites are observed (Watt et al., 2000). A later phase of cordierite-rich granites crystallized in boudin necks and small shear zones, suggesting a minor amount of decompression melting (Watt et al., 2000).

Southern Segment: Upper Plate on Middle Plate

The upper splay of the Fjord region detachment puts the Nathorst Land Group clastic rocks, i.e., the lower part of the Eleonore Bay Supergroup, atop the Krummedal migmatites. The Neoproterozoic Nathorst Land Group is a 9-km-thick package of shallow-marine sandstones and mudstones (Smith and Robertson, 1999). Locally, it can be very difficult to distinguish from the Mesoproterozoic Krummedal rocks because the basal units reach garnet zone, are strongly deformed, and are intruded by granitoids (Caby, 1976; Smith and Robertson, 1999; White et al., 2002). The base of the Nathorst Land Group is everywhere truncated by an extensional detachment or obscured by granites that lie along the Fjord region detachment.

In Forsblad Fjord, the upper splay is known as the Tinderne detachment, and it puts the little deformed Klosterbjerg granite over the sheet-like Caledoniaø granite (White et al., 2002). The

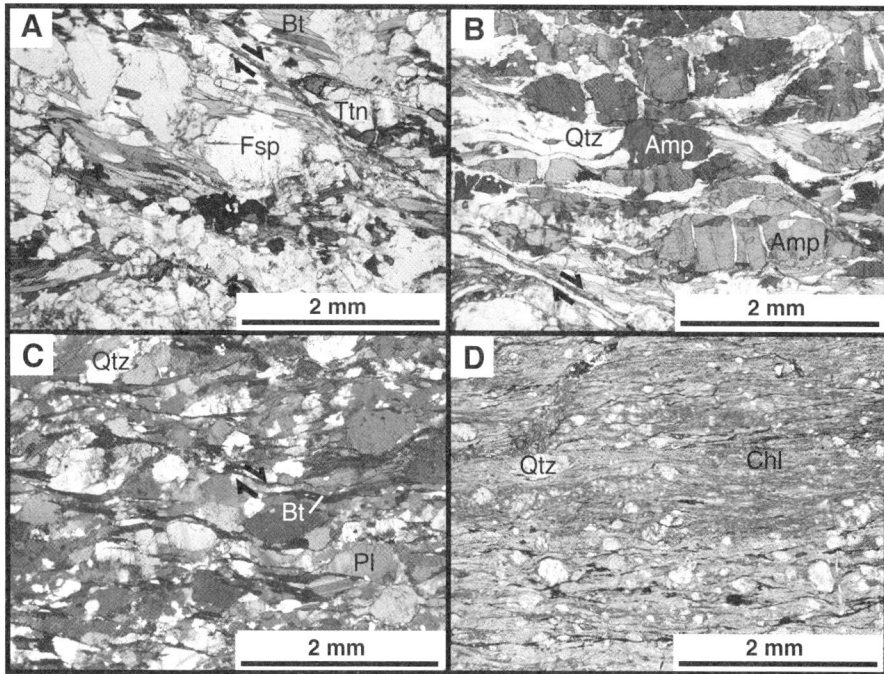

Figure 6. Photomicrographs of typical tectonites from the Fjord region detachment zone at Murgangsdal. All samples are oriented with east on the right. See cross section (Fig. 4B) for sample locations. (A) Protomylonitic quartzofeldspathic orthogneiss (sample 422850) with biotite (Bt) and titanite (Ttn) fish located between shear bands defined by fine-grained biotite and chlorite, all of which give a top-to-the-east sense of displacement. Note lack of asymmetric tails on feldspar (Fsp) porphyroclasts. (B) Banded calc-silicate sample 422857. Green amphibole (Amp) rods lie in the foliation and form a strong mineral lineation, but they are also boudinaged and filled with quartz (Qtz), forming a prominent stretching lineation. Biotite-rich shear bands give a top-to-the-east sense of shear. (C) Protomylonitic quartz-rich metasandstone sample 422856. Biotite fish and some poorly developed σ-porphyroclasts of plagioclase (Pl) give a top-to-the-east sense of shear. The quartz exhibits surprisingly little evidence of dynamic recrystallization. (D) Foliated cataclasite sample 422870 with abundant chlorite (Chl) in the dark splotchy areas and quartz porphyroclasts.

Caledoniaø granite is interpreted to be pre- to syntectonic, but evidence of solid-state extensional deformation suggests the former, and the tabular shape is probably due to deformation. White et al. (2002) described a 700-m-thick, footwall mylonite zone with top-to-the-east displacement capped by 2 m of cataclasite in Schaffhauserdalen. Synthetic and antithetic normal faults occur in the hanging wall and are asymptotic to the detachment.

The southward continuation of the Tinderne detachment is a major, subhorizontal detachment at the top of the Stauning Alper migmatite zone that truncates kilometer-scale isoclinal folds (Watt and Kinny, 1998; Watt and Friderichsen, 1999; Watt et al., 2000). Watt et al. (2000) examined the detachment between Nathorst Land Group phyllites and Krummedal mylonitic migmatites on Fangsthytte Gletscher, east of Alpefjord. At this locality, a 500-m-thick, foliated granite sheet lies along the contact. The granite is cut by numerous protomylonitic to mylonitic shear zones that are parallel to the foliation. A well-developed E-W, downdip lineation is also present. These observations support the idea that the extensional deformation is later than the granite emplacement.

The Cryptic Northern Segments

On the north side of Kejser Franz Joseph Fjord, near Junctiondal, the Fjord region detachment branches into two splays. The splay that puts the upper plate on the middle plate continues northward to Waltershausen Gletscher in Strindberg Land (Hartz and Andresen, 1995; Andresen et al., 1998). In Andrée Land, the upper splay is locally called the Grejsdalen fault and is described as a steep, brittle extensional structure (Andresen et al., 2007). Escher and Jones (1999) and Andresen et al. (2007) recognized a lower extensional detachment, the Rendal detachment, in a NW-trending valley called Junctiondal. This shear zone separates our middle plate from the lower plate, and it contains evidence for top-up-to-the-west thrust sense kinematics (Leslie and Higgins, 1998) overprinted by top-down-to-the-east extensional shear sense indicators (Escher and Jones, 1999; Smith and Robertson, 1999). We interpret this entire contact as another lower splay of the Fjord region detachment (Fig. 1) that may be a reworked segment of the Hagar Bjerg thrust. In contrast, the 1:1,000,000 tectonic map (Henriksen, 2003) portrays the contact in Strindberg Land as a thrust, in keeping with the orogenic architecture described by Higgins et al. (2004). This is one of a number of places where kinematics are unresolved, but the present geometry suggests an extensional detachment.

Petermann Bjerg Detachment and Boyd Bastion Fault

Between 73°N and 74°N, the Neoproterozoic sedimentary section overlies the Krummedal units of the Hagar Bjerg thrust sheet (Fig. 1; Higgins et al., 2004). Although Andresen et al. (1998) argued that this is a deformed unconformity, Escher and Jones (1998, 1999) observed several associated west-dipping, low-angle extensional shear zones, which they referred to collectively as the Petermann Bjerg detachment. The Neoproterozoic rocks in the hanging wall are little deformed and retain sedimentary structures but show an increase in metamorphism

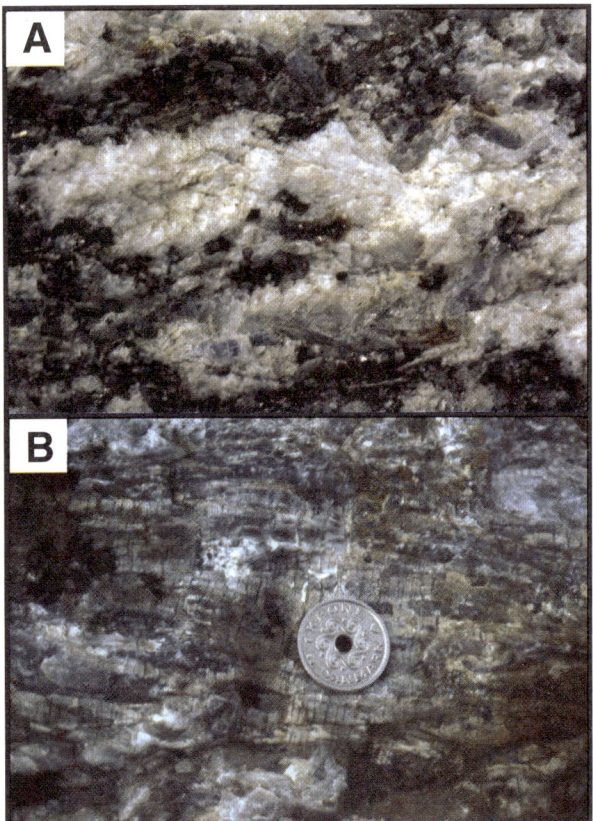

Figure 7. (A) Kyanite + garnet–bearing leucosomes in the lower plate at Kap Bayard (Fig. 1) are similar to the well-documented high-pressure crystallized melts in Payer Land. Field of view is approximately 15 cm wide. (B) Massive sillimanite pull-aparts with feldspar + quartz in the Fjord region detachment at Murgangsdal are interpreted to be the deformed and retrogressed equivalents of the high-pressure leucosomes above.

to garnet grade at the contact. The Petermann Bjerg detachment cuts different stratigraphic levels in the hanging wall. Complexly folded, migmatitic metapelites containing garnet and kyanite occur beneath the detachment, and they exhibit similar relationships to those observed in the middle plate along the upper splay of the Fjord region detachment. The displacement sense in the mylonites is top-to-the-NNW. We interpret the Petermann Bjerg detachment as an extensional detachment, recognizing that the NNW displacement represents a complexity within the extensional system.

A N-S–striking, brittle normal fault, the Boyd Bastion fault (Higgins et al., 2004; Higgins and Leslie, this volume), bounds the Neoproterozoic outcrops to the west (Wenk and Haller, 1953; Larsen and Bengaard, 1991). The fault dips steeply to the east and has significant offset with top-down-to-the-east displacement. East-plunging lineations with well-developed shear sense indicators are found along the fault (Escher and Jones, 1999). The Boyd Bastion fault is interpreted as a later feature relative to the Petermann Bjerg detachment.

TIMING OF MOVEMENT ON EXTENSIONAL DETACHMENTS

The absolute timing of ductile extension in the Greenland Caledonides is poorly known. Current estimates are based on (1) the age of high-pressure metamorphism in the lower plate, (2) ages from peraluminous granites within the middle and upper plates, and (3) metamorphic and cooling ages from all plates (Fig. 9). Two main periods of ductile extension are recognized, i.e., older displacement between the middle and upper plate after ca. 420 Ma, and younger movement after 400 Ma on the detachments involving the lower plate.

Extension between the Middle and Upper Plates

Estimates for the timing of extension between the middle and upper plates are based on zircon and monazite U-Pb data from migmatites developed in the Mesoproterozoic metasedimentary rocks of the middle plate and from peraluminous granitic bodies emplaced into the middle and upper plates and spatially associated with the extensional faults (Fig. 9; Hartz et al., 2000, 2001; Watt et al., 2000; Strachan et al., 2001; White et al., 2002). Geochronological data sets come from Ardencaple Fjord (Hansen et al., 1994; Strachan et al., 2001), Payer Land (Gilotti and McClelland, 2005), Nordenskiöld Gletscher (Kalsbeek et al., 2001a), Andrée Land (Andresen et al., 2007), Kejser Franz Joseph Fjord (Hartz et al., 2000, 2001), Forsblad Fjord (White et al., 2002; White and Hodges, 2002), Stauning Alper (Kalsbeek et al., 2001a; Watt et al., 2000), and Renland (Leslie and Nutman, 2003) (Fig. 1), and the majority of the reported U-Pb zircon and monazite ages range from 435 to 415 Ma. Although a number of these authors propose that extension was synchronous with granite emplacement, Gilotti and McClelland (2005) argued that leucogranite ages are of limited use for directly dating extension due to the presence of inherited components and a lack of specific crosscutting relationships that tie the dated rocks to major extensional structures. Many of the granites that lie along the upper detachment surface contain solid-state, extensional deformation fabrics and thus clearly predate extension (White et al., 2002; Watt et al., 2000). In addition, P-T paths suggest that prograde dehydration melting of micas, rather than decompressional melting, produced the peraluminous migmatites and granites in the middle plate (Gilotti et al., this volume). The best candidates for decompression melts are the late cordierite-bearing leucosomes in Stauning Alper (Watt et al., 2000), but these have not been dated.

The timing of extension between the middle and upper plates is inferred from their thermal history, assuming that exhumation and cooling are due to extension. For example, displacement along the Tinderne detachment is interpreted to have occurred just prior to 423 Ma on the basis of $^{40}Ar/^{39}Ar$ muscovite cooling ages above and below the detachment (White and Hodges, 2002). The constant values of both muscovite and biotite cooling ages in a transect across the Tinderne detachment provide another

Figure 8. Steep, conjugate normal faults cut the upper part of the Eleonore Bay Supergroup in the hanging wall of the Fjord region detachment, Kempe Fjord. View is to the north. The cliff is ~1000 m high.

line of evidence that the ductile extension is Silurian (White and Hodges, 2002). A range of $^{40}Ar/^{39}Ar$ muscovite ages from undeformed granites within the upper and middle plates at Ardencaple Fjord gives similar plateau ages of ca. 405–423 Ma, while muscovite from a foliated granite in the Smallefjord unit (our middle plate) is ca. 385 Ma (Dallmeyer et al., 1994). These ages are difficult to interpret in the context of extension because the samples are not from the detachment zone mylonites. The muscovite ages from granites in the upper plate may simply reflect cooling of the magma and may have little bearing on extension.

Extension Involving the Lower Plate

The timing of displacement along extensional faults that separate the structurally lowest plate from the overlying plates is limited by the age of high-pressure–high-temperature granulite-facies metamorphism observed in the lower plate. Age relationships are best documented in Payer Land, where a U-Pb sensitive high-resolution ion microprobe (SHRIMP) age of 403 ± 5 Ma was obtained from metamorphic zircon rims in high-pressure– high-temperature metapsammites (McClelland and Gilotti, 2003). Kyanite + K-feldspar + garnet–bearing leucosomes derived from high-pressure granulite-facies melting of metapelites crystallized at 404 ± 4 Ma (McClelland and Gilotti, 2003) and are deformed within the mylonitic extensional fabrics of the Payer Land detachment. Extensional deformation associated with the Payer Land detachment must be younger than ca. 400 Ma (Gilotti and McClelland, 2005; Fig. 9).

The $^{40}Ar/^{39}Ar$ cooling ages are consistent with extension after 400 Ma. Displacement above the lower plate of the Fjord region detachment system is inferred to have occurred between 400 and 370 Ma on the basis of $^{40}Ar/^{39}Ar$ cooling ages on muscovite and biotite from the structurally lowest plate (Hartz et al., 2000, 2001; White and Hodges, 2002). White and Hodges (2002) pointed out that biotite K-Ar ages (Rex and Higgins, 1985) across the Højedal detachment show a >15 m.y. difference, with the young ages (ca. 385 Ma) occurring in the footwall; they also inferred two periods of extension. Dallmeyer et al. (1994) obtained a bulk muscovite age of ca. 381 Ma from a mylonitic felsic pegmatite in the Bessel Fjord shear zone, which also falls within this range.

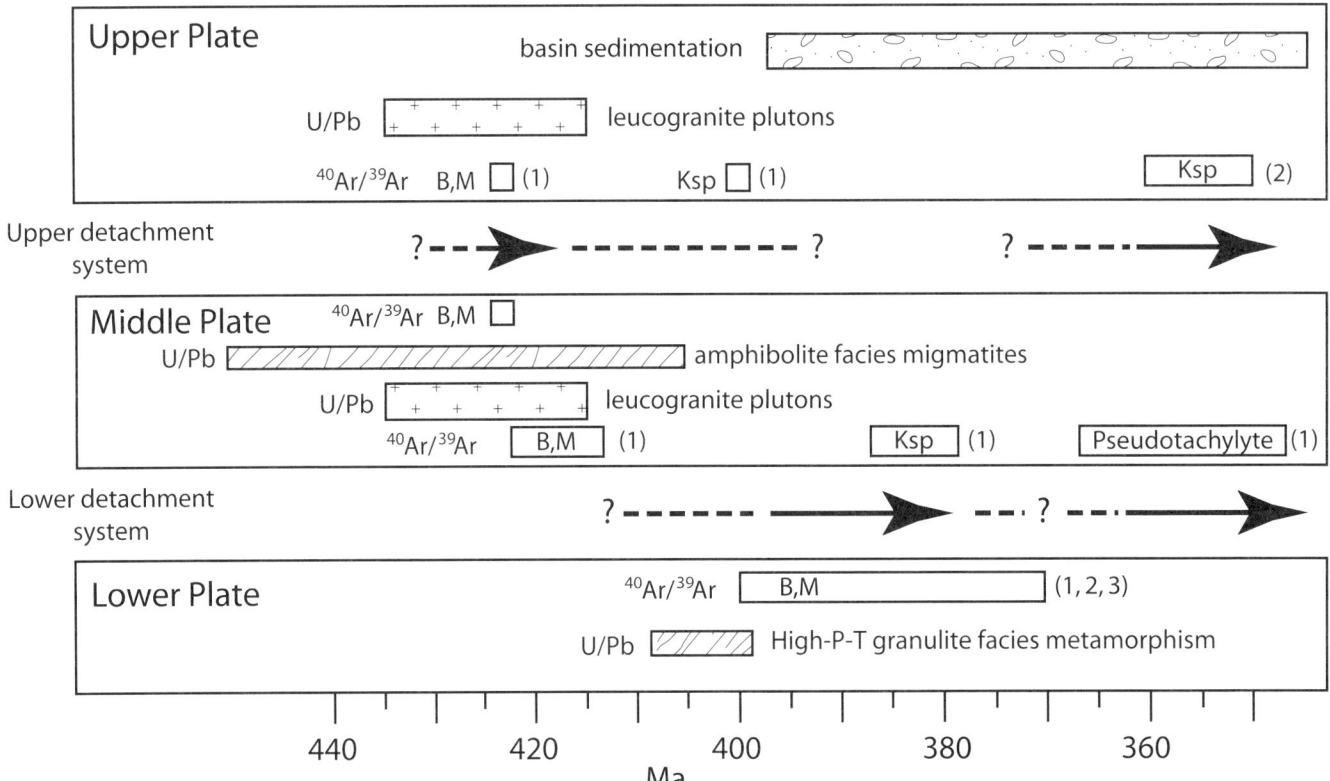

Figure 9. Summary of the timing relationships in the Fjord region, showing the variation in basin evolution, magmatism, metamorphism, and detachment faulting as a function of structural level. The U-Pb age ranges depicted are based on data discussed and referenced in the text. The $^{40}Ar/^{39}Ar$ age ranges are based on data from (1) White and Hodges (2002); (2) Hartz et al. (2000, 2001); and (3) Dallmeyer et al. (1994). Timing of basin sedimentation is from Larsen et al. (this volume). Mineral abbreviations: B—biotite, M—muscovite, Ksp—K-feldspar.

Reactivated or continued displacement at shallow crustal levels on both the upper and lower detachment system is recorded by brittle deformation. Part of the brittle history is dated at ca. 360–350 Ma on the basis of $^{40}Ar/^{39}Ar$ ages on K-feldspar and pseudotachylyte (Hartz et al., 2000; White and Hodges, 2002).

Devonian Basin–Bounding Faults and Later Brittle Structures

Steep, N-S–striking brittle faults bound the main outcrop area of the Devonian basins and cut the younger Carboniferous to Cretaceous basins (1:1,000,000 map; Henriksen, 2003). Their complete geometry and kinematics are not well understood, but their very existence raises the question of whether the high-angle faults interacted with the low-angle extensional detachments. The Devonian to Cretaceous basins sit atop the upper plate, since they unconformably overlie the Ordovician and older sedimentary rocks of the Franz Joseph allochthon. It is conceivable that the steep, basin-bounding, brittle faults sole into the low-angle detachments that extend beneath the Devonian and younger basins at some unknown depth (e.g., McClay et al., 1986; Hartz and Andresen, 1995).

The Western and Eastern fault zones (WFZ and EFZ, Fig. 1) that bound the Devonian basin are prominent, through-going structures. Larsen et al. (this volume) show that the sinistral Western fault zone developed in the Middle Devonian, while the sinistral Eastern fault zone appears in the Late Devonian to form a wrench system. In detail, the fault zones are segmented and complicated (Hartz, 2000), and, overall, the strike-slip component of displacement on the Western fault zone is thought to be <20 km (Larsen and Bengaard, 1991). Sedimentation patterns reflect different periods of motion on these faults (Larsen et al., this volume). The entire Devonian to Permian basin fill is deformed by regional, N-S–striking, low-amplitude, upright folds that have puzzled geologists since Bütler (1935). These structures and locally observed thrust faults within the Devonian strata (e.g., Geographical Society Ø, Henriksen, 2003) suggest that the region experienced a prolonged period of contractional deformation. Smaller faults with other orientations, such as 10–20-km-long E-W–striking normal faults (Larsen and Bengaard, 1991), are common in the Devonian basins. These E-W normal faults have stratigraphic throws up to 1500 m and truncate the N-S folds.

Abundant steep, brittle faults cut through the Carboniferous to Cretaceous strata to Precambrian basement gneisses in Jameson Land and along the continental shelf (Hamann et al.,

2005). Although periods of strike slip may have alternated with dip-slip movement, normal separation is prominent in seismic profiles and is not surprising since a main rifting event in the North Atlantic took place in the Jurassic and Early Cretaceous (Hamann et al., 2005). Paleogene sedimentary rocks onlap the normal faults in seismic sections. Movement on steep, brittle faults cutting the upper plate and the Devonian and younger basins can be as late as Cretaceous.

Styles of Extensional Detachments in East Greenland

The extensional detachments in Greenland share common characteristics with those observed in other orogens. Low-angle extensional detachments bound three major tectonostratigraphic units, described here as a lower, middle, and upper plate, and are largely responsible for the regional map pattern we see today. The detachments are defined by up to 1-km-thick mylonitic zones that are primarily developed in the footwalls. Many of the rocks in the high-strain zones are protomylonitic and do not show tremendous grain-size reduction in comparison to the gneissic protoliths, perhaps because much of the deformation is amphibolite facies. Mylonitization commonly begins in the amphibolite facies and progresses toward lower grade with increasing exhumation. The thick mylonitic zones are capped by relatively thin cataclastic layers, suggesting that extension was active from deep to shallow crustal levels and that displacement became more and more localized along brittle boundaries through time. Ductile mylonites are not thick or well developed in the hanging walls. Where found (e.g., Payer Land), deformed upper-plate rocks are commonly greenschist-facies mylonites <100 m thick.

The evolution and overprinting relationships in the tectonites along the Greenland detachments are similar to those described for metamorphic core complexes, where material is exhumed from depth (e.g., Armstrong, 1982; Davis and Lister, 1988; Lister and Davis, 1989). However, the tectonic setting is fundamentally different. The Greenland detachments formed in an overall syncollisional setting like the Himalayas and the Alps, rather than in an intracontinental extensional setting characterized by lithospheric-scale thinning as seen in the Cordilleran core complexes of North America (e.g., Wernicke, 1981; Coney, 1987).

The Greenland Caledonides experienced some form of extensional tectonics from the Silurian (ca. 420 Ma) at the earliest through the Cretaceous. The long-lived extensional history is complex in detail but generally developed in a setting that changed from a collisional orogen to a continental rift environment. Extension can be thought of as syncollisional at least through the Mississippian, because ultrahigh-pressure (UHP) metamorphism at 365–350 Ma at 78°N (Gilotti et al., 2004; McClelland et al., 2006) requires plate convergence through this time (Gilotti and McClelland, 2007). Later in the Carboniferous, Visean and younger extension characterized by high-angle faulting is postcollisional with respect to Baltica-Laurentia convergence, and it eventually led to continental rifting (Stemmerik and Worsley, 2005).

The geochronological studies reviewed in this paper suggest that there were at least two periods of regional ductile extension in East Greenland. The detachment between the middle and upper plate (e.g., the upper splays of the Fjord region detachment, the Petermann Bjerg detachment, and the detachments beneath the upper plate in Ardencaple Fjord) may have been initiated shortly after leucogranite intrusion around 420 Ma, while detachments that juxtapose the lower plate with either overlying plate must be younger than high-pressure metamorphism at ca. 400 Ma. The early period of extension was superimposed on fabrics associated with crustal thickening, juxtaposing midcrustal migmatites of the middle plate with the upper plate at shallow crustal levels. Leucogranites derived from midcrustal levels are interpreted to have intruded the upper plate as discrete plutons during crustal thickening, implying that the Franz Joseph allochthon already existed above the Mesoproterozoic sequences, either as an original integral part of the Hagar Bjerg thrust sheet (Higgins et al., 2004) or as a dominant thrust sheet where the thrust is now excised. The early displacement along the upper detachments may be analogous to the South Tibetan detachment system, which places midcrustal migmatites and peraluminous granites against shallow-level rocks, perhaps above a coeval thrust in an extrusive flow scenario (e.g., Searle, 1999; Beaumont et al., 2001; Vannay and Grasemann, 2001; Grujic et al., 2002). A significant difference between the Greenland and Himalayan examples, which may simply reflect varying structural level, is that leucogranites rarely if ever lie above the South Tibetan detachment (Searle and Godin, 2003). In addition, earlier structures have been severely modified during continued or younger periods of extension on the structurally lowermost detachment. This second episode of ductile extension must be younger than ca. 400 Ma, but the actual displacement age is not known. Thrust emplacement of the lower plate onto the foreland must be younger than high-pressure metamorphism, as well. This raises the possibility that extensional displacement on the lower detachment may have been synchronous with structurally lower thrusts framing the foreland windows.

The extensional fault systems in East Greenland show complex linkages. In some cases, for example, along the Lerbugt shear zone (Fig. 3), the connectivity of splays defines an extensional duplex. The high level of shear zone connectivity is an artifact of formation at relatively deep crustal levels in a progressive, but singular, extensional event. For the most part, however, linkages are between the two different levels of detachment and are seen where branch lines join the upper-middle plate detachment and the middle-lower plate detachment faults (e.g., at Kap Hedlund, Junctiondal, and around Ardencaple Fjord). Places where the lower plate is in extensional contact with the upper plate (e.g., Payer Land detachment, Fjord region detachment main segment, northwest segment in Ardencaple Fjord) formed when the lower level of detachment captured and excised the extensional structures formed earlier between the middle and upper plates at higher crustal levels. Davis and Lister (1988) introduced the concept of excisement splays for single detachment systems in the Whipple and Sacramento Mountains metamorphic core complexes. If the inference that there have been two main periods of extensional deformation in East Greenland is correct, then the younger episode of extension involving structurally deeper rocks truncated and excised the older, higher extensional structures between the middle and upper plates after 400 Ma. The complex pattern of excisement in East Greenland reflects superposition of two major extensional systems during collisional orogenesis, and it increases the difficulty of unraveling the extensional history of the Greenland Caledonides.

The orientation of the detachments and their displacement sense vary around the region (Fig. 1). The east-dipping, top-down-to-the-east Fjord region detachment system dominates the central part of the East Greenland Caledonides. However, the SE-dipping Payer Land detachment and the SW-dipping Bessel Fjord detachment both have top-to-the-SW kinematics, while the Petermann Bjerg detachment is a west-dipping, top-to-the-west shear zone. Kinematics do not vary systematically with structural level. Extension probably did not result from gravitational collapse of a single crustal welt because the displacement sense is not consistently outwardly radial and did not occur in a single event (e.g., Selverstone, 2005). The variable kinematics demonstrate that shear sense alone cannot be used to distinguish between thrusts and extensional structures. Complex patterns in Ardencaple Fjord, where the Kildedal and Bessel Fjord detachments dip toward each other, and both have downdip displacements (Strachan, 1994), are difficult to explain without the superposition of multiple shear zones. The Kildedal and Bessel Fjord detachments could have evolved as simultaneous opposed-dip extensional structures, such as those modeled by Rosenbaum et al. (2005) and observed for continental extensional settings, and then subsequently been linked by capture and excisement along the lower detachment.

Our maximalist interpretation of extension in East Greenland (Fig. 1) shows an extensional detachment between the middle and lower plate everywhere along the Hagar Bjerg thrust of Higgins et al. (2004). Unlike Andresen et al. (2007), who believe that the only significant thrust in the entire East Greenland hinterland is the Niggli Spids thrust, we think that thrusting had to have played an important role in the contraction of the entire hinterland in order to produce the complex deformation and regional Caledonian metamorphic patterns (Gilotti et al., this volume). The Hagar Bjerg thrust is obscured by extension south of 73°N, but in northern Andrée Land, Leslie and Higgins (this volume) show that this thrust puts migmatitic Krummedal metasedimentary rocks over lower-grade psammitic sections of the same unit with bedding preserved in the Niggli Spids thrust sheet. Although Leslie and Higgins (this volume) do not document the top-to-the-west kinematics in the thrust zone itself, the high-grade on low-grade relationship is, indeed, indicative of thrusting. Elsewhere in the Niggli Spids thrust sheet (our lower plate), relics of high-pressure granulites are found, e.g., in Payer Land and Gletscherland. These discrepancies point to heterogeneities within the large thrust sheets and our incomplete understanding of the detailed geology of the region. The inferred extensional detachments on our Figure 1 are places where future workers should concentrate their efforts. Our prediction is that detailed work will show complex relationships with the Hagar Bjerg thrust excised by extensional detachments.

Numerous models for Devonian basin initiation have been proposed (e.g., Friend et al., 2000; Larsen et al., this volume), and most hinge on the significance of the Western and Eastern fault zones (Fig. 1). The most plausible models fall into two categories: models that involve a sinistral, wrench component (Friend et al., 1983; Larsen and Bengaard, 1991) to form a pull-apart basin versus those that call on extensional tectonics to open space for the evolving terrestrial sedimentation (e.g., Hossack, 1984; McClay et al., 1986). Documentation of extensional detachments in East Greenland led Hartz and Andresen (1995) to build on the latter conceptual models and propose that basin development coincided with collapse of the hanging wall above the Fjord region detachment. Our view is that the end-member models are not mutually exclusive. Regardless of whether the steep, brittle faults bounding the basins were dominantly wrench or normal in nature, there must have been an active, subhorizontal detachment separating the upper crust and evolving supradetachment basin from the lower crustal crystalline complexes that underwent high-pressure metamorphism in the Middle Devonian and were exhumed shortly thereafter. Although folds are possible in extension (e.g., Harris et al., 2002, and references therein), the folds in the Devonian strata were most likely formed in a contractional environment, perhaps transpressional, due to ongoing plate convergence into the Carboniferous.

Steep, brittle faults, such as the Boyd Bastion fault, Western fault zone, Eastern fault zone, and others along the Fjord region detachment, display a Devonian and later displacement history at shallow crustal levels. The switch from extension along low-angle detachments to development of high-angle structures roughly corresponds to the transition from plate convergence to plate divergence in the Mississippian. The upright folds in Carboniferous and Permian strata may have formed via wrench motion on the Western fault zone and Eastern fault zone or as warps associated with normal displacement. From the surface geology, it is clear that brittle faults were active through the Cretaceous (Fig. 1), while seismic sections show that the faults cut Precambrian basement with a horst and graben–type geometry (Larsen, 1990; Hamann et al., 2005). Carboniferous through Cretaceous brittle displacement on steep faults was probably episodic based on the sedimentary record (Stemmerik, 2000).

Orogen-wide Extension in the Caledonides

Extension in the Greenland Caledonides should be viewed in the context of the whole orogen. The Greenland Caledonides represent a portion of the Laurentian margin that evolved as the overriding plate in the continent-continent collision with Baltica (Gee, 1975; Hossack and Cooper, 1986; Roberts, 2003). Migmatite complexes and leucogranite intrusions, not unlike the High Himalayan leucogranites (e.g., Searle et al., 2003), formed from melted metasedimentary rocks during crustal thickening and were probably exhumed during a Silurian phase of extension along ductile detachments at midcrustal levels. Continued convergence produced greater crustal thickening seen as widespread high-pressure metamorphism in the Precambrian gneisses that comprise the Laurentian basement. Intermontane Devonian basins developed shortly after high-pressure metamorphism at depth and were necessarily separated by a low-angle detachment. This second detachment system became active sometime after high-pressure metamorphism at 405 Ma and localized at a lower crustal level. The lower detachment excised portions of the older, upper system as it exhumed the high-pressure rocks.

In contrast, extension in the Scandinavian Caledonides developed in the upper parts of the subducting slab of Baltica. Low-angle detachments floor the Devonian basins in Norway (e.g., Hossack, 1984; Norton, 1987), and extension has been shown to be synchronous with deposition (Osmundsen et al., 1998). The extensional detachments and basins in Norway are affected by kilometer-scale upright folds with axes parallel to the E-W or NE-SW displacement directions (Krabbendam and Dewey, 1998; Osmundsen et al., 1998, 2006). Similar large-amplitude, extension-parallel folds are not common in Greenland. The Norwegian detachments separate the Old Red Sandstone basins from the underlying high- and ultrahigh-pressure continental crust of Baltica that was exhumed out of the subduction zone from depths ≤150 km. It is unlikely that the detachments were the sole exhumation mechanism for the ultrahigh-pressure rocks; for example, Andersen and Jamtveit (1990) proposed that a large component of vertical pure shear thinned the crust and partially exhumed the eclogites prior to extension in the middle crust.

Extension must have been younger than the ultrahigh-pressure metamorphism of the subducted Baltica margin that culminated between 410 and 400 Ma (Terry et al., 2000; Carswell et al., 2003; Root et al., 2004), at approximately the same time as high-pressure metamorphism on the Laurentian margin. Rapid exhumation of the Western Gneiss Region to midcrustal levels was accomplished by 395 Ma, based on U-Pb titanite geochronology (Tucker et al., 2004) and ^{40}Ar/^{39}Ar thermochronometry (Chauvet and Dallmeyer, 1992; Eide et al., 1999; Kendrick et al., 2004; Root et al., 2005). Significant sedimentation and movement on the detachment flooring the basins occurred in the Middle Devonian (Andersen, 1998; Friend et al., 2000) into the Late Devonian and Mississippian (Eide et al., 2005), and it happened after the early stage of eclogite exhumation. Large-magnitude listric detachments also cut some of the thrust sheets in the Scandinavian Caledonides (Bergman and Sjöström, 1997; Fossen and Dunlap, 1998; Fossen, 2000) and underlie major tectonostratigraphic domains, such as the Helgeland Nappe Complex (Osmundsen et al., 2003). These detachments were active from the Middle Devonian onward (Fossen and Dunlap, 1998; Eide et al., 2002), synchronous with orogenwide extension associated with the development of the Old Red Sandstone basins. The easternmost faults of the west-dipping detachment system in Scandinavia form an exposed rift shoulder that bounds the extension that led to the growth of the North Atlantic (Mosar, 2003).

The early period of extension postulated between the middle and upper plates in Greenland is not recognized in Scandinavia.

Strike-slip faulting concurrent with extension plays a role on both the Laurentian and Baltica margins. Models for extension and basin development between 400 and 360 Ma in the Norwegian Caledonides build on a variety of data suggestive of a sinistral transtensional setting (e.g., Krabbendam and Dewey, 1998; Terry and Robinson, 2003; Osmundsen et al., 1998, 2006). Sinistral transpression has been suggested as a mode of deformation in Greenland (Holdsworth and Strachan, 1991; Larsen and Bengaard, 1991; Soper et al., 1992); however, the presence of dextral shear zones in North-East Greenland (Hull and Gilotti, 1994; Sartini-Rideout et al., 2006) indicates that the strike-slip history is complex. Dewey and Strachan (2003) called for a change in plate motion from transpression to transtension between 410 and 395 Ma, but such a switch must have occurred later in order to allow for eclogite-facies metamorphism on both margins at this time (Gilotti et al., 2004; Gilotti and McClelland, 2007). Transtension may have played a role in the late extension in the Fjord region, the strike-slip deformation in North-East Greenland, and exhumation of ultrahigh-pressure rocks in the North-East Greenland eclogite province after 350 Ma (Gilotti et al., 2004; Sartini-Rideout et al., 2006; McClelland et al., 2006), but this remains to be demonstrated.

SUMMARY

A synthesis of observations from the North-East Greenland Caledonides suggests that extensional structures involving upper-, middle-, and lower-crustal rocks may control the final map pattern in the hinterland of collisional orogens. The development of multiple detachment systems at different structural levels at different times during the evolution of a orogen leads to complex geometric and temporal relationships between preserved structures. In the case of the Greenland Caledonides, early-formed extensional structures involving midcrustal migmatites and leucogranites may be analogous to the South Tibetan detachment system. Younger detachments that involved exhumed high-pressure rocks and developed below Devonian supradetachment basins modified the older extensional systems. They were broadly syncollisional due to overlap in age with ultrahigh-pressure metamorphism of Laurentian continental crust in the north, and they may have been synchronous with thrusts that emplaced high-pressure lower-plate rocks over the foreland, as well as with strike-slip faulting in the hinterland. Conversion to extension accommodated by high-angle faulting in the Carboniferous marks the final transition to plate divergence that ultimately led to continental rifting.

ACKNOWLEDGMENTS

Gilotti is grateful to the Geological Survey of Denmark and Greenland (GEUS) for the invitation to participate in the 1997 and 1998 mapping campaigns, and particularly to Niels

Henriksen for organizing the expeditions. This manuscript would not be possible without the contributions of the many people who participated in the GEUS field work, particularly Synnøve Elvevold, Jan Escher, the late J.D. Friderichsen, Kevin Jones, and Gordon Watt. A University of Iowa Central Investment Fund for Research Enhancement to Gilotti and a University of Idaho Seed Grant to McClelland supported a 2001 season working on the Payer Land detachment. U.S. National Science Foundation grants EAR-0208236 and EAR-0208158 to Gilotti and McClelland, respectively, supported some of the geochronology discussed in this paper. We appreciate the efforts of Andrew Stolba for his senior thesis work on the Murgangsdal thin sections. Discussions with Synnøve Elvevold, A.K. Higgins, and Kevin Jones, and careful reviews by Niels Henriksen, Tekla Harms, and Per Terje Osmundsen helped us improve the manuscript. We thank A.K. Higgins for his helpful editorial work on this paper.

REFERENCES CITED

Andersen, T.B., 1998, Extensional tectonics in the Caledonides of southern Norway, an overview: Tectonophysics, v. 285, p. 333–351, doi: 10.1016/S0040-1951(97)00277-1.

Andersen, T.B., and Jamtveit, B., 1990, Uplift of deep crust during orogenic extensional collapse: A model based on field studies in the Sogn-Sunnfjord region of western Norway: Tectonics, v. 9, p. 1097–1111, doi: 10.1029/TC009i005p01097.

Andresen, A., and Hartz, E.H., 1998, Basement-cover relationships and orogenic evolution in the Central East Greenland Caledonides: GFF, v. 120, p. 191–198.

Andresen, A., Hartz, E.H., and Vold, J., 1998, A late orogenic extensional origin for the infracrustal gneiss domes of the East Greenland Caledonides (72°–74°N): Tectonophysics, v. 285, p. 353–369, doi: 10.1016/S0040-1951(97)00278-3.

Andresen, A., Rehnström, E.F., and Holte, M., 2007, Evidence for simultaneous contraction and extension at different crustal levels during the Caledonian orogeny in NE Greenland: Journal of the Geological Society of London, v. 164, p. 869–880, doi: 10.1144/0016-76492005-056.

Armstrong, R.L., 1982, Cordilleran metamorphic core complexes—From Arizona to southern Canada: Annual Review of Earth and Planetary Sciences, v. 10, p. 129–154, doi: 10.1146/annurev.ea.10.050182.001021.

Beaumont, C., Jamieson, R.A., Nguyen, M.H., and Lee, B., 2001, Himalayan tectonics explained by extrusion of a low viscosity crustal channel coupled to focused surface denudation: Nature, v. 414, p. 738–742, doi: 10.1038/414738a.

Bergman, S., and Sjöström, H., 1997, Accretion and lateral extension in an orogenic wedge: Evidence from a segment of the Seve-Köli terrane boundary, central Scandinavian Caledonides: Journal of Structural Geology, v. 19, p. 1073–1091, doi: 10.1016/S0191-8141(97)00028-X.

Burchfiel, B.C., and Royden, L.H., 1985, North-south extension within the convergent Himalayan region: Geology, v. 13, p. 679–682, doi: 10.1130/0091-7613(1985)13<679:NEWTCH>2.0.CO;2.

Burchfiel, B.C., Zhiliang, C., Hodges, K.V., Yuping, L., Royden, L.H., Changrong, D., and Jiene, X., 1992, The South Tibetan Detachment System, Himalayan Orogen: Extension Contemporaneous with and Parallel to Shortening in a Collisional Mountain Belt: Geological Society of America Special Paper 269, 41 p.

Burg, J.P., and Vanderhaeghe, O., 1993, Structures and way-up criteria in migmatites, with application to the Velay dome (French Massif Central): Journal of Structural Geology, v. 15, p. 1293–1301, doi: 10.1016/0191-8141(93)90103-H.

Burg, J.P., Guiraud, M., Chen, G.M., and Li, G.C., 1984, Himalayan metamorphism and deformation in the north Himalayan belt (southern Tibet, China): Earth and Planetary Science Letters, v. 69, p. 391–400, doi: 10.1016/0012-821X(84)90197-3.

Bütler, H., 1935, Some New Investigations of the Devonian Stratigraphy and Tectonics of East Greenland: Meddelelser om Grønland, v. 102, 35 p.

Caby, R., 1976, Investigations of the lower Eleonore Bay Group in the Alpefjord region, central East Greenland: Rapport Grønlands Geologiske Undersøgelse, v. 80, p. 102–106.

Carswell, D.A., Tucker, R.D., O'Brien, P.J., and Krogh, T.E., 2003, Coesite micro-inclusions and the U/Pb age of zircons from the Hareidland eclogite in the Western Gneiss Region of Norway: Lithos, v. 67, p. 181–190, doi: 10.1016/S0024-4937(03)00014-8.

Chauvet, A., and Dallmeyer, R.D., 1992, $^{40}Ar/^{39}Ar$ mineral dates related to Devonian extension in the southwestern Scandinavian Caledonides: Tectonophysics, v. 210, p. 155–177, doi: 10.1016/0040-1951(92)90133-Q.

Chen, Z., Liu, Y., Hodges, K.V., Burchfiel, B.C., Royden, L.H., and Deng, C., 1990, The Kangmar Dome: A metamorphic core complex in southern Xizang (Tibet): Science, v. 250, p. 1552–1556, doi: 10.1126/science.250.4987.1552.

Coney, P.J., 1980, Cordilleran metamorphic core complexes: An overview, in Crittenden, M.D., Coney, P.J., and Davis, G.H., eds., Cordilleran Metamorphic Core Complexes: Geological Society of America Memoir 153, p. 7–31.

Coney, P.J., 1987, The regional tectonic setting and possible causes of Cenozoic extension in the North American Cordillera, in Coward, M.P., Dewey, J.F., and Hancock, P.L., eds., Continental Extensional Tectonics: Geological Society of London Special Publication 28, p. 177–186.

Coney, P.J., and Harms, T.A., 1984, Cordilleran metamorphic core complexes; Cenozoic extensional relics of Mesozoic compression: Geology, v. 12, p. 550–554, doi: 10.1130/0091-7613(1984)12<550:CMCCCE>2.0.CO;2.

Dallmeyer, R.D., Strachan, R.A., and Henriksen, N., 1994, $^{40}Ar/^{39}Ar$ mineral age record in NE Greenland: Implications for tectonic evolution of the North Atlantic Caledonides: Journal of the Geological Society of London, v. 151, p. 615–628, doi: 10.1144/gsjgs.151.4.0615.

Davis, G.A., and Lister, G.S., 1988, Detachment faulting in continental extension: Perspectives from the Southwestern U.S. Cordillera, in Clark, S.P., Jr., Burchfiel, B.C., and Suppe, J., eds., Processes in Continental Lithospheric Deformation: Geological Society of America Special Paper 218, p. 133–159.

Dewey, J.F., 1988, Extensional collapse of orogens: Tectonics, v. 7, p. 1123–1139, doi: 10.1029/TC007i006p01123.

Dewey, J.F., and Strachan, R.A., 2003, Changing Silurian-Devonian relative plate motion in the Caledonides: Sinistral transpression to sinistral transtension: Journal of the Geological Society of London, v. 160, p. 219–229.

Eide, E.A., Torsvik, T.H., Andersen, T.B., and Arnaud, N.O., 1999, Early Carboniferous unroofing in western Norway and alkali feldspar thermochronology: The Journal of Geology, v. 107, p. 353–374, doi: 10.1086/314351.

Eide, E.A., Osmundsen, P.T., Meyer, G.B., and Kendrik, M.A., 2002, The Nesna shear zone, north-central Norway—An $^{40}Ar/^{39}Ar$ record of Early Devonian–Early Carboniferous ductile extension and unroofing: Norwegian Journal of Geology, v. 82, p. 317–339.

Eide, E.A., Haabesland, N.E., Osmundsen, P.T., Andersen, T.B., Roberts, D., and Kendrick, M.A., 2005, Modern techniques and Old Red problems—Determining the age of continental sedimentary deposits with $^{40}Ar/^{39}Ar$ provenance analysis in west-central Norway: Norwegian Journal of Geology, v. 85, p. 133–149.

Elvevold, S., and Gilotti, J.A., 1998, Metamorphic and structural studies in the Caledonian fold belt of East Greenland (72°30′–73°N): Danmarks og Grønlands Geologiske Undersøgelse Rapport, v. 1998/28, p. 43–53.

Elvevold, S., Thrane, K., and Gilotti, J.A., 2003, Metamorphic history of high-pressure granulites in Payer Land, Greenland Caledonides: Journal of Metamorphic Geology, v. 21, p. 49–63, doi: 10.1046/j.1525-1314.2003.00419.x.

Escher, J.C., and Jones, K.A., 1998, Caledonian thrusting and extension in Frænkel Land, East Greenland (73°–73°30′N): Preliminary results: Danmarks og Grønlands Geologiske Undersøgelse Rapport, v. 1998/28, p. 29–42.

Escher, J.C., and Jones, K.A., 1999, Caledonian geology of Frænkel Land and adjacent areas (73°00′–73°30′N), East Greenland: Danmarks og Grønlands Geologiske Undersøgelse Rapport, v. 1999/19, p. 27–36.

Fossen, H., 2000, Extensional tectonics in the Caledonides: Synorogenic or post-orogenic?: Tectonics, v. 19, p. 213–224, doi: 10.1029/1999TC900066.

Fossen, H., and Dunlap, W.J., 1998, Timing and kinematics of Caledonian thrusting and extensional collapse, southern Norway: Evidence from

^{40}Ar/^{39}Ar thermochronology: Journal of Structural Geology, v. 20, p. 765–781, doi: 10.1016/S0191-8141(98)00007-8.

Fränkl, E., 1953, Geologische Untersuchungen in Ost-Andrées Land (NE Grønland): Meddelelser om Grønland, v. 113, 160 p.

Friderichsen, J.D., 1999, Hard rock geology of Th. Thomsen land, East Greenland Caledonides: Danmarks og Grønlands Geologiske Undersøgelse Rapport, v. 1999/19, p. 111–119.

Friderichsen, J.D., and Thrane, K., 1998, Caledonian and pre-Caledonian geology of the crystalline complexes of the Stauning Alper, Nathorst Land and Charcot Land, East Greenland: Danmarks og Grønlands Geologiske Undersøgelse Rapport, v. 1998/28, p. 55–71.

Friderichsen, J.D., Henriksen, N., and Strachan, R.A., 1994, Basement-cover relationships and regional structure in the Grandjean Fjord–Bessel Fjord region (75–76°N), North-East Greenland Caledonides: Rapport Grønlands Geologiske Undersøgelse, v. 162, p. 17–23.

Friend, P.F., Alexander-Marrack, P.D., Allen, K.C., Nicholsen, J., and Yeats, A.K., 1983, Devonian Sediments of East Greenland: VI. Review of Results: Meddelelser om Grønland, v. 206, no. 6, 96 p.

Friend, P.F., Williams, B.P.J., Ford, M., and Williams, E.A., 2000, Kinematics and dynamics of Old Red Sandstone basins, in Friend, P.F., and Williams, B.P.J., eds., New Perspectives on the Old Red Sandstone: Geological Society of London Special Publication 180, p. 29–60.

Frotzheim, N., Schmid, S.M., and Conti, P., 1994, Repeated change from crustal shortening to orogen-parallel extension in the Austroalpine units of Graubunden: Eclogae Geologische Helvetica, v. 87, p. 559–612.

Gee, D.G., 1975, A tectonic model for the central part of the Scandinavian Caledonides: American Journal of Science, v. 275a, p. 468–515.

Gilotti, J.A., and Elvevold, S., 2002, Extensional exhumation of a high-pressure granulite terrane in Payer Land, Greenland Caledonides: Structural, petrologic and geochronologic evidence from metapelites: Canadian Journal of Earth Sciences, v. 39, p. 1169–1187, doi: 10.1139/e02-019.

Gilotti, J.A., and McClelland, W.C., 2005, Leucogranites and the time of extension in the East Greenland Caledonides: The Journal of Geology, v. 113, p. 399–417, doi: 10.1086/430240.

Gilotti, J.A., and McClelland, W.C., 2007, Characteristics of, and a tectonic model for, ultrahigh-pressure metamorphism in the overriding plate of the Caledonian orogen: International Geology Review, v. 49, p. 777–797.

Gilotti, J.A., Nutman, A.P., and Brueckner, H.K., 2004, Devonian to Carboniferous collision in the Greenland Caledonides: U-Pb zircon and Sm-Nd ages of polyphase high-pressure and ultrahigh-pressure metamorphisms: Contributions to Mineralogy and Petrology, v. 148, p. 216–235, doi: 10.1007/s00410-004-0600-4.

Gilotti, J.A., Jones, K.A., and Elvevold, S., 2008, this volume, Caledonian metamorphic patterns in Greenland, in Higgins, A.K., Gilotti, J.A., and Smith, M.P., eds., The Greenland Caledonides: Evolution of the Northeast Margin of Laurentia: Geological Society of America Memoir 202, doi: 10.1130/2008.1202(08).

Grujic, D., Hollister, L.S., and Parrish, R.R., 2002, Himalayan metamorphic sequence as an orogenic channel: Insight from Bhutan: Earth and Planetary Science Letters, v. 198, p. 177–191, doi: 10.1016/S0012-821X(02)00482-X.

Haller, J., 1971, The Geology of the East Greenland Caledonides: New York, Interscience Publishers, 413 p.

Hamann, N.E., Whittaker, R.C., and Stemmerik, L., 2005, Structural and geological development of the North-East Greenland Shelf, in Doré, A.G., and Vining, B., eds., Petroleum Geology: North-West Europe and Global Perspectives: Proceedings of the 6th Petroleum Conference: London, Geological Society of London, p. 887–902.

Hansen, B.T., Henriksen, N., and Kalsbeek, F., 1994, Age and origin of Caledonian granites in the Grandjean Fjord–Bessel Fjord region (75–76°N), North-East Greenland: Rapport Grønlands Geologiske Undersøgelse, v. 162, p. 139–151.

Harris, L.B., Koyi, H.A., and Fossen, H., 2002, Mechanisms of folding high-grade rocks in extensional tectonic settings: Earth-Science Reviews, v. 59, p. 163–210, doi: 10.1016/S0012-8252(02)00074-0.

Hartz, E., 2000, Early syndepositional tectonics of East Greenland's Old Red Sandstone basin, in Friend, P.F., and Williams, B.P.J., eds., New Perspectives on the Old Red Sandstone: Geological Society of London Special Publication 180, p. 537–555.

Hartz, E., and Andresen, A., 1995, Caledonian sole thrust of central East Greenland: A crustal-scale Devonian extensional detachment?: Geology, v. 23, p. 637–640, doi: 10.1130/0091-7613(1995)023<0637:CSTOCE>2.3.CO;2.

Hartz, E.H., Andresen, A., Martin, M.W., and Hodges, K.V., 2000, U-Pb and ^{40}Ar/^{39}Ar constraints on the Fjord region detachment zone: A long-lived extensional fault in the central East Greenland Caledonides: Journal of the Geological Society of London, v. 157, p. 795–809.

Hartz, E.H., Andresen, A., Hodges, K.V., and Martin, M.W., 2001, Syncontractional extension and exhumation of deep crustal rocks in the east Greenland Caledonides: Tectonics, v. 20, p. 58–77, doi: 10.1029/2000TC900020.

Henriksen, N., 2003, Geological Map of the Caledonian Orogen, East Greenland, 70–82°N: Copenhagen, Geological Survey of Denmark and Greenland, scale 1:1,000,000.

Henriksen, N., and Higgins, A.K., 2008, this volume, Caledonian orogen of East Greenland 70°N–82°N: Geological map at 1:1,000,000—Concepts and principles of compilation, in Higgins, A.K., Gilotti, J.A., and Smith, M.P., eds., The Greenland Caledonides: Evolution of the Northeast Margin of Laurentia: Geological Society of America Memoir 202, doi: 10.1130/2008.1202(14).

Higgins, A.K., 1988, The Krummedal supracrustal sequence in East Greenland, in Winchester, J.A., ed., Later Proterozoic Stratigraphy of the Northern Atlantic Regions: Glasgow, Blackie and Son, p. 86–96.

Higgins, A.K., and Leslie, A.G., 2000, Restoring thrusting in the East Greenland Caledonides: Geology, v. 28, p. 1019–1022, doi: 10.1130/0091-7613(2000)28<1019:RTITEG>2.0.CO;2.

Higgins, A.K., and Leslie, A.G., 2008, this volume, Architecture and evolution of the East Greenland Caledonides—An introduction, in Higgins, A.K., Gilotti, J.A., and Smith, M.P., eds., The Greenland Caledonides: Evolution of the Northeast Margin of Laurentia: Geological Society of America Memoir 202, doi: 10.1130/2008.1202(02).

Higgins, A.K., Friderichsen, J.D., and Thyrsted, T., 1981, Precambrian metamorphic complexes in the East Greenland Caledonides (72°–74°N): Their relationships to the Eleonore Bay Group and Caledonian orogenesis: Rapport Grønlands Geologiske Undersøgelse, v. 104, p. 4–46.

Higgins, A.K., Henriksen, N.H., Jepsen, H.F., Kalsbeek, F., Thrane, K., Elvevold, S., Escher, J.C., Frederiksen, K.S., Gilotti, J.A., Jones, K., Leslie, A.G., Smith, M.P., Kinny, P.D., and Watt, G.R., 2004, The foreland-propagating architecture of the East Greenland Caledonides 72°–75°N: Journal of the Geological Society of London, v. 161, p. 1009–1026, doi: 10.1144/0016-764903-141.

Hodges, K.V., 2000, Tectonics of the Himalaya and southern Tibet from two perspectives: Geological Society of America Bulletin, v. 112, p. 324–350, doi: 10.1130/0016-7606(2000)112<0324:TOTHAS>2.3.CO;2.

Holdsworth, R.E., and Strachan, R.A., 1991, Interlinked system of ductile strike-slip and thrusting formed by Caledonian sinistral transpression in northeastern Greenland: Geology, v. 19, p. 510–513, doi: 10.1130/0091-7613(1991)019<0510:ISODSS>2.3.CO;2.

Hossack, J.R., 1984, The geometry of listric growth faults in the Devonian basins of Sunnfjord, W Norway: Journal of the Geological Society of London, v. 141, p. 629–637, doi: 10.1144/gsjgs.141.4.0629.

Hossack, J.R., and Cooper, M.A., 1986, Collision tectonics of the Scandinavian Caledonides, in Coward, M.P., and Ries, A.C., eds., Collision Tectonics: Geological Society of London Special Publication 19, p. 287–304.

Hull, J.M., and Gilotti, J.A., 1994, The Germania Land deformation zone and related structures, North-East Greenland: Rapport Grønlands Geologiske Undersøgelse, v. 162, p. 113–127.

Jones, K.A., and Escher, J.C., 1999, Thickening and collapse history of the Caledonian crust, Clavering Ø, East Greenland: Danmarks og Grønlands Geologiske Undersøgelse Rapport, v. 1999/19, p. 101–109.

Jones, K.A., and Escher, J.C., 2002, Near-isothermal decompression within a clockwise P-T evolution recorded in migmatitic mafic granulites from Clavering Ø, NE Greenland: Implications for the evolution of the Caledonides: Journal of Metamorphic Geology, v. 20, p. 365–378, doi: 10.1046/j.1525-1314.2002.00375.x.

Jones, K.A., and Strachan, R.A., 2000, Crustal thickening and ductile extension in the NE Greenland Caledonides: A metamorphic record from anatectic pelites: Journal of Metamorphic Geology, v. 18, p. 719–735, doi: 10.1046/j.1525-1314.2000.00282.x.

Kalsbeek, F., Thrane, K.T., Nutman, A.P., and Jepsen, H.P., 2000, Late Mesoproterozoic to early Neoproterozoic history of the East Greenland Caledonides: Evidence for Grenvillian orogenesis?: Journal of the Geological Society of London, v. 157, p. 1215–1225.

Kalsbeek, F., Jepsen, H.P., and Nutman, A.P., 2001a, From source migmatites to plutons: Tracking the origin of the ca. 435 Ma S-type granites in the East Greenland Caledonian orogen: Lithos, v. 57, p. 1–21, doi: 10.1016/S0024-4937(00)00071-2.

Kalsbeek, F., Jepsen, H.F., and Jones, K.A., 2001b, Geochemistry and petrogenesis of S-type granites in the East Greenland Caledonides: Lithos, v. 57, p. 91–109, doi: 10.1016/S0024-4937(01)00038-X.

Kalsbeek, F., Higgins, A.K., Jepsen, H.F., Frei, R., and Nutman, A.P., 2008, this volume, Granites and granites in the East Greenland Caledonides, in Higgins, A.K., Gilotti, J.A., and Smith, M.P., eds., The Greenland Caledonides: Evolution of the Northeast Margin of Laurentia: Geological Society of America Memoir 202, doi: 10.1130/2008.1202(09).

Kendrick, M.A., Eide, E.A., Roberts, D., and Osmundsen, P.T., 2004, The Mid–Late Devonian Høybakken detachment, central Norway: $^{40}Ar/^{39}Ar$ evidence for prolonged late/post-Scandian extension and uplift: Geological Magazine, v. 141, p. 329–344, doi: 10.1017/S0016756803008811.

Koch, L., 1955, Report on the Expeditions to Central East Greenland 1926–39 Conducted by Lauge Koch, Part II: Meddelelser om Grønland, v. 143, no. 2, 642 p.

Krabbendam, M., and Dewey, J.F., 1998, Exhumation of UHP rocks by transtension in the Western Gneiss Region, Scandinavian Caledonides, in Holdsworth, R.E., Strachan, R.A., and Dewey, J.F., eds., Continental Transpressional and Transtensional Tectonics: Geological Society of London Special Publication 135, p. 159–181.

Larsen, H.C., 1990, The East Greenland shelf, in Grant, A., Johnson, L., and Sweeney, J.F., eds., The Arctic Ocean region: Boulder, Colorado, Geological Society of America, Geology of North America, v. L, p. 185–210.

Larsen, P.-H., and Bengaard, H.J., 1991, Devonian basin initiation in East Greenland: A result of sinistral wrench faulting and Caledonian extensional collapse: Journal of the Geological Society of London, v. 148, p. 355–368, doi: 10.1144/gsjgs.148.2.0355.

Larsen, P.-H., Olsen, H., and Clack, J.A., 2008, this volume, The Devonian basin in East Greenland—Review of basin evolution and vertebrate assemblages, in Higgins, A.K., Gilotti, J.A., and Smith, M.P., eds., The Greenland Caledonides: Evolution of the Northeast Margin of Laurentia: Geological Society of America Memoir 202, doi: 10.1130/2008.1202(11).

Leslie, A.G., and Higgins, A.K., 1998, On the Caledonian geology of Andrée Land, Eleonore Sø and adjacent nunataks (73°30′–74°N), East Greenland: Danmarks og Grønlands Geologiske Undersøgelse Rapport, v. 1998/28, p. 11–27.

Leslie, A.G., and Higgins, A.K., 2008, this volume, Foreland-propagating Caledonian thrust systems in East Greenland, in Higgins, A.K., Gilotti, J.A., and Smith, M.P., eds., The Greenland Caledonides: Evolution of the Northeast Margin of Laurentia: Geological Society of America Memoir 202, doi: 10.1130/2008.1202(07).

Leslie, A.G., and Nutman, A.P., 2003, Evidence of Neoproterozoic orogenesis and early high temperature Scandian deformation events in the southern East Greenland Caledonides: Geological Magazine, v. 140, p. 309–333, doi: 10.1017/S0016756803007593.

Lister, G.S., and Davis, G.A., 1989, The origin of metamorphic core complexes and detachment faults during Tertiary continental extension in the northern Colorado River region: Journal of Structural Geology, v. 11, p. 65–94, doi: 10.1016/0191-8141(89)90036-9.

Mancktelow, N.S., and Pavlis, T.L., 1994, Fold–fault-relationships in low-angle detachment systems: Tectonics, v. 13, p. 668–685, doi: 10.1029/93TC03489.

Mattauer, M., Brunel, M., and Matte, P., 1988, Failles normales ductiles et grands chevauchements: Une nouvelle analogie dans l'Himalaya et la Chaîne Hercynienne du Massif Central Français: Comptes Rendus de l'Académie des Sciences [Paris], v. 306, p. 671–676.

McClay, K.R., Norton, M.G., Coney, P., and Davis, G.H., 1986, Collapse of the Caledonian orogen and the Old Red Sandstone: Nature, v. 323, p. 147–149, doi: 10.1038/323147a0.

McClelland, W.C., and Gilotti, J.A., 2003, Late-stage extensional exhumation of high-pressure granulites in the Greenland Caledonides: Geology, v. 31, p. 259–262, doi: 10.1130/0091-7613(2003)031<0259:LSEEOH>2.0.CO;2.

McClelland, W.C., Power, S.E., Gilotti, J.A., Mazdab, F.K., and Wopenka, B., 2006, U-Pb SHRIMP geochronology and trace-element geochemistry of coesite-bearing zircons, North-East Greenland Caledonides, in Hacker, B., McClelland, W.C., and Liou, J.G., eds., Ultrahigh-Pressure Metamorphism: Deep Continental Subduction: Geological Society of America Special Paper 403, p. 23–43.

Mezger, K., van der Pluijm, B.A., Essene, E.J., and Halliday, A.N., 1991, Synorogenic collapse; a perspective from the middle crust, the Proterozoic Grenville orogen: Science, v. 254, p. 695–698, doi: 10.1126/science.254.5032.695.

Mosar, J., 2003, Scandinavia's North Atlantic passive margin: Journal of Geophysical Research, v. 108, no. 8, 2360, doi: 10.1029/2002JB002134.

Norton, M.G., 1987, The Nordfjord–Sogn detachment, W. Norway: Norsk Geologisk Tidsskrift, v. 67, p. 93–106.

Osmundsen, P.T., Andersen, T.B., Markussen, S., and Svendby, A.K., 1998, Tectonics and sedimentation in the hanging wall of a regional extensional detachment: The Devonian Kvamshesten basin, W. Norway: Basin Research, v. 10, p. 213–234, doi: 10.1046/j.1365-2117.1998.00064.x.

Osmundsen, P.T., Braathen, A., Nordgulen, Ø., Roberts, D., Meyer, G.B., and Eide, E.A., 2003, The Devonian Nesna shear zone and adjacent gneiss-cored culminations, north-central Norwegian Caledonides: Journal of the Geological Society of London, v. 160, p. 137–150, doi: 10.1144/0016-764901-173.

Osmundsen, P.T., Eide, E.A., Haabesland, N.E., Roberts, D., Andersen, T.B., Kendrick, M., Bingen, B., Braathen, A., and Redfield, T.F., 2006, Kinematics of the Høybakken detachment zone and the Møre-Trøndelag fault complex, central Norway: Journal of the Geological Society of London, v. 163, p. 303–318, doi: 10.1144/0016-764904-129.

Peucat, J.J., Tisserant, D., Caby, R., and Clauer, N., 1985, Resistance of zircons to U-Pb resetting in a prograde metamorphic sequence of Caledonian age in East Greenland: Canadian Journal of Earth Sciences, v. 22, p. 330–338.

Platt, J.P., and Vissers, R.L.M., 1989, Extensional collapse of thickened continental lithosphere: A working hypothesis for the Alboran Sea and Gibraltar arc: Geology, v. 17, p. 540–543, doi: 10.1130/0091-7613(1989)017<0540:ECOTCL>2.3.CO;2.

Rasmussen, T.V., and Andresen, A., 1998, Kinematic indicators in Højedal relative to the Fjord region detachment: Danmarks og Grønlands Geologiske Undersøgelse Rapport, v. 1998/46, p. 55.

Rex, D.C., and Gledhill, A.R., 1981, Isotopic studies in the East Greenland Caledonides (72°–74°N)—Precambrian and Caledonian ages: Rapport Grønlands Geologiske Undersøgelse, v. 104, p. 47–72.

Rex, D.C., and Higgins, A.K., 1985, Potassium-argon mineral ages from the East Greenland Caledonides between 72° and 74°N, in Gee, D.G., and Sturt, B.A., eds., The Caledonide Orogen—Scandinavia and Related Areas, Volume 2: New York, John Wiley, p. 1115–1124.

Roberts, D., 2003, The Scandinavian Caledonides: Event chronology, paleogeographic setting and likely modern analogues: Tectonophysics, v. 365, p. 283–299, doi: 10.1016/S0040-1951(03)00026-X.

Root, D.B., Hacker, B.R., Mattinson, J.M., and Wooden, J.L., 2004, Zircon geochronology and ca. 400 Ma exhumation of Norwegian ultrahigh-pressure rocks: An ion microprobe and chemical abrasion study: Earth and Planetary Science Letters, v. 228, p. 325–341, doi: 10.1016/j.epsl.2004.10.019.

Root, D.B., Hacker, B.R., Gans, P.B., Ducea, M.N., Eide, E.A., and Mosenfelder, J.L., 2005, Discrete ultrahigh-pressure domains in the Western Gneiss Region, Norway: Implications for formation and exhumation: Journal of Metamorphic Geology, v. 23, p. 45–61, doi: 10.1111/j.1525-1314.2005.00561.x.

Rosenbaum, G., Regenauer-Lieb, K., and Weinberg, R., 2005, Continental extension: From core complexes to rigid block faulting: Geology, v. 33, p. 609–612, doi: 10.1130/G21477.1.

Sartini-Rideout, C., Gilotti, J.A., and McClelland, W.C., 2006, Geology and timing of dextral strike-slip shear zones in Danmarkshavn, North-East Greenland Caledonides: Geological Magazine, v. 143, p. 431–446, doi: 10.1017/S0016756806001968.

Schlindwein, V., and Jokat, W., 1999, Structure and evolution of the continental crust of northern East Greenland from integrated geophysical studies: Journal of Geophysical Research, v. 104, p. 15,227–15,245, doi: 10.1029/1999JB900101.

Searle, M.P., 1999, Extensional and compressional faults in the Everest-Lhotse massif, Khumbu Himalaya, Nepal: Journal of the Geological Society of London, v. 156, p. 227–240, doi: 10.1144/gsjgs.156.2.0227.

Searle, M.P., and Godin, L., 2003, The South Tibetan detachment and Manaslu leucogranite: 431–446 Ma structural reinterpretation and restoration of the Annapurna-Manaslu Himalaya, Nepal: The Journal of Geology, v. 111, p. 505–523, doi: 10.1086/376763.

Searle, M.P., Simpson, R.F., Law, R.D., Parrish, R.R., and Waters, D., 2003, The structural geometry, metamorphic and magmatic evolution of the Everest Massif, High Himalaya of Nepal–South Tibet: Journal of the Geological Society of London, v. 160, p. 345–366.

Selverstone, J., 1988, Evidence for east-west crustal extension in the eastern Alps: Implications for the unroofing history of the Tauern Window: Tectonics, v. 7, p. 87–105, doi: 10.1029/TC007i001p00087.

Selverstone, J., 2005, Are the Alps collapsing?: Annual Review of Earth and Planetary Sciences, v. 33, p. 113–132, doi: 10.1146/annurev.earth.33.092203.122535.

Smith, M.P., and Robertson, S., 1999, The Nathorst Land Group (Neoproterozoic) of East Greenland—Lithostratigraphy, basin geometry and tectonic history: Danmarks og Grønlands Geologiske Undersøgelse Rapport, v. 1999/19, p. 127–143.

Sønderholm, M., and Tirsgaard, H., 1993, Lithostratigraphic framework of the Upper Proterozoic Eleonore Bay Supergroup of East and North-East Greenland: Bulletin Grønlands Geologiske Undersøgelse, v. 167, p. 1–38.

Soper, N.J., and Higgins, A.K., 1993, Basement-cover relationships in the East Greenland Caledonides: Evidence from the Eleonore Bay Supergroup at Ardencaple Fjord: Transactions of the Royal Society of Edinburgh, Earth Sciences, v. 84, p. 103–115.

Soper, N.J., Strachan, R.A., Holdsworth, R.E., Gayer, R.A., and Greiling, R.O., 1992, Sinistral transpression and the Silurian closure of Iapetus: Journal of the Geological Society of London, v. 149, p. 871–880, doi: 10.1144/gsjgs.149.6.0871.

Stemmerik, L., 2000, Late Paleozoic evolution of the North Atlantic margin of Pangea: Palaeogeography, Palaeoclimatology, Palaeoecology, v. 161, p. 95–126, doi: 10.1016/S0031-0182(00)00119-X.

Stemmerik, L., and Worsley, D., 2005, 30 years on—Arctic Upper Palaeozoic stratigraphy, depositional evolution and hydrocarbon prospectivity: Norwegian Journal of Geology, v. 85, p. 151–168.

Stouge, S., Boyce, W.D., Christiansen, J.L., Harper, D.A.T., and Knight, I., 2002, Lower–Middle Ordovician stratigraphy of North-East Greenland: Geology of Greenland Survey Bulletin, v. 191, p. 117–125.

Strachan, R.A., 1994, Evidence in North-East Greenland for Late Silurian–Early Devonian regional extension during the Caledonian orogeny: Geology, v. 22, p. 913–916, doi: 10.1130/0091-7613(1994)022<0913:EINEGF>2.3.CO;2.

Strachan, R.A., Martin, M.S., and Friderichsen, J.D., 2001, Evidence for contemporaneous yet contrasting styles of granite magmatism during extensional collapse of the northeast Greenland Caledonides: Tectonics, v. 20, p. 458–473, doi: 10.1029/2000TC001206.

Terry, M.P., and Robinson, P., 2003, Evolution of amphibolite-facies structural features and boundary conditions for deformation during exhumation of high- and ultrahigh-pressure rocks, Nordøyane, Western Gneiss Region, Norway: Tectonics v. 22, p. 10-1–10-23.

Terry, M.P., Robinson, P., Hamilton, M.A., and Jercinovic, M.J., 2000, Monazite geochronology of UHP and HP metamorphism, deformation and exhumation, Nordøyane, Western Gneiss Region, Norway: The American Mineralogist, v. 85, p. 1651–1664.

Thrane, K., 2002, Relationships between Archean and Paleoproterozoic crystalline basement complexes in the southern part of the East Greenland Caledonides: An ion microprobe study: Precambrian Research, v. 113, p. 19–42, doi: 10.1016/S0301-9268(01)00198-X.

Tucker, R.D., Robinson, P., Solli, A., Gee, D.G., Thorsnes, T., Krogh, T.E., Nordgulen, Ø., and Bickford, M.E., 2004, Thrusting and extension in the Scandian hinterland, Norway: New U-Pb ages and tectonostratigraphic evidence: American Journal of Science, v. 304, p. 477–532, doi: 10.2475/ajs.304.6.477.

Vannay, J.C., and Grasemann, B., 2001, Himalayan inverted metamorphism and syn-convergence extension as a consequence of general shear extrusion: Geological Magazine, v. 138, p. 253–276, doi: 10.1017/S0016756801005313.

Vold, J., 1997, Structural and Metamorphic Evolution of the Fjord Region Detachment Zone at Kap Hedlund, Central East Greenland [Cand. Scient. thesis]: Oslo, University of Oslo, 122 p.

Watt, G., and Friderichsen, J.D., 1999, Further remarks on Caledonian anatexis in the Stauning Alper migmatite zone, East Greenland Caledonides: Danmarks og Grønlands Geologiske Undersøgelse Rapport, v. 1999/19, p. 59–69.

Watt, G., and Kinny, P.D., 1998, Caledonian migmatisation and granite formation in the northern Gåsefjord–Stauning Alper migmatite and granite zone, East Greenland: Danmarks og Grønlands Geologiske Undersøgelse Rapport, v. 1998/28, p. 117–133.

Watt, G., Kinny, P.D., and Friderichsen, J.D., 2000, U-Pb geochronology of Neoproterozoic and Caledonian tectonothermal events in the East Greenland Caledonides: Journal of the Geological Society of London, v. 157, p. 1031–1048.

Wawrzyniec, T.F., Selverstone, J., and Axen, G.J., 2001, Styles of footwall uplift along the Simplon and Brenner normal fault systems, central and eastern Alps: Tectonics, v. 20, p. 748–770, doi: 10.1029/2000TC001253.

Wenk, E., and Haller, J., 1953, Geological explorations in the Petermann Bjerg region, western part of Frænkels Land, East Greenland: Meddelelser om Grønland, v. 111, p. 1–48.

Wernicke, B., 1981, Low-angle normal faults in the Basin and Range Province: Nappe tectonics in an extending orogen: Nature, v. 291, p. 645–648, doi: 10.1038/291645a0.

White, A.P., and Hodges, K.V., 2002, Multistage extensional evolution of the central East Greenland Caledonides: Tectonics, v. 22, p. 34–45.

White, A.P., and Hodges, K.V., 2003, Pressure-temperature-time evolution of the central East Greenland Caledonides: Quantitative constraints on crustal thickening and synorogenic extension: Journal of Metamorphic Geology, v. 21, p. 875–897.

White, A.P., Hodges, K.V., Martin, M.W., and Andresen, A., 2002, Geologic constraints on middle-crustal behavior during broadly synorogenic extension in the central East Greenland Caledonides: International Journal of Earth Sciences, v. 91, p. 187–208, doi: 10.1007/s005310100227.

MANUSCRIPT ACCEPTED BY THE SOCIETY 14 JANUARY 2008

The Geological Society of America
Memoir 202
2008

The Devonian basin in East Greenland—Review of basin evolution and vertebrate assemblages

Poul-Henrik Larsen*
Mærsk Olie og Gas AS, Esplanaden 50, DK-1263 Copenhagen K, Denmark

Henrik Olsen
COWI AS, Parallelvej 2, DK-2800 Kongens Lyngby, Denmark

Jennifer A. Clack
University Museum of Zoology, University of Cambridge, Downing Street, Cambridge CB2 3EJ, UK

ABSTRACT

From Middle Devonian times, the continental Old Red Sandstone Basin in North-East Greenland has accumulated more than 8 km of mainly coarse clastic sediments. These sediments have been studied for more than 100 yr, and they became world famous prior to the Second World War for the discovery of the earliest four-legged vertebrates, the tetrapods later assigned to the genus *Ichthyostega*. Basin initiation in East Greenland was caused mainly by extensional collapse of an overthickened Caledonian crustal welt accommodated by SE-NW–oriented dip-slip faulting and, subordinately, by N-S–oriented sinistral wrench faulting due to late Caledonian shear displacements along plate boundaries. Four main tectonostratigraphic basin stages have been recognized in the succession. The stages of basin development are separated by subregional to basinwide unconformities and represent depositional episodes punctuated by major tectonic events. Each basin stage is built up of one or several depositional complexes that share roughly similar drainage patterns, measured on a basinwide scale. The four basin stages indicate initial eastward drainage, followed by southward drainage, northward drainage, and finally southwestward drainage.

In this paper, we review previous research on the dynamics of Devonian basin initiation, its filling, and the tectonic and climatic controls on sedimentary processes. The successive vertebrate faunal assemblages of the different basin stages are also reviewed, with some consideration of the preservational, ecological, and wider faunal contexts of the components of those faunas. Some remaining problems of correlation and precise dating are noted, and suggestions are made for further work.

Keywords: Caledonides, Greenland, Devonian, vertebrate paleontology, structural geology, sedimentology.

*phljo@os.dk

INTRODUCTION TO PREVIOUS RESEARCH

The presence of Devonian rocks in East Greenland was first noted by Nathorst (1901), who discovered fish remains in sandstones on an expedition in 1899. Woodward (1900) interpreted these as Late Devonian in age. Many scientists (Orvin, 1930, 1931; Heintz, 1930, 1932; Koch, 1929a, 1929b; Kulling, 1930, 1931; Säve-Söderbergh, 1932a, 1932b) subsequently worked on the sedimentology and vertebrate fossils of the area. Säve-Söderbergh (1932a, 1933, 1934) divided the Devonian sequences into several stratigraphic series based on the fish fossils and thus established the first biostratigraphy of the Devonian Old Red Sandstone succession in East Greenland. The lithostratigraphy and structural development of the basin were extensively investigated by Bütler (1935a, 1935b, 1954, 1955, 1957, 1959, 1961), and later by Haller (1971). Subsequent work by Friend and coworkers extended Bütler's work to demonstrate the fluvial nature of the Old Red Sandstone and establish the architecture, sediment, and dispersal patterns in the basin (Friend et al., 1976a, 1976b, 1983; Alexander-Marrack and Friend, 1976; Yeats and Friend, 1978; Nicholson and Friend, 1976).

Later studies, including structural geology and mapping (Marcussen et al., 1987, 1988; P.H. Larsen et al., 1989; Larsen and Olsen, 1991; Larsen, 1990a, 1990b, 1900c, 1990d), led to a new geological map of the basin by Bengaard and Larsen (1991) and a new tectonic model for basin initiation (Larsen and Bengaard, 1991). Olsen and Larsen (Olsen, 1993a; Olsen and Larsen, 1993a) integrated sedimentologic studies and mapping of stratigraphic units to produce a comprehensively revised and formal lithostratigraphy for the Devonian succession. Other aspects of the interplay of climate, tectonics, and sedimentation have also been published (Christiansen et al., 1990; Kelly and Olsen, 1993; Olsen and Larsen, 1993b).

More recently, Clack, Bendix-Almgreen, and colleagues (Bendix-Almgreen, 1976; Bendix-Almgreen et al., 1988, 1990; Clack and Neininger, 2000) collected more vertebrate material. Based on this and earlier collected material, Clack was able to describe the tetrapods from East Greenland in detail (e.g., Clack, 1998, 2003; Clack et al., 2003) and address problems of the fish to tetrapod transition (Clack, 2002, 2005, 2006). Vertebrate studies have been complemented by palynologic and ecologic discoveries that throw light on the stratigraphic, chronologic, and ecologic context of the sediments and the fossils they contain (Marshall and Astin, 1996; Marshall et al., 1999; Marshall and Hemsley, 2003). Reviews of the Devonian vertebrate fauna have recently been compiled by Blom and colleagues (Blom et al., 2005, 2007). These reviews brought together all the records of East Greenland Devonian vertebrate fossils from the literature, museum collections, and unpublished field notes, producing the first comprehensive documentation of the fossils and their distribution in time and space within the Devonian basin.

The Cambridge Arctic Shelf Programme (CASP) East Greenland Project in 1992 (e.g., Marshall and Stephenson, 1997) and the Alfred Wegener Institute for Polar and Marine Research (e.g., Schlindwein, 1998), and several expeditions led by Andresen (University of Oslo) have resulted in many papers that offer further insight into the complexities of the Devonian basin initiation and its structural development (e.g., Andresen et al., 1998; Hartz, 2000; Hartz and Andresen, 1995, 1997; Hartz et al., 1996, 2000). In addition, controversial radiometric dating results have formed the basis for discussion of the correct age of the sediments (e.g., Hartz et al., 1997, 1998; Stemmerik and Bendix-Almgreen, 1998; Marshall et al., 1999).

THE DEVONIAN OF EAST GREENLAND

From late Middle Devonian time, the continental Old Red Sandstone Basin in North-East Greenland has accumulated more than 8 km of mainly coarse clastic sediments. The sediments generally rest unconformably on Cambrian-Ordovician carbonates and, to a lesser extent, on older rocks in the northern part of the basin (see map in Plate 1 (Henriksen and Higgins, this volume)) (Larsen, 1990a; Bengaard and Larsen, 1991; Sønderholm and Tirsgaard, 1993; Tirsgaard and Sønderholm, 1997). Toward the east, the sediments are generally in fault contact with Upper Paleozoic and Mesozoic rocks topped by Tertiary plateau basalts (Fig. 1 and Plate 1, Henriksen and Higgins, this volume).

The main part of the exposed Devonian basin is situated in North-East Greenland, although small outliers occur in Canning Land and on Wegener Halvø in East Greenland (Fig. 1). Seismic studies in Jameson Land have revealed Devonian sediments indicating a much larger paleogeographic extension of the basin than is seen today (H.C. Larsen et al., 1989; Larsen and Marcussen, 1992; Schlindwein, 1998). The Old Red Sandstone sediments crop out over more than 300 km in a N-S–trending direction from Canning Land in the south to Wordie Gletscher in the north (71°30′N to 74°15′N) (Fig. 1). The widest E-W extent of the outcrops occurs in the Kejser Franz Joseph Fjord area. Devonian sediments from Kap Franklin in the east to the mouth of Geologfjord in the west show that the outcrops cover a land area of ~5600 km^2.

BASIN INITIATION

Several tectonic models have been proposed to explain the initiation and development of the East Greenland Devonian basin.

Bütler (1959) was the first to suggest that the deposition of the Old Red Sandstone occurred in a N-S–trending intramontane basin or depression created after the late Caledonian Orcadian orogenic phase followed by renewal of the relief and accumulation in the Devonian trough. He envisaged that the syn- to post-depositional deformation of the basin should be regarded as local manifestations of wider Middle and Late Devonian late Caledonian orogenic processes elsewhere.

This model was further developed by Haller (1970, 1971), who suggested that central East Greenland could be subdivided into major late Caledonian structural provinces, in which a downflexed graben-like tract oriented NW-SE became filled

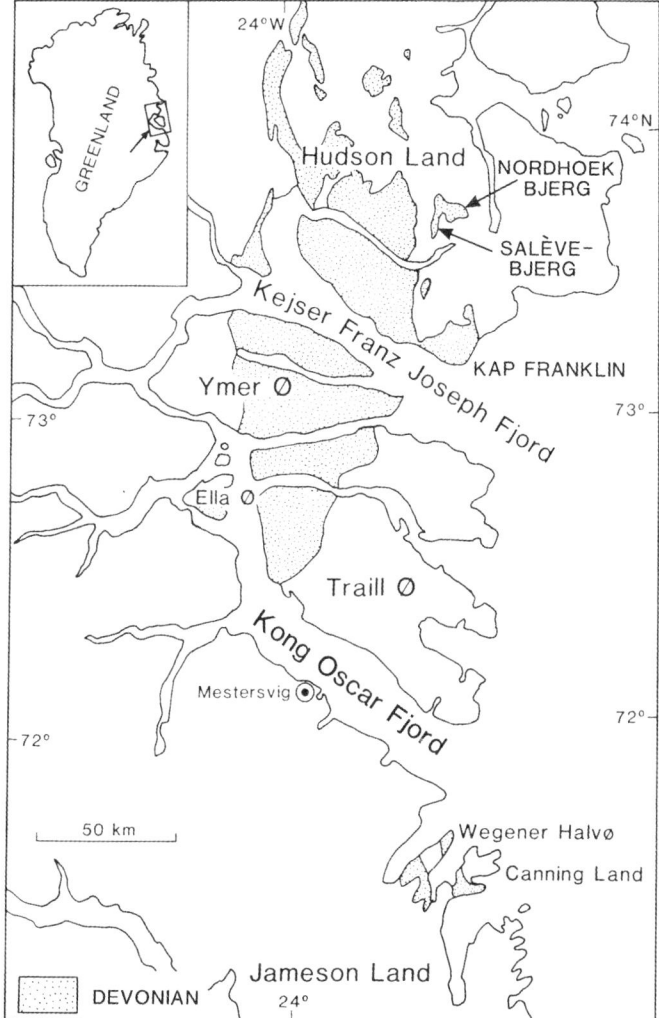

Figure 1. Map showing the extent of the Devonian outcrops in North-East Greenland.

with continental Devonian deposits. Sedimentologic studies led Friend et al. (1983) to suggest a different model in which the tectonic situation was dominated by four N-S–trending fracture zones. The positions of these fractures were established using the presence of aligned granite intrusions and other geological lineaments (Friend et al., 1983, p. 88). They proposed that these were probably largely generated by wrench stresses, although the fractures were thought to include both vertical and horizontal components of motion. McClay et al. (1986) and Norton et al. (1987) proposed that the Devonian basins in Norway and Scotland had developed due to collapse of the overthickened crustal welt, which had resulted from Caledonian compressional tectonics. These authors extended their models to include East Greenland and postulated that a comparable Devonian basin evolution might have taken place there.

Based on their own work (e.g., Bengaard, 1985, 1989; Larsen, 1990a), Larsen and Bengaard (1991) concluded that basin initiation in East Greenland was caused mainly by extensional collapse of an overthickened Caledonian crustal welt accommodated by SE-NW–oriented dip-slip faulting and, subordinately, by N-S–oriented, sinistral wrench faulting as a result of late Caledonian shear displacements along plate boundaries. The extension was suggested to have been accommodated along two major NNE-SSW–trending east-dipping normal faults (the Fjord zone fault and the Nunatak zone fault) that transect the crystalline basement and Caledonian fold structures west of the present-day outcrop of the Devonian basin (Fig. 2). Based on K-Ar ages in the basement of the exhumed footwall block to the west, the extensional collapse was suggested to have taken place prior to Middle Devonian times. However, recent radiometric dating seems to indicate that the large extensional structures, especially the Fjord zone fault, are long-lived structures that may have been active from Late Silurian to Devonian time (e.g., Andresen et al., 1998; Hartz et al., 2000, 2001).

In addition to the extensional faults, a N-S–trending sinistral wrench zone developed, which marks the western margin (the Western fault zone) of the present-day outcrop of the Devonian basin (Fig. 2). Both the extensional faults and wrench fault can be followed as regional structures or lineaments through much of East Greenland and can be compared to similar Devonian structures elsewhere in the Caledonian orogen. No clear separation in time between the extensional collapse and the initial wrenching was found by Larsen and Bengaard (1991), which therefore led these authors to suggest a transtensional kinematic history for the early stages of the basin formation.

After initiation and during deposition of the 8 km of basin fill, the basin underwent a series of deformation events that continued into the Early Carboniferous [Mississippian] (the Hudson Land and Ymer Ø phases of Bütler, 1935a, 1935b). These deformation events were suggested by Larsen and Olsen (1991) to have been caused by left-lateral strike-slip movements along the eastern basin margin, so the basin during this period could be viewed as undergoing transpressional deformation between overstepping strike-slip faults (Fig. 2).

The major extensional collapse faults suggested by Larsen and Bengaard (1991) inspired Hartz and Andresen (1995) and Andresen et al. (1998) to reexamine the Fjord zone fault. Their work clearly indicated the extensional nature of this feature on the basis of kinematic indicators. They renamed the Fjord zone fault as the Fjord region detachment and upgraded it to the main overall extensional feature within the region. The Western fault zone along the Devonian basin was referred to as a subordinate but major normal fault within the hanging wall of the extensional detachment. The Western fault zone may have been reactivated later as a wrench fault. The Devonian basin itself was interpreted to have developed in the stretched hanging-wall block above the Fjord region detachment, and some of the Middle Devonian unconformities within the basin were suggested to be the result of crustal extension and block rotation (Hartz and Andresen, 1995). The other main extensional collapse feature mentioned by Larsen and Bengaard (1991), the Nunatak zone fault, was not examined

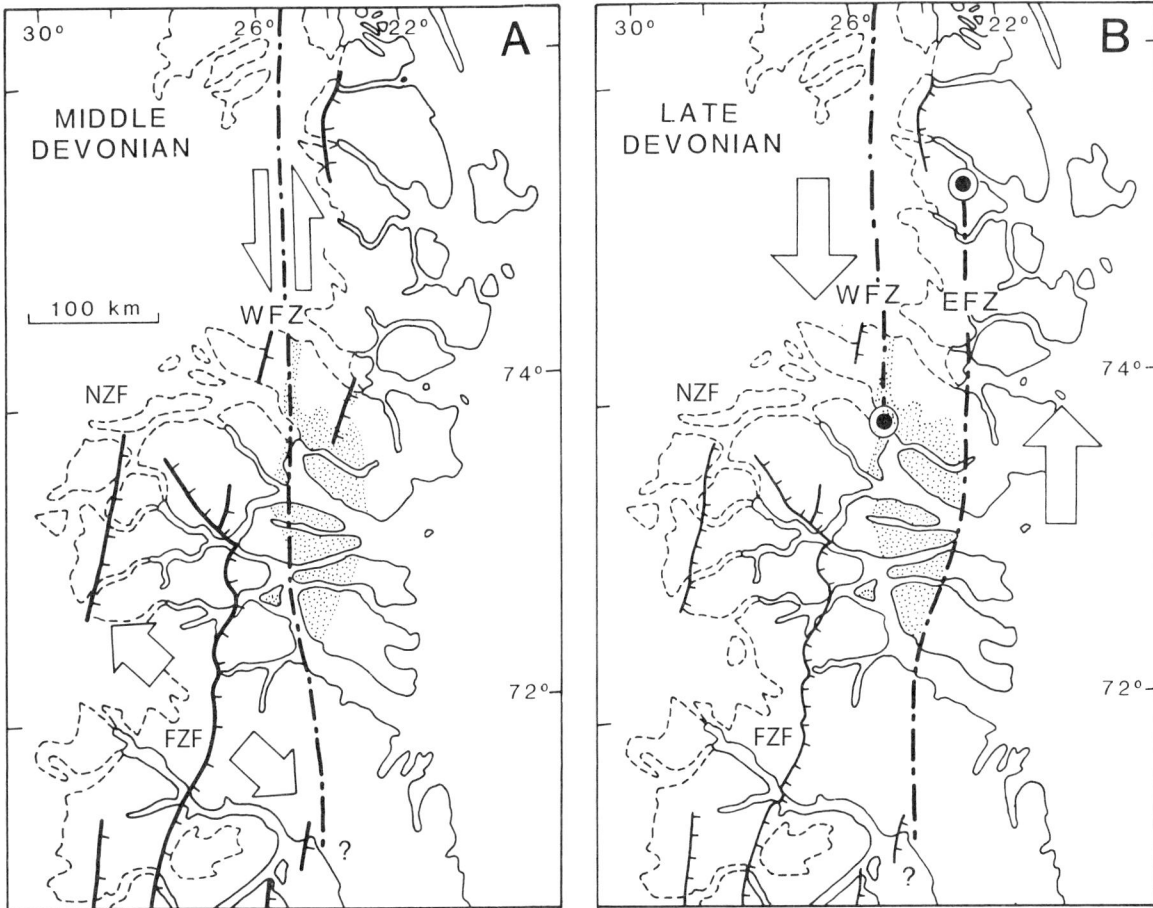

Figure 2. Tectonic setting of the East Greenland Devonian basin in (A) Middle and (B) Late Devonian time (after Larsen and Bengaard, 1991; Larsen and Olsen, 1991; Olsen, 1993a). NZF—Nunatak Zone Fault; FZF—Fjord Zone Fault; WFZ—Western Fault Zone; EFZ—Eastern Fault Zone. Circles indicate tip lines for the Eastern and Western fault zone structures.

by Hartz and Andresen (1995), but they suggested this fault to be subordinate to the Fjord region detachment. However, later publications (e.g., Andresen et al., 1998; Hartz et al., 2001) included the Nunatak zone fault in the extensional structures and thus fully support the proposed extensional collapse fault pattern of Larsen and Bengaard (1991) (Fig. 2).

STRATIGRAPHY OF THE DEVONIAN BASIN FILL

Orvin (1930, 1931) was the first to group the strata according to the color of the beds. Unfortunately, since the color of the succession changed in both vertical and horizontal directions (a problem that later stratigraphers also had to face), he was not able to correlate his units regionally. In addition, the fjord valleys represented broad areas where the beds could not be followed, and he therefore failed to correlate the fossil horizons with certainty. Kulling (1930, 1931) formulated a similar preliminary stratigraphy based on several measured stratigraphic sections and study of the unconformity underlying the sedimentary sequence on Ymer Ø. On the eastern tip of Ymer Ø, at Celsius Bjerg, he found the first Devonian tetrapod fossils.

From studies of the vertebrate fossils and especially the fish fauna, Säve-Söderbergh (1932a, 1933, 1934) established the first biostratigraphy of the Old Red Sandstone sequence, dividing it into several stratigraphic series. Unfortunately, only the uppermost series were rich in fossils, and these were often confined to small areas. The lower part of the succession was far thicker and had a more regional distribution, but it did not yield fossil material suitable for stratigraphic purposes. Therefore, the lower series was designated as the Devonian Basal Conglomerate, and its age within the Devonian was left open. An Upper Devonian *Phyllolepis* series (also termed "Lower Sandstone Complex") was correlated with the youngest beds in other Old Red Sandstone areas in Europe. It was suggested that the highest series (*Remigolepis* series and *Arthrodire* sandstone) were younger than all the Old Red Sandstone formations in other areas known at that time.

Bütler (1935a) established a Devonian lithostratigraphy based on angular unconformities in the Hudson Land–Moskusoksefjord area and divided the succession into five "Orogene Serien," which he tried to correlate with the biostratigraphy of Säve-Söderbergh (1934). The stratigraphy of the Devonian succession was summarized by Bütler (1959, 1961) (Fig. 3). Jarvik (1961) correlated the previous biostratigraphic divisions with some of Bütler's orogenic series, but it was the latter's allostratigraphic subdivisions that formed the basis for the Devonian part of the geological map of North-East Greenland produced by Koch and Haller (1971) (Fig. 3).

Friend and colleagues attempted to disentangle this confusing mix of bio- and lithostratigraphy in studies from 1968 to 1970 using sedimentology and computer data handling techniques. They produced a series of papers (Friend et al., 1976a, 1976b, 1983; Alexander-Marrack and Friend, 1976; Yeats and Friend, 1978; Nicholson and Friend, 1976) that integrated their stratigraphic studies with names used by Bütler. They erected a new series of lithostratigraphic units (Fig. 3), formalized in Friend et al. (1983, p. 10–15). Much new fossil vertebrate material was collected associated with a palynologic investigation that supported a Givetian (Middle Devonian) age for some of the oldest sediments (Allen, 1972).

More recent lithostratigraphic work, between 1986 and 1988, led to a revised and formal lithostratigraphy (in accordance with the lithostratigraphic code of NACSN, 1983) for the main parts of the Devonian Old Red Sandstone succession in East Greenland (Olsen and Larsen, 1993a) (Figs. 3–5). The work was carried out as an integrated structural and sedimentologic study based on detailed facies logging plus mapping of facies associations and lithostratigraphic units. Recently, new vertebrate studies complemented by palynologic discoveries have clarified some of the stratigraphic, chronologic, and ecologic contexts of the sediments and fossils (Marshall and Astin, 1996; Marshall et al., 1999; Marshall and Hemsley, 2003; Blom et al., 2005, 2007). In Figure 4, the most recent biostratigraphic evidence has been added to the formal lithostratigraphy proposed by Olsen and Larsen (1993a).

The lithostratigraphic units established by Olsen and Larsen (1993a) have been mapped within the larger part of the Devonian outcrop areas in East Greenland (Fig. 5) (Larsen, 1990a; Bengaard and Larsen, 1991). The areas around Kap Franklin, Nordhoek Bjerg, and Salévebjerg have been visited only briefly by the present authors (Larsen and Olsen). Based on previous work by Alexander-Marrack and Friend (1976) and Friend et al. (1983), the lithostratigraphy was extended to include the Kap Franklin area, while the areas around Nordhoek Bjerg and Salévebjerg were excluded. The Devonian rocks in Wegener Halvø and Canning Land in northern Jameson Land were likewise not visited by Olsen and Larsen (1993a), and they suggested that the existing lithostratigraphy presented in Alexander-Marrack and Friend (1976) and Friend et al. (1983) should be maintained for this area (Fig. 4).

The lithologic characteristics of the Devonian succession are summarized next, together with a summary of the sedimentary basin evolution and paleogeography. For a more detailed review of the lithostratigraphy and sedimentologic aspects, the reader is referred to Olsen and Larsen (1993a) and Olsen (1993a).

DEVONIAN BASIN OF NORTH-EAST GREENLAND

Development of the Devonian basin has been controlled by an interplay of climate and tectonics; this is reflected in its paleogeography, sedimentary processes, and deposits. Olsen (1993a) recognized four main tectonostratigraphic basin stages that individually are separated by subregional (tens of kilometers) to basinwide unconformities and represent depositional episodes punctuated by major tectonic events (Fig. 6). Each basin stage is built up of one or several depositional complexes that share roughly similar drainage patterns, measured on a basinwide scale. These complexes (numbered successively 1–11 in Fig. 6) are bounded by event-stratigraphic surfaces, and each complex is composed of one, or more commonly several, depositional system. In the absence of high-resolution biostratigraphy, the definition of event-stratigraphic surfaces was based solely on sedimentologic criteria, i.e., marked changes in depositional systems in the entire basin. Each depositional system was defined from three-dimensional assemblages of process-related facies associations that recorded major paleogeomorphic basin elements. A total of eight facies associations formed the basis for this subdivision of the Devonian succession: eolian dune, eolian sand sheet/interdune, flood basin, ephemeral stream, meandering river, braided river, gravelly stream, and lacustrine association.

In the following, sedimentary basin evolution is briefly illustrated by eleven established depositional complexes associated with a summary of the most recent vertebrate paleontologic observations (Fig. 4).

Vilddal Basin Stage: Eastward-Draining Alluvial Plains

The Vilddal basin stage (Eifelian–Givetian) constitutes one of the oldest sediments in the Devonian basin and is represented by the 2500-m-thick alluvial Vilddal and Nathorst Fjord Groups (Fig. 4), which accumulated in a generally eastward-draining fluvial system. The deposits unconformably overlie pre-Devonian sediments, mainly of Cambrian-Ordovician age. Two depositional complexes build up this stage. The lowest, complex 1, is dominated by gravelly basinwide braid-plain systems, draining eastward, represented by the Solstrand Formation. Small alluvial fans may have developed locally in connection with local tectonism (Kap Bull Formation) (Figs. 4 and 7A).

Depositional complex 2 is composed entirely of the sandy and silty alluvial systems of the Ankerbjergselv Formation, which is mainly built up of tabular, red and green sedimentary units, alternating on a 10 m scale. These are interpreted as reflecting short-lived changes between ephemeral stream systems (red) and perennial alluvial systems (green). The perennial systems were dominated by braid plains in the proximal (western) areas, whereas meandering rivers dominated in the distal parts, grading into lacustrine systems.

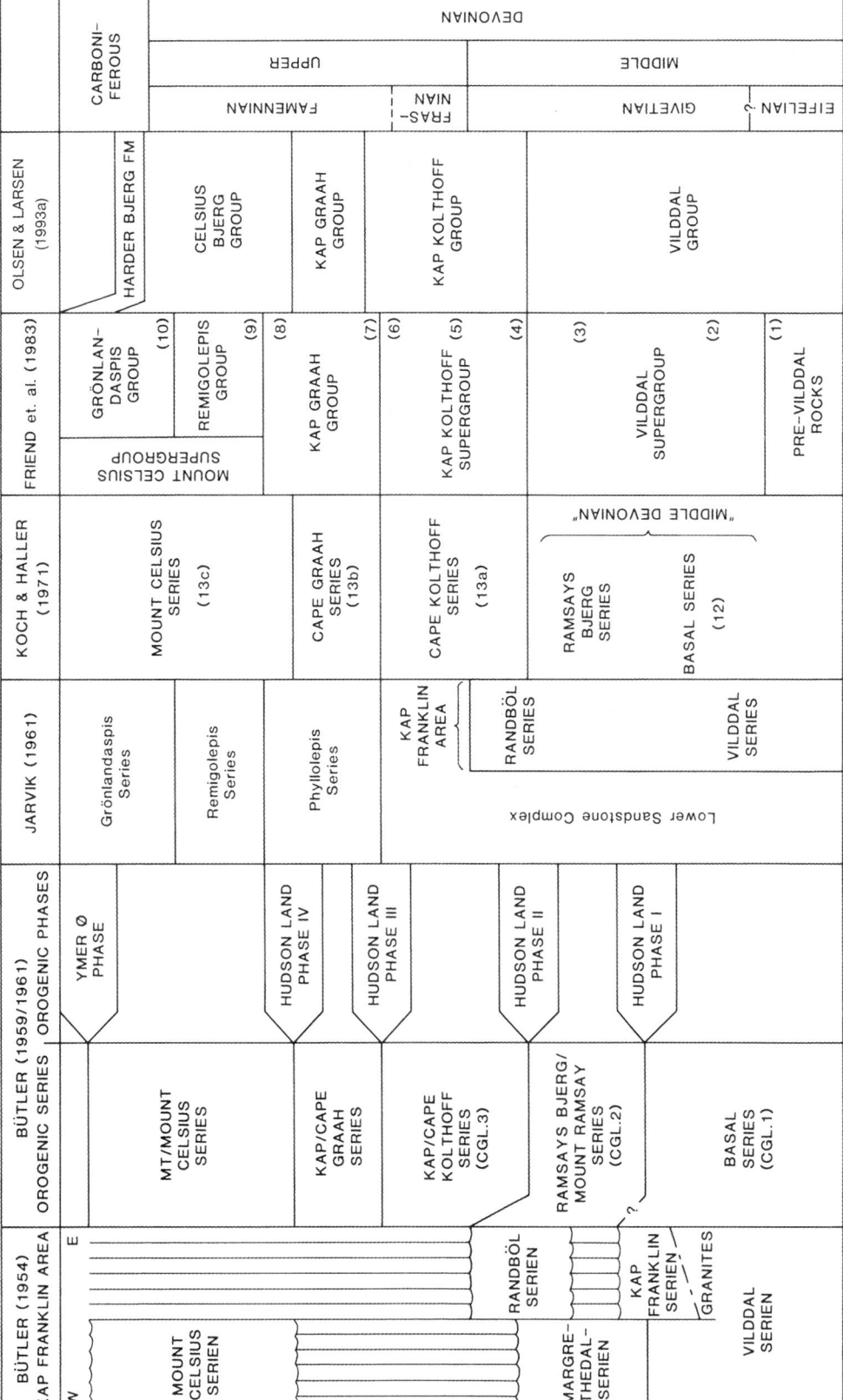

Figure 3. Stratigraphic schemes from 1954 to 1983 covering the Devonian sediments of North-East Greenland, and correlation to the main lithostratigraphic subdivisions proposed by Olsen and Larsen (1993a). The studies of Alexander-Marrack and Friend (1976), Nicholson and Friend (1976), and Yeats and Friend (1978) are represented by the formalized lithostratigraphy of Friend et al. (1983). Figure is based on Olsen and Larsen (1993a).

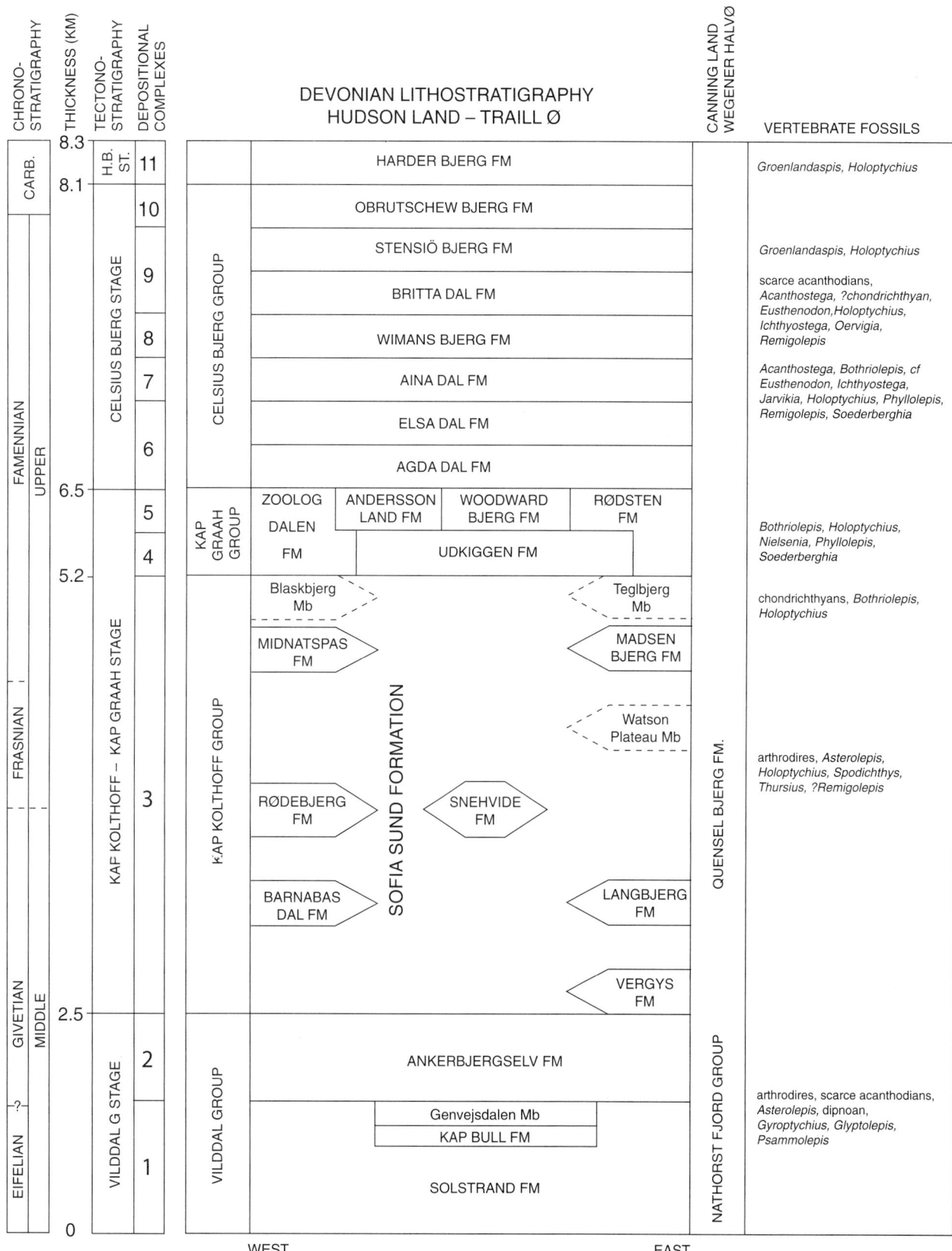

Figure 4. Stratigraphic subdivision of the Devonian succession in North-East Greenland, showing the distribution of vertebrates after Olsen and Larsen (1993a) and Olsen (1993a) and the latest biostratigraphic evidence from Blom et al. (2007).

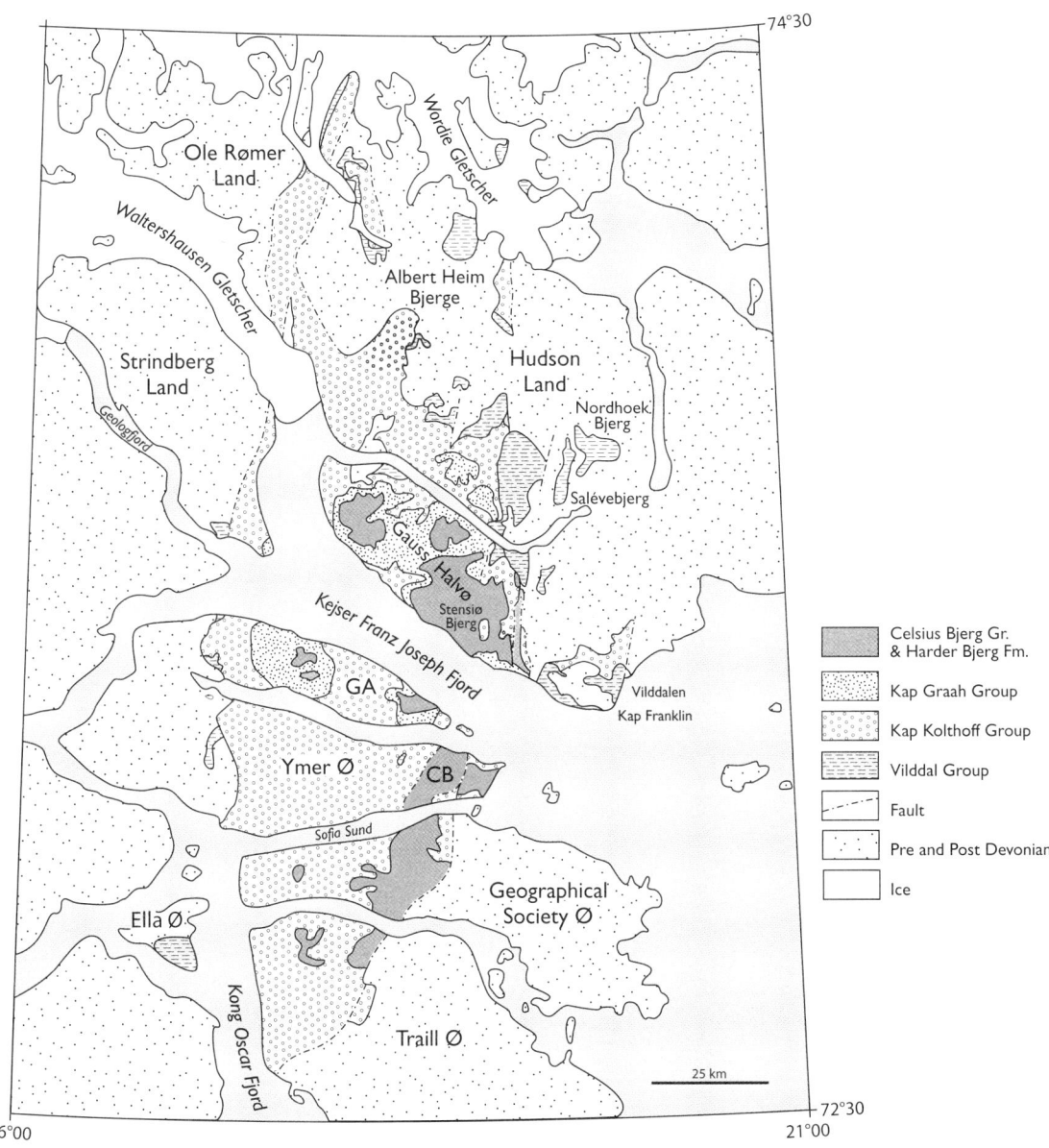

Figure 5. Geological map of the Devonian basin based on Larsen (1990a), showing the areal distribution of lithostratigraphic groups proposed by Olsen and Larsen (1993a). GA—Gunnar Anderssons Land; CB—Celsius Bjerg.

The transition from depositional complex 1 to 2 was probably related to a general change in climatic regime toward more arid conditions. Moreover, during complex 2, cyclic variations in climate apparently persisted, resulting in alternating red and green deposits laid down during relatively arid and humid conditions, respectively.

Vertebrate paleontology (complexes 1 and 2). The fossils contained in these earliest deposits are Eifelian to Givetian in age (though the Eifelian–Givetian boundary is not well-defined in these sediments). In these relatively early deposits, some of the most common vertebrate fossils are placoderm fishes: the often large, predatory arthrodires *Heterostius* and *Homostius* and the bottom-feeding antiarch *Asterolepis*. The jawless vertebrates, usually common in Devonian deposits, are represented only by a single fragment of the heterostrachan *Psammolepis*. As yet unidentified "spiny sharks" (acanthodians) are represented by a few spines and scales, though it is perhaps surprising that these animals are so poorly represented. Also present in these early deposits are the lobe-finned fishes, and the most common of these is the osteolepidid genus *Gyroptychius*. This was a medium-sized fish with no obvious predatory specializations, in contrast to the rarer porolepiform *Glyptolepis*, which is also found in these deposits. Porolepiforms are often some of the largest predators in the fauna of the East Greenland deposits (Figs. 8H–8K). Lung-

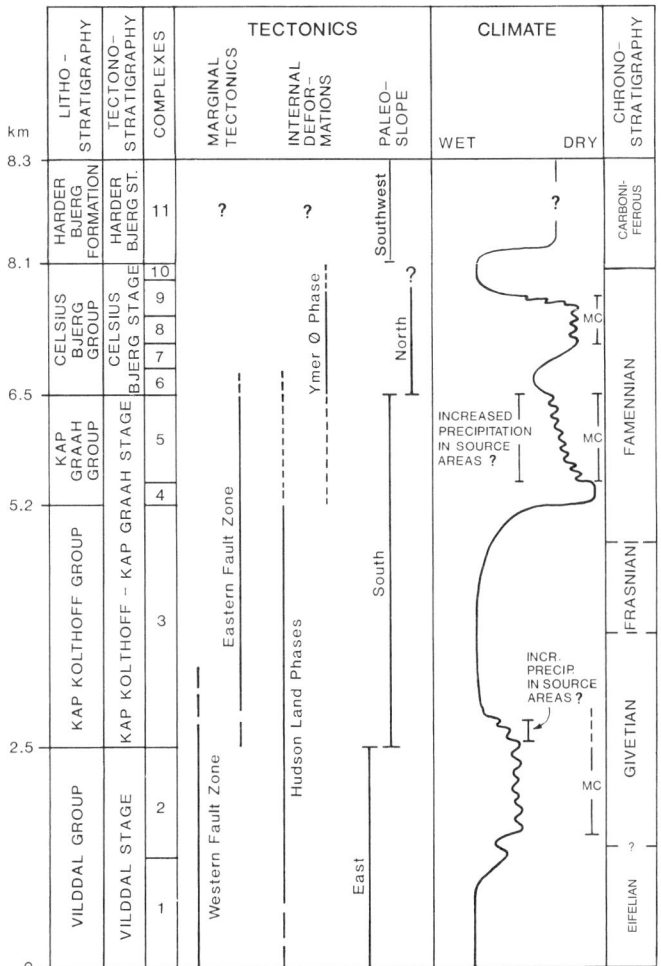

Figure 6. Summary of external controls on the sedimentary basin evolution. See text and Olsen (1993a) for discussion. MC—Milankovitch cyclicity.

fishes are represented by a single bone from a *Dipterus*-like form, comparable with contemporary forms from Scotland. For further details, see Blom et al. (2007) and references therein.

Kap Kolthoff–Kap Graah Basin Stage: Southward-Draining Alluvial Plains and Associated Eolian Systems

A basinwide unconformity and a general change in drainage toward the south (Fig. 7B) indicate the transition from the Vilddal to the succeeding Kap Kolthoff–Kap Graah basin stage of Givetian–Famennian age. The Kap Kolthoff–Kap Graah stage is composed of the deposits of the Kap Kolthoff Group, up to 2700 m thick, and the Kap Graah Group, which is up to 1300 m thick (Fig. 4). The Kap Kolthoff–Kap Graah basin stage is distinguished from the Vilddal basin stage by the shift to a southward-directed tributary drainage pattern along a N-S axis (Olsen and Larsen, 1993a). The dominant depositional systems are alluvial, but major eolian systems also occur.

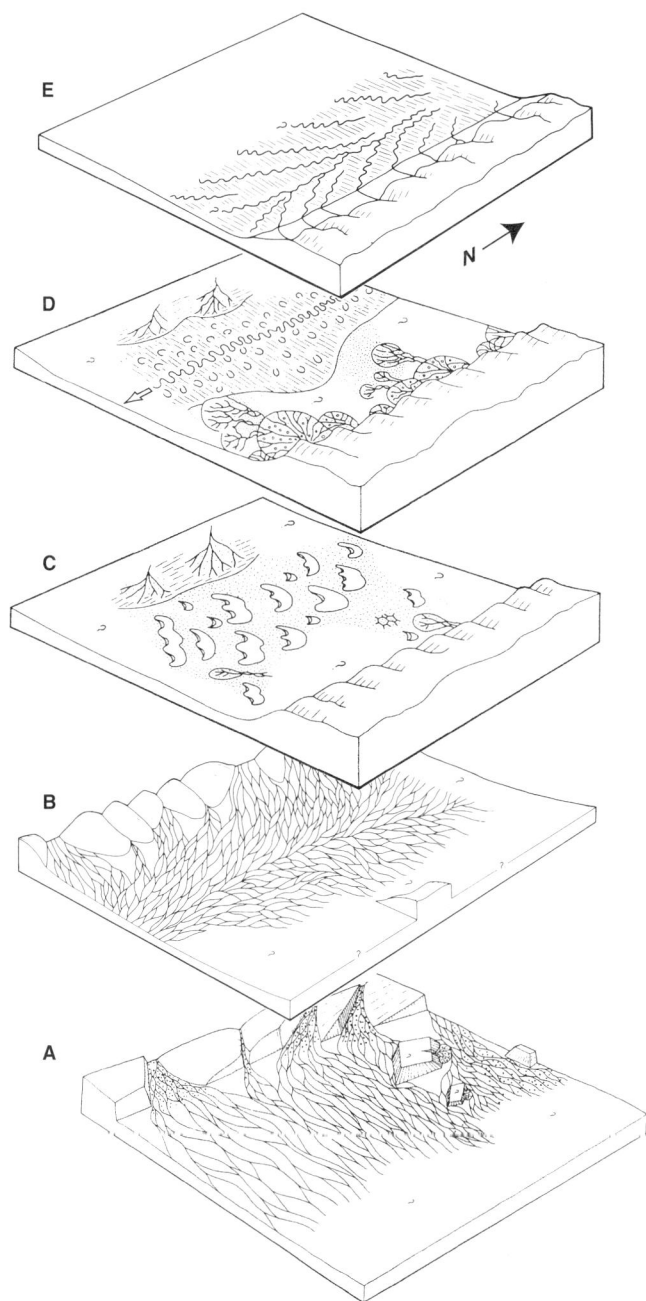

Figure 7. Paleogeography in five snapshots. (A) Vilddal basin stage, depositional complex 2 (eastward-draining alluvial plains). (B, C, and D) Early, middle, and late in the Kap Kolthoff–Kap Graah basin stage, depositional complex 3 (southward-draining alluvial plains), depositional complex 4 (dominated by an erg), and depositional complex 5 (characterized by southward-draining axial meander belt). (E) Celsius Bjerg basin stage, depositional complex 7 (northward-draining alluvial plains).

The Kap Kolthoff–Kap Graah basin stage is subdivided into three complexes, of which the lowermost is represented by the Kap Kolthoff Group (complex 3) (Figs. 4 and 6). Complex 3 is mainly fluvial, though locally eolian systems dominate (Fig. 7B). The Sofia Sund Formation forms the dominant

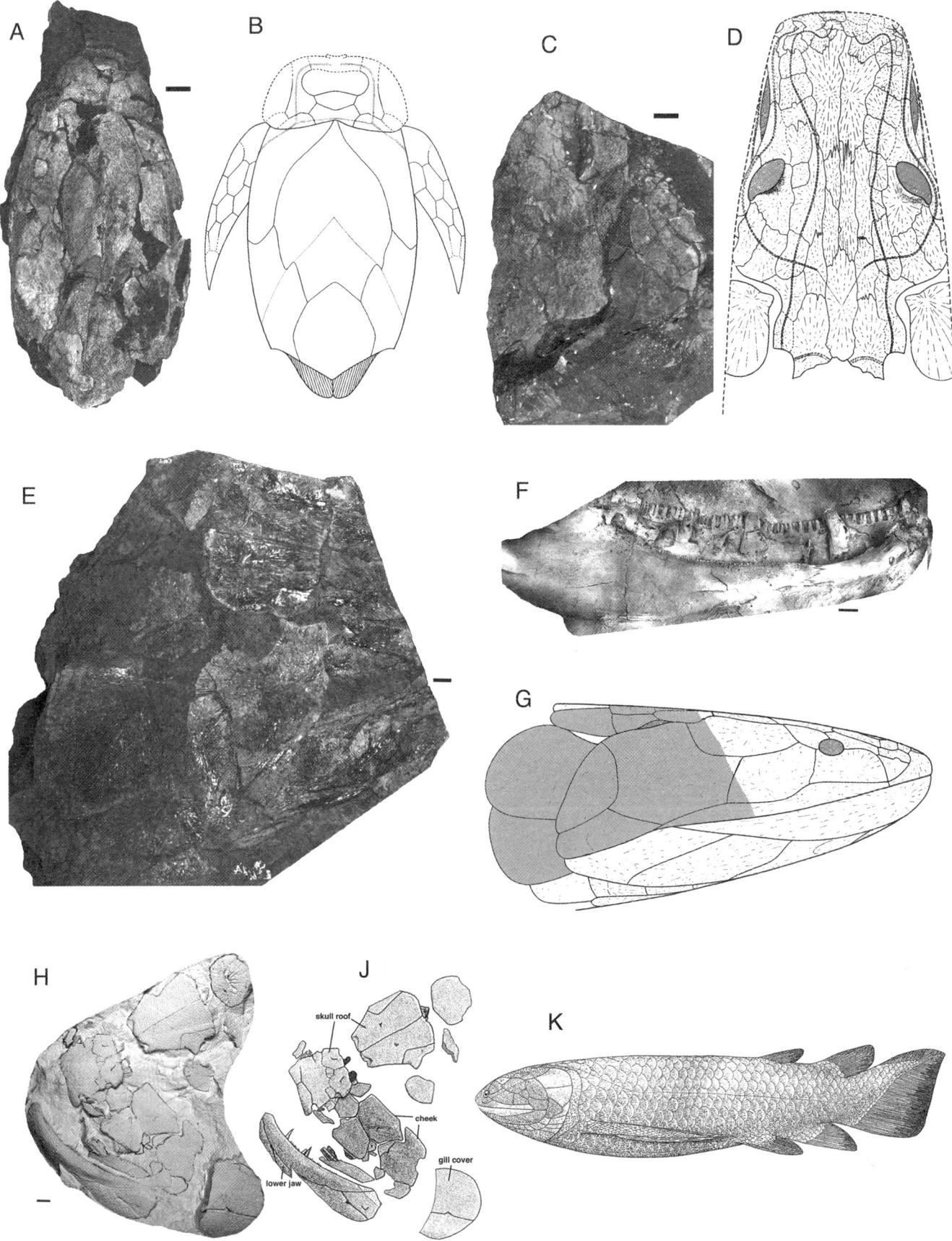

Figure 8. (A) Specimen of *Remigolepis* collected in 1987 from the north side of Celsius Bjerg. (B) Reconstruction of the carapace of *Remigolepis* from Stensiö (1931). (C) *Soederberghia* partial skull roof specimen collected in 1987 from the Aina Dal Formation of Stensiö Bjerg. (D) Reconstruction of the skull of *Soederberghia* in dorsal view from Lehman (1959). (E) Cf. *Eusthenodon* partial skull specimen collected in 1987 expedition from the Aina Dal Formation of Stensiö Bjerg. (F) Latex peel coated with ammonium chloride showing the internal view of the lower jaw of cf. *Eusthenodon* collected in 1987 from Celsius Bjerg. (G) Reconstruction of the skull of *Eusthenodon* from Jarvik (1952); shaded areas indicate parts preserved on the specimen in E. (H) Silastomer peel of a partial skull of *Holoptychius* collected in 1987 from the Aina Dal Formation on Gauss Halvø. (J) Interpretation of the skull in H by P.E. Ahlberg (1989, personal commun.). (K) Reconstruction of the porolepiform *Glyptolepis* by P.E. Ahlberg (1989, personal commun.). Scale bars represent 10 mm in all cases. B, D, and G appear courtesy of Meddelelser om Grønland.

depositional system of the complex (~90%) and is almost entirely composed of sandy braid-plain deposits. At least seven local depositional systems form the remaining part of the complex, and these occur as large-scale sedimentary prisms within the Sofia Sund Formation. These smaller systems are composed of eolian dune, ephemeral stream, meandering river, and braided river deposits. Some of these systems form so-called terminal fan systems (Kelly and Olsen, 1993).

Complex 4 is dominantly eolian (Fig. 7C) and is succeeded by the alluvial and eolian complex 5 (Fig. 7D), and together these two complexes form the Kap Graah Group (Olsen and Larsen, 1993a). Complex 4 is dominated by the Udkiggen Formation, which represents an erg that covered almost the entire basin and a dominant wind direction toward the southwest. Terminal fans fringed the eastern and western basin margins, depositing the Rødsten and Zoologdalen Formations, respectively. Complex 4 accordingly represents a pronounced arid interval in comparison with complex 5.

The succeeding complex 5 is composed of four depositional systems, each represented by lithostratigraphic formations (Fig. 4). Initially, the erg of the Udkiggen Formation is almost entirely replaced by a southward-draining axial meandering river belt of the Anderson Land Formation. The terminal fan systems of the Zoologdalen and Rødsten Formations still existed, and the latter expanded into the basin. The establishment of the axial meandering river system is explained by tectonism in the borderlands north of the basin (Olsen and Larsen, 1993b). The meander belt deposits are restricted to the western part of the basin in the upper part of complex 5, and eolian deposits of the Woodward Bjerg Formation gradually occupied a larger part of the basin. This gradual change was interpreted by Olsen and Larsen (1993b) in terms of syndepositional folding.

Vertebrate paleontology (complexes 3, 4, and 5). The Kap Kolthoff Group (complex 3) spans the interval from rocks of undoubted Givetian age through to the Famennian, although the position of the Frasnian boundary within it is still uncertain. As well as being probably the longest and most complex series of sequences in the Devonian basin, it is also so far one of the least fossiliferous. Fishes that have been recovered from the Kap Kolthoff deposits include poorly preserved arthrodires, the antiarch placoderm *Asterolepis* (apparently the same species as found in the earlier deposits), and potentially the earliest example of the antiarch *Remigolepis* (Figs. 8A and 8B) found in Greenland (the taxonomic affinity of the specimen remains uncertain). Toward the top of the Kap Kolthoff Group, the first appearance in Greenland of the antiarch *Bothriolepis*, a very common and cosmopolitan genus, is recorded. *Thursius* (an osteolepidid), *Spodichthys* (a tristichopterid), and *Holoptychius* (a porolepiform) are found, along with a few chondrichthyan remains. For further details, see Blom et al. (2007) and references therein.

The Kap Graah Group (complexes 4 and 5) appears to be lowest Famennian in age, although the boundary with the Frasnian is again unclear. Some parts of it, notably the Zoologdalen Formation (Fig. 4), are relatively richly fossiliferous. The placoderms *Bothriolepis* and *Phyllolepis* are common, the lungfishes *Nielsenia* and *Soederberghia* (Figs. 8C and 8D) are present, and *Holoptychius* is found throughout the sequence.

Celsius Bjerg Basin Stage: Northward-Draining Alluvial Plains and Lakes

The Celsius Bjerg basin stage (Famennian) is represented by the up to 1550-m-thick Celsius Bjerg Group, and it succeeds the Kap Kolthoff–Kap Graah basin stage with subregional unconformities. It is distinguished by northward drainage, in contrast to the southward drainage of the preceding basin stage. The basin fill is built up of five depositional complexes (6–10), of which several are composed of only one depositional system (Figs. 4 and 6).

Depositional complex 6 is composed of two systems, represented by the Agda Dal Formation and succeeding Elsa Dal Formation (Figs. 4 and 6). The Agda Dal Formation forms the basal part of the Celsius Bjerg Group. In some parts of the basin, the unit fills out a very irregular topography with an erosional relief of up to ~100–200 m (Olsen and Larsen, 1993a). The formation is entirely fluvial, and paleocurrents are generally toward the north, with local variations toward east and west. The formation is interpreted as an alluvial plain of mainly ephemeral streams. The Elsa Dal Formation forms a rapid transitional contact to the underlying Agda Dal Formation (Fig. 4). The Elsa Dal Formation is also entirely fluvial, and paleocurrents are in general identical to those in the Agda Dal Formation. The Elsa Dal Formation is interpreted as an alluvial plain of braided rivers that mainly carried sandy material.

Depositional complex 7 is composed of the Aina Dal Formation, which conformably overlies the Elsa Dal Formation (Figs. 4 and 6) and is entirely fluvial. The lower half of the formation is interpreted as small meandering river deposits with associated overbank deposits. The meandering rivers were perennial and probably characterized by slow discharge. The upper half of the formation was also deposited in small meandering rivers, but in contrast to the rivers in the lower half, the rivers in the upper part were characterized by ephemeral discharge. The uppermost sediments of the Aina Dal Formation indicate a gradual transition to the overlying depositional complex 8. Paleocurrents in the Aina Dal Formation are directed northward (Fig. 7E).

Depositional complex 8 (Wimans Bjerg Formation) conformably overlies the Aina Dal Formation with a few exceptions (Figs. 4 and 6). The deposits are composed almost entirely of laminated siltstones (lacustrine association) and brecciated to massive mudstones (flood basin association). These deposits are interpreted as an ephemeral lake–dry mud flat complex. The brecciated mudstones indicate repeated exposed mud-flat conditions. The stacked laminated siltstone beds record the repeated expansions and contractions of ephemeral lakes on the mud flat.

Depositional complex 9 is composed of two depositional systems, represented by the Britta Dal and the succeeding Stensiö Bjerg Formations (Figs. 4 and 6). The Britta Dal Formation conformably overlies the Wimans Bjerg Formation and is composed of ephemeral stream and flood basin deposits. The base of the formation is a rapid transition from the lacustrine deposits of the Wimans Bjerg Formation. Paleocurrents from the ephemeral stream deposits are rather sparse, but a general NE direction is apparent (Nicholson and Friend, 1976). The Britta Dal Formation as a whole is interpreted as an alluvial plain composed of meandering rivers with ephemeral discharge associated with flood basin areas. The Stensiö Bjerg Formation is composed of meandering river, ephemeral stream, flood basin, and lacustrine facies. The lower boundary is transitional to the underlying Britta Dal Formation. In the lower ~100 m of the Stensiö Bjerg Formation, sandstone bodies of perennial meandering river origin are associated with ephemeral meandering stream and flood basin deposits similar to the upper part of the underlying Britta Dal Formation. The ephemeral meandering stream deposits gradually decrease in importance upward and multistory sandstone bodies of the meandering river association dominate in the rest of the formation. Paleocurrents are toward the NE.

Depositional complex 10 is the uppermost depositional complex in the Celsius Bjerg basin stage. It is entirely built up of the Obrutschew Bjerg Formation (Figs. 4 and 6), which is a lacustrine sequence of black shales and limestones with local gray silty mudstones. The general lack of coarse clastic sediments indicates that the coastline was situated at some distance from the outcrop localities. The lake, accordingly, seems to have been developed as a basinwide feature.

The sediments of the Kap Graah and Celsius Bjerg Groups in North-East Greenland, as defined by Olsen and Larsen (1993a) (Fig. 4), have generally been considered to be Late Devonian (Famennian) in age since Säve-Söderbergh (1934) described fossil vertebrate assemblages from these rocks of the early tetrapods *Ichthyostega* and *Acanthostega*. Recently, there has been some controversy over the dating of these sequences (Hartz et al., 1997; Stemmerik and Bendix-Almgreen, 1998; Hartz et al., 1998). Paleomagnetic and geochronologic studies (Hartz et al., 1997) have suggested a radiometric date (^{40}Ar/^{39}Ar) of ca. 336 Ma for a basalt in the Kap Graah Group, which would place the group within the late Early Carboniferous, i.e., about middle Visean in age.

More recently, palynological samples encompassing the Celsius Bjerg Group (Marshall et al., 1999) have indicated a Late Devonian (Famennian) age. These show the distinctive changes in the palynologic microflora that are internationally accepted as defining the Devonian–Carboniferous boundary. This boundary was placed with confidence within the Obrutschew Bjerg Formation, in the uppermost part of the Celsius Bjerg Group, and thus fully supports Olsen and Larsen (1993a), who also placed this group at the top of the Famennian (Fig. 4).

Vertebrate paleontology (complexes 6–10). For further details, see Blom et al. (2007) and references therein. Once into the Famennian Celsius Bjerg Group, the vertebrate fauna becomes generally rich, except for the Elsa Dal and Wimans Bjerg Formations. Antiarch placoderms are by far the most common component, including *Bothriolepis* and *Phyllolepis* in the lower parts of the sequence, and three described species of *Remigolepis*. Higher in the sequence, in the Britta Dal Formation, two more named species of *Remigolepis* occur (Figs. 8A and 8B). At face value, that would make *Remigolepis* the most speciose type of vertebrate in the sequence, but further study is required to clarify its systematics.

The lungfishes *Jarvikia* and particularly *Soederberghia* (Figs. 8C and 8D) are known from good material from the Aina Dal Formation, and the large predators *Holoptychius* (Figs. 8H and 8J) and the tristichopterid *Eusthenodon* (Figs. 8E–8G) are found through the Aina and Britta Dal Formations. These latter two fish grew up to two meters long: *Holoptychius* is known from body scales up to 120 mm in diameter (J. Clack, 1998, personal observation). In contrast to *Remigolepis*, *Holoptychius* is usually identified only as "*Holoptychius* sp." throughout the whole Devonian sequence in East Greenland. As well as being of biological interest, *Holoptychius* and *Remigolepis* species could be extremely valuable for stratigraphic correlation if correctly diagnosed and identified.

Smaller components of the Celsius Bjerg Group fauna are represented by a few acanthodian scales and spines and a possible chondrichthyan spine. In contrast to the earlier parts of the sequence, only a single arthrodire specimen has been found in the major part of the Celsius Bjerg Group. The arthrodire *Groenlandaspis*, ubiquitous and diverse in the Famennian, occurs at the top of the sequence in the Stensiö Bjerg Formation (Daeschler et al., 2003).

The Wimans Bjerg Formation is generally described as unfossiliferous, although one or two specimens from the expeditions of the 1930s are described as occurring near its base (see Blom et al., 2005). Otherwise, only trace fossils, including *Cruziana* and arthropods trackways, have been found there. The only other record of invertebrate fossils from the whole sequence is bivalves from the Celsius Bjerg Group on the south side of Celsius Bjerg itself, although the formation from which they derive is unknown (Clack and Neininger, 2000).

It is the Celsius Bjerg Group that has yielded the most significant component of the vertebrate fauna of East Greenland, those of the early tetrapods *Ichthyostega* (Figs. 9A–9C) and *Acanthostega* (Figs. 9D and 9E). Both genera appear in the Aina and Britta Dal Formations, though they have not been found together in the same fossiliferous lenses or horizons (they may occur in the same levels of the Aina Dal Formation, but this is not certain). *Ichthyostega* has been collected in quantity since the 1930s, and more than 300 specimens are known. It is relatively common in both formations. *Acanthostega*, on the other hand, is known mainly from a single lens in the Britta Dal Formation, and only a few scattered specimens have been found elsewhere (see Blom et al. [2007] for a full listing). Recent work on *Ichthyostega* has suggested that rather than the five species originally named, there are no more than three, but interestingly, the Aina Dal and Britta Dal Formations each have their own characteristic species, at least on Gauss Halvø, where the two formations are clearly defined (see below and Blom, 2005). The sample of *Acanthostega* specimens is inadequate to make comparable distinctions between species.

The tetrapods are very different from each other in body form and presumably in habits, and they appear to have occupied quite separate environments or habitats. Unfortunately, most of the specimens derive from flood deposits, so we know little about their place or environment of origin. Most *Acanthostega* specimens came from one site that represents a point-bar in an active river channel where a series of burial events took place, resulting in deposition of several well-articulated specimens. These are interpreted as having been deposited during a flooding event. The good preservation of these specimens suggests that they had not been carried far, and thus that the animals inhabited these channels (Bendix-Almgreen et al., 1990). The skeletal morphology of these two genera is nonetheless well known, and they have contributed to a radical reassessment of the origin of tetrapods (see Clack [2005, 2006] for recent reviews).

At the top of the Celsius Bjerg Group lies the Obrutschew Bjerg Formation, in which the Devonian–Carboniferous boundary occurs. Marshall et al. (1999) considered a black-shale layer within this formation to mark the boundary. This shale horizon has yielded a single vertebrate species; a ray-finned fish has recently been described (Friedman and Blom, 2006).

Harder Bjerg Basin Stage: Southwestward-Draining Alluvial Plains and Associated Eolian Systems

The Celsius Bjerg basin stage is succeeded by the Harder Bjerg basin stage, which is represented by the up to 200 m thick Harder Bjerg Formation, referred to depositional complex 11 (Figs. 4 and 6). This was considered by Olsen (1993a) and Olsen and Larsen (1993a) to be the final stage in the evolution of the Devonian basin. Because of the lack of age-diagnostic fossil material, the age of the Harder Bjerg Formation was considered to be either latest Devonian or earliest Carboniferous by Olsen and Larsen (1993a). Based on the work of Marshall et al. (1999), the Harder Bjerg Formation may now be considered as early Tournaisian (Carboniferous) in age. The formation consists of a lower part, a few tens to 100 m thick, of cross-bedded sandy braided river deposits overlain by alternating eolian, ephemeral stream, and braided river deposits. Sparse paleocurrent data indicate a drainage toward the southwest, in contrast to the general northward drainage of the Celsius Bjerg stage.

Vertebrate paleontology (complex 11). According to the literature, the Lower Carboniferous Harder Bjerg Formation has yielded both *Groenlandaspis* and *Holoptychius* specimens, which is anomalous. These genera are not known elsewhere beyond the end of the Devonian. They may have been misidentified, the boundary may have been wrongly defined, they may represent unique occurrences, or they may mean that our preconceptions about their distribution in time are mistaken, and that they may eventually be found in the Carboniferous elsewhere. Further investigations are needed to clarify this anomaly. For further details, see Blom et al. (2007) and references therein.

EXTERNAL CONTROLS OF THE SEDIMENTARY BASIN EVOLUTION

Development of the basin and its sediments has been strongly controlled by an interplay of tectonic and climatic events (see Olsen, 1990; Olsen, 1993a, 1993b; Olsen and Larsen, 1993b). This interplay is summarized in Figure 6 and will briefly be discussed next.

Initiation of the Devonian basin and subsequent development of the Vilddal basin stage was governed by Caledonian extensional collapse and the establishment of the Western fault zone as a normal fault with sinistral wrench components (Fig. 2) (Larsen and Bengaard, 1991; Larsen and Olsen, 1991). During deposition of the Vilddal basin stage, marginal faulting and internal deformation of the basin fill occurred, related to the Hudson Land phases of Bütler (1935a, 1959). The paleoslope was toward the east.

Establishment of the Eastern fault zone, associated with waning activity along the Western fault zone, characterized the onset of the Kap Kolthoff–Kap Graah basin stage. The Eastern fault zone was probably also a sinistral wrench fault zone or oblique-slip fault zone (Larsen and Olsen, 1991). During deposition of complex 3, the Hudson Land deformation phases resulted in several internal unconformities, mainly along the Eastern fault zone (Larsen, 1990d). The general paleoslope was southward, and sediment input occurred from both east and west. A gradual change to a more humid climate is observed in early complex 3 time.

Depositional complex 4 marks a rapid aridification that resulted in desert conditions and development of a basinwide erg. Relatively dry climatic conditions also persisted during deposition of complex 5. Uplift east of the Eastern fault zone occurred during deposition of complex 5, and tectonic movements apparently also occurred north of the basin. The tectonism caused the eastern transverse alluvial system to expand into the basin and also resulted in the establishment of an axial river system with

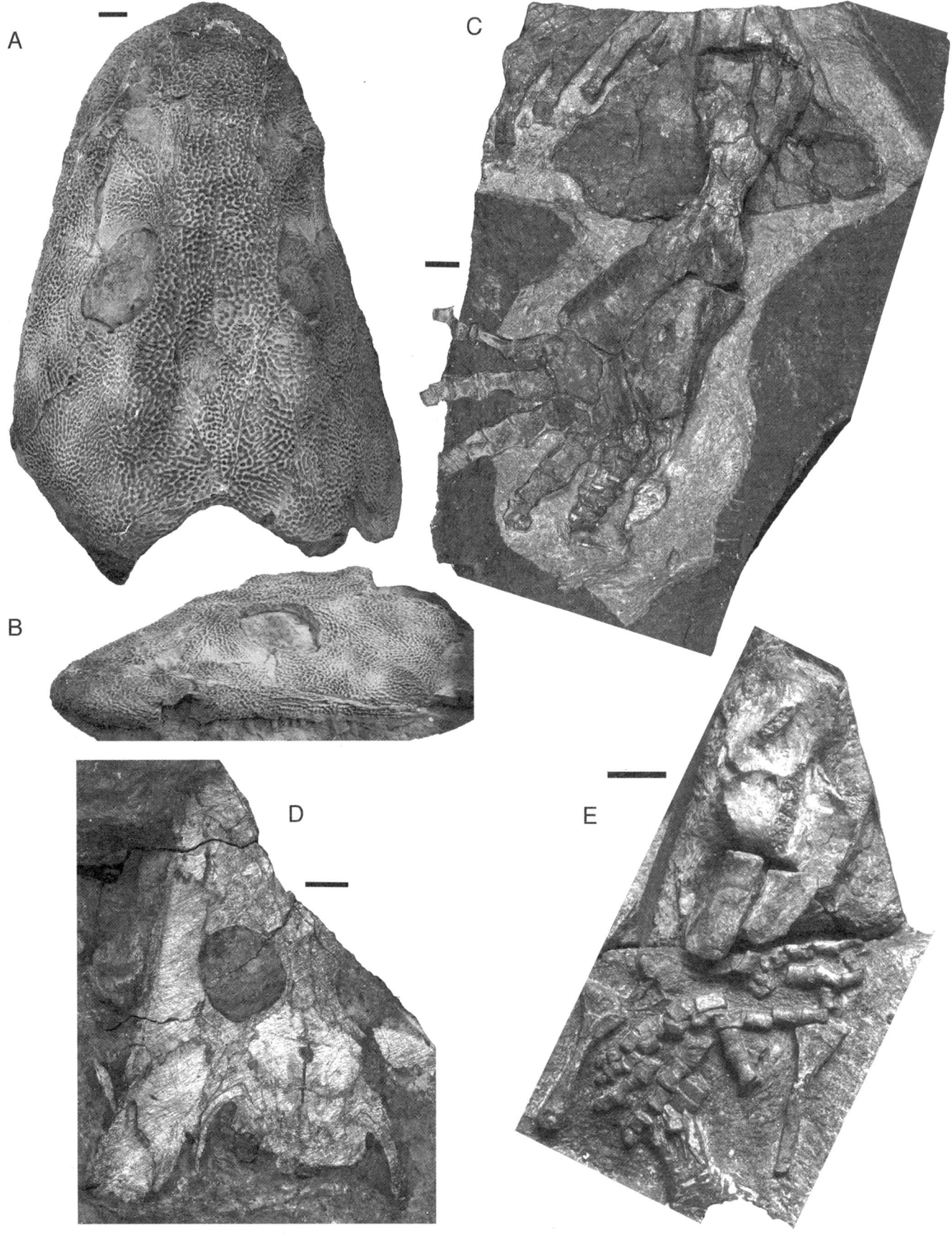

Figure 9. (A) Latex peel of the skull roof of *Ichthyostega* coated with ammonium chloride to enhance the contrast. This skull was collected in 1931 from the base of the Britta Dal Formation on Gauss Halvø. (B) Oblique lateral view of the same skull. (C) Hind limb and tail specimen of *Ichthyostega* collected in 1987 from the Aina Dal Formation on Stensiö Bjerg. (D) Skull roof of *Acanthostega* collected in 1987 from Stensiö Bjerg. (E) Forearm of *Acanthostega* collected in 1987 from the Britta Dal Formation on Stensiö Bjerg. Scale bars are all 10 mm.

headwaters in the north. The Ymer Ø deformation phase (Bütler, 1935a) gradually replaced the Hudson Land phases and resulted in synsedimentary open folds locally associated with thrusts or normal faults (Larsen, 1990d; Olsen and Larsen, 1993b). The axial drainage was southward also during deposition of complex 5, but the axis was structurally forced toward the west during deposition of the complex.

The Celsius Bjerg basin stage is characterized by a northward paleoslope. The change from the underlying Kap Kolthoff–Kap Graah basin stage was also associated with a transgression of the Eastern fault zone, which apparently exhibited limited activity during deposition of the basin fill. The Ymer Ø deformation phase, however, continued throughout this basin stage and greatly influenced the valley slope and position of depocenters. The climate changed gradually to more humid conditions and finally, due to tectonic movements, a closed basin was established and a basinwide perennial lake developed (depositional complex 10). A humid climate with sufficient water supply or limited evaporation may have supported the formation of the lake.

This event marks the final episode in the development of the basin in the Devonian. However, it continued into the Harder Bjerg stage, which now is known to be of early Tournaisian (Carboniferous) age (see previous section). The Harder Bjerg basin stage exhibits southwestward drainage. Tectonism is interpreted as having controlled the development of this new drainage direction.

The Devonian basin accordingly underwent five major events of basin-scale influence: (1) the establishment of the Western fault zone and initiation of the Vilddal basin stage; (2) the establishment of the Eastern fault zone and the Kap Kolthoff–Kap Graah basin stage; (3) the aridification halfway through the Kap Kolthoff–Kap Graah basin stage; (4) the establishment of the Celsius Bjerg basin stage associated with an inversion of paleoslope; and (5) the change in paleoslope toward the southwest in connection with the transition to the Harder Bjerg basin stage in the early Tournaisian (Carboniferous). No signs of the termination of the basin are preserved.

PALEOCLIMATE

The North-East Greenland Old Red Sandstone Basin was situated at a paleolatitude of ~5°N–10°N during the Middle to Late Devonian, and it formed part of the equatorial Laurasia continent (Smith et al., 1981; Van der Voo, 1988; Ziegler, 1988; Olsen, 1993a, 1993b; Hartz et al., 1997). The basin was probably located in a trade wind belt, though it experienced a monsoonal climate in the summer (Olsen, 1990). The two general paleowind directions measured within the sedimentary succession are toward the southwest and northeast. According to Olsen (1993a), this probably represents the trade wind pattern north of the equator and monsoonal summer winds, respectively.

Variations in Earth's rotation and orbit, Milankovitch cyclicity, may cause cyclic changes in the geographic and seasonal distribution of solar radiation, resulting in climatic cyclicity (e.g., Hays et al., 1976, Berger, 1977). Increased seasonal variation in solar radiation results in a higher temperature gradient between the ocean and the continent. The increased temperature gradient would cause a northward progradation of the Intertropical Convergence zone during summer in the Northern Hemisphere and thus an intensified monsoonal circulation and increased summer precipitation (e.g., Kutzbach and Street-Parrott, 1985).

When the Intertropical Convergence zone summer position was close to the Devonian basin in North-East Greenland, the local climate would be expected to be very sensitive to Milankovitch cyclicity. Accordingly, a cyclic pattern should be reflected in the deposits as seen in the Andersson Land and Wimans Bjerg Formations (Olsen, 1990; 1993a, 1993b), and probably also the Ankerbjergselv and Britta Dal Formations (Fig. 6) (Olsen, 1993a). When the zone was continuously north or south of the basin during the summer, the climate would be comparatively stable.

The extremely arid climate during deposition of complex 4 in the Kap Kolthoff–Kap Graah basin stage (Fig. 6) may have corresponded to a period when the Intertropical Convergence zone was continuously situated south of the basin, preventing humid monsoonal winds from reaching the basin. Such a situation would occur when the orbit of Earth was extremely elliptical for a prolonged period. However, relatively arid conditions were also maintained during deposition of complex 5 and the main part of the Celsius Bjerg basin stage, thus involving the main part of the Famennian (Fig. 6). This implies a duration of dry climate on the order of 10^6 yr, which is much longer than that suggested by the Milankovitch cycle theory (e.g., Berger et al., 1989). In addition, Milankovitch cyclicity is very well developed in the Upper Kap Graah Group and lower Celsius Bjerg Group, indicating a summertime position of the Intertropical Convergence zone close to the basin and not constantly south of the basin. Accordingly, the aridity must have been caused by some other factor.

During the Famennian, extensive inland ice sheets developed in Gondwana associated with a major regression (Caputo and Crowell, 1985). The combined effect was a cooling of the ocean and an increased distance from the ocean to the Devonian basin in Greenland. A cooler ocean probably resulted in decreased evaporation and thus drier air blowing from the ocean toward the continent. Moreover, the increased distance from the ocean to the basin may have resulted in increased loss of water due to precipitation prior to reaching the basin in Greenland. The aridification in the upper Kap Kolthoff–Kap Graah and lower–middle Celsius Bjerg basin stages accordingly may be explained by the glacially induced Famennian regression.

DEVONIAN VERTEBRATES OF EAST GREENLAND AND THEIR PRESERVATIONAL AND ECOLOGIC CONTEXT

It was during the Late Devonian that creatures with limbs and digits, colloquially described as tetrapods, emerged from the lobe-finned group of fishes and founded the roots of the terrestrial vertebrate groups that have populated the land from the Carboniferous to the present. All land vertebrates living today owe their existence to these earliest representatives; the most completely known and the first discovered of these tetrapods were found in East Greenland.

The Devonian basin of East Greenland, which was the first region to yield tetrapod fossils representing the earliest evolution of limbed vertebrates, understandably became the focus of interest for interpreting the fish to tetrapod transition. For many decades, however, the anatomy of the best known Devonian tetrapod, *Ichthyostega*, remained enigmatic, and neither it, nor the fauna in which it was found, threw much light on the environments in which an early tetrapod might have lived. The so-called Red Beds of East Greenland and elsewhere indeed led to a view that they evolved in arid conditions, acquiring legs to transport them overland from a drying pool to find another pool. This idea is no longer generally held (Clack, 2002). In recent years, studies of Devonian tetrapods and their associated faunas has advanced apace, such that those of East Greenland can now be placed in a wider context.

It remains true that the East Greenland Devonian vertebrate fauna is not particularly informative about the development of the basin as a whole; however, some conclusions can be drawn in the light of evidence from other localities and sources. The recent survey by Blom et al. (2007) has set out a comprehensive table of vertebrate taxa and their distribution through the sequence. One of the problems that has emerged is the poor quality and/or uncertain identity of many of the taxa, which makes comparisons with other horizons in East Greenland or with other Devonian vertebrate localities rather uncertain. This is particularly true of the lower parts of the sequence from the Eifelian and Givetian. Nonetheless, Blom et al. (2007) recognized five faunal assemblages, associated with the Vilddal Group, Kap Kolthoff Group, Kap Graah Group, Celsius Bjerg Group, and Harder Bjerg Formation. Each of these has a complement of placoderms and sarcopterygians and various other vertebrates, such as acanthodians, chondrichthyans or actinopterygians, as occasional occurrences. Most also contain at least one species of lungfish. Only the Celsius Bjerg Group has yielded tetrapods.

The most common placoderms are the antiarch genera *Bothriolepis*, *Remigolepis*, and *Phyllolepis*; each species is probably endemic to East Greenland. *Bothriolepis* is the most common antiarch genus worldwide in the Middle and Late Devonian, and it is found in a wide range of environments from fully marine to fully freshwater. *Remigolepis* seems confined to freshwater environments, but it is also found worldwide, not only on the Old Red Sandstone continent in the Catskill Formation in Pennsylvania, but also in Gondwanan deposits in Australia. In many cases, it seems to be associated with tetrapods. *Phyllolepis* is likewise found in these regions, sometimes alongside tetrapods, but sometimes not (for a review of the vertebrate faunas of Devonian tetrapod sites, see Clack, 2006). The wide-ranging distribution of these genera thus gives little clue to the environments of the East Greenland Devonian basin.

The lungfish fauna of the basin is quite varied. The earliest is a possible dipterid from the Eifelian Vilddal Group. *Dipterus valenciennesi* is the best-known dipterid, from the Eifelian/Givetian of Scotland, and recent work has suggested that this is one of the earliest lungfishes to show the beginnings of adaptations for air-breathing (den Blaauwen et al., 2005). The environments of the Orcadian Basin where it is found suggest that it lived in occasionally deoxygenated waters (Trewin, 1986; Long, 1993). If the East Greenland specimen is confirmed as a species of *Dipterus*, it could suggest the existence of eutrophic or otherwise oxygen-poor conditions at the time.

Later lungfishes from the East Greenland basin are also thought to show air-breathing adaptations, such as *Soederberghia* and *Jarvikia* (Campbell and Barwick, 1988). These genera occur alongside the tetrapods in the Celsius Bjerg Formation. *Soederberghia* is a genus found in North American and Gondwanan deposits, where it is exclusively associated with freshwater environments.

One ubiquitous genus is the porolepiform *Holoptychius*, which, as well as occurring throughout the vertebrate-bearing sequences of East Greenland, also occurs in almost all Devonian tetrapod localities worldwide. This large predator is often represented only by scales, which are problematic to identify to species, so that its use as an environmental or stratigraphic indicator is extremely limited.

There appears to be no geological evidence to suggest that the later stages of the Devonian basin in East Greenland were under any marine influence, and the fish fauna of the Celsius Bjerg Group does not contradict this idea. For that reason, it has for many years been assumed that Devonian tetrapods evolved in freshwater environments. This and other evidence was used to support the view that tetrapods were confined to fresh water for most of the Paleozoic. More recently, this idea has been refuted by finds of Devonian tetrapods in marginal marine and even fully marine environments (Clack, 2006).

The tetrapods *Ichthyostega* and *Acanthostega* of the Late Devonian in East Greenland represent highly divergent morphologies, suggesting that they inhabited different environments and different ecological niches from one another. *Acanthostega* is thought to have been primarily aquatic; it has a large oar-shaped, finned tail, multidigited paddle-like limbs unsuitable for bearing weight on land, and short, straight ribs. Its numerous small teeth combined with larger fangs on the palate suggest a diet of small invertebrates (Clack, 2002). The preservation of articulated specimens of *Acanthostega* in point-bar deposits indicates carcass flotation rapidly followed by stranding and burial within a fluvial channel (Bendix-Almgreen et al., 1990), perhaps as the result of a flood event. *Acanthostega* is thought to have inhabited active fluvial channels. The fluvial channels were ephemeral and probably mainly active during the monsoonal rainfall season, during which *Acanthostega* may have inhabited the whole fluvial system. In the drought season, smaller channels on the fluvial plain would have dried out, and *Acanthostega* may have been forced to seek larger, active channels. The forelimbs do not seem to have been capable of lifting the body from the ground, so that *Acanthostega* may have moved partly as a belly crawler along the drying-out fluvial channels.

Because of the preservation of *Ichthyostega* in allochthonous deposits, little information about its original habitat can be inferred. *Ichthyostega*, in contrast to *Acanthostega*, was more obviously a predator, and it had fewer but larger and recurved marginal teeth. It had a large and robust shoulder girdle and muscular pectoral limbs. The hind limbs, multidigited paddles not unlike those of *Acanthostega*, were held with the feet vertically positioned and posteriorly directed when at rest. The vertebral column and ribs were uniquely modified to suggest specialized locomotion involving differentiated shoulder girdle and axial musculature that could have been involved in dorsoventral movements (Ahlberg et al., 2005). The tail, though finned, was much shorter and less deep than in *Acanthostega*. The overall morphology of *Ichthyostega* suggests some ability to haul itself out onto shallow banks or overbank mud flats.

Three species of *Ichthyostega* have recently been distinguished and associated with different formations in the Celsius Bjerg Group. Distinguished by skull proportions, they suggest changing environments between the Aina Dal and Britta Dal Formations, though the way in which the skull morphologies relate to environmental changes is unknown (Blom, 2005).

A third genus of Devonian tetrapod has recently been described and named. Based on two partial skull specimens, it has been found only in deposits from the south side of Celsius Bjerg, probably from the Britta Dal Formation. There are now five recognized species of Devonian tetrapods in East Greenland, which is a greater diversity than that known from any other Devonian tetrapod locality (Clack et al., 2008).

As far as present knowledge goes, both *Ichthyostega* and *Acanthostega* are endemic to East Greenland. However, some recent discoveries hint that they may have relatives elsewhere. Clement et al. (2004) recognized an ichthyostegid-like lower jaw in deposits at Strud in Belgium. From this, they inferred that at least ichthyostegids were able to travel along shorelines and estuaries to reach what is now Europe. The North American and European regions of Euramerica, though part of the same continent, were separated by 1500 km of seaway in the Late Devonian.

Work in progress on the late Famennian tetrapod *Ventastega*, from Latvia, has revealed postcranial remains that resemble those of *Acanthostega* quite closely. The skull of *Ventastega* shows a number of features in which it is more primitive than *Acanthostega*, and several in which it appears apomorphic. It is slightly later in time than *Acanthostega* and occurs in marginal marine environments. Present analyses show no direct relationship between the two—*Ventastega* is always more primitive than *Acanthostega* in cladistic analyses (e.g., Ahlberg and Clack, 1998)—but there remains the possibility that the two could be shown to form a clade in the light of further discoveries. The alternative possibility is that both show what is essentially the primitive bauplan for tetrapods.

The flora of the East Greenland Devonian basin is poorly known because oxidation of the deposits has for the most part converted plant material into indeterminate coalified cortices or natural molds containing little anatomical information. These nonetheless indicate at least that the plant material associated with tetrapods could reach quite large sizes, with stems of up to 10 cm across. However, work in progress by C. Berry (2008, personal commun.) reports a single specimen, collected by one of us (Clack) in 1998, that shows the presence of a fern-like pinnate branching system. Unfortunately, no leaves are preserved, and details of the exact branching pattern have yet to be established. Though it has not been determined whether it comes from an archaeopteridalean progymnosperm or a *Rhacophyton*-like fern, this is one of the only known plant specimens from the tetrapod-bearing localities to show possibly identifiable morphology.

Recent work on climates, atmospheres, and environments of the late Paleozoic has shown events that could be related to the origin of tetrapods and the tetrapodomorph stem group (Clack, 2007). Scotese (2002) produced Late Devonian paleomaps showing an arid band stretched across the continents from the equator to ~30°S. East Greenland fitted into the northerly part of this band, and the suggested aridity accords with the known sedimentology. However, the tetrapods of the period are found around the continental margins, in river basins and estuaries, where the climate may have been ameliorated somewhat. The most recent model of atmospheric composition during the Middle and Late Devonian has suggested a precipitous decline in oxygen concentration in the atmosphere, reaching a low point of only 13% in the mid-Frasnian (Berner, 2006). Contemporaneously, terrestrial plant cover showed its most rapid increase in size, diversity, and area of land cover during this time (Algeo et al., 2001). The consequences of this have been suggested to be potentially dramatic, causing runoff of nutrients, decaying plant matter, and increased bacterial activity causing eutrophication and oxygen depletion in most water systems. This has further been suggested as a significant factor in the development of air-breathing specializations in the tetrapod stem group and possibly a parallel development in lungfishes, with the emergence of limbed tetrapods themselves as an outcome (Clack, 2007). The black shale situated at the top of the Late Devonian sequences may be one of many such black shales found throughout the world during the Late Devonian indicative of anoxic events.

FUTURE RESEARCH

The Devonian basin in East Greenland is still a challenge to structural geologists, sedimentologists, and paleontologists. The timing of overall basin initiation is still not fully understood. Likewise the distribution of initial sediments is unknown, but possible future offshore hydrocarbon activities on the East Greenland shelf will reveal more information to solve these aspects. The lack of macrofossils in some of the thick sedimentary successions makes stratigraphic correlations difficult and sequence stratigraphic studies (sensu stricto) almost impossible. Furthermore, the link between radiometric dating and paleontologic dating of the sedimentary succession is still a puzzle. The paleomagnetic and geochronologic studies of Hartz et al. (1997) suggested a radiometric date ($^{40}Ar/^{39}Ar$) of ca. 336 Ma for a basalt in the Kap Graah Group, which would place the group within the late Early Carboniferous, i.e., about middle Visean in age. If geochronologic studies within the sedimentary sequence do not match with the paleontologic evidence, some doubts can be cast on the timing of the extensional collapse, i.e., as published recently by Gilotti and McClelland (2005) based on geochronologic studies in the surrounding basement for the basin.

As for the vertebrates, among the outstanding problems is the need to correlate the vertebrate-bearing strata more precisely, if detailed studies of faunal changes are to be undertaken. There are also unexploited areas of Frasnian and Famennian outcrops that could be explored for new near-tetrapod and fully tetrapod material, respectively. The existing tetrapod fossil material, could, where amenable to the use of computed tomography (CT) scanning techniques, be the subject of biomechanical studies of cranial and locomotory function.

ACKNOWLEDGMENTS

We wish to thank Niels (Oscar) Henriksen for inviting us to contribute to the present publication on the geological development of North-East Greenland; his enthusiasm and encouragement through the writing process and help with the figures are greatly appreciated. Finally, we would like to thank the reviewers within the Geological Survey of Denmark and Greenland as well as the two external reviewers, Peter Friend and Geoff Manby, for comments and suggestions about earlier versions of the present manuscript.

REFERENCES CITED

Ahlberg, P.E., and Clack, J.A., 1998, Lower jaws, lower tetrapods—A review based on the Devonian genus *Acanthostega*: Transactions of the Royal Society of Edinburgh, Earth Sciences, v. 89, p. 11–46.

Ahlberg, P.E., Clack, J.A., and Blom, H., 2005, The axial skeleton of the Devonian tetrapod *Ichthyostega*: Nature, v. 437, p. 137–140, doi: 10.1038/nature03893.

Alexander-Marrack, P.D., and Friend, P.F., 1976, Devonian Sediments of East Greenland: III. The Eastern Sequence, Vilddal Supergroup and Part of the Kap Kolthoff Supergroup: Meddelelser om Grønland, v. 206, no. 3, 122 p.

Algeo, T.J., Scheckler, S.E., and Maynard, J.B., 2001, Effects of the Middle to Late Devonian spread of vascular land plants on weathering regimes, marine biotas, and global climate, *in* Gensel, P.G., and Edwards, D., eds., Plants Invade the Land—Evolutionary and Environmental Perspectives: New York, Columbia University Press, p. 213–236.

Allen, K.C., 1972, Devonian megaspores from East Greenland: Their bearing of the development of certain trends: Review of Palaeobotany and Palynology, v. 14, p. 7–17, doi: 10.1016/0034-6667(72)90004-8.

Andresen, A., Hartz, E.H., and Vold, J., 1998, A late orogenic extensional origin for the infracrustal gneiss domes of the East Greenland Caledonides (72–74°N): Tectonophysics, v. 285, p. 353–369, doi: 10.1016/S0040-1951(97)00278-3.

Bendix-Almgreen, S.E., 1976, Palaeovertebrate faunas of Greenland, *in* Escher, A., and Watt, W.S., eds., Geology of Greenland: Copenhagen: Geological Survey of Greenland, p. 536–573.

Bendix-Almgreen, S.E., Clack, J.A., and Olsen, H., 1988, Upper Devonian and Upper Permian vertebrates collected in 1987 around Kejser Franz Joseph Fjord, central East Greenland: Rapport Grønlands Geologiske Undersøgelse, v. 140, p. 95–102.

Bendix-Almgreen, S.E., Clack, J.A., and Olsen, H., 1990, Upper Devonian tetrapod palaeoecology in the light of new discoveries in East Greenland: Terra Nova, v. 2, p. 131–137, doi: 10.1111/j.1365-3121.1990.tb00053.x.

Bengaard, H.-J., 1985, Beskrivelse til et geologisk, et metamorft og et strukturelt kort i målestokken 1:1,000,000 over Østgrønland 70°–82°N [M.S. thesis]: Copenhagen, University of Copenhagen.

Bengaard, H.-J., 1989, Geometrical and geological analysis of photogrammetrically measured deformed sediments of the fjord zone, central east Greenland: Open-File Series Grønlands Geologiske Undersøgelse 89/6.

Bengaard, H.-J., and Larsen, P.-H., 1991, Upper Proterozoic (Eleonore Bay Supergroup) to Devonian, Central Fjord Zone, East Greenland: Copenhagen, Grønlands Geologiske Undersøgelse, scale 1:250,000.

Berger, A.L., 1977, Support for the astronomical theory of climatic change: Nature, v. 269, p. 44–45, doi: 10.1038/269044a0.

Berger, A., Loutre, M.F., and Dehant, V., 1989, Astronomical frequencies for pre-Quaternary palaeoclimate studies: Terra Nova, v. 1, p. 474–479, doi: 10.1111/j.1365-3121.1989.tb00413.x.

Berner, R.A., 2006, GEOCARBSULF: A combined model for Phanerozoic atmospheric O_2 and CO_2: Geochimica et Cosmochimica Acta, v. 70, p. 5653–5664, doi: 10.1016/j.gca.2005.11.032.

Blom, H., 2005, Taxonomic revision of the Late Devonian tetrapod *Ichthyostega* from Greenland: Palaeontology, v. 48, p. 111–134, doi: 10.1111/j.1475-4983.2004.00435.x.

Blom, H., Clack, J.A., and Ahlberg, P.E., 2005, Localities, distribution and stratigraphical context of the Late Devonian tetrapods of East Greenland: Meddelelser om Grønland: Geoscience, v. 43, p. 1–50.

Blom, H., Clack, J.A., Ahlberg, P.E., and Friedman, M., 2007, Devonian vertebrates from East Greenland: A review of faunal composition and distribution: Geodiversitas, v. 29, p. 119–141.

Bütler, H., 1935a, Die Mächtigkeit der kaledonischen Molasse in Ostgrönland: Mitteilungen der Naturforschenden Gesellschaft Shaffhausen, v. 12, p. 17–33.

Bütler, H., 1935b, Some New Investigations of the Devonian Stratigraphy and Tectonics of East Greenland: Meddelelser om Grønland, v. 103, no. 2, 35 p.

Bütler, H., 1954, Die Stratigraphische Gliederung der Mitteldevonischen Serien im Gebiet von Kap Franklin in Zentral-Ostgrönland: Meddelelser om Grønland, v. 116, no. 7, 126 p.

Bütler, H., 1955, Das Variscish Gefaltete Devon zwischen Dusén Fjord und Kongeborgen in Zentral-Ostgrönland: Meddelelser om Grønland, v. 155, no. 1, 131 p.

Bütler, H., 1957, Beobachtungen an der Hauptbruchzone der Küste von Zentral-Ostgrönland: Meddelelser om Grønland, v. 160, no. 1, 79 p.

Bütler, H., 1959, Das Old-Red Gebiet am Moskusoksefjord (Attempt at a correlation of the series of various Devonian areas in Central East Greenland): Meddelelser om Grønland, v. 160, no. 5, 88 p.

Bütler, H., 1961, Devonian deposits of central East Greenland, *in* Raasch, G.O., ed., Geology of the Arctic, Volume 1: Toronto, Toronto University Press, p. 188–196.

Campbell, K.S.W., and Barwick, R.E., 1988, Geological and palaeontological information and phylogenetic hypotheses: Geological Magazine, v. 125, p. 207–227.

Caputo, M.V., and Crowell, J.C., 1985, Migration of glacial centers across Gondwana during the Paleozoic era: Geological Society of America Bulletin, v. 96, p. 1020–1036, doi: 10.1130/0016-7606(1985)96 <1020:MOGCAG>2.0.CO;2.

Christiansen, F.G., Olsen, H., Piasecki, S., and Stemmerik, L., 1990, Organic geochemistry of Upper Palaeozoic lacustrine shales in the East Greenland basin, *in* Durand, B., and Behar, F., ed., Advances in Organic Geochemistry 1989: Organic Geochemistry, v. 16, p. 287–294, doi: 10.1016/0146-6380(90)90048-5.

Clack, J.A., 1998, The neurocranium of *Acanthostega gunnari* and the evolution of the otic region in tetrapods: Zoological Journal of the Linnean Society, v. 122, p. 61–97, doi: 10.1006/zjls.1997.0114.

Clack, J.A., 2002, Gaining Ground: The Origin and Early Evolution of Tetrapods; Life of the Past: Bloomington, Indiana University Press, 369 p.

Clack, J.A., 2003, A revised reconstruction of the dermal skull roof of *Acanthostega*, an early tetrapod from the Late Devonian: Transactions of the Royal Society of Edinburgh, v. 93, p. 163–165.

Clack, J.A., 2005, Making headway and finding a foothold: Tetrapods come ashore, *in* Briggs, D.E.G., ed., Evolving Form and Function: Fossils and Development: New Haven, Connecticut, Yale University, Peabody Museum of Natural History Special Publications, p. 223–244.

Clack, J.A., 2006, The emergence of early tetrapods: Palaeogeography, Palaeoclimatology, Palaeoecology, v. 232, p. 167–189, doi: 10.1016/j.palaeo.2005.07.019.

Clack, J.A., 2007, Devonian climate change, breathing, and the origin of the tetrapod stem group: Integrative and Comparative Biology, v. 47, p. 510–523, doi: 10.1093/icb/icm055.

Clack, J.A., and Neininger, S.L., 2000, Fossils from the Celsius Bjerg Group, Upper Devonian sequence, East Greenland: Significance and sedimentological distribution, *in* Friend, P.F., and Williams, B., eds., New Perspectives on the Old Red Sandstone: London, Geological Society Symposium Volume, p. 557–566.

Clack, J.A., Ahlberg, P.E., Finney, S.M., Dominguez Alonso, P., Robinson, J., and Ketcham, R.A., 2003, A uniquely specialised ear in a very early tetrapod: Nature, v. 425, p. 65–69, doi: 10.1038/nature01904.

Clack, J.A., Ahlberg, P.E., Blom, H., and Finney, S.M., 2008, A new genus of Devonian tetrapod from East Greenland with new information on the lower jaw of *Ichthyostega*, *in* Elliott, D.K., Maisey, J.G., Yu, X., and Miao, D., eds., Morphology, Phylogeny, and Biogeography of Fossil Fishes Honoring Meemann Chang: Munich, Verlag der Friedrich Pfeil (in press).

Clement, G., Ahlberg, P.E., Blieck, A., Blom, H., Clack, J.A., Poty, E., Thorez, J., and Janvier, P., 2004, A Devonian tetrapod from western Europe: Nature, v. 427, p. 412–413, doi: 10.1038/427412a.

Daeschler, E.B., Frumes, A.C., and Mullinson, F., 2003, Groenlandaspid placoderm fishes from the Late Devonian of North America: Records of the Australian Museum, v. 55, p. 45–60.

den Blaauwen, J.L., Barwick, R.E., and Campbell, K.S.W., 2005, Structure and function of the tooth plates of the Devonian lungfish *Dipterus valenciennesi* from Caithness and the Orkney Islands: Records of the Australian Museum, v. 23, p. 91–113.

Friedman, M., and Blom, H., 2006, A new actinopterygian from the Famennian of East Greenland and the interrelationships of Devonian ray-finned fishes: Journal of Paleontology, v. 80, p. 1186–1204, doi: 10.1666/0022-3360(2006)80[1186:ANAFTF]2.0.CO;2.

Friend, P.F., Alexander-Marrack, P.D., Nicholson, J., and Yeats, A.K., 1976a, Devonian Sediments of East Greenland: I. Introduction, Classification of Sequences, Petrographic Notes: Meddelelser om Grønland, v. 206, no. 1, 56 p.

Friend, P.F., Alexander-Marrack, P.D., Nicholson, J., and Yeats, A.K., 1976b, Devonian Sediments of East Greenland: II. Sedimentary Structures and Fossils: Meddelelser om Grønland, v. 206, no. 2, 91 p.

Friend, P.F., Alexander-Marrack, P.D., Allen, K.C., Nicholson, J., and Yeats, A.K., 1983, Devonian sediments of East Greenland: VI. Review of Results: Meddelelser om Grønland, v. 206, no. 6, 96 p.

Gilotti, J.A., and McClelland, W.C., 2005, Leucogranites and the time of extension in the East Greenland Caledonides: The Journal of Geology, v. 113, p. 399–417, doi: 10.1086/430240.

Haller, J., 1970, Tectonic Map of East Greenland (1:500,000): An Account of Tectonism, Plutonism, and Volcanism in East Greenland: Meddelelser om Grønland, v. 171, no. 5, 286 p.

Haller, J., 1971, Geology of the East Greenland Caledonides: New York, Interscience Publishers, 415 p.

Hartz, E.H., 2000, Early syndepositional tectonics of East Greenland's Old Red Sandstone Basin, *in* Friend, P.F., and Williams, B.P.J., eds., New Perspectives in the Old Red Sandstone: Geological Society of London Special Publication 180, p. 537–555.

Hartz, E.H., and Andresen, A., 1995, Caledonian sole thrust of central East Greenland: A crustal-scale Devonian extensional detachment?: Geology, v. 23, no. 7, p. 637–640, doi: 10.1130/0091-7613(1995)023<0637:CSTOCE>2.3.CO;2.

Hartz, E.H., and Andresen, A., 1997, From collision to collapse: Complex strain permutations in the hinterland of the Scandinavian Caledonides: Journal of Geophysical Research, v. 102, no. B11, p. 24,697–24,711, doi: 10.1029/97JB02275.

Hartz, E.H., Osmundsen, P.T., and Andresen, A., 1996, Structural control of the Devonian basin in the East Greenland Caledonides: Geologiska Föreningens Stockholm Förhandlingar, v. 118, p. 37–38.

Hartz, E.H., Torsvik, T.H., and Andresen, A., 1997, Carboniferous age for the East Greenland "Devonian" basin: Palaeomagnetic and isotopic constraints on age, stratigraphy, and plate reconstructions: Geology, v. 25, no. 8, p. 675–678, doi: 10.1130/0091-7613(1997)025<0675:CAFTEG>2.3.CO;2.

Hartz, E.H., Torsvik, T.H., and Andresen, A., 1998, Reply to comment: Carboniferous age for the East Greenland "Devonian" basin: Paleomagnetic and isotopic constraints on age, stratigraphy, and plate reconstructions: Geology, v. 26, p. 285–286.

Hartz, E.H., Andresen, A., Hodges, K., and Martin, M.W., 2000, The Fjord region detachment zone, a long-lived extensional fault in the East Greenland Caledonides: Journal of the Geological Society of London, v. 157, p. 795–809.

Hartz, E.H., Andresen, A., Hodges, K.V., and Martin, M.W., 2001, Syncontractional extension and exhumation of deep crustal rocks in the East Greenland Caledonides: Tectonics, v. 20, p. 58–77, doi: 10.1029/2000TC900020.

Hays, J.D., Imbrie, J., and Schackleton, N.J., 1976, Variations in the Earth's orbit: Pacemaker of the ice ages: Science, v. 194, p. 1121–1132, doi: 10.1126/science.194.4270.1121.

Heintz, A., 1930, Oberdevonische Fischreste aus Ostgrönland: Skrifter om Svalbard og Ishavet, v. 30, p. 31–46.

Heintz, A., 1932, Beitrag zur Kenntnis der oberdevonischen Fischfauna Ostgrönlands: Skr: Skrifter om Svalbard og Ishavet, v. 42, p. 1–27.

Henriksen, N., and Higgins, A.K., 2008, this volume, Caledonian orogen of East Greenland 70°N–82°N: Geological map at 1:1,000,000—Concepts and principles of compilation, *in* Higgins, A.K., Gilotti, J.A., and Smith, M.P., eds., The Greenland Caledonides: Evolution of the Northeast Margin of Laurentia: Geological Society of America Memoir 202, doi: 10.1130/2008.1202(14).

Jarvik, E., 1952, On the Fish-Like Tail in the Ichthyostegid Stegocephalians, with Descriptions of a New Crossopterygian from the Upper Devonian of East Greenland: Meddelelser om Grønland, v. 114, no. 12, 90 p.

Jarvik, E., 1961, Devonian vertebrates, *in* Raasch G.O., ed., Geology of the Arctic, Volume 1: Toronto, Toronto University Press, p. 197–204.

Kelly, S.B., and Olsen, H., 1993, Terminal fans—A review with reference to Devonian examples, *in* Fielding, C.R., ed., Current Research in Fluvial Sedimentology: Sedimentary Geology, v. 85, p. 339–374.

Koch, L., 1929a, The geology of East Greenland: Meddelelser om Grønland, v. 73, no. 2, p. 1–204.

Koch, L., 1929b, Stratigraphy of Greenland: Meddelelser om Grønland, v. 73, no. 2, p. 205–320.

Koch, L., and Haller, J., 1971, Geological Map of East Greenland 72°–76°N Lat. (1:250,000): Meddelelser om Grønland, v. 183, 26 p.

Kulling, O., 1930, Stratigraphic studies of the geology of Northeast Greenland (preliminary report): Meddelelser om Grønland, v. 74, no. 13, p. 317–346.

Kulling, O., 1931, An Account of the Localities of the Upper Devonian Vertebrate Finds in East Greenland in 1929: Meddelelser om Grønland, v. 86, no. 2, 14 p.

Kutzbach, J.E., and Street-Parrott, F.A., 1985, Milankovitch forcing of fluctuations in the level of tropical lakes from 18 to 0 kyr BP: Nature, v. 317, p. 130–134, doi: 10.1038/317130a0.

Larsen, H.-C., and Marcussen, C., 1992, Sill-intrusion, flood basalt emplacement and deep crustal structure of the Scoresby Sund region, East Greenland *in* Storey, B.C., Alabaster, T., and Pankhurst, R.J., eds., Magmatism and the Causes of Continental Break-up: Geological Society of London Special Publication 68, p. 365–386.

Larsen, H.-C., Armstrong, G., Marcussen, C., Moore, S., and Stemmerik, L., 1989, Deep seismic data from the Jameson Land basin, East Greenland (from IFP/ILP/IUGS [Institut Français de Pétrole/International Lithosphere Program/International Union of Geological Sciences] Exploration Research Conference: The Potential of Deep Seismic Profiling for Hydrocarbon Research, Arles, France, June 1989): Terra Abstracts, v. 1, p. 10–11.

Larsen, P.-H., 1990a, Geological Map (1:100,000) of the Devonian Basin, North-East Greenland: Grønlands Geologiske Undersøgelse internal report, 6 p., 3 maps.

Larsen, P.-H., 1990b, Geological, Structural Contour and Isopach Maps (1:50,000) of the Devonian Celsius Bjerg Group on Eastern Gauss Halvø, North-East Greenland: Grønlands Geologiske Undersøgelse internal report, 6 p., 12 maps.

Larsen, P.-H., 1990c, Structural Contour and Isopach Maps (1:50,000) of the Upper Devonian Kap Graah Group on Western Gauss Halvø, North-East Greenland: Grønlands Geologiske Undersøgelse internal report, 6 p., 6 maps.

Larsen, P.-H., 1990d, The Devonian Basin in East Greenland: Status of Structural Studies, June 1990: Grønlands Geologiske Undersøgelse internal report, 54 p.

Larsen, P.-H., and Bengaard, H.-J., 1991, The Devonian basin initiation in East Greenland: A result of sinistral wrench faulting and Caledonian extensional collapse: Journal of the Geological Society of London, v. 148, p. 355–368, doi: 10.1144/gsjgs.148.2.0355.

Larsen, P.-H., and Olsen, H., 1991, The Devonian basin project, North-East Greenland—A summary: Rapport Grønlands Geologiske Undersøgelse, v. 152, p. 17–20.

Larsen, P.-H., Olsen, H., Rasmussen, F.O., and Wilken, U.G., 1989, Sedimentological and structural investigations of the Devonian basin, East Greenland: Rapport Grønlands Geologiske Undersøgelse, v. 145, p. 108–113.

Lehman, J.-P., 1959, Les Dipneustes du Dévonien Supérieur du Groenland: Meddelelser om Grønland, v. 160, no. 4, 58 p. (21 plates).

Long, J.A., 1993, Cranial ribs in Devonian lungfish and the origin of dipnoan air-breathing: Memoirs of the Association of Australasian Palaeontologists, v. 15, p. 199–209.

Marcussen, C., Christiansen, F.G., Larsen, P.-H., Olsen, H., Piasecki, S., Stemmerik, L., Bojesen-Koefoed, J., Jepsen, H.F., and Nøhr-Hansen, H., 1987, Studies of the onshore hydrocarbon potential in East Greenland 1986–1987: Fieldwork from 72° to 74°N: Rapport Grønlands Geologiske Undersøgelse, v. 135, p. 72–81.

Marcussen, C., Larsen, P.-H., Olsen, H., Piasecki, S., and Stemmerik, L., 1988, Studies of the onshore hydrocarbon potential in East Greenland 1986–1987: Fieldwork from 73° to 76°N: Rapport Grønlands Geologiske Undersøgelse, v. 140, p. 89–95.

Marshall, J.E.A., and Astin, T.R., 1996, An ecological control on the distribution of the Devonian fish *Asterolepis*: Newsletter on Stratigraphy, v. 33, p. 133–144.

Marshall, J.E.A., and Hemsley, A.R., 2003, A mid-Devonian seed megaspore from East Greenland and the origin of seed plants: Palaeontology, v. 46, p. 647–670, doi: 10.1111/1475-4983.00314.

Marshall, J.E.A., and Stephenson, B.J., 1997, Sedimentological responses to basin initiation in the Devonian of East Greenland: Sedimentology, v. 44, p. 407–419, doi: 10.1046/j.1365-3091.1997.d01-29.x.

Marshall, J.E.A., Astin, T.R., and Clack, J.A., 1999, The East Greenland tetrapods are Devonian in age: Geology, v. 27, p. 637–640, doi: 10.1130/0091-7613(1999)027<0637:EGTADI>2.3.CO;2.

McClay, K.R., Norton, M.G., Coney, P., and Davis, G.H., 1986, Collapse of the Caledonian orogen and the Old Red Sandstone: Nature, v. 323, p. 147–149, doi: 10.1038/323147a0.

NACSN, 1983, North American Stratigraphic Code: American Association of Petroleum Geology Bulletin, v. 67, p. 841–875.

Nathorst, A.G., 1901, Bidrag til nordöstra Grönlanfs Geologi (with petrographic descriptions by H. Bäckström): Geologiska Föreningens Stockholm Förhandlingar, v. 23, no. 207, p. 275–306.

Nicholson, J., and Friend, P.F., 1976, Devonian sediments of East Greenland: V. The Central Sequence, Kap Graah Group and Mount Celsius Supergroup: Meddelelser om Grønland, v. 206, no. 5, 117 p.

Norton, M.G., McClay, K.R., and Way, N.A., 1987, Tectonic evolution of Devonian basins in northern Scotland and southern Norway: Norsk Geologisk Tidsskrift, v. 67, p. 323–338.

Olsen, H., 1990, Astronomical forcing of meandering river behaviour: Milankovitch cycles in Devonian of East Greenland: Palaeogeography, Palaeoclimatology, Palaeoecology, v. 79, p. 99–115, doi: 10.1016/0031-0182(90)90107-I.

Olsen, H., 1993a, Sedimentary Basin Analysis of the Continental Devonian Basin in North-East Greenland: Bulletin Grønlands Geologiske Undersøgelse, v. 168, 80 p.

Olsen, H., 1993b, Orbital forcing on continental depositional systems—Lacustrine and fluvial cyclicity in the Devonian of East Greenland, *in* De Boer, P.L., and Smith, D.G., eds., Orbital Forcing and Cyclicity Sequences: International Association of Sedimentologists Special Publication 19, p. 429–438.

Olsen, H., and Larsen, P.-H., 1993a, Lithostratigraphy of the Continental Devonian Sediments in North-East Greenland: Bulletin Grønlands Geologiske Undersøgelse, v. 165, 111 p.

Olsen, H., and Larsen, P.-H., 1993b, Structural and climatic control on fluvial depositional systems—Devonian, North-East Greenland, *in* Marzo, M., and Puigdefàbregas, C., ed., Alluvial Sedimentation: International Association of Sedimentologists Special Publication 17, p. 401–423.

Orvin, A.K., 1930, Beitrage zur Kenntnis des Oberdevons Ostgrönlands: Skrifter om Svalbard og Ishavet, v. 30, p. 1–30.

Orvin, A.K., 1931, A fossil river bed in East-Greenland: Norsk Geologisk Tidsskrift, v. 12, p. 469–474.

Säve-Söderbergh, G., 1932a, Notes on the Devonian Stratigraphy of East Greenland: Meddelelser om Grønland, v. 94, no. 4, 40 p.

Säve-Söderbergh, G., 1932b, Preliminary Note on Devonian Stegocephalians from East Greenland: Meddelelser om Grønland, v. 94, no. 7, 107 p.

Säve-Söderbergh, G., 1933, Further Contributions to the Devonian Stratigraphy of East Greenland: I. Results from the Summer Expeditions 1932: Meddelelser om Grønland, v. 96, no. 1, 40 p.

Säve-Söderbergh, G., 1934, Further Contributions to the Devonian Stratigraphy of East Greenland: II. Investigations on Gauss Peninsula during the Summer of 1933. With an Appendix: Notes on the Geology of the Passage Hills (East Greenland): Meddelelser om Grønland, v. 96, no. 2, 74 p.

Schlindwein, V., 1998, Architecture and Evolution of the Continental Crust of East Greenland from Integrated Geophysical Studies: Berichte zur Polarforschung, v. 270, 148 p.

Scotese, C., 2002, Paleomap Project: www.scotese.com, accessed February 2, 2003.

Smith, A.G., Hurley, A.M., and Briden, J.C., 1981, Phanerozoic Paleocontinental World Maps: Cambridge, Cambridge University Press, 102 p.

Sønderholm, M., and Tirsgaard, H., 1993, Lithostratigraphic Framework of the Upper Proterozoic Eleonore Bay Supergroup of East and North-East Greenland: Bulletin Grønlands Geologiske Undersøgelse, v. 167, 38 p.

Stemmerik, L., and Bendix-Almgreen, S.E., 1998, Comment on: Carboniferous age for the East Greenland "Devonian" basin: Paleomagnetic and isotopic constraints on age, stratigraphy, and plate reconstructions: Geology, v. 26, p. 284, doi: 10.1130/0091-7613(1998)026<0284:CAFTEG>2.3.CO;2.

Stensiö, E.A., 1931, Upper Devonian Vertebrates from East Greenland, Collected by the Danish Greenland Expeditions in 1929 and 1930: Meddelelser om Grønland, v. 86, no. 1, 212 p.

Tirsgaard, H., and Sønderholm, M., 1997, Lithostratigraphy, Sedimentary Evolution and Sequence Stratigraphy of the Upper Proterozoic Lyell Land Group (Eleonore Bay Supergroup) of East and North-East Greenland: Geological Survey of Greenland Bulletin, v. 178, 60 p.

Trewin, N.H., 1986, Palaeoecology and sedimentology of the Achanarras fish bed of the Middle Old Red Sandstone, Scotland: Transactions of the Royal Society of Edinburgh, Earth Sciences, v. 77, p. 21–46.

Van der Voo, R., 1988, Paleozoic paleogeography of North America, Gondwana, and intervening displaced terranes: Comparisons of paleomagnetism with paleoclimatology and biogeographical patterns: Geological Society of America Bulletin, v. 100, p. 311–324, doi: 10.1130/0016-7606(1988)100<0311:PPONAG>2.3.CO;2.

Woodward, A.S., 1900, Notes on some Upper Devonian fish-remains discovered by Prof. A. G. Nathorst in East Greenland: Bihang till Kungliga Vetenskapsakademiens Handlingar, v. 26, no. 4, p.1–10.

Yeats, A.K., and Friend, P.F., 1978, Devonian Sediments of East Greenland: IV. The Western Sequence, Kap Kolthoff Supergroup of the Western Areas: Meddelelser om Grønland, v. 206, no. 4, 112 p.

Ziegler, P.A., 1988, Laurussia—The Old Red Continent, *in* McMillan, N.J., Embry, A.F., and Glass, D.J., eds., Devonian of the World, Volume I: Regional Syntheses: Canadian Society of Petroleum Geologists Memoir 14, p. 15–48.

MANUSCRIPT ACCEPTED BY THE SOCIETY 14 JANUARY 2008

The Geological Society of America
Memoir 202
2008

Mineral occurrences in central East Greenland (70°N–75°N) and their relation to the Caledonian orogeny— A Sr-Nd-Pb isotopic study of scheelite

Henrik Stendal*
Geological Survey of Denmark and Greenland, Øster Voldgade 10, DK-1350 Copenhagen K, Denmark

Robert Frei
Department of Geography and Geology, University of Copenhagen, Øster Voldgade 10, DK-1350 Copenhagen K, Denmark

ABSTRACT

The Caledonian orogen of North-East Greenland hosts numerous mineral occurrences related to (1) pre-Caledonian crystalline complexes (Pb-Zn skarn type); (2) Neoproterozoic basins (strata-bound copper); (3) Caledonian granites (vein-type gold, silver, tungsten, arsenic, and antimony); and (4) late Caledonian extensional structures (vein base metal ± silver). Sr, Pb, and Sm-Nd isotope analyses of scheelite ($CaWO_4$) indicate a heterogeneous, probably local, source for tungsten, and Sr isotopic data support a genetic link to Caledonian magmatic activity. Pb isotopes indicate mixing of Pb derived from late waning-stage fluids from the granites and from interaction with wall rocks. Sm-Nd isotopic data for the investigated scheelites indicate that a portion of the rare earth elements was derived from fluids that had interacted with both Archean-Paleoproterozoic crystalline basement and Mesoproterozoic-Neoproterozoic sedimentary rocks.

Mineral occurrences associated with fault zones and late Caledonian veins all show a genetic relationship with Caledonian granite emplacement. Sm-Nd isotopic data from scheelite define an errorchron with a slope corresponding to 382 ± 39 Ma (mean square of weighted deviates [MSWD] = 2.6) and an initial $^{143}Nd/^{144}Nd$ value of 0.511642 ± 0.000049. This indicates emplacement during the latest stages or even subsequent to emplacement of most Caledonian granites around 425 Ma. The initial Nd isotopic ratio defined by the scheelite Sm-Nd isotopic correlation line is identical within error to the values of S-type granitoids. The multi-isotope studies indicate that tungsten may have been deposited from fluids associated with Caledonian granites, which provided heat sources for local hydrothermal circulation cells. Forced into faults, thrusts, and fractures, the fluids were trapped by dominantly Ca-rich sediments.

Keywords: scheelite, Caledonian orogeny, Neoproterozoic, isotopes, East Greenland.

*hdal@gh.gl

Stendal, H., and Frei, R., 2008, Mineral occurrences in central East Greenland (70°N–75°N) and their relation to the Caledonian orogeny—A Sr-Nd-Pb isotopic study of scheelite, *in* Higgins, A.K., Gilotti, J.A., and Smith, M.P., eds., The Greenland Caledonides: Evolution of the Northeast Margin of Laurentia: Geological Society of America Memoir 202, p. 293–306, doi: 10.1130/2008.1202(12). For permission to copy, contact editing@geosociety.org. ©2008 The Geological Society of America. All rights reserved.

INTRODUCTION

Mineral occurrences in central East Greenland and North-East Greenland occur in Archean to Paleoproterozoic crystalline basement, Mesoproterozoic-Neoproterozoic metasediments, and in Paleogene volcanic rocks. This paper focuses on mineral occurrences in the southern half of the Caledonian orogen (70°N–75°N, 22°W–28°W) that are genetically related to the Caledonian orogenic processes and hosted by Mesoproterozoic-Neoproterozoic sedimentary rocks. In the Eleonore Bay Supergroup (Neoproterozoic clastic sediments) strata-bound copper-bearing mineral occurrences are common. The mode of mineral occurrence related to Caledonian events suggests a close relationship to crustal faults and thrust zones. These zones acted as pathways for fluids, and they have been research targets for mineralization investigations in central East Greenland (Jensen, 1993, 1994a, 1998; Pedersen, 1997; Pedersen and Stendal, 1999; Stendal and Wendorff, 1998; Stendal et al., 1999). This paper places the various types of Caledonian mineral occurrences into a regional, structural, and time contexts based on the most recent results gained from multi-isotopic studies. We present Sr, Pb, and Nd-Sm isotopic data on scheelite ($CaWO_4$) from various occurrences in the central East to North-East Greenland metallogenetic province, and we discuss them within the framework of previously published Pb isotopic data of sulfides from mineral occurrences to assess the framework for Caledonian mineral occurrences.

Most of the numerous mineral occurrences of the southern half of the East Greenland Caledonides were discovered during Lauge Koch's geological expeditions (1926–1958; Koch, 1963; Haller, 1971) and during the later extensive mineral exploration by Nordisk Mineselskab A/S up to 1984. The mineral exploration history of East Greenland is outlined in detail by Harpøth et al. (1986), who also provided a thorough overview of the individual occurrences. The earliest reports of mineral occurrences relate to the exploration period from 1822 to 1900 (for details and references, see Harpøth et al., 1986). From 1979 to 1984, Nordisk Mineselskab carried out a tungsten exploration program supported by the Commission of the European Communities (Hallenstein et al., 1981; Hallenstein and Pedersen, 1983; Pedersen and Stendal, 1987). During this campaign, several vein-type mineral occurrences were located, including: base metal, precious and base metal, tungsten-antimony, antimony-gold, and bismuth-gold mineralizations. In 1990 and 1997–1998, the Geological Survey of Denmark and Greenland undertook mineralization investigations during geological expeditions to central East and North-East Greenland, with particular focus on major regional faults, thrusts, and detachment zones (Jensen, 1993, 1994a, 1998; Pedersen, 1997; Stendal and Wendorff, 1998; Pedersen and Stendal, 1999; Stendal, 1999; Stendal et al., 1999).

GEOLOGICAL SETTING

The crystalline basement rocks of the Caledonian fold belt are overprinted by a Mesoproterozoic to early Neoproterozoic "Grenvillian" event. During a tectono-stratigraphic event around 950 Ma, the supracrustal rocks (Krummedal supracrustal sequence) and underlying basement rocks were reworked to form a migmatite and paragneiss complex (Henriksen et al., 2000). The Krummedal supracrustal sequence consists of metamorphosed sandstones, mudstones, and calcareous mudstones. This metasedimentary package makes up a major part of the Caledonian Hagar Bjerg thrust sheet, which hosts both Caledonian (435–425 Ma) and older 940–910 Ma granites (Kalsbeek et al., 2001; Higgins et al., 2004).

The Neoproterozoic Eleonore Bay Supergroup, the Tillite Group, and the lower Paleozoic sequence (Neoproterozoic–Middle Ordovician sedimentary to low-grade metamorphic rocks in Fig. 1) comprise a sedimentary succession up to 18.5 km thick. This composite unit makes up the Franz Joseph allochthon, the uppermost segment of the Caledonian thrust pile (Higgins et al., 2004), which is widely exposed between latitudes 72°N and 76°N, with a small outlier in Canning Land (71°50′N) (Higgins and Soper, 1994; Sønderholm and Tirsgaard, 1993). This succession is dominated by shallow-water siliciclastic and carbonate deposits that were laid down in a major sedimentary basin.

The Eleonore Bay Supergroup is unconformably overlain by the Tillite Group, which consists of a 700–800-m-thick sedimentary succession of Vendian age (610–570 Ma) and includes two glaciogenic diamictite formations (Hambrey and Spencer, 1987). The Lower Paleozoic of East Greenland is composed of 4000 m of Cambrian-Ordovician sediments (Higgins et al., 2004). Later in the Caledonian orogeny, a period of extensional faulting led to the initiation of the Devonian sedimentary basin in central East Greenland. The Devonian sediments unconformably overlie Ordovician and older rocks (Fig. 1) and make up more than 8 km of continental siliciclastic sediments locally intercalated with volcanic rocks (Larsen et al., this volume).

The faults, thrusts, and detachment zones in central East and North-East Greenland are important structural features that host many of the known mineral occurrences. Most of the structural zones have NNE-SSW trends, parallel to the length of the East Greenland Caledonian orogen (Fig. 1). In general, the structures can be divided into (1) structures that formed during the main Caledonian collisional orogeny, and (2) late Caledonian extensional features (Pedersen and Stendal, 2000). In the region between latitudes 72°N and 74°N, Caledonian faults are found that cut the Precambrian crystalline complexes as well as Neoproterozoic to Paleozoic sedimentary rocks. Late Caledonian faults developed in relation to the formation of the Devonian sedimentary basins. The Fjord region detachment (FRD; Fig. 1) separates the Neoproterozoic sediments of the Eleonore Bay Supergroup from the Paleoproterozoic to Mesoproterozoic infracrustal complexes in Andrée Land, Suess Land, and Lyell Land (Hartz and Andresen, 1995; Andresen et al., 1998; Higgins et al., 2004; Gilotti and McClelland, this volume). During the Late Devonian, the so-called "eastern fault zone" (EFZ; Fig. 1) was an important tectonic feature in the region (Larsen and Olsen, 1991).

Granitic intrusions are prominent in various units of the Caledonian fold belt in East Greenland. The earliest intrusive

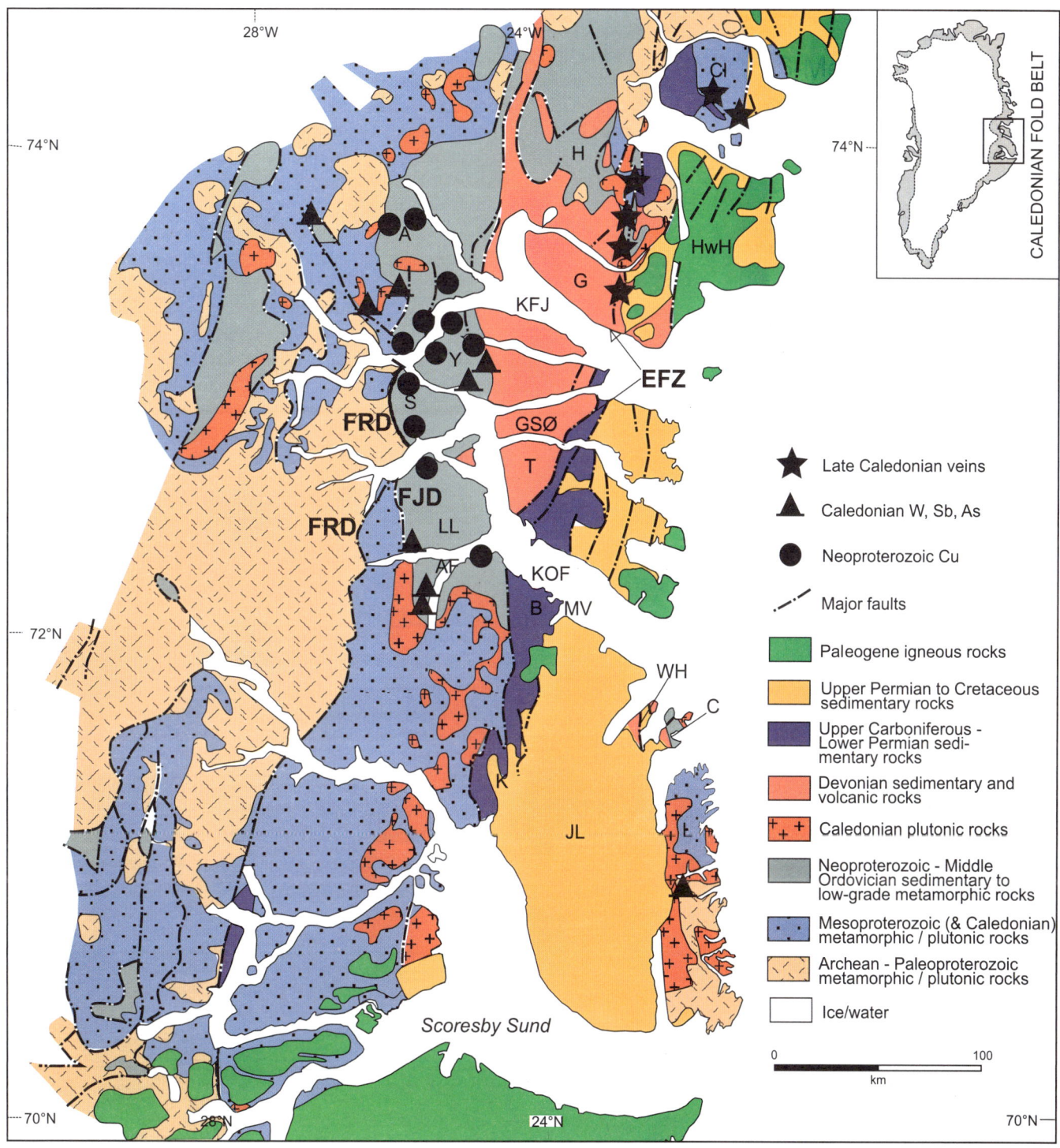

Figure 1. Simplified geological map of central East and North-East Greenland, showing the lineaments and localities mentioned in the text. A—Andrée Land; AF—Alpefjord; B—Blyklippen; C—Canning Land; Cl—Clavering Ø; EFZ—eastern fault zone; FJD—Franz Joseph detachment; FRD—Fjord region detachment; G—Gauss Halvø; GSØ—Geographical Society Ø; H—Hudson Land; HwH—Hold with Hope; JL—Jameson Land; K—Karstryggen; KFJ—Kejser Franz Joseph Fjord; KOF—Kong Oscar Fjord; L—Liverpool Land; LL—Lyell Land; MV—Mesters Vig; S—Suess Land; T—Traill Ø; WH—Wegener Halvø; Y—Ymer Ø. Geological map was modified from Henriksen (2003), and mineral occurrences are from Harpøth et al. (1986) and Pedersen and Stendal (2000).

bodies and associated basic dikes in the crystalline basement are probably of Archean age and have a minimum age of 2500 Ma (e.g., Rex and Gledhill, 1974; Steiger et al., 1979). An early phase of intrusions emplaced into the Krummedal supracrustal sequence have yielded 940–910 Ma ages (Kalsbeek et al., 2000, 2001), while Caledonian plutonic intrusions were emplaced between ca. 435 and 425 Ma (Kalsbeek et al., this volume).

MINERAL OCCURRENCES

Mineral occurrences located within the present Caledonian orogen can be related to (1) pre-Caledonian crystalline complexes (e.g., Pb-Zn skarn occurrences), (2) Neoproterozoic basins (e.g., strata-bound copper-type occurrences), (3) Caledonian granites (e.g., gold, tungsten, arsenic, and antimony deposits), and (4) late Caledonian extensional detachments and faults (e.g., mainly vein base metal occurrences ± silver).

Pre-Caledonian Crystalline Complexes

Pre-Caledonian mineral occurrences are scarce, and their structural setting is not always clear. Several areas with mineral occurrences in Archean to Paleoproterozoic regions are mentioned by Harpøth et al. (1986). Mafic to ultramafic rocks in different parts of the crystalline basement are host to nickel-copper and gold, chromium, and iron-titanium mineral associations. In addition, supracrustal enclaves in the gneisses may host massive iron-sulfides and strata-bound copper occurrences. Minor uranium-bearing occurrences are widespread in the infracrustal gneiss complexes. In the Mesoproterozoic metasedimentary sequences, small occurrences of skarn mineralization related to granitoids include lead, zinc, silver, copper, and uranium minerals. The most prominent occurrences are a lead-zinc skarn-type deposit in the Alpefjord region and strata-bound copper mineralization in quartz-rich sediments (Krummedal sequence) in northern Suess Land and southern Andrée Land.

Neoproterozoic Basins

In the Neoproterozoic basins, strata-bound copper occurrences (Fig. 1) are located in the Eleonore Bay Supergroup (Stendal, 1979, 1980; Ghisler et al., 1980a, 1980b; Stendal and Hock, 1981; Stendal and Ghisler, 1984). The strata-bound copper occurrences (0.2–2 m thick) are found at different stratigraphic levels in shale and quartzite and extend over a distance of 275 km (Fig. 1). Eight levels of strata-bound occurrences have been recorded (Table 1). The two lowest levels are found at the top of the Nathorst Land Group in metamorphosed quartz-rich shales (1 and 2, Table 1). The Lyell Land Group is dominated by sandstone and quartz-rich shale, and it hosts three levels with copper mineralization. The first level is in the middle of the group, and the two other levels are at the top of the group (3–5, Table 1). Epigenetic overprinting resulted in the addition of antimony to the upper copper-bearing layers on Strindberg Land (Fig. 2; 5 in Table 1). Three additional strata-bound copper-bearing layers are found in the lower part of Ymer Ø Group, which is dominated by red and green shales at the base, followed by dolomitic shales (6–8, Table 1).

The mineral assemblages in the shales are dominated by an iron-poor paragenesis represented by chalcocite-(bornite) and an iron-rich paragenesis consisting of pyrite-chalcopyrite. Chalcopyrite is dominant in the coarse-grained laminae, together with pyrite ± pyrrhotite. In addition, strata-bound sulfide-bearing veins occur together with chalcopyrite-pyrite-(tetrahedrite). The grade of copper in the mineralized zones is generally low (Table 1). The strata-bound copper sulfides were probably concentrated mainly by diagenetic processes and then later modified by metamorphic mobilization and local enrichment from hydrothermal solutions. In the sandstones, chalcopyrite occurs interstitially to the detrital quartz grains.

The Vendian Tillite Group consists of diamictites, sandstones, mudstones, and carbonates (1 km thick) and the Cambrian-Ordovician mainly carbonates (4.5 km thick) with intercalated siliciclastic sediments. No mineralization is recorded in the Vendian sedimentary rocks.

TABLE 1. STRATIGRAPHY OF THE NEOPROTEROZOIC ELEONORE BAY SUPERGROUP (EXCEPT THE YOUNGEST GROUP) WITH THE COPPER MINERALIZED LEVELS INDICATED

Period	Lithostratigraphy	Age (Ma)	Lithology	No.	Thickness (m)	Sulfide mineralization (Cu grade, %)
Riphean	Ymer Ø Group (b)	700	1.3 km sandstones, mudstones	8	0.1–0.25	Pyrite-chalcopyrite-bornite-chalcocite
				7	0.1–0.25	Chalcocite (<0.1)
			and carbonates	6	1.0–2.0	Chalcocite (0.1–0.5; max. 6)
				5	0.2–1.0	Pyrite-chalcopyrite-tetrahedrite (0.1–1.35)
	Lyell Land Group (b)		2.8 km sandstones and mudstones	4	0.2–2.0	Pyrite-chalcopyrite-bornite-chalcocite (0.1–1.25)
				3	0.2–2.0	Pyrite-chalcopyrite (0.1)
	Nathorst Land Group (a)	900	9 km sandstones, mudstones	2	0.2–1.5	Pyrite-chalcopyrite-(pyrrhotite)-cobaltite (0.25–2.5)
			and carbonates	1	0.2–5.0	Pyrrhotite-pyrite-chalcopyrite (0.1)

Note: (a)—Lower Eleonore Bay Supergroup; (b)—Upper Eleonore Bay Supergroup.

Figure 2. (A) Copper occurrence in white quartzite, Holmesø, Strindberg Land. Inserts: (B) Strata-bound copper-bearing layers (chalcocite, bornite, chalcopyrite); the scale is in centimeters. (C) Veinlets with chalcocite, bornite, and tetrahedrite in white quartzite; the sample is 10 cm long.

Caledonian Granites and Related Hydrothermal Activity

Mineralization related to Caledonian intrusions includes arsenic, tungsten, and stibnite mineral occurrences. Arsenic as arsenopyrite is found at several places and is particularly concentrated in the Alpefjord region (Stendal, 1981). The arsenopyrite was hydrothermally formed and is distributed in fracture and vein systems closely related to the contact between the base of the Eleonore Bay Supergroup and older metamorphic crystalline complexes (Stendal, 1981). Fluid inclusions in the arsenopyrite-bearing veins indicate accumulation at 225–260 °C from CO_2-bearing water with moderate salinity at a pressure of ~1 kbar (Stendal and Ghisler, 1984).

Tungsten mineralization occurs in the form of scheelite (Hallenstein et al., 1981; Hallenstein and Pedersen, 1983; Pedersen and Stendal, 1987). Hallenstein and Pedersen (1983) grouped the scheelite occurrences into three main geological settings. The first ore setting is scheelite in metasedimentary rocks, often associated with granodiorite or granite intrusions. In this type, scheelite occurs within calcareous quartzite (garnet-hornblende quartzite), also called "skarnoid" by Hallenstein and Pedersen (1983). This type is commonly associated with minor tin occurrences. The second ore style is scheelite associated with granite, pegmatite, and quartz veins. This type is only found in the Lower Eleonore Bay Supergroup. The third group is scheelite and stibnite in sedimentary rocks without direct magmatic association, and these occurrences are found on Ymer Ø (Fig. 1; Pedersen and Stendal, 1987). This third type occurs distributed along the extensional fault zone within the Neoproterozoic–Middle Ordovician sedimentary basin close to the border of the Devonian sedimentary basin, where several levels of Ymer Ø Group carbonates are mineralized. The scheelite is associated with stibnite (Sb_2S_3) and located in brecciated carbonates in minor fault zones; the grade of tungsten (Fig. 3) and antimony can be high. One of the scheelite-stibnite veins carries 42,000 Mt with 0.7% tungsten and 108,000 Mt with 3.5% antimony (Harpøth et al., 1986). The hydrothermal system on Ymer Ø reflects a crude zonation with scheelite and stibnite at a higher stratigraphic level. At lower levels, pyrite, galena, sphalerite, chalcopyrite, and locally arsenopyrite with gold occur; gold contents can be up to a few parts per million (Pedersen, 1993).

Late Caledonian Extensional Structures

Mineral occurrences related to late Caledonian faults include vein-type occurrences that are particularly widespread along the eastern fault zone (EFZ). The occurrences consist of base metals, arsenic, bismuth, antimony, minor silver, gold, and uranium. The occurrences are epigenetic and spatially related to complex conjugate fault sets associated with the main N-S faults (e.g., in Hudson Land). The main fault zones acted as pathways for basinal brines, which precipitated minor vein-type

Figure 3. (A) Scheelite-bearing limestone breccia from Ymer Ø. (B) Massive scheelite crystals (white) in limestone breccia (in ultraviolet light).

mineralization on a regional scale. The mineral occurrences are brecciated and hydrothermally altered. The veins typically contain quartz + fluorite + calcite as the main mineral assemblage, which occurs in both Proterozoic and Paleozoic sediments. The variety of the mineral occurrences is illustrated by the following three descriptions adapted from Harpøth et al. (1986) and Pedersen and Stendal (2000):

(1) On Gauss Halvø, several quartz + fluorite + calcite veins occur subparallel to the main fault. Veins and veinlets with base metals grade up to 2.5% Cu, 0.35% Pb, 250 ppm Ag, and 150 ppm Bi.

(2) In Hudson Land, base metal mineralization occurs generally in the form of centimeter- to decimeter-thick quartz veins with galena, sphalerite, and minor pyrite, locally accompanied by Bi and Ag. At one locality close to the eastern fault zone, a quartz vein (up to 6 m wide) rich in sulfides (galena, sphalerite, and arsenopyrite) yielded 663 ppb Au, 790 ppm Ag, 1.3% As, 79 ppm Sb, 2.6% Cu, 2.5% Pb, and 7% Zn.

(3) Clavering Ø was intensively explored for Pb-Zn and Au in the 1930s. A pyrite vein (90% pyrite) was said to be traceable for 1500 m, with an average thickness of 1.3 m, but new investigations could not confirm the described dimensions (Stendal et al., 1999). The vein appears not to be continuous but to be point-shaped due to crosscutting faults. Brecciation, silicification, and kaolinization are related to the quartz-pyrite mineralization. The pyrite ore contains up to 0.5 ppm Au and 25 ppm Ag.

ISOTOPE STUDIES

Methods

Scheelite-bearing hand samples representing different tungsten occurrences hosted by Mesoproterozoic to Neoproterozoic metasediments were chosen for analysis (Table 2). The tungsten-mineralized rock samples included the following: (1) two skarn samples from the Krummedal supracrustal sequence of Andrée Land; (2) four skarnoid samples from Lyell Land and Alpefjord; (3) one quartz vein from Alpefjord; and (4) six samples of carbonate-bearing breccia from Ymer Ø (Fig. 3; Table 1).

Mineral separates of scheelite were obtained from crushed, washed, and sieved (<0.23 mm) powder aliquots using magnetic separation techniques, heavy liquids, and subsequent purification by handpicking under binocular microscope. Scheelite is difficult to recognize, but because scheelite fluoresces, the handpicking was carried out under short-waved ultraviolet light.

For the Sm-Nd and Sr isotopic analyses, scheelite separates were powdered in an agate mortar. About 200 mg of powdered scheelite sample was spiked with a ^{150}Nd-^{149}Sm mixed tracer solution and then treated with concentrated aqua regia in a Teflon vial on a hot plate at 130 °C for 2 d. Chemical separation of Sr and bulk rare earth elements (REEs) was carried out on conventional cation exchange columns, followed by Sm-Nd separation using Hydroxyl-Bis (2-ethylhexyl)-phospate (HDEHP)–coated beads (BIO-RAD) charged in 6 mL quartz glass columns. The Sr fraction was subsequently purified over 600 μL glass columns using SrSpec™ resin. Average total procedural blanks for Sr and REE were small enough not to significantly affect the isotopic analyses, and therefore no blank corrections were applied.

Scheelite powder aliquots (100 mg) were dissolved for separate Pb isotopic analyses. Pb was separated via a conventional HCl-HBr elution recipe over 0.5 mL glass columns charged with BIO RAD AG-1X8 anion resin. Purification of the Pb concentrates was achieved over 300 μL Teflon columns. Procedural blanks for Pb remained below 87 pg (an amount which insignificantly affected the isotopic data of the analyzed samples).

REEs, Sr, and Pb were analyzed on a VG Sector 54 IT mass spectrometer at the Geological Institute, University of Copenhagen, Denmark. Both static and multidynamic routines were used for collection of the isotopic ratios. Nd ratios were normalized

TABLE 2. SCHEELITE SAMPLES GROUPED ACCORDING TO THE DESCRIPTION OF THE HOST ROCKS, HOST UNIT, AND THE SETTING OF THE SCHEELITES

Sample no.	Locality	Description	Host unit	Scheelite
9502	Ymer Ø	Light carbonate-breccia	Upper Eleonore Bay Supergroup - N	Breccia filling
9503	Ymer Ø	Light carbonate-breccia	Upper Eleonore Bay Supergroup - N	Breccia filling
9025	Ymer Ø	Light carbonate-breccia	Upper Eleonore Bay Supergroup - N	Breccia filling
9026	Ymer Ø	Dark carbonate-breccia	Upper Eleonore Bay Supergroup - N	Breccia filling
9027	Ymer Ø	Light carbonate-breccia	Upper Eleonore Bay Supergroup - N	Breccia filling
8007004	Ymer Ø	Dark carbonate-breccia	Upper Eleonore Bay Supergroup - N	Breccia filling
7822/2	Alpefjord	Quartz veins in quartzites	Lower Eleonore Bay Supergroup - N	Disseminated
5858	Alpefjord	Garnet quartzite (skarnoid)	Lower Eleonore Bay Supergroup - N	Disseminated
5814	Alpefjord	Garnet quartzite (skarnoid)	Lower Eleonore Bay Supergroup - N	Disseminated
12095/5	Lyell Land	Garnet quartzite (skarnoid)	Lower Eleonore Bay Supergroup - N	Disseminated
12079/3	Lyell Land	Garnet quartzite (skarnoid)	Lower Eleonore Bay Supergroup - N	Disseminated
9501	Andrée Land	Carbonate-scapolite breccia (skarn)	Krummedal supracrustal sequence - M	Breccia filling
9075	Andrée Land	Biotite-zoizite schist (skarn)	Krummedal supracrustal sequence - M	Disseminated

Note: N—Neoproterozoic; M—Mesoproterozoic.

to ^{146}Nd/^{144}Nd = 0.7219. The mean value of ^{143}Nd/^{144}Nd for our internal JM Nd standard (referenced against La Jolla) during the period of measurement was 0.511105 ± 0.000013 (2σ; n = 3). Fractionation for Pb was controlled by repeated analysis of the NBS 981 standard (Todt et al., 1996) and amounted to 0.103 ± 0.007%/a.m.u. (2σ; n = 5). The mean ^{87}Sr/^{86}Sr value of the NBS 987 Sr standard was 0.710234 ± 0.000011 (2σ; n = 7).

Correction algorithms, appropriate error propagation modes, and correlation line–isochron calculations are based on those of Ludwig, using the ISOPLOT toolkit for Microsoft Excel (Ludwig, 2003). In order to deduce the timing and assess the Caledonian mineral occurrences, we chose to compare the scheelite data with other available isotopic data from the study area. The available published data are Pb isotopic data from sulfides (e.g., Jensen, 1993, 1994a, 1994b, 1998; Pedersen, 2000).

Results

The isotopic analyses of 13 scheelite samples are presented in Tables 2–4. Previous Pb isotopic results of the strata-bound copper occurrences and of several vein-type occurrences were used for comparative purposes (Jensen, 1993, 1994a, 1994b, 1998; Pedersen, 2000).

Sr isotopic results are presented in Table 3. Sr is highly concentrated in scheelites relative to the extremely small amounts of Rb, and therefore the ^{87}Sr/^{86}Sr values can be directly used for source tracing. The ^{87}Sr/^{86}Sr values vary from 0.7298 to 0.7429 and basically cover the entire range of initial Sr isotopic values for Caledonian granites in East Greenland reported by Kalsbeek et al. (2001), with the exception of two

TABLE 3. MEASURED ^{87}Sr/^{86}Sr RATIOS OF SCHEELITE SAMPLES GROUPED AS IN TABLE 2

Sample	Measured ^{87}Sr/^{86}Sr	±1σ (ppm)	±1σ (abs)
9502	0.734789	9	0.000007
9503	0.734819	8	0.000006
9025	0.739322	8	0.000006
9026	0.736809	16	0.000012
9027	0.736769	10	0.000007
8007004	0.734826	9	0.000007
7822/2	0.730792	10	0.000007
5858	0.729802	9	0.000007
5814	0.728264	10	0.000007
12095/5	0.740810	10	0.000007
12079/3	0.742937	10	0.000007
9501	0.734914	10	0.000007
9075	0.738495	10	0.000007

granites that have lower initial Sr isotopic values of 0.712 and 0.724. Scheelites formed within carbonate breccias have quite homogeneous ^{87}Sr/^{86}Sr values (0.7348–0.7368), except for one sample (9025), which has a slightly elevated value of 0.7393. The skarnoid samples and scheelites from the biotite-zoisite schist can be divided into two groups, one with values lower than the carbonate-hosted scheelites (0.7283–0.7308), and the other with higher initial values (0.7385–0.7429).

Pb isotopic results of the scheelites are listed in Table 4, and data are plotted in Figure 4. The scheelite samples are characterized by rather uniform $^{208}Pb/^{204}Pb$ ratios (Fig. 4A), while samples 9502, 8007004, and 9501 in the $^{207}Pb/^{204}Pb$ versus $^{206}Pb/^{204}Pb$ diagram (Fig. 4B) have strongly uranogenic Pb ($^{206}Pb/^{204}Pb$ ratios exceeding 136). These samples belong to the carbonate-hosted group within the types that occur as breccia-fillings (Fig. 3). This either indicates that these scheelites are characterized by high U/Pb ratios, and/or that these particular separates were contaminated by a U-rich phase (e.g., zircon). Disseminated scheelite from skarnoids and from biotite-zoisite schists is characterized by less radiogenic values. All 13 samples define a correlation line in uranogenic Pb isotope space with a slope corresponding to an age of 420 ± 16 Ma (mean square of weighted deviates [MSWD] = 28; Fig. 4B). This correlation line intercepts the Stacey and Kramers (1975) Pb growth curve at ca. 700 Ma, which corresponds roughly to the depositional age of the Ymer Ø Group and/or the Andrée Land Group of the upper Eleonore Bay Supergroup. The uranogenic versus thorogenic isotopic pattern (Fig. 4A) is disperse and does not add to a better understanding of the uranogenic Pb isotopic data.

Figure 5 is a common Pb diagram that shows the lower end of the scheelite trend line from Figure 4. In this diagram, Pb isotopic results are plotted from various sulfide occurrences from central East Greenland (data from Jensen, 1994a), and they are compared to a field including whole-rock Pb isotope data of Caledonian granites (from Hansen and Friderichsen, 1989; Jensen, 1994a; Pedersen, 2000). The correlation line through the scheelite data from this study passes through the granite field, suggesting a common or similar source of Pb. With the exception of sulfide data from Hudson Land and from the Alpefjord area (Fig. 5), sulfide Pb from the other occurrences in central East Greenland is compatible with a Pb source similar or identical to that derived from the granitoids and/or from the metasedimentary rocks. This attests to a close genetic link between the sulfides and scheelite mineralization in this province. Sulfides from the Hudson Land and Alpefjord areas show less radiogenic and more primitive Pb isotopic ratios, pointing to mixing of granite-type Pb with some old Archean-Paleoproterozoic source(s), as indicated by the steep stippled tie line (2) in Figure 5.

Sm-Nd isotopic results of the scheelites are presented in Table 5, and data are plotted in an isochron diagram (Fig. 6). All 13 samples define an errorchron of 373 ± 85 Ma (MSWD = 19), which, within error, is identical to the intrusion ages of the Caledonian granites (Kalsbeek et al., 2000, 2001). If three outlier samples are omitted, the correlation slope indicates an age corresponding to 382 ± 39 Ma, which is apparently younger than the intrusion age of the granites. The lack of corresponding data from granites in this area, from the Krummedal metasedimentary rocks, and from the Eleonore Bay Supergroup prevents a direct comparison between the initial $^{143}Nd/^{144}Nd$ ratio of 0.51164 ± 0.00011 defined by the scheelites and these "reservoirs." Sm and Nd concentrations in the analyzed scheelites vary strongly, and values are in the lower ppm to tens of ppm range. The $^{147}Sm/^{144}Nd$ ratios of the East Greenland scheelites compare well with suprachondritic ratios of scheelite in many tungsten (W)-Au deposits worldwide (e.g., those classified as "type I" from Mt. Charlotte, Australia; Brugger et al., 2002). However, the concentrations of Nd in the studied scheelites are low compared to scheelite from: (1) Archean epigenetic or lode Au deposits, such as Val d'Or, Quebec, Canada (Nd = 30–385 ppm; Anglin et al., 1996), the Hollinger-McIntyre-

TABLE 4. Pb ISOTOPIC DATA OF SCHEELITE SAMPLES GROUPED AS IN TABLE 2

Sample	$^{206}Pb/^{204}Pb$	±2σ*	$^{207}Pb/^{204}Pb$	±2σ*	$^{208}Pb/^{204}Pb$	±2σ*	r_1[†]	r_2[§]
9502	136.466	0.539	22.041	0.088	39.979	0.160	0.997	0.994
9503	46.013	0.037	17.136	0.015	38.553	0.038	0.974	0.934
9025	18.321	0.009	15.596	0.009	37.998	0.028	0.961	0.929
9026	18.689	0.012	15.669	0.012	38.217	0.034	0.953	0.876
9027	18.682	0.011	15.628	0.011	38.122	0.032	0.957	0.917
8007004	210.361	0.477	26.213	0.062	38.912	0.096	0.974	0.936
7822/2	18.587	0.012	15.653	0.012	38.029	0.038	0.910	0.808
5858	25.089	0.015	15.931	0.011	43.654	0.036	0.967	0.933
5814	28.983	0.022	16.153	0.013	44.172	0.041	0.971	0.943
12095/5	19.258	0.011	15.666	0.011	39.831	0.033	0.962	0.930
12079/3	24.655	0.024	15.893	0.017	41.528	0.050	0.961	0.909
9501	164.684	0.975	23.717	0.141	41.275	0.247	0.996	0.990
9075	19.225	0.012	15.672	0.011	38.701	0.034	0.945	0.881

*Errors are two standard deviations absolute (Ludwig, 2003).
[†]r_1 = $^{206}Pb/^{204}Pb$ vs. $^{207}Pb/^{204}Pb$ error correlation (Ludwig, 2003).
[§]r_2 = $^{206}Pb/^{204}Pb$ vs. $^{208}Pb/^{204}Pb$ error correlation (Ludwig, 2003).

Coniaurum gold deposit, Ontario, Canada (Nd = 51–151 ppm; Bell et al., 1989), Mount Charlotte, Kalgoorlie, Western Australia (Nd = 58–280 ppm; Brugger et al., 2002; Kent et al., 1995); and (2) younger hydrothermal deposits, such as the Paleoproterozoic Omai gold deposit, Guinea Shield (Nd = 23–111 ppm; Voicu et al., 2000). Some published data that show a large spread in REE concentrations in scheelite are from within a single deposit, for example, the Felbertal tungsten deposit, central Alps, Austria (Nd = 0.3–276 ppm; Eichhorn et al., 1997).

DISCUSSION

The Neoproterozoic Eleonore Bay Supergroup sedimentary basin contains various sedimentary copper mineral occurrences (Fig. 2). These were evidently formed by diagenetic processes, as supported by several observations such as: (1) the copper sulfides occur interstitially in the cement to detrital quartz, (2) the sulfides were deposited as coarse-grained laminae in shaly to sandy lithologies, (3) mineralization is associated with

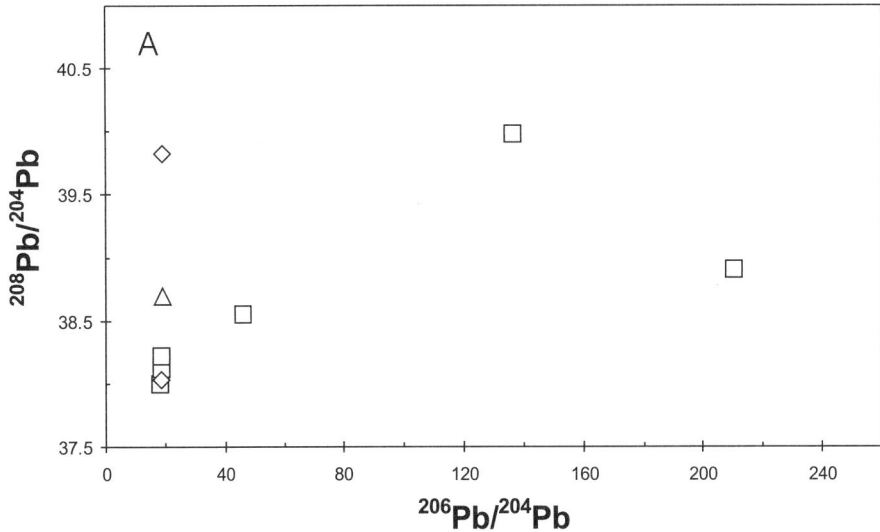

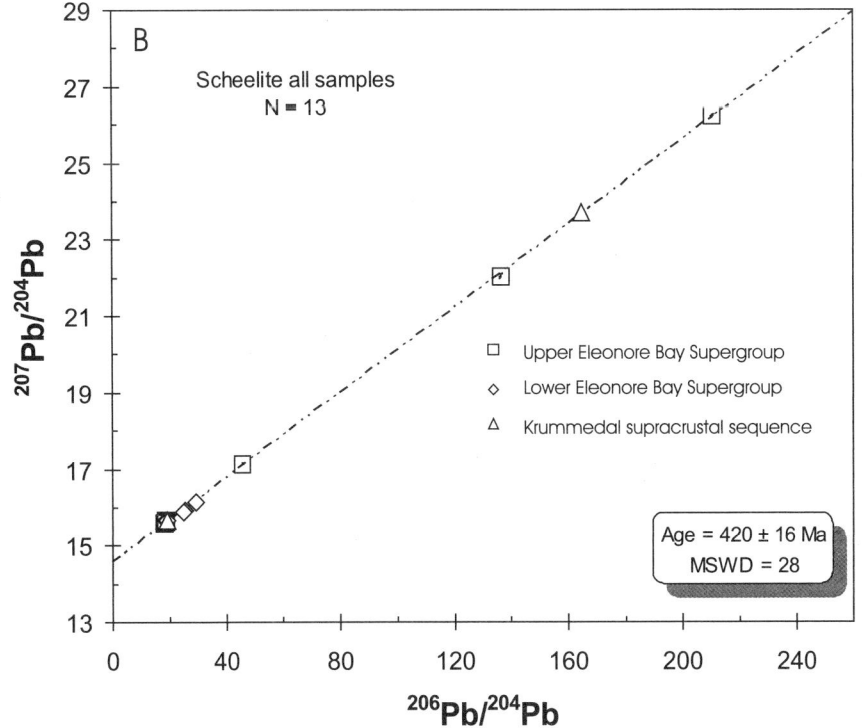

Figure 4. (A) $^{208}Pb/^{204}Pb$ vs. $^{206}Pb/^{204}Pb$ and (B) $^{207}Pb/^{204}Pb$ vs. $^{206}Pb/^{204}Pb$ lead isotope compositions of scheelite separates. MSWD—mean square of weighted deviates.

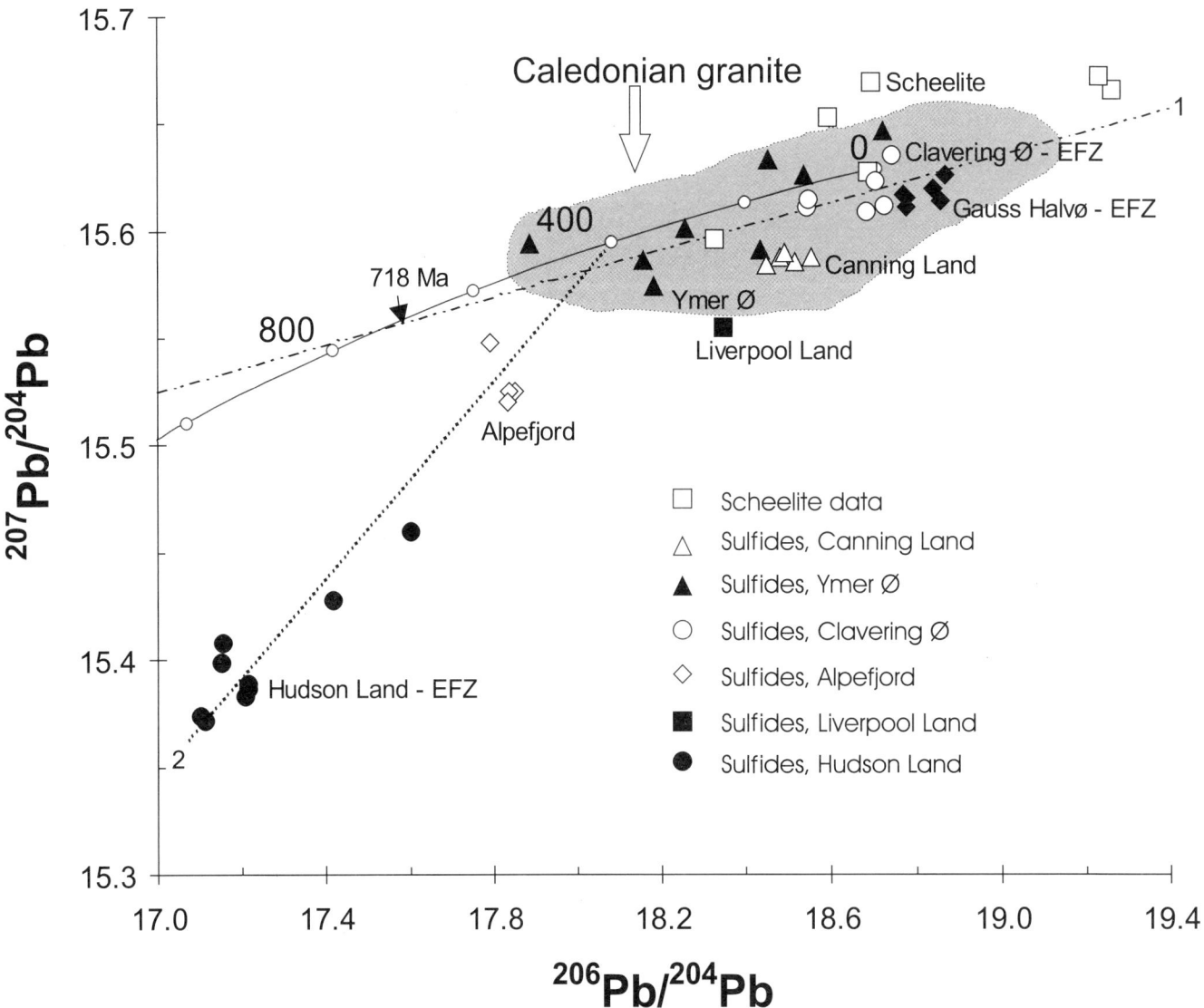

Figure 5. ^{206}Pb/^{204}Pb and ^{207}Pb/^{204}Pb ratios of scheelite and sulfides. 1—trend line through scheelite data from Figure 4; 2—mixing line; the shaded area is the outline of Pb isotopic data of Caledonian granites. Scheelite data are from this study, and sulfide and granite data are from Hansen and Friderichsen (1989), Jensen (1994a), and Pedersen (2000). EFZ—eastern fault zone.

"green" lithologies (due to reduced iron) forming part of alternating green and red shales, and (4) the mineralization is often found together with organic matter and/or graphite, and with framboidal pyrite in shales.

The types of copper occurrence in the clastic sediments of the Eleonore Bay Supergroup have many similarities with typical strata-bound copper mineralization, especially from the Central African copper belt. The deposits in the study area are subeconomic.

Pb isotopic ratios of strata-bound copper sulfides from the Eleonore Bay Supergroup (chalcocites) were analyzed by Jensen (1994a). The Pb isotopic ratios are radiogenic, and ^{206}Pb/^{204}Pb ranges from 18.5 to 26.4. A Pb-Pb isochron of 680 ± 65 (MSWD = 0.25) was defined by twelve chalcocite separates (Jensen, 1993). The age is compatible with the depositional age of the upper part of Eleonore Bay Supergroup. The unconformity at the base of the overlying Tillite Group has a Vendian age of 610–570 Ma (Hambrey and Spencer, 1987).

Mineral deposits associated with the fault zone that juxtaposes sedimentary rocks of the Eleonore Bay Supergroup and Devonian continental clastic rocks have some economic potential. This is particularly so on Ymer Ø, where hydrothermal fluids percolating along E-W– and NW-SE–trending high-angle faults have precipitated tungsten-, antimony-, and arsenic-bearing minerals and gold. Harpøth et al. (1986) suggested that the W-Sb mineralization on Ymer Ø was deposited by fluids from a deep-seated Caledonian granite. However, the Pb isotope composition of the Ymer Ø sulfides has been interpreted to result from mixing

TABLE 5. Sm-Nd ISOTOPIC DATA AND CONCENTRATIONS OF SCHEELITE SAMPLES GROUPED AS IN TABLE 2

Sample	Sm (ppm)	Nd (ppm)	$^{147}Sm/^{144}Nd$	2σ abs.	$^{143}Nd/^{144}Nd$	±2σ$_m$ abs.
9502	2.06	6.74	0.1754	0.0035	0.512087	0.000014
9503	4.03	11.11	0.2078	0.0042	0.512166	0.000008
9025	0.80	3.01	0.1533	0.0031	0.511925	0.000015
9026	0.33	1.52	0.1239	0.0025	0.511917	0.000013
9027	0.68	3.00	0.1292	0.0026	0.511930	0.000011
80007004	11.48	17.43	0.3775	0.0075	0.512569	0.000013
7822/2	9.04	23.58	0.2197	0.0044	0.512044	0.000009
5858	11.50	45.85	0.1437	0.0029	0.512027	0.000011
5814	13.59	53.15	0.1465	0.0029	0.512027	0.000012
12095/5	4.37	18.29	0.1368	0.0027	0.511994	0.000007
12079/3	4.99	9.34	0.3062	0.0061	0.512405	0.000023
9501	2.70	7.77	0.1993	0.0040	0.512162	0.000017
9075	10.85	36.54	0.1702	0.0034	0.512172	0.000011

Subscript m ($_m$) signifies mean.

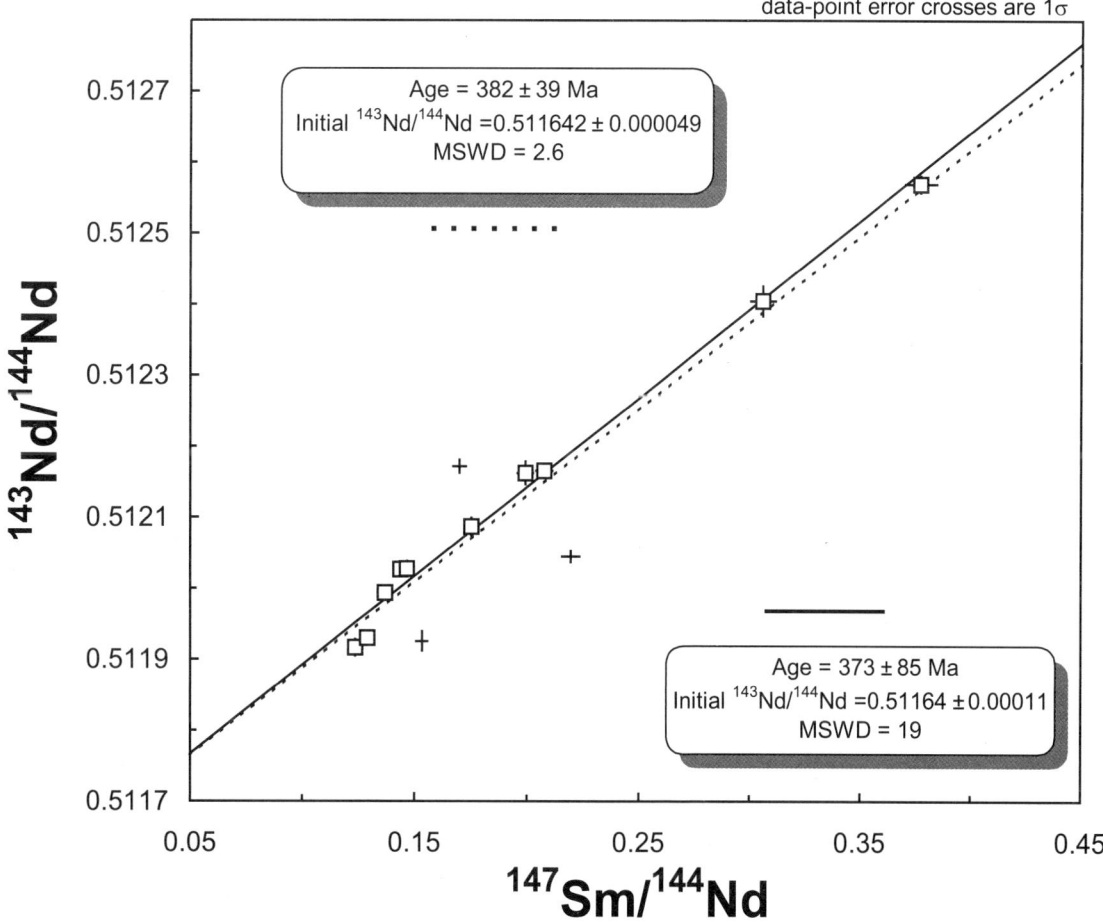

Figure 6. $^{143}Nd/^{144}Nd$ vs. $^{147}Sm/^{144}Nd$ isochron diagram of the scheelite samples. Full line represents all 13 scheelite samples defining an errorchron of 373 ± 85 Ma. The stippled line represents omission of three outlier samples; the correlation slope indicates an age corresponding to 382 ± 39 Ma. MSWD—mean square of weighted deviates.

of Pb from the Eleonore Bay Supergroup host rocks and Pb from the Mesoproterozoic and older basement rocks (Jensen, 1994a). If this interpretation holds, then a deep crustal structure acted as a plumbing system for the mineral fluids. Pb isotopic trends, as in the case of the Ymer Ø sulfides (Fig. 5), are merely mixing arrays and probably have no geochronological significance.

Sr Isotopes

Scheelite $^{87}Sr/^{86}Sr$ ratios (ranging from 0.728 to 0.743) are compatible with the initial $^{87}Sr/^{86}Sr$ range of Caledonian granites (0.723–0.742; Kalsbeek et al., 2001), as well as those values from the Krummedal sequence calculated at the age of Caledonian granites (Sr_{430} values ranging between 0.726 and 0.746; Kalsbeek et al., 2001). This attests to a genetic link between the formation of the Caledonian granites and/or Pb derived from the Mesoproterozoic and Neoproterozoic metasedimentary rocks. The range of observed initial $^{87}Sr/^{86}Sr$ ratios is also compatible with that defined by similar rocks from the southern part of Lyell Land in the same area, where tin and tungsten mineralization is found (Rex et al., 1976).

Based on Sr isotopic results, Kalsbeek et al. (2001) provided convincing evidence that the Caledonian leucogranites formed through anatexis of the pelitic lithologies in the Krummedal sequence. There was no significant contribution to suggest that the leucogranites were formed at depth from the crystalline basement. Inherited zircons in the Caledonian granites show a close similarity in sensitive high-resolution ion microprobe (SHRIMP) U-Pb ages to those of the detrital zircon populations in the Krummedal metasediments. Based on Sr isotopes alone, it is not possible to distinguish a granitic source from a metasedimentary source for Sr in the scheelites, but Sr isotopic data clearly support a Caledonian depositional age for the scheelite in the studied area.

Pb Isotopes

Pb isotopic data can be regressed along a correlation line with a slope corresponding to an age of 420 ± 16 Ma, which is in perfect accordance with the majority of intrusion ages of Caledonian granites in this area (Kalsbeek et al., this volume). The correlation line also passes through the field defined by whole-rock Pb isotopic data for these granites, and it intersects the Stacey and Kramers (1975) Pb growth line at ca. 700 Ma. Taken together, this is interpreted to indicate a close genetic relationship between the emplacement of anatectic granites and tungsten mineralization in the sediments. As is the case for the Sr isotopes, the overlap of Pb isotopes from the granites with those of the metasedimentary rocks means that the ultimate source of Pb (and by inference that of tungsten) in scheelite cannot be determined. However, a combination of trace-element and isotope data points to mixing of material from both sources. With respect to the Pb isotopic signatures of sulfides from a number of occurrences in East Greenland, the Pb data are broadly compatible with the range of whole-rock Pb from the granites. The scatter and subtle differences between the occurrences may be the result of mixing of Pb from the different possible reservoirs, i.e., the crystalline basement, the Neoproterozoic and Mesoproterozoic sedimentary rocks, and the granitoids themselves, as is also implied by the Ymer Ø Pb isotopic data (Jensen, 1994a). Enhanced scatter of scheelite data around the 420 Ma correlation line can be similarly interpreted.

Sm-Nd Isotopes

The errorchron age of 373 ± 85 Ma defined by all scheelite samples (Fig. 6) is within error compared to the intrusion ages of the Caledonian granites. The fact that scheelite Sm-Nd isotopic data from various occurrences define a linear array in an isochron diagram implies that REEs were derived from a rather homogeneous source region. Thus, the REE systematics simply reflect the results from the Sr and Pb isotopes, and the Sm-Nd isotopic data imply a Caledonian age of tungsten mineralization. Enhanced scatter of data points around the correlation line may be due to admixture of variable but local REE sources (i.e., granites, metasediments, and basement rocks). The lack of Nd isotopic data from the metasediments prevents further speculation on the source of REEs in the investigated scheelites. If preliminary Sm-Nd isotopic data for Caledonian granites in the East Milne Land plutonic complex (Kalsbeek et al., this volume) and from the granitoid basement rocks are taken as representative of those in the granitoids from the studied area, then the initial $^{143}Nd/^{144}Nd$ ratios of 0.51164 ± 0.00004 defined by the scheelites would be fully compatible with an average Nd_{430} (average Caledonian granite intrusion age) value of 0.51181 ± 0.00013 for such granitoids. As suggested for the granitoids, Nd isotopes also imply the possibility of an older crustal Nd source mixed into the fluids from which the scheelite was deposited. Such sources may well be compatible with the Archean-Paleoproterozoic basement (Nd_{430} values of 0.51120 ± 0.00024; Kalsbeek et al., this volume), but a significant contribution can be expected from the Neoproterozoic and Mesoproterozoic sedimentary rocks for which data do not exist.

CONCLUSIONS

This multi-isotopic study of scheelite, together with previously published Pb data on sulfides, indicates deposition of mineral occurrences from fluids with mixed sources, mainly from metasedimentary rocks (Mesoproterozoic and Neoproterozoic).

Based on Sr isotopes alone, it is not possible to distinguish granitic from metasedimentary sources for the Sr in the scheelites; however, Sr isotopic data clearly support a Caledonian depositional age for the scheelites in the studied region.

Pb isotopes in scheelite have a similar composition to that of the granites and Mesoproterozoic and Neoproterozoic metasediments. It is not possible to elaborate on the ultimate source of Pb (and by inference of tungsten) in scheelite, although, in summary, the isotope data point to admixing of fluids from both sources during emplacement of the Caledonian granites.

Sm-Nd isotopic data for scheelite define a correlation line indicating an age of 373 ± 85 Ma, which is broadly comparable, within error, to the age of emplacement of Caledonian granites in East Greenland. As for Nd in the granites, the isotopic data indicate the occurrence of an older REE component in the mineralizing fluid, possibly derived from the Archean-Paleoproterozoic crystalline basement or from the Mesoproterozoic or Neoproterozoic metasediments.

The Caledonian granites were formed by anatexis of metasedimentary rocks (Kalsbeek et al., this volume). The multi-isotope data of scheelite presented here are compatible with this scenario and indicate that tungsten may have been mobilized from metasedimentary rocks, and mineralization was driven by the thermal activity associated with the Caledonian granite formation, which also seems to have initiated hydrothermal fluid circulation. The fluids preferentially circulated along faults and fractures and were apparently trapped by Ca-rich sedimentary rocks, which mainly host the scheelite occurrences in the studied region.

ACKNOWLEDGMENTS

The authors acknowledge comments and suggestions from colleagues at the Geological Survey of Denmark and Greenland for comments during the compilation of this paper. The authors would especially like to acknowledge Sven Monrad Jensen for permission to quote Pb isotope data from his thesis (Jensen, 1994a), and Mikael Pedersen for the initial draft of Figure 1. The authors are grateful to David Peate and Marian Lupulescu for their helpful review of the manuscript. John Bailey kindly improved the use of English herein. This paper is published with permission of the Geological Survey of Denmark and Greenland.

REFERENCES CITED

Andresen, A., Hartz, E.H., and Vold, J., 1998, A late orogenic extensional origin for the infracrustal gneiss domes of the East Greenland Caledonides (72–74°N): Tectonophysics, v. 285, p. 353–369, doi: 10.1016/S0040-1951(97)00278-3.

Anglin, C.D., Jonasson, I.R., and Franklin, J.M., 1996, Sm-Nd dating of scheelite and tourmaline: Implications for the genesis of Archean gold deposits, Val d'Or, Canada: Economic Geology and the Bulletin of the Society of Economic Geologists, v. 91, p. 1372–1382.

Bell, K., Anglin, C.D., and Franklin, J.M., 1989, Sm-Nd and Sr-Rb isotope systematics of scheelites: Possible implications for the age and genesis of vein-hosted gold deposits, Western Australia: Geology, v. 17, p. 500–504, doi: 10.1130/0091-7613(1989)017<0500:SNARSI>2.3.CO;2.

Brugger, J., Maas, R., Lahaye, Y., McRae, C., Ghaderi, M., Costa, S., Lambert, D., Bateman, R., and Prince, K., 2002, Origin of Nd-Sr-Pb isotopic variations in single scheelite grains from Archaean gold deposits, Western Australia: Chemical Geology, v. 182, p. 203–225, doi: 10.1016/S0009-2541(01)00290-X.

Eichhorn, R., Höll, R., Jagoutz, E., and Schärer, U., 1997, Dating scheelite stages: A strontium, neodymium, lead approach from the Felbertal tungsten deposit, central Alps, Austria: Geochimica et Cosmochimica Acta, v. 61, p. 5005–5022, doi: 10.1016/S0016-7037(97)00349-9.

Ghisler, M., Jensen, A., Stendal, H., and Urban, H., 1980a, Stratabound copper mineralization in the late Precambrian of the East Greenland Caledonides, in Jankovic, S., and Sillitoe, R.H., eds., European Copper Deposits, Society for Geology Applied to Mineral Deposits Special Publication 1, p. 160–165.

Ghisler, M., Jensen, A.A., Stendal, H., and Urban, H., 1980b, Stratabound scheelite, arsenopyrite and copper sulphide mineralization in the late Precambrian sedimentary succession of the East Greenland Caledonides, in Vokes, F., and Zachrisson, E., eds., Review of Caledonian-Appalachian Stratabound Sulphides: Contribution to IGCP (International Geological Correlation Programme) Project 60, Correlation of Caledonian Stratabound Sulphides.

Gilotti, J.A., and McClelland, W.C., 2008, this volume, Geometry, kinematics, and timing of extensional faulting in the Greenland Caledonides—A synthesis, in Higgins, A.K., Gilotti, J.A., and Smith, M.P., eds., The Greenland Caledonides: Evolution of the Northeast Margin of Laurentia: Geological Society of America Memoir 202, doi: 10.1130/2008.1202(10).

Hallenstein, C.P., and Pedersen, J.L., 1983, Scheelite mineralisation in central East Greenland: Mineralium Deposita, v. 18, p. 315–333, doi: 10.1007/BF00206482.

Hallenstein, C.P., Pedersen, J.L., and Stendal, H., 1981, Exploration for scheelite in East Greenland—A case study: Journal of Geochemical Exploration, v. 15, p. 381–392, doi: 10.1016/0375-6742(81)90074-1.

Haller, J., 1971, Geology of the East Greenland Caledonides: London, Interscience Publishers, 413 p.

Hambrey, M.J., and Spencer, A.M., 1987, Late Precambrian Glaciation of Central East Greenland: Meddelelser om Grønland, v. 19, 50 p.

Hansen, B.T., and Friderichsen, J.D., 1989, The influence of recent lead loss on the interpretation of disturbed U-Pb systems in zircons from igneous rocks in East Greenland: Lithos, v. 23, p. 209–223, doi: 10.1016/0024-4937(89)90006-6.

Harpøth, O., Pedersen, J.L., Schønwandt, H.K., and Thomassen, B., 1986, The Mineral Occurrences of Central East Greenland: Meddelelser om Grønland, Geoscience, v. 17, 139 p.

Hartz, E., and Andresen, A., 1995, Caledonian sole thrust of central East Greenland: A crustal-scale Devonian extensional detachment?: Geology, v. 23, p. 637–640, doi: 10.1130/0091-7613(1995)023<0637:CSTOCE>2.3.CO;2.

Henriksen, N., 2003, Caledonian Orogen, East Greenland 70°–82°N: Geological Map: Copenhagen, Geological Survey of Denmark and Greenland, scale 1:1,000,000.

Henriksen, N., Higgins, A.K., Kalsbeek, F., and Pulvertaft, T.C.R., 2000, Greenland from Archaean to Quaternary: Descriptive Text to the Geological Map of Greenland, 1:2,500,000: Geology of Greenland Survey Bulletin, v. 185, 93 p.

Higgins, A.K., and Soper, N.J., 1994, Structure of the Eleonore Bay Supergroup at Ardencaple Fjord, North-East Greenland, in Higgins, A.K., ed., Geology of North-East Greenland: Rapport Grønlands Geologiske Undersøgelse, v. 162, p. 91–101.

Higgins, A.K., Elvevold, S., Escher, J.C., Frederiksen, K.S., Gilotti, J., Henriksen, N., Jepsen, H.F., Jones, K.A., Kalsbeek, F., Kinny, P.D., Leslie, A.G., Smith, M.P., Thrane, K., and Watt, G.R., 2004, The foreland-propagating thrust architecture of the East Greenland Caledonides: Journal of the Geological Society of London, v. 161, p. 1–18, doi: 10.1144/0016-764903-141.

Jensen, S.M., 1993, Lead isotope composition of stratabound Cu-Pb-Zn-Ba occurrences in Upper Palaeozoic–Mesozoic sediments in East Greenland, in Hach-Alí, F., Torres-Ruiz, J., and Gervilla, J., eds., Current Research in Geology Applied to Ore Deposits: Granada, Universidad de Granada, p. 327–330.

Jensen, S.M., 1994a, Lead Isotope Systematics of Ore Deposits and Mineral Occurrences in East and North-East Greenland [Ph.D. thesis]: Aarhus, University of Aarhus, 161 p.

Jensen, S.M., 1994b, Reconnaissance for mineral occurrences in North-East Greenland (76–78°N): Rapport Grønlands Geologiske Undersøgelse, v. 162, p. 163–s168.

Jensen, S.M., 1998, Tertiary mineralization and magmatism, East Greenland: Lead isotope evidence for remobilization of continental crust: Chemical Geology, v. 150, p. 119–144, doi: 10.1016/S0009-2541(98)00063-1.

Kalsbeek, F., Thrane, K., Nutman, A.P., and Jepsen, H.F., 2000, Late Mesoproterozoic to early Neoproterozoic history of the East Greenland Caledonides: Evidence for Grenvillian orogenesis?: Journal of the Geological Society of London, v. 157, p. 1215–1225.

Kalsbeek, F., Jepsen, H.F., and Nutman, A., 2001, From source migmatites to plutons: Tracking the origin of c. 435 Ma granites in the East Greenland Caledonian orogen: Lithos, v. 57, p. 1–21, doi: 10.1016/S0024-4937(00)00071-2.

Kalsbeek, F., Higgins, A.K., Jepsen, H.F., Frei, R., and Nutman, A.P., 2008, this volume, Granites and granites in the East Greenland Caledonides, in Higgins, A.K., Gilotti, J.A., and Smith, M.P., eds., The Greenland Caledonides: Evolution of the Northeast Margin of Laurentia: Geological Society of America Memoir 202, doi: 10.1130/2008.1202(09).

Kent, A.J.R., Campell, I.H., and McCulloch, M.T., 1995, Sm-Nd systematics of hydrothermal scheelite from the Mount Charlotte mine, Kalgoorlie, Western Australia: An isotopic link between gold mineralization and komatiites: Economic Geology and the Bulletin of the Society of Economic Geologists, v. 90, p. 2329–2335.

Koch, L., 1963, Rapport om den geologiske kortlægning af Østgrønland fra 1926–1958: Betænkning, Minelovskommision for Grønland: Bilag, nr. 10, 340, p. 104–115.

Larsen, P.-H., and Olsen, H., 1991, The Devonian basin project, North-East Greenland—A summary: Rapport Grønlands Geologiske Undersøgelse, v. 152, p. 17–20.

Larsen, P.-H., Olsen, H., and Clack, J.A., 2008, this volume, The Devonian basin in East Greenland—Review of basin evolution and vertebrate assemblages, in Higgins, A.K., Gilotti, J.A., and Smith, M.P., eds., The Greenland Caledonides: Evolution of the Northeast Margin of Laurentia: Geological Society of America Memoir 202, doi: 10.1130/2008.1202(11).

Ludwig, K.R., 2003, User's Manual for Isoplot 3.00: A Geochronological Toolkit for Microsoft Excel: Berkeley Geochron Center, 70 p.

Pedersen, J.L., 1993, Gold-Antimony Exploration at Noa Dal on Ymer Ø, East Greenland: Copenhagen, Internal Report Nunaoil A/S, 22 p.

Pedersen, J.L., and Stendal, H., 1987, Geology and geochemistry of tungsten-antimony vein mineralization on Ymer Ø, East Greenland: Transactions of the Institution of Mining and Metallurgy, Section B–Applied Earth Science, v. 96, p. B31–B36.

Pedersen, M., 1997, Sources of Metals and Mineralising Fluids in Base Metal and Barite Occurrences in the Jameson Land Basin, Ventral East Greenland [Ph.D. thesis]: Copenhagen, University of Copenhagen, 172 p.

Pedersen, M., 2000, Lead isotope signature related to sediment-hosted base metal and barite occurrences in the Jameson Land Basin, central East Greenland: Transactions of the Institution of Mining and Metallurgy, Section B–Applied Earth Science, v. 109, p. B49–B59.

Pedersen, M., and Stendal, H., 1999, Ground Check of Airborne Geophysical Anomalies in Northern Jameson Land, Central East Greenland: Danmarks og Grønlands Geologiske Undersøgelse Rapport, v. 1999/38, 76 p.

Pedersen, M., and Stendal, H., 2000, Mineral occurrences in relation to lineaments and intrusions in Central and North East Greenland—New possibilities?: Transactions of the Institution of Mining and Metallurgy, Section B–Applied Earth Science, v. 109, p. B42–B48.

Rex, D.C., and Gledhill, A., 1974, Reconnaissance geochronology of the infracrustal rocks of Flyvefjord, Scoresby Sund, East Greenland: Bulletin of the Geological Society of Denmark, v. 23, p. 49–54.

Rex, D.C., Gledhill, A.R., and Higgins, A.K., 1976, Precambrian Rb-Sr isochron ages from the crystalline complexes of inner Forsblad Fjord, East Greenland fold belt: Rapport Grønlands Geologiske Undersøgelse, v. 85, p. 122–126.

Sønderholm, M., and Tirsgaard, H., 1993, Lithostratigraphic Framework of the Upper Proterozoic Eleonore Bay Supergroup of East and North-East Greenland: Bulletin Grønlands Geologiske Undersøgelse, v. 167, 38 p.

Stacey, J.S., and Kramers, J.D., 1975, Approximation of terrestrial lead isotope evolution by a two-stage model: Earth and Planetary Science Letters, v. 26, p. 207–221, doi: 10.1016/0012-821X(75)90088-6.

Steiger, R.H., Hansen, B.T., Schuler, C., Bär, M.T., and Henriksen, N., 1979, Polyorogenic nature of the southern Caledonian fold belt in East Greenland: An isotopic age study: The Journal of Geology, v. 87, p. 475–495.

Stendal, H., 1979, Geochemical copper prospecting by use of inorganic drainage sediment sampling in central East Greenland: Transactions of the Institution of Mining and Metallurgy, Section B–Applied Earth Science, v. 88, p. B1–B4.

Stendal, H., 1980, Distribution of metals and heavy minerals in drainage sediments in late Precambrian rocks of central East Greenland (72–74°N): Bulletin Grønlands Geologiske Undersøkelse, v. 360, p. 99–111.

Stendal, H., 1981, Geochemistry and genesis of arsenopyrite mineralization in late Precambrian sediments in central East Greenland: Transactions of the Institution of Mining and Metallurgy, Section B–Applied Earth Science, v. 91, p. B187–B191.

Stendal, H., 1999, Mineralisation follow-up in fault zones, Hudson Land, Gauss Halvø and Steno Land, North-East Greenland, in Higgins, A.K., and Frederiksen, K.S., eds., Geology of East Greenland 72°–75°, Mainly Caledonian: Preliminary Reports from the 1998 Expedition: Danmarks og Grønlands Geologiske Undersøgelse Rapport, v. 1999/19, p. 201–205.

Stendal, H., and Ghisler, M., 1984, Strata-bound copper sulfide and nonstratabound arsenopyrite and base metal mineralization in the Caledonides of East Greenland; a review: Economic Geology and the Bulletin of the Society of Economic Geologists, v. 79, p. 1574–1584.

Stendal, H., and Hock, M., 1981, Geochemical prospecting for stratabound mineralization in late Precambrian sediments of East Greenland (72–74°N): Journal of Geochemical Exploration, v. 15, p. 261–269, doi: 10.1016/0375-6742(81)90067-4.

Stendal, H., and Wendorff, M.B., 1998, Mineral resource investigations in the Caledonian crystalline complexes between Andrée Land and Lyell Land, East Greenland, in Higgins, A.K., and Frederiksen, K.S., eds., Caledonian Geology of East Greenland 72°–74°N: Preliminary Reports from the 1997 Expedition: Danmarks og Grønlands Geologiske Undersøgelse Rapport, v. 1998/28, p. 95–105.

Stendal, H., Pedersen, M.R., and Sørensen, L.L., 1999, Mineralisation studies on Clavering Ø, North-East Greenland, in Higgins, A.K., and Frederiksen, K.S., eds., Geology of East Greenland 72°–75°N, Mainly Caledonian: Preliminary Reports from the 1998 Expedition: Danmarks og Grønlands Geologiske Undersøgelse Rapport, v. 1999/19, p. 195–200.

Todt, W., Cliff, R.A., Hanser, A., and Hofmann, A.W., 1996, Re-calibration of NBS lead standards using a $^{202}Pb+ ^{205}Pb$ double spike, in Basu, A., and Hart, S., eds., Earth Processes: Reading the Isotopic Code: American Geophysical Union Geophysical Monograph 95, p. 429–437.

Voicu, G., Bardoux, M., Stevenson, R., and Jebrak, M., 2000, Nd and Sr isotope study of hydrothermal scheelite and host rocks at Omai, Guiana shield: Implications for ore fluid source and flow path during the formation of orogenic deposits: Mineralium Deposita, v. 35, p. 302–314, doi: 10.1007/s001260050243.

MANUSCRIPT ACCEPTED BY THE SOCIETY 14 JANUARY 2008

Laurentian margin evolution and the Caledonian orogeny— A template for Scotland and East Greenland

A. Graham Leslie*
Martin Smith
British Geological Survey, Murchison House, Edinburgh, EH9 3LA, UK

N.J. Soper
Gams Bank, Threshfield, North Yorkshire, BD23 5NP, UK

ABSTRACT

The orthotectonic Scottish Caledonides constitute only a small fragment of the Neoproterozoic to Paleozoic margin of Laurentia, albeit one which lies at a prominent bend in that margin. Sequences exposed in the Scottish outcrop include Mesoproterozoic, Neoproterozoic, and Cambrian-Ordovician strata that record sedimentation, volcanism, and deformation related to the latter stages of the amalgamation of Rodinia, the subsequent breakout of Laurentia, and growth of the Iapetus Ocean. Metamorphic and tectonic overprints then record the destruction of that ocean through Ordovician arc accretion and mid-to-late Silurian collision of Laurentia, Baltica, and Avalonia and the final closure of Iapetus by end-Silurian time. New isotopic data and recent advances in the understanding of the late Mesoproterozoic (Stenian) to Cambrian-Ordovician stratigraphic framework now better constrain the sequence and timing of events across the "Scottish Corner" and invite a dynamic comparison with the current research into the East Greenland Caledonides summarized in this volume. Although many broad similarities exist, the comparisons described here reveal for the first time a number of significant contrasts in the spatial arrangement of depocenters, location of rifting, and patterns and timing of magmatism, metamorphism, and contractional deformation. This expanded understanding of the late Neoproterozoic evolution of these adjacent sectors of Laurentia provides an important basis for reconstructions of the subsequent early Paleozoic Caledonian orogenic evolution of the present North Atlantic region.

Keywords: Laurentia, Caledonian, Scotland, Greenland, orogenic evolution.

INTRODUCTION AND REGIONAL COMPARISONS

The Caledonides of East Greenland disappear southward beneath Paleogene flood basalts at Scoresby Sund (70°N). Thereafter, the most proximal sector of the preserved Laurentian margin at the onset of Silurian (Scandian) collision of East Greenland with Baltica is the Scottish Highlands and northern parts of Ireland. Palinspastic reconstructions (Dickin, 1992; Cambridge PalaeoMap, 1998) indicate that during early Paleozoic time, Scotland may have lain as little as 500 km to the south of central East Greenland (Fig. 1).

With a present-day along-strike section of ~600 km, and as one of the most intensively studied orogens in the world (Strachan et al., 2002), Scotland shares many aspects of Laurentian geology

*agle@bgs.ac.uk

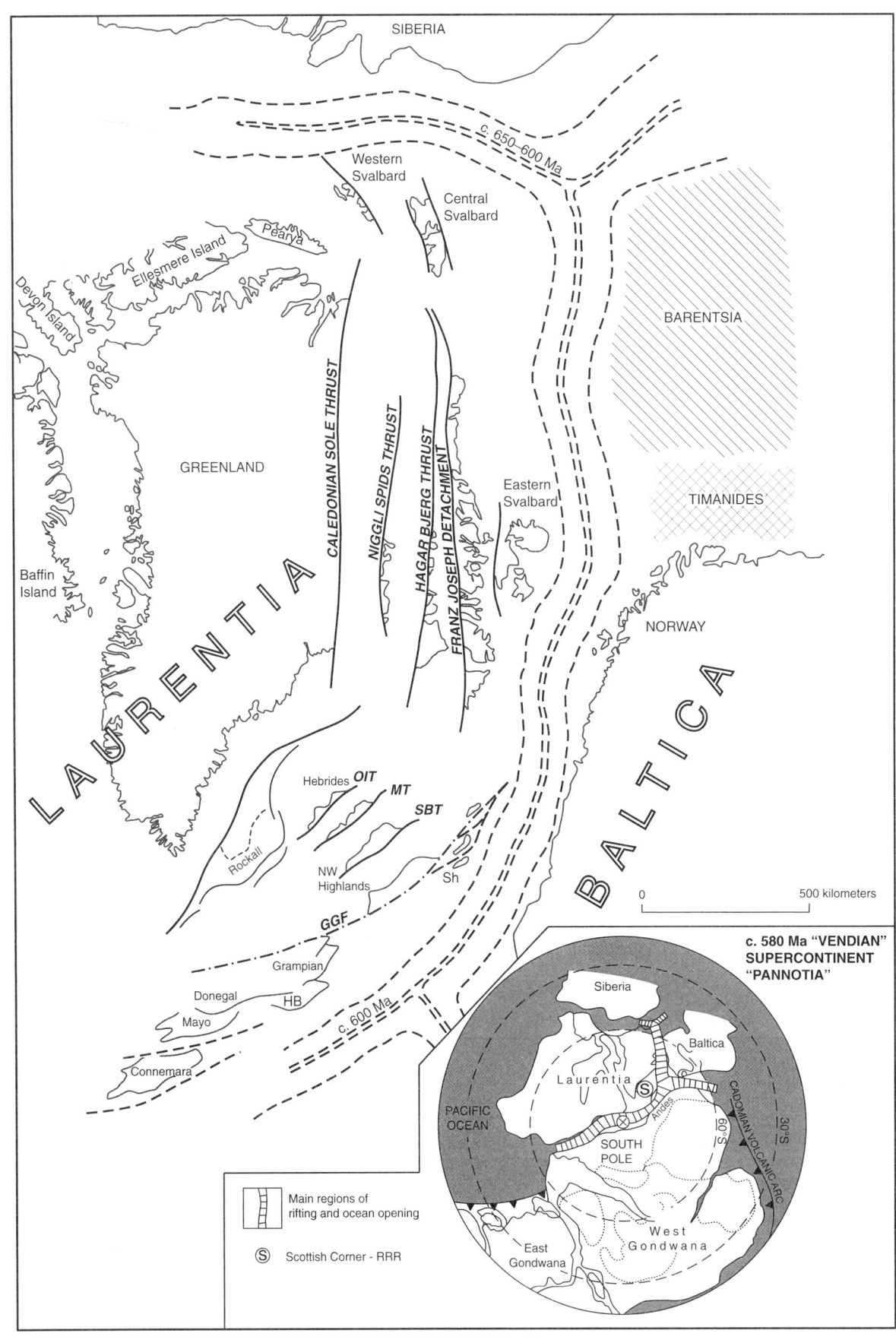

Figure 1. Latest Neoproterozoic template for the various segments of the Laurentian margin discussed in this text. The line of Iapetan separation is shown, along with suggested positions for Baltica, Barentsia, and Siberia. The locations of western, central, and eastern Svalbard are modified after Soper et al. (1992) and Gee and Teben'kov (2004). The inset projection shows location of the Scottish "Promontory" (S) within a wider global continental plate reconstruction for Pannotia; note the alternative orientation shown for Baltica. GGF—Great Glen fault; HB—Highland Border; MT—Moine thrust; OIT—Outer Isles thrust; SBT—Sgurr Beag thrust; Sh—Shetland.

(Fig. 2) with the 1300-km-long East Greenland Caledonides. Nevertheless, several fundamental problems still lack a definitive interpretation despite thousands of publications detailing decades of field and laboratory investigations into the perplexing architecture and history of the Scottish Caledonides. The basin architecture accommodating each of the Stoer and Torridon Groups (the "Torridonian") and the Moine Supergroup has not been resolved; whether or not parts of these sequences might be correlated within a single depositional system is still a matter of debate. The cause and spatial extent of the Knoydartian tectonothermal event (or events) are likewise unresolved, and we cannot as yet define the age and nature of the base of the Dalradian Supergroup. The glacigenic deposits within the Dalradian Supergroup (tillites) are not yet age-constrained or definitively placed within a global sequence. These problems arise, at least in part, from the difficulty in dealing with the lack of critical exposure in an upland glaciated terrain where there is an extensive cover of superficial deposits.

Thus, there is much to envy in the geological vistas of the fjords and mountains of East Greenland. With the completion of the 1:500,000 mapping and research program by the Geological Survey of Denmark and Greenland (GEUS), it is therefore timely to explore the similarities and contrasts between these two sectors and their roles in the Precambrian to early Paleozoic evolution of the Laurentian margin. We note, however, that while with our broadened perspective we may have moved on from the vestige of a beginning in our understanding of this part of the Laurentian margin, there seems to be no prospect of an end in view!

In this account, we take the stance that the East Greenland and Scottish Caledonides record a shared Neoproterozoic to lower Paleozoic tectono-stratigraphic and tectono-metamorphic Laurentian geological history, albeit with a diachroneity and difference in detail that reflect their individual locations on the margin. An alternative "mobilistic" model, which is not further discussed here but involves a protracted history of major lateral movements and the amalgamation of terranes of quite separate affinities, has been proposed for the Scottish Caledonides (Bluck et al., 1997).

The late Neoproterozoic (ICS [International Commission on Stratigraphy] time scale, Gradstein et al., 2004) position of Scotland has been previously interpreted as occupying a stable promontory in the Archean to Paleoproterozoic Hebridean Shield (Dalziel and Soper, 2001) and ultimately close to an Ediacaran (Vendian) RRR junction (inset to Fig. 1) (Soper, 1994a). The most recent analysis of the paleomagnetic and geological constraints argues that the western Scandinavian margin of Baltica faced the eastern Greenland margin of Laurentia in its right-way-up orientation (Cawood and Pisarevsky, 2006). The reconstruction in Figure 1 adopts this configuration and includes a schematic restoration on the numerous major thrusts identified in the Caledonides of East Greenland and in the north of Scotland. On this basis, Scotland does lie on a corner in the Laurentian margin and potentially in the vicinity of a RRR junction, but the concept of a "Scottish promontory" is not sustained.

In contrast, a general absence of volcanic activity or other evidence of extensive or rapid upper Neoproterozoic extension suggests that the East Greenland and Eastern Svalbard margin of Laurentia lay distant from such an Iapetan RRR locus. East Greenland was also apparently isolated from the active spreading junction that affected northern Greenland, western/central Svalbard, Scandinavia, and Siberia as recorded by ca. 650–600 Ma igneous mafic activity (Gromet and Gee, 1997; Bingen et al., 1998) (Fig. 1). Earlier attempts at rifting, possibly signaling the initiation of continental breakup, are recorded in the ca. 720 Ma Coronation mafic dike swarm of Baffin Island and West Greenland (Shellnutt et al., 2004) and ca. 700 Ma mafic sills in NE Svalbard (Johansson et al., 2004). Abundant mafic volcanics also mark extension and rifting on the Appalachian sector at this time (Bailey and Tollo, 1998; Tollo and Hutson, 1996; Tollo et al., 2004). Such activity is absent from central East Greenland. Thus, over many millions of years, the Neoproterozoic successions of Greenland perhaps record relatively quiescent deposition on a continental margin lying between migratory RRR junctions throughout the evolution of the Iapetus Ocean.

For reference we delineate, in the inset to Figure 2, those fault-bounded crustal terranes identified in Scotland as having a distinctive geological history (cf. Strachan et al., 2002). Where practicable, we will refer to geographical regions (Fig. 1) rather than specific terranes; fault-bounded terranes are not defined in the East Greenland Caledonides, and the full status of some of the Scottish "terrane boundaries" outlined in the inset map (Fig. 2) is still a matter of debate. In general terms, the Neoproterozoic–Ordovician rocks of the northwest Scottish Highlands preserve a record of the Laurentian autochthonous-parautochthonous foreland (Figs. 2 and 3A). These successions correspond stratigraphically and chronologically with rocks of the Laurentian foreland succession preserved in the nunatak region of central East Greenland, the parautochthonous belt of western Kronprins Christian Land in eastern North Greenland, and the parautochthonous successions exposed in tectonic windows along the length of the East Greenland Caledonides (Figs. 3B and 3C). Elements of the geology of the Northern and Grampian Highlands of Scotland may be compared with similar sequences in the Niggli Spids thrust sheet, Hagar Bjerg thrust sheet, and Franz Joseph allochthon of central East Greenland, to the eastern hinterland of Dronning Louise Land, and to the Vandredalen thrust sheet of Kronprins Christian Land (Figs. 3A, 3B, and 3C).

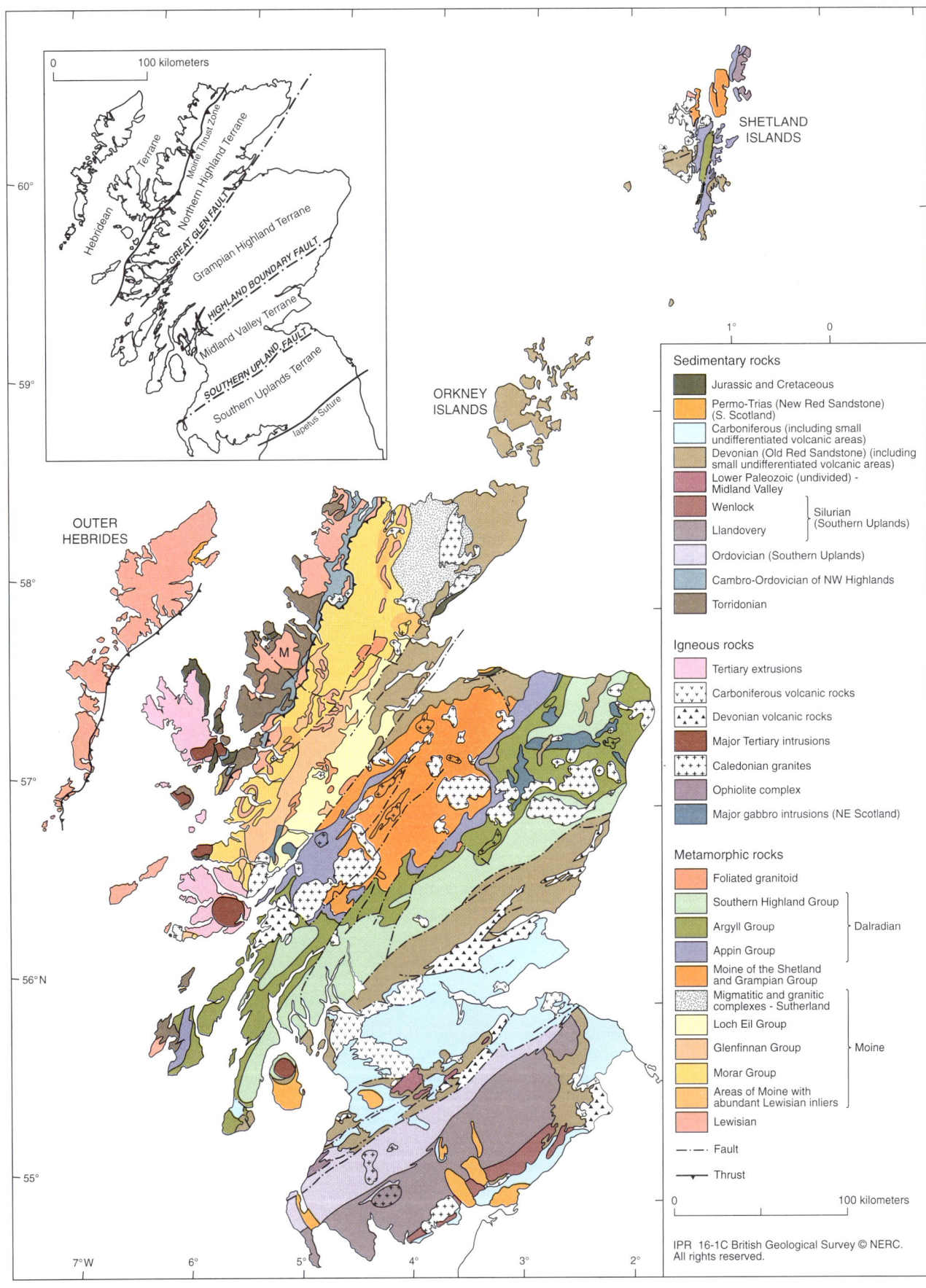

Figure 2. Simplified geological map of Scotland; small areas of outcrop and dike swarms omitted. The inset shows the major terranes of Scotland after Strachan et al. (2002). M—Loch Maree Group. Used with permission from the British Geological Survey.

Bearing in mind the relative positions and regional structural trends (Fig. 1), it is then tempting to link the individual major fault structures in Scotland (e.g., Moine thrust, Great Glen fault, Highland Boundary fault) with the major bounding structures of the East Greenland thrust sheets (e.g., Caledonian sole thrust, Fjord region fault, Western fault zone). That said, the differences in present knowledge and understanding of the history of these fault zones mean that such correlations remain speculative.

While such broad similarities invite comparison, intriguing contrasts also exist between Scotland and East Greenland. For example, the thick siliciclastic contemporaneous (early Neoproterozoic) sequences of the Krummedal succession of central East Greenland (Higgins, 1988) and the Moine Supergroup of the Northern Highlands in Scotland (Figs. 3A and 3B) share a similar provenance (Friend et al., 2003; Kalsbeek et al., 2000; Watt et al., 2000) and together, are remarkable for evidence of repeated high-pressure, high-temperature metamorphism. While the metasedimentary rocks from the Northern and Grampian Highlands record evidence for a series of Neoproterozoic tectonothermal events between 820 Ma and 730 Ma (Fig. 4), these are unknown in Greenland. Instead, the Krummedal succession was there affected by a single high-grade metamorphic event that culminated in the generation of voluminous S-type augen granite, at ca. 910 Ma (Leslie and Nutman, 2003).

The younger and well-preserved mid- to late Neoproterozoic and early Cambrian sequences of the Grampian Highlands reveal a history of repeated uplift, rifting, and complex internal basin architecture. Mafic volcanic rocks developed at several levels in the evolving depositional pile, and the transition from rift to drift in the developing Iapetus Ocean occurred at ca. 610–600 Ma (Figs. 3A and 4). In contrast, the succession of mid- to late Neoproterozoic sediments in central East Greenland, the Franz Joseph allochthon, although extremely thick, records sustained subsidence but no active rifting. An apparent depositional thickness of over 14 km of sediment is assigned to the Eleonore Bay Supergroup alone, and a further 1 km is assigned to the Ediacaran Tillite Group (Sønderholm et al., this volume); mafic volcanic rocks are conspicuous by their absence (Fig. 3B). While there may be some limited evidence of stretching and rift shoulder uplift farther inboard on the restored margin (Leslie and Higgins, 1998, this volume), major extension and rifting are only clearly evident much farther north in eastern North Greenland where the late Neoproterozoic Hekla Sund Basin rift-sag sequence dominates the geology (Fig. 3C; Higgins et al., 2001b). Neoproterozoic tillites, if originally deposited, are now absent on the foreland in Scotland, whereas they are present in significant thicknesses, in *both* the parautochthonous foreland windows and the fjord region allochthon in East Greenland.

The ensuing Neoproterozoic and early Paleozoic deformation and metamorphism across Scotland and Greenland record the final amalgamation then breakup of the ancient supercontinent of Rodinia, followed by convergence, and eventual collision of Baltica with the East Greenland sector of the Laurentian margin in the Caledonian orogeny (Fig. 4). Grampian (Ordovician) orogenesis and arc accretion dominate the structural framework of the Grampian and parts of the Northern Highlands in Scotland but are apparently only expressed in the southernmost extremity of the East Greenland Caledonides. Conversely, the subsequent Scandian (Silurian) orogenesis, which documents the final collision and docking of Baltica with Laurentia, dominates the structure in only the Northern Highlands in Scotland but is the pervasive control on Caledonian structural chronology and architecture throughout East Greenland.

In this chapter, we present a synopsis of the Archean to lower Paleozoic geology of the "Scottish Corner" of Laurentia as a series of time slices summarized in the tectono-stratigraphic template presented in Figures 3 and 4. In this framework, we explore and test the key comparisons and contrasts between Scotland and East Greenland, drawing particular attention to the locations of rifting and spatial arrangements of depocenters, as well as to the key orogenic events. We conclude by presenting a dynamic synthesis of Iapetan rifting and Caledonian orogenesis in the Scottish Caledonides, and, while recognizing that no consensus can currently be reached, we contend that the combined geology of Scotland and Greenland provides real constraints upon the Neoproterozoic to early Paleozoic evolution of Laurentia.

ASSEMBLY OF A STABLE BASEMENT: PRE-STENIAN (1200 Ma) HISTORY

The oldest rocks in Scotland are the crystalline basement rocks of the Lewisian Complex in the Hebrides and Northwest Highlands (Fig. 2; Plates 1A and 1B). This complex is exposed along the mainland coastal strip in the footwall of the Moine thrust zone (Fig. 2; Plate 1B) and extends westwards across the Outer Hebrides and to a broad region (Rockall) on the edge of the UK continental shelf (Dickin, 1992). East of the Moine thrust zone, geophysical data (Trewin and Rollin, 2002) imply that Lewisian, or similar, rocks underlie the Moine rocks in Scotland at least as far as the trace of the Great Glen fault (Fig. 2). The Lewisian Complex is, and has been, the subject of intense scrutiny. Only a brief summary is provided here, and the reader is referred to the comprehensive summary of Park et al. (2002) for further details and the recent review of Park (2005).

The oldest rocks are Archean-age granulite-facies orthogneiss (the Scourian), which had already been reworked by two tectonothermal events prior to the intrusion of the basic and ultrabasic magmas of the Scourie dike suite in mid-Paleoproterozoic time (Park et al., 2002; Park, 2005). The Scourian rocks are typically gray, banded trondhjemite-tonalite-granodiorite (TTG) orthogneisses with rare metasedimentary enclaves; basic enclaves may represent relict oceanic crust (Plate 1B). Studies of

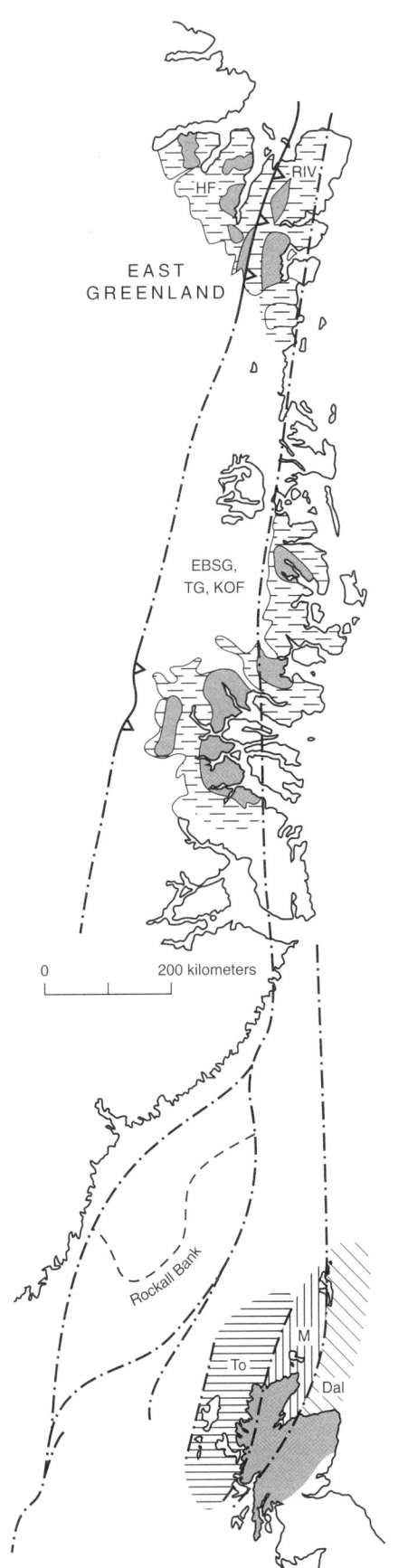
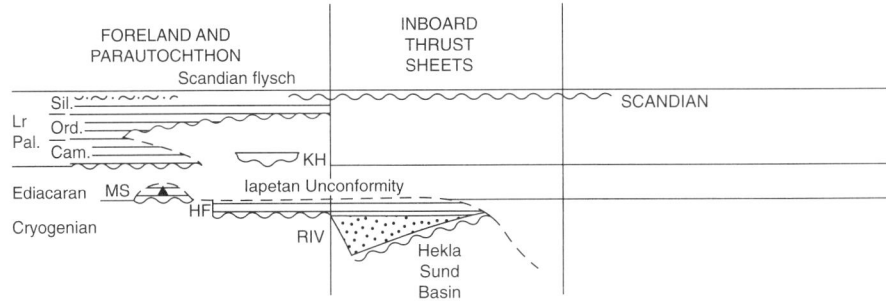

Figure 3. Comparison of the Mesoproterozoic to Ordovician successions of (A) Scotland and (B–C) East Greenland (shaded areas on the map). The successions are shown with reference to their relative position on the inboard to outboard location in the evolving continental margin of Laurentia. The timing of orogenesis (Scandian, Grampian, etc.) in each sector is shown schematically. Ap—Appin Group; Arg—Argyll Group; D—Durness Group; Dal—Dalradian; D, GB—Dava Glen Banchor succession; EBSG—Eleonore Bay Supergroup; Gr—Grampian Group; HF—Hagen Fjord Group; KH—Kap Holbæk Formation; KOF—Kong Oscar Fjord Group; Kr—Krummedal; M—Moine Supergroup; MF—Målebjerg Formation; MS—Moræneso Formation; RIV—Rivieradal Group; S—Slottet Formation; SH—Southern Highland Group; St—Stoer Group; TG—Tillite Group; To—Torridon Group; ZS—Zebra "series."

the tectonothermal history in the Lewisian Complex have previously assumed that these rocks are broadly correlatable across the Hebrides and Northwest Highlands. However, recent work by Friend and Kinny (2001), Kinny et al. (2005), and Park (2005) has proposed a series of discrete Archean blocks that amalgamated during the Paleoproterozoic, each with a more complicated, but as yet unresolved, early history. Orthogneisses of the mainland central block yield 3.03–2.96 Ga protolith ages and record granulite-facies metamorphism (the Badcallian event) at ca. 2.5 Ga (Whitehouse, 1989; Friend and Kinny, 1995; Corfu et al., 1998). Protolith ages in the northern block range from 2.84 to 2.68 Ga, and in the southern block, ages range from 2.82 to 2.73 Ga (Kinny and Friend, 1997; Corfu et al., 1998).

The ensuing Inverian event postdated a suite of pegmatites (2.49–2.48 Ga) and is marked by retrogression to amphibolite facies and the formation of shear zones that predate the earliest Scourian dikes (Corfu et al., 1998). The Scourian dikes were intruded, predominantly as quartz-dolerites, over a considerable period of time, and the main swarm was emplaced ca. 2.42 Ga (Heaman and Tarney, 1989).

Younger supracrustal strata are represented in the Loch Maree Group, which consists of two structural belts of metasedimentary and meta-igneous rocks (M on Fig. 2). The Loch Maree succession accumulated at ca. 2.0 Ga and includes graywacke, quartzite, carbonate rock, and banded iron formation, along with sheets of amphibolite. Park et al. (2001) interpreted these as graywackes that accumulated close to a continental source; the rare earth element (REE) chemistry of the amphibolite suggests an origin as oceanic plateau lava and subsidiary primitive island-arc basalt. The supracrustal rocks are cut by orthogneiss that yields U-Pb zircon ages of 1.9 Ga, which are interpreted to be the age of emplacement (Park et al., 2001). These crosscutting gneisses have a primitive arc signature and thus provide further evidence of a Paleoproterozoic magmatic arc and subduction of oceanic material (Park et al., 2001).

The Laxfordian refers to tectonothermal events that modify the Scourie dikes. An early phase, dated at 1.86–1.63 Ga, may be related to a network of low-angle transpressional shear zones (Coward and Park, 1987). Later-phase, steep shear zones and refolding in upright structures were accompanied by retrogression to greenschist-facies conditions.

The rocks of the Lewisian Complex exposed on the islands of the Outer Hebrides (Fig. 2) are broadly similar to the northern and southern mainland blocks but do show some significant differences. There are, for example, far fewer Scourie dikes, and although the supracrustal rocks on South Harris have similar lithologies to the Loch Maree Group, they yield younger U-Pb ages, younger than ca. 1.9 Ga (Whitehouse and Bridgwater, 1999). Granulite-facies metamorphism dated at 1.9–1.8 Ga is dominant. These differences led Friend and Kinny (2001) to argue that the Outer Isles basement has more affinity with East Greenland than with the adjacent Scottish mainland.

The pattern of protolith ages in the basement complexes in East Greenland also reflects this widespread Paleoproterozoic activity. Archean protolith (2.8–2.7 Ga) is also evident in the gneisses in Gletscherland and Nathorst Land, which are broadly equivalent, therefore, to the Archean gneisses of the Ammassalik region (Rex and Gledhill, 1981; Thrane, 2002; F. Kalsbeek, 2005, personal commun.) (Fig. 1 in Higgins and Leslie, this volume). Paleoproterozoic (ca. 1.9 Ga) gneissose and granitoid rocks that occur farther north in East Greenland (eastern Frænkel Land and Suess Land; Fig. 1 in Higgins and Leslie, this volume) were accreted around 2.0 Ga (Kalsbeek et al., 1993) and are in many respects similar to the somewhat younger (and better preserved) Ketilidian orogenic belt in South Greenland.

Reconstructions of the Paleoproterozoic belts and Archean cratons of the North Atlantic region (Buchan et al., 2000) show the Lewisian Complex as part of the Paleoproterozoic internal belt linking the eastern Churchill province of the Canadian Shield, the Nagssugtoqidian of Greenland, and the Lapland-Kola belt of Scandinavia (Fig. 5). Wright et al. (1973), Myers (1987), Kalsbeek et al. (1993), and Park (1994) have highlighted the many similarities of the Lewisian Complex to the Nagssugtoqidian, or Ammassalik belt, of South-East Greenland. The latter consists of reworked Archean granitoid gneisses cut by mafic dikes and other intrusions (Kalsbeek, 1989). The central part of the Ammassalik belt contains abundant metasediments with a depositional age of ca. 2.0–1.9 Ga (Bridgwater et al., 1996) cut by subduction-related, 1.9 Ga calc-alkaline granitoid intrusions (Kalsbeek et al., 1993) and by deformed and metamorphosed intrusions of the 1.89 Ga Ammassalik Complex. The Charcot Land and Eleonore Sø supracrustal successions (Higgins et al., 2001a) of central East Greenland share many similarities with the Loch Maree Group rocks of Scotland and should be regarded as directly comparable.

By ca. 1.9 Ga, the ancient basement of Scotland (Lewisian Complex), Greenland (Nagssugtoqidian), and the coeval Lapland-Kola belt formed one continuous accretionary belt composed of various Archean cratonic components welded together during the assembly of Laurentia and Baltica (Fig. 5; Buchan et al., 2000; Dickin, 1992). By ca. 1.84 Ga, calc-alkaline magmatism was concentrated along a new active margin represented by the ca. 1.9–1.85 Ga Makkovik/Ketilidian belt of Labrador and

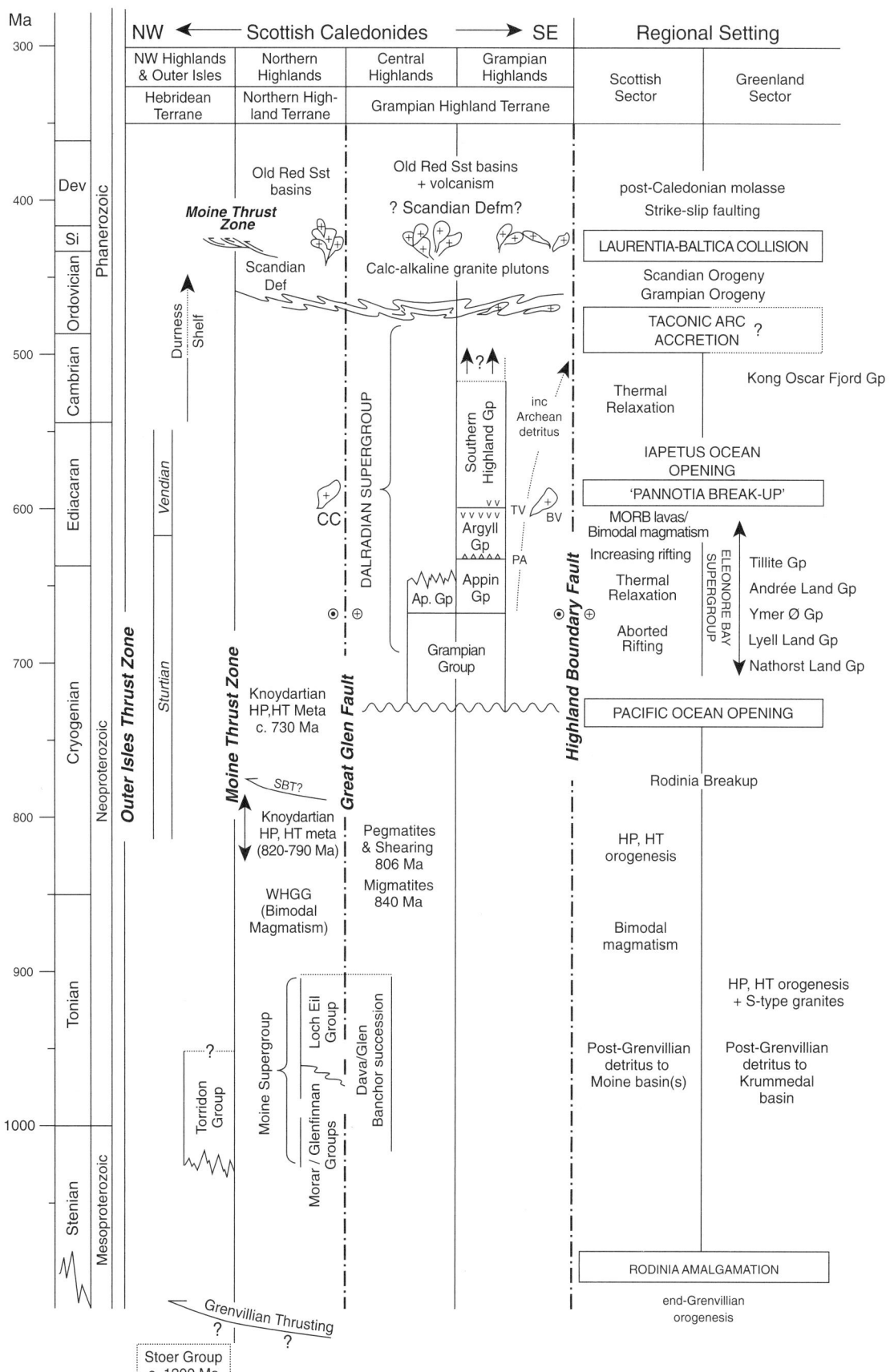

Figure 4. Record of the geological evolution of the Scottish sector of Laurentia from the post-Grenvillian amalgamation of Rodinia until post-Caledonian time. The principal comparable events recorded on the East Greenland sector are included here for reference, but see greater detail included in Figure 6. BV—Ben Vuirich Granite; CC—Carn Chuinneag Granite; PA—Port Askaig Tillite; SBT—Sgurr Beag thrust; TV—Tayvallich volcanics; WHGG—West Highland Granite Gneiss; MORB—mid-ocean-ridge basalt; HP, HT—high pressure, high temperature. Sst—Sandstone.

South Greenland and the younger (1.85–1.50 Ga) Labradorian-Gothian belt of NE Canada and SW Scandinavia. In Scotland, the Labradorian-Gothian belt is represented by the Rhinns Complex of Islay (Muir et al., 1994), part of a largely submerged area (the Malin block) of juvenile Proterozoic crust that formed ca. 1.78 Ga (Marcantonio et al., 1988).

MESOPROTEROZOIC BASIN EVOLUTION (Ca. 1200–900 Ma)

There is a paucity of evidence to define the regional extent of the Mesoproterozoic to early Neoproterozoic Grenvillian global orogenic event (ca. 1200–950 Ma) in Scotland and East Greenland. Evidence has yet to be determined in Greenland and, in the Western Highlands of Scotland, is only locally recorded in the Glenelg Inlier, where eclogite- and amphibolite-facies metamorphism is dated at ca. 1000 Ma (Sanders et al., 1984; Storey et al., 2004). This time interval is marked in both East Greenland and Scotland by the accumulation of thick siliciclastic sedimentary successions, and the degree to which syndepositional crustal extension and active rifting were involved in these accumulations remains unclear.

The Stoer and Torridon Groups ("Torridonian")

In Scotland, the Lewisian Complex had by late Proterozoic time been deeply eroded and buried by an unconformable succession of red arkosic sandstones, informally referred to as the "Torridonian" (Figs. 2 and 3; Plate 1C). These sediments are facies equivalents of similar red-bed successions in Labrador and the Great Lakes area, and they are interpreted to have occupied rifts that developed peripherally to the eroding and maturing Grenville orogenic belt that extended across Rodinia (Winchester, 1988; Turnbull et al., 1996; Gower, 1988). While a number of lines of sedimentologic evidence support a rift-dominated setting for these deposits in Scotland (Stewart, 2002), evidence of related volcanism is restricted to undersaturated mafic volcaniclastic detritus present in the Mesoproterozoic (ca. 1200 Ma) Stoer Group (Stewart, 1991). The early Neoproterozoic (ca. 1000–900 Ma) Torridon Group lacks any volcanic association, is unconformable on the Stoer Group, and the relative ages are in accord with paleomagnetic evidence that shows a 90° change in polarity across the unconformity (Smith et al., 1983). This unconformity broadly coincides with the climactic Grenvillian orogenic deformation in North America.

Active syndepositional faulting cannot unequivocally be demonstrated in the Torridon Group since east-facing faults reactivated during Iapetan thrusting (Butler, 1997) may signify Iapetan rather than early Neoproterozoic extension. An alternative model is that the Torridon Group sediments were deposited in fluvial environments by major river systems that drained the foreland to the Grenville orogen (Rainbird et al., 2001). Detrital zircons from the Torridon Group have yielded a minimum U-Pb age of 1060 ± 18 Ma (Rainbird et al., 2001), coeval with the later stages of Grenville orogenic activity. No comparable sediments are preserved in East Greenland.

The Moine Supergroup

In areal terms, the Moine Supergroup dominates the rocks of the Northern Highlands of Scotland (Figs. 2 and 3A). Inliers of similar strata occur southeast of the Great Glen fault (the Dava and Glen Banchor successions) and form the basement to the younger Dalradian Supergroup in the Grampian Highlands (Smith et al., 1999).

The monotonous siliciclastic lithology of the Moine Supergroup provides few distinctive horizons that can be easily correlated over great distance. Three formal lithostratigraphic groups are recognized (Fig. 4): Morar (oldest), Glenfinnan, and Loch Eil (youngest) (Holdsworth et al., 1994). The Morar Group is a 5-km-thick tripartite psammite-pelite-psammite succession. The Glenfinnan Group is characterized by striped units of thinly interbanded psammites, semipelites, pelites, and quartzites (Plate 1D) together with thick pelite formations; estimates of thickness vary from 1 to 4 km, largely as a result of the high levels of ductile strain. The largely psammitic Loch Eil Group may be up to 5 km thick.

Detrital and inherited zircon grains constrain a source age for detritus between ca. 1.8 and 0.9 Ga; Archean sources (ca. 2.9 Ga) are only prominent in basal units just above the inliers of Lewisian basement rocks (Friend et al., 2003; Cawood et al., 2004). The Moine rocks were thus probably derived from erosion of the assembled domains of the ca. 1.1–1.0 Ga Grenville deformed basement and deposited in a distal foreland setting with respect to the relict mountainous hinterland in the core of the orogen. The youngest Moinian detrital zircons detected so far (ca. 900 Ma) have been found in samples from the Loch Eil Group in the eastern part of the Northern Highlands, whereas the minimum-age detrital zircons from the western Morar Group rocks are more in accord with the minimum-age samples from the Torridon Group, at ca. 1.06–1.00 Ga (Rainbird et al., 2001; Cawood et al., 2004).

Determination of the depositional environment of parts of the Morar (regressive tidal shelf) and Loch Eil (shallow marine) Groups has been possible where low strain permits sedimentologic study; paleocurrents in both areas indicate a general flow direction from south to north (Glendinning, 1988; Strachan, 1986). Soper et al. (1998) proposed deposition in two major half-graben basins that were bounded to the west by inferred

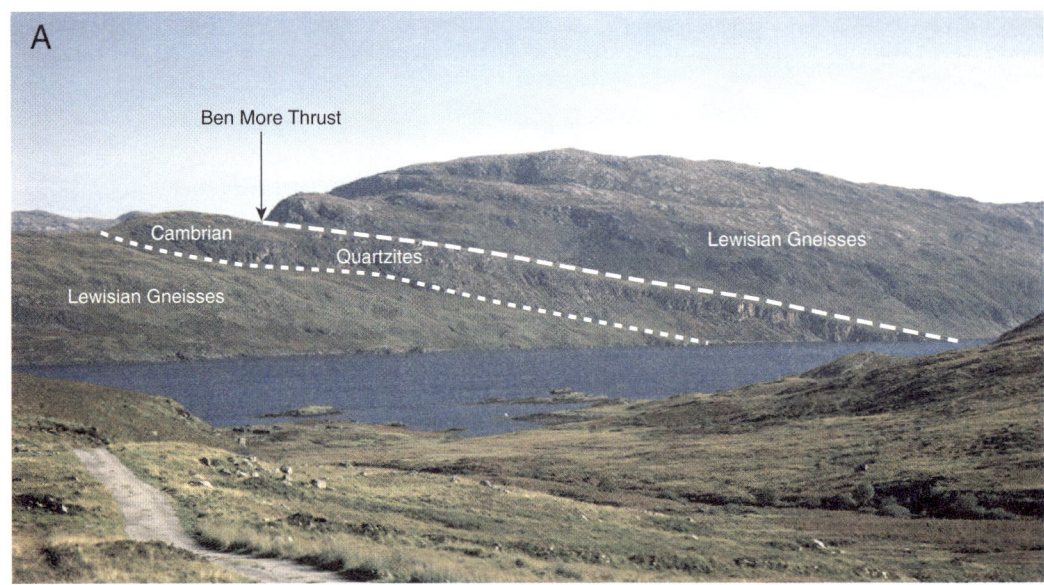

Plate 1. (A) View north across Loch Glencoul, northwest Scottish Highlands. The Ben More thrust is one of the principal thrusts developed within the Assynt culmination in the Moine thrust zone (Caledonian front) and here places Lewisian gneisses in the Ben More thrust sheet onto Cambrian quartzites and Lewisian gneisses of the foreland. The Moine thrust (sensu stricto) lies structurally above and to the right in the view shown. The summit is 450 m above the loch. British Geological Survey (BGS) photograph no. P000965. (B) Typical banded Scourian (Paleoproterozoic) Lewisian tonalite-trondhjemite-granodiorite (TTG) orthogneiss, Achmelvich Bay, northwest Scottish Highlands. The mafic enclave identified in the photograph has a longest dimension of ~80 cm. BGS photograph no. P513590. (C) View east-northeast across Suilven Mountain in the northwest Scottish Highlands. The unconformity between the horizontally bedded early Neoproterozoic Torridonian sandstones, which make up the mountain, and the underlying "cnoc and lochan" terrain of the Paleoproterozoic Lewisian gneisses is clearly demonstrated. The near summit (Caisteal Liath, 731 m) lies some 550 m above the surrounding plain. BGS photograph no. P000827. (D) Interlayered cross-bedded psammitic and striped semipelitic rocks belonging to the Glenfinnan Group of the early Neoproterozoic Moine Supergroup, Loch Cluanie, Wester Ross, Northern Highlands. Hammer shaft is ~35 cm long. Photograph taken by J.R. Mendum, BGS. Photographs used with permission of the British Geological Survey © NERC. All rights reserved.

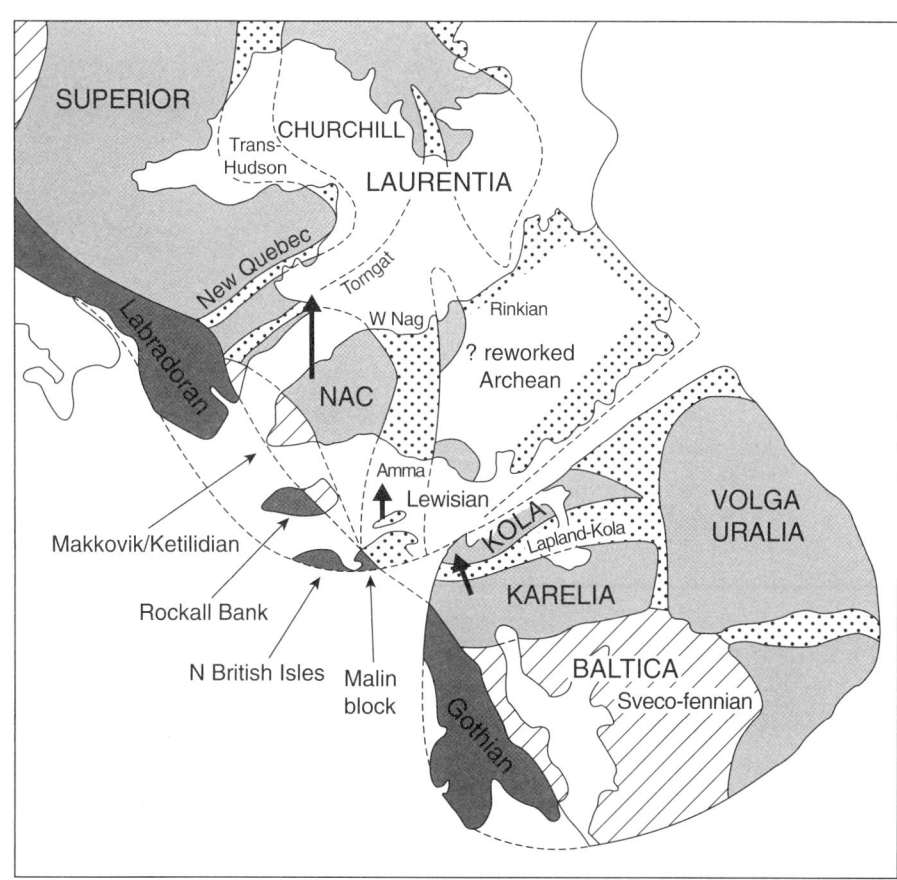

Figure 5. Reconstruction of the Paleoproterozoic belts and Archean cratons of the North Atlantic region (after Buchan et al., 2000). Amma—Ammassalik; NAC—North Atlantic craton; Nag—Nagssugtoqidian.

east-dipping normal faults, schematically represented in Figure 3A. An initial phase of rifting is proposed to have accommodated deposition of the Morar and Glenfinnan Groups; the former displays marked westward thickening in its upper part, which is consistent with deposition in a half-graben (Glendinning, 1988). The same formations appear to become progressively more distal eastward, and the striped and pelitic rocks of the Glenfinnan Group may represent a distal facies of the Morar Group (Soper et al., 1998). In this model, the Loch Eil Group marks the onset of renewed rifting; asymmetrical facies distribution and westward thickening are again consistent with deposition in a second half-graben bounded by an east-dipping normal fault (Strachan, 1986). Dalziel and Soper (2001) suggested that the Moine rocks were deposited within an aborted rift zone that formed during the early stages of Laurentian break-out from the supercontinent of Rodinia as East Gondwana separated from West Laurentia to form the Pacific Ocean.

Metasediments of the Glenfinnan and Loch Eil Groups are cut by the West Highland Granite Gneiss and intruded by metabasic intrusions of tholeiitic affinities both dated at ca. 870 Ma (Friend et al., 1997; Millar, 1999; Fig. 2; Plate 2A). The chemistry of the basic rocks is consistent with intrusion into thinned continental crust. Ryan and Soper (2001) proposed emplacement of the basic intrusions at depth to provide the heat necessary to locally melt the underlying basement and Moine sediments. Granitic melts then migrated through the sedimentary pile to higher levels.

Geikie (1893) first proposed that the "Torridonian" and Moine rocks could represent different elements of the same sedimentary succession foreshortened by Caledonian thrusting. The available isotopic constraints on the age of these deposits (Turnbull et al., 1996), and on the provenance of detritus, suggest that parts of these two extended sequences were broadly contemporaneous, and so it remains a plausible hypothesis—the Torridon Group may have been deposited by a major fluvial

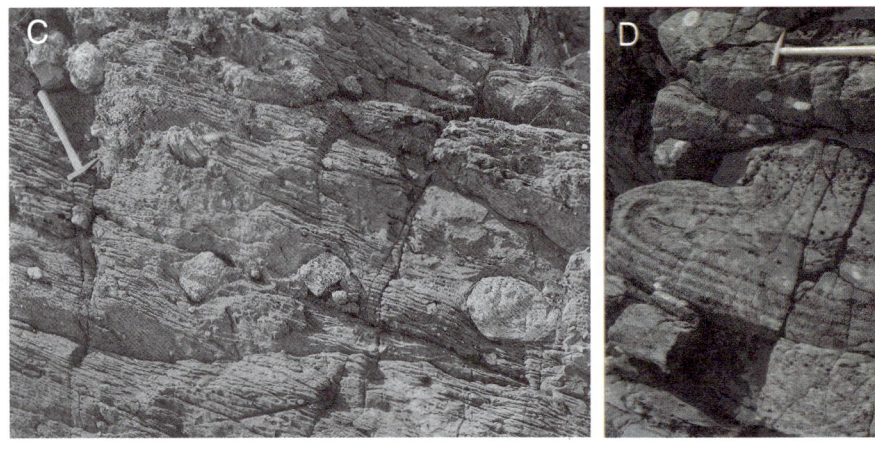

Plate 2. (A) Roadside exposure of the Neoproterozoic (ca. 870 Ma) West Highland Granite Gneiss along with the mafic sheets intruded into the granite prior to Knoydartian deformation and metamorphism at ca. 820 Ma. Invermoriston, Wester Ross, Northern Highlands. The thickest of the mafic sheets is ~40 cm thick. British Geological Survey (BGS) photograph no. P579984. The inset shows the strongly foliated granite gneiss. Coin is 2 cm in diameter. BGS photograph no. P580522. (B) Prominently ribbed and graded late Neoproterozoic psammitic turbidites; the more micaceous uppermost parts of each layer preferentially erode giving a distinctive "sawtooth" profile at outcrop. Grampian Group Dalradian, Ben Alder, central Grampian Highlands. Hammer shaft is ~35 cm long. BGS photograph no. P605180. (C) Diamictite in the Port Askaig Tillite Formation, cleaved and containing large clasts, mainly of granite. Argyll Group Dalradian, Islay, southwest Scottish Highlands. The hammer shaft is ~40 cm long. BGS photograph no. P215702. (D) Basaltic pillow lavas of the Vendian Tayvallich Volcanic Formation. Argyll Group Dalradian, coast southwest of Tayvallich, Argyll, southwest Scottish Highlands. The hammer shaft is ~40 cm long. BGS photograph no. P219459. Photographs used with permission of the British Geological Survey © NERC. All rights reserved.

system that flowed eastward into a marine setting for the dispersal of the Morar Group sediments (M. Krabbendam, 2006, personal commun.). This cycle of deposition probably spanned the period from ca. 1000 to 900 Ma; accumulation of the youngest parts of the sedimentary pile (Loch Eil Group; Soper et al., 1998) was followed relatively soon after by ca. 870 Ma intrabasinal acid and basic magmatism (Millar, 1999).

In East Greenland, comparable sequences of thick Mesoproterozoic metasedimentary rocks are represented by the Krummedal succession (Higgins, 1988; Leslie and Higgins, this volume), which is widely distributed in both the Niggli Spids and Hagar Bjerg thrust sheets. The Krummedal succession and equivalent successions are exposed over a N-S distance of at least 600 km in East Greenland; the reconstruction of Higgins et al. (2004) suggests that the original depositional basin was at least 300 km wide (with the eastern limit undefined). Ion microprobe analyses of detrital zircon (Kalsbeek et al., 2000; Watt et al., 2000) and electron microprobe analyses of monazite (Gilotti and Elvevold, 2002) are available from Krummedal succession metasediments of both the Niggli Spids and Hagar Bjerg thrust sheets. Like the Moine Supergroup, deposition occurred after ca. 1050 Ma, while Archean and mid-Paleoproterozoic (1800–2000 Ma) detrital zircon grains are rare and indicate that the older rocks of the western foreland region to the Grenville belt were not a significant source region. Watt et al. (2000) and Watt and Thrane (2001) advocated a distant source area.

The widespread occurrence of these marine siliciclastic successions in Greenland, NW Scotland, and in the eastern province of Spitsbergen (Brennevinsfjorden Group, Helvetsflya Formation; Gee and Teben'kov, 1996; Gee et al., 1995) indicates that late Mesoproterozoic sedimentary basins extended over several thousand square kilometers, perhaps in a number of interconnected depocenters. The general lack of Archean detritus in the Moine and Krummedal sequences implies that detritus was derived from more recently active interior areas of the Laurentian hinterland (e.g., the Grenville belt) rather than more parochial Archean or Paleoproterozoic basement. Those sediments were then deposited in marginal settings floored by Archean to Paleoproterozoic continental crust. Since no equivalent of the early Neoproterozoic Torridon Group fluvial red-bed system seems to have been present at the paleolatitude represented by East Greenland, we conclude that it is possible that the Krummedal succession may represent a northward continuation of marine dispersal of parts of the Moine succession.

NEOPROTEROZOIC OROGENESIS (Ca. 900–730 Ma)

In Scotland, the nature and timing of Neoproterozoic tectonothermal activity affecting the Moine Supergroup are problematical and highly contentious. Current debate is focused on whether or not the Moine rocks were affected by one or more high-pressure, high-temperature granulite-facies orogenic event between ca. 820 and 730 Ma. These events both led to extensive migmatization of the Moine metasediments but did not generate significant volumes of granitic melt in the form of discrete plutonic bodies. This contrasts markedly with central East Greenland, where high-pressure, high-temperature granulite-facies metamorphism culminated in the generation of significant volumes of S-type granite at ca. 910–930 Ma.

For Scotland, Ryan and Soper (2001) proposed that the protolith to the 870 Ma West Highland Granite Gneiss (and the contemporaneous basic sheets) was generated in response to *crustal thinning and extension* affecting an undeformed sedimentary pile. The East Greenland granites were intruded by basic sheets and subsequently affected by sillimanite-grade ductile shear zones, but no precise age constraints are available for these later events (Leslie and Nutman, 2003).

Evidence for a ca. 820 Ma orogenic (Knoydartian) event exists in the Moine rocks of the Northern Highlands (Fig. 4). Bodies of pegmatite generated in situ in localized zones of high strain yield precise U-Pb zircon and monazite ages of 827 ± 2 Ma, and Sm-Nd ages of ca. 820–790 Ma obtained from post-D_1 Morar Group garnets apparently date the early metamorphism (Rogers et al., 1998; Vance et al., 1998). Peak pressures of 12.5–14.5 kbar indicate *crustal thickening* at this time. A further episode of high-pressure, high-temperature metamorphism (ca. 10 kbar, 800 °C) occurred at 730 Ma (Fig. 4) but was seemingly restricted to the eastern part of the Moine outcrop (Tanner and Evans, 2003; Emery, 2005).

Combined, these data from the Scottish and Greenland sectors apparently indicate three periods of orogenesis (ca. 930 Ma, ca. 820 Ma, ca. 730 Ma) in this sector of Laurentia, each separated by ~100 m.y. The wider significance of these orogenic events is unclear, not least because they occurred at a time when the Rodinian supercontinent was undergoing late-stage amalgamation, even the initiation of fragmentation in some locations (Hoffman, 1991; Dalziel, 1992). It has been proposed that intracratonic movements accommodating extension and compressional deformation may account for these early to mid-Neoproterozoic orogenic events (e.g. Cawood et al., 2004). Repeated high-pressure, high-temperature conditions, however, are largely inconsistent with models of intracratonic (low-pressure, high-temperature) orogenic belts (cf. Petermann and Alice Springs orogenies of Australia; Sandiford and Hand, 1998; Scrimgeour and Close, 1999; Shaw et al., 1992).

We note here that ca. 950 Ma calc-alkaline volcanics have been recorded in the eastern Svalbard Nordaustlandet terrane Kapp Hansteen Group along with 930–960 Ma augen granites that were emplaced synchronously with an episode of deformation and high-pressure, high-temperature crustal anatexis (Johansson et al., 2000). An alternative hypothesis might propose a more marginal continental setting, in which a continental block actively overrode oceanic crust as a consequence of active spreading on an opposing plate margin. Thus, repeated Neoproterozoic tectonothermal events on the eastern Laurentian margin may have some analogy to the patterns of high heat flow and repeated deformation found in subduction-zone continental back arcs (Hyndman et al., 2005).

MID-NEOPROTEROZOIC TO EARLY PALEOZOIC SEDIMENTATION

During the late Neoproterozoic, post–Pan African rifting of eastern Rodinia culminated with the formation of the Iapetus Ocean as Baltica and Amazonia drifted away from Laurentia (inset to Fig. 1, Fig. 4; Soper, 1994b; Cawood et al., 2001, 2003, 2004). These rift and rift-drift events were recorded by the formation of widespread progradational passive-margin sedimentary sequences, igneous activity, and oceanic sedimentation along the eastern side of Laurentia that lasted from ca. 700 Ma until the mid-Ordovician.

In Scotland, this geological evolution is recorded in two separate successions, the Cambrian to mid-Ordovician carbonate shelf succession of the Northwest Highlands and the late Neoproterozoic to mid-Cambrian, shallow- and deep-water sediments and volcanic rocks of the Dalradian Supergroup in the Grampian Highlands. The former readily correlates with the foreland succession in East Greenland, and the latter is comparable in age to the Eleonore Bay Supergroup in East Greenland (Soper, 1994b). Thus, we can interpret the Hebridean Cambrian-Ordovician and Grampian Dalradian rocks and their Laurentian correlatives as forming two subparallel sedimentary belts located along the eastern continental margin, where the shelf carbonate rocks were positioned on the landward side of the generally deeper-water marine lithologies. Oceanic crust is presumed to have developed to the southeast of the present Dalradian outcrop Scotland, where the youngest sediments ultimately prograded onto Iapetan oceanic crust.

The Pre-Dalradian Basement

In Scotland, southeast of the Great Glen fault, the stratigraphically, and structurally, oldest known strata within the Grampian Highlands comprise the recrystallized and mainly gneissose to locally migmatitic psammite and semipelite of the Glen Banchor and Dava successions (Fig. 4; Piasecki, 1980; Smith et al., 1999; see also review by Strachan et al., 2002).

Rocks of the Glen Banchor and Dava successions are invariably intensely recrystallized, do not preserve sedimentary structures, and commonly contain intrafolial isoclinal folds that deform the first foliation (usually a gneissosity) (Smith et al., 1999). In contrast, the overlying Grampian Group rocks are variably recrystallized, structurally less complex, and commonly preserve sedimentary structures.

In the Dava and Glen Banchor succession, locally close to the contact with younger rocks, late Neoproterozoic syntectonic blastesis and pegmatite segregation are present within ductile shear zones (Hyslop, 1992; Hyslop and Piasecki, 1999). U-Pb monazite analyses from such pegmatites have provided high-precision ages of 808 +11/–9 Ma and 806 ± 3 Ma, and a concordant age of 804 +13/–12 Ma from the host mylonite matrix (Noble et al., 1996). Recent U-Pb dating of single zircon grains within kyanite-bearing migmatites yielded an age of 840 ± 11 Ma (Highton et al., 1999). These data are interpreted as the effects, in the Dava–Glen Banchor succession rocks, of high-grade metamorphism and migmatization associated with Knoydartian orogenesis recognized farther west in the Moine Supergroup.

In contrast, isotopic evidence for Neoproterozoic tectonothermal events has yet to be recorded in the Dalradian rocks. There is evidence of progressive overstep of various lithologies onto the older strata (Smith et al., 1999; Robertson and Smith, 1999; British Geological Survey, 2002, 2004), and this, combined with $^{87}Sr/^{86}Sr$ isotopic signatures from the lowermost Dalradian strata (Thomas et al., 2004, see following), provides evidence for a significant stratigraphic and tectonothermal break at the base of the Grampian Group (Smith et al., 1999).

In Greenland, the very thick Neoproterozoic–Ordovician succession recognized within the Franz Joseph allochthon is dominated by the Neoproterozoic Eleonore Bay Supergroup (Higgins et al., 2004). The lower contact of the Eleonore Bay Supergroup against the older gneissose metasedimentary rocks of the Krummedal succession in the Hagar Bjerg thrust sheet is likely to be an unconformity modified during westward Caledonian transport (Leslie and Higgins, this volume).

The Dava and Glen Banchor succession rocks are interpreted to form a Moine-like metasedimentary basement that was affected by a Neoproterozoic Knoydartian tectonothermal event prior to deposition of the overlying Grampian Group. The relationship is thus analogous to that observed at the base of the Eleonore Bay Supergroup in Greenland.

Dalradian Sedimentation (Ca. 730–470 Ma)

The Dalradian Supergroup is a relatively well-differentiated progradational passive-margin sedimentary sequence dominated by marine metasandstones, siltstones, mudstones, and carbonate rocks. They are subdivided into a Cryogenian (Grampian, Appin Groups) and an overlying Ediacaran to mid-Cambrian (Argyll Group and Southern Highland Group) succession (Harris et al., 1978, 1994) (Figs. 4 and 6). The Dalradian Supergroup has an apparent total thickness of at least ~25 km, but this is unlikely to have been deposited in a single continuous succession. Significant volumes of the middle to upper parts of the succession (e.g., in the southwest Scottish Highlands) are affected only by low-grade metamorphism and/or relatively weak deformation, and here the sedimentologic and basin evolution history can be interpreted with some confidence. When a restoration is made of the distribution of lithostratigraphy prior to Ordovician (Grampian) folding, it seems reasonable to presume that deposition would have migrated broadly southeastward with time.

Figure 6. Correlation of the Neoproterozoic to Lower Paleozoic Laurentian stratigraphy of Scotland and the central fjord region of East Greenland. The correlation is structured around the dated global record of Neoproterozoic glacials and a tentative but chronologically unconstrained set of important flooding surfaces identified in the Neoproterozoic depositional record. The timing of the principal orogenic events is shown schematically. B.B and ▲ indicate boulder bed.

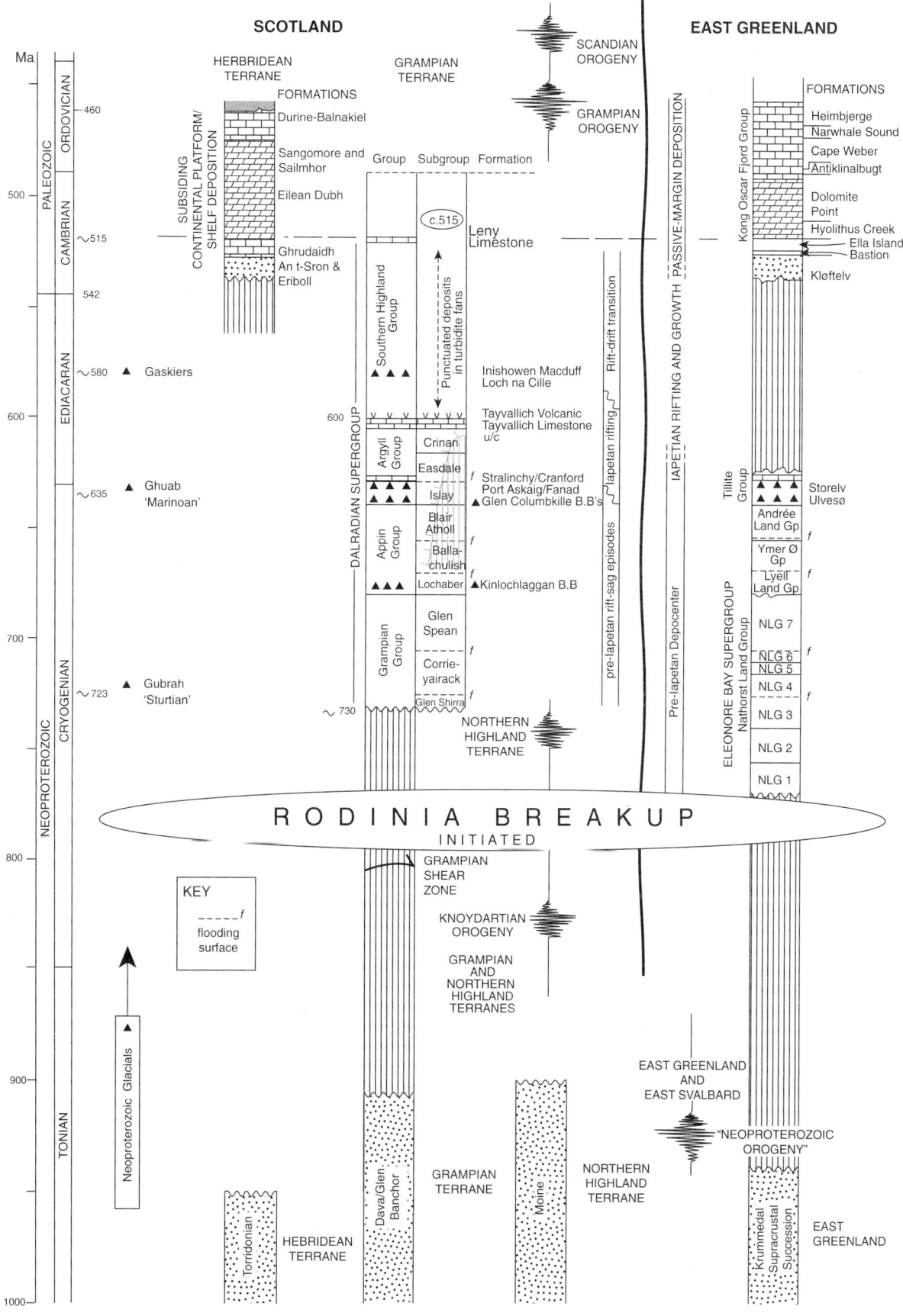

One key to unraveling the stratigraphy is a number of regionwide events that were probably broadly synchronous across Scotland and Ireland. These include transgressive flooding surfaces (e.g., base of the Ballachulish and Easdale subgroups; Fig. 6), Neoproterozoic glaciation events (e.g., Port Askaig Tillite, base of the Argyll Group; Plate 2C; McCay et al., 2006), and rift-related magmatism (as represented by A_2-group granitoids and large volumes of basic volcanic rocks; e.g., Tayvallich Volcanics Formation; Plate 2D) in the Argyll Group (Tanner et al., 2006). These key horizons are used to help constrain the suggested correlation of Figure 6.

Modern sequence stratigraphic concepts have only been applied to the lowermost Grampian Group (Glover and McKie, 1996; Banks, 2005). Previously undetected intrabasinal unconformities and periods of nondeposition are now recognized in the southern and central Scottish Highlands (A.G. Leslie for the British Geological Survey [BGS], unpublished data). One such important unconformity that affects the Appin Group and Argyll Group succession has been added to the detail on Figure 6. Alternative evidence for major stratigraphic *and* tectonic orogenic breaks (Prave, 1999; Alsop et al., 2000; Hutton and Alsop, 2004) has not been regionally validated and remains speculative.

Age constraints on the initiation of Dalradian sedimentation are poor. The Grampian Group must be younger (<800 Ma) than the pegmatites contained within the basement Dava and Glen Banchor successions, and the youngest deformation event recorded within the equivalent Moine rocks to the northwest of the Great Glen fault is now ca. 730 Ma (Tanner and Evans, 2003; Emery, 2005). The youngest Dalradian detrital zircons yield ages of 900 Ma (Cawood et al., 2003), and $^{87}Sr/^{86}Sr$ whole-rock isotope data from the lowermost metacarbonate rocks of the Grampian Group are consistent with a global late Neoproterozoic strontium seawater signature younger than 800 Ma and possibly as young as ca. 670 Ma (Thomas et al., 2004). Thus, Grampian Group sedimentation occurred after 800 Ma and could have initiated as late as ca. 700–730 Ma.

Neoproterozoic augen granites such as the Ben Vuirich pluton (BV on Fig. 4) were intruded into Appin-Argyll Group Dalradian strata at 590 Ma (Rogers et al., 1989; Pidgeon and Compston, 1992). These rift-related intrusions have an A_2-group chemistry (Tanner et al., 2006) and are probably genetically linked with the Argyll Group mafic volcanism (Tayvallich volcanics) dated at 595 Ma (Dempster et al., 2002). This magmatism was contemporaneous with a second pulse of bimodal magmatism throughout the Appalachians (e.g., Badger and Sinha, 1988; Rankin et al., 1989; Aleinikoff et al., 1995), and it was related to continental breakup of Laurentia (Cawood et al., 2001). Reliable information critical to the biostratigraphic age of the Dalradian Supergroup is only preserved in the uppermost parts of the Southern Highland Group, where locally developed metacarbonate rocks (Leny Limestone) contain topmost Lower Cambrian *Pagetia* trilobites, indicating an approximate age of ca. 515 Ma for the upper limits of deposition (Fig. 6; Pringle, 1940; Tanner, 1995).

An age of ca. 730 Ma for the base of the Grampian Group has important implications for the ages of the tillite formations in the mid- to upper Dalradian (Port Askaig and Inishowen-Macduff). Brasier and Shields (2000), Condon and Prave (2000), and McCay et al. (2006) proposed that the Argyll Group tillites correlate with the Ghubrah (Sturtian) glacial dated at 723 +16/–10 Ma (Brasier et al., 2000), but deposits of this age would mean that accumulation of both the Grampian and Appin Groups would have overlapped the high-pressure, high-temperature tectonothermal event recorded in the eastern parts of the Moine (Tanner and Evans, 2003; Emery, 2005), which is contradicted by the $^{87}Sr/^{86}Sr$ isotope data from the Grampian Group metacarbonate rocks (Thomas et al., 2004). Displacements on the Great Glen fault would permit some room for maneuver here, but even the larger estimates of displacement (e.g., Dewey and Strachan, 2003) seem unlikely to solve the space problems inferred from these overlapping ages. The scenario favored here (Fig. 6) envisages two principal preserved glacigenic intervals in the Dalradian. We equate the younger Southern Highland Group glacigenic deposits in Inishowen (Donegal, Eire) and Macduff (NE Scotland) with the Gaskiers Formation (ca. 580 Ma; Bowring et al., 2003) and the Varangerian tillites of Norway (620–590 Ma; Gorokhov et al., 2001; Bingen et al., 2005). The older Port Askaig Tillite Formation (Argyll Group) and the Storelv and Ulvesø Formations (Tillite Group) of East Greenland (Hambrey and Spencer, 1987) are then equated with the Marinoan-Ghaub glacial (ca. 635 Ma; Hoffmann et al., 2004). The Kinlochlaggan Boulder Bed is an isolated set of occurrences restricted to the central Scottish Highlands and assigned to the Lochaber Subgroup (British Geological Survey, 2002) (Fig. 6). It must, on that basis, presumably be significantly younger than 700 Ma and cannot correlate with the Ghubrah glaciation. The Kinlochlaggan Boulder Bed is interpreted as a mature sandy deposit containing glacially rafted dropstones (J.R. Mendum, 2006, personal commun.) rather than a subglacial till and may thus still mark the earliest record of glacial influence in the Dalradian.

Provenance studies summarized here (Fig. 7; Cawood et al., 2003, 2004) provide useful data that track the evolution and denudation history of the hinterland. Probability density distributions of concordant detrital zircon ages show that Grampian Group detrital zircons maintain the earliest Neoproterozoic to late Paleoproterozoic spectrum of the Torridon Group and Moine Supergroup (Cawood et al., 2003, 2004; Friend et al., 2003). However, quite different distributions are apparent in the data currently available for the remainder of the Dalradian Supergroup. After an early (Grampian Group) rifting episode in the Scottish sector of Laurentia, Appin Group shallow-marine-shelf sedimentary facies associations developed during postrift thermal subsidence (Stephenson and Gould, 1995). Flooding at the base of the Ballachulish Subgroup introduced, or just preceded, the arrival of Archean zircons in detritus and an apparent absence of Grenvillian ones. The onset of Argyll Group deposition coincided with supply of a broad spectrum of earliest Neoproterozoic to Archean detritus to an increasingly unstable and volcanically

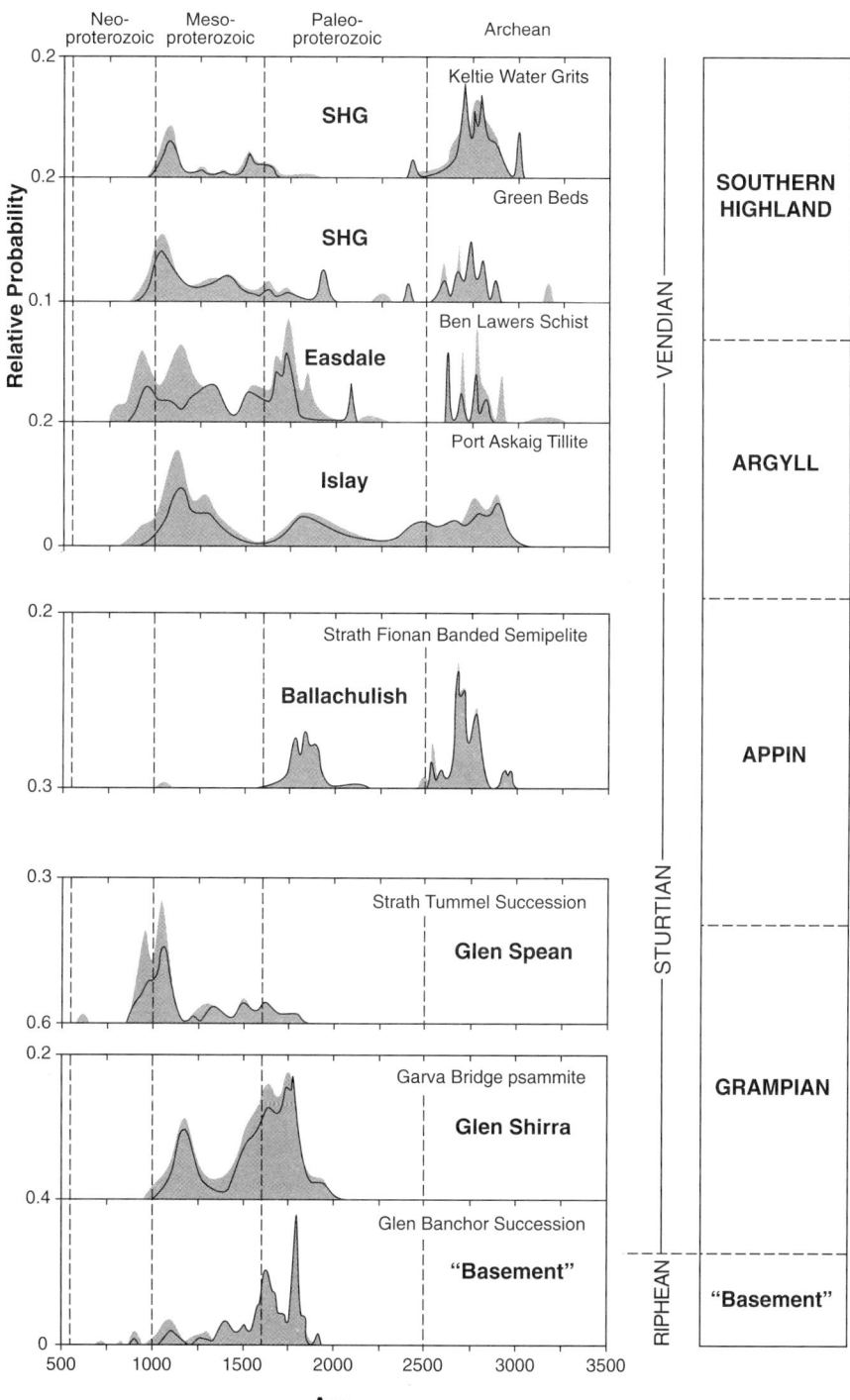

Figure 7. Selected frequency distribution diagrams for detrital zircon ages determined for the Dalradian Supergroup of Scotland and its basement (after Cawood et al., 2003). SHG—Southern Highland Group.

active margin (Stephenson and Gould, 1995). These successions were ultimately overstepped as the margin foundered in response to Iapetan rifting and was inundated with immature siliciclastic detritus from Crinan Subgroup time onward (upper Argyll Group). We see apparently less abundant Paleoproterozoic zircons in the sediment load upward from the base of the Crinan Subgroup, and these data might suggest that any ca. 1800 Ma Rhinnian source became increasingly isolated from the Dalradian basins as Archean supply increased. Iapetan rifting may have removed, or more likely submerged, that sector of the Laurentian margin that had been supplying the Paleoproterozoic detritus, presumably as Amazonia and Baltica broke away.

Thus, the picture emerges in Scotland of a Dalradian succession deposited in an evolving pericontinental environment over a period of ~180–200 m.y. Fluctuations in water depth accompanied active rifting and the development of second- and third-order subbasins on this sector of the continental margin.

In Greenland, the broadly time-equivalent Eleonore Bay Supergroup (apparently 14.5 km thick) and overlying Tillite Group (0.8 km thick) and Kong Oscar Fjord Group (4.5 km thick) are, by comparison, more poorly constrained chronologically. Here, there is no evidence of contemporaneous volcanic activity, and the available age data loosely bracket deposition in the period ca. 940 to ca. 460 Ma (Sønderholm and Tirsgaard, 1993; Smith et al., 2004). These East Greenland strata record shelf and ramp sedimentation, which was probably punctuated by significant periods of nondeposition on a slowly subsiding passive margin, a setting in stark contrast to the linked Neoproterozoic rift-basin architecture of Scotland. Farther north on the Laurentian margin, the Neoproterozoic Hekla Sund Basin (Higgins et al., 2001b) represents another locus of rifting activity in eastern North Greenland. In such a setting, and based upon the data currently available, we propose that a generalized correlation is possible between the Greenland and Scottish sectors in Laurentia (Figs. 3 and 6).

Cryogenian to Early Ediacaran Basin Evolution (Ca. 730–610 Ma)

Grampian Group

The Grampian Group records the initiation of middle to late Neoproterozoic extension and basin development and consists of three main lithofacies associations (subgroups), which are interpreted as representing distinct phases of early and synrift extension followed by a protracted period of postrift thermal subsidence (Fig. 6). Deposition occurred within a series of linked NE-trending rift basins bounded by major crustal lineaments (Glover and Winchester, 1989; Smith et al., 1999; Banks, 2005). Despite regional deformation and metamorphism to amphibolite-facies conditions, the stratigraphic integrity and overall geometry of these basins has been preserved largely intact.

The actual base to the Grampian Group is unexposed, but the oldest unit is the spatially restricted and fault-bounded Glenshirra Subgroup, which is nowhere observed in primary undisturbed contact with the underlying Glen Banchor or Dava succession rocks (Fig. 6). With a maximum exposed thickness of ~2 km, the Glenshirra Subgroup is composed of stacked shoaling sequences of geochemically distinct, immature arkosic psammite and beds of metaconglomerate (Banks and Winchester, 2004). Banks and Winchester (2004) interpreted the sediments as alluvial-fan and shallow-water sediments deposited within a SE-thinning fan-delta clastic wedge. Progressive thickening and coarsening of the strata toward the west may imply the presence of a basin margin to the west or northwest (approximately coincident with the present trace of the Great Glen fault). The clastic wedge was supplied from an exposed hinterland of mature crust beyond the basin margin (Banks and Winchester, 2004), which, based upon clast populations, was predominantly composed of quartzofeldspathic gneiss and granitic rock. Detrital zircon populations are dominated by 1.8 Ga detritus, with subsidiary 1.2 Ga detritus (Fig. 7), suggesting that "Rhinnian-type" basement was an important source area (Cawood et al., 2003).

The Glenshirra Subgroup is abruptly but conformably overlain by a distinctive and regionally widespread succession of psammite and semipelite assigned to the Corrieyairack Subgroup (Fig. 6). This change records a basinwide flooding event that heralded a period of subsidence and rift-related extension (Banks, 2005). A near-complete sequence is preserved through the main rift cycle in 4–5 km of siliciclastic deposits deposited by prograding turbidite complexes (Banks, 2005). Variations in sediment supply and source area are indicated by changes in the proportions of plagioclase and K-feldspar, whereas variations in bed thickness and form reflect depositional processes. Bouma cycles are well represented, but bottom structures are extremely rare (Banks, 2005). A reduction in sand-grade sediment supply and development of shelf conditions along the tectonically active basin margins and intrabasinal highs are recorded by lateral thickness and facies changes to striped semipelite and psammite. This was followed by a renewed influx of sand-dominated turbidites (Plate 2B) deposited by fan-lobe systems derived from the northwest, passing south and eastward into shelf environments (Glover et al., 1995; Robertson and Smith, 1999; Banks, 2005).

The turbidites of the upper Corrieyairack Subgroup are overlain by shallow-marine sediments of the Glen Spean Subgroup, which prograded into the basin from the northwest and southeast (Fig. 6) after a flooding event (Banks, 2005). Reduced subsidence and relative tectonic stability at this time are interpreted to represent a postrift thermal subsidence phase (Glover et al., 1995). The lithological associations of the Glen Spean Subgroup, combined with well-preserved sedimentary structures, indicate deposition in shallow-marine (tidally influenced) shelf environments with intensive sediment recycling and winnowing of the underlying turbiditic rocks (Banks, 2005). Analysis of detrital zircon populations shows a marked absence of any Archean detritus, peaks at 2.0 Ga and 1.4 Ga in the Corrieyairack Subgroup, and progressive dilution by 1.1–0.9 Ga Grenvillian detritus in the Glen Spean Subgroup (Fig. 7; Cawood et al., 2003; Banks, 2005).

In Greenland, the Nathorst Land Group (Fig. 6; Sønderholm and Tirsgaard, 1993) consists of up to 9 km thickness of siliciclastic sediment that rests in sheared or unexposed contact on the Hagar Bjerg thrust sheet (Higgins et al., 2004). The Nathorst Land succession is informally subdivided into seven formations (NLG1–NLG7; Smith and Robertson, 1999), all of which record persistent fine-grained and shallow-marine shelf sedimentation. Several lithofacies associations of sandstone–dolostone–quartz arenite alternating with heterolithic fine sandstones, siltstones, and mudstone are identified. Carbonate deposits with parallel lamination of possible algal origin are present in the upper part of the group. Beautifully preserved delicate sedimentary structures including desiccation cracks, ripple-lamination, cross-lamination, and heavy mineral bands are characteristic at several levels. Depositional environments include outer-shelf storm-influenced and inner-shelf to tidally influenced shoreface environments. Two major flooding surfaces are identified at the junctions between the Nathorst Land Group formations 3 and 4 and between formations 6 and 7 (Smith and Robertson, 1999), and we tentatively match these with the flooding events that bracket the Corrieyairack Subgroup in the Scottish sector in order to help constrain the suggested correlation in Figure 6.

Appin Group

In Scotland, siliciclastic sedimentation continued up into the lowermost Appin Group (Lochaber Subgroup, see Fig. 6). Although locally conformable, the Lochaber Subgroup has a markedly diachronous base at the basin scale, and, with an overall decrease in thickness of the subgroup to the southwest of its crop, it is interpreted to have been deposited with considerable lateral facies variation during marine transgression (Key et al., 1997). Similarly, above the Nathorst Land Group in East Greenland, the base of the Lyell Land Group is marked by transgression, marked locally in places by an angular unconformity (Smith and Robertson, 1999). The Lyell Land Group is composed of 3 km of siliciclastic shelf and coastal-plain tidal sediments dominated by storm and wave events. Cyclical changes in sea level and shoreward reorganization of facies are linked to large-scale regressions that can be traced along 300 km of inferred paleocoastline (Tirsgaard and Sønderholm, 1997).

Across Scotland, we interpret the base of the succeeding Ballachulish Subgroup as transgressive, and maximum flooding is likely to have coincided with deposition of the Ballachulish Slate Formation in the locally anoxic environments that characterize the lower part of the subgroup (Fig. 6). Subsequent progradation is marked by progressive development of extensive shallow, tidally influenced, shelf sedimentation and a period of stability (Anderton, 1985). A fourfold subdivision at formation level can be traced with remarkable continuity for some 300 km across the Grampian Highlands in Scotland and northwest Ireland. Limestone pelite quartzite facies associations are characteristic and mark a significant break from the siliciclastic-dominated record of the Grampian Group and Lochaber Subgroup as presently defined. Continuity at this regional scale, almost on a bed-for-bed basis, attests to the widespread stability and relatively uniform nature of the subsidence. Interestingly, available data indicate that Archean detrital zircon grains become evident in the sediment load at this juncture (Fig. 7; Cawood et al., 2003), further emphasizing the change that occurs at the base of the Ballachulish Subgroup.

In Greenland, the base of the Ymer Ø Group may record the same event; there is a sharp break in sedimentary facies association from heterolithic sandstones in the Lyell Land Group below to fine-grained mudstone above. A wide variation in lithology comparable to the Ballachulish Subgroup is also evident—siliciclastic mudstone and sandstone pass upward into black limestone and dolomite with algal biostromes (Sønderholm and Tirsgaard, 1993). A wide range of environments is indicated at this time, including basinal and slope deposits, inner-shelf environments, and horizons of evaporitic sulfate deposition.

The Blair Atholl Subgroup marks the continued diversification of Dalradian lithologies in Scotland with renewed flooding and a change to deeper deoxygenated marine conditions in the Scottish sector (Fig. 6; Stephenson and Gould, 1995). In the type area, the basal slates and phyllites are conformable with the underlying Ballachulish Subgroup, but, importantly, some hints of volcaniclastic detritus and minor tuff horizons are also recorded, pointing to the earliest signs of basin instability on this sector of the Laurentian margin.

In East Greenland, the Ymer Ø Group is succeeded by the Andrée Land Group, which should thus be broadly equivalent to the Blair Atholl Subgroup. The Andrée Land Group is composed of 1275 m of algal limestone and dolomite deposited on a NE-facing storm-influenced carbonate ramp (Frederiksen, 2001). Laterally extensive facies form cyclical stacking patterns in response to sea-level fluctuations and are the basis for subdivision into seven formations. Changes in ramp geometry and transgression have been linked to an episode of extensional faulting that marked incipient stretching on this part of the margin (Frederiksen, 2001).

Lower Argyll Group

Argyll Group Dalradian sedimentation in Scotland as a whole records the rapid onset of instability in the mid- to late Neoproterozoic and the replacement of widespread shallow-marine conditions of the Appin Group by cycles of rapid basin deepening. Initially, the distinctive successions of black graphitic pelite, metacarbonate rock, and quartzite of the Appin Group are succeeded by an equally distinctive glaciomarine tillite (Plate 2C), the Port Askaig Tillite Formation and other correlatives. Above, deeper-water psammites and quartzites comprise the remainder of the Islay Subgroup. The tillite formation is a prominent marker horizon across Scotland and Ireland and marks the onset of cold-climate glaciomarine sedimentation equated here with the Marinoan glacial period at ca. 635 Ma (Fig. 6). The top of the subgroup is located in Ireland by a cap carbonate (the Cranford Limestone Formation; McCay et al., 2006), after which cold-climate conditions apparently ameliorated with no further sign of glaciogenic deposits in the Argyll Group. Thereafter, variable lithofacies of psammite and quartzite and locally thick accumulations indicate that sediment input kept pace with extension in a series of NE-trending basins (Stephenson and Gould, 1995). While detrital zircons from the tillite units are comparable to the enclosing Appin and Argyll Group rocks (Cawood et al., 2003), this change in sedimentation is also marked by increasing volumes of Archean grains above the level of the tillite formations (Fig. 7; Cawood et al., 2003).

The Tillite Group in Greenland is likewise marked by units of diamictite sandstone, carbonate, and shale (Fig. 6). Here, massive bedded diamictites and cross-bedded sandstones of eolian origin mark the base and are overlain by shales and sandstones formed by debris-flow and turbidite events. There is a second horizon of diamictite below tidally influenced dolomites and shales at the top of the group (Hambrey and Spencer, 1987; Moncrieff and

Hambrey, 1988). We follow the assessment of Sønderholm et al. (this volume) for a "Marinoan" age for these deposits and thus make the ca. 635 Ma chronostratigraphic correlation with the Dalradian Islay Subgroup used in the construction of Figure 6.

In Scotland and Ireland, the Easdale Subgroup sees a return to a wide range of finer-grained lithologies, including graphitic black pelite, calcareous semipelite, and metacarbonate rock, commonly associated with pebbly quartzite and sheets of basic meta-igneous rock of varying abundance (Stephenson and Gould, 1995). Exhalative saline brines gave rise to a laterally persistent bed of strata-bound sulfide, barite, and vein mineralization (Hall et al., 1991). Taken together with the greater abundance of mafic meta-igneous rocks, these occurrences point to an increased extension in this sector of the Laurentian margin at the end of the Cryogenian.

In Greenland, an erosional break encompassing the later Ediacaran and earliest Cambrian separates the sediments of the Tillite Group from the Cambrian-Ordovician succession of the Franz Joseph allochthon in the central fjord region of East Greenland (Fig. 6). Uplift and erosion in Greenland at this time coincided with the onset of enhanced rifting and mafic volcanism leading up to continental rupture in the Scottish sector.

Ediacaran Basin Evolution (Ca. 610–542 Ma)

Upper Argyll Group

In Scotland, individually thick formations of often immature sediment and deep-water turbiditic facies characterize the succeeding Crinan Subgroup. This pronounced change in sedimentary facies association coincides with regional overstep in the southern and northeast Grampian Highlands, and we suggest that it is this change that indicates the onset of rift-drift transition in the Scottish sector of Laurentia as Iapetan rupture expanded northward. The overlying Tayvallich Subgroup is dominated by carbonates, which are locally accompanied by thick extrusive mafic volcanic rocks (including pillow lavas, Plate 2D), and subvolcanic sills that mark, perhaps for the first time, rupture of the continental crust during rifting. Felsic tuffs within the Tayvallich Volcanic Formation have yielded U-Pb zircon ages of 601 ± 4 Ma (Dempster et al., 2002). Rapid lateral variations in facies and thickness associations with unconformities, overstep relations, and pebbly beds typify this part of the succession.

Southern Highland Group

The uppermost unit of the Dalradian Supergroup is characterized by an ~4-km-thick pile of coarse-grained turbiditic siliciclastic and volcaniclastic strata that lie immediately above the Tayvallich Subgroup (Fig. 6). These sediments mark rapid basin deepening that persistently stayed ahead of the sedimentary and volcanic fill. The coarse-grained sediments were probably laid down in slope apron or ramp settings with channels on the lower slopes and inner zones of deep-water submarine fans with overbank deposits or as outer-fan facies (Burt, 2003). No apparent match for these immature turbidite-dominated sedimentary facies associations occurs anywhere along the Greenland sector of Laurentia.

In Scotland, volcaniclastic units are a conspicuous component of the Southern Highland Group. They are most prevalent in the lowermost 1 km and are interpreted as recording, in part, the erosion of the underlying basic volcanics, but they may also have resulted from contemporaneous volcanism and ash fall on the hinterland (Pickett et al., 2006). This interpretation is not, however, strongly supported by the detrital zircon data, which are dominated by Archean detritus in both the volcaniclastic "Green Beds" and their siliciclastic Southern Highland Group counterparts. Ages younger than 900 Ma are generally absent, although one grain yielded an age of 553 ± 24 Ma (Cawood et al., 2003). An important glaciogenic deposit is recognized in Inishowen in the north of Ireland (Condon and Prave, 2000). We follow those authors in correlating the Inishowen occurrences with others in the Southern Highland Group Dalradian section at Loch na Cille and Macduff in Scotland and then on a global scale with the Gaskiers glacial at ca. 580 Ma (Fig. 6).

Cambrian-Ordovician Sedimentation (Ca. 540–470 Ma)

No mid- to late Neoproterozoic tillite deposits are preserved on the foreland of the NW Highlands, so the early Neoproterozoic Torridon Group rocks are overlain unconformably by the Lower Cambrian Ardvreck Group (Eriboll and An t-Sron Formations; Fig. 6). This siliciclastic succession is dominated by feldspathic to quartzitic sandstones and subsidiary siltstone and is interpreted as a transgressive sequence passing upward into storm-dominated calcareous siltstones and regressive sands (McKie, 1990). The Ardvreck Group sediments are conformably but sharply overlain by 900 m of Durness Group dolostone with limestone and minor chert (Ghrudaidh to Durine Formations; Fig. 6) that accumulated on a low-energy shelf (Park et al., 2002). From this change onward, sedimentation in peri- and subtidal environments continued into the Middle Ordovician (Wright and Knight, 1995), and, thus, we envision Greenland-style passive subsidence and a broad platformal shelf extending across inboard parts of the Scottish sector (cf. Higgins et al., 2001a). The lower siliciclastic formations contain distinctive *Skolithos* burrows (Piperock) and pass up into dolomitic siltstone and minor limestone containing diverse macrofaunal assemblages and *Planolites* burrows (Park et al., 2002).

In the central fjord region of East Greenland (Fig. 6), the Cambrian-Ordovician Kong Oscar Fjord Group is separated from the Tillite Group by a disconformity or erosional unconformity with cross-bedded quartz arenites and rippled bed tops passing upward into shales, thin sandstones, and limestones (Smith et al., 2004). Limestones and dolostones then increase upward to become the dominant facies. The base of the Ordovician is marked by a transgression, above which subtidal carbonate environments dominate (Smith et al., 2004). The Cambrian-Ordovician lithostratigraphy can be traced continuously over many hundreds of kilometers N-S along strike and demonstrates systematic thickening of the succession from inboard to outboard positions on the original depositional margin (Higgins et al., 2001a).

In summary, the NW Highlands Cambrian-Ordovician succession of Scotland apparently represents an intermediate position on the slowly subsiding Laurentian carbonate platform that was located between more inboard and more outboard environments, each of which is represented by Cambrian-Ordovician sedimentary rocks in East Greenland (Smith et al., 2004).

The lower parts of the shallow-water carbonate shelf succession represented by the NW Highlands Durness Group and the Kong Oscar Fjord Group of East Greenland are thought to be contemporaneous with the younger elements of the deep-marine turbidite basins of the southern Highlands of Scotland (Fig. 6; Wright and Knight, 1995; Park et al., 2002). Tanner (1995) made it clear that the lower Paleozoic Leny Limestone Formation occurs in stratigraphic continuity with the uppermost parts of the Southern Highland Group, and thus we have the only reliable biostratigraphic age for the Dalradian Supergroup. These metacarbonate rocks preserve *Pagetia* trilobites (Pringle, 1940) and constrain the uppermost Dalradian to be topmost Lower Cambrian, i.e., ca. 515 Ma. Acritarchs from the Leny Limestone Formation have been correlated with Greenland but are long-ranging (Downie, 1982).

Fault-bound slivers preserved locally along the Highland Boundary fault and assigned to the Highland Border Complex preserve Arenig carbonate sediments and black shale and pillow lava of Arenig age along with remnants of a fragmented pre-Arenig ophiolite (Tanner and Sutherland, 2007; but see also review in Bluck, 2002). While the provenance of these fault-bound slivers is undoubtedly Laurentian, and their stratigraphic ages overlap with part of the Ordovician Durness Group, their affinity with the Dalradian succession with which they are now juxtaposed has remained equivocal until now. Tanner and Sutherland (2007) reappraised the paleontologic and stratigraphic evidence and argued for a largely autochthonous Highland Border Complex in stratigraphic continuity with the Dalradian, which was overridden by a Highland Border ophiolite early in the arc-accretion process.

THE CALEDONIAN OROGENY

Ordovician Arc Accretion: Grampian Orogenesis (470–460 Ma)

By mid-Ordovician time, the Grampian phase of orogenesis halted passive-margin sedimentation (Fig. 6; Lambert and McKerrow, 1976; Soper et al., 1999; McKerrow et al., 2000). This phase records the convergence of the Laurentian continental margin with an intra-oceanic subduction zone and volcanic arc. The paleogeography of the margins of the Iapetus Ocean is likely to have been complex, and the potential for preservation is low; much of the evidence remaining is fragmentary. Parts of an early Cambrian to Early Ordovician continent-facing mafic to silicic arc and suprasubduction ophiolites are exposed in western Ireland, where it has proved possible to determine the sequence of events in a short-lived continent-arc collision orogeny (Dewey and Ryan, 1990; Dewey and Mange, 1999). There is indirect evidence that such an arc is buried beneath the Devonian-Carboniferous sedimentary cover in the Midland Valley of Scotland (Bluck, 1983, 1984).

Accretion is thought to have resulted in an overthrust ophiolite nappe, perhaps analogous to the Shetland ophiolite complex, which structurally overlies Dalradian rocks on Unst in NE Shetland. However, this particular fragment of Iapetan crust was apparently obducted at ca. 490 Ma (Flinn et al., 1991), some 20 m.y. prior to the peak of Grampian regional deformation and Barrovian metamorphism in Dalradian and Moine sediments. It is this later stage (ca. 470 Ma) that probably represents major collision and arc accretion, although oblique convergence resulted in some diachroneity of development in the regional structural architecture. Geochronologic constraints for Grampian orogenesis indicate that deformation, magmatism, and regional metamorphism and migmatization occurred within an interval of ~10 m.y. in the Middle Ordovician, between ca. 471 Ma and 462 Ma (Friedrich et al., 1999). Switching subduction polarity at about this time initiated a south-facing arc and an Ordovician-Silurian subduction-accretion complex (Dewey and Ryan, 1990; Dewey and Mange, 1999). The Tyrone ophiolite of Northern Ireland may represent an Arenig-Llanvirn back-arc basin overthrust by the Sperrins nappe, part of the SE-facing and southerly directed nappe complex that includes the Tay nappe in the southern part of the Grampian Highlands (Fig. 8; Krabbendam et al., 1997).

Grampian orogenesis affected all of the Dalradian and older rocks of the Grampian Highlands. Isotopic evidence proves that parts of the Northern Highlands were also affected by a ca. 470 Ma tectonothermal event, which has been correlated with Grampian orogenesis (Kinny et al., 1999; Rogers et al., 2001; Emery, 2005). The effects of this event are most evident in east Sutherland and eastern Inverness-shire, where the effects of the later (Silurian) Scandian reworking are weak. Peak Grampian deformation culminated in the formation of major fold stacks or nappe complexes and associated zones of structural attenuation. Although deformation was superimposed upon a complex stratigraphic template, the gross lateral continuity of the Dalradian lithostratigraphy precludes the existence of any large-scale thrusting at the present exposure level in the Grampian Highlands (see the cross section in Fig. 8). Illustration of the gross architecture of the deformation is best portrayed in the three-dimensional (3D) block diagram reproduced by Stephenson and Gould (1995, after Thomas, 1979).

Several suites of Ordovician plutonic rocks were intruded into the Dalradian rocks of the northeast Grampian Highlands during regional deformation and metamorphism at ca. 470 Ma (Kneller and Aftalion, 1987; Dempster et al., 2002). These include a syn- to late tectonic suite of basic and ultramafic plutons and two suites of syn- to late tectonic diorites and granites (Fig. 8).

There is scant evidence in the East Greenland Caledonides (or in Svalbard; Harland et al., 1997) of these short-lived early

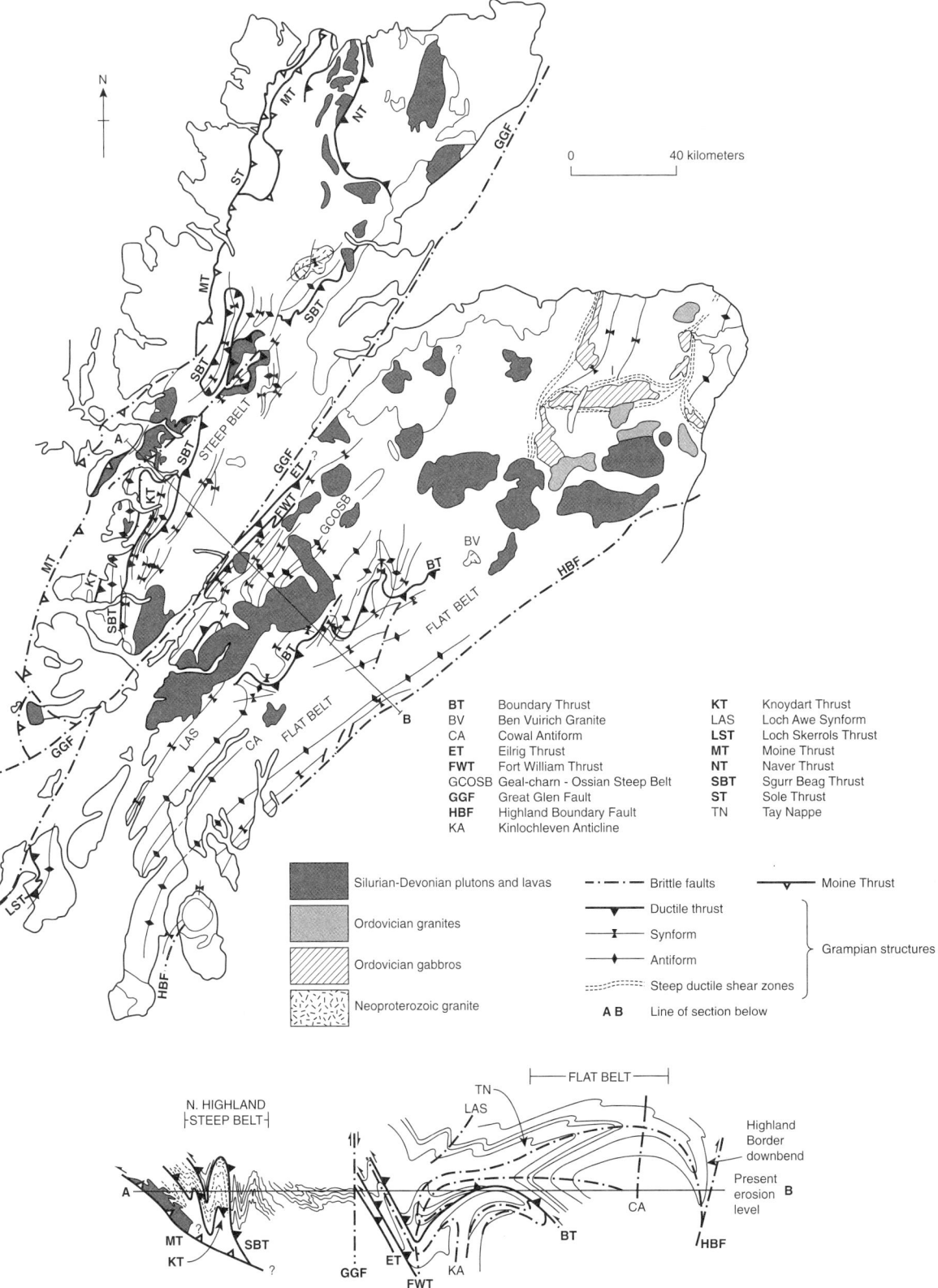

Figure 8. Map of the principal Grampian and Scandian structural features in the Northern and Grampian Highlands of Scotland along with the principal Ordovician and Silurian-Devonian intrusions (cf. Fig. 2). The cross section along the line A–B illustrates the gross structural architecture of each region and overall pattern of structural divergence that results in this part of the Caledonian orogen.

Paleozoic marginal arcs and basins that must have accommodated subduction of Iapetan oceanic crust and convergence with Baltica. Rather, the evidence now available suggests that these distinctive orogenic elements were mainly incorporated as thrust sheets into the higher structural levels of the Scandinavian Caledonides (Roberts et al., 2001; Yoshinobu et al., 2002; Andréasson et al., 2003).

The earliest known (Ordovician and Silurian) granitoids in the East Greenland Caledonides are I-type calc-alkaline granodiorite and quartz-diorite intrusions in the Scoresby Sund region (70°N–72°N), dated by sensitive high-resolution ion microprobe (SHRIMP) U-Pb analyses of zircons to between 466 ± 9 Ma and 432 ± 10 Ma (F. Kalsbeek, 2005, personal commun.). The older date is close to that of the youngest (Middle–Late Ordovician, ca. 460 Ma) sediments preserved in the Franz Joseph allochthon, suggesting that a tectonic control may have brought sediment accumulation to a close in this sector of the Laurentian margin. These I-type granitoids are only known in the southeastern portions of the Hagar Bjerg thrust sheet, which perhaps indicates that this part of the Laurentian continental margin was closest to the site of collision during the Grampian phase of arc accretion on the Laurentian margin. No similar rocks are known in the whole of the East Greenland orogen farther north.

Laurentia-Baltica Scandian Collision (Ca. 430 Ma)

Continued closure of the Iapetus Ocean in the Scottish sector after the Grampian orogenic event was achieved by reversal of the polarity of oceanic subduction (Dewey and Ryan, 1990). The paleogeographic details of these latter stages of convergence and collision are complex and result from the collision and interaction of three continental blocks, namely Laurentia, Baltica, and Avalonia (Soper and Hutton, 1984; Soper et al., 1992; van Staal et al., 1998).

Baltica-Laurentia collision is expressed in Scotland as the Scandian orogeny in the Northern Highlands. Regionally significant ductile thrusting and folding of the Moine rocks and associated basement inliers culminated in the development of the Moine thrust zone at ca. 430 Ma, marking the boundary with the autochthonous-parautochthonous foreland rocks of the Northwest Highlands. The Moine thrust zone is therefore a comparable structure to the Caledonian sole thrust of East Greenland (Higgins and Leslie, 2000).

Scandian Ductile Thrusting and Folding of the Moine Supergroup

Scandian thrust-related folding and fabric development were pervasive within the Moine rocks of west Sutherland but were restricted to localized reworking of migmatites and structures in the east above the Naver thrust (Strachan et al., 2002) (NT on Fig. 8). By analogy, and in the absence of any reliable isotopic evidence to the contrary, a Scandian age may be inferred for some of the movement on similar structures farther south in Ross-shire and Inverness-shire, including the Sgurr Beag thrust (SBT on Fig. 8). The total displacement along these thrusts is uncertain, but it is likely to be at least tens of kilometers and conceivably >100 km in the case of the Sgurr Beag thrust (Powell et al., 1981). This places this structure in the same order of magnitude as the Hagar Bjerg thrust in East Greenland, with which it shares a similar structural level in a foreland-propagating system (cf. Higgins et al., 2004). In Scotland, intense upright folding followed internal ductile thrusting and resulted in the structure referred to as the Northern Highland steep belt (Fig. 9; Strachan et al., 2002). Regional deformation was accompanied by amphibolite-facies Barrovian metamorphism (Strachan et al., 2002).

This regional-scale foreland-propagating thrust system was responsible for the major interleaving of Moine rocks with Lewisian-type basement in Sutherland (Strachan et al., 2002). Many basement inliers occupy the cores of sheath and isoclinal folds along the trace of many of the major thrusts (Fig. 8). The trace of the Sgurr Beag thrust through Ross-shire and northern Inverness-shire to Loch Hourn is commonly marked by allochthonous slices of basement (Tanner et al., 1970); these may have been derived from a rift shoulder within the Moine sedimentary basin (Tanner et al., 1970; Soper et al., 1998) in the same style as the modified rift shoulder of the Hekla Sund Basin in eastern North Greenland (Higgins et al., 2001b).

Moine Thrust Zone

The Moine thrust zone is the westernmost and youngest of the system of Scandian thrusts on the Scottish mainland (Fig. 8). Although localized Caledonian displacements may also have occurred along the Outer Isles fault zone farther to the west, the Moine thrust zone is generally taken to define the northwest edge of the Caledonian orogenic belt (Strachan et al., 2002). In this regard, it is most likely to correlate in style and structural level with the Caledonian sole thrust in East Greenland (Higgins et al., 2004; Leslie and Higgins, this volume), providing a suitable analogue for the margin of the orogen in that region.

The Moine thrust zone varies from a relatively simple planar structure to a complex array of interconnected thrust sheets (Plate 1A; Krabbendam and Leslie, 2004). Detailed analysis has shown that the thrusts generally developed in a foreland-propagating sequence, and successively younger and lower thrusts transported older and higher thrusts to the WNW in piggyback fashion (Elliott and Johnson, 1980; McClay and Coward, 1981; Butler, 1982). Early-formed thrusts within the foreland-propagating sequence are commonly folded as a result of the accretion of underlying thrust sheets. This simple pattern is complicated in some areas by later, low-angle "out-of-sequence" faults that cut through previously thrust-and-folded strata (Holdsworth et al., 2006).

Rb-Sr and K-Ar dating of recrystallized micas within Moine mylonites suggests that emplacement of the Moine rocks onto the foreland occurred ca. 435–430 Ma (Johnson et al., 1985; Kelley, 1988; Freeman et al., 1998). This is consistent with the U-Pb zircon age of 430 ± 4 Ma obtained from the syntectonic Loch Borralan Complex within the Moine thrust zone in the

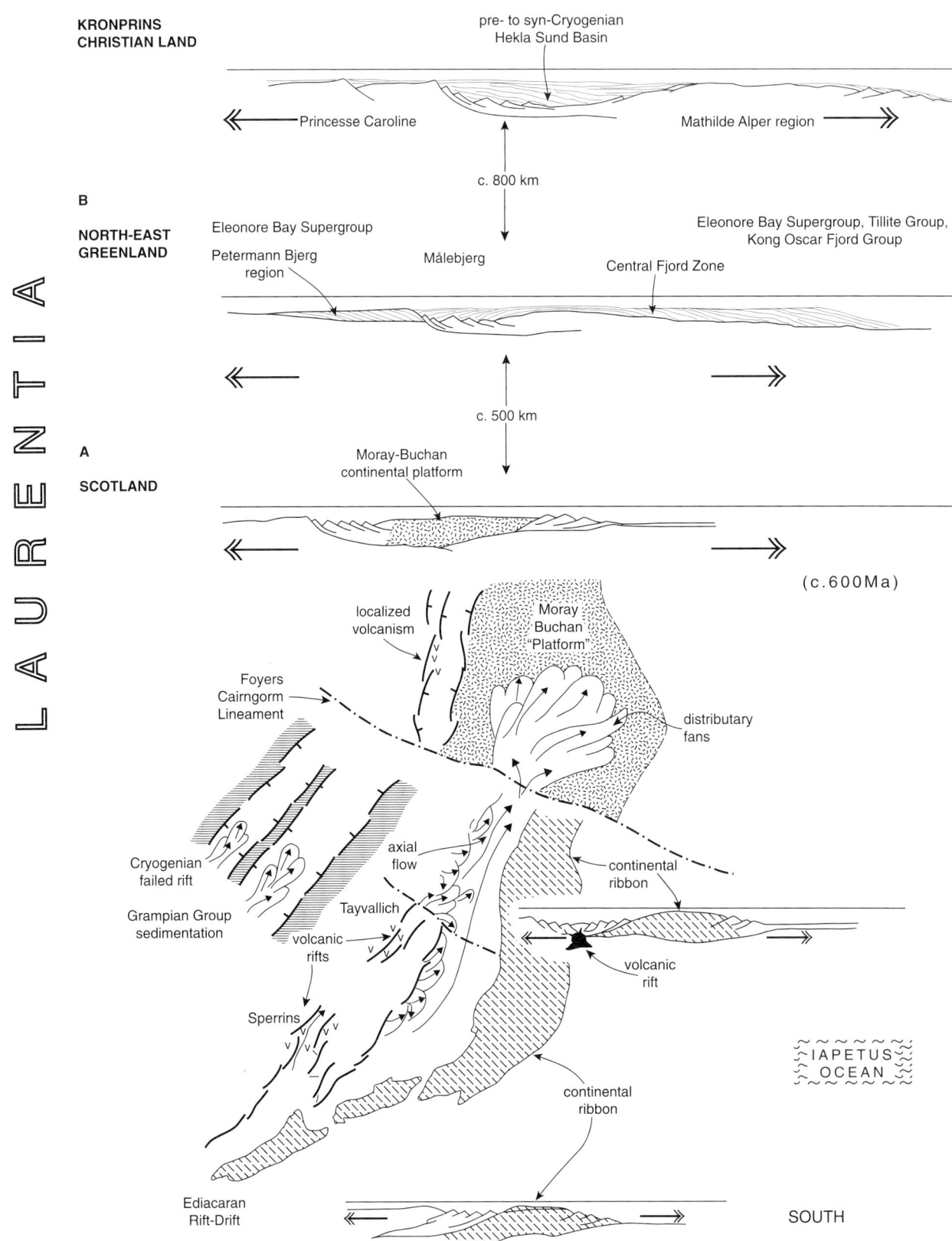

Figure 9. Schematic early Ediacaran paleogeography (A) for the Laurentian "Scottish Corner," along with possible crustal cross sections at this time. Note the suggestion for a change in the dominant polarity of Iapetan faulting in the onboard parts of the margin across the Foyers Cairngorm lineament (cf. Smith et al., 1999). Schematic crustal cross sections for the fjord region of central East Greenland (B) and for Kronprins Christian Land (C) are added for comparison. See discussion in text.

Assynt area (van Breemen et al., 1979). However, isotopic ages as young as ca. 408 Ma have been obtained from mylonites in the Dundonell area, leading to the suggestion that, locally at least, thrusting may have continued into the Early Devonian after the main Scandian collision (Freeman et al., 1998).

The direction of regional thrusting was toward 290°N (McClay and Coward, 1981). It is difficult to estimate the displacement on the Moine thrust itself, but its association with a very thick belt of mylonites (up to 100 m) suggests that it is a major displacement zone with a minimum offset of many tens of kilometers. The construction of balanced sections drawn parallel to the direction of thrusting demonstrates a minimum slip across the thrust zone of 77 km (Elliott and Johnson, 1980), and Butler and Coward (1984) showed that the Cambrian shelf sequence can be restored for ~54 km to the ESE. A total minimum displacement for the Moine thrust zone of around 100 km is therefore commonly accepted.

The Caledonian orogeny in East Greenland was the result of Silurian collision of Baltica with the margin of Laurentia. The structural record and architecture of that part of the orogen preserved onshore record collision in a system of foreland-propagating thrust sheets, which were derived from the Laurentian margin and translated westward across the orogenic foreland (Higgins and Leslie, 2000, this volume; Higgins et al., 2004; Leslie and Higgins, this volume). Restoration of thrusting indicates that the site of collision was probably several hundred kilometers east of the present-day onshore preserved part of the orogen. The thickened orogen was already recording the effects of E-directed thinning and collapse toward the core of the orogen as latest Silurian–Early Devonian orogen-parallel strike-slip deformation began to dominate, dissecting the orogenic welt along its axis.

Silurian-Devonian Magmatism

Mid-Silurian to Early Devonian (ca. 430–400 Ma) subduction-related magmatism is recognized throughout the Highlands of Scotland (Stephenson et al., 1999, and references therein). Recent geochronologic research indicates that emplacement of the majority of these granites in the Northern and Grampian Highlands was focused around 427–425 Ma (Rogers and Dunning, 1991; Oliver, 2001; Fraser et al., 2004), an interval more or less synchronous with the final closure of the Iapetus Ocean in the late Wenlockian (e.g., Stone et al., 1987, 1993; Soper et al., 1992; Torsvik et al., 1992).

Magmatism overlapped Scandian-age deformation, and the emplacement mechanism of many intrusions was structurally controlled either by ductile thrusts or by the later strike-slip brittle faulting. Two main groups are recognized (Stephenson et al., 1999) (Fig. 8): the first is represented by a series of small alkaline intrusions that occur in the northwest of Scotland, mainly in the Assynt area. The second, and volumetrically more important, is commonly referred to as the "Newer Granites" (Read, 1961), although it includes a range of related rock types, including diorite, tonalite, and granodiorite. Members of this group are present on both sides of the Great Glen fault but are particularly common in the Grampian Highlands. The magmatism is represented by mainly I-type, high-K calc-alkaline rocks, some of which are shoshonitic (high-K and high-Mg) in nature. Early Devonian intrusions may have acted as feeders to volcanic sequences (Stephenson et al., 1999).

A subduction-zone setting, where of magmas were derived from the melting of mantle and/or lower-crustal sources, has been considered appropriate for both the alkaline and Newer Granite suites (Stephenson et al., 1999; Strachan et al., 2002). This is reinforced by the calc-alkaline nature of the Newer Granite suite, in particular, the Devonian volcanic rocks (Thirlwall, 1981, 1982). The isotope characteristics of the Newer Granite suite (Halliday, 1984; Stephens and Halliday, 1984; Thirlwall, 1988), as well as the presence of inherited zircons that can only have been derived from older continental basement, indicate some crustal recycling (O'Nions et al., 1983; Harmon, 1983). However, it is also clear that a proportion of magma was derived from the subcontinental lithospheric mantle (Stephens and Halliday, 1984; Tarney and Jones, 1994; Fowler et al., 2001). This mantle-derived magma is represented by the mafic enclaves and appinites associated with some plutons, and also by the calc-alkaline lamprophyre dike swarms. Thus, we conclude that the Newer Granites were derived mainly from the melting of lithospheric mantle and lower-crustal sources and that melting was probably initiated by the introduction of fluids derived from a northward-subducting oceanic slab into an overlying mantle wedge.

It should be noted, however, that the emplacement of the Newer Granites lagged some 30 m.y. after the commencement of NW-directed subduction in the Middle–Late Ordovician at ca. 460 Ma (Oliver, 2001). Onset of plutonism in the late Llandovery at ca. 430 Ma coincided with the time when tectonics in the Southern Uplands of Scotland changed from orthogonal underthrusting to sinistrally oblique underthrusting (Stone, 1995). This change in plate kinematics may ultimately have resulted in development of the crustal-scale sinistral wrench faults that are believed to have acted as fundamental controls on the locus of magma emplacement (Hutton and Reavy, 1992; Jacques and Reavy, 1994).

Kalsbeek et al. (this volume) relate the voluminous Silurian S-type Caledonian granites of the central fjord region of East Greenland to partial fusion of fertile lithologies within a thickening Krummedal supracrustal sequence, most obviously during foreland-directed translation of the Hagar Bjerg thrust sheet.

Formation of this granite magma occurred prior to and during emplacement of the major thrust units and subsequent collapse of the thickened orogen between 435 and 425 Ma. Orogenic collapse followed rapidly after foreland-propagating thrusting linked to Laurentia-Baltica collision (Gilotti and McClelland, this volume; Leslie and Higgins, this volume) and may reflect the same change in plate kinematics that triggered emplacement of the Scottish Newer Granites at ca. 425 Ma.

Sinistral Transtensional Faulting in the Northern and Grampian Highlands

The main phase of mid-Silurian Scandian ductile thrusting was followed by sinistral strike-slip displacements along an array of NE-trending structures that dissect the Northern and Grampian Highlands (Fig. 2). Strike-slip faulting and ductile shear zones developed prior to and during the oblique collision of Avalonia and Baltica with Laurentia throughout the Late Silurian to Early Devonian (Soper et al., 1992). Most of these structures developed prior to the onset of post-Caledonian Old Red Sandstone (Devonian) deposition (Watson, 1984; Mykura, 1991). The most prominent structures are the Great Glen–Walls Boundary and Highland Boundary faults, along which hundreds of kilometers of displacement may have occurred (Dewey and Strachan, 2003).

Seismic-reflection studies show that the Great Glen fault is coincident with a subvertical structure that extends to at least 40 km depth (Hall et al., 1984). Silurian mantle-derived lamprophyre dikes appear to have different isotopic signatures on either side of the fault, suggesting that this structure has some expression in the upper mantle (Canning et al., 1996, 1998). Stewart et al. (1999) argued that blastomylonitic rocks preserved in the core of the fault zone may reflect the presence of an exhumed positive flower structure that formed during sinistral transpression along the same zone of weakness. Relationships among fault-zone structures, dated igneous intrusions, and postorogenic sedimentary rocks constrain the main sinistral displacement along the Great Glen fault to the period between ca. 428 Ma and ca. 400 Ma (Stewart et al., 1999). Newer Granite plutonism was thus initiated broadly concurrently with the onset of major sinistral transtensional displacement. A lower age limit of ca. 400 Ma is indicated by the low-strain nature of Old Red Sandstone (upper Emsian?) sedimentary rocks within the fault zone. Post–Old Red Sandstone structures along the fault zone are invariably brittle in style, and fault products are typically incohesive, consisting of clay fault gouge and poorly consolidated fault breccia (May and Highton, 1997; Stewart et al., 1999).

Although the timing of late Caledonian sinistral transtension is relatively well constrained, the magnitude of early displacement along the Great Glen fault is less certain because there is no unambiguous correlation of pre-Devonian features across the fault. The general consensus (Strachan et al., 2002) is that sinistral strike-slip displacements along the Great Glen fault are unlikely to have exceeded 200–300 km, although Dewey and Strachan (2003) later argued that a bare minimum of 700 km displacement is required if no Scandian deformation should be identified in the Grampian Highlands. The lower value is, however, consistent with the most reliable paleomagnetic evidence (Briden et al., 1984), the inferred offset of reflectors within the mantle lithosphere (Snyder and Flack, 1990), correlation between the Moine Supergroup and the Dava–Glen Banchor successions, and similarities in the timing of Neoproterozoic and Ordovician tectonothermal events on either side of the fault (Bluck, 1995; Stewart et al., 1999; Highton et al., 1999; Rogers et al., 2001). The possibility remains that the Great Glen fault does not after all represent a terrane boundary (sensu stricto), and that a greater apparent discontinuity (including a contrast in crystalline basement properties) exists between the Midland Valley and Southern Uplands of Scotland along the trace of the present Southern Uplands fault.

The present Highland Boundary fault is a high-angle reverse fault that emplaced Dalradian rocks onto the Highland Border Complex and Old Red Sandstone rocks of the Midland Valley (Anderson, 1946; Bluck, 1984). Geophysical studies have shown that the fault is broadly coincident with a change in lower crustal structure (Bamford et al., 1978; Barton, 1992; Rollin, 1994), and this implies that the present structure may have reactivated an older and more fundamental structure. Various workers have speculated that this may correspond to the edge of the Laurentian craton (e.g., Soper and Hutton, 1984). Although Late Silurian to Early Devonian sinistral displacements comparable with the Great Glen fault have been commonly assumed (e.g., Harte et al., 1984; Soper and Hutton, 1984; Hutton, 1987; Soper et al., 1992), other workers have argued against major displacements (e.g., Hutchison and Oliver, 1998), and, thus, the regional tectonic significance of this fault is uncertain.

In central East Greenland, mid-Devonian continental sediments were deposited on the eroded Caledonian orogen; sedimentation was controlled, in part, by sinistral displacements along the Western fault zone (Larsen and Bengaard, 1991; Olsen, 1993). Larsen and Bengaard (1991) tentatively linked the Western fault zone with the Storstrømmen shear zone to the north and with the Great Glen fault to the south. Dewey and Strachan (2003) argued that relative motion between Laurentia and Avalonia-Baltica changed from sinistrally transpressive collision at ca. 425 Ma to more orogen-parallel sinistrally transtensional movements, which persisted until ca. 400 Ma and were terminated, in Britain, by the brief Acadian orogeny (Soper and Woodcock, 2003). Dewey and Strachan (2003) argued that the Great Glen fault would have been the principal structure on which sinistral displacement was accommodated in the Scottish sector of the orogen, linked northward via the Walls Boundary fault in Shetland into East Greenland as the Western fault zone.

Similar translations have been proposed to explain the juxtaposition of the separate terranes identified in Svalbard (e.g., Soper et al., 1992; Harland et al., 1997). Gee and Teben'kov (2004) proposed an alternative interpretation which minimizes the scale of strike-slip displacement and, if proven, might argue that displacement on the major strike-slip fault systems transecting the Scottish Caledonides should be of a similar dimension, i.e., <200 km.

A TEMPLATE FOR COMPARISON OF IAPETAN RIFTING AND CALEDONIAN OROGENESIS IN THE SCOTLAND-GREENLAND SECTOR OF LAURENTIA

The Iapetan geological record from Scotland constrains the establishment of the Laurentian passive margin and culmination of the two-stage breakout of Laurentia from the supercontinent Rodinia. The uppermost Neoproterozoic to lower Paleozoic Scottish successions record proximity to a RRR junction (Soper, 1994a), and by late Neoproterozoic (Ediacaran) time, the "Scottish Corner" may have resembled a patchwork of marginal plateaus or continental ribbons extending some 1000 km or more along the margin (Fig. 9), similar in style to the paleogeography proposed by Waldron and van Staal (2001) and Cawood et al. (2001) for the Newfoundland sector. The geological map of the United Kingdom–Faroes sector of the present-day continental shelf presents a framework of interlocking depocenters (Hatton, Rockall, and Porcupine Basins) and structural basement highs (Rockall and Porcupine highs) and provides a good modern analogue for the late Neoproterozoic paleogeography of the "Scottish Corner" of Laurentia.

In contrast, the lithostratigraphy of East Greenland can be traced continuously over considerable distances along strike along the original depositional margin. A general absence of volcanic activity suggests that the East Greenland sector may have lain (dependent upon stretching rates) at some distance from any focus of Iapetan volcanism (Fig. 1). There is systematic thickening of the successions from inboard to outboard positions in the sector, and the pattern of N-S facies belts shapes a slowly subsiding but relatively stable continental margin that must have extended beyond and included northeastern Svalbard at this time (Gee and Teben'kov, 2004).

In the following sections, we provide a dynamic synthesis of the geological evolution of the Scottish sector of Laurentia (the "Scottish Corner") through a complete Wilson cycle that encompasses the opening and closure of the Iapetus Ocean. The synthesis reflects the authors' stance in regard to the wealth of new and archived data available to geologists studying the evolution of the Scottish and East Greenland Caledonides.

Cycle I: Early (Failed) Cryogenian Rifting

Mid-Cryogenian rifting on the "Scottish Corner" probably initiated after 730 Ma and then lasted for 60–70 m.y., with post-rift thermal subsidence extending for another ~50 m.y. or so toward the end of the Cryogenian. Turbiditic sands and muds of the Grampian Group, derived from a hinterland to the west and northwest, infilled a series of basins (Banks, 2005), the distribution and depositional geometry of which are constrained by the position, and uplift history, of discrete intrabasinal highs, e.g., Robertson and Smith (1999). These localized basins were infilled by the time progradational shelf sands extended from the south and east across these early (presumably failed) rifts and their margins in the upper Grampian Group. The continued sand- and mud-dominated sedimentary facies associations of this first cycle culminated with diminished sediment input into marginal off-shelf to lagoonal, locally emergent, even evaporitic environments as recorded in the lowermost Appin Group (Lochaber Subgroup) (Stephenson and Gould, 1995).

In Greenland, the contemporaneous Nathorst Land Group appears to reflect a more gradual and widely distributed subsidence, and there is no sign of localized depocenters (Sønderholm et al., this volume). The correlation proposed here at significant flooding surfaces (Fig. 6) may signify short-lived enhanced stretching that affected all of this part of Laurentia. Active stretching probably advanced northward from the vicinity of the "Scottish Corner" toward East Greenland, perhaps linking with an active rift system that was encroaching from the north.

Cycle I: Sag

A major marginwide transgression across a flooding surface marks the base of the Ballachulish Subgroup in Scotland and the Ymer Ø Group in East Greenland (Fig. 6). Flooding began the onset and development of wide-ranging and uniform sedimentary lithofacies. Typically, deposition of dark anoxic limestone and mud is followed by shallowing-upward cycles of progradational clean-washed sands and shallow-water muds and limestones (Stephenson and Gould, 1995). Basins at this time were probably wide and shallow, as evidenced by remarkably similar successions along strike extending across Ireland and Scotland and with tentative correlations possible at formation level across Greenland into the Hekla Sund Basin of eastern North Greenland (Fig. 6). Renewed flooding (eustatic change?) heralded the deposition of further muds and limestones (Blair Atholl Subgroup in Scotland and the Andrée Land Group in East Greenland) before "Marinoan" diamictites were deposited during a major lowstand at ca. 635 Ma. In the Southern Grampian Highlands, some parts of these basins were interspersed with sediment starved areas, which are now expressed as major stratigraphic omission and subsequent overstep (A.G. Leslie for BGS, unpublished data).

Cycle II: Renewed Rifting to Rift-Drift Transition in the Ediacaran

Renewed and vigorous extension is recorded by rifting in the Scottish Dalradian in early Ediacaran time. Contemporaneous mafic volcanism marks the transition to Iapetan rift-drift. Sharply defined thickness changes become apparent in the Islay Subgroup depositional record and are most likely fault controlled (Anderton, 1985). The upward-shallowing cyclical behavior recorded previously begins to repeat again in the Easdale Subgroup, albeit in an increasingly unstable, but still essentially shallow-marine, on-shelf environment, with deposition of quartzitic sands, limestones, and muds. Instability is recorded in influxes of pebbly sands (Stephenson and Gould, 1995). Localized volcanogenic centers become a feature of this sector of the margin, with punctuated episodes of mafic volcanism (Goodman and Winchester, 1993; Macdonald et al., 2005).

Sediment-starved sectors in Scotland are only overstepped at this later stage (A.G. Leslie for BGS, unpublished data), and a more rapidly foundering rift system (Crinan Subgroup) then evolved, accumulating debris flows and slumps (Stephenson and Gould, 1995). Incipient volcanic spreading centers located mainly in the southwest (Tayvallich Subgroup, Fig. 6) gave way to organized turbiditic submarine fans (Southern Highland Group) as final rupture occurred and this sector of Iapetus began to widen. The major change to more immature sediment, increasingly dominated by Archean detritus, occurs at the base of the Crinan Subgroup and persists on through the Southern Highland Group as separation occurred (rift-drift transition) and the foundering margin evolved toward a stable passive margin by Cambrian-Ordovician time. Laurentian fauna (*Pagetia*) contained within the uppermost, youngest Dalradian strata (Tanner, 1995) and conodont assemblages in the Durness Group (Wright and Knight, 1995) place the "Scottish Corner" firmly on the Laurentian margin.

Erosion of the upper formations of the Eleonore Bay Supergroup to provide clasts in the latest Cryogenian to early Ediacaran (Marinoan) Tillite Group indicates pre–Tillite Group uplift, but it is not clear whether the absence of any younger Ediacaran deposits merely records lack of supply and nondeposition or if extensional rifting and block tilting removed any sediment accumulation prior to deposition of the Cambrian succession (Smith et al., 2004; Sønderholm et al., this volume; Smith and Rasmussen, this volume). Any accumulations may have been scavenged from this relatively stable, nonvolcanic marginal platform into the active and volcanic rifts on the "Scottish Corner."

Figure 9 presents a summary of how the Scottish sector may have looked at this time. Rifting leading to separation occurred ca. 600 Ma; fragments of the continental margin then broke away, and Iapetan spreading ridges became active (off to the lower right of the cartoon). Relative to the central parts of the Grampian and the southwest Scottish Highlands, we speculate that Connemara in western Ireland may have constituted a marginal plateau or ribbon, while the Moray-Buchan area of northeast Scotland may have constituted a subdued marginal platform. The newly established peri-Laurentian trough received a deluge of submarine debris flows (Crinan Subgroup); localized volcanic centers built out volcaniclastic deltas. Carbonate buildups were reworked from the shelf into deeper water as redeposited metacarbonate rocks in the Tayvallich Subgroup (Thomas et al., 2004). Submarine fans prograded from the margin into the trough and may have extended along the trough axis, spilling out onto adjacent marginal platforms. Deltaic volcaniclastic debris was reworked along the margin as "green beds" in the Southern Highland Group (Pickett et al., 2006). The extensional geometry of the various components of this architecture must of necessity be speculative; possible cross-section configurations are incorporated in the model of Figure 9. We can suggest, however, that these geometries subsequently exerted control on the collisional geometry and acted as nuclei for deformation structures during arc accretion in the Grampian orogeny.

The analogue model of Figure 9 can be extended to include the East Greenland sector of the Laurentian margin. Although late Neoproterozoic extension may locally have affected the inboard portion of North-East Greenland (e.g., in the Målebjerg region; Leslie and Higgins, 1998, this volume), the lack of volcanism and persistent shallow-marine shelf depositional environments imply a broad, gradually subsiding continental platform where subsidence rate was broadly matched by sediment influx, at least until the late Cryogenian. The Hekla Sund Basin of eastern North Greenland preserves evidence of the northward change into to a more active, but nonvolcanic, extensional setting on Laurentia, in the Cryogenian at least.

Evolution to a Cambrian-Ordovician Passive Margin

By Cambrian-Ordovician time, a broad carbonate platform was established along the continental margin of Laurentia facing the Iapetus Ocean. Orthoquartzite/carbonate successions that accumulated in this period are recognized all along the length of the Caledonian fold belt in East Greenland (Higgins et al., 2001a), across the far northwest of Scotland, and through Newfoundland and the Appalachians (see compilation in Dewey and Mange, 1999). Whether or not Cambrian-Ordovician shelf, shelf-slope, and rise successions either did (Dewey, 1969) or did not (Bluck et al., 1997) continuously extend across Scotland from the Laurentian foreland into the deeper-water environments of the Cambrian upper Dalradian of the Southern Grampian Highlands has been a matter of debate. Some support for the former position comes from reconstructions of the Newfoundland margin (Waldron and van Staal, 2001) and from East Greenland (Leslie and Higgins, 1998, this volume), and the new work of Tanner and Sutherland (2007) has resolved the matter in the Scottish sector. Analogues for the Cambrian-Ordovician successions of the Highland Border Series are perhaps to be found in later pelagic passive-margin sedimentation on a fragmented continental margin (Stoker et al., 2001), which would have accumulated in the troughs featured in the model of Figure 9.

Grampian Arc Accretion

Figure 10A illustrates our Early Ordovician (Tremadoc–Arenig) reconstruction of the Scottish sector of Laurentia. At this point, Cryogenian siliciclastic detritus had built out a progradational sedimentary prism, drowning in the process such intrabasinal structural features as the Glen Banchor "basement high" (Robertson and Smith, 1999; GB on Fig. 10A). Ediacaran (ca. 600 Ma) intrusion of extension-related bimodal magmatism (Tanner et al., 2006) into the expanding sediment prism is represented by the Ben Vuirich Granite Pluton and the Tayvallich Volcanics Formation (BV and TV on Fig. 10A, respectively).

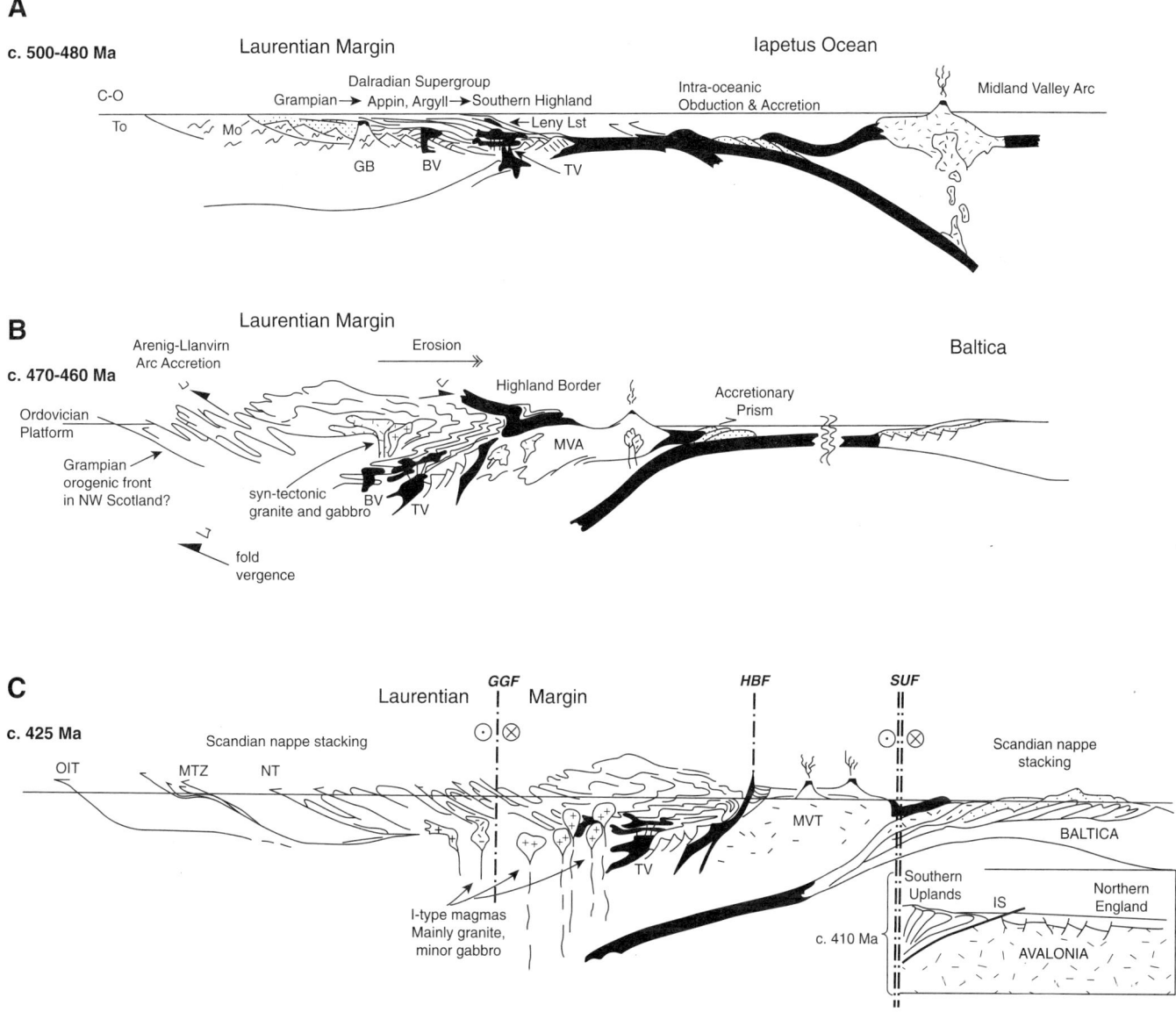

Figure 10. Schematic sections (A–C) illustrating the Ordovician to Silurian plate-tectonic evolution of the "Scottish Corner" on the Laurentian continental margin. In the inset addition to section C, end-Silurian (ca. 410 Ma) juxtaposition of Avalonia against the Midland Valley was achieved along a major terrane boundary now expressed as the Southern Upland fault. BV—Ben Vuirich Granite; C-O—Cambrian-Ordovician shelf succession; GB—Glen Banchor "High"; GGF—Great Glen fault; HBF—Highland Boundary fault; IS—Iapetus suture; Mo—Moine Supergroup; MTZ—Moine thrust zone; MVA—Midland Valley arc; MVT—Midland Valley terrane; NT—Naver thrust; OIT—Outer Isles thrust; SUF—Southern Upland fault; To—Torridonian; TV—Tayvallich volcanics.

The Cambrian-Ordovician Durness Group represents major marine transgression onto the subsiding continental margin, while deeper-water contemporaneous successions accumulated on the Laurentian continental slope and rise (upper Dalradian Leny Limestone). Intra-oceanic obduction and accretion was under way by this time (Bluck, 2001), and the Midland Valley Arc began to encroach upon the Laurentian continental margin.

By Arenig-Llanvirn time, the Midland Valley Arc (MVA on Fig. 10B) had been accreted onto the continental margin. The arc was partly underthrust on the margin such that a zone of top-to-the-south or -southeast noncoaxial simple shear formed in the lower structural levels of the Tay nappe, as recorded in the southern Scottish Highlands (Krabbendam et al., 1997). SE-facing extensional fault blocks on the ruptured continental margin (Fig. 9) would have rotated and steepened toward the interior of the orogenic wedge underneath the developing nappe (HB on Fig. 10A). Syntectonic mafic and calc-alkaline felsic magmas emplaced into the thickening orogenic pile at ca. 470 Ma (Friedrich et al., 1999) probably formed as a consequence of the suborogenic heating as the subducting oceanic crust detached southeastward beneath the accreting magmatic arc (Dewey and Mange, 1999). The Grampian front clearly extended into the Moinian rocks of the Northern Highlands of Scotland (Dallmeyer et al., 2001), and crustal flexure at this time (ca. 460 Ma) would have ended shallow-water carbonate deposition on the continental platform.

No widespread evidence of Ordovician arc accretion survives in East Greenland; the terranes involved in collision have been transported eastward and incorporated into the higher structural levels of the Scandinavian Caledonides (Yoshinobu et al., 2002).

With rapid uplift and erosion of the thickened orogenic wedge postdating the Grampian metamorphic peak (ca. 465 Ma), Grampian-age metamorphic detritus was dispersed across the accreted arc (or arcs) and supplied to an accretionary wedge facing the narrowing Iapetus Ocean. In Scotland, such detritus appears in the sedimentary record of the Southern Uplands accretionary prism in the Caradoc (Oliver, 2001). However, the system of large-scale strike-slip displacements that affected this sector of the closing Iapetus in the Silurian-Devonian (ca. 425–400 Ma) means that the Late Ordovician–Early Silurian sediment dispersal pathways cannot now be determined but are likely to have incorporated considerable distances along the orogenic belt.

Laurentia-Baltica Collision

Baltica collided with the East Greenland sector of Laurentia by late Llandoverian–early Wenlockian time. The reconstruction in Figures 10B and 10C speculates that the Scottish sector would also have experienced oblique collision with parts of Baltica. Post-metamorphic peak exhumation of the Grampian orogenic wedge was largely complete, and the lower inverted limb of the Tay nappe had by now formed the flat belt identified in the cross section in Figure 8. In contrast, Scandian nappe stacking dominated the structural architecture of the Northern Highlands of Scotland and the East Greenland Caledonides. One possible view is that the Grampian Highlands may thus have formed a relatively rigid block entrained between the contractional deformation zones at the leading edges of the colliding Laurentian and Baltican plates. At ca. 425 Ma, in the mid-to-late Silurian, large volumes of felsic magma were intruded into the Grampian and Northern Highlands, and partitioned sinistral transtensional stresses began to replace Early Silurian oblique convergence and transpression, as evidenced by movements along the Great Glen (Stewart et al., 1999), Highland Boundary, and Southern Upland faults. Juxtaposition of the present-day Southern Uplands region with the Midland Valley and Scottish Highlands was largely achieved at this time as "Scottish" Laurentia and Avalonia were juxtaposed along WSW-ESE strike-slip systems, expressed today in the Southern Uplands fault (see ca. 410 Ma inset on Fig. 10C). Major pull-apart and extension in the more N-S–trending tracts of the Caledonian suture between "East Greenlandic" Laurentia and western Baltica may have initiated the Early Devonian eclogite exhumation processes in the Western Gneiss Region of Norway and the eclogite-bearing terrain of East Greenland (Krabbendam and Dewey, 1998; cf. Gilotti et al., this volume). Evidence for orogenic collapse in the Scottish sector is possibly represented by the depositional architecture and deformation of the Old Red Sandstone (Middle Devonian) Orcadian Basin in Orkney and Shetland (Seranne, 1992), and the top-to-the-NE–directed shear fabrics recorded in the Shetland ophiolite complex and the metasedimentary successions of Unst and Fetlar in NE Shetland (Cannat, 1989). NE-vergent extension in Shetland is opposite to the SW-vergent extension and exhumation of the Western Gneiss Region in Norway (Krabbendam and Dewey, 1998), which suggests perhaps an internal zone of collapse toward the interior of the Caledonian orogen.

CONCLUSIONS AND FUTURE RESEARCH

A number of key observations and conclusions emerge from these correlations and discussion.

1. By ca. 1.9 Ga, the ancient basement of Scotland and Greenland lay within a continuous accretionary belt made up of various Archean cratonic components welded together during the assembly of Laurentia and Baltica. Calc-alkaline magmatism concentrated along a new active margin is represented by the ca. 1.9–1.85 Ga Makkovik-Ketilidian belt of Labrador and South Greenland, juvenile Proterozoic crust forming at ca. 1.78 Ga in the Rhinns Complex in Scotland, and the 1.85–1.50 Ga Labradorian-Gothian belt of NE Canada and SW Scandinavia.

2. Early Neoproterozoic fluvial red-bed successions (Torridon Group) buried deeply eroded Archean-Paleoproterozoic basement rocks in the foreland to the Caledonian orogen in Scotland; no comparable sediments are known in East Greenland.

3. Late Mesoproterozoic to early Neoproterozoic siliciclastic successions are widespread in Greenland (Krummedal supracrustal succession), NW Scotland (Moine Supergroup), and in the eastern province of Spitsbergen (Brennevinsfjorden Group). These successions are almost entirely marine deposits and indicate that a depocenter, or interconnected depocenters, may have extended over several thousand square kilometers along the Laurentian margin during this time. Since no equivalent of the early Neoproterozoic Torridon Group fluvial red-bed system seems to have been present at the paleolatitude represented by East Greenland, we conclude that it is possible that the Krummedal succession may represent a northward continuation of marine dispersal of parts of the Moine succession.

4. Three episodes of early to mid-Neoproterozoic orogenesis (ca. 930 Ma, ca. 820 Ma, ca. 730 Ma) are recorded in the Scottish and Greenland sectors of Laurentia, and each is separated by ~100 m.y. The Rodinian supercontinent was undergoing late-stage amalgamation during this time, and so intracratonic movements accommodating extension and compressional deformation may be required to account for these orogenic events (e.g., Cawood et al., 2004).

5. In Scotland, the mid-Neoproterozoic to mid-Cambrian Dalradian Supergroup was deposited in an evolving pericontinental environment over a period of ~180–200 m.y. Fluctuations in water depth accompanied active rifting and the development of second- and third-order subbasins on this sector of the Laurentian

margin. In East Greenland, farther north on that same margin, the broadly time-equivalent Eleonore Bay Supergroup and overlying Tillite Group and Kong Oscar Fjord Group record no active rifting and contain no evidence of contemporaneous volcanic activity. East Greenland strata record shelf and ramp sedimentation, which was probably punctuated by significant periods of non-deposition on a slowly subsiding passive margin. Farther north still, the Neoproterozoic Hekla Sund Basin represents another locus of rifting activity in eastern North Greenland. The youngest units of the Dalradian Supergroup in Scotland record rapid basin deepening that persistently stayed ahead of the sedimentary and volcanic basin fill. No apparent match for these immature turbidite-dominated sedimentary facies associations occurs anywhere along the Greenland sector of Laurentia.

6. The NW Highlands Cambrian-Ordovician succession of Scotland apparently represents an intermediate position on the slowly subsiding Laurentian carbonate platform that was located between more inboard and more outboard environments, each of which is represented by the Cambrian-Ordovician succession in East Greenland.

7. Mid-Ordovician Grampian orogenesis affected all of the Dalradian and older rocks of the Grampian Highlands as well as parts of the Moine Supergroup in the Northern Highlands. This phase records the convergence of the Laurentian continental margin with an intra-oceanic subduction zone and volcanic arc. There is, in contrast, scant evidence in the East Greenland Caledonides (or in Svalbard) of these short-lived early Paleozoic marginal arcs and basins that must have accommodated subduction of Iapetan oceanic crust and convergence with Baltica.

8. Baltica-Laurentia collision is expressed in Scotland and East Greenland as the Scandian orogeny. Regional-scale ductile thrusting and folding of the Moine rocks in the Northern Highlands of Scotland culminated in the development of the Moine thrust zone at ca. 430 Ma, marking the boundary with the autochthonous-parautochthonous foreland rocks of the Northwest Highlands. The Moine thrust zone is therefore a comparable structure to the Caledonian sole thrust of East Greenland. Restoration of the system of foreland-propagating thrust sheets derived from the Laurentian margin and translated westward across the orogenic foreland in East Greenland indicates that the site of collision was probably several hundred kilometers east of the present-day onshore preserved part of the orogen.

9. Mid-Silurian to Early Devonian (ca. 430–400 Ma) subduction-related magmatism is recognized throughout the Highlands of Scotland (Stephenson et al., 1999, and references therein). Kalsbeek et al. (this volume) relate the voluminous Silurian S-type Caledonian granites of this age in the central fjord region of East Greenland to partial fusion of fertile lithologies as a consequence of crustal thickening, most obviously during foreland-directed translation of the Hagar Bjerg thrust sheet.

10. The present distribution of Caledonian domains across the northern Atlantic region is in part a consequence of the transtensional shearing that resulted from relative lateral motion of the two large continental segments (Laurentia and Baltica) after their respective continental margins were in contact and subduction of oceanic crust had ceased. Sinistral wrench faults dissect the Caledonian orogen of both Scotland and East Greenland.

Together, East Greenland and Scotland preserve a partial 350 m.y. record of deposition and rifting leading toward opening and spreading of the Iapetus Ocean, and then of the arc accretion and continent-continent collision that consumed Iapetus. While the "Scottish Corner" is undoubtedly geologically fascinating, and perplexing, we have found it hugely rewarding to work on the bigger scale. We present this overview as a comparative synopsis of our present understanding of the Scottish and East Greenland orthotectonic sectors of the Caledonian orogen and as an encouragement to future researchers to address issues of diachroneity in otherwise similar events and strive for better understanding of the "big picture." The tentative (tectono)stratigraphic correlations proposed here must be tested further. Many of those answers will lie in systematic analysis of Caledonian geology around the North Atlantic region, linking Scandinavia, Svalbard, Greenland, Scotland, Labrador, and Norway.

ACKNOWLEDGMENTS

Smith and Leslie publish with the permission of the Director of the British Geological Survey. Constructive comments and criticism were gratefully received from P. Cawood and from an anonymous referee.

REFERENCES CITED

Aleinikoff, J.N., Zartman, R.E., Walter, M., Rankin, D.W., Lyttle, P.T., and Burton, W.C., 1995, U-Pb ages of metarhyolites of the Catoctin and Mount Rogers Formations, Central and Southern Appalachians: Evidence for two pulses of Iapetan rifting: American Journal of Science, v. 295, p. 428–454.

Alsop, G.I., Prave, A.R., Condon, D.J., and Phillips, C.A., 2000, Cleaved clasts in Dalradian conglomerates: Possible evidence for Neoproterozoic compressional tectonism in Scotland and Ireland?: Geological Journal, v. 35, p. 87–98, doi: 10.1002/1099-1034(200004/06)35:2<87::AID-GJ842>3.0.CO;2-7.

Anderson, J.G.C., 1946, The geology of the Highland Border: Stonehaven to Arran: Transactions of the Royal Society of Edinburgh, Earth Sciences, v. 61, p. 479–515.

Anderton, R., 1985, Sedimentation and tectonics in the Scottish Dalradian: Scottish Journal of Geology, v. 21, p. 407–436.

Andréasson, P.G., Gee, D.G., Whitehouse, M.J., and Schoberg, H., 2003, Subduction-flip during Iapetus Ocean closure and Baltica-Laurentia collision, Scandinavian Caledonides: Terra Nova, v. 15, p. 362–369, doi: 10.1046/j.1365-3121.2003.00486.x.

Badger, R.L., and Sinha, A.K., 1988, Age and Sr isotopic signature of the Catoctin volcanic province: Implications for subcrustal mantle evolution: Geology, v. 16, p. 692–695, doi: 10.1130/0091-7613(1988)016<0692:AASISO>2.3.CO;2.

Bailey, C.M., and Tollo, R.P., 1998, Late Neoproterozoic extension-related magma emplacement in the Central Appalachians: An example from the Polly Wright Cove pluton: The Journal of Geology, v. 106, p. 347–359.

Bamford, D., Nunn, K., Proedehl, C., and Jacob, B., 1978, LISPB-IV. Crustal studies of northern Britain: Geophysical Journal of the Royal Astronomical Society, v. 54, p. 43–60.

Banks, C.J., 2005, Neoproterozoic Basin Analysis: A Combined Sedimentological and Provenance Study in the Grampian Group, Central Highlands, Scotland [Ph.D. thesis]: Keele, University of Keele, 520 p.

Banks, C.J., and Winchester, J.A., 2004, Sedimentology and stratigraphic affinities of Neoproterozoic coarse clastic successions, Glenshirra Group, Inverness-shire, Scotland: Scottish Journal of Geology, v. 40, p. 159–174.

Barton, P.J., 1992, LISPB revisited: A new look at the Caledonides of northern Britain: Geophysical Journal International, v. 110, p. 371–391, doi: 10.1111/j.1365-246X.1992.tb00881.x.

Bingen, B., Demaiffe, D., and Van Breemen, O., 1998, The 616 Ma old Egersund basaltic dyke swarm, SW Norway, and late Neoproterozoic opening of the Iapetus Ocean: The Journal of Geology, v. 106, p. 565–574.

Bingen, B., Griffin, W.L., Torsvik, T.H., and Saeed, A., 2005, Timing of late Neoproterozoic glaciation on Baltica constrained by detrital zircon geochronology in the Hedmark Group, south-east Norway: Terra Nova, v. 17, p. 250–258, doi: 10.1111/j.1365-3121.2005.00609.x.

Bluck, B.J., 1983, Role of the Midland Valley of Scotland in the Caledonian orogeny: Transactions of the Royal Society of Edinburgh, Earth Sciences, v. 74, p. 275–295.

Bluck, B.J., 1984, Pre-Carboniferous history of the Midland Valley of Scotland: Transactions of the Royal Society of Edinburgh, Earth Sciences, v. 75, p. 275–295.

Bluck, B.J., 1995, W.Q. Kennedy, the Great Glen fault and strike-slip motion, in Le Bas, M.J., ed., Milestones in Geology: Geological Society of London Memoir 16, p. 57–65.

Bluck, B.J., 2001, Caledonian and related events in Scotland: Transactions of the Royal Society of Edinburgh, Earth Sciences, v. 91, p. 375–404.

Bluck, B.J., 2002, The Midland Valley terrane, in Trewin, N.H., ed., The Geology of Scotland (4th edition): London, Geological Society of London, p. 149–166.

Bluck, B.J., Dempster, T.J., and Rogers, G., 1997, Allochthonous metamorphic blocks on the Hebridean passive margin, Scotland: Journal of the Geological Society of London, v. 154, p. 921–924, doi: 10.1144/gsjgs.154.6.0921.

Bowring, S., Myrow, P., Landing, E., Ramezani, J., and Grotzinger, J., 2003, Geochronological constraints on terminal Neoproterozoic events and the rise of metazoans: Geophysical Research Abstracts, v. 5, p. 13,219.

Brasier, M.D., and Shields, G., 2000, Neoproterozoic chemostratigraphy and correlation of the Port Askaig glaciation, Dalradian Supergroup of Scotland: Journal of the Geological Society of London, v. 157, p. 909–914.

Brasier, M.D., McCarron, G., Tucker, R., Leather, J., Allen, P., and Shields, G., 2000, New U-Pb zircon dates for the Neoproterozoic Ghubrah glaciation and for the top of the Huqf Supergroup, Oman: Geology, v. 28, p. 175–178, doi: 10.1130/0091-7613(2000)28<175:NUZDFT>2.0.CO;2.

Briden, J.C., Turnell, H.B., and Watts, D.R., 1984, British paleo-magnetism, Iapetus Ocean and the Great Glen fault: Geology, v. 12, p. 428–431, doi: 10.1130/0091-7613(1984)12<428:BPIOAT>2.0.CO;2.

Bridgwater, D., Campbell, L., Mengel, F., Marker, M., and Scott, D., 1996, The Nagssugtoqidian of West Greenland in the light of comparative studies of juvenile components in the Palaeoproterozoic Torngat, SE Greenland Nagssugtoqidian, and Lapland-Kola 'collisional' belt, in Proceedings of the 2nd Danish Lithosphere Centre Workshop on Nagssugtoquidian Geology: Copenhagen, Danish Lithosphere Centre, p. 8–19.

British Geological Survey, 2002, Solid Geological Map, Sheet 63E (Dalwhinnie, Scotland): Keyworth, Nottingham, British Geological Survey, scale 1:50,000.

British Geological Survey, 2004, Solid Geological Map, Sheet 74W (Tomatin, Scotland): Keyworth, Nottingham, British Geological Survey, scale 1:50,000.

Buchan, K.L., Mertanen, S., Park, R.G., Pesonen, L.J., Elming, S.A., Abrahamsen, N., and Bylund, G., 2000, Comparing the drift of Laurentia and Baltica in the Proterozoic: The importance of key palaeomagnetic poles: Tectonophysics, v. 319, p. 167–198, doi: 10.1016/S0040-1951(00)00032-9.

Burt, E.C., 2003, Sedimentology, Provenance and Basin Evolution of the Upper Dalradian: Tayvallich Subgroup and the Southern Highland Group [Ph.D. thesis]: London, Kingston University, 377 p.

Butler, R.W.H., 1982, The terminology of structures in thrust belts: Journal of Structural Geology, v. 4, p. 239–245, doi: 10.1016/0191-8141(82)90011-6.

Butler, R.W.H., 1997, Late Proterozoic rift faults and basement-cover relationships within the Ben More thrust sheet, NW Scotland: Journal of the Geological Society of London, v. 154, p. 761–764, doi: 10.1144/gsjgs.154.5.0761.

Butler, R.W.H., and Coward, M.P., 1984, Geological constraints, structural evolution and deep geology of the NW Scottish Caledonides: Tectonics, v. 3, p. 347–365, doi: 10.1029/TC003i003p00347.

Cambridge Paleomap Services, 1998, Timetrek v. 3.11: Cambridge, Cambridge Paleomap Services.

Cannat, M., 1989, Late Caledonian northeastward ophiolite thrusting in the Shetland Islands, UK: Tectonophysics, v. 169, p. 257–270, doi: 10.1016/0040-1951(89)90090-5.

Canning, J.C., Henney, P.J., Morrison, M.A., and Gaskarth, J.W., 1996, Geochemistry of late Caledonian minettes from northern Britain: Implications for the Caledonian sub-continental lithospheric mantle: Mineralogical Magazine, v. 60, p. 221–236, doi: 10.1180/minmag.1996.060.398.15.

Canning, J.C., Henney, P.J., Morrison, M.A., Van Calsteren, P.W.C., Gaskarth, J.W., and Swarbrick, A., 1998, The Great Glen fault: A major vertical lithospheric boundary: Journal of the Geological Society of London, v. 155, p. 425–428, doi: 10.1144/gsjgs.155.3.0425.

Cawood, P.A., and Pisarevsky, S.A., 2006, Was Baltica right-way-up or upside-down in the Neoproterozoic?: Journal of the Geological Society of London, v. 163, p. 753–759, doi: 10.1144/0016-76492005-126.

Cawood, P.A., McCausland, P.J.A., and Dunning, G.R., 2001, Opening Iapetus: Constraints from the Laurentian margin in Newfoundland: Geological Society of America Bulletin, v. 113, p. 443–453, doi: 10.1130/0016-7606 (2001)113<0443:OICFTL>2.0.CO;2.

Cawood, P.A., Nemchin, A.A., Smith, M., and Loewy, S., 2003, Source of the Dalradian Supergroup constrained by U-Pb dating of detrital zircon and implications for the East Laurentian margin: Journal of the Geological Society of London, v. 160, p. 231–246.

Cawood, P.A., Nemchin, A.A., Strachan, R.A., Kinny, P.D., and Loewy, S., 2004, Laurentian provenance and an intracratonic tectonic setting for the upper Moine Supergroup, Scotland, constrained by detrital zircons from the Loch Eil and Glen Urquhart successions: Journal of the Geological Society of London, v. 161, p. 861–874, doi: 10.1144/16-764903-117.

Condon, D.J., and Prave, A.R., 2000, Two from Donegal: Neoproterozoic glacial episodes on the northeast margin of Laurentia: Geology, v. 28, p. 951–954, doi: 10.1130/0091-7613(2000)28<951:TFDNGE>2.0.CO;2.

Corfu, F., Crane, A., Moser, D., and Rogers, G., 1998, U-Pb zircon systematics at Gruinard Bay, northwest Scotland: Implications for the early orogenic evolution of the Lewisian Complex: Contributions to Mineralogy and Petrology, v. 133, p. 329–345, doi: 10.1007/s004100050456.

Coward, M.P., and Park, R.G., 1987, The role of mid-crustal shear zones in the early Proterozoic evolution of the Lewisian, in Park, R.G., and Tarney, J., eds., Evolution of the Lewisian and Comparable Precambrian High-Grade Terrains: Geological Society of London Special Publication 27, p. 127–138.

Dallmeyer, R.D., Strachan, R.A., Rogers, G., Watt, G.R., and Friend, C.R.L., 2001, Dating deformation and cooling in the Caledonian thrust nappes of north Sutherland, Scotland: Insights from ^{40}Ar/^{39}Ar and Rb-Sr chronology: Journal of the Geological Society of London, v. 158, p. 501–512.

Dalziel, I.W.D., 1992, On the organization of American plates in the Neoproterozoic and the breakout of Laurentia: GSA Today, v. 2, no. 11, p. 237, 240–241.

Dalziel, I.W.D., and Soper, N.J., 2001, Neoproterozoic extension on the Scottish promontory of Laurentia: Paleogeographic and tectonic implications: The Journal of Geology, v. 109, p. 299–317, doi: 10.1086/319974.

Dempster, T.J., Rogers, G., Tanner, P.W.G., Bluck, B.J., Muir, R.J., Redwood, S.D., Ireland, T.R., and Paterson, B.A., 2002, Timing of deposition, orogenesis and glaciation within the Dalradian rocks of Scotland: Constraints from U-Pb zircon ages: Journal of the Geological Society of London, v. 157, p. 909–914.

Dewey, J.F., 1969, Evolution of the Caledonian/Appalachian orogen: Nature, v. 222, p. 124–129, doi: 10.1038/222124a0.

Dewey, J.F., and Mange, M., 1999, Petrography of Ordovician and Silurian sediments in the western Irish Caledonides: Tracers of a short-lived Ordovician continent-arc collision orogeny and the evolution of the Laurentian Appalachian–Caledonian margin, in MacNiocaill, C., and Ryan, P.D., eds., Continental Tectonics: Geological Society of London Special Publication 164, p. 55–107.

Dewey, J.F., and Ryan, P.D., 1990, The Ordovician evolution of the South Mayo trough, Western Ireland: Tectonics, v. 9, p. 887–903, doi: 10.1029/TC009i004p00887.

Dewey, J.F., and Strachan, R.A., 2003, Changing Silurian-Devonian relative plate motion in the Caledonides: Sinistral transpression to sinistral transtension: Journal of the Geological Society of London, v. 160, p. 219–229.

Dickin, A.P., 1992, Evidence for an early Proterozoic crustal province in the North Atlantic region: Journal of the Geological Society of London, v. 149, p. 483–486, doi: 10.1144/gsjgs.149.4.0483.

Downie, C., 1982, Lower Cambrian acritarchs from Scotland, Norway, Greenland and Canada: Transactions of the Royal Society of Edinburgh, Earth Sciences, v. 72, p. 257–285.

Elliott, D., and Johnson, M.R.W., 1980, Structural evolution in the northern part of the Moine thrust belt, NW Scotland: Transactions of the Royal Society of Edinburgh, Earth Sciences, v. 71, p. 69–96.

Emery, M., 2005, Polyorogenic history of the Moine rocks of Glen Urquhart, Inverness-shire [Ph.D. thesis]: Portsmouth, University of Portsmouth, 232 p.

Flinn, D., Miller, J.A., and Roddom, D., 1991, The age of the Norwick hornblende schists of Unst and Fetlar and the obduction of the Shetland ophiolite: Scottish Journal of Geology, v. 27, p. 11–19.

Fowler, M.B., Henney, P.J., Darbyshire, D.P.F., and Greenwood, P.B., 2001, Petrogenesis of high Ba-Sr granites: The Rogart Pluton, Sutherland: Journal of the Geological Society of London, v. 158, p. 521–534.

Fraser, G.L., Pattison, D.R.M., and Heaman, L.M., 2004, Age of the Ballachulish and Glencoe igneous complexes (Scottish Highlands), and paragenesis of zircon, monazite and baddeleyite in the Ballachulish aureole: Journal of the Geological Society of London, v. 161, p. 447–462.

Frederiksen, K.S., 2001, A Neoproterozoic Carbonate Ramp and Base-of-Slope Succession, the Andrée Land Group, Eleonore Bay Supergroup, North-East Greenland: Sedimentary Facies, Stratigraphy and Basin Evolution [Ph.D. thesis]: Copenhagen, University of Copenhagen, 250 p.

Freeman, S.R., Butler, R.W.H., Cliff, R.A., and Rex, D.C., 1998, Direct dating of mylonite evolution: A multi-disciplinary geochronological study from the Moine thrust zone, NW Scotland: Journal of the Geological Society of London, v. 155, p. 745–758, doi: 10.1144/gsjgs.155.5.0745.

Friedrich, A.M., Hodges, K.V., Bowring, S.A., and Martin, M.W., 1999, Geochronological constraints on the magmatic, metamorphic and thermal evolution of the Connemara Caledonides, western Ireland: Journal of the Geological Society of London, v. 156, p. 1217–1230, doi: 10.1144/gsjgs.156.6.1217.

Friend, C.R.L., and Kinny, P.D., 1995, New evidence for protolith ages of Lewisian granulites, northwest Scotland: Geology, v. 23, p. 1027–1030, doi: 10.1130/0091-7613(1995)023<1027:NEFPAO>2.3.CO;2.

Friend, C.R.L., and Kinny, P.D., 2001, A reappraisal of the Lewisian Gneiss Complex: Geochronological evidence for its tectonic assembly from disparate terranes in the Proterozoic: Contributions to Mineralogy and Petrology, v. 142, p. 198–218.

Friend, C.R.L., Kinny, P.D., Rogers, G., Strachan, R.A., and Paterson, B.A., 1997, U-Pb zircon geochronological evidence for Neoproterozoic events in the Glenfinnan Group (Moine Supergroup): The formation of the Ardgour Granite Gneiss, north-west Scotland: Contributions to Mineralogy and Petrology, v. 128, p. 101–113, doi: 10.1007/s004100050297.

Friend, C.R.L., Strachan, R.A., Kinny, P.D., and Watt, G.R., 2003, Provenance of the Moine Supergroup of NW Scotland: Evidence from geochronology of detrital and inherited zircons from (meta)sedimentary rocks, granites and migmatites: Journal of the Geological Society of London, v. 160, p. 247–257.

Gee, D.G., and Teben'kov, A.M., 1996, Two major unconformities beneath the Neoproterozoic Murchisonfjorden Supergroup in the Caledonides of central Nordaustlandet, Svalbard: Polar Research, v. 15, p. 81–91, doi: 10.1111/j.1751-8369.1996.tb00460.x.

Gee, D.G., and Teben'kov, A.M., 2004, Svalbard: A fragment of the Laurentian margin, in Gee, D.G., and Pease, V., eds., The Neoproterozoic Timanide Orogen of Eastern Baltica: Geological Society of London Memoir 30, p. 191–206.

Gee, D.G., Johansson, Å., Ohta, Y., Teben'kov, A.M., Krasil'schikov, A.A., Balashov, Yu.A., Larionov, A.N., Gannibal, L.A., and Ryungenen, G.I., 1995, Grenvillian basement and a major unconformity within the Caledonides of Nordaustlandet, Svalbard: Precambrian Research, v. 70, p. 215–234, doi: 10.1016/0301-9268(94)00041-O.

Geikie, A., 1893, On the pre-Cambrian rocks of the British Isles: The Journal of Geology, v. 1, p. 1–14.

Gilotti, J.A., and Elvevold, S., 2002, Extensional exhumation of a high-pressure granulite terrane in Payer Land, Greenland Caledonides: Structural, petrologic and geochronologic evidence from metapelites: Canadian Journal of Earth Sciences, v. 39, p. 1169–1187, doi: 10.1139/e02-019.

Gilotti, J.A., and McClelland, W.C., 2008, this volume, Geometry, kinematics, and timing of extensional faulting in the Greenland Caledonides—A synthesis, in Higgins, A.K., Gilotti, J.A., and Smith, M.P., eds., The Greenland Caledonides: Evolution of the Northeast Margin of Laurentia: Geological Society of America Memoir 202, doi: 10.1130/2008.1202(10).

Gilotti, J.A., Jones, K.A., and Elvevold, S., 2008, this volume, Caledonian metamorphic patterns in Greenland, in Higgins, A.K., Gilotti, J.A., and Smith, M.P., eds., The Greenland Caledonides: Evolution of the Northeast Margin of Laurentia: Geological Society of America Memoir 202, doi: 10.1130/2008.1202(08).

Glendinning, N.R.W., 1988, Sedimentary structures and sequences within a late Proterozoic tidal shelf deposit: The upper Morar Psammite Formation of northwestern Scotland, in Winchester, J.A., ed., Later Proterozoic Stratigraphy of the Northern Atlantic Regions: Glasgow, Blackie, p. 14–31.

Glover, B.W., and McKie, T.L., 1996, A sequence stratigraphic approach to the understanding of basin history in orogenic Neoproterozoic successions: An example from the central Highlands of Scotland, in Hesselbo, S.P., and Parkinson, D.N., eds., Sequence Stratigraphy in British Geology: Geological Society of London Special Publication 103, p. 257–269.

Glover, B.W., and Winchester, J.A., 1989, The Grampian Group: A major late Proterozoic clastic sequence in the Central Highlands of Scotland: Journal of the Geological Society of London, v. 146, p. 85–97, doi: 10.1144/gsjgs.146.1.0085.

Glover, B.W., Key, R.M., May, F., Clark, G.C., Phillips, E.R., and Chacksfield, B.C., 1995, A Neoproterozoic multi-phase rift sequence: The Grampian and Appin Groups of the southwestern Monadhliath Mountains of Scotland: Journal of the Geological Society of London, v. 152, p. 391–406, doi: 10.1144/gsjgs.152.2.0391.

Goodman, S., and Winchester, J.A.W., 1993, Geochemical variations within metavolcanic rocks of the Dalradian Farragon beds and adjacent formations: Scottish Journal of Geology, v. 29, p. 131–141.

Gorokhov, I.M., Siedlecka, A., Roberts, D., Melnikov, N.N., and Turchenko, T.L., 2001, Rb-Sr dating of diagenetic illite in Neoproterozoic shales, Varanger Peninsula, northern Norway: Geological Magazine, v. 138, p. 541–562, doi: 10.1017/S001675680100574X.

Gower, C.F., 1988, The Double-Mer Formation, in Winchester, J.A., ed., Later Proterozoic Stratigraphy of the Northern Atlantic Regions: Glasgow, Blackie, p. 113–118.

Gradstein, F.M., Ogg, J.G., Smith, A.G., Agterberg, F.P., Bleeker, W., Cooper, R.A., Davydov, V., Gibbard, P., Hinnov, L.A., House, M.R., Lourens, L., Luterbacher, H.P., McArthur, J., Melchin, M.J., Robb, L.J., Shergold, J., Villeneuve, M., Wardlaw, B.R., Ali, J., Brinkhuis, H., Hilgen, F.J., Hooker, J., Howarth, R.J., Knoll, A.H., Laskar, J., Monechi, S., Plumb, K.A., Powell, J., Raffi, I., Röhl, U., Sadler, P., Sanfilippo, A., Schmitz, B., Shackleton, N.J., Shields, G.A., Strauss, H., Van Dam, J., van Kolfschoten, T., Veizer, J., and Wilson, D., 2004, A Geologic Time Scale 2004: Cambridge, Cambridge University Press, 589 p.

Gromet, L.P., and Gee, D.G., 1997, Age of high-pressure metamorphism in the high arctic Caledonides: U-Pb results from Biskayerhalvoya, MW Svalbard: Geological Society of America Abstracts with Programs, v. 29, no. 6, p. 49.

Hall, A.J., Boyce, A.J., Fallick, A.E., and Hamilton, P.J., 1991, Isotopic evidence of the depositional environment of late Proterozoic stratiform barite mineralisation, Aberfeldy, Scotland: Chemical Geology, v. 87, p. 99–114.

Hall, J., Brewer, J.A., Matthews, D.H., and Warner, M.R., 1984, Crustal structure across the Caledonides from the WINCH seismic reflection profile: Influences on the evolution of the Midland Valley of Scotland: Transactions of the Royal Society of Edinburgh, Earth Sciences, v. 75, p. 97–109.

Halliday, A.N., 1984, Coupled Sm-Nd and U-Pb systematics in late Caledonian granites and basement under northern Britain: Nature, v. 307, p. 229–233, doi: 10.1038/307229a0.

Hambrey, M.J., and Spencer, A.M., 1987, Late Precambrian glaciation of central East Greenland: Meddelelser om Grønland: Geoscience, v. 19, p. 1–50.

Harland, W.B., Anderson, L.M., and Manasrah, D., 1997, The Geology of Svalbard: Geological Society of London Memoir 17, 521 p.

Harmon, R.S., 1983, Oxygen and strontium isotope evidence regarding the role of continental crust in the origin and evolution of the British Caledonian granites, in Atherton, M.P., and Gribble, C.D., eds., Migmatites, Melting and Metamorphism: Orpington, Shiva, p. 62–79.

Harris, A.L., Baldwin, C.T., Bradbury, H.J., Johnson, H.D., and Smith, R.A., 1978, Ensialic basin sedimentation: The Dalradian Supergroup, in Bowes,

D.R., and Leake, B.E., eds., Crustal Evolution in Northwestern Britain: Geological Journal, special issue, no. 10, p. 115–138.

Harris, A.L., Haselock, P.J., Kennedy, M.J., and Mendum, J.R., 1994, The Dalradian Supergroup in Scotland, Shetland and Ireland, in Gibbons, W., and Harris, A.L., eds., A Revised Correlation of Precambrian Rocks in the British Isles: Geological Society of London Special Report 22, p. 33–53.

Harte, B., Booth, J.E., Dempster, T.J., Fettes, D.J., Mendum, J.R., and Watts, D., 1984, Aspects of post-depositional evolution of Dalradian and Highland Border Complex rocks in the Southern Highlands of Scotland: Transactions of the Royal Society of Edinburgh, Earth Sciences, v. 75, p. 151–163.

Heaman, L., and Tarney, J., 1989, U-Pb baddeleyite ages for the Scourie dyke swarm, Scotland: Evidence for two distinct intrusion events: Nature, v. 340, p. 705–708, doi: 10.1038/340705a0.

Higgins, A.K., 1988, The Krummedal supracrustal sequence in East Greenland, in Winchester, J.A., ed., Later Proterozoic Stratigraphy of the Northern Atlantic Regions: Glasgow, Blackie, p. 86–96.

Higgins, A.K., and Leslie, A.G., 2000, Restoring thrusting in the East Greenland Caledonides: Geology, v. 28, p. 1019–1022, doi: 10.1130/0091-7613 (2000)28<1019:RTITEG>2.0.CO;2.

Higgins, A.K., and Leslie, A.G., 2008, this volume, Architecture and evolution of the East Greenland Caledonides—An introduction, in Higgins, A.K., Gilotti, J.A., and Smith, M.P., eds., The Greenland Caledonides: Evolution of the Northeast Margin of Laurentia: Geological Society of America Memoir 202, doi: 10.1130/2008.1202(02).

Higgins, A.K., Leslie, A.G., and Smith, M.P., 2001a, Neoproterozoic–Lower Paleozoic stratigraphical relationships in the marginal thin-skinned thrust belt of the East Greenland Caledonides: Comparisons with the foreland of Scotland: Geological Magazine, v. 138, p. 143–160, doi: 10.1017/S0016756801005076.

Higgins, A.K., Smith, M.P., Soper, N.J., Leslie, A.G., Rasmussen, J.A., and Sønderholm, M., 2001b, The Neoproterozoic Hekla Sund Basin, eastern North Greenland: A pre-Iapetan extensional sequence thrust across its rift shoulders during the Caledonian orogeny: Journal of the Geological Society of London, v. 158, p. 487–499.

Higgins, A.K., Elvevold, S., Escher, J.C., Frederiksen, K.S., Gilotti, J.A., Henriksen, N., Jepsen, H.F., Jones, K.A., Kalsbeek, F., Kinny, P.D., Leslie, A.G., Smith, M.P., Thrane, K., and Watt, G.R., 2004, The foreland-propagating thrust architecture of the East Greenland Caledonides 72°–75°N: Journal of the Geological Society of London, v. 161, p. 1009–1026, doi: 10.1144/0016-764903-141.

Highton, A.J., Hyslop, E.K., and Noble, S.R., 1999, U-Pb zircon geochronology of migmatization in the northern Central Highlands: Evidence for pre-Caledonian (Neoproterozoic) tectonometamorphism in the Grampian block, Scotland: Journal of the Geological Society of London, v. 156, p. 1195–1204, doi: 10.1144/gsjgs.156.6.1195.

Hoffman, P.F., 1991, Did the breakout of Laurentia turn Gondwanaland inside-out?: Science, v. 252, p. 1409–1412, doi: 10.1126/science.252.5011.1409.

Hoffmann, K.H., Condon, D.J., Bowring, S.A., and Crowley, J.L., 2004, U-Pb zircon date from the Neoproterozoic Ghaub Formation, Namibia: Constraints on Marinoan glaciation: Geology, v. 32, p. 817–820, doi: 10.1130/G20519.1.

Holdsworth, R.E., Strachan, R.A., and Harris, A.L., 1994, Pre-Cambrian rocks in northern Scotland east of the Moine thrust: The Moine Supergroup, in Gibbons, W., and Harris, A.L., eds., A Revised Correlation of Precambrian Rocks in the British Isles: Geological Society of London Special Report 22, p. 23–32.

Holdsworth, R.E., Strachan, R.A., Alsop, G.I., Grant, C.J., and Wilson, R.W., 2006, Thrust sequences and the significance of low-angle, out-of-sequence faults in the northernmost Moine nappe and Moine thrust zone, NW Scotland: Journal of the Geological Society of London, v. 163, p. 801–814, doi: 10.1144/0016-76492005-076.

Hutchison, A.R., and Oliver, G.J.H., 1998, Garnet provenance studies, juxtaposition of Laurentian marginal terranes and timing of the Grampian orogeny in Scotland: Journal of the Geological Society of London, v. 155, p. 541–550, doi: 10.1144/gsjgs.155.3.0541.

Hutton, D.H.W., 1987, Strike-slip terranes and a model for the evolution of the British and Irish Caledonides: Geological Magazine, v. 124, p. 405–425.

Hutton, D.H.W., and Alsop, G.I., 2004, Evidence for a major Neoproterozoic orogenic unconformity within the Dalradian Supergroup of NW Ireland: Journal of the Geological Society of London, v. 161, p. 629–640, doi: 10.1144/0016-764903-094.

Hutton, D.H.W., and Reavy, R.J., 1992, Strike-slip tectonics and granite petrogenesis: Tectonics, v. 11, p. 960–967, doi: 10.1029/92TC00336.

Hyndman, R.D., Currie, C.A., and Mazotti, S.P., 2005, Subduction zone backarcs, mobile belts, and orogenic heat: GSA Today, v. 15, no. 2, p. 4–10.

Hyslop, E.K., 1992, Strain-Induced Metamorphism and Pegmatite Development in the Moine Rocks of Scotland [Ph.D. thesis]: Hull, University of Hull, 216 p.

Hyslop, E.K., and Piasecki, M.A.J., 1999, Mineralogy, geochemistry and the development of ductile shear zones in the Grampian slide zone of the Scottish Central Highlands: Journal of the Geological Society of London, v. 156, p. 577–589, doi: 10.1144/gsjgs.156.3.0577.

Jacques, J.M., and Reavy, R.J., 1994, Caledonian plutonism and major lineaments in the SW Scottish Highlands: Journal of the Geological Society of London, v. 151, p. 955–969, doi: 10.1144/gsjgs.151.6.0955.

Johansson, Å., Larionov, A.N., Tebenkov, A.M., Gee, D.G., Whitehouse, M.J., and Vestin, J., 2000, Grenvillian magmatism of western and central Nordaustlandet, northeastern Svalbard: Transactions of the Royal Society of Edinburgh, Earth Sciences, v. 90, p. 221–254.

Johansson, Å., Larionov, A.N., Gee, D.G., Ohta, Y., Tebenkov, A.M., and Sandelin, S., 2004, Grenvillian and Caledonian tectono-magmatic activity in northeasternmost Svalbard, in Gee, D.G., and Pease, V., eds., The Neoproterozoic Timanide Orogen of Eastern Baltica: Geological Society of London Memoir 30, p. 207–232.

Johnson, M.R.W., Kelley, S.P., Oliver, G.J.H., and Winter, D.A., 1985, Thermal effects and timing of thrusting in the Moine thrust zone: Journal of the Geological Society of London, v. 142, p. 863–874, doi: 10.1144/gsjgs.142.5.0863.

Kalsbeek, F., ed., 1989, Geology of the Ammassalik Region, South-East Greenland: Rapport Grønlands Geologiske Undersøgelse, v. 146, 106 p.

Kalsbeek, F., Nutman, A.P., and Taylor, P.N., 1993, Palaeoproterozoic basement province in the Caledonian fold belt of North-East Greenland: Precambrian Research, v. 63, p. 163–178, doi: 10.1016/0301-9268(93)90010-Y.

Kalsbeek, F., Thrane, K., Nutman, A.P., and Jepsen, H.F., 2000, Late Mesoproterozoic to early Neoproterozoic history of the East Greenland Caledonides: Evidence for Grenvillian orogenesis?: Journal of the Geological Society of London, v. 57, p. 1215–1225.

Kalsbeek, F., Higgins, A.K., Jepsen, H.F., Frei, R., and Nutman, A.P., 2008, this volume, Granites and granites in the East Greenland Caledonides, in Higgins, A.K., Gilotti, J.A., and Smith, M.P., eds., The Greenland Caledonides: Evolution of the Northeast Margin of Laurentia: Geological Society of America Memoir 202, doi: 10.1130/2008.1202(09).

Kelley, S.P., 1988, The relationship between K-Ar mineral ages, mica grain sizes and movement on the Moine thrust zone, NW Highlands, Scotland: Journal of the Geological Society of London, v. 145, p. 1–10, doi: 10.1144/gsjgs.145.1.0001.

Key, R.M., Clark, G.C., May, F., Phillips, E.R., Chacksfield, B.C., and Peacock, J.D., 1997, Geology of the Glen Roy District: Memoir of the British Geological Survey, Her Majesty's Stationery Office, Sheet 63 W (Scotland), 127 p.

Kinny, P.D., and Friend, C.R.L., 1997, U/Pb isotopic evidence for the accretion of different crustal blocks to form the Lewisian complex of northwest Scotland: Contributions to Mineralogy and Petrology, v. 129, p. 326–340, doi: 10.1007/s004100050340.

Kinny, P.D., Friend, C.R.L., Strachan, R.A., Watt, G.R., and Burns, I.M., 1999, U-Pb geochronology of regional migmatites in East Sutherland, Scotland: Evidence for crustal melting during the Caledonian orogeny: Journal of the Geological Society of London, v. 156, p. 1143–1152, doi: 10.1144/gsjgs.156.6.1143.

Kinny, P.D., Friend, C.R.L., and Love, G.J., 2005, Proposal for a terrane-based nomenclature for the Lewisian Gneiss Complex of NW Scotland: Journal of the Geological Society of London, v. 162, p. 175–186, doi: 10.1144/0016-764903-149.

Kneller, B.C., and Aftalion, M., 1987, The isotopic and structural age of the Aberdeen Granite: Journal of the Geological Society of London, v. 144, p. 717–721, doi: 10.1144/gsjgs.144.5.0717.

Krabbendam, M., and Dewey, J.F., 1998, Exhumation of UHP rocks by transtension in the Western Gneiss Region, Scandinavian Caledonides, in Holdsworth, R.E., Strachan, R.A., and Dewey, J.F., eds., Continental Transpressional and Transtensional Tectonics: Geological Society of London Special Publication 135, p. 159–181.

Krabbendam, M., and Leslie, A.G., 2004, Lateral ramps and thrust terminations: An example from the Moine thrust zone, NW Scotland: Jour-

nal of the Geological Society of London, v. 161, p. 551–554, doi: 10.1144/0016-764904-015.
Krabbendam, M., Leslie, A.G., Crane, A., and Goodman, S., 1997, Generation of the Tay nappe, Scotland, by large-scale SE-directed shearing: Journal of the Geological Society of London, v. 154, p. 15–24, doi: 10.1144/gsjgs.154.1.0015.
Lambert, R.S.J., and McKerrow, W.S., 1976, The Grampian orogeny: Scottish Journal of Geology, v. 12, p. 271–292.
Larsen, P.H., and Bengaard, H.J., 1991, Devonian basin initiation in East Greenland: A result of sinistral wrench faulting and Caledonian extensional collapse: Journal of the Geological Society of London, v. 148, p. 355–368, doi: 10.1144/gsjgs.148.2.0355.
Leslie, A.G., and Higgins, A.K., 1998, The Caledonian geology of Andrée Land, Eleonore Sø and adjacent nunataks, East Greenland: Danmarks og Grønlands Geologiske Undersøgelse Rapport, v. 1998/28, p. 11–27.
Leslie, A.G., and Higgins, A.K., 2008, this volume, Foreland-propagating Caledonian thrust systems in East Greenland, in Higgins, A.K., Gilotti, J.A., and Smith, M.P., eds., The Greenland Caledonides: Evolution of the Northeast Margin of Laurentia: Geological Society of America Memoir 202, doi: 10.1130/2008.1202(07).
Leslie, A.G., and Nutman, A.P., 2003, Evidence for Neoproterozoic orogenesis and early high temperature Scandian deformation events in the southern East Greenland Caledonides: Geological Magazine, v. 140, p. 309–333, doi: 10.1017/S0016756803007593.
Macdonald, R.L., Fettes, D.J., Stephenson, D., and Graham, C.M., 2005, Basic and ultrabasic volcanic rocks from the Argyll Group (Dalradian) of NE Scotland: Scottish Journal of Geology, v. 41, p. 159–174.
Marcantonio, F., Dickin, A.P., McNutt, R.H., and Heaman, L.M., 1988, A 1800-million-year-old Proterozoic gneiss terrane in Islay with implications for the crustal structure evolution of Britain: Nature, v. 335, p. 62–64, doi: 10.1038/335062a0.
May, F., and Highton, A.J., 1997, Geology of the Invermoriston district: Memoir of the British Geological Survey, Her Majesty's Stationery Office, Sheet 73W (Scotland), 77 p.
McCay, G.A., Prave, A.R., Alsop, G.I., and Fallick, A.E., 2006, Glacial trinity: Neoproterozoic Earth history within the British-Irish Caledonides: Geology, v. 34, p. 909–912, doi: 10.1130/G22694A.1.
McClay, K.R., and Coward, M.P., 1981, The Moine thrust zone: An overview, in McClay, K.R., and Price, N.J., eds., Thrust and Nappe Tectonics: Geological Society of London Special Publication 9, p. 241–260.
McKerrow, W.S., MacNiocaill, C., and Dewey, J.F., 2000, The Caledonian orogen redefined: Journal of the Geological Society of London, v. 157, p. 1149–1154.
McKie, T., 1990, Tidal and storm-influenced sedimentation from a Cambrian transgressive passive margin sequence: Journal of the Geological Society of London, v. 147, p. 785–794, doi: 10.1144/gsjgs.147.5.0785.
Millar, I.L., 1999, Neoproterozoic extensional basic magmatism associated with emplacement of the West Highland Granite Gneiss in the Moine Supergroup of NW Scotland: Journal of the Geological Society of London, v. 156, p. 1153–1162, doi: 10.1144/gsjgs.156.6.1153.
Moncrieff, A.C.M., and Hambrey, M.J., 1988, Late Precambrian glacially-related grooved and striated surfaces in the Tillite Group of central East Greenland: Palaeogeography, Palaeoclimatology, Palaeoecology, v. 65, p. 183–200, doi: 10.1016/0031-0182(88)90023-5.
Muir, R.J., Fitches, W.R., and Maltman, A.J., 1994, The Rhinns Complex: Proterozoic basement on Islay and Colonsay, Inner Hebrides, Scotland, and on Inishtrahull, NW Ireland: Transactions of the Royal Society of Edinburgh, Earth Sciences, v. 85, p. 77–90.
Myers, J.S., 1987, The East Greenland Nagssuqtoqidian mobile belt compared with Lewisian complex, in Park, R.G., and Tarney, J., eds., Evolution of the Lewisian and Comparable Precambrian High-Grade Terrains: Geological Society of London Special Publication 27, p. 235–246.
Mykura, W., 1991, Old Red Sandstone, in Craig, G.Y., ed., Geology of Scotland (3rd edition): London, Geological Society of London, p. 297–346.
Noble, S.R., Hyslop, E.K., and Highton, A.J., 1996, High precision U-Pb monazite geochronology of the c. 806 Ma Grampian shear zone and the implications for the evolution of the Central Highlands of Scotland: Journal of the Geological Society of London, v. 153, p. 511–514, doi: 10.1144/gsjgs.153.4.0511.
Oliver, G.J.H., 2001, Reconstruction of the Grampian episode in Scotland: Its place in the Caledonian orogeny: Tectonophysics, v. 332, p. 23–49, doi: 10.1016/S0040-1951(00)00248-1.
Olsen, H., 1993, Sedimentary Basin Analysis of the Continental Devonian Basin in North-East Greenland: Bulletin Grønlands Geologiske Undersøgelse, v. 168, 80 p.
O'Nions, R.K., Hamilton, P.J., and Hooker, P.J., 1983, A Nd isotope investigation of sediments related to crustal development in the British Isles: Earth and Planetary Science Letters, v. 63, p. 229–240, doi: 10.1016/0012-821X(83)90039-0.
Park, R.G., 1994, Early Proterozoic tectonic overview of the northern British Isles and neighbouring terrains in Laurentian and Baltica: Precambrian Research, v. 68, p. 65–79, doi: 10.1016/0301-9268(94)90065-5.
Park, R.G., 2005, The Lewisian terrane model: A review: Scottish Journal of Geology, v. 41, p. 105–118.
Park, R.G., Tarney, J., and Connelly, J.N., 2001, The Loch Maree Group: Palaeoproterozoic subduction–accretion complex in the Lewisian of NW Scotland: Precambrian Research, v. 105, p. 205–226, doi: 10.1016/S0301-9268(00)00112-1.
Park, R.G., Stewart, A.D., and Wright, D.T., 2002, The Hebridean terrane, in Trewin, N.H., ed., The Geology of Scotland (4th edition): London, Geological Society of London, p. 45–80.
Piasecki, M.A.J., 1980, New light on the Moine rocks of the Central Highlands of Scotland: Journal of the Geological Society of London, v. 137, p. 41–59, doi: 10.1144/gsjgs.137.1.0041.
Pickett, E.A., Hyslop, E.K., and Petterson, M.G., 2006, The Green Beds of the SW Highlands: Deposition and origin of a basic igneous-rich sedimentary sequence in the Dalradian Supergroup of Scotland: Scottish Journal of Geology, v. 42, p. 43–57.
Pidgeon, R.T., and Compston, W., 1992, A SHRIMP ion microprobe study of inherited and magmatic zircons from four Scottish Caledonian granites, in Brown, P.E., and Chappell, B.W., eds., The Second Hutton Symposium on the Origin of Granites and Related Rocks: Geological Society of America Special Paper 272, p. 473–483.
Powell, D., Baird, A.W., Charnley, N.R., and Jordon, P.J., 1981, The metamorphic environment of the Sgurr Beag Slide, a major crustal displacement zone in Proterozoic, Moine rocks of Scotland: Journal of the Geological Society of London, v. 138, p. 661–673, doi: 10.1144/gsjgs.138.6.0661.
Prave, A.R., 1999, The Neoproterozoic Dalradian Supergroup of Scotland: An alternative hypothesis: Geological Magazine, v. 136, p. 609–617, doi: 10.1017/S0016756899003155.
Pringle, J., 1940, The discovery of Cambrian trilobites in the Highland Border rocks near Callander, Perthshire (Scotland): Report of the British Association for the Advancement of Science, v. 1, p. 252.
Rainbird, R.H., Hamilton, M.A., and Young, G.M., 2001, Detrital zircon geochronology and provenance of the Torridonian, NW Scotland: Journal of the Geological Society of London, v. 158, p. 15–27.
Rankin, D.W., Drake, A.A., Jr., Glover, L., III, Goldsmith, R., Hall, L.M., Murray, D.P., Ratcliffe, N.M., Read, J.F., Secor, D.T., Jr., and Stanley, R.S., 1989, Pre-orogenic terranes, in Hatcher, R.D., Jr., Thomas, W.A., and Viele, G.W., eds., The Appalachian-Ouachita Orogen in the United States: Boulder, Colorado, Geological Society of America, Geology of North America, v. F-2, p. 7–100.
Read, H.H., 1961, Aspects of the Caledonian magmatism in Britain: Liverpool and Manchester Geological Journal, v. 2, p. 653–683.
Rex, D.C., and Gledhill, A.R., 1981, Isotopic studies in the East Greenland Caledonides (72°–74°N): Precambrian and Caledonian ages: Rapport Grønlands Geologiske Undersøgelse, v. 104, p. 47–72.
Roberts, D., Melezhik, V.M., and Heldal, T., 2001, Carbonate formations and early NW-directed thrusting in the highest allochthons of the Norwegian Caledonides: Evidence of a Laurentian ancestry: Journal of the Geological Society of London, v. 159, p. 117–120.
Robertson, S., and Smith, M., 1999, The significance of the Geal Charn-Ossian steep belt in basin development in the central Scottish Highlands: Journal of the Geological Society of London, v. 156, p. 1175–1182, doi: 10.1144/gsjgs.156.6.1175.
Rogers, G., and Dunning, G.R., 1991, Geochronology of appinitic and related granitic magmatism in the W Highlands of Scotland: Constraints on the timing of transcurrent fault movement: Journal of the Geological Society of London, v. 148, p. 17–27, doi: 10.1144/gsjgs.148.1.0017.
Rogers, G., Dempster, T.J., Bluck, B.J., and Tanner, P.W.G., 1989, A high-precision age for the Ben Vuirich granite: Implications for the evolution of the Scottish Dalradian Supergroup: Journal of the Geological Society of London, v. 146, p. 789–798, doi: 10.1144/gsjgs.146.5.0789.

Rogers, G., Hyslop, E.K., Strachan, R.A., Peterson, B.A., and Holdsworth, R.E., 1998, The structural setting and U-Pb geochronology of Knoydartian pegmatites in W Inverness-shire: Evidence for Neoproterozoic tectonothermal events in the Moine of NW Scotland: Journal of the Geological Society of London, v. 155, p. 685–696, doi: 10.1144/gsjgs.155.4.0685.

Rogers, G., Kinny, P.D., Strachan, R.A., Friend, C.R.L., and Patterson, B.A., 2001, U-Pb geochronology of the Fort Augustus Granite Gneiss; constraints on the timing of Neoproterozoic and Paleozoic tectonothermal events in the NW Highlands of Scotland: Journal of the Geological Society of London, v. 158, p. 7–14.

Rollin, K.E., 1994, Geophysical correlation of Precambrian rocks in northern Britain, in Gibbons, W., and Harris, A.L., eds., A Revised Correlation of Precambrian Rocks in the British Isles: Geological Society of London Special Report 22, p. 65–74.

Ryan, P.D., and Soper, N.J., 2001, Modelling anatexis in intra-cratonic rift basins: An example from the Neoproterozoic rocks of the Scottish Highlands: Geological Magazine, v. 138, p. 577–588.

Sanders, I.S., van Calsteren, P.W.C., and Hawkesworth, C.J., 1984, A Grenville Sm-Nd age for the Glenelg eclogite in north-west Scotland: Nature, v. 312, p. 439–440, doi: 10.1038/312439a0.

Sandiford, M., and Hand, M., 1998, Controls on the locus of Phanerozoic intraplate deformation in central Australia: Earth and Planetary Science Letters, v. 162, p. 97–110, doi: 10.1016/S0012-821X(98)00159-9.

Scrimgeour, I., and Close, D., 1999, Regional high pressure metamorphism during intracratonic deformation: The Petermann orogeny, central Australia: Journal of Metamorphic Geology, v. 17, p. 557–572, doi: 10.1046/j.1525-1314.1999.00217.x.

Seranne, M., 1992, Devonian extensional tectonics versus Carboniferous inversion in the northern Orcadian basin: Journal of the Geological Society of London, v. 149, p. 27–37, doi: 10.1144/gsjgs.149.1.0027.

Shaw, R.D., Zeitler, P.K., McDougall, I., and Timgate, P., 1992, The Paleozoic history of an unusual intracratonic thrust belt in central Australia based on ^{40}Ar-^{39}Ar, K-Ar and fission track dating: Journal of the Geological Society of London, v. 149, p. 937–954, doi: 10.1144/gsjgs.149.6.0937.

Shellnutt, J.G., Dostal, J., and Keppie, J.D., 2004, Petrogenesis of the 723 Ma Coronation sills, Amundsen Basin, Arctic Canada: Implications for the break-up of Rodinia: Precambrian Research, v. 129, p. 309–324, doi: 10.1016/j.precamres.2003.10.006.

Smith, M., Robertson, S., and Rollin, K.E., 1999, Rift basin architecture and stratigraphical implications for basement-cover relationships in the Neoproterozoic Grampian Group of the Scottish Caledonides: Journal of the Geological Society of London, v. 156, p. 1163–1173, doi: 10.1144/gsjgs.156.6.1163.

Smith, M.P., and Rasmussen, J.A., 2008, this volume, Cambrian–Silurian development of the Laurentian margin of the Iapetus Ocean in Greenland and related areas, in Higgins, A.K., Gilotti, J.A., and Smith, M.P., eds., The Greenland Caledonides: Evolution of the Northeast Margin of Laurentia: Geological Society of America Memoir 202, doi: 10.1130/2008.1202(06).

Smith, M.P., and Robertson, S., 1999, The Nathorst Land Group (Neoproterozoic) of East Greenland—Lithostratigraphy, basin geometry and tectonic history: Danmarks og Grønlands Geologiske Undersøgelse Rapport, v. 1999/19, p. 127–143.

Smith, M.P., Rasmussen, J.A., Robertson, S., Higgins, A.K., and Leslie, A.G., 2004, Lower Palaeozoic stratigraphy of the East Greenland Caledonides: Geological Survey of Denmark and Greenland Bulletin, v. 6, p. 5–28.

Smith, R.L., Sterans, J.E.F., and Piper, J.D.A., 1983, Palaeomagnetic studies of the Torridonian sediments, NW Scotland: Scottish Journal of Geology, v. 19, p. 29–45.

Snyder, D.B., and Flack, C.A., 1990, A Caledonian age for reflectors within the mantle lithosphere north and west of Scotland: Tectonics, v. 9, p. 903–922, doi: 10.1029/TC009i004p00903.

Sønderholm, M., and Tirsgaard, H., 1993, Lithostratigraphic Framework of the Upper Proterozoic Eleonore Bay Supergroup of East and North-East Greenland: Bulletin Grønlands Geologiske Undersøgelse, v. 167, 38 p.

Sønderholm, M., Frederiksen, K.S., Smith, M.P., and Tirsgaard, H., 2008, this volume, Neoproterozoic sedimentary basins with glacigenic deposits of the East Greenland Caledonides, in Higgins, A.K., Gilotti, J.A., and Smith, M.P., eds., The Greenland Caledonides: Evolution of the Northeast Margin of Laurentia: Geological Society of America Memoir 202, doi: 10.1130/2008.1202(05).

Soper, N.J., 1994a, Was Scotland a Vendian RRR junction?: Journal of the Geological Society of London, v. 151, p. 579–582, doi: 10.1144/gsjgs.151.4.0579.

Soper, N.J., 1994b, Neoproterozoic sedimentation on the northeast margin of Laurentia and the opening of Iapetus: Geological Magazine, v. 131, p. 291–299.

Soper, N.J., and Hutton, D.H.W., 1984, Late Caledonian sinistral displacements in Britain: Implications for a three-plate model: Tectonics, v. 3, p. 781–794, doi: 10.1029/TC003i007p00781.

Soper, N.J., and Woodcock, N.H., 2003, The lost Lower Old Red Sandstone of England and Wales: A record of post-Iapetan flexure or Early Devonian transtension?: Geological Magazine, v. 140, p. 627–647, doi: 10.1017/S0016756803008380.

Soper, N.J., Strachan, R.A., Holdsworth, R.E., Gayer, R.A., and Greiling, R.O., 1992, Sinistral transpression and the Silurian closure of Iapetus: Journal of the Geological Society of London, v. 149, p. 871–880, doi: 10.1144/gsjgs.149.6.0871.

Soper, N.J., Harris, A.L., and Strachan, R.A., 1998, Tectonostratigraphy of the Moine Supergroup: A synthesis: Journal of the Geological Society of London, v. 155, p. 13–24, doi: 10.1144/gsjgs.155.1.0013.

Soper, N.J., Ryan, P.D., and Dewey, J.F., 1999, Age of the Grampian orogeny in Scotland and Ireland: Journal of the Geological Society of London, v. 156, p. 1231–1236, doi: 10.1144/gsjgs.156.6.1231.

Stephens, W.E., and Halliday, A.N., 1984, Geochemical contrasts between late Caledonian granitoid plutons of northern, central and southern Scotland: Transactions of the Royal Society of Edinburgh, Earth Sciences, v. 75, p. 259–273.

Stephenson, D., and Gould, D., 1995, British Regional Geology: The Grampian Highlands (4th edition): London, Her Majesty's Stationery Office for the British Geological Survey, 261 p.

Stephenson, D., Bevins, R.E., Millward, D., Highton, A.J., Parsons, I., Stone, P., and Wadsworth, W.J., 1999, Caledonian igneous rocks of Great Britain: Geological Conservation Review Series, No. 17, Joint Nature Conservation Committee, 648 p.

Stewart, A.D., 1991, Geochemistry, provenance and climate of the Upper Proterozoic Stoer Group in Scotland: Scottish Journal of Geology, v. 26, p. 89–97.

Stewart, A.D., 2002, The Later Proterozoic Torridonian Rocks of Scotland: Their Sedimentology, Geochemistry and Origin: Geological Society of London Memoir 24, 130 p.

Stewart, M., Strachan, R.A., and Holdsworth, R.E., 1999, Structure and early kinematic history of the Great Glen fault zone, Scotland: Tectonics, v. 18, p. 326–342, doi: 10.1029/1998TC900033.

Stoker, M.S., van Weering, T.C.E., and Svaerdborg, T., 2001, A mid-late Cenozoic tectonostratigraphic framework for the Rockall trough, in Shannon, P.M., Haughton, P.D.W., and Corcoran, D.V., eds., The Petroleum Exploration of Ireland's Offshore Basins: Geological Society of London Special Publication 188, p. 411–438.

Stone, P., 1995, Geology of the Rhins of Galloway: Memoir of the British Geological Survey: Her Majesty's Stationery Office, sheets 1 and 3 (Scotland), 102 p.

Stone, P., Floyd, J.D., Barnes, R.P., and Lintern, B.C., 1987, A sequential back-arc and foreland basin thrust duplex model for the Southern Uplands of Scotland: Journal of the Geological Society of London, v. 144, p. 753–764, doi: 10.1144/gsjgs.144.5.0753.

Stone, P., Green, P.M., Lintern, B.C., Simpson, P.R., and Plant, J.A., 1993, Regional geochemical variation across the Iapetus suture zone: Tectonic implications: Scottish Journal of Geology, v. 29, p. 113–121.

Storey, C.D., Brewer, T.S., and Parrish, R.R., 2004, Late Proterozoic tectonics in northwest Scotland: One contractional orogeny or several?: Precambrian Research, v. 134, p. 227–247, doi: 10.1016/j.precamres.2004.06.004.

Strachan, R.A., 1986, Shallow-marine sedimentation in the Proterozoic Moine succession, northern Scotland: Precambrian Research, v. 32, p. 17–33, doi: 10.1016/0301-9268(86)90027-6.

Strachan, R.A., Harris, A.L., Fettes, D.J., and Smith, M., 2002, The Highland and Grampian terranes, in Trewin, N.H., ed., The Geology of Scotland (4th edition): London, Geological Society of London, p. 81–148.

Tanner, P.W.G., 1995, New evidence that the Lower Cambrian Leny Limestone at Callander, Perthshire, belongs to the Dalradian Supergroup, and a reassessment of the 'exotic' status of the Highland Border Complex: Geological Magazine, v. 132, p. 473–483.

Tanner, P.W.G., and Evans, J.A., 2003, Late Precambrian U-Pb titanite age for peak regional metamorphism and deformation (Knoydartian orogeny) in the western Moine, Scotland: Journal of the Geological Society of London, v. 160, p. 555–564, doi: 10.1144/0016-764902-080.

Tanner, P.W.G., and Sutherland, S., 2007, The Highland Border Complex, Scotland: A paradox resolved: Journal of the Geological Society of London, v. 164, p. 111–116, doi: 10.1144/0016-76492005-188.

Tanner, P.W.G., Johnstone, C.S., Smith, D., and Harris, A.L., 1970, Moinian stratigraphy and the problem of the Central Ross-shire inliers: Geological Society of America Bulletin, v. 81, p. 299–306, doi: 10.1130/0016-7606 (1970)81[299:MSATPO]2.0.CO;2.

Tanner, P.W.G., Leslie, A.G., and Gillespie, M.R., 2006, Structural setting and petrogenesis of the Ben Vuirich Granite Pluton of the Grampian Highlands: A pre-orogenic, rift-related intrusion: Scottish Journal of Geology, v. 42, p. 113–136.

Tarney, J., and Jones, C.E., 1994, Trace element geochemistry of orogenic igneous rocks and crustal growth models: Journal of the Geological Society of London, v. 151, p. 855–868, doi: 10.1144/gsjgs.151.5.0855.

Thirlwall, M.F., 1981, Implications for Caledonian plate tectonic models of chemical data from volcanic rocks of the British Old Red Sandstone: Journal of the Geological Society of London, v. 138, p. 123–138, doi: 10.1144/gsjgs.138.2.0123.

Thirlwall, M.F., 1982, Systematic variation in chemistry and Nd-Sr isotopes across a Caledonian calc-alkaline volcanic arc: Implications for source materials: Earth and Planetary Science Letters, v. 58, p. 27–50, doi: 10.1016/0012-821X(82)90101-7.

Thirlwall, M.F., 1988, Geochronology of late Caledonian magmatism in northern Britain: Journal of the Geological Society of London, v. 145, p. 951–967, doi: 10.1144/gsjgs.145.6.0951.

Thomas, C.W., Graham, C.M., Ellam, R.M., and Fallick, A.E., 2004, $^{87}Sr/^{86}Sr$ chemostratigraphy of Neoproterozoic Dalradian limestones of Scotland and Ireland: Constraints on depositional ages and time scales: Journal of the Geological Society of London, v. 161, p. 229–242, doi: 10.1144/0016-764903-001.

Thomas, P.R., 1979, New evidence for a Central Highland root zone, in Harris, A.L., Holland, C.H., and Leake, B.E., eds., The Caledonides of the British Isles—Reviewed: Geological Society of London Special Publication 8, p. 205–211.

Thrane, K., 2002, Relationships between Archean and Palaeoproterozoic crystalline basement complexes in the southern part of the East Greenland Caledonides: An ion microprobe study: Precambrian Research, v. 113, p. 19–42, doi: 10.1016/S0301-9268(01)00198-X.

Tirsgaard, H., and Sønderholm, M., 1997, Lithostratigraphy, Sedimentary Evolution and Sequence Stratigraphy of the upper Lyell Land Group (Eleonore Bay Supergroup) of East and North-East Greenland: Geology of Greenland Survey Bulletin, v. 178, 60 p.

Tollo, R.P., and Hutson, E.H., 1996, 700 Ma rift event in the Blue Ridge province of Virginia: A unique time constraint on pre-Iapetan rifting of Laurentia: Geology, v. 24, p. 59–62, doi: 10.1130/0091-7613(1996)024 <0059:MREITB>2.3.CO;2.

Tollo, R.P., Aleinikoff, J.N., Bartholomew, M.J., and Rankin, D.W., 2004, Neoproterozoic A-type granitoids of the central and southern Appalachians: Intraplate magmatism associated with episodic rifting of the Rodinian supercontinent: Precambrian Research, v. 128, p. 3–38, doi: 10.1016/j.precamres.2003.08.007.

Torsvik, T.H., Smethurst, M.A., Van der Voo, R., Trench, A., Abrahamsen, N., and Halvorsen, E., 1992, Baltica: A synopsis of Vendian-Permian palaeomagnetic data and their palaeotectonic implications: Earth-Science Reviews, v. 33, p. 133–152, doi: 10.1016/0012-8252(92)90023-M.

Trewin, N.H., and Rollin, K., 2002, Geological history and structure of Scotland, in Trewin, N.H., ed., The Geology of Scotland (4th edition): London, Geological Society of London, p. 1–25.

Turnbull, M.J.M., Whitehouse, M.J., and Moorbath, S., 1996, New isotope age determinations for the Torridonian, NW Scotland: Journal of the Geological Society of London, v. 153, p. 955–964, doi: 10.1144/gsjgs.153.6.0955.

van Breemen, O., Aftalion, M., Pankhurst, R.J., and Richardson, S.W., 1979, Age of the Glen Dessary syenite, Inverness-shire: Diachronous Paleozoic metamorphism across the Great Glen: Scottish Journal of Geology, v. 15, p. 49–62.

Vance, D., Meier, M., and Oberli, F., 1998, The influence of high U-Th inclusions on the U-Th-Pb systematics of almandine-pyrope garnet: Results of a combined bulk dissolution, stepwise-leaching, and SEM study: Geochimica et Cosmochimica Acta, v. 62, p. 3527–3540, doi: 10.1016/S0016-7037(98)00252-X.

van Staal, C.R., Dewey, J.F., MacNiocall, C., and McKerrow, W.S., 1998, The Cambrian-Silurian tectonic evolution of the northern Appalachian and British Caledonides: History of a complex west and southwest Pacific-type segment of Iapetus, in Blundell, D.J., and Scott, A.C., eds., Lyell: The Past Is the Key to the Present: Geological Society of London Special Publication 143, p. 199–142.

Waldron, J.W.F., and van Staal, C.R., 2001, Taconian orogeny and the accretion of the Dashwoods block: A peri-Laurentian microcontinent in the Iapetus Ocean: Geology, v. 29, p. 811–814, doi: 10.1130/0091-7613(2001)029 <0811:TOATAO>2.0.CO;2.

Watson, J.V., 1984, The ending of the Caledonian orogeny in Scotland: Journal of the Geological Society of London, v. 141, p. 193–214, doi: 10.1144/gsjgs.141.2.0193.

Watt, G.R., and Thrane, K., 2001, Early Neoproterozoic events in East Greenland: Precambrian Research, v. 110, p. 165–184, doi: 10.1016/S0301-9268(01)00186-3.

Watt, G.R., Kinny, P.D., and Friderichsen, J.D., 2000, U-Pb geochronology of Neoproterozoic and Caledonian tectonothermal events in the East Greenland Caledonides: Journal of the Geological Society of London, v. 157, p. 1031–1048.

Whitehouse, M.J., 1989, Sm-Nd evidence for diachronous crustal accretion in the Lewisian complex of NW Scotland: Tectonophysics, v. 161, p. 245–256, doi: 10.1016/0040-1951(89)90157-1.

Whitehouse, M.J., and Bridgwater, D., 1999, Palaeoproterozoic evolution of the Outer Hebridean Lewisian Complex, northwest Scotland: Constraints from ion microprobe zircon geochronology: European Union of Geosciences conference abstracts; EUG 10, Journal of Conference Abstracts 4; 1, p. 129.

Winchester, J.A., 1988, Later Proterozoic environments and tectonic evolution in the northern Atlantic lands, in Winchester, J.A., ed., Later Proterozoic Stratigraphy of the Northern Atlantic Region: Glasgow, Blackie, p. 253–270.

Wright, A.E., and Tarney, J., Palmer, K.F., Moorlock, B.S.P., and Skinner, A.C., 1973, The geology of the Angmassalik area, East Greenland, and possible relationships with the Lewisian of Scotland, in Park, R.G., and Tarney, J., eds., The Early Precambrian of Scotland and Related Rocks of Greenland: Keele, University of Keele, p. 157–177.

Wright, D.T., and Knight, I., 1995, A revised chronostratigraphy for the lower Durness Group: Scottish Journal of Geology, v. 31, p. 11–22.

Yoshinobu, A.S., Barnes, C.G., Nordgulen, Ø., Prestvik, T., Fanning, M., and Pedersen, R.B., 2002, Ordovician magmatism, deformation, and exhumation in the Caledonides of central Norway: An orphan of the Taconic orogeny?: Geology, v. 30, p. 883–886, doi: 10.1130/0091-7613(2002)030 <0883:OMDAEI>2.0.CO;2.

Manuscript Accepted by the Society 14 January 2008

The Geological Society of America
Memoir 202
2008

Caledonian orogen of East Greenland 70°N–82°N: Geological map at 1:1,000,000—Concepts and principles of compilation

Niels Henriksen
A.K. Higgins*
Geological Survey of Denmark and Greenland, Øster Voldgade 10, DK-1350 Copenhagen K, Denmark

ABSTRACT

A geological map of the Caledonian orogen in East Greenland at the scale of 1:1,000,000 accompanies this volume. The map sheet is a compilation of lithostructural data, and it includes cross sections and inset synoptic tectonic maps with profiles. The ~1300 km length of the N-S–trending Caledonian orogen in East Greenland is divided into three lithostructural domains—the Caledonian foreland, which is partly exposed in the west, a western marginal thrust belt with foreland windows exposed in anticlinal culminations, and an eastern thick-skinned thrust belt that incorporates major segments of reworked Laurentian gneiss basement in major thrust sheets. Caledonian migmatites and granite intrusions are widespread in the southern part of the orogen. Transport directions of the major thrusts are to the west-northwest, and restoration indicates total displacements on the order of 200–400 km, with estimated shortening of 40%–60%.

Archean and Paleoproterozoic gneiss complexes, reworked during Caledonian orogenesis, are widespread. In the south, they are overlain by late Mesoproterozoic metasedimentary rocks. Overlying Neoproterozoic and Lower Paleozoic sediments laid down at the margin of Iapetus reach ~20 km in thickness. In the north, the Paleoproterozoic basement gneisses are overlain by late Paleoproterozoic to early Mesoproterozoic quartzites interbedded with basaltic rocks and cut by doleritic dikes and sills. Overlying Neoproterozoic and Lower Paleozoic sedimentary rocks relate to developments on the south side of the Franklinian Basin, which extends across North Greenland and into Arctic Canada.

Keywords: Caledonides, Greenland, geological map.

INTRODUCTION

A colored geological map of the East Greenland Caledonian orogen accompanies this volume (Plate 1 [on loose insert, on CD-ROM acompanying this volume, or in the GSA Data Repository[1]]). The map covers the region between 70°N and 82°N at a scale of 1:1,000,000 (1:1 million) and is based on five geological maps at a scale of 1:500,000 published between 1984 and 2001 (Fig. 1). The regional mapping project that produced the series of maps at 1:500,000 scale was initiated by the former Geological Survey of Greenland (GGU) and completed by the present Geological Survey of Denmark and Greenland (GEUS). The five 1:500,000 geological maps are standard lithostratigraphic maps,

*akh@geus.dk

[1]GSA Data Repository Item 2008168, Plate 1, is available at www.geosociety.org/pubs/ft2008.htm, or on request from editing@geosociety.org, Documents Secretary, GSA, P.O. Box 9140, Boulder, CO 80301-9140, USA.

Henriksen, N., and Higgins, A.K., 2008, Caledonian orogen of East Greenland 70°N–82°N: Geological map at 1:1,000,000—Concepts and principles of compilation, *in* Higgins, A.K., Gilotti, J.A., and Smith, M.P., eds., The Greenland Caledonides: Evolution of the Northeast Margin of Laurentia: Geological Society of America Memoir 202, p. 345–368, doi: 10.1130/2008.1202(14). For permission to copy, contact editing@geosociety.org. ©2008 The Geological Society of America. All rights reserved.

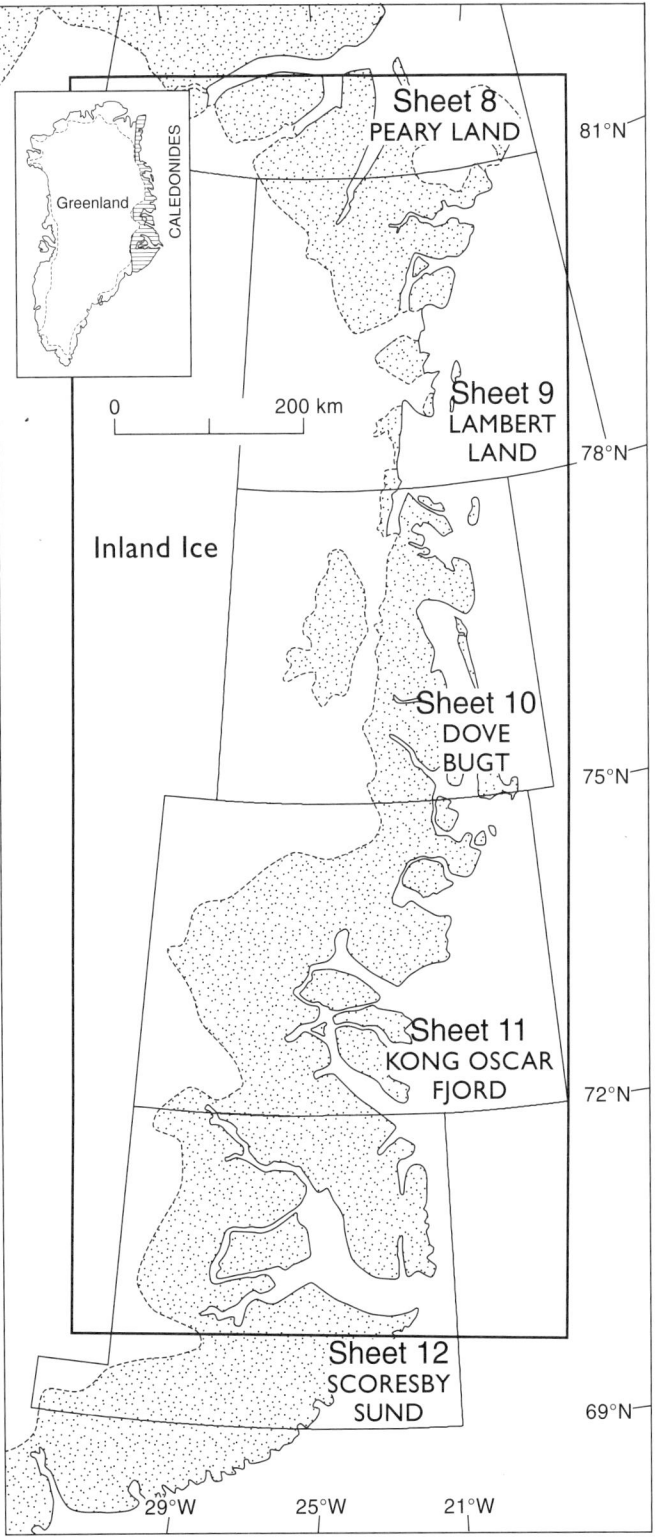

Figure 1. Index map showing the coverage of the Geological Survey of Denmark and Greenland (GGU/GEUS) published 1:500,000 geological maps of the Caledonian orogen in East Greenland (Henriksen, 1984; Bengaard and Henriksen, 1986; Henriksen, 1997; Jepsen, 2000; Escher, 2001). The area covered by the 1:1,000,000 geological sheet is shown by the rectangular box.

and they present the full range of geological units that occur in the mapped region, spanning a time period from Archean to Quaternary. These maps were compiled from field mapping, mainly at a scale of 1:100,000, and there is only a limited amount of interpretation. In contrast, the 1:1,000,000 map is a lithostructural interpretation, designed to provide a geological overview of current knowledge, as well as a structural interpretation of the East Greenland Caledonian orogen. Successions formed in the foreland are distinguished from those found in parautochthonous and allochthonous settings within the Caledonian orogen. The post-Caledonian sequences that occur mainly in the eastern coastal regions of the map sheet are presented with a chronostratigraphic division; lithologies are not depicted, with the exception of the Paleogene igneous province, where a basic distinction is made between intrusive complexes, dikes and sills, and plateau basalt sequences.

The term "Caledonide orogen" is employed on the map sheet to encompass structural events and related rock units up to the end of Silurian time, whereas Devonian and later structures and rock units are classified as "post-Caledonian." However, McKerrow et al. (2000) redefined the Caledonian orogen to include Devonian tectonic events, and this volume essentially adopts the broader view of the Caledonian orogeny: Devonian sedimentation and related structural events are here interpreted as pertaining to late Caledonian extension and exhumation.

The map depicts only the geology of the ice-free land areas. To the west, the Inland Ice largely conceals the transition from the orogen to the foreland. The offshore shelf region, to the east of the coastline with the North Atlantic Ocean, is up to 300 km wide. The subsurface geology of the shelf is not shown on the map but is well known from a variety of geophysical surveys mainly investigating the potential for oil and gas (e.g., Hamann et al., 2005).

The 1:1,000,000 geological map is the first at this scale depicting the northern part of East Greenland. The topographic base used was photogrammetrically plotted from wide-angle aerial photographs (flown in 1978, 1985, and 1987 for the Danish Geodetic Institute [GI]), and a network of new fixed points was also established by the Danish Geodetic Institute (now incorporated into Kort & Matrikelstyrelsen [KMS], the National Survey and Cadastre). All place names used are listed in an appendix to this article with accompanying latitudes and longitudes. A version of the map has been prepared on CD-ROM. This CD-ROM includes the entire map sheet with all topographic and geological data.

The term "East Greenland" officially refers to the entire east-facing coastal region between 59°30′N and 82°N. The former Geological Survey of Greenland (GGU) introduced standard geographic subdivisions of Greenland (Ghisler, 1990), partly based on informal usage by earlier expeditionary activity, and these continue to be used by the present Geological Survey of Denmark and Greenland (GEUS). The East Greenland Caledonian orogen thus extends through part of the survey's "central East Greenland" (70°N–72°N) and all of "North-East

Greenland" (72°N–79°30′N), and the northernmost orogenic segment (79°30′N–82°N) forms part of "eastern North Greenland." These GGU subdivisions are widely used by contributors to the chapters of this volume, but where there is no confusion, they have not been rigidly applied. East Greenland north of ~71°30′N and all of North Greenland form part of the North-East Greenland National Park.

Cartography

In 1968, when GGU began the geological mapping program at 1:500,000 scale in East Greenland, the existing topographic maps were limited to 1:250,000 maps of the region from 72°N to 76°N and a few map sheets from the Scoresby Sund region (70°N–72°N). These maps were based on a ground theodolite survey by the Geodetic Institute (GI) dating from the 1930s, supplemented by photogrammetric plotting using oblique aerial photographs. The contour interval was 100 m, but in many places, contours were omitted, and other areas were blank, because of blind angles on the oblique photographs. North of 76°N, the only existing maps of East Greenland were those of the United States Army Map Service (AMS) at 1:250,000, based on combined vertical and oblique aerial photographs flown from an altitude of ~6000 m. The data from these maps were adapted for the United States Air Force Aeronautical Chart and Information Center (ACIC), and their aeronautical maps were used for geological reconnaissance work in the northernmost parts of East Greenland by geologists working with Lauge Koch's 1926–1958 Danish Expeditions to East Greenland (Koch and Haller, 1971). Since existing ground control points were then widely spaced, the American maps were by modern standards not very accurate, lacked topographic details, and were inadequate for the larger-scale geological mapping planned by the GGU.

In the region around Scoresby Sund (70°N–72°N), where the new geological mapping program began in 1968, new topographic maps were constructed by the Geodetic Institute using existing data and new 1:50,000 vertical aerial photographs; these provided an adequate topographic base for the geological mapping. These new topographic maps were then used as a base for the set of 16 geological map sheets at 1:100,000 published from this region (Henriksen, 1986).

Following the 1968–1972 work in the Scoresby Sund region, GGU planned to extend the geological mapping program to both North Greenland and North-East Greenland, a vast region for which modern topographic base maps were lacking (Henriksen and Higgins, 1991). In cooperation with the Geodetic Institute in Copenhagen, a new topographic mapping project was initiated and subsequently extended to all of Greenland. Beginning in 1978, new super wide-angle aerial photographs at a scale of ~1:150,000 were obtained for the ice-free land areas of all of Greenland, and a network of new ground control points was established using a Doppler-satellite-position system (Bengtsson, 1983). The black-and-white aerial photographs were taken from a flight altitude of 14 km with 40% lateral overlap and 80% overlap in the flight direction. The control points were defined with an absolute precision of a few meters in all three dimensions.

New topographic maps of North and North-East Greenland were compiled photogrammetrically in the survey's (GGU/GEUS) photogeological laboratory at a scale of 1:100,000 with 100 m contours (Figs. 2 and 3). Topographic map compilation and photogeological interpretation were carried out as a joint process by technicians and geologists (see, e.g., Hougaard et al., 1991; Jepsen et al., 1994). The new 1:100,000 topographic maps were used by geologists in the field, and they provided the base for the published 1:500,000 geological map sheets as well as the 1:1,000,000 map that accompanies this volume.

GEUS reprocessed all the topographic data for North and North-East Greenland into a Geographic Information System (GIS) database in 2003, as part of a Greenland Home Rule project to produce maps of the North-East Greenland National Park. One of the products is a set of 48 new 1:250,000 topographic maps with 100 m contours that covers all North Greenland and North-East Greenland north of 70°N. This new topographic GIS database, "G/250vector," is available from Kort & Matrikelstyrelsen (KMS) in Copenhagen (KMS incorporates the former Geodetic Institute). One of the features of the GIS database is that the original compilation data can be utilized to print out topographic maps of selected areas at scales between 1:250,000 and 1:100,000.

Place Names

The responsibility for administration and approval of place names in Greenland was transferred from the Place Name Committee for Greenland in Copenhagen (Denmark), to Nunat Aqqinik Aalajangiisartut–Grønlands Stednavnenævn in Nuuk (Greenland), following the establishment of Greenland Home Rule in 1979. The use of names with European origins has declined in the populated areas of West and South-East Greenland as a consequence of the implementation of Home Rule. This is particularly noticeable with respect to the former double-naming of many towns; the former Danish names for towns have now almost completely disappeared in daily usage, while the Greenland town names have achieved widespread acceptance. The Danish town names can, however, still be found on many maps and have been widely used in the scientific literature.

Place names in thinly populated and unpopulated regions such as North and North-East Greenland still largely reflect the European exploration and discovery voyages. The house ruins of several phases of Eskimo cultures are widespread in North and North-East Greenland, but any place names they may have used have been lost. The only Greenlandic names within the part of East Greenland covered by the 1:1,000,000 map are a small group in the vicinity of Illoqqortoormiut/Scoresbysund, a small town founded in 1924 with settlers mainly from Ammassalik in South-East Greenland. While a few of the large fjords in the Scoresby Sund region have alternative Greenlandic names, these have not been used in the scientific literature, and they have been

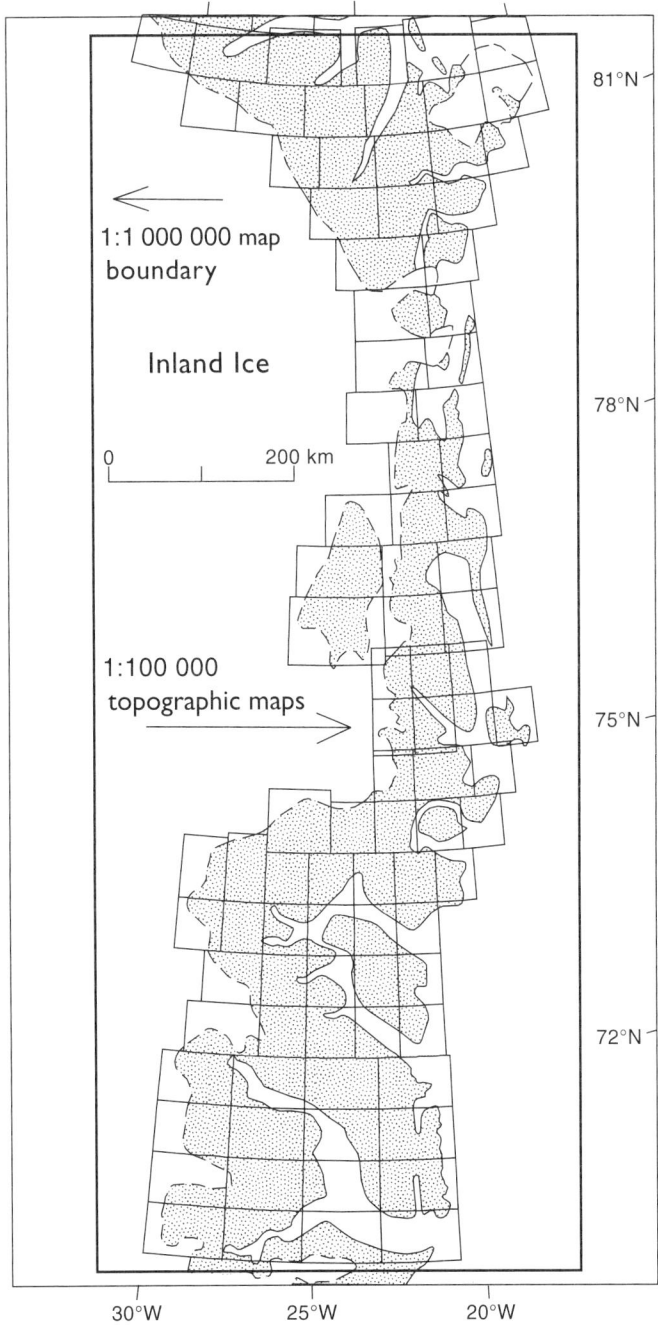

Figure 2. Index to the Geological Survey of Denmark and Greenland (GGU/GEUS) 1:100,000 topographic base maps. The maps show 100 m contours and portray basic physiographic features such as coasts, rivers, lakes, glaciers, and permanent ice cover. The 16 maps in the Scoresby Sund region (70°N–72°N) correspond to the printed 1:100,000 geological map sheets. The topographic maps were produced at the GEUS photogrammetric laboratory using ground control data supplied by Kort & Matrikelstyrelsen, Copenhagen.

omitted from the map here to avoid confusion. The only Greenlandic place name that is used on the map is that for the settlement Illoqqortoormiut/Scoresbysund (misspelled "Illoqqortormiut" on the map), but in this case, the official West Greenland spelling "Illoqqortoormiut" differs from the local East Greenland dialect form "Ittoqqortoormiit" used by the inhabitants.

Out of ~3150 officially authorized place names for features in the northern half of East Greenland, 486 appear on the printed 1:1,000,000 map sheet, and they are listed in an appendix to this article with their longitudes and latitudes. The earliest names were given by Dutch and British whalers in the 1600s (e.g., Hold with Hope, Lambert Land, Gael Hamkes Bugt), but the first recorded European landings on the coast were those of William Scoresby in 1822 and Douglas Clavering and Edward Sabine in 1823. The names for features given on their charts (Scoresby, 1823; Clavering, 1830) are still in use in Danish form. In the late 1800s and early 1900s, a series of discovery voyages by German, French, Swedish, and Danish shipborne expeditions mapped and explored the coast, several of which overwintered and continued their explorations on boat and sledge journeys (Koldewey, 1873–1874; Ryder, 1895; Nathorst, 1901). In the early and mid-1900s, scientific expeditions became increasingly common (e.g., Amdrup, 1913; Koch, 1913; Knuth, 1942), and commercial activities were pioneered by Norwegian and Danish fox hunters (Mikkelsen, 1995). The most intense phase of geological investigations included the long series of Danish expeditions to East Greenland led by Lauge Koch, which, with a break for the war years, lasted from 1924 to 1958 (Haller, 1971). Mining has so far been limited to the exploitation of lead deposits near Mestersvig, discovered by Lauge Koch's expeditions in 1948, and described by Bondam and Brown (1955). All these activities have resulted in new place names being applied to various features, of which the majority of names now have approved status.

Physiography

The GGU/GEUS geological investigations associated with the regional mapping of North-East Greenland extended from south of Scoresby Sund (69°N) to the northern point of Kronprins Christian Land (~82°N), a distance of more than 1400 km. The strip of land between the Inland Ice to the west and the North Atlantic Ocean varies in width from ~100 km to 300 km, and the region depicted on the 1:1,000,000 map has an ice-free area of ~127,000 km^2.

The Inland Ice covers 81% of Greenland's land area, is the second largest body of ice in the world, and its highest point is above 3000 m. In East Greenland, the margin of the Inland Ice reaches down to altitudes of 1500–1000 m in the south and generally 800–600 m in the north, although in the Jøkelbugten region (78°N–79°N) it is close to sea level. The numerous glaciers draining from the Inland Ice that reach the heads of the fjords range from a few kilometers to more than 30 km in width. The most productive glacier is Daugaard-Jensen Gletscher, which drains into Nordvestfjord in the northwest part of the Scoresby Sund

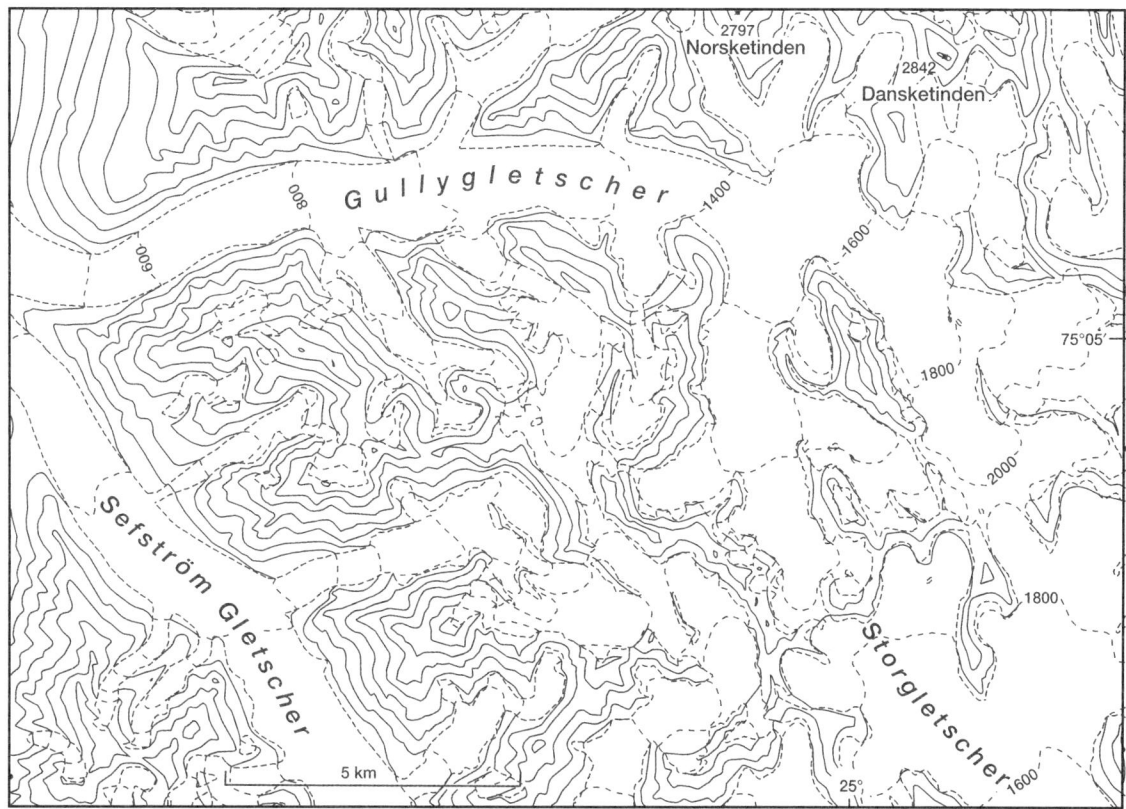

Figure 3. Extract from one of the topographic base maps, reproduced at 1:100,000 scale. It shows part of the northern Stauning Alper, which is characterized by rugged alpine topography, is dissected by glaciers, and has a highest summit above 2800 m. Contours are at 200 m intervals. See also Figure 5.

region, and it has a mean velocity of ~10 m per 24 h and calf-ice production of ~10 km^3 annually (Olesen and Reeh, 1969).

Large, independent local ice caps and networks of alpine glaciers are prominent in the inland areas at altitudes above ~1500 m. In Kronprins Christian Land (79°N–82°N), several ice caps locally reach sea level, of which the largest is more than 120 km long and 75 km wide and terminates with Flade Isblink. The topography of the ice-free areas of North-East Greenland generally reflects the influence of the former ice cover during the Last Glacial Maximum (Funder, 1989; Funder et al., 1998). Glacier-eroded U-shaped valleys cut through the landscape, radiating out from the Inland Ice or local glaciation centers, with a relief varying from a few hundred meters to over 2000 m. Moraines and other glacial deposits are widespread, whereas fluviatile Quaternary deposits are confined to valley floors and lowland areas.

Areas in the Caledonian orogen dominated by gneisses and granites often preserve relics of a dissected peneplain at 1500–2000 m altitude. The numerous fjords and valleys often have an extremely well-exposed steep relief (Fig. 4). Late uplift of parts of the region and subsequent erosion by numerous independent valley glaciers have resulted in high alpine relief, notably in the Stauning Alper (71°30′N–72°15′N; Fig. 5) and western Frænkel Land (73°10′N), where the highest summits exceed 2900 m.

Post-Caledonian, often flat-lying, sedimentary rocks are widespread in eastern coastal regions, and they are characterized by gentle undulating forms and summits usually below 1500–1000 m. Steep-sided fjords and valleys cut these areas, but since the sedimentary rocks are more readily eroded, exposure is not as good as in the crystalline areas. Paleogene basalts are most widespread on the south side of Scoresby Sund, where they form spectacular cliffs more than 2000 m high.

Climate

The northern part of East Greenland lies in the high arctic zone, and average temperatures are 3–5 °C in July, the warmest summer month. Winter temperatures are lowest in February, with mean temperatures of –16 °C at Illoqqortoormiut/Scoresbysund (71°29′N), –24 °C at Danmarkshavn (76°46′N), and –30 °C at Station Nord (81°37′N). In the high arctic climatic zone, the sparse vegetation consists of scattered low bushes and plants.

Annual precipitation, mainly as snow, is low in the whole region: 300–400 mm water equivalent in the south (70°N–72°N), 200–300 mm in the central sector (72°N–75°N), and 100–200 mm farther north (75°N–82°N). The snow that

Figure 4. View eastward across the 5-km-wide inner part of Kejser Franz Joseph Fjord. The snow-covered summit at right is Payer Tinde (2368 m high), the northern flanks of which are formed of dark metasedimentary rocks. The cliff at left center is Ättestupan, showing a 1200-m-high near-vertical face carved in leucocratic 1800 Ma orthogneisses. At the top of the cliff, a light-colored dolomite unit occurs near the base of a thick succession of dark siliciclastic sedimentary rocks.

Figure 5. View southeast toward the two highest summits in the northern Stauning Alper. Dansketinden at left is 2842 m high, and Norsketinden at center right is 2797 m high (see map in Fig. 3). The rocks are Caledonian and early Neoproterozoic granites and migmatitic gneisses.

falls after early September remains over the winter and first begins to melt in May–June. Geological fieldwork is usually only practical in July and August.

Long periods of calm and sunny weather are frequent during the summer months. Winds are mainly katabatic, and they are strongest near the Inland Ice margin. Temperature contrasts between the land and the ice-covered sea create sea breezes, and coastal fog is common. The occasional passage of low-pressure areas may give rise to storms with rain, and sometimes snow even in summer; these conditions usually only last a few days.

Conditions for fieldwork in the summer season are generally outstanding. The summer season, with 24 h of daylight, low precipitation, and many hours of sunshine, has led to the central fjord region being characterized as the "Arctic Riviera" (Hofer, 1957).

FIELDWORK

The region depicted on the 1:1,000,000 map is uninhabited apart from the small town of Illoqqortoormiut/Scoresbysund and outlying settlements (total population 550 in 2004), a few military outposts, and the weather station at Danmarkshavn. The lack of supporting infrastructure and logistics means that any large-scale activity, such as geological fieldwork, must be carried out as self-supporting expeditions. All personnel, provisions, and camp equipment must be transported in and out of the region each field season; an independent local transport and communication system also must be established.

Drifting sea ice from the Polar Basin prevents ship access to the coastal regions for a large part of the year. Ice-strengthened ships can normally only reach the coast between mid-July and early October south of ~76°N. Fjords in southern areas are usually ice-free from mid-July to September. Reliable access to the region is thus only possible by air, using one of the long gravel airstrips capable of handling aircraft such as the Hercules C-130 (Constable Pynt at 70°44′N, Mestersvig at 72°14′N, Station Nord at 81°37′N). Many unmanned, rough, airstrips where STOL (short take-off and landing) aircraft can land have been established throughout the region by the GGU/GEUS and the Danish military sledge patrol "Sirius."

The main fieldwork, which formed the basis for the published 1:500,000 map sheets and the 1:1,000,000 map sheet, took place over ~15 field seasons between 1968 and 1998. In addition to the major mapping expeditions, a variety of local and specialized activities were undertaken in various parts of the region, and relevant results have been integrated into the maps.

The major mapping projects were designed for an expedition group of between 30 and 50 participants each season, including scientific personnel, base camp support staff, and pilots and mechanics of helicopters and STOL aircraft. The two-person geological parties worked in the field from small field camps (Fig. 6) that were resupplied and moved during periodic helicopter visits.

The first three seasons of geological fieldwork in the Scoresby Sund region (1968–1970) were based on ice-strengthened expedition ships fitted with helicopter platforms (Fig. 7) that transported the entire group of scientific and support personnel to and from the working region. During each of the three field seasons, the ships functioned as floating bases. In all the later seasons, transport to and from East Greenland was by charted aircraft, most commonly, military C-130 (Hercules) transport aircraft, and tent base camps were established by STOL aircraft (Pilatus Porter and Twin-Otter)

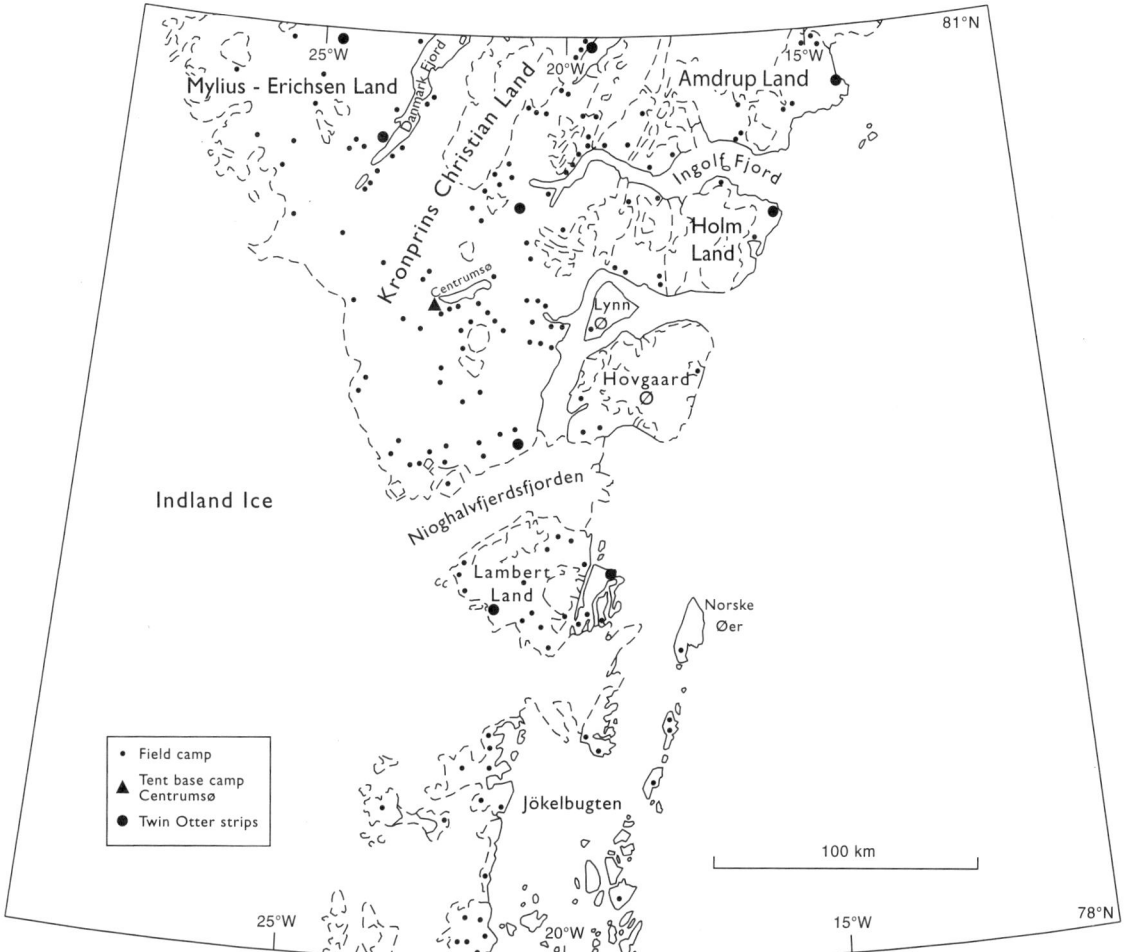

Figure 6. An indication of the intensity of field activities is given for the region 78°N–81°N, corresponding to geological map sheet 9 (Lambert Land). The location of the base camp at Centrumsø and rough landing strips used by the STOL Twin Otter aircraft are shown, together with the field camp sites used by different two-person field parties during the two and a half field seasons.

at convenient locations. The five major fieldwork projects were carried out during 1968–1972, 1979–1980, 1988–1990, 1993–1995, and 1997–1998.

1968–1972: Scoresby Sund Region (69°N–72°N)

The Scoresby Sund region was mapped during five field seasons. The expedition groups consisted of between 30 and 45 participants each season, traveling from Denmark to the Scoresby Sund region by ship (Fig. 7) the first three summers, and working from tent base camps the remaining two summers. The 12–16 two-person field teams operated mainly from field camps moved by helicopter, but some work was carried out using small boats. In each field season, a region of 7000–9500 km² was covered. The geological regional mapping work covered the southernmost segment of the East Greenland Caledonian orogen, post-Caledonian Upper Paleozoic–Mesozoic sedimentary rocks of the Jameson Land Basin, and large areas of Paleogene plateau basalts on the south side of Scoresby Sund. Various special studies were carried out at the same time, including Quaternary geology, glaciology, and paleomagnetic and geochronologic investigations (see overview in Henriksen, 1986). In addition to geological and Quaternary map sheets at 1:500,000, this project produced 16 geological maps at a scale of 1:100,000.

1979–1980: Northern Kronprins Christian Land, Eastern North Greenland (81°N–82°N)

Between 1978 and 1985, the GGU mapped the whole of North Greenland in a total of five field seasons (Henriksen and Higgins, 1991). In 1979 and 1980, a few field teams worked in the northern parts of Kronprins Christian Land and areas south of Independence Fjord that appear on the 1:1,000,000 map. The areas north of 81°N form the southern part of the 1:500,000

Peary Land map sheet, which was printed in 1986 (Bengaard and Henriksen, 1986; Henriksen, 1992). The geological work included mapping of the northernmost segment of the Caledonian orogen, the transition with the foreland areas to the west, and the post-Caledonian sedimentary rocks of the Upper Paleozoic–Mesozoic Wandel Sea Basin. The 1978–1980 expeditions were flown by military C-130 into Station Nord, and field activities were operated from a tent base camp at Fastelavnssø (82°12′N, 30°00′W) in southern Peary Land.

1988–1990: Dove Bugt Region (75°N–78°N)

The region between Bessel Fjord in the south and Jøkelbugten was the main focus of activity during three summer field seasons. The expeditions were flown into Mestersvig by military C-130 aircraft and transported into the field area by Twin Otter (STOL) aircraft. The first summer, a tent base camp was established on the west side of Fligely Fjord (74°53′N), and it was moved to Hvalrosodden (76°55′N) for the two subsequent summers. The number of participants ranged from 31 in 1988 to 47 in 1990, including the geological teams and all support personnel. As in previous years, the two-person field teams were supported by helicopter. In 1988, the activities of five field teams extended south of 75°N (Henriksen, 1989). The geology of the Dove Bugt region is dominated by Precambrian crystalline basement complexes reworked during the Caledonian orogeny, but a few teams undertook special studies of the economic geology, the limited areas of Mesozoic sediments, the northernmost representatives of the Paleogene basalt province, as well as Quaternary geology and glaciology studies at the margin of the Inland Ice. During the 1989 summer season, activities were concentrated in the network of large nunataks making up Dronning Louise Land, where strong katabatic winds and low temperatures periodically hindered the fieldwork.

1993–1995: Lambert Land Region (78°N–81°N)

The region covered by this three-year project extended from Jøkelbugten in the south through Lambert Land to central Kronprins Christian Land in the north. Access to the region was by military C-130 aircraft to the small Danish military base at Station Nord in northernmost Kronprins Christian Land, where the expedition personnel and equipment were transferred by Twin Otter aircraft to a tent base camp at Centrumsø (Fig. 8). The first year of the project was a short reconnaissance season, with 19 participants and only one supporting helicopter. The two remaining seasons were full-scale expeditions with 46 and 47 participants, respectively, including 15 two-person scientific teams, supporting base camp staff, and helicopter crews. Most geological parties worked on the structure and petrology of the crystalline rocks of the Caledonian orogen, and on the sedimentology and stratigraphy of the Mesoproterozoic to Lower Paleozoic successions that are widely exposed in the foreland and extend eastward into the marginal fold-and-thrust belt of the Caledonian orogen. A few parties worked in the post-Caledonian sedimentary rocks of the Wandel Sea Basin, and two parties undertook glaciologic investigations at the margin of the Inland Ice and at the major outlet glacier from the Inland Ice that fills Nioghalvfjerdsfjorden.

1997–1998: Kong Oscar Fjord Region (72°N–75°N)

The region extending from Mestersvig and the Stauning Alper in the south to Grandjean Fjord and Hochstetterbugten in the north is ~330 km from south to north and up to 300 km from east to west. This extensive and largely ice-free region is cut by numerous major fjords that extend westward to the margin of the Inland Ice, and they provide spectacular sections through the geo-

Figure 7. The ice-strengthened expedition ship *Magga Dan*, which functioned as a floating base during mapping in the Scoresby Sund region during the 1969 summer. Two Bell 3 GB helicopters operated from the helicopter platforms. Photograph was taken in the outer part of Nordvestfjord.

Figure 8. The 1994 tent base camp established on river terraces at the west end of Centrumsø, Kronprins Christian Land, the expedition's two small helicopters (Hughes 500E Notar), and Twin Otter aircraft. Large tents and small temporary huts provided equipment storage and radio and kitchen facilities, while the 20 small tents housed personnel.

logical structure. Geological investigations began in the late nineteenth century in this region, which was generally accessible by ship except in the worst ice years, and the region became the main focus of the long series of Danish geological expeditions to East Greenland led by Lauge Koch from 1926 to 1958 (see, e.g., Haller, 1971). A major result of this work was a set of geological maps at a scale of 1:250,000 (Koch and Haller, 1971). A limited amount of work was carried out in the region by GGU field teams from ~1973 to 1980, but they were focused on specific investigations of stratigraphy, structure, and mineralization rather than mapping. A wealth of geological information of all kinds thus existed prior to systematic mapping of this region by GEUS in 1997–1998.

The expedition groups in 1997 and 1998 were flown into Mestersvig by military C-130 aircraft. This airport in the southern part of the mapping region was originally constructed for the former lead mine near Mestersvig, and it functioned as a civil airport until 1985. Its continued use for expeditions visiting the national park was dependent on a small military group, and the Danish Polar Center, which maintained a minimum service related to resupply of the Zackenberg research station (74°29′N). The housing facilities at Mestersvig served as a center for the GEUS field operations and as a base for the GEUS two chartered helicopters and Twin Otter aircraft. In 1997, the expedition group numbered 38 persons, and in 1998, there was a total of 43 participants. The majority of the ~30 geoscientists worked in the Caledonian orogenic belt, and a few teams worked in the post-Caledonian sedimentary basins (Henriksen, 1999). Although the major geological features of the region were known from the extensive earlier investigations, significant new detail was added, and discoveries at key localities in the Caledonian orogen led to new interpretations of the thrust architecture (Higgins and Leslie, 2000; Higgins et al., 2001a, 2004b).

Arild Andresen, of the University of Oslo, began a major project to study aspects of the collisional and extensional history of the Caledonides and the post-Caledonian sedimentation linked to orogenic collapse. Geological groups, under the leadership of Andresen, were active throughout the region between Nordvestfjord (71°50′N) and Ardencaple Fjord (75°30′N) between 1995 and 2003, and they have included participants from Norway, Denmark, and the United States. Significant work has focused on the N-S–trending extensional fault system known as the Fjord region detachment zone (Hartz and Andresen, 1995; Hartz et al., 2000).

PREVIOUS GEOLOGICAL MAPS

The greater part of the geological information on North-East Greenland resulting from the activities of the "Danish Expeditions to East Greenland 1926–1958" is published in the scientific monograph series *Meddelelser om Grønland*, issued by the Scientific Commission for Greenland, Copenhagen. Many of the volumes are accompanied by colored geological maps at a scale of 1:250,000, which were compiled by John Haller into the set of map sheets covering the region 72°N–76°N (Koch and Haller, 1971; Fig. 9). A reconnaissance map of the region north of 75°N at a scale of 1:1,000,000 was also published (Haller, 1983).

The Danish expeditions to East Greenland led by Lauge Koch were mostly funded by the Danish State on the basis of annual grants; however, the work did not have the status of an official survey. Lauge Koch initiated and organized the work programs with the aid of a very small administrative staff in Copenhagen, made short-term contracts with the participating scientists and technical personnel, and carried out the logistic arrangements. In the postwar years, the participating geologists were mainly from Swiss and British universities, who returned to their institutions after the field seasons to work on their data and rock samples. Their reports on their work, submitted to Lauge Koch as part of their contract, often formed the basis for the monographs published in *Meddelelser om Grønland*.

The geological maps produced by the "Danish Expeditions to East Greenland" were excellent for their time, an era essentially predating the use of helicopters. Geological mapping was mainly carried out by long foot traverses, with small boats used for transport from area to area, and the occasional use of ponies in suitable terrain. Lauge Koch was at the forefront in adapting new means of transport for expedition purposes, and early versions of fixed-wing Heinkel seaplanes were used for transport and aerial reconnaissance in the 1930s. After the Second World War, Catalina and DC-3 aircraft were used for transport to and from Greenland instead of ships, and Norseman seaplanes were extensively used for aerial reconnaissance and aerial photography. The Norseman seaplanes could only operate from ice-free fjords and lakes, but their long range meant that many otherwise inaccessible areas could be overflown, and more than 10,000 low-altitude aerial photographs were taken by professional Swiss photographers for use by the mapping geologists. The reconnaissance map of northern East Greenland at 1:1,000,000 (Haller, 1983) was largely based on photogeological interpretations of photographs taken from Norseman seaplanes and vertical aerial photographs of the region north of latitude 76°N flown for the Geodetic Institute from 1959 to 1963; ground observations were few and widely scattered.

The complex of large nunataks that make up Dronning Louise Land was first investigated scientifically by the 1952–1954 British North Greenland expedition, a military joint services expedition that overwintered in northern Dronning Louise Land. Their geological maps at a scale of ~1:250,000 (Peacock, 1956, 1958), supplemented by John Haller's photogeological interpretations, were incorporated into regional descriptions and map compilations (Haller, 1970, 1983).

John Haller's involvement with Lauge Koch's East Greenland geological expeditions began in 1949 with the mapping of the crystalline complexes of Andrée Land (Haller, 1955). Mapping of additional large areas of mainly crystalline gneisses and granites in the Stauning Alper, Nathorst Land, Frænkel Land, and parts of the nunatak region occupied subsequent summers. All these investigations were published, accompanied by geological maps and cross sections (Haller, 1956, 1958; Wenk and Haller, 1953). Lauge Koch's funding from the Danish government was suspended after the 1958 field season, although further geological expeditions to East Greenland had been planned. Short-term

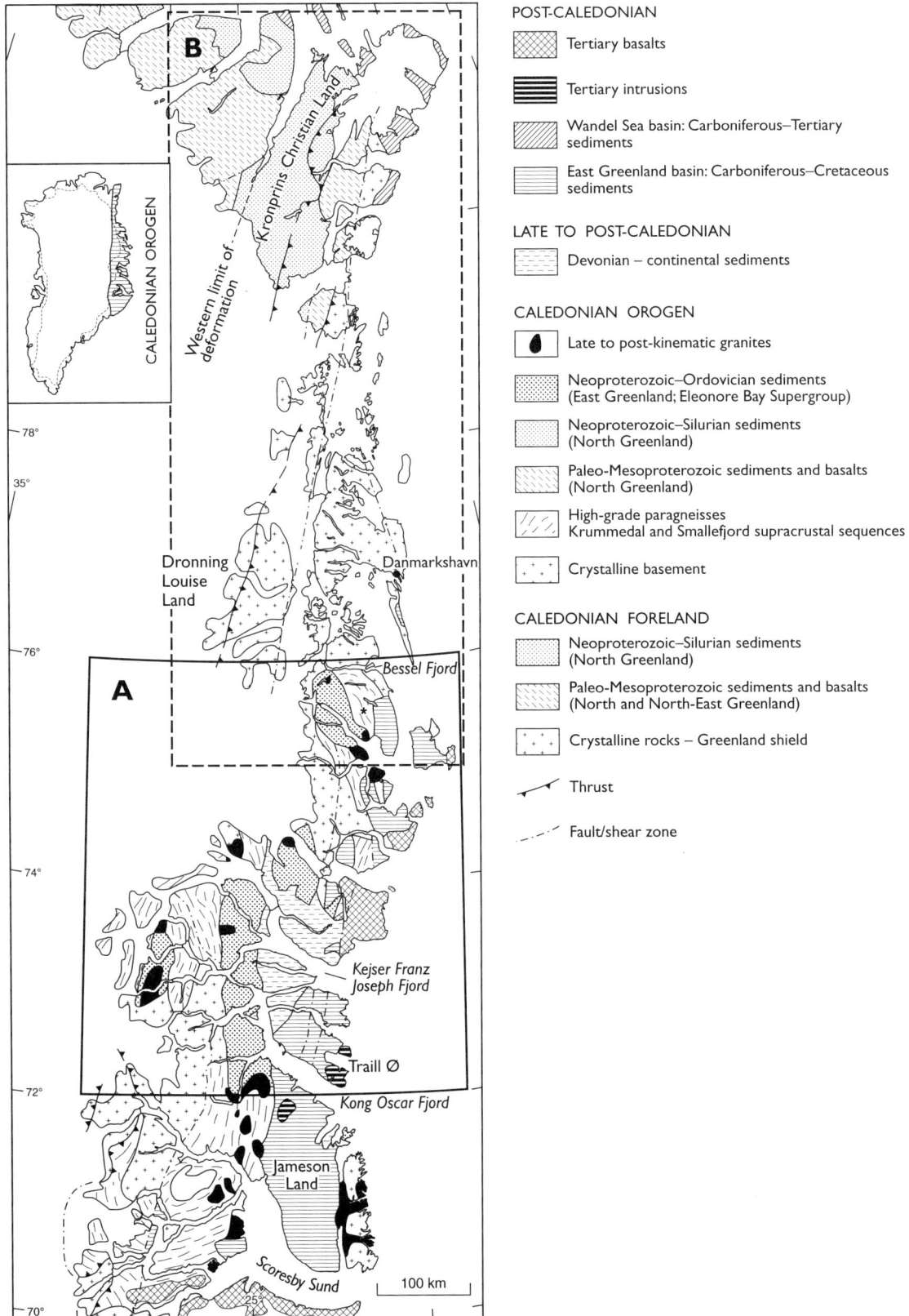

Figure 9. Simplified geological map of northern East Greenland, showing the coverage of maps by Lauge Koch's 1926–1958 Danish Expeditions to East Greenland. (A) Region covered by the published 1:250,000 geological maps of Koch and Haller (1971). (B) Region covered by the 1:1,000,000 geological reconnaissance map of Haller (1983).

funding was made available to John Haller, as chief scientist, to undertake compilation of the set of geological maps from his home base at Basel University, Switzerland. The compilation work was continued with funding from Swiss foundations. In 1964, he was appointed professor of geology at Harvard University, Cambridge, Massachusetts, and many of Haller's major East Greenland publications were completed during his tenure at Harvard. The 13 sheets of the 1:250,000 *Geological Map of East Greenland 72°N–76°N*, which had been printed in 1965, were published with a short description by Koch and Haller (1971). An extensive geological description to accompany a tectonic map of East Greenland at 1:500,000 scale in three map sheets was published in *Meddelelser om Grønland* in 1970 (Haller, 1970). His most significant work, the *Geology of the East Greenland Caledonides*, was published the following year (Haller, 1971). The reconnaissance map at a scale of 1:1,000,000 of northern East Greenland was finally published in 1983 (Haller, 1983), just one year before his untimely death in 1984.

COMPILATION OF THE 1:1,000,000 GEOLOGICAL MAP

The geological map at a scale of 1:1,000,000 that accompanies this volume is based on the five 1:500,000 geological map sheets published by the GGU/GEUS between 1984 and 2001. Geological information and the implications of geochronological data that have become available since the compilation and printing of the 1:500,000 map sheets have been incorporated into the new map. These additional data sources are listed at the foot of the legend of the map sheet.

The printed 1:500,000 geological maps are lithostratigraphic interpretations, with units divided according to age and rock type. Precambrian gneissic and metamorphic rocks are usually divided into lithologic-age categories, whereas Proterozoic and Phanerozoic sedimentary rocks are divided into formal groups and formations. Paleogene basalts and intrusive rocks are distinguished by rock type and age, and, where formally defined, by group or formation. The former Proterozoic subdivisions, Lower, Middle, and Upper are used on all sheets except the most recent, sheet 11, Kong Oscar Fjord. The latter sheet, and the 1:1,000,000 map, use the newer terminology Paleo- (Palaeo-), Meso-, and Neoproterozoic. Similarly, the usage of Tertiary on the older 1:500,000 map sheets is replaced by Paleogene (Palaeogene) on sheet 11 and the 1:1,000,000 map. All the GGU/GEUS map sheets employ "English" rather than "American" spelling in the legend: e.g., "Palaeogene" and "Palaeozoic" are used for Paleogene and Paleozoic, and "dyke" is used rather than "dike." Structural information on the maps depicts observed features, although the ages given for the features are in part interpretative. The content and regional setting depicted on each of the five 1:500,000 geological maps are briefly summarized here, from north to south (Fig. 1). Description of the Caledonian geology within the five 1:500,000 maps and the area depicted on the 1:1,000,000 map is given in articles in this volume and will, therefore, not be treated here.

A short geological description of the Peary Land map sheet (sheet 8) was given by Henriksen (1992). Many aspects of the regional geology are outlined in a description of the sedimentary basins of North Greenland (Peel and Sønderholm, 1991). The only parts of the Peary Land sheet used for compilation of the 1:1,000,000 geological map of the Caledonian orogen are in the southeastern part between latitudes 81°N and 82°N, including northern Kronprins Christian Land and the areas west of Danmark Fjord and south of Independence Fjord. The Caledonian foreland region west of Danmark Fjord was partly reinvestigated after the Peary Land map was published, and new regional interpretations and geochronologic studies have provided better control on ages of some units. Thus, the Kap Holbæk Formation is now considered to be Early Cambrian in age rather than Neoproterozoic, and it is excluded from the Hagen Fjord Group (Smith et al., 2004). Furthermore, the age range of the Independence Fjord Group has been extended back to the Paleoproterozoic (1740 Ma), following the dating of rhyolitic lavas interbedded with the lower part of the sandstone succession (Kalsbeek et al., 1999; Pedersen et al., 2002; Collinson et al., this volume).

The Lambert Land map (sheet 9) covers the region between central Kronprins Christian Land (81°N) and Jøkelbugten (78°N), and it includes the most completely exposed transition between the Caledonian foreland and the Caledonian orogenic belt. The Lambert Land map sheet contains two cross sections through the Caledonian orogenic belt in southern Kronprins Christian Land, and one cross section through the foreland. These cross sections are also reproduced on the 1:1,000,000 geological map. The Lambert Land map sheet was mapped in the years 1993–1995 and printed in 2000 (Jepsen, 2000). The compilation was based on field maps at 1:100,000 scale, supplemented by considerable photogeological interpretation. This was one of the earliest digital compilations by GEUS, and all data are in a vectorized digital format. The geological data of the 1:500,000 map compilation were adapted for the 1:1,000,000 map by simplifying the detailed subdivisions of the legend. The 25 foreland units on the 1:500,000 map have been included in 12 units on the 1:1,000,000 map, and the 22 units of the Caledonian orogen are included in eight units. The overall structural geometry and the boundary lines have, however, been retained essentially unchanged.

The Dove Bugt map (sheet 10) covers the region between Grandjean Fjord in the south and Hertugen af Orlèans Land and the Danske Øer in the north (75°N–78°N). The border zone of the Caledonian orogen against the Precambrian Greenland shield is well exposed in the extensive nunatak area of Dronning Louise Land. The fieldwork on which the map sheet is based was carried out in 1988–1990. The 1:500,000 geological map was compiled from a 1:250,000 draft map, prepared on the basis of original field maps and photogeological maps at 1:100,000. A collection of 19 articles on the geology of the Dove Bugt region was published in the GGU *Report* series (Higgins, 1994). The geological data on the 1:500,000 Dove Bugt map sheet were adapted for the 1:1,000,000 map by simplification of the details. The 39 units on the legend of the Dove Bugt map sheet have been included in 18 units on the 1:1,000,000 map. The overall structural geometry and contacts are, however, essentially unchanged.

The Kong Oscar Fjord map (sheet 11) covers the geologically best known part of East Greenland (72°N–75°N). This region, which was the main focus of Lauge Koch's long series of geological expeditions, is covered by the 1:250,000 geological maps of Koch and Haller (1971), and it has become a classic area for Caledonian studies (e.g., Haller, 1971). The new 1:500,000 map sheet is based on new fieldwork during the summers of 1997 and 1998 (Henriksen, 1999). The map sheet was printed in 2001 (Escher, 2001). Previously published geological data (Bengaard and Larsen, 1992) were combined with new mapping, supplemented in some parts of the region by photogeological interpretation. All available information was initially compiled into a draft map at 1:250,000, transferred to a digital format, and stored in a vectorized form using ARC/INFO. For the purposes of the 1:1,000,000 map, the 73 subdivisions on the legend on the 1:500,000 map were combined into 28 units. Contacts and structures depicted were essentially retained on the smaller-scale sheet. The two E-W cross sections of the Kong Oscar Fjord map sheet have been reproduced on the 1:1,000,000 map.

The Scoresby Sund map (sheet 12) was the first of the 1:500,000 geological maps of the Caledonian orogen to be published (Henriksen, 1984), following a five-year period of fieldwork from 1968 to 1972. The region covered by the 1:500,000 map sheet extends from latitude 69°N to 72°N. A descriptive text to the Scoresby Sund map sheet was also published (Henriksen, 1986). The compiled map sheet was based on 16 geological maps at 1:100,000 scale, published between 1975 and 1986, with simplification of details to accommodate the smaller scale. The 1:1,000,000 map sheet represents a further simplification of detail. The 69 units on the legend of the 1:500,000 Scoresby Sund map sheet are represented by 29 units on the corresponding part of the 1:1,000,000 map.

PHOTOGEOLOGICAL WORK

New ground control points were established throughout East Greenland by the Geodetic Institute (GI; since 1988 part of Kort & Matrikelstyrelsen, the National Survey and Cadastre), for use in conjunction with the super wide-angle vertical aerial photographs that cover the region at a scale of ~1:150,000. Photogeological interpretation was carried out for selected areas of the East Greenland Caledonian orogen using the same photogrammetric instrument used for preparation of the 1:100,000 topographic maps.

From the early 1970s until the end of the 1990s, the GGU photogrammetric laboratory used a Kern PG 2-D stereoplotter instrument, initially equipped with analogue drawing facilities, and later with a digital recording system. In 2000, GEUS replaced this instrument with a Digital Photogrammetric Workstation (BAE Systems), which has a flat-bed plotter and links to color-plotter facilities. Expanded computer facilities for the new instrument allow the automatic generation of contour lines. Geological interpretations with both instruments have been supported by specially developed programs that, in addition to accurate plotting of geological boundaries, permit the geologist to calculate and plot dips and strikes, fault displacements, and bed thickness. Calculated dip and strike measurements can also be combined to calculate local fold axes and extrapolate bedrock geological boundaries from well-exposed areas into poorly exposed and covered areas (Hougaard et al., 1991). The photogrammetric plotting of boundary lines, with the 1:150,000 photographs, is estimated to be accurate to a few meters.

Photogeological interpretations were mainly carried out in areas where the geological setting favored such a working method. Thus, areas with Proterozoic stratiform successions gave good results where beds or groups of beds could be distinguished on the black-and-white aerial photographs. Large parts of Kronprins Christian Land and the foreland areas west of Danmark Fjord (map sheets 8 and 9) were interpreted photogeologically, and subsequent field control was done in critical areas. The 1:250,000 map of the Upper Proterozoic (Eleonore Bay Supergroup) to Devonian succession in the Kong Oscar Fjord region (72°N–74°N) was entirely constructed using photogrammetric methods (Bengaard and Larsen, 1992). Photogeological interpretation was largely limited to plotting geological boundaries between contrasting rock types and drawing structural trend lines in areas dominated by late Paleoproterozoic and Mesoproterozoic metasedimentary rocks and Proterozoic or Archean gneiss complexes.

THE 1:1,000,000 GEOLOGICAL MAP—CONCEPTS OF COMPILATION

The 1:1,000,000 map of the Caledonian orogen differs from the standard regional 1:500,000 geological maps in its specific focus on the East Greenland Caledonides. The map links lithostructural units across the boundaries of the original 1:500,000 map sheets and, thus, provides a regional overview. Geological divisions in the post-Caledonian areas have been converted from lithostratigraphic to chronostratigraphic units.

Three main divisions are applied on the map and legend: (1) post-Caledonian; (2) Caledonian foreland; and (3) Caledonian orogen. The post-Caledonian succession is divided into chronostratigraphic volcanic units and sedimentary (basinal) units. Units in the Caledonian orogen are arranged chronologically within the overall structural framework.

Post-Caledonian Rock Units

Quaternary deposits are only shown where they cover extensive areas, mainly in coastal lowlands and major valley systems. No subdivisions are made on the map. Descriptions of Quaternary investigations in the Scoresby Sund region can be found in Funder (1987, 1990), while studies in eastern North Greenland were reported by Funder and Hjort (1980).

A regional overview of the Quaternary of Greenland has been presented by Funder (1989).

Three units are distinguished for the Paleogene volcanic province: basaltic plateau lavas (58–54 Ma), basaltic sills and dikes (ca. 52 Ma), and intrusive complexes (48–28 Ma). The basaltic plateau lavas are extensively exposed south of Scoresby Sund (Larsen et al., 1989, 1999), and they occupy smaller areas in a northern region between Hold with Hope and Shannon (Upton et al., 1984, 1995; Watt, 1994). Basaltic dikes and sills are locally large and conspicuous, notably where emplaced into the Paleozoic and Mesozoic sedimentary successions between Jameson Land and Geographical Society Ø (Upton and Hald, 1980; Price et al., 1997). Felsic and intermediate to mafic intrusions form large complexes in Scoresby Land and southeast Traill Ø (Nielsen, 1987).

The Upper Paleozoic and Mesozoic successions are divided chronostratigraphically. Devonian–Lower Permian continental sedimentary rocks with scattered Devonian intrusions are widely exposed in southern areas (Olsen, 1993; Olsen and Larsen, 1993; Stemmerik et al., 1993). These are overlain by Upper Permian–Cretaceous rift-basin deposits, which are dominated by marine sediments (Stemmerik et al., 1993; Stemmerik and Piasecki, 2004; Surlyk, 1990, 2003). Two divisions of the Permian are made on the map to distinguish the Lower Permian continental rocks from the Upper Permian marine sedimentary rocks. The rocks of the Wandel Sea Basin, which are Carboniferous to Paleogene in age, are confined to the region north of 80°N (Stemmerik et al., 1998; Pedersen and Håkansson, 2001).

Caledonian Foreland

Autochthonous foreland successions, unaffected by the Caledonian orogeny, are well exposed in the region north of latitude 79°N and west of Danmark Fjord. In areas farther south, the foreland is almost completely concealed beneath the Inland Ice; the foreland areas exposed in western Dronning Louise Land and in windows eroded through thrust sheets nearly all exhibit some Caledonian disturbance, and are viewed as parautochthonous foreland; only the foreland exposed around inner Hamberg Gletscher and that in the Gåseland window are viewed as essentially autochthonous. The autochthonous foreland west of Danmark Fjord includes the 2-km-thick sandstone succession of the Paleoproterozoic–Mesoproterozoic Independence Fjord Group (Collinson, 1980; Sønderholm and Jepsen, 1991), the basalts of the Mesoproterozoic Zig-Zag Dal Formation (probably ca. 1380 Ma; Upton et al., 2005), and the correlative Midsommersø Dolerites (Kalsbeek and Jepsen, 1984). A substantial hiatus of ~600 m.y. is followed by the ~1000-m-thick succession of siliciclastic and carbonate shallow-water shelf sedimentary rocks of the Neoproterozoic Hagen Fjord Group (Sønderholm and Jepsen, 1991; Clemmensen and Jepsen, 1992). The upper part of the foreland succession is composed of Cambrian–Silurian, dominantly carbonate, sedimentary rocks of the Franklinian Basin (Peel and Sønderholm, 1991).

Caledonian Orogen

The subdivisions of the Caledonian orogen are based on structural levels and geographic position. The four principal subdivisions are: parautochthonous units north of 79°15′N; parautochthonous-allochthonous units north of ~78°N; allochthonous Caledonian thrust sheets (70°N–81°N); and Caledonian tectonic windows and western parautochthonous foreland.

Parautochthonous Units North of 79°15′N

The parautochthonous thin-skinned fold-and-thrust belt of western Kronprins Christian Land marks the transition between the intact foreland to the west and the allochthonous thrust sheets to the east. The units in this zone include the Neoproterozoic Hagen Fjord Group and Cambrian to Silurian sedimentary rocks of the Franklinian Basin, which are identical to their counterparts on the foreland to the west. On the map, the parautochthonous units are distinguished from the autochthonous units by slightly darker-color tones.

Parautochthonous-Allochthonous Units North of ~78°N

A series of thrust sheets structurally overlies the thin-skinned fold-and-thrust belt of Kronprins Christian Land. These sheets preserve Paleoproterozoic to Neoproterozoic sedimentary and volcanic successions, of which the older sedimentary rocks are invaded by abundant dikes and sills. The lowest thrust sheet, the Vandredalen thrust sheet, contains the youngest rocks; these include a Neoproterozoic clastic succession known as the Rivieradal Group, which is not represented on the foreland and can be shown to have accumulated in a rift basin that was overstepped by the later Neoproterozoic Hagen Fjord Group (Higgins et al., 2001b, 2004a). The Rivieradal Group deposits were thrust westward across the shoulders of the rift basin during the Caledonian orogeny and make up the Vandredalen thrust sheet. Continued shortening was taken up on splays off the Vandredalen thrust that propagated westward into the Ordovician platform carbonates to produce the thin-skinned parautochthonous fold-and-thrust belt. The thickness of the overburden was calculated from the thermal effects on conodonts (color alteration indices) in Ordovician platform carbonates (Rasmussen and Smith, 2001). The higher-level thrust sheets that occupy central Kronprins Christian Land and extend southward through Lambert Land to the nunatak region at ~78°N consists of strongly folded Paleoproterozoic to Mesoproterozoic siliciclastic sedimentary rocks of the Independence Fjord Group, with abundant dikes and sills. The lower part of the Independence Fjord Group succession is interbedded with basaltic rocks of the Hekla Sund and Aage Berthelsen Gletscher Formations (Pedersen et al., 2002; Collinson et al., this volume), of which the former formation has yielded a sensitive high-resolution ion microprobe (SHRIMP) zircon age of 1740 Ma (Kalsbeek et al., 1999). The magnitude of displacement of these higher thrust units is unknown but is probably on the order of 50–100 km. Gneissic units believed to form the basement underlying the Independence Fjord Group are found as isolated occurrences in Lambert Land and on nunataks west of Jøkelbugten.

Allochthonous Caledonian Thrust Sheets (70°N–81°N)

The rock units making up the allochthonous thrust sheets throughout the length of the Caledonian orogen are included within this broad subdivision. A regional thrust architecture of two major thrusts sheets, the Niggli Spids thrust sheet and the Hagar Bjerg thrust sheet, overlying foreland windows can be distinguished south of latitude 76°N (Higgins et al., 2004b; Leslie and Higgins, this volume). The allochthonous thrust sheets north of latitude 76°N are dominated by orthogneiss complexes (Fig. 10) that contain abundant eclogite enclaves over a wide region; these demonstrate that the gneisses have been exhumed from a great depth and carried westward across the foreland by thrusting at a late stage in the Caledonian orogeny (Gilotti et al., this volume). The orthogneisses both south and north of latitude 76°N have Paleoproterozoic protolith ages and are shown as one unit on the map. Future detailed work will probably reveal a complex interleaving of structural elements in both regions.

The Niggli Spids thrust sheet south of 76°N consists of Archean and Paleoproterozoic gneisses overlain by thick late Mesoproterozoic metasedimentary successions (Fig. 11). There is a diffuse boundary between Archean gneisses to the south and Paleoproterozoic gneisses to the north at ~72°50′N (Thrane, 2002). The overlying metasedimentary successions, distinguished as the Krummedal supracrustal sequence, have yielded SHRIMP ages on detrital zircons that show the sequence must have been deposited after ca. 1050 Ma (Kalsbeek et al., 2000; Watt et al., 2000; Watt and Thrane, 2001). The structurally overlying Hagar Bjerg thrust sheet consists of Paleoproterozoic gneisses overlain by metasedimentary successions correlated with the Krummedal supracrustal sequence. The Krummedal sequence metasedimentary rocks of the Hagar Bjerg thrust sheet are, however, extensively migmatized and invaded by two generations of granites. A significant thermal event at ca. 930 Ma led to extensive melting of the pelitic siliciclastic sedimentary rocks of the Krummedal sequence to produce abundant migmatitic neosome veins. Larger accumulations of melt produced granite sheets and bodies, commonly with large feldspar phenocrysts (Kalsbeek et al., 2001a), the "augen granites" of many descriptions (Steiger et al., 1979; Henriksen, 1986; Kalsbeek et al., 2000; Leslie and Nutman, 2003). On the map, the Krummedal metasedimentary rocks of the Niggli Spids thrust sheet are grouped as Mesoproterozoic, whereas the migmatitic varieties of the same Krummedal metasedimentary rocks of the Hagar Bjerg thrust sheet were formed in the early Neoproterozoic.

Figure 10. Major isoclinal fold in basement orthogneisses, refolded by a N-S–trending open structure. Black unit at left side of photograph is amphibolite. Cliff height is ~1200 m. View is north side of innermost Grandjean Fjord (~75°N), looking northward.

Figure 11. View southeast across Kejser Franz Joseph Fjord to Payer Tinde (2368 m). The photograph shows the two part division of the Niggli Spids thrust sheet, where light-colored, well-foliated granitic gneisses and augen granites (1800 Ma) are conformably overlain by dark-colored metasedimentary rocks (Krummedal supracrustal sequence ca. 1000 Ma).

Renewed melting of Krummedal metasedimentary rocks produced a younger generation of granites (435–425 Ma) during the Caledonian orogeny. At about the same time, major thrusting displaced the rock units that became part of the Hagar Bjerg thrust sheet westward above the rock units that were to become part of the Niggli Spids thrust sheet. The 930 Ma granites are often indistinguishable in the field from the younger 435–425 Ma Caledonian granites; the two granite generations can only be reliably separated on the basis of geochronological investigations (Kalsbeek et al., 2000, 2001a, 2001b). On the legend of the map sheet, three categories of granite are therefore distinguished, including one category for granites with as-yet undetermined ages.

The S-type Caledonian granites produced by melting of the Krummedal metasedimentary rocks of the Hagar Bjerg thrust sheet migrated upward to form dikes and granite bodies in the overlying Eleonore Bay Supergroup. The Neoproterozoic Eleonore Bay Supergroup (up to 13.5 km thick), together with the Vendian Tillite Group (800–1000 m; Sønderholm et al., this volume) and the Cambrian-Ordovician Kong Oscar Fjord Group (up to 4 km; Smith and Rasmussen, this volume), make up the 18.5-km-thick Neoproterozoic to Ordovician succession described as the Franz Joseph allochthon. This allochthonous unit is always in tectonic contact with the underlying Krummedal supracrustal sequence, typically as a bedding-parallel detachment that may be a reworked unconformity (Leslie and Higgins, this volume).

North of latitude 76°N, the allochthonous thrust sheets are dominated by orthogneiss complexes that contain eclogite enclaves testifying to deep burial and exhumation. Metasedimentary rocks are uncommon; equivalents to the late Mesoproterozoic (Krummedal sequence) and Neoproterozoic (Eleonore Bay Supergroup) successions have not been recognized. Caledonian granites, largely generated by melting of metasediments south of 76°N, have not been recorded in the onshore part of the orogen north of 76°N.

Caledonian Tectonic Windows and Western Parautochthonous Foreland

The foreland is continuously exposed west of Kronprins Christian Land (79°N–82°N). Farther south, the Inland Ice conceals much of the western margin of the East Greenland Caledonian orogen. The largest area of near-autochthonous Caledonian foreland crops out in western Dronning Louise Land (76°N–77°30′N), and smaller areas occur in nunataks in the upper reaches of Hamberg Gletscher (73°30′N–74°N) and in western Gåseland (70°N–70°30′N). Other foreland rocks crop out in large and small anticlinal windows eroded through the thrust sheet pile, and the most important windows occur in Nørreland (78°42′N), around Eleonore Sø (73°N–74°25′N), at Målebjerg (73°40′N), and in Charcot Land (71°N–72°30′N). The rock units exposed in these windows differ significantly from those found in the structurally overlying thrust sheets. All these foreland areas have, however, been affected to some degree by Caledonian deformation or shearing that can be related to displacement of the overlying major Caledonian thrust sheets on basal shear zones, or the Caledonian deformation that exposed the windows in anticlinal culminations; they are thus arranged in the map legend as parautochthonous foreland units within the Caledonian orogen.

The gneiss complexes represented in the foreland windows, where dated, appear to be Paleoproterozoic in age. In the Charcot Land window, a succession of volcanic and sedimentary rocks of Paleoproterozoic age overlies the gneiss complexes, and a comparable volcanic and sedimentary succession occurs in the Eleonore Sø window. Volcanic rocks of this age are not known within the thrust sheets structurally overlying the windows. In western Dronning Louise Land, the Trekant "series" sandstones overlying the gneisses are correlated with the Independence Fjord Group of the foreland west of Danmark Fjord (~80°N; Kalsbeek et al., 1999). Late Mesoproterozoic successions comparable to the Krummedal sequence, and Neoproterozoic successions similar to the Eleonore Bay Supergroup, are not present.

Diamictites correlated with the diamictites of the Vendian Tillite Group have been recorded in the Gåseland, Charcot Land, and Målebjerg windows, but only as lenticular deposits occupying depressions in the eroded gneiss surface. In contrast to the ~4-km-thick Cambrian-Ordovician succession preserved in the structurally overlying Franz Joseph allochthon, the Lower Paleozoic rocks of the foreland windows are restricted to a 200–350-m-thick Early Cambrian sandstone unit and ~50 m of possible Ordovician carbonates; these have been formally defined by Smith et al. (2004) as the Slottet Formation and Målebjerg Formation, respectively (see also Smith and Rasmussen, this volume).

The thrust contacts that outline the windows are for the most part sharply defined shear zones. However, the shear zone above the Gåseland window is a several-hundred-meter-thick package of highly deformed calcareous sandstones and carbonates. In western Dronning Louise Land, a complex, imbricate thrust zone, several kilometers thick, forms the western margin of the Caledonian orogen. Within the thrust zone, gneisses and sedimentary rocks of the Zebra series and Trekant series are interleaved, and where they cannot be distinguished on the scale of the map, the sedimentary rocks are indicated as "Proterozoic metasediments, mixed rocks in thrust zone."

STRUCTURAL SYMBOLS AND BOUNDARIES

The structural symbols and boundaries shown on the 1:1,000,000 map are adapted from the 1:500,000 maps; locally, additional data have been taken from published descriptions or unpublished field maps. Distinction is made between established and inferred boundaries. Inferred boundaries are projected in areas of ice or sea to facilitate reading of the map.

Three categories of thrusts are defined according to their structural and lithological type and relative size. The western border thrust of the Caledonian orogen, or the Caledonian sole thrust, has the most prominent (thickest) line boundary. Other major thrusts, such as those separating thrust sheets, are depicted with slightly thinner lines; the third category, depicted with the thinnest lines, is used for less-important features. Thrusts related to the end-Cretaceous events in the Wandel Sea Basin succession in Kronprins Christian Land are indicated with blue lines (Pedersen and Håkansson, 2001).

Faults are distinguished according to age, relative size, and variety. Major steep-angle faults within the orogen often have a significant strike-slip component. Other important steep faults may have normal or wrench displacement. Major extensional faults related to late Caledonian orogenic collapse (Gilotti and McClelland, this volume) are also distinguished, as are detachment zones separating thrust units, which often have both contractional and extensional movements. Two categories of post-Caledonian normal faults are distinguished where they can be seen to cut post-Caledonian units (shown in black on the map), whereas faults of Mesozoic and Cenozoic age affecting the Wandel Sea Basin succession in Kronprins Christian Land are colored blue.

Axial traces of major fold structures are distinguished by age. In the basement gneiss complexes, most of the fold structures are thought to relate to Caledonian reactivation, but in some places, structures cut by amphibolitic dikes can be shown to be pre-Caledonian. Folds that deform the Neoproterozoic–Lower Paleozoic succession are classified as Caledonian, and these are indicated with a red color. Open upright structures that affect the Devonian sediments are shown in brown on the map and relate to several phases of moderate Devonian to Carboniferous deformation (Bütler, 1959; Haller, 1971; Bengaard and Larsen, 1992). The folds affecting the Wandel Sea Basin succession are shown in blue and are taken from Pedersen and Håkansson (2001).

Major Caledonian strike-slip shear zones are depicted with red symbols, and the sense of relative displacement is indicated where known. For example, the sinistral Storstrømmen shear zone can be traced for several hundred kilometers from Lambert Land, southward through Jøkelbugten and Storstrømmen, to L. Bistrup Bræ (Holdsworth and Strachan, 1991). The dextral Germania Land shear zone can be traced for several tens of kilometers (Hull and Gilotti, 1994).

Ore Deposits

Only a few localities with major mineralization are indicated on the map. These include the site of the lead-zinc mine at Mestersvig, and the intensively investigated molybdenum-wolfram deposits at Malmbjerg, ~35 km south of Mestersvig. Other mineral occurrences within the region covered by the map have been described by Harpøth et al. (1986) and Jensen and Stendal (1994); see also Stendal and Frei (this volume).

Profiles

Five approximately WNW-ESE profiles are reproduced below the map, at a scale of 1:500,000. These profiles were taken from the published 1:500,000 map sheets Lambert Land (sheet 9, Jepsen, 2000) and Kong Oscar Fjord (sheet 11, Escher, 2001), with slight modifications of legend colors. The 1:500,000 profiles are drawn without vertical exaggeration, and they illustrate the thrust-sheet architecture in the northern and south-central parts of orogen.

Synoptic Tectonic Maps

Two interpretative small-scale synoptic tectonic maps, with profiles, present a general overview of the structure of the East Greenland Caledonian orogen. The two maps depict the northern and southern halves of the orogen at the same scale, but the profiles are in different scales. The northern and southern maps and profiles have separate legends.

MAP ON CD—EXPLANATION AND DATA FORMATS

The entire map sheet with all topographic and geological data is stored on the CD-ROM that accompanies this volume. The full map sheet image, including profiles, legend, and other margin information, is stored in two scaleable graphics formats: postscript and pdf. The printed map was produced using ESRI (Environmental Systems Research Institute, Inc.) Arc/Info data. Additional information and last-minute changes are described in a readme file. This can be opened using, for example, the Windows program Notepad or the Mac program TextEdit.

SUMMARY

The 1:1,000,000 map presents the first regional overview of the entire East Greenland Caledonian orogen. The lithostructural interpretation is based on published 1:500,000 geological maps resulting from the GGU/GEUS regional geological mapping between 1968 and 1998.

The topographic base maps used for the fieldwork were at a scale of 1:100,000, with 100 m contours. The maps were produced in the GGU/GEUS photogrammetric laboratory from super wide-angle aerial photographs taken from ~14 km altitude, with a print scale of ~1:150,000. New ground control points, established using a Doppler satellite-position system combined with barometric measurements, were supplied by GI/KMS (National Survey and Cadastre), Copenhagen. The control points are defined with an absolute accuracy of 1 m in all three dimensions, and the resulting maps have a precision down to a few meters.

The 1:1,000,000 map lists 486 place names. They are recorded in an appendix to this article, together with latitude and longitude.

Geological fieldwork was carried out during independent GGU/GEUS expeditions, and all provisions, equipment, and personnel were brought in from Denmark for each field season. The main fieldwork occupied 15 field seasons of 7–8 weeks each. The expedition groups numbered 30–50 participants each year, of which 75% were scientific personnel;

the remaining 25% made up a support group of base camp personnel, together with helicopter and aircraft pilots and mechanics. The geoscientists generally worked as two-person parties from small field camps, supported with helicopter camp moves and reconnaissance flights. Approximately half of the scientific group consisted of specialists from geological institutes and universities in Europe and North America who participated as guests of GGU/GEUS, and often worked in partnership with GGU/GEUS staff. The contribution of non-GGU/GEUS geoscientific personnel has been considerable, and the work has included experts with experience in other Caledonian orogenic belts, as well as specialists in Precambrian and Phanerozoic geological problems.

Vertical aerial photographs of substantial parts of the region were studied photogeologically at GGU/GEUS prior to fieldwork using photographs taken with super wide-angle cameras. The exceptional degree of exposure makes the high arctic particularly suited to photogeological interpretation, especially in regions dominated by well-stratified sequences. Photogrammetric geological interpretation programs facilitate accurate plotting of boundaries and calculations of dip and strike, fault displacements, and bed thicknesses.

The legend of the 1:1,000,000 map is divided into three main parts: post-Caledonian units, Caledonian foreland, and Caledonian orogen. The foreland geology and the representation of post-Caledonian units have been simplified to focus on the Caledonian orogenic features. Divisions of the Caledonian orogen into structural groups reflect their tectono-stratigraphic position in the orogen.

The northern part the orogen in Kronprins Christian Land includes a thin-skinned thrust belt developed in an Ordovician–Silurian sedimentary succession, at the transition between the foreland in the west and the major allochthonous thrust sheets in the east. The lowest major thrust sheet here is dominated by Neoproterozoic siliciclastic sedimentary rocks overlain by a late Neoproterozoic dominantly carbonate succession. The thrust sheet has a westward displacement of ~40 km.

The southern half of the orogen (70°N–76°N) is divided into three principal lithostructural domains. Foreland windows are overridden by two major thrust sheets, both of which incorporate substantial units of crystalline Archean and Paleoproterozoic orthogneisses, together with a late Mesoproterozoic metasedimentary cover. The highest part of the upper thrust sheet consists of an 18.5-km-thick Neoproterozoic–Ordovician sedimentary succession, the Franz Joseph allochthon.

The late Mesoproterozoic metasedimentary rocks of the upper (Hagar Bjerg) thrust sheet host S-type granites of two major age groups: ca. 930 and ca. 430 Ma. These granites are not found in the lower (Niggli Spids) thrust sheet. A younger generation of Caledonian granites (435–425 Ma), generated in the late Mesoproterozoic metasedimentary rocks of the upper thrust sheet, also occurs in the basal part of the Neoproterozoic sediments of the Franz Joseph allochthon.

Transport directions of the major thrust sheets are toward the west-northwest. Total westward displacement of the thrust sheets is estimated to be several hundred kilometers, with shortening across the orogen estimated at 40%–60%.

Contractional deformation structures are overprinted by extensional structures of late Silurian to mid-Devonian age. In the northern half of the orogen, major units of basement gneiss lithologies containing eclogitic enclaves have been exhumed from deep crustal levels, and they form parts of thrust sheets displaced westward across the foreland. Crustal extension led to formation of orogen-parallel intermontane continental basins in the Middle to Upper Devonian, in which >8 km of siliciclastic sediments were deposited.

The fieldwork for the 1:500,000 scale regional mapping project by GGU/GEUS between 1968 and 1998 has provided a general geological overview of the entire East Greenland Caledonian orogen. The orogen extends for more than 1300 km between latitudes 70°N and ~81°30′N. In the course of 15 field seasons, spread over 30 years, more than 70 geologists have participated in one or more field seasons each. The scientific results are documented in numerous publications, both in international journals and in the GGU/GEUS *Bulletins* and *Reports*. The present volume, *The Greenland Caledonides: Evolution of the Northeast Margin of Laurentia*, presents a general summary and conclusions of this work.

ACKNOWLEDGMENTS

The original field mapping was carried out during 15 field seasons by more than 70 geologists, too numerous to list individually here. The names of all participating geologists are given in the legends of the individual GGU/GEUS published geological map sheets: the 16 sheets at 1:100,000 that cover the Scoresby Sund region (69°N–72°N), and the four 1:500,000 sheets that cover the region 72°N–82°N. Valuable suggestions and comments have been provided during the compilation of the 1:1,000,000 map by J.A. Gilotti, F. Kalsbeek, A.G. Leslie, S.A.S. Pedersen, M.P. Smith, H. Stendal, and W.S. Watt. Special thanks are due to the staff in the GEUS photogeological laboratory, who adapted the topographic base for the purposes of the map. The geological map compilation would not have been possible without the skilled and enthusiastic cooperation of Margareta Christoffersen at all stages of the project. The systematic management of the data for the compilation and the transference of the digital data to the CD were carefully and competently undertaken by Willy L. Weng. A proof copy of the map was externally refereed by David G. Gee, University of Uppsala, Sweden, and his valuable comments and suggestions have been incorporated into the printed map sheet. F. Kalsbeek is thanked for a careful first internal review of this paper and for his suggestions for improvements. The two external reviewers, Winfried Dallman (Norwegian Polar Institute) and Robert H. Fakundiny (New York State Geological Survey), are thanked for their numerous constructive comments and suggestions.

APPENDIX. PLACE NAMES USED ON THE 1:1,000,000 MAP

Place	Location
A.B. Drachmann Gletscher	76°10′N, 24°30′W
A.P. Olsen Land	74°38′N, 21°30′W
Academy Gletscher	81°30′N, 32°30′W
Achton Friis Ø	78°58′N, 19°14′W
Ad Astra Iskappe	77°00′N, 23°55′W
Ad. S. Jensen Land	76°06′N, 21°00′W
Admiralty Gletscher	77°02′N, 24°20′W
Adolf Hoel Gletscher	73°57′N, 27°30′W
Æbeltoft Vig	72°30′N, 22°09′W
Alabama	75°18′N, 17°52′W
Alabama Nunatak	77°45′N, 24°00′W
Albert Heim Bjerge	74°06′N, 23°08′W
Alfabet Nunatakker	71°53′N, 30°05′W
Alpefjord	72°15′N, 25°25′W
Ambolten	78°18′N, 19°13′W
Amdrup Højland	80°27′N, 24°55′W
Amdrup Land	80°45′N, 16°00′W
Andrée Land	73°42′N, 26°25′W
Ankerbjerg	73°35′N, 22°32′W
Annekssøen	77°18′N, 21°06′W
Antarctic Bugt	80°55′N, 14°00′W
Antarctic Havn	72°01′N, 23°08′W
Antarctic Sund	73°07′N, 25°30′W
Ardencaple Fjord	75°19′N, 21°00′W
Arnold Escher Land	74°00′N, 28°15′W
Bach Dal	76°52′N, 23°33′W
Bagfjorden	76°34′N, 22°22′W
Bartholin Land	74°24′N, 25°00′W
Berg Fjord	76°34′N, 18°58′W
Bessel Fjord	75°59′N, 21°00′W
Bildsøe Nunatakker	78°05′N, 23°38′W
Bjørneøer	71°06′N, 25°30′W
Blæsebræ	78°13′N, 21°24′W
Blåsø	79°35′N, 22°30′W
Blomsterbugten	73°20′N, 25°18′W
Bontekoe Ø	73°07′N, 21°20′W
Borgfjorden	76°41′N, 22°00′W
Borgjøkelen	76°40′N, 23°35′W
Bourbon Ø	78°46′N, 18°15′W
Boyd Bastion	73°26′N, 28°35′W
Bræfjorden	76°25′N, 22°00′W
Bræøerne	76°45′N, 22°10′W
Bredefjord	75°34′N, 21°40′W
Bregnepynt	70°55′N, 25°24′W
Bristol Plateau	80°40′N, 24°10′W
Britannia Gletscher	77°12′N, 24°00′W
Britannia Sø	77°09′N, 23°20′W
Brogetdalen	73°45′N, 24°45′W
Brønlunds Grav	79°12′N, 19°04′W
Brune Nunatakker	71°10′N, 29°45′W
Budolfi Isstrøm	76°20′N, 24°38′W
C.F. Mourier Fjord	77°21′N, 20°17′W
C.H. Ostenfeld Land	75°14′N, 21°30′W
C.H. Ostenfeld Nunatak	74°17′N, 22°55′W
C. Hofmann Halvø	70°58′N, 27°50′W
Campbell Sund	71°18′N, 21°44′W
Canning Land	71°41′N, 22°15′W
Canongletscher	75°28′N, 22°40′W
Carlsberg Fjord	71°22′N, 22°25′W
Carlsbergfondet Land	76°25′N, 25°00′W

Place	Location
Cecilia Nunatak	72°30′N, 27°55′W
Celcius Bjerg	73°08′N, 23°15′W
Centrumsø	80°10′N, 22°00′W
Charcot Havn	70°47′N, 25°24′W
Charcot Land	72°00′N, 29°00′W
Clavering Ø	74°18′N, 21°00′W
Cloos Klippe	76°48′N, 24°53′W
Constable Pynt	70°44′N, 22°36′W
Daneborg	74°20′N, 20°14′W
Daniel Bruun Land	76°53′N, 21°52′W
Danmark Fjord	81°15′N, 21°30′W
Danmark Ø	70°30′N, 26°15′W
Danmarkshavn	76°46′N, 18°41′W
Danske Øer	78°07′N, 19°00′W
Daugaard-Jensen Gletscher	71°50′N, 28°50′W
Davy Sund	72°02′N, 22°30′W
Depotfjeld	80°10′N, 16°50′W
Depotnæsset	77°34′N, 18°54′W
Dickson Fjord	72°51′N, 26°50′W
Dijmphna Sund	80°02′N, 19°20′W
Djævlekløften	73°30′N, 26°30′W
Dove Bugt	76°35′N, 20°00′W
Dronning Louise Land	76°30′N, 24°30′W
Dronning Margrethe II Land	75°40′N, 21°00′W
Dusén Fjord	73°14′N, 24°00′W
Edvard Ø	76°36′N, 21°20′W
Eielson Gletscher	71°10′N, 28°00′W
Eigil Sø	76°43′N, 25°05′W
Eigtvedsund	75°56′N, 20°20′W
Eleonore Bugt	73°26′N, 25°25′W
Eleonore Sø	74°00′N, 28°10′W
Ella Ø	72°51′N, 25°00′W
Emilia Bjerg	72°53′N, 27°27′W
Engdalen	73°13′N, 27°17′W
Erik S. Henius Land	81°30′N, 11°47′W
Eskimonæs	74°05′N, 21°15′W
Evers Gletscher	73°42′N, 29°20′W
F. Graae Gletscher	72°10′N, 28°45′W
F. Toula Plateau	77°05′N, 18°45′W
Falkonerklippe	76°38′N, 26°22′W
Fegin Elv	71°13′N, 23°45′W
Femte Maj Sø	81°15′N, 25°52′W
Finsch Øer	74°00′N, 21°00′W
Firndalen	80°20′N, 18°00′W
Fladebugt	74°24′N, 19°09′W
Fladebugt	77°15′N, 19°16′W
Flade Isblink	81°30′N, 14°30′W
Flakkerhuk	70°28′N, 23°21′W
Fleming Fjord	71°45′N, 22°50′W
Fligely Fjord	74°53′N, 20°40′W
Flyversø	77°45′N, 20°30′W
Fønfjord	70°28′N, 27°00′W
Forsblad Fjord	72°25′N, 25°35′W
Foster Bugt	73°15′N, 21°30′W
Frænkel Land	73°18′N, 27°35′W
Franske Øer	78°40′N, 18°20′W
Frederiksdal	71°45′N, 26°50′W
Freeden Bugt	75°00′N, 18°00′W
Furesø	72°01′N, 26°00′W
Fyn Sø	80°32′N, 24°10′W

(continued)

APPENDIX. PLACE NAMES USED ON THE 1:1,000,000 MAP (continued)

Place	Location	Place	Location
Gael Hamke Bugt	74°00'N, 20°20'W	Hochstetterbugten	74°50'N, 18°45'W
Gamle Jim Øer	79°21'N, 19°25'W	Højedal	72°25'N, 26°30'W
Gamma Ø	77°50'N, 19°52'W	Holbæk Bugt	80°36'N, 23°45'W
Gammel Hellerup Gletscher	78°34'N, 21°45'W	Hold with Hope	73°45'N, 21°00'W
Garde Nunatakker	78°28'N, 23°06'W	Holger Danskes Briller	71°25'N, 25°05'W
Gåseelv	70°47'N, 22°50'W	Holm Bugt	72°30'N, 24°05'W
Gåsefjord	70°07'N, 27°30'W	Holm Land	80°20'N, 17°00'W
Gåseland	70°15'N, 28°00'W	Horsens Fjord	70°48'N, 21°48'W
Gåsepynt	70°22'N, 26°18'W	Hovgaard Ø	79°55'N, 18°30'W
Gauss Halvø	73°30'N, 23°00'W	Hudson Land	73°55'N, 23°00'W
Geikie Plateau	69°55'N, 25°15'W	Hurry Inlet	70°40'N, 22°35'W
Geographical Society Ø	72°55'N, 23°30'W	Hvalrosodden	76°55'N, 20°06'W
Geologfjord	73°45'N, 25°15'W	Île de France	77°43'N, 17°45'W
Gerard de Geer Gletscher	73°35'N, 27°17'W	Illoqqortoormiut / Ittoqqortoormiit	70°29'N, 21°58'W
Germania Havn	74°32'N, 18°50'W	Independence Fjord	81°50'N, 32°00'W
Germania Land	77°05'N, 19°00'W	Inderbredningen	76°15'N, 21°40'W
Giesecke Bjerge	74°29'N, 22°08'W	Ingolf Fjord	80°33'N, 17°30'W
Gletscherland	72°40'N, 27°00'W	Isfjord	73°21'N, 27°00'W
Godfred Hansen Ø	76°26'N, 20°55'W	J.C. Christensen Land	81°35'N, 30°00'W
Godthåb Golf	74°08'N, 21°50'W	J.L. Mowinckel Land	73°51'N, 28°35'W
Gog	73°14'N, 28°24'W	J.P. Koch Fjeld	70°40'N, 22°56'W
Goodenough Land	72°55'N, 28°20'W	Jackson Ø	73°55'N, 20°10'W
Graben Land	71°08'N, 28°55'W	Jættegletscher	73°25'N, 27°50'W
Græselv	80°03'N, 23°03'W	Jameson Land	71°00'N, 23°15'W
Grandjean Fjord	74°58'N, 21°45'W	Jøkelbugten	78°30'N, 20°00'W
Grejsdalen	73°35'N, 26°00'W	Jomfrudal	72°04'N, 27°09'W
Grønlandshavet	79°30'N, 13°30'W	Junctiondal	73°16'N, 25°58'W
Grønne Nunatak	78°32'N, 23°20'W	Jyllandselv	70°45'N, 23°50'W
Grottedal	80°23'N, 21°40'W	Jyske Ås	81°32'N, 23°30'W
Gullygletscher	72°06'N, 25°20'W	Kaj Munk Iskappe	81°54'N, 28°00'W
Gunnar Andersson Land	73°20'N, 24°25'W	Kap Aage Bertelsen	76°40'N, 23°04'W
Gurreholm	71°14'N, 24°36'W	Kap Achton Friis	76°46'N, 23°05'W
Gurreholm Dal	71°28'N, 24°46'W	Kap Alf Trolle	75°56'N, 18°29'W
Hagar Bjerg	72°53'N, 27°44'W	Kap Amélie	77°32'N, 19°14'W
Hagen Bræ	81°28'N, 27°00'W	Kap Anna Bistrup	79°42'N, 18°13'W
Hagen Fjord	81°45'N, 24°15'W	Kap Bellevue	77°07'N, 23°14'W
Hall Bredning	70°54'N, 25°45'W	Kap Biot	71°55'N, 22°32W
Hamberg Gletscher	73°34'N, 29°00'W	Kap Bismarck	76°42'N, 18°32'W
Hammeren	78°17'N, 19°34'W	Kap Brewster	70°09'N, 22°03'W
Hanseeraq Fjord	80°18'N, 16°17'W	Kap Broer Ruys	73°32'N, 20°23'W
Hareelv	70°42'N, 22°46'W	Kap Buch	75°08'N, 20°30'W
Harefjord	70°55'N, 28°00'W	Kap Fletcher	71°37'N, 22°06'W
Haystack	75°44'N, 19°24'W	Kap Franklin	73°15'N, 22°08'W
Heinkel Gletscher	75°10'N, 22°40'W	Kap Freuchen	76°21'N, 23°40'W
Hekla Havn	70°27'N, 26°11'W	Kap Georg Cohn	80°16'N, 26°15'W
Hekla Sund	80°14'N, 19°19'W	Kap Godfred Hansen	71°27'N, 21°42'W
Helgenæs	70°22'N, 25°03'W	Kap Greg	70°57'N, 21°38'W
Helgoland	76°23'N, 26°14'W	Kap Gustav Rasmussen	80°50'N, 28°15'W
Hellefjord	76°52'N, 21°15'W	Kap Hedlund	72°43'N, 26°11'W
Henrik Kröyer Holme	80°38'N, 13°45'W	Kap Holbæk	80°43'N, 23°28'W
Hermelintop	70°26'N, 27°57'W	Kap Hooker	70°28'N, 23°24'W
Herthadal	72°10'N, 27°00'W	Kap Hope	70°28'N, 22°21'W
Hertugen af Orléans Land	77°50'N, 21°50'W	Kap Jones	71°07'N, 21°44'W
Hesteelv	70°29'N, 22°58'W	Kap Jungersen	80°37'N, 16°05'W
Hinks Land	71°40'N, 28°30'W	Kap Kolthoff	73°43'N, 24°02'W
Hisinger Gletscher	72°47'N, 27°45'W	Kap Lagerberg	72°31'N, 24°40'W
Hjelmbjergene	73°23'N, 23°00'W	Kap Leslie	70°39'N, 25°17'W
Hjørnegletscher	80°40'N, 19°30'W	Kap McClintock	72°41'N, 21°56'W
Hobbs Land	74°02'N, 29°00'W	Kap Mackenzie	72°54'N, 21°54'W
Hochstetter Forland	75°30'N, 19°50'W	Kap Mohn	73°11'N, 25°45'W

(continued)

APPENDIX. PLACE NAMES USED ON THE 1:1,000,000 MAP (continued)

Place	Location
Kap Montpensier	77°51′N, 17°38′W
Kap Negri	75°03′N, 20°39′W
Kap Oswald Heer	75°33′N, 19°25′W
Kap Ovibos	73°33′N, 24°24′W
Kap Pansch	75°09′N, 17°25′W
Kap Peschel	76°15′N, 19°58′W
Kap Philip Broke	74°56′N, 17°37′W
Kap Rink	75°08′N, 19°38′W
Kap Seaforth	71°48′N, 22°48′W
Kap Simpson	72°08′N, 22°12′W
Kap Skt. Jacques	77°37′N, 18°08′W
Kap Stevenson	70°24′N, 25°12′W
Kap Stewart	70°27′N, 22°38′W
Kap Stop	76°38′N, 21°40′W
Kap Stosch	74°03′N, 21°43′W
Kap Syenit	72°03′N, 23°06′W
Kap Tobin	70°25′N, 21°57′W
Kap Wardlaw	71°44′N, 21°55′W
Kap Weber	73°30′N, 24°43′W
Kap Young	72°16′N, 22°02′W
Karupelv	72°33′N, 23°40′W
Kejser Franz Joseph Fjord	73°13′N, 26°00′W
Kempe Fjord	72°52′N, 25°35′W
Kilen	81°12′N, 13°30′W
Kjerulf Fjord	73°04′N, 27°22′W
Klitdal	70°58′N, 22°30′W
Knækket	70°16′N, 26°40′W
Knebel Vig	72°17′N, 22°18′W
Koefoed-Hansen Bræ	77°30′N, 21°45′W
Kong Oscar Fjord	72°22′N, 24°06′W
Kongeborgen	72°39′N, 24°20′W
Korsgletscher	74°18′N, 25°13′W
Kronprins Christian Land	80°40′N, 22°00′W
Krumme Langsø	74°04′N, 23°45′W
Krummedal	71°24′N, 29°00′W
Kuhn Ø	74°50′N, 20°10′W
L. Bistrup Bræ	76°19′N, 23°10′W
Lady Øer	77°59′N, 20°25′W
Lægervallen	79°14′N, 19°00′W
Lambert Land	79°15′N, 20°30′W
Langsø	75°49′N, 20°45′W
Lille Koldewey	76°39′N, 18°40′W
Lille Pendulum	74°40′N, 18°28′W
Lillefjord	70°38′N, 21°40′W
Lindhard Ø	76°31′N, 22°10′W
Liverpool Land	71°00′N, 22°00′W
Loch Fyne	73°47′N, 21°48′W
Lodin Elv	71°23′N, 24°00′W
Lollandselv	70°57′N, 23°45′W
Louise Boyd Land	73°35′N, 28°00′W
Lyell Land	72°40′N, 25°35′W
Lynn Ø	80°08′N, 19°14′W
Mackenzie Bugt	73°27′N, 21°30′W
Målebjerg	73°34′N, 27°09′W
Mallemukfjeld	80°12′N, 16°40′W
Månevig	80°33′N, 20°30′W
Maria Ø	72°57′N, 24°53′W
Mariager Fjord	70°39′N, 21°51′W
Marie Sophie Gletscher	81°50′N, 33°00′W
Marmorvigen	80°05′N, 20°05′W
Martin Knudsen Nunatakker	73°15′N, 29°00′W
Mestersvig	72°14′N, 23°54′W
Micardbu	77°05′N, 18°12′W
Milne Land	70°45′N, 26°30′W
Mørkefjord	76°57′N, 21°02′W
Moskusoksefjord	73°35′N, 22°59′W
Moskusokselandet	73°47′N, 23°15′W
Mountnorris Fjord	72°22′N, 22°20′W
Mudderbugt	70°35′N, 25°48′W
Myggbukta	73°29′N, 21°34′W
Mylius-Erichsen Land	81°18′N, 24°30′W
Nakkehoved	81°42′N, 13°10′W
Nanok Ø	76°20′N, 20°35′W
Narhvalsund	72°47′N, 25°10′W
Nathorst Fjord	71°40′N, 22°30′W
Nathorst Land	71°50′N, 26°30′W
Niggli Spids	73°15′N, 26°40′W
Nioghalvfjerdsfjorden	79°33′N, 21°00′W
Nordbugten	71°36′N, 26°30′W
Nordenskiöld Bugt	75°14′N, 18°06′W
Nordenskiöld Gletscher	73°08′N, 28°00′W
Nordfjord	73°40′N, 24°18′W
Nordhoek Bjerg	73°49′N, 22°04′W
Nordmarken	77°45′N, 21°00′W
Nordostrundingen	81°21′N, 11°30′W
Nordredepot Ø	78°12′N, 20°28′W
Nordvestfjord	71°31′N, 26°00′W
Nordvestklint	79°23′N, 21°31′W
Nørlund Alper	73°59′N, 22°29′W
Nørre Mellemland	78°20′N, 21°15′W
Nørreland	78°42′N, 21°12′W
Norsemandal	80°48′N, 25°00′W
Norske Øer	79°04′N, 17°50′W
Øfjord	71°00′N, 26°12′W
Okselandet	77°12′N, 21°30′W
Ole Rømer Land	74°10′N, 24°10′W
Olsen Nunatakker	76°38′N, 26°50′W
Orienteringsøerne	76°47′N, 19°45′W
Orléans Sund	77°49′N, 20°00′W
Palnatoke Bjerg	74°34′N, 20°32′W
Pariserøerne	78°24′N, 19°00′W
Påskenæsset	76°10′N, 19°47′W
Pasterze	74°41′N, 22°35′W
Paul Stern Land	70°25′N, 29°30′W
Payer Land	74°30′N, 22°30′W
Payer Tinde	73°08′N, 26°22′W
Pendulum Øer	74°37′N, 18°40′W
Penthievre Fjord	77°39′N, 20°00′W
Petermann Bjerg	73°05′N, 28°37′W
Peters Bugt	75°18′N, 20°15′W
Polhem Dal	72°36′N, 25°18′W
Pony Gletscher	76°29′N, 25°03′W
Port Arthur	76°46′N, 21°13′W
Poulsen Nunatakker	76°51′N, 26°55′W
Prins Frederik Øer	81°37′N, 19°45′W
Prinsesse Dagmar Ø	81°40′N, 17°45′W
Prinsesse Ingeborg Halvø	81°35′N, 16°30′W
Prinsesse Thyra Ø	81°55′N, 19°00′W
Rathbone Ø	70°40′N, 21°28′W
Ravnebjerg	73°35′N, 21°20′W
Rechnitzer Land	76°19′N, 22°00′W
Renbugten	73°20′N, 26°28′W
Renland	71°15′N, 26°30′W
Rhedin Fjord	72°40′N, 26°20′W

(continued)

APPENDIX. PLACE NAMES USED ON THE 1:1,000,000 MAP (continued)

Place	Location	Place	Location
Rivieradal	80°04′N, 21°10′W	Suzanne Bræ	77°19′N, 24°20′W
Rødefjord	70°44′N, 27°48′W	Svedenborg Bjerg	72°57′N, 24°26′W
Röhss Fjord	72°44′N, 26°35′W	Sydbræ	70°08′N, 26°22′W
Rolige Bræ	70°32′N, 29°00′W	Sydgletscher	71°56′N, 26°23′W
Romer Sø	80°58′N, 19°13′W	Sydkap	71°18′N, 25°05′W
Rosio	77°17′N, 18°21′W	Syttenkilometernæsset	76°50′N, 18°16′W
Roslin Gletscher	71°50′N, 24°55′W	Tærskelsø	72°21′N, 26°35′W
Royston Nunatakker	71°23′N, 29°40′W	Terrasseodde	70°19′N, 24°51′W
Ruth Ø	73°00′N, 24°53′W	Teufelkap	76°23′N, 20°10′W
Rypefjord	71°00′N, 27°40′W	Th. Sørensen Land	71°22′N, 28°15′W
Sabine Ø	74°36′N, 18°58′W	Th. Thomsen Land	74°52′N, 21°30′W
Sælsøen	77°05′N, 20°30′W	Tillit Nunatak	71°55′N, 29°45′W
Sanddal	78°05′N, 21°34′W	Tobias Dal	73°46′N, 21°00′W
Savoia Halvø	70°05′N, 22°15′W	Tobias Gletscher	80°46′N, 17°26′W
Schnauder Ø	78°47′N, 19°30′W	Tobias Øer	79°20′N, 15°48′W
Schuchert Flod	71°30′N, 24°25′W	Torvgletscher	70°04′N, 23°20′W
Scoresby Land	72°00′N, 24°30′W	Trækpasset	76°10′N, 18°40′W
Scoresby Sund	70°17′N, 23°00′W	Traill Ø	72°40′N, 23°40′W
Scoresbysund	70°29′N, 21°58′W	Trangsund	76°17′N, 20°46′W
Segelsällskapet Fjord	72°26′N, 25°00′W	Trefork Sø	76°56′N, 24°18′W
Sengstacke Bugt	75°20′N, 18°15′W	Trekanten	76°51′N, 25°22′W
Shannon	75°07′N, 18°30′W	Trekroner	77°00′N, 20°12′W
Shannon Sund	75°13′N, 19°15′W	Troldehaven	77°55′N, 18°50′W
Sjælland Fjelde	80°45′N, 22°30′W	Tuborgfondet Land	78°26′N, 22°15′W
Sjællandselv	70°42′N, 23°25′W	Tuppiat Qeqertai	79°20′N, 15°48′W
Skærfjorden	77°25′N, 19°30′W	Tyrolerfjord	74°28′N, 21°12′W
Skallingen	79°50′N, 22°00′W	Ugleelv	70°53′N, 22°48′W
Skeldal	72°15′N, 24°17′W	V. Clausen Fjord	77°27′N, 20°35′W
Skjoldungeelv	80°21′N, 25°00′W	Valdemar Glückstadt Land	81°50′N, 22°00′W
Skrænterne	70°05′N, 24°35′W	Valhal	75°05′N, 22°25′W
Snenæs	76°49′N, 19°21′W	Vandreblokken	70°39′N, 24°00′W
Snespurvefjeld	79°43′N, 20°49′W	Vandredalen	80°25′N, 20°56′W
Snesund	70°44′N, 27°30′W	Vega Sund	72°50′N, 23°00′W
Sofia Sund	73°02′N, 24°00′W	Vestre Borggletscher	70°06′N, 23°51′W
Solgletscher	70°13′N, 24°30′W	Vestfjord	70°29′N, 28°38′W
Solitærbugt	72°53′N, 25°06′W	Vestfjord Gletscher	70°18′N, 29°29′W
Sønderland	77°52′N, 21°52′W	Vibeke Gletscher	74°14′N, 23°55′W
Søndermarken	77°18′N, 20°45′W	Victor Madsen Gletscher	73°15′N, 28°54′W
Søndre Mellemland	78°07′N, 21°30′W	Vikingebugt	70°20′N, 25°12′W
Sophus Muller Næs	80°47′N, 14°09′W	Vildtland	81°36′N, 33°00′W
Soranerbræen	76°11′N, 21°54′W	Vindseløen	76°49′N, 20°17′W
Sortebakker	80°10′N, 17°20′W	Vindue Gletscher	71°15′N, 28°50′W
Station Nord	81°36′N, 16°40′W	Violingletscher	72°16′N, 26°48′W
Stauning Alper	72°00′N, 25°00′W	Volquart Boon Kyst	70°06′N, 23°15′W
Steno Land	74°17′N, 23°40′W	Wahlenberg Gletsher	72°30′N, 27°00′W
Stigbøjlen	78°14′N, 19°05′W	Waltershausen Gletscher	73°55′N, 24°25′W
Stordal	73°48′N, 22°25′W	Wandel Hav	81°40′N, 10°00′W
Store Bælt	76°20′N, 19°30′W	Wegener Halvø	71°41′N, 22°45′W
Store Koldewey	76°25′N, 18°45′W	Wegener Øer	80°34′N, 16°45′W
Store Sødal	74°31′N, 21°00′W	Weinschenck Ø	77°54′N, 21°13′W
Storefjord	71°05′N, 22°00′W	Wollaston Forland	74°26′N, 19°35′W
Storgletscher	71°57′N, 24°45′W	Wordie Bugt	74°04′N, 22°20′W
Stormbugt	76°48′N, 19°00′W	Wordie Gletscher	74°14′N, 23°00′W
Stormgletscher	75°41′N, 22°45′W	Ymer Nunatak	77°25′N, 24°17′W
Stormlandet	77°40′N, 19°40′W	Ymer Ø	73°15′N, 24°30′W
Storø	70°49′N, 27°35′W	Young Sund	74°20′N, 20°20′W
Storøen	78°04′N, 19°00′W	Zachariae Isstrøm	78°54′N, 20°30′W
Storstrømmen	77°00′N, 22°30′W	Zackenberg	74°29′N, 20°55′W
Strindberg Land	73°50′N, 25°00′W	Zebra Klippe	77°13′N, 24°50′W
Suess Land	73°00′N, 26°00′W	Zig-Zag Dal	81°15′N, 24°00′W

REFERENCES CITED

Amdrup, G., 1913, Report on the Danmark Expedition to the North-East Coast of Greenland 1906–08: Meddelelser om Grønland, v. 41, no. 1, 270 p.

Bengaard, H.-J., and Henriksen, N., 1986, Geological Map of Greenland, Peary Land, Sheet 8: Copenhagen, Geological Survey of Greenland, scale 1:500,000.

Bengaard, H.-J., and Larsen, P.-H., 1992, Upper Proterozoic (Eleonore Bay Supergroup) to Devonian, Central Fjord Zone, East Greenland: Geological Map: Copenhagen, Geological Survey of Greenland, scale 1:250,000.

Bengtsson, T., 1983, The mapping of northern Greenland: The Photogrammetric Record, v. 11, p. 135–150.

Bondam, J., and Brown, H., 1955, The Geology and Mineralisation of the Mesters Vig Area, East Greenland: Bulletin Grønlands Geologiske Undersøgelse, v. 11, 40 p.

Bütler, H., 1959, Das Old Red Gebiet am Moskusoksefjord. Attempt at a Correlation of the Series of Various Devonian Areas in Central East Greenland: Meddelelser om Grønland, v. 160, no. 5, 188 p.

Clavering, D.C., 1830, Journal of a voyage to Spitzbergen and the east coast of Greenland in His Majesty's ship *Griper*: Edinburgh New Philosophical Journal, v. 9, p. 1–30.

Clemmensen, L.B., and Jepsen, H.F., 1992, Lithostratigraphy and Geological Setting of Upper Proterozoic Shelf Deposits, Hagen Fjord Group, Eastern North Greenland: Rapport Grønlands Geologiske Undersøgelse, v. 157, 27 p.

Collinson, J.D., 1980, Stratigraphy of the Independence Fjord Group (Proterozoic) of eastern North Greenland: Rapport Grønlands Geologiske Undersøgelse, v. 99, p. 7–23.

Collinson, J.D., Kalsbeek, F., Jepsen, H.F., Pedersen, S.A.S., and Upton, B.G.J., 2008, this volume, Paleoproterozoic and Mesoproterozoic sedimentary and volcanic successions in the northern parts of the East Greenland Caledonian orogen and its foreland, *in* Higgins, A.K., Gilotti, J.A., and Smith, M.P., eds., The Greenland Caledonides: Evolution of the Northeast Margin of Laurentia: Geological Society of America Memoir 202, doi: 10.1130/2008.1202(04).

Escher, J.C., 2001, Geological Map of Greenland, Kong Oscar Fjord, Sheet 11: Copenhagen, Geological Survey of Denmark and Greenland, scale 1:500,000.

Funder, S., 1987, Kvartærgeologisk kort Jameson Land: Copenhagen, Geological Survey of Greenland, scale 1:125,000.

Funder, S., coordinator, 1989, Quaternary geology of the ice-free areas and adjacent shelves of Greenland, *in* Fulton, R.J., ed., Quaternary Geology of Canada and Greenland: Boulder, Colorado, Geological Society of America, Geology of North America, v. K-1, p. 741–792 (also Geology of Canada, Volume 1, Geological Survey of Canada).

Funder, S., 1990, Quaternary Map of Greenland, 1:500 000, Sheet 12, Scoresby Sund: Descriptive Text: Copenhagen, Geological Survey of Greenland, 24 p. (+ 1 map).

Funder, S., and Hjort, C., 1980, A reconnaissance of the Quaternary geology of eastern North Greenland: Rapport Grønlands Geologiske Undersøgelse, v. 99, p. 99–105.

Funder, S., Hjort, C., Landvik, J.Y., Seung-Il Nam, Reeh, N., and Stein, R., 1998, History of a stable ice margin—East Greenland during the Middle and Upper Pleistocene: Quaternary Science Reviews, v. 17, p. 77–123.

Ghisler, M., 1990, Geographical subdivision of Greenland: Rapport Grønlands Geologiske Undersøgelse, v. 148, p. 8–10.

Gilotti, J.A., and McClelland, W.C., 2008, this volume, Geometry, kinematics, and timing of extensional faulting in the Greenland Caledonides—A synthesis, *in* Higgins, A.K., Gilotti, J.A., and Smith, M.P., eds., The Greenland Caledonides: Evolution of the Northeast Margin of Laurentia: Geological Society of America Memoir 202, doi: 10.1130/2008.1202(10).

Gilotti, J.A., Jones, K.A., and Elvevold, S., 2008, this volume, Caledonian metamorphic patterns in Greenland, *in* Higgins, A.K., Gilotti, J.A., and Smith, M.P., eds., The Greenland Caledonides: Evolution of the Northeast Margin of Laurentia: Geological Society of America Memoir 202, doi: 10.1130/2008.1202(08).

Haller, J., 1955, Der "Zentrale Metamorphe Komplex" von NE-Grönland. Teil I. Der geologische Karte von Suess Land, Gletscherland und Goodenoughs Land: Meddelelser om Grønland, v. 73, I, no. 3, 174 p.

Haller, J., 1956, Geologie der Nunatakker Region von Zentral-Ostgrönland: Meddelelser om Grønland, v. 154, no. 1, 172 p.

Haller, J., 1958, Der "Zentrale Metamorphe Komplex" von Nordostgrönland. Teil II. Die geologische Karte des Stauning Alper und des Forsblads Fjordes: Meddelelser om Grønland v. 154, no. 3, 153 p.

Haller, J., 1970, Tectonic Map of East Greenland (1:500,000): An Account of Tectonism, Plutonism and Volcanism in East Greenland: Meddelelser om Grønland, v. 171, no. 5, 286 p.

Haller, J., 1971, Geology of the East Greenland Caledonides: London, Interscience, 413 p.

Haller, J., 1983, Geological Map of Northeast Greenland 75°–82°N Lat. 1:1,000,000: Meddelelser om Grønland, v. 200, no. 5, 22 p. (+ 1 map sheet).

Hamann, N.E., Whittaker, R.C., and Stemmerik, L., 2005, Structural and geological development of the North-East Greenland shelf, *in* Doré, A.G., and Vining, B.A., eds., Petroleum Geology: North-West Europe and Global Perspectives—Proceedings of the 6th Petroleum Geology Conference: London, Geological Society of London, p. 887–902.

Harpøth, O., Pedersen, J.L., Schønwandt, H.K., and Thomassen, B., 1986, The Mineral Occurrences of Central East Greenland: Meddelelser om Grønland, Geoscience, v. 17, 139 p.

Hartz, E., and Andresen, A., 1995, Caledonian sole thrust of central East Greenland: A crustal-scale Devonian extensional detachment: Geology, v. 23, p. 637–640, doi: 10.1130/0091-7613(1995)023<0637:CSTOCE>2.3.CO;2.

Hartz, E.H., Andresen, A., Martin, M.W., and Hodges, K.V., 2000, U-Pb and ^{40}Ar/^{39}Ar constraints on the Fjord region detachment zone: A long-lived extensional fault in the central East Greenland Caledonides: Journal of the Geological Society of London, v. 157, p. 795–809.

Henriksen, N., 1984, Geological Map of Greenland, Scoresby Sund, Sheet 12: Copenhagen, Geological Survey of Greenland, scale 1:500,000.

Henriksen, N., 1986, Geological Map of Greenland, Sheet 12, Scoresby Sund: Descriptive Text: Copenhagen, Geological Survey of Greenland, scale 1:500,000, 27 p.

Henriksen, N., 1989, Regional geological investigations and 1:500,000 mapping in North-East Greenland: Rapport Grønlands Geologiske Undersøgelse, v. 145, p. 88–90.

Henriksen, N., 1992, Geological Map of Greenland, Nyeboe Land, sheet 7, Peary Land, Sheet 8: Descriptive Text: Copenhagen, Geological Survey of Greenland, scale 1:500,000, 40 p. (+2 maps).

Henriksen, N., 1997, Geological Map of Greenland, Dove Bugt, Sheet 10: Copenhagen, Geological Survey of Greenland, scale 1:500,000.

Henriksen, N., 1999, Conclusion of the 1:500,000 mapping project in the Caledonian fold belt in North-East Greenland: Geology of Greenland Survey Bulletin, v. 183, p. 10–22.

Henriksen, N., and Higgins, A.K., 1991, The North Greenland Project, *in* Peel, J.S., and Sønderholm, M., eds., Sedimentary Basins of North Greenland: Bulletin Grønlands Geologiske Undersøgelse, v. 160, p. 9–24.

Higgins, A.K., ed., 1994, Geology of North-East Greenland: Rapport Grønlands Geologiske Undersøgelse, v. 162, 209 p.

Higgins, A.K., and Leslie, A.G., 2000, Restoring thrusting in the East Greenland Caledonides: Geology, v. 28, p. 1019–1022, doi: 10.1130/0091-7613 (2000)28<1019:RTITEG>2.0.CO;2.

Higgins, A.K., Leslie, A.G., and Smith, M.P., 2001a, Neoproterozoic–Lower Palaeozoic stratigraphical relationships in the marginal thin-skinned thrust belt of the East Greenland Caledonides: Comparisons with the foreland of Scotland: Geological Magazine, v. 138, p. 143–160, doi: 10.1017/S0016756801005076.

Higgins, A.K., Smith, M.P., Soper, N.J., Leslie, A.G., Rasmussen, J.A., and Sønderholm, M., 2001b, The Neoproterozoic Hekla Sund Basin, eastern North Greenland: A pre-Iapetan extensional sequence thrust across its rift shoulders during the Caledonian orogeny: Journal of the Geological Society of London, v. 158, p. 487–499.

Higgins, A.K., Soper, N.J., Smith, M.P., and Rasmussen, J.A., 2004a, The Caledonian thin-skinned thrust belt of Kronprins Christian Land, eastern North Greenland: Geological Survey of Denmark and Greenland Bulletin, v. 6, p. 41–56.

Higgins, A.K., Elvevold, S., Escher, J.C., Frederiksen, K.S., Gilotti, J.A., Henriksen, N., Jepsen, H.F., Jones, K.A., Kalsbeek, F., Kinny, P.D., Leslie, A.G., Smith, M.P., Thrane, K., and Watt, G.R., 2004b, The foreland-propagating thrust architecture of the East Greenland Caledonides 72°–75°N: Journal of the Geological Society of London, v. 161, p. 1009–1026, doi: 10.1144/0016-764903-141.

Hofer, E., 1957, Arctic Riviera, North-East Greenland: Berne, Kümmerly and Frey, Geographical Publishers, 128 p.

Holdsworth, R.E., and Strachan, R.A., 1991, Interlinked system of ductile strike-slip and thrusting formed by Caledonian sinistral transpression in northeastern Greenland: Geology, v. 19, p. 510–513, doi: 10.1130/0091-7613(1991)019<0510:ISODSS>2.3.CO;2.

Hougaard, G., Jepsen, H.F., and Neve, J.K., 1991, GGU's photogeological laboratory: Aerial photogrammetry—A valuable geological mapping tool in Greenland: Rapport Grønlands Geologiske Undersøgelse, v. 152, p. 29–32.

Hull, J.M., and Gilotti, J.A., 1994, The Germania Land deformation zone and related structures, North-East Greenland: Rapport Grønlands Geologiske Undersøgelse, v. 162, p. 113–127.

Jensen, S.M., and Stendal, H., 1994, Reconnaissance for mineral occurrences in North-East Greenland (76°–78°N), in Higgins, A.K., ed., Geology of North-East Greenland: Rapport Grønlands Geologiske Undersøgelse, v. 162, p. 163–168.

Jepsen, H.F., 2000, Geological Map of Greenland, Lambert Land, Sheet 9: Copenhagen, Geological Survey of Denmark and Greenland, scale 1:500,000.

Jepsen, H.F., Escher, J.C., Friderichsen, J.D., and Higgins, A.K., 1994, The geology of the north-east corner of Greenland—Photogeological studies and 1993 field work: Rapport Grønlands Geologiske Undersøgelse, v. 161, p. 21–33.

Kalsbeek, F., and Jepsen, H.F., 1984, The late Proterozoic Zig-Zag Dal Basalt Formation of eastern North Greenland: Journal of Petrology, v. 25, p. 644–664.

Kalsbeek, F., Nutman, A.P., Escher, J.C., Friderichsen, J.D., Hull, J.M., Jones, K.A., and Pedersen, S.A.S., 1999, Geochronology of granitic and supracrustal rocks from the northern part of the East Greenland Caledonides: Ion microprobe U-Pb zircon ages: Geology of Greenland Survey Bulletin, v. 184, p. 31–48.

Kalsbeek, F., Thrane, K., Nutman, A.P., and Jepsen, H.F., 2000, Late Mesoproterozoic to early Neoproterozoic history of the East Greenland Caledonides: Evidence for Grenvillian orogenesis?: Journal of the Geological Society of London, v. 157, p. 1215–1225.

Kalsbeek, F., Jepsen, H.F., and Nutman, A.P., 2001a, From source migmatites to plutons: Tracking the origin of c. 435 Ma granites in the East Greenland Caledonian orogen: Lithos, v. 57, p. 1–21, doi: 10.1016/S0024-4937(00)00071-2.

Kalsbeek, F., Jepsen, H.F., and Jones, K.A., 2001b, Geochemistry and petrogenesis of S-type granites in the East Greenland Caledonides: Lithos, v. 57, p. 91–109, doi: 10.1016/S0024-4937(01)00038-X.

Knuth, E., 1942, Report on the expedition and on subsequent work at the Mørkefjord Station: Meddelelser om Grønland, v. 126, no. 1, 159 p.

Koch, J.P., 1913, Gennem den hvide Ørken, Gyldendalske Boghandel: Copenhagen, Nordisk Forlag, 286 p.

Koch, L., and Haller, J., 1971, Geological Map of East Greenland 72°–76°N. Lat.: Meddelelser om Grønland, v. 183, scale 1:250,000, 26 p. (+ 13 map sheets).

Koldewey, K., 1873–1874, Die zweite deutsche Nordpolarfahrt in den Jahren 1869 und 1870 unter Führung des Kapitän Karl Koldewey: Leipzig, F.A. Brockhaus, 699 and 962 p.

Larsen, L.M., Watt, W.S., and Watt, M., 1989, Geology and Petrology of the Lower Tertiary Plateau Basalts of the Scoresby Sund Region, East Greenland: Bulletin Grønlands Geologiske Undersøgelse, v. 157, 164 p.

Larsen, L.M., Waagstein, R., Pedersen, A.K., and Storey, M., 1999, Trans-Atlantic correlation of the Palaeogene volcanic successions in the Faeroe Islands and East Greenland: Journal of the Geological Society of London, v. 156, p. 1081–1095, doi: 10.1144/gsjgs.156.6.1081.

Leslie, A.G., and Higgins, A.K., 2008, this volume, Foreland-propagating Caledonian thrust systems in East Greenland, in Higgins, A.K., Gilotti, J.A., and Smith, M.P., eds., The Greenland Caledonides: Evolution of the Northeast Margin of Laurentia: Geological Society of America Memoir 202, doi: 10.1130/2008.1202(07).

Leslie, A.G., and Nutman, A.P., 2003, Evidence for Neoproterozoic orogenesis and early high temperature Scandian deformation events in the southern East Greenland Caledonides: Geological Magazine, v. 140, p. 309–333, doi: 10.1017/S0016756803007593.

McKerrow, W.S., Mac Niocaill, C., and Dewey, J.F., 2000, The Caledonian Orogeny redefined: Journal of the Geological Society of London, v. 157, p. 1149–1154.

Mikkelsen, P.S., 1995, Nordøstgrønland 1908–1960. Fangstmandsperioden: Copenhagen, Dansk Polarcenter, 408 p.

Nathorst, A.G., 1901, On the map of King Oscar Fjord and Kaiser Franz Josef Fjord in North-Eastern Greenland: The Geographical Journal, v. 17, p. 48–63, doi: 10.2307/1775853.

Nielsen, T.F.D., 1987, Tertiary alkaline magmatism in East Greenland: A review, in Fitton, J.G., and Upton, B.G.J., eds., Alkaline Igneous Rocks: Geological Society of London Special Publication 30, p. 489–515.

Olesen, O.B., and Reeh, N., 1969, Preliminary report on glacier observations in Nordvestfjord, East Greenland: Rapport Grønlands Geologiske Undersøgelse, v. 21, p. 41–53.

Olsen, H., 1993, Sedimentary Basin Analysis of the Continental Devonian Basin in East Greenland: Bulletin Grønlands Geologiske Undersøgelse, v. 168, 80 p.

Olsen, H., and Larsen, P.-H., 1993, Lithostratigraphy of the Continental Devonian Sediments in North-East Greenland: Bulletin Grønlands Geologiske Undersøgelse, v. 165, 108 p.

Peacock, J.D., 1956, The Geology of Dronning Louise Land, N.E. Greenland: Meddelelser om Grønland, v. 137, no. 7, 38 p.

Peacock, J.D., 1958, Some Investigations into the Geology and Petrography of Dronning Louise Land, N.E. Greenland: Meddelelser om Grønland, v. 157, no. 4, 139 p.

Pedersen, S.A.S., and Håkansson, E., 2001, Kronprins Christian Land orogeny: Deformation styles of the end Cretaceous transpressional mobile belt in eastern North Greenland, in Tessensohn, F., and Roland, N.W., eds., ICAM (International Conference on Arctic Margins) III International Conference on Arctic margins: Polarforschung, v. 69, p. 117–130.

Pedersen, S.A.S., Craig, L.E., Upton, B.G.J., Rämö, O.T., Jepsen, H.F., and Kalsbeek, F., 2002, Palaeoproterozoic (1740 Ma) rift-related volcanism in the Hekla Sund region, eastern North Greenland: Field occurrence, geochemistry and tectonic setting: Precambrian Research, v. 114, p. 327–346, doi: 10.1016/S0301-9268(01)00234-0.

Peel, J.S., and Sønderholm, M., eds., 1991, Sedimentary Basins of North Greenland: Bulletin Grønlands Geologiske Undersøgelse, v. 160, 164 p.

Price, S., Brodie, J., Whitham, A., and Kent, R., 1997, Mid-Tertiary rifting and magmatism in the Traill Ø region, East Greenland: Journal of the Geological Society of London, v. 154, p. 419–434, doi: 10.1144/gsjgs.154.3.0419.

Rasmussen, J.A., and Smith, M.P., 2001, Conodont geothermometry and tectonic overburden in the northernmost East Greenland Caledonides: Geological Magazine, v. 138, p. 687–698.

Ryder, C., 1895, Beretning om den østgrønlandske Expedition 1891–92: Meddelelser om Grønland, v. 17, p. 1–159.

Scoresby, W., Jr., 1823, Journal of a Voyage to the Northern Whale Fishery, including Researches and Discoveries on the Eastern Coast of West Greenland, Made in Summer of 1922, in the ship Baffin of Liverpool: Edinburgh, A. Constable, 472 p.

Smith, M.P., and Rasmussen, J.A., 2008, this volume, Cambrian–Silurian development of the Laurentian margin of the Iapetus Ocean in Greenland and related areas, in Higgins, A.K., Gilotti, J.A., and Smith, M.P., eds., The Greenland Caledonides: Evolution of the Northeast Margin of Laurentia: Geological Society of America Memoir 202, doi: 10.1130/2008.1202(06).

Smith, M.P., Rasmussen, J.A., Robertson, S., Higgins, A.K., and Leslie, A.G., 2004, Lower Palaeozoic stratigraphy of the East Greenland Caledonides: Geological Survey of Denmark and Greenland Bulletin, v. 6, p. 5–28.

Sønderholm, M., and Jepsen, H.F., 1991, Proterozoic basins of North Greenland: Bulletin Grønlands Geologiske Undersøgelse, v. 160, p. 49–69.

Sønderholm, M., Frederiksen, K.S., Smith, M.P., and Tirsgaard, H., 2008, this volume, Neoproterozoic sedimentary basins with glacigenic deposits of the East Greenland Caledonides, in Higgins, A.K., Gilotti, J.A., and Smith, M.P., eds., The Greenland Caledonides: Evolution of the Northeast Margin of Laurentia: Geological Society of America Memoir 202, doi: 10.1130/2008.1202(05).

Steiger, R.H., Hansen, B.T., Schuler, C., Bär, M.T., and Henriksen, N., 1979, Polyorogenic nature of the southern Caledonian fold belt in East Greenland: The Journal of Geology, v. 87, p. 475–495.

Stemmerik, L., and Piasecki, S., 2004, Isotopic evidence for the age of the Røde Ø Conglomerate, inner Scoresby Sund, East Greenland: Bulletin of the Geological Society of Denmark, v. 51, p. 137–140.

Stemmerik, L., Christiansen, F.G., Piasecki, S., Jordt, B., Marcussen, C., and Nøhr-Hansen, H., 1993, Depositional history and petroleum geology of the Carboniferous to Cretaceous sediments in the northern part of East Greenland, in Vorren, T.O., et al., eds., Arctic Geology and Petro-